ESKAPE Pathogens

Siddhardha Busi • Ram Prasad
Editors

ESKAPE Pathogens

Detection, Mechanisms and Treatment Strategies

Springer

Editors
Siddhardha Busi
Department of Microbiology
School of Life Sciences
Pondicherry University
Puducherry, India

Ram Prasad
Department of Botany
Mahatma Gandhi Central University
Motihari, Bihar, India

ISBN 978-981-99-8798-6 ISBN 978-981-99-8799-3 (eBook)
https://doi.org/10.1007/978-981-99-8799-3

This Springer imprint is published by the registered company Springer Nature Singapore Pte Ltd.
The registered company address is: 152 Beach Road, #21-01/04 Gateway East, Singapore 189721, Singapore

Preface

ESKAPE pathogens—the 'escapers' from the biocidal action of antibiotics—comprise *Enterococcus faecium, Staphylococcus aureus, Klebsiella pneumoniae, Acinetobacter baumannii, Pseudomonas aeruginosa,* and *Enterobacter* spp. Infectious diseases due to this group of bacteria continue to be one of the most alarming health issues that cost millions of lives globally. Also, this group of bacteria is the most encountered nosocomial pathogens and is highly resistant to commonly used antibiotics. Noticeably, now reports on strains developing resistance to last-resort antibiotics are also frequent. Not surprisingly, this alarming situation of antimicrobial resistance (AMR) has been declared a threat to human and animal health by the WHO and CDC. According to the current rate of emergence and spread of antibiotic resistance, it is expected that the annual loss of life will reach around ten million deaths with an estimated economic loss of 100 trillion dollars by the year 2050. In the US and Europe, AMR infections account for three million cases costing over 6.2 billion USD annually. Estimated figures are not available for the LMIC countries, although, undoubtedly, they also highly contribute to harbouring MDR pathogens.

The lack of rapid diagnosis and the usage of broad-spectrum antibiotics have accelerated the development of antibiotic resistance. Hence, the detection of antibiotic resistance and virulence factors is of utmost importance and to adapt the right course of action upon infection. Recently, numerous studies have been reported on determining virulence factors through protein-protein interaction, biochemical, and molecular techniques. However, a compilation of the comprehensive list of such techniques remains unreported. It is important and highly demanding to understand the ESKAPE pathogens, their medical relevance, detection techniques, resistant patterns, and current strategies against these pathogens to further study, progress, and thereby overcome the frightening situation of antimicrobial resistance.

The contents of this book are arranged systematically, ensuring complete insights into ESKAPE pathogens. The medical relevance, basic to high-end detection techniques, mechanisms of antibiotic resistance, antimicrobial-resistant gene detection, etc., cover a portion of the book that delivers introductory knowledge to the readers. Elaborate elucidation of each ESKAPE pathogen is provided followed by the introductory chapters. These chapters will equip the readers with a comprehensive understanding of the synthesis, structure, and function of each pathogen's virulence factors and their 'defensive fort'—biofilm. The chapters will also discuss the inhibition approaches and modalities against the virulence factors and biofilm besides available antibiotics. Final chapters of the book deal with novel strategies, including therapy with under-explored and novel phytochemicals, antibiotic adjuvant therapy, antimicrobial photodynamic therapy, therapy using antimicrobial peptides, nanomaterial and composites, and bacteriophage therapy for combating ESKAPE pathogens and to fight against these microorganisms in the 'post-antibiotic era'. All the techniques are discussed in detail with a significant number of recently studied reports. The book ends with the chapter dealing with in silico/computational approaches for the inhibition of ESKAPE pathogen, which reflects one of the most dependable future aspects and ways to the battle against ESKAPE pathogens.

Puducherry, India Siddhardha Busi
Motihari, Bihar, India Ram Prasad

Contents

Editors and Contributors

About the Editors

Siddhardha Busi, M.Sc., Ph.D. is Assistant Professor in the Department of Microbiology, Pondicherry University, Pondicherry, India. Dr. Siddhardha Busi worked in the Biology Division, CSIR-IICT, Hyderabad, India, for his Ph.D. His research interest includes antimicrobial drug discovery, host-pathogen interactions, photodynamic therapy, and nanotechnology-based drug discovery and drug delivery. His research interests focus on quorum sensing and biofilm inhibition in the pathogens of medical importance. He has more than 12 years of teaching and research experience. Dr. Siddhardha Busi published more than 90 publications (total citations 3292 with an h-index 30, i10-index 74), including research papers and review articles in peer-reviewed international journals. He authored or co-authored numerous book chapters. He edited four books from his areas of expertise. Dr. Siddhardha Busi also served as the principal investigator of two project grants funded by the Science & Engineering Research Board (SERB), India. He serves as an editorial board member for several reputed journals and is a member of many national and international scientific societies

Ram Prasad, Ph.D. is Associate Professor, Department of Botany, Mahatma Gandhi Central University, Motihari, Bihar, India. His research interest includes applied and environmental microbiology, plant-microbe interactions, nanobiotechnology, sustainable agriculture, and chemistry of biology. Dr. Prasad has more than 250 publications (total citations 14,275 with an h-index 60, i10-index 189) to his credit, including research papers, review articles and book chapters and seven patents issued or pending, and edited or authored several books. Dr. Prasad has 14 years of teaching experience and has been awarded the Young Scientist Award and Prof. J.S. Datta Munshi Gold Medal by the International Society for Ecological Communications, Sir CV Raman Scientist Award, Fellow of Biotechnology Research Society of India; Fellow of Agricultural Technology Development Society, India; Fellow of the Society for Applied Biotechnology; the American Cancer Society UICC International Fellowship for

Beginning Investigators, USA; Outstanding Scientist Award, BRICPL Science Investigator Award, and Research Excellence Award, etc. He has been serving as an editorial board member of *BMC Microbiology*, *BMC Biotechnology*, *Current Microbiology*, *Archives of Microbiology*, *Annals of Microbiology*, *Nanotechnology for Environmental Engineering*, *SN Applied Sciences*, etc., including series editor of Nanotechnology in the Life Sciences, Springer Nature, USA. Previously, Dr. Prasad served as Assistant Professor at Amity University Uttar Pradesh, India; Visiting Assistant Professor, Whiting School of Engineering at Johns Hopkins University, Baltimore, United States; and Research Associate Professor at School of Environmental Science and Engineering, Sun Yat-sen University, Guangzhou, China

Contributors

Ankita Agrawal Department of Biotechnology and Bioinformatics, Sambalpur University, Jyoti Vihar, Sambalpur, Odisha, India

Dinakara Rao Ampasala Department of Bioinformatics, School of Life Sciences, Pondicherry University, Puducherry, India

V. T. Anju Department of Biochemistry and Molecular Biology, School of Life Sciences, Pondicherry University, Pondicherry, India

Sayak Bhattacharya Department of Microbiology, Bijoykrishna Howrah Girls' College, Howrah, West Bengal, India

Indranil Chattopadhyay Department of Biotechnology, School of Life Sciences, Central University of Tamil Nadu, Thiruvarur, Tamil Nadu, India

Srujana Chitipothu Centre for Bioinformatics, Kamalnayan Bajaj Institute for Research in Vision and Ophthalmology,, Vision Research Foundation, Chennai, Tamil Nadu, India

Central Research Instrumentation Facility, Kamalnayan Bajaj Institute for Research in Vision and Ophthalmology, Vision Research Foundation, Chennai, Tamil Nadu, India

Madhu Dyavaiah Department of Biochemistry and Molecular Biology, School of Life Sciences, Pondicherry University, Pondicherry, India

Nishel Forgia Department of Microbiology, School of Life Sciences, Pondicherry University, Puducherry, India

Pitchaipillai Sankar Ganesh Department of Microbiology, Centre for Infectious Diseases, Saveetha Dental College and Hospitals, Saveetha Institute of Medical and Technical Sciences (SIMATS), Chennai, India

Rajesh Kanna Gopal Department of Microbiology, Centre for Infectious Diseases, Saveetha Dental College and Hospitals, Saveetha Institute of Medical and Technical Sciences (SIMATS), Chennai, India

Archana Priyadarshini Jena Department of Biotechnology, School of Life Sciences, Pondicherry University, Pondicherry, India

Jeyakanthan Jeyaraman Structural Biology and Bio-Computing Lab, Department of Bioinformatics, Science Block, Alagappa University, Karaikudi, Tamil Nadu, India

Bicky Jerin Joseph School of Biosciences, Mahatma Gandhi University, Kottayam, Kerala, India

Srujana Kathi UGC-Human Resource Development Centre, Pondicherry University, Puducherry, India

Anitha Katta Department of Biosciences and Biotechnology, Krishna University, Machilipatnam, Andhra Pradesh, India

G. Kishore Department of Biosciences and Biotechnology, Krishna University, Machilipatnam, Andhra Pradesh, India

Mounika Kovvali Department of Biosciences and Biotechnology, Krishna University, Machilipatnam, Andhra Pradesh, India

Suseela Lanka Department of Biosciences and Biotechnology, Krishna University, Machilipatnam, Andhra Pradesh, India

Jung-Kul Lee Department of Chemical Engineering, Konkuk University, Seoul, Republic of Korea

Jyothis Mathew School of Biosciences, Mahatma Gandhi University, Kottayam, Kerala, India

Maya Mathew School of Biosciences, Mahatma Gandhi University, Kottayam, Kerala, India

Monika Mishra Department of Biotechnology and Bioinformatics, Centre of Excellence in Natural Products and Therapeutics, Sambalpur University, Sambalpur, Odisha, India

Mahima S. Mohan Department of Microbiology, School of Life Sciences, Pondicherry University, Puducherry, India

Raman Muthusamy Department of Microbiology, Centre for Infectious Diseases, Saveetha Dental College and Hospitals, Saveetha Institute of Medical and Technical Sciences (SIMATS), Chennai, India

Hemavathy Nagarajan Centre for Bioinformatics, Kamalnayan Bajaj Institute for Research in Vision and Ophthalmology, Vision Research Foundation, Sankara Nethralaya, Chennai, Tamil Nadu, India

Pradeep Kumar Naik Department of Biotechnology and Bioinformatics, Centre of Excellence in Natural Products and Therapeutics, Sambalpur University, Sambalpur, Odisha, India

P. Naji Naseef Department of Biotechnology, Infection and Inflammation, Central University of Tamil Nadu, Thiruvarur, Tamil Nadu, India

Gantala Sarva Sai Nikhilesh Andhra University, Visakhapatnam, Andhra Pradesh, India

Santhilatha Pandrangi Department of Biochemistry and Bioinformatics, GITAM School of Sciences, GITAM (Deemed to be) University, Visakhapatnam, India

Paramanantham Parasuraman Department of Chemical Engineering, Konkuk University, Seoul, Republic of Korea

Amiya Kumar Patel Department of Biotechnology and Bioinformatics, Sambalpur University, Jyoti Vihar, Sambalpur, Odisha, India

Subhaswaraj Pattnaik Department of Biotechnology and Bioinformatics, Centre of Excellence in Natural Products and Therapeutics, Sambalpur University, Sambalpur, Odisha, India

Sudhakar Pola Department of Biotechnology, College of Science and Technology, Andhra University, Visakhapatnam, Andhra Pradesh, India

Riya Rachel St. Berchmans College, Kottayam, Kerala, India

E. K. Radhakrishnan School of Biosciences, Mahatma Gandhi University, Kottayam, Kerala, India

Sampathkumar Ranganathan Department of Chemical Engineering, Konkuk University, Seoul, Republic of Korea

Department of Bioinformatics, School of Life Sciences, Pondicherry University, Puducherry, India

Simi Asma Salim Department of Microbiology, School of Life Sciences, Pondicherry University, Puducherry, India

Sathish Sankar Department of Microbiology, Centre for Infectious Diseases, Saveetha Dental College and Hospitals, Saveetha Institute of Medical and Technical Sciences (SIMATS), Chennai, India

Vemuri Venkateswara Sarma Department of Biotechnology, School of Life Sciences, Pondicherry University, Pondicherry, India

Esaki Muthu Shankar Department of Biotechnology, Infection and Inflammation, Central University of Tamil Nadu, Thiruvarur, Tamil Nadu, India

Siddhardha Busi Department of Microbiology, School of Life Sciences, Pondicherry University, Puducherry, India

Zarin Taj Department of Biotechnology, School of Life Sciences, Central University of Tamil Nadu, Thiruvarur, Tamil Nadu, India

Medical Importance of ESKAPE Pathogens

1

Simi Asma Salim, Mahima S. Mohan, Nishel Forgia, and Siddhardha Busi

Abstract

Being the 'escapers' from the action of antibiotics, the ESKAPE pathogens—*Enterococcus faecium, Staphylococcus aureus, Klebsiella pneumoniae, Acinetobacter baumannii, Pseudomonas aeruginosa* and *Enterobacter* spp.—a group of nosocomial pathogens are presently considered greatly by the health sector because of their highly infective nature and influence on mortality and morbidity. It is crucial to know their priority ranks as pathogens, resistant mechanisms, transmission of resistance and strategies to compete against these pathogens. All of these pathogens fall under the critical or high priority tier according to WHO's Global Priority Pathogens List (GPP) and are characterized by high resistance to different first-line to last-hope antibiotic classes. Resistant strains of these organisms are isolated frequently from clinical samples. In general, ESKAPE pathogens are associated with hospital-acquired infections including skin and soft tissue infections (SSTIs), pneumonia, endocarditis, and other bloodstream infections (BSI), surgical site infections (SSI) and urinary tract infections (UTIs). Each of these organisms has unique mechanisms of pathogenesis and resistance. This chapter discusses in detail how and why ESKAPE pathogens are medically relevant, what is their role in antibiotic resistance, mechanism and transmission and their current status as infectious pathogens as per various priority lists and reports.

S. A. Salim · M. S. Mohan · N. Forgia · S. Busi (✉)
Department of Microbiology, School of Life Sciences, Pondicherry University, Puducherry, India

S. Busi, R. Prasad (eds.), *ESKAPE Pathogens*,
https://doi.org/10.1007/978-981-99-8799-3_1

Keywords

ESKAPE pathogens · Antibiotic resistance · Nosocomial infections · Antibiotics

1.1 Introduction

1.1.1 ESKAPE Pathogens and Nosocomial Infections

The Infectious Diseases Society of America classified a cluster of bacteria adept of withstanding antibiotic lethality, representing a new standard in pathogenesis, transmission and resistance as ESKAPE pathogens (*Enterococcus faecium, Staphylococcus aureus, Klebsiella pneumoniae, Acinetobacter baumannii, Pseudomonas aeruginosa* and *Enterobacter* spp.) (Pendleton et al. 2013). The mainstream nosocomial infections produced by ESKAPE pathogens (Navidinia 2016) cause high mortality and morbidity, and trouble in the initiation of empirical treatment due to the high level of resistance shown by them (Ma et al. 2020). To tackle the rising dilemma of antibiotic resistance, WHO published a catalogue of pathogenic bacteria that urgently require new antibiotics to treat them. The catalogue of 12 'priority pathogens' that cause serious hazards to human health is categorized into three—critical, high and medium priority. *Acinetobacter*, *Pseudomonas* and various Enterobacteriaceae constituting the critical group accountable for various lethal infections such as bloodstream infections and pneumonia. While *E. faecium* and *S. aureus* come under the list of high-priority category causing common infections such as urinary tract infections (UTIs) (WHO 2017). Soil, food, various water sources, plants and manure or domestic waste are the common environmental pools of the ESKAPE pathogens, act as drivers of antimicrobial resistance (Denissen et al. 2022). The multidrug resistance mechanism shown by ESKAPE pathogens is generally classified into three—enzyme-mediated drug inactivation, modification of the target site, and reduced penetrability or elevated drug outflow. They are also capable of forming biofilms to prevent immune cell response and antibiotic penetration (Mulani et al. 2019).

Nosocomial infections (also called hospital-acquired infection) are systemic or localized infections that arise in patients under medical care within 48 h of admission; this is one of the chief communal healthcare issues throughout the world (Tolera et al. 2018). About 10–25% of the hospital waste is hazardous and serves as a potential source of pathogens. Nosocomial infections are ground for 7% and 10%, respectively, in developed and developing nations. It can be classified into central line-associated bloodstream infections (CLABSIs), catheter-associated urinary tract infections (CAUTIs), surgical site infections (SSIs) and ventilator-associated pneumonia (VAPs). CLABSIs are deadly nosocomial infections that cause 12–25% of mortality. CAUTI is the most commonly occurring nosocomial infection, causing orchitis, epididymitis and prostatitis in males, and pyelonephritis, cystitis and meningitis in all patients. Around 2–5% of patients subjected to surgery have a high mortality risk due to SSIs. It is the second utmost frequent kind of nosocomial illness, principally caused by *S. aureus*. Another nosocomial pneumonia, VAP cases were noticed in 9–27% of patients on artificial ventilation (Khan et al. 2017).

Although bacteria, viruses and fungi are associated with nosocomial infections, incidence and complications due to bacteria are of major concern (Khan et al. 2017). Nosocomial infections due to *Enterococci* in the hospital turned out to increase widely due to their inherent resistance and resistance acquired through either gene alteration or by transfer of plasmids and transposons. Likewise, *Enterococci* can transmit vancomycin resistance determinants to other organisms like *S. aureus* which has been also reported in the laboratory. About 30% of the skin or noses of healthy individuals are inhabited by *S. aureus*. Such strains isolated from the medical environment are termed as hospital-acquired *S. aureus* (HASA) and can cause atopic dermatitis (AD) and causes skin inflammation. Diseased individuals and healthcare acquaintances act as noteworthy carriers for transmission. CAUTI and lower respiratory tract infections are more often caused by *K. pneumoniae.* Stress tolerance and advancement in resistance through antibiotic selective pressure led to the expansion of *A. baumannii* in hospital while *Pseudomonas* cause high morbidity and mortality rate in infected patients. *Enterobacter* spp. is also responsible for the cause of some vital nosocomial infections, by acquiring multi-drug-resistant (MDR) through plasmid-encoded varieties of β-lactamase enzymes (Navidinia 2016).

Recent studies show that patients diseased with COVID-19 are highly infected with bacterial and fungal nosocomial infections. Bardi et al. (2021) showed the prevalence of *E. faecium* over *E. faecalis* in COVID-19 patients with bloodstream infections (BSI). Recent studies also proved the presence of Enterobacterales, *A. baumannii* and *P. aeruginosa* in patients with COVID-19 (Langford et al. 2020; Rawson et al. 2020). The pathogenic bacteria, *Acinetobacter*, is responsible for the occurrence of 80% of infections reported in ICUs. Methicillin-resistant *S. aureus* (MRSA) enters open wounds, can travel through organs or bloodstream and causes sepsis, pneumonia and SSI (Khan et al. 2017). Tolera et al. assessed the manifestation of bacterial nosocomial illness among hospital patients in Eastern Ethiopia. They reported that *S. aureus* (18.5%) was the regular pathogen present in them succeeded by *Escherichia coli* (16.7%) and typical multidrug-resistant isolates identified were *P. aeruginosa* (30.4%) and *S. aureus* (21.7%) (Tolera et al. 2018). Feleke et al. acknowledged that nosocomial illnesses were commonly due to Gram-negative (53.2%) than Gram-positive bacteria (46.8%). *S. aureus* (35.6%) and *E. coli* (15.3%) were the most isolated strains. Gram-negative bacteria such as *E. coli* (15.3%) and *Klebsiella* sp. (13.4%) were commonly isolated and showed high resistance against antibiotics (Feleke et al. 2018).

1.2 ESKAPE Organisms and Pathogenic Concerns

1.2.1 *Enterococcus faecium*

It is a Gram-positive commensal bacterium normally dwelling in the gastrointestinal tract of humans and animals. This opportunistic pathogen is amongst the top pathogens causing nosocomial and multiple-resistant infections and is found mostly related to infections like various bloodstream infections, urinary tract infections and endocarditis. World Health Organization (WHO) enlisted vancomycin-resistant *E. faecium* as a priority 2 pathogen. According to the Indian Pathogen Priority List (IPPL), vancomycin, linezolid-resistant and daptomycin non-susceptible *Enterococcus* species are listed in the high-priority pathogen category. In various animal-origin antimicrobial resistance (AMR) surveillance programmes, such as Canada's—Canadian Integrated Program for Antimicrobial Resistance Surveillance (CIPARS), European Union's—European Antimicrobial Susceptibility Surveillance in Animals (EASSA), USA's—National Antimicrobial Resistance Monitoring System for Enteric Bacteria (NARMS) and The Japanese veterinary antimicrobial resistance monitoring systems (JVARM) of Japan, *E. faecium* is considered as a Gram-positive indicator bacteria (Xuan et al. 2021). The highest increase in positive blood culture was observed with *E. faecium* compared to several other pathogens as per the 2002–2008 data of the European Antimicrobial Resistance Surveillance System (EARSS). The rise in bloodstream infections (BSI) was also followed by investigation data from the University Medical Center Groningen (UMCG, The Netherlands). Most European countries, countries outside Europe and the United States, also reported the rise of *E. faecium* BSIs during the last 20 years. The Australian Enterococcal Sepsis Outcome Program (AESOP) identified in 2014 that *E. faecium* serves as the causative agent of 39.9% of enterococcal bacteraemia (Zhou et al. 2020; Coombs et al. 2014). The European Center for Diseases and Control (ECDC) reports of 2021–2023 show different proportion rates of vancomycin-resistant *E. faecium* among different countries varying from less than 1 to more than 46% (ECDC). According to 2011–2014 data on antibiotic-resistant hospital-acquired infections from the Centers for Disease Control and Prevention (CDC), the prevalence of vancomycin-resistant *E. faecium* in the US was 80.5% in 2011 and 75.6% in 2014. 'Antibiotic Resistance Threats in the United States'—a CDC report of 2019, in which 18 organisms are classified into one among three categories: urgent, serious and concerning. In that vancomycin-resistant Enterococcus (VRE) comes under the serious threat category and the report states that most VRE infections occur in patients with long-term healthcare exposures such as those who received treatment for cancer and other cases like organ transplantations or in patients with weakened immune systems. Moreover, as stated by CDC's National Healthcare Safety Network, in units of solid organ transplant, *E. faecium* was the frequent pathogen associated with CLABSIs, and over 70% of this pathogen was resistant to vancomycin and it is of concern that VRE is progressively gaining resistance to other antibiotics, raising fear that the other persisting medications for the management of VRE infections may become ineffective (CDC 2019).

1.2.2 *Staphylococcus aureus*

Methicillin-resistant *Staphylococcus aureus* (MRSA) is classified as a high-priory pathogen in both WHO's global priority pathogens list and Indian priority pathogen list (Asokan et al. 2019; DBT 2021). It is the commonest multidrug-resistant pathogen related to HAI, causing a high disease and death rate (Kramer et al. 2019). The hospital-acquired strains of MRSA are associated with severe aggressive infections like bloodstream infection (BSI), pneumonia, and skin and soft tissue infections (Choo 2017). Nosocomial or healthcare-associated pneumonia is frequently caused by MRSA (Torre-Cisneros et al. 2018), it is the second most dominant pathogen responsible for almost 17.8% of infection among all the known cases (Cabrera et al. 2020). Ventilator-associated pneumonia (VAP) is common among significantly sick COVID-19 patients. In an investigation among COVID-19 patients with VAP, the most persistent deep respiratory bacteria isolated was *S. aureus* (Giacobbe et al. 2021).

S. aureus can survive harsh environments for longer periods and allowing it to spread throughout the hospital through contact with permanently or transiently colonized people (Matta et al. 2018). The increasing concern of infectious outbursts in neonatal intensive care units (NICUs) is mostly associated with the incidence of *S. aureus*. The bacteriological profile causing HAI in a NICU study group showed a high proportion of Gram-positive pathogens where *S. aureus* constitutes 65%, out of which MRSA accounted for 40%. The mean frequency of isolation of *S. aureus* was 15% and 9% in the high-risk and low-risk NICU, respectively (Kumar et al. 2018).

Staphylococcus aureus is a predominant bloodstream pathogen. Recent studies report that it is responsible for 14.2% of BSI in Korea, 13.0% in Japan and 6.5% in Taiwan, respectively (Chiang et al. 2018). In surgical wards, HAI is one of the major difficulties, increasing the rate of death incidence, morbidity and extent of the hospitalization period (Mateescu et al. 2023). About 23.6% of sepsis in the surgical section is due to *S. aureus* (Voidazan et al. 2020). A recent study testified the occurrence of *S. aureus* as the dominant pathogen among 973 samples tested at hospitalization, constituting 64.7% of the total isolated strains (Mateescu et al. 2023). It is responsible for 20.7% of surgical site infections in the USA (O'Toole 2021).

1.2.3 *Klebsiella pneumoniae*

Klebsiella pneumoniae is an extremely versatile Gram-negative pathogen associated with multi-drug-resistant (MDR) healthcare-associated and hyper-virulent community-acquired infections (Meatherall et al. 2009). Ominously, over the last 10 years, the acquisition of antimicrobial resistance determining factors among *K. pneumoniae* is upsurging relentlessly, predominantly resistance to third-generation antibiotics such as carbapenems and cephalosporins. According to the Indian Pathogen Priority List reviewed in 2019, *K. pneumoniae* is categorized as a critical priority pathogen along with *E. coli* (DBT 2021). As per the study of Global Research on Antimicrobial Resistance (GRAM), one of the six pathogens which caused more than 250,000 AMR-related deaths in 2019 were *K. pneumoniae* (Murray et al. 2022). Also, it is estimated that *K. pneumoniae* BSI has an approximately 20% mortality rate (Jung et al. 2012). Predictably, this pathogen was included in the WHO's highest category of priority pathogens, which demands novel antimicrobial agents or strategies urgently (Foster-Nyarko et al. 2023). Frequently, *K. pneumoniae* related to nosocomial infections in South Asia are extensively drug resistant (XDR) (Kaza et al. 2021). In a 5-year retrospective survey carried out in a tertiary care teaching hospital, in Hungary, the second most prevalent genus isolated was *Klebsiella*, in which *K. pneumoniae* was the most frequently occurring species (Benkő et al. 2020). Coming to the Indian scenario, in a five-month prospective study of the samples received from the ICU, in order to estimate the incidence of carbapenem-resistant Gram-negative bacilli in the ICU setups, *K. pneumoniae* constituted 36.7%, the highest proportion (Rout et al. 2021). *K. pneumoniae* is associated with both endemic infections and epidemic strain outbreaks in hospital sites. It causes infections like bacteremia, infections of the respiratory, and urinary tract, invasive liver abscesses, endophthalmitis and endocarditis, and it also has been found associated with catheter-associated urinary tract infection owing to their proficiency to develop biofilm in invasive medical devices. The pathogen is also the predominant agent of neonatal sepsis (Mukherjee et al. 2021). According to data from ICMR-AMRSN published in 2019, the second most isolated pathogen (34%) was from the genus *Klebsiella,* and *K. pneumoniae* showed 73–77% resistance to cephalosporins. It also showed more than 65% resistance to the fluroquinolone class of antibiotics (Walia et al. 2019). As per the 2021–2023 European Centre for Disease Prevention and Control (ECDC) antimicrobial surveillance data of Europe, resistance to third-generation cephalosporins and carbapenems was found generally higher in *K. pneumoniae* (ECDC 2023).

1.2.4 *Acinetobacter baumannii*

A. baumannii accounts for more than 20% of nosocomial infections (Ayobami et al. 2019). Carbapenem-resistant *A. baumannii* (CRAB) is considered as 'priority 1' group by WHO and the Indian Priority Pathogen List (IPPL) (Asokan et al. 2019; DBT 2021). Sultan and Seliem 2018 analysed NICU patients having HAIs due to *A. baumannii* and reported that 73.4% of infection was caused by CRAB. *A. baumannii* cause ICU-acquired pneumonia and account for more than 12% of bloodstream infections (BSI) in ICU, thus ICUs have been considered as the epicentres of *A. baumannii* infections. Zhang et al. reported that around 55.6% of COVID-19 ICU patients were coinfected with CRAB (Zhang et al. 2020). It is liable for more than 36% of HAP cases reported in Asia (Garnacho-Montero and Timsit 2019). *A. baumannii* coinfection secondary to SARS-CoV-2 has also been testified in different countries including Wuhan (China), France, Spain, Iran, Egypt, New York (USA), Italy and Brazil. The 2009–2010 report of the National Healthcare Safety Network (NHSN) specifies the presence of *A. baumannii* in CAUTI, CLABI, SSI and VAP (Rangel et al. 2021). In India, the incidence of 32.1% of *Acinetobacter* sp. is one of the most common isolated organisms among VAP patients (Divatia and Abraham 2018).

Acinetobacter baumannii has an innate ability to resist many antibiotics, making them difficult to treat and increasing mortality, morbidity and extent of the period of hospitalization. The global rate of mortality caused by multidrug-resistant *A. baumannii* (MDRAB) in HAP and VAP ranges between 38% and 48% (Mohd Sazlly Lim et al. 2019). MDRAB can also be responsible for infectious outbreaks in burn units. A recent study reported the occurrence of 17% methicillin-resistant-A. *baumannii* nosocomial infection among patients admitted with burn injuries (Munier et al. 2019).

A. baumannii can withstand the dry atmospheres in hospitals for weeks, and enables its spread through contaminants, thus hospitals are thought to be the key source of *A. baumannii* infections (Al-Gethamy et al. 2017). A survey based on the healthcare-associated infections at a University Hospital in Turkey reported that 67.5% of the isolated nosocomial representatives were Gram-negative bacteria where *A. baumannii* accounts for 11.3% of it. In this study, *A. baumannii* is the commonly isolated organism causing VAP (Erdem et al. 2022). In another study *A. baumannii* was one of the frequently isolated organisms from the ICU and 30% of HAIs were caused by them. Also, they are the most common pathogens present in blood samples and respiratory tract samples account for 21.5% of all organisms isolated from patients with BSI (Alfouzan et al. 2021).

1.2.5 *Pseudomonas aeruginosa*

The list of global priority pathogens (GPP) classified carbapenem-resistant *P. aeruginosa* as a critical pathogen that urgently requires novel antibiotics for therapy (Asokan et al. 2019). According to the report of the Taiwan Nosocomial Infection Surveillance System (TNIS), 2014, the CDC showed that 7.5–9.2% of HAI is induced by *P. aeruginosa*. According to the annual report of TNIS, carbapenem-resistant *P. aeruginosa* (CRPA) was accountable for nearly 15.5–18.8% of HAI during early 2014 and the rate of CRPA in ICU patients increased since 2013 and reached 33.3% during the late 2014 (Tsao et al. 2018).

The Indian priority pathogen list also placed *P. aeruginosa* in the critical category (DBT 2021). It is the most prevalent potentially pathogenic microorganism (PPM) in patients with CF and causes 7% of entire nosocomial infections including pneumonia, urinary tract infections, surgical site infections and bacteraemia (Garcia-Clemente et al. 2020; Ng et al. 2023). Several recent international studies also reported the increased occurrence of *P. aeruginosa* bacteraemia during the COVID-19 pandemic (Ng et al. 2023). It is one of the principal organisms causing ventilator-associated pneumonia (VAP) in the USA and Europe and is frequently identified in paediatric JIS (Labovská 2021). Hospital-acquired strains of *P. aeruginosa* tend to be more resistant than community-acquired strains. Research by Reza et al. found that during the years 2014–2015, ESBL *P. aeruginosa* causes 36.5% of VAP and is responsible for 16.6% of sepsis due to VAP at ICUs of 18 hospitals in the north of Iran (Rezai et al. 2018).

P. aeruginosa is liable for elevated morbidity–mortality in intensive care units (ICUs) (Hoang et al. 2018). Wang et al. 2019 investigated a total of 15,588 patients, out of which 7579 have incidence with HAI. About 42.5% of HAI infection was caused by MDR isolates, whereas 19.5% was MDR *Pseudomonas aeruginosa*. In another study by Onifade et al. 2019, among 240 samples from three hospitals in Makurdi metropolis, 71 bacterial pathogens were isolated of which 35.2% of the isolates were Gram-negative and 12.68% was *P. aeruginosa*. Healthcare-associated infection (HAI) is a chief potential concern in surgical wards. Mateescu et al. stated the occurrence of *P. aeruginosa* in surgical departments of emergency hospitals. Out of the 973 samples collected during hospitalization, *P. aeruginosa* represents 5.63% of the total isolated strains (Mateescu et al. 2023). *P. aeruginosa* is also a common cause of catheter-associated UTI (CAUTI) responsible for ~10% of all CAUTIs, and up to 16% of UTIs in patients in ICU (Reynolds and Kollef 2021).

1.2.6 *Enterobacter* Species

Enterobacter is a Gram-negative genus which includes various species known to cause many bloodstream infections, urinary and respiratory tract infections. It is known for its wide range of antibiotic-resistant mechanisms, although most important one is drug inactivation or modification via Ambler A and B class enzymes such as metallo-β-lactamases, extended-spectrum β-lactamases (ESBLs), broad-spectrum β-lactamases and carbapenemases (Pendleton et al. 2013; Santajit and Indrawattana 2016). In 2017, WHO enlisted carbapenem-resistant and ESBL-producing Enterobacteriaceae in the critical priority category which urgently demands the development of novel antimicrobials (WHO 2017). As we earlier discussed about the category in which vancomycin-resistant *Enterococcus* is placed according to the report—'Antibiotic Resistance Threats in the United States', the Enterobacteriaceae that are resistant to carbapenem and which produce ESBL are placed under critical threat category and serious threat category, respectively. The estimated cases of hospitalized patients by these resistant organisms also increased from 2012 to 2017 (CDC 2019). As per the annual report (2021) of the Antimicrobial Resistance Research and Surveillance Network program of the Indian Council of Medical Research (ICMR-AMRSN), almost half proportion (49.5%) of the positive culture isolates were from Enterobacterales except *Salmonella* and *Shigella*. Furthermore, *Enterobacter cloacae* from the *Enterobacter* genus was among the top 10 pathogens (1.7%) (AMRSN 2021) According to CARSS data, although the infections due to hospital-acquired carbapenemase producing Enterobacterales are less in acute care hospitals of Canada, a rise in its frequency was witnessed during 2016–2019, followed by a decrease in 2020. Of note, during this period the all-cause mortality per 100 hospital-acquired patients infected with carbapenemase-producing enterobacterales was 18.02% (CARSS 2022). *Enterobacter* infections are characterized by generally high mortality rates. In a research led by Kang et al. in which they analysed mortality rates due to *Enterobacter* bacteremia within 30 days, 24.6% died of those who received proper antibiotic medication, whereas 34.7% of patients died as they got infected with cephalosporin-resistant strains (Kang et al. 2004).

1.3 Major Nosocomial Infections by ESKAPE Pathogens

Every pathogen possesses distinctive risk factors and different effects on populations. Several species of *Enterococcus* cause various types of fatal infections, including the bloodstream, surgical site and urinary tract infections. The majority of *S. aureus* infections are superficial, such as infections of skin wounds and lesions. Particularly, among patients having lesions and wounds as part of the surgery, there is a higher chance of developing life-threatening *S. aureus* infections. *K. pneumoniae* commonly inhabiting the human gastrointestinal system develops infections, especially in patients receiving care in hospital settings and those who necessitate medical device interventions. *A. baumannii* is also frequently found associated with the infections of the bloodstream, urinary tracts or skin lesions and is comparatively resistant to respiratory tract infection. *P. aeruginosa* can cause severe infections, especially in immunocompromised individuals and individuals having chronic lung illnesses including asthma, cystic fibrosis and chronic obstructive pulmonary diseases. Like all others, pathogenic members of the genus *Enterobacter* cause infections in the circulatory, respiratory and urinary systems, and are more susceptible to individuals having pre-existing health issues including diabetes, cancer malignancy and lupus. In general, ESKAPE organisms associated with major HAIs like bloodstream infections, ventilator-associated pneumonia, urinary tract infections and surgical site infections are described briefly below.

1.3.1 Bloodstream Infection (BSI)

Presently, the prevalence of bloodstream infection (BSI) due to ESKAPE pathogens has rapidly increased, leading to prolonged stays in hospitals, higher economic costs and a high incidence of death (Peng et al. 2021). In one study based on the US data set of 1.1 million and more patient encounters, they found that 42.2% of the total isolated species were ESKAPE pathogens and the most dominant among them were *S. aureus*, *K. pneumoniae* and *P. aeruginosa*, which comprised 21.9%, 7.5% and 7.2% of the total, respectively. Enterobacter species such as *E. cloacae* and *E. aerogenes* represented 0.81% and 2.1% respectively, whereas *A. baumannii* constituted 0.83% among total positive culture (Marturano and Lowery 2019). Two-year study in China among hospitalized children with BSI showed that 76.3% of total BSI was due to ESKAPEEc (ESKAPE pathogens and *E. coli*) and mostly (73.8%) MDR ESKAPEEc (Peng et al. 2021). Studies in the USA among children showed that ESKAPE pathogens accounted for 20% of incidents. They also found that methicillin-resistant *S. aureus* prevalence increased over time (Larru et al. 2016). Of note, according to the CDC, as a consequence of Central line-associated bloodstream infections (CLABSIs), thousands of deaths are reported to occur each year (CDC 2014).

1.3.2 Ventilator Associated Pneumonia (VAP)

Ventilator-associated pneumonia (VAP) due to ESKAPE pathogens are continuously increasing in occurrence and resistance profile in ICUs throughout the world. Except for *E. faecium* all other ESKAPE organisms are among the first six VAP-causing pathogens according to the data of the 2004–2008 SENTRY Antimicrobial Surveillance Program. Estimates of various other multinational, national or local clinical studies are concordant with these data. Although *E. faecium* has been regularly recognized as the third most frequent cause of hospital-acquired BSI in the USA, its association with VAP is not that threatening. Out of all ICU pneumonias due to *S. aureus*, 25% were due to methicillin-resistant strains according to previous estimations of a Spanish yearly ICU infection surveillance study—ENVIN. Moreover, this proportion was higher in the Latin VAP study and EU-VAP study (Sandiumenge and Rello 2012). In one study where 107 VAP patients participated, one-third of participant's infection was due to MRSA (Shorr et al. 2006). Nosocomial pneumonia due to *A. baumannii* represented more than 19% of all incidents of ICU nosocomial pneumonia in Europe (Koulenti et al. 2009). Even though a study by Chan et al. showed that carbapenem-resistant *Acinetobacter* VAP can be successfully treated with second-line agents like colistin and tigecycline, higher nephrotoxicity because of colistin and decreased susceptibility to tigecycline appeared on treatment indicating the boundaries with these second-line drugs (Chan et al. 2010).

1.3.3 Urinary Tract Infection (UTI)

Currently, urinary tract infections (UTIs) are increasingly encountered with antibiotic-resistant ESKAPE uropathogens. A recent study that evaluated the antibiotic resistance among ESKAPE uropathogens in paediatric UTIs in Southeast Gabon reported that *K. pneumoniae* (34%) constituted among the major ESKAPE implicated in UTIs followed by *Enterococcus* spp. (8%) and *S. aureus* (6%) (Mouanga-Ndzime et al. 2023). In another recent one-year retrospective study among Jordanian patients by Alsharedeh et al. to evaluate the frequency of ESKAPE pathogens in infections of the urinary tract and to understand their pattern of antibiotic susceptibility, they found that 87.4% of the UTIs were due to ESKAPE pathogens, and all ESKAPE organisms except *A. baumannii* were identified in the urine samples (Alsharedeh et al. 2023). An antimicrobial resistance pattern study in South India among community-acquired UTI patients demonstrated that although *E. coli* continues as the major associated pathogen, the incidence of ESBL producers and enterococcal UTI has increased (Venkat Ramanan et al. 2014). According to CDC reports, ~75% of UTIs acquired in the hospital are associated with a urinary catheter and are one of the most common hospital-acquired infections (CDC 2014).

1.3.4 Surgical Site Infections (SSI)

Surgical site infection (SSI) explained as an infection that follows after surgery in the body part where the surgery happened and it represents one of the most widespread hospital-associated infections worldwide (CDC 2014). The average occurrence of SSI is 7–15 cases/100 hospitalized patients, whereas in lower-income nations, a higher frequency is observed. The most SSI frequent pathogen includes skin flora-associated *S. aureus* (30%) and coagulase-negative staphylococci (12%). Enterobacteriaceae members—*Escherichia coli, Enterobacter* spp. and *K. pneumoniae* as well as *E. faecalis and P. aeruginosa* were the other identified pathogens belonging to the multi-drug resistant ESKAPE group (Deslouches and Di 2017). In a 5-year prospective study of surgical site infection carried out in southeastern USA, the authors concluded that the incidence frequency of SSIs was higher after colon surgery, peripheral vascular bypass surgery and small bowel surgery, and the major isolated pathogen was *S. aureus* (Baker et al. 2016). Metallo-beta lactamases (MBLs) producing *P. aeruginosa* is frequently found to be associated with SSIs. In a study performed to evaluate the incidence of MBL production by Raouf et al., *P. aeruginosa* had a prevalence rate of 35%. Among the isolated *P. aeruginosa*, imipenem resistance was found in 28.57% and the incidence of MBL-producing isolates among imipenem-resistant *P. aeruginosa* was 85% (Raouf et al. 2018). Helal et al. found that the major pathogen associated with orthopaedic SSI is *S. aureus* with a prevalence percentage of 44.4 and all *S. aureus* isolates were MRSA, followed by *K. pneumoniae* (24.44%). Finally, the occurrence of *Acinetobacter* was 16.67% (Helal et al. 2015).

1.4 Prevalence of ESKAPE Pathogens in Environment

The frequency of clinically relevant ESKAPE pathogens in the milieu is a great concern. These pathogens are resistant to most of the clinical drugs, i.e. antimicrobial-resistant bacteria (AMR). The occurrence of these pathogens in the milieu is due to the illegitimate dumping of hospital waste, sewage spills, dumping of human solid waste and other activities such as bathing, swimming and agricultural waste (Hrenovic et al. 2017; Denissen et al. 2022). Furthermore, the incidence of antimicrobial-resistant organisms and antibiotic resistance genes (ARG) was obtained from the beaches, wastewater system, raw and ready-to-eat foods, water systems including groundwater, drinking water, irrigation water and so on. Moreover, it is found that about 46.4% of the bacteria isolated from sewage, hospitals and pharmaceutical industries have shown resistance towards several antimicrobial drugs (Denissen et al. 2022). Soil is an ideal platform for gene transfer, especially the exchange of resistance genes from the milieu or the animals to the humans. Besides, the contamination of surface water causes the spread of resistant microorganisms through food webs, water supplies and food trade routes across the globe. Through the international food trade web, pathogenic organisms as well as resistant genes were spreading extensively, which are not traceable through visual observation (Pendleton et al. 2013). Precautions should be taken in order to avoid the dissemination of such resistance and resistant organisms. In addition, wild animals were reported as the source of pathogenic microorganisms and resistance, especially birds, which are able to survive and adapt in rural, semi-urban or urban areas. It is believed that resistant organisms were transmitted from human or veterinary origin to the wild birds via contaminated food or water. Birds fly and the bird droppings can act as a carrier for the transmission of resistant bacteria and genetic factors to the environment (Russo et al. 2022).

The advent of resistance causes chronic losses to human health and the economy. Therapeutic failure towards antimicrobial-resistant pathogenic infections causes chronic life-threatening infections. Furthermore, this causes annual economic losses of up to 1.5 billion euros in Europe and five billion dollars in the USA (Pendleton et al. 2013). Due to AMR, more than 700,000 deaths are reported each year globally. In 2050, it was predicted that more than 33,000 mortalities will happen due to antimicrobial drug resistance in the European Union (Arbune et al. 2021). The incidence of antimicrobial drug resistance enterococci was detected in municipal wastewater (including hospital wastewater) even after mechanical and biological treatment in wastewater treatment plants. Anthropogenic activities such as wastewater treatment, agriculture and tourism also take part in the spread of antibiotic-resistant organisms in and outside the hospital milieu and also from the effluent release spot into rivers. Studies revealed that numerous MDR enterococci were released into the river system from the treatment plants (Gotkowska-Płachta 2021). Environmental isolates of *P. aeruginosa* isolated from the sink were suspected as a reservoir of infections in hospitalized patients which may result in outbreaks (Buhl et al. 2015).

Human well-being is closely connected to environmental and animal health. In order to manage the health threat that arises from the human–animal–environment crosslink worldwide, it is necessary to contemplate the One Health approach. This approach could correspondingly reduce the dissemination of antimicrobial resistance across the globe (Russo et al. 2022; Denissen et al. 2022). One health approach is described by the American Veterinary Medical Association (AVMA) as the 'collaborative effort of multiple disciplines, working locally, nationally and globally to attain optimal health for people, animals and environment' (Rubin et al. 2013). Previously, it was thought that zoonosis was a major concern to humans and animals. Later, in 2008, the position of the environment in these infections was highlighted in the One Health Approach, which was documented by different associations like the Food and Agriculture Organization (FAO), World Organization for Animal Health (OIE) and World Health Organization (WHO) (Kim and Cha 2021).

1.5 Antibiotic Resistance in ESKAPE Organisms

There is an increasing concern and curiosity to identify the predisposal factors that promote the advent of antimicrobial drug resistance in ESKAPE organisms. The emergence of such resistant organisms is due to the indiscriminate practice of using antimicrobial agents in humans, animals, agriculture, etc. One of the crucial aspects that steered the upsurge of antimicrobial drug-resistant bacteria is the overexploited use of antimicrobial drugs. Besides human usage, antibiotics were widely used for veterinary and farming purposes. Studies reported the existence of antibiotic resistance strains in animals and foods as well as the presence of antibiotic resistance in the milieu clearly depicted the chance of such resistance in humans (Russo et al. 2022; Nishiyama et al. 2021). Globally, between 2000 and 2010, the usage of antibiotics was improved by 36%, and went up to 45% in the case of carbapenems. In 2030, the number of drugs that are used in animal foods would be estimated around 200,235 tons (Marutescu et al. 2023).

In order to deal with antibiotics, ESKAPE pathogens have acquired several resistance mechanisms, which are detailed below (Fig. 1.1):

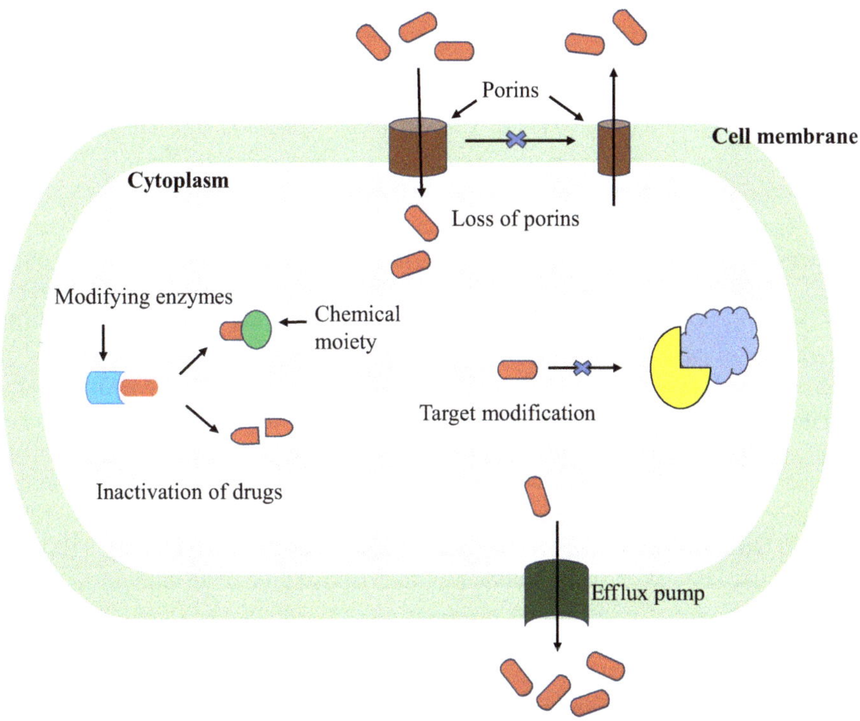

Fig. 1.1 Several resistance mechanisms developed by ESKAPE pathogens to overcome the effects of antibiotics

1.5.1 Alteration/Modification of Drugs

Alteration of drugs by enzymes is one of the most crucial resistance mechanisms shown by antimicrobial-resistant ESKAPE organisms. Enzymes irreversibly alter, inactivate or neutralize the drug molecule. Modifying enzymes are more prevalent in Gram-negative organisms and which include: (1) the enzymes that destroy the active site of antibiotics (hydrolase enzymes) and (2) the enzymes that covalently modify the key sites of the drug molecule (aminoglycoside modifying enzymes) (De Oliveira et al. 2020).

1.5.1.1 Hydrolase Enzymes

β-Lactamases are the enzymes widely produced by pathogens, which can cleave the amide bond of the β-lactam ring. β-Lactam antibiotics include penicillin, cephamycin, carbapenem, cephalosporin and monobactam. They were known shortly after the discovery and purification of penicillin. These enzymes were commonly seen in Gram-negative ESKAPE pathogens, which serve as a key resistance mechanism against antibiotics. β-Lactamase is concentrated in the periplasmic region of the bacterial cell and can cleave the antibiotics without binding to the penicillin-binding proteins (PBP) present in the cell wall (De Oliveira et al. 2020).

β-Lactamases were classified into two: based on the (1) molecular structure (Ambler scheme) and (2) function (Bush–Jacoby system). According to the Ambler scheme, lactamases are categorized into four classes, namely A, B, C and D. In class A enzymes, the active site contains a serine group, which includes penicillinases, narrow or broad spectrum β-lactamases, extended spectrum β-lactamases (ESBLs), cephalosporins and carbapenemases. These enzymes act on drugs such as penicillins, early cephalosporins, third-generation oxyimino-cephalosporins, monobactams, cephamycins and carbapenems (De Oliveira et al. 2020; Jadimurthy et al. 2022). Class A were described in Gram-negative and positive pathogens. Subsequently, the application of penicillin to treat various infections led to the emergence of *blaZ*-encoded penicillinases and was identified in almost 85% of *S. aureus* clinical isolates and in some *Enterococcus* sp. Some of the ESBL family lactamases are TEM (Temoniera), SHV (sulphydral reagent variable), CTX (cefotaximase from Munich), GES (Guiana extended spectrum β-lactamase), VEB (Vietnam extended-spectrum β-lactamase), etc. All Gram-negative ESKAPE organisms produce CTX-M, PER (Pseudomonas extended resistant), GES and VEB. Most of the ESBL enzymes can be repressed by clavulanic acid excluding TEM-30, SHV-10 and TEM-50. *K. pneumoniae* was described as producing inhibitor-resistant lactamase enzyme, *Klebsiella pneumonaiae* carbapenemase (KPC) is able to deteriorate all β-lactams including carbapenems (Jadimurthy et al. 2022; De Oliveira et al. 2020; Santajit and Indrawattana 2016).

The Class B Ambler group consists of metallo-β-lactamase (MBL) with Zn^{+} ions in their active sites, which is predominant in Gram-negative ESKAPE organisms. Microorganisms such as *P. aeruginosa, K. pneumoniae, E. cloacae* and *Acinetobacter* sp. possess IMP (Imipenemase)- and VIM (Verona integron-encoded MBL)-type

MBLs. Another type of group B lactamases found in the majority of Gram-negative bacteria are NDM (New Delhi metallo β-lactamase) type MBLs; these are incorporated into the transposons and encode genes accountable for resistance to other antimicrobial drugs (Jadimurthy et al. 2022).

Class C Ambler group includes enzymes such as penicillinase and cephalosporinase (for instance, AmpC β-lactamase), which show resistance towards narrow to intermediate spectrum cephalosporins and aztreonam. This type was recognized in *Enterobacter* sp., *P. aeruginosa* and *Acinetobacter* sp. The Class D group composed of oxacillin-hydrolysing enzymes (OXA) was reported to inactivate oxacillin, also its derived products, which portrays properties similar to ESBL and shows resistance against β-lactam inhibitors. The prevalent members belonging to this class are OXA-11, OXA-14 and OXA-16, which are commonly seen in *P. aeruginosa*. Besides, OXA-type carbapenemases are generally detected in *Acinetobacter* sp. (Jadimurthy et al. 2022; De Oliveira et al. 2020; Santajit and Indrawattana 2016).

1.5.1.2 Aminoglycoside-Modifying Enzymes

Resistance to aminoglycosides in ESKAPE pathogens is due to the secretion of aminoglycoside-modifying enzyme (AME). Aminoglycosides have broad-spectrum activity by inhibiting protein synthesis via binding to subunits of ribosomes. Genes encoded for AME production were located on the plasmid, transposons and rarely in the DNA of some bacteria. AMEs are grouped according to their function (chemical alterations) and they are aminoglycoside acetyl-, phospho-, and nucleotidyl-transferases. Aminoglycoside acetyltransferases (AACs) represent the largest class of modifying enzymes, linked in the acetylation of NH_2 groups of antibiotics, and were divided into four sub-classes. AAC (1) and AAC (2) subclasses participated in the addition of the acetyl group of 2-deoxystreptamine of aminoglycosides at first and third positions, respectively. AAC (2′) and AAC (6′) subclasses aid in the acetylation of the 2,6-dideoxy-2,6-diamino-glucose group at 2′ and 6’ NH_2 groups of the antibiotics. According to epidemiological studies in Europe, the USA and Asia, the prevalence of AAC (3) and AAC (6′) modifying enzymes were reported in Gram-negative ESKAPE pathogens, and provide resistance towards antibiotics such as gentamicin, amikacin and tobramycin. Aminoglycoside phosphotransferases (APHs) being the second major class of AMEs take part in the phosphorylation of the -OH group and inhibit hydrogen bond formation between antibiotics and the target. It has been further divided into seven sub-classes, among them, APH (3′) phosphorylates amikacin and was found in *S. aureus* and *Enterococcus* sp. Aminoglycoside nucleotidyltransferases (ANTs) represent the third largest class of modifying enzymes, aid in the addition of nucleotide monophosphate at 2″, 3″, 4′,6 and ninth positions of the OH moiety of aminoglycosides. Out of these, ANT (2″) and ANT (4′) were secreted by *K. pneumoniae*; the former showed resistance towards 4,6-di-substituted aminoglycoside while the latter showed resistance towards kanamycin A/B/C, amikacin, gentamicin A, tobramycin and neomycin B/C (Jadimurthy et al. 2022; De Oliveira et al. 2020).

1.5.2 Alteration of Drug Target Sites

It is another important strategy to overcome the binding of drugs to target sites. Here, the drug target site is modified and decreases the binding affinity of the antibiotics. It includes (1) the alteration of the target enzyme, (2) the modification of the ribosomal target site and (3) modifications in cell wall precursors (De Oliveira et al. 2020).

1.5.2.1 Target Enzyme Alteration

Alteration in the penicillin-binding proteins is one of the resistance mechanisms to overcome β-lactam antibiotics. These antibiotics bind to the PBP present in the cell wall and inhibit bacteria. *mecA* encodes for PBP2a, which has minimal affinity towards the β-lactam class. MRSA possesses PBP2a type of PBP, where the organism showed resistance against methicillin and β-lactam antibiotics. Thus, rendering the survival of MRSA in high concentrations of β-lactam drugs or methicillin. Around 90% of the nosocomial pathogen, *E. faecium* showed resistance towards ampicillin at higher levels due to the overexpression of PBP5 (an orthologue of PBP2a in MRSA, confers resistance to β-lactams from low to moderate levels) or polymorphism in PBP5 (De Oliveira et al. 2020; Santajit and Indrawattana 2016).

1.5.2.2 Modification of Ribosomal Target

Several antibiotics can target bacterial ribosomal subunits such as 50S and 30S. Antibiotic classes such as aminoglycoside binds with 16S rRNA of the 30S unit and blocks the tRNA interaction with anticodon, which finally leads to the inhibition of translation. In order to overcome the bactericidal effects of aminoglycosides, methylation modification at the N7-guanine residue or N1-adenine of 16S rRNA by bacterial 16S rRNA methyltransferases will take place (Jadimurthy et al. 2022). Recently, studies reported that the methylation of 16S rRNA by enzymes provides resistance to all aminoglycosides at higher levels, an attained mechanism in Gram-negative pathogens (De Oliveira et al. 2020).

1.5.2.3 Modifications in Cell Wall Precursors

Glycopeptides can hinder cell wall synthesis in Gram-positive bacteria. They aim for the acyl-D-alanyl-D-alanine (acyl-D-Ala-D-Ala) moieties of the peptidoglycan. In *E. faecium* and *E. faecalis*, change in the peptidoglycan crosslinks such as D-Ala-D-Ala to D-Ala-D-Lac or D-Ala-D-Ser encoded by cluster of genes such as *Van-A, Van-B, Van-D, Van-C, Van-E* and *Van-G* can enhance the resistance to glycopeptides (Santajit and Indrawattana 2016). Alteration in the Lipopolysaccharide (in lipid A residue) causes polymyxin resistance in bacteria, due to the genes present in PmrA/PmrB, PhoP/PhoQ, ParR/ParS, ColR/ColS or the CprR/CprS two-component pathway or plasmid-facilitated *mcr* genes (Zhu et al. 2022).

1.5.3 Inhibition of Drug Influx and Accumulation

Maximum efficacy of the drug molecule is attained only when a certain quantity of the molecule is present inside the bacterial cell. Microorganisms have tolerance to most antibiotics by modifying the channels and through efflux pumps. In Gram-negative bacteria, channels called porins span the outer membrane and aid in the transportation of substances in and out of the cell. In *A. baumanii*, loss of outer membrane (OM) porin proteins (29 kDa) makes them resistant towards carbapenem drugs such as imipenem and meropenem. Loss of OM proteins, OmpK35 and OmpK36 in *K. pneumoniae* causes decreased susceptibility towards β-lactams, for instance, cephalosporins and carbapenems. The loss or modification of porin, OprD, gives rise to carbapenem-resistant *P. aeruginosa* (Santajit and Indrawattana 2016; De Oliveira et al. 2020).

1.5.3.1 Efflux Pumps

These are the membrane transport channels intended for the pumping of antibiotics from the cells. These proteins eject the antimicrobial agents at a higher rate from the cells, causing inappropriate amounts of antibiotics to elicit antagonistic effects. Most of them are multidrug transporters and include: the ATP-binding cassette (ABC) family, small multidrug resistance family (SMR), major facilitator superfamily (MFS), resistance nodulation division (RND) family, multidrug and toxic compound extrusion family (MATE) and proteobacterial antimicrobial compound efflux (PACE). Some of the important genes involved in each efflux pump family are: *acrB, mdtF* in the RND family; *bcr, cmr* in MFS; *mdtK, yeeO* in MATE; *emrE, ydgE* in SMR and *macB* in ABC. In Gram-negative bacteria, RNDs are the most principal efflux pump family intricated in drug resistance. RND includes MexAB-OprM, MexCD-OprJ, MexEF-OprN, MexXY and MuxBC-OpmB. In *P. aeruginosa*, these efflux pumps were implicated in the exclusion of antibiotics such as β-lactams, aminoglycosides, chloramphenicol, fluoroquinolones, tetracyclines, quinolones, novobiocin and macrolides. *P. aeruginosa* is relatively resistant to tigecycline, but a deficit of MexXY (OprM) causes improved susceptibility to tigecycline (Santajit and Indrawattana 2016; Zhu et al. 2022; Jadimurthy et al. 2022). *A. baumanii* shows the MDR phenomenon through the overexpression of the RND pump, AdeABC and elicits resistance towards a wide spectrum of drugs including fluoroquinolones, β-lactams, tetracyclines, macrolides/lincosamides, chloramphenicol and aminoglycosides. RND-efflux pumps in all bacterial species including AdeABC, AdeDE, AdeFGH and AdeIJK showed resistance towards aminoglycosides, fluoroquinolones, tetracycline, chloramphenicol and erythromycin (Santajit and Indrawattana 2016).

1.5.4 Transmission of Resistant Strains and ARGs

Antimicrobial resistance is currently a global concern. Resistance is due to spontaneous mutation, or via gene transfer including conjugation, transformation and transduction. Bacteria acquire resistance to antibiotics through the constant usage of antimicrobials, which leads to the advent of antibiotic-resistant bacteria (ARB) (Russo et al. 2022). Antimicrobials were extensively used by humans as well as for veterinary and agricultural purposes. Therefore, there is an abundance of ARB and antibiotic-resistant genes (ARG) in human and animal faeces. The human microbiome can be modified and augmented by ARB, hence, considered as a reservoir of ARB and ARGs. This is adapted via hospital- and community-based pathways. In hospital-based routes, the resistant bacteria present in the hospitals are primarily linked with nosocomial infections. These pathogens mainly belong to the ESKAPE family, consisting of both Gram-negative and -positive bacteria, which are highly resistant to most antibiotics and cause life-threatening illness. Bacteria shows resistance to last resort drugs such as fluoroquinolone, carbapenems, glycopeptides and third-generation cephalosporins, which have been reported in both hospital-attained and community-attained infections (Nishiyama et al. 2021). The transmission routes of ARGs and ARBs are shown in Fig. 1.2.

High sequence similarities were reported in environmental ARGs with that of human faecal microbiota. Soil became the largest environmental source of antibiotic resistance, due to agricultural methods, animal husbandry and manures. The

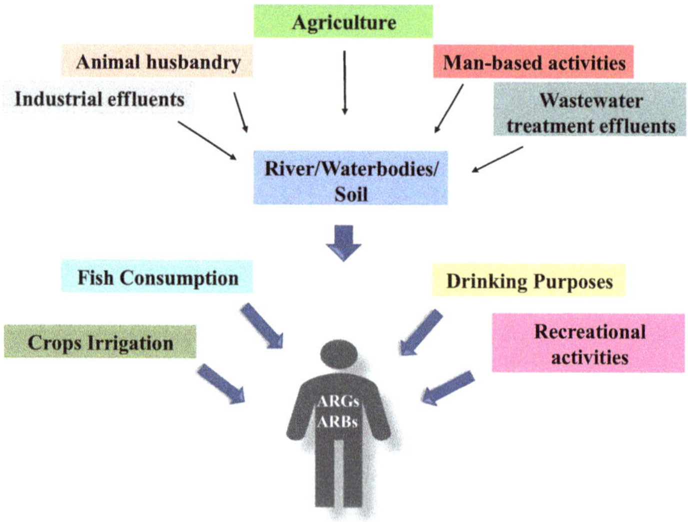

Fig. 1.2 Various transmission routes of ARGs and ARBs to humans. Through industrial effluents, animal rearing, agriculture, man-based activities and wastewater ARBs and ARGs enter into aquatic systems and soil. From there, water used for drinking, farming, washing foods, and recreational activities reaches humans

application of manure for agricultural purposes into the soil is an entry route for ARGs into the environment. Studies reported that the process of composting manures can reduce antibiotic resistance considerably, but still, there is a chance of spreading resistome from manure to the environment. Indigenous microorganisms acquire resistance genes from the manure containing ARBs through horizontal gene transfer. The diverseness of ARGs and transposons in the manure causes the spread of ARGs in agricultural ecosystems (Zhang et al. 2019).

Similar to soil, ARGs and ARBs were reported in aquatic ecosystems, showing a higher movement of ARBs and transposons. Disinfection using chlorine in water treatments or sanitation can reduce the abundance of ARGs. However, exposure to chlorine enhances cell membrane permeability and facilitates the transfer of resistance genes. The routes of transmission of AMR in humans are via drinking, bathing, foods, washing foods by running water, swimming pool, etc. A survey of antibiotic resistance in tap water collected from 25 cities in 7 countries reported the presence of 181 ARG subtypes belonging to 16 ARG types (Ma et al. 2022). The load of ARGs in human faeces, skin and effluents from household sewage were nearly 23, 2 and 7 times more than the plenitude in rivers. Around 53 ARGs and 28 bacteria found in the human faeces were obtained from the influent and effluent of the rural sewage system and from the downstream effluent discharge site (Zhou et al. 2018). In a study, the river water samples obtained from upstream and downstream of wastewater discharge points revealed the existence of multidrug-resistant *Enterococci*. Strains obtained from wastewater and the downstream water contained *van* genes resistant to vancomycin (57%), and also *van*C1 genes (27.6%) observed in treated municipal wastewater. The release of pathogenic *Enterococci* into the environment causes critical environmental issues for public health (Gotkowska-Płachta 2021). The data obtained from 3482 isolates of carbapenem-resistant *Klebsiella* through the One Health Approach across Italy showed that the milieu is accountable for 0.21% of human ARB illness (Leonard et al. 2022).

Whole genome sequencing (WGS), a method to recognize the causes of resistance in microorganisms using in silico methods, includes read-based (Genefinder), de Bruijn graph-based (Mykrobe) and BLAST-based (Typewriter). Through this approach, the surveillance of AMR is possible with less time and low cost. Some of the genes intricated in the resistance towards β-lactam drugs are $bla_{\mathrm{OXA\text{-}51}}$, $bla_{\mathrm{OXA\text{-}23}}$, $bla_{ADC\text{-}25}$, $bla_{ADC\text{-}73}$, $bla_{TEM\text{-}1}$ and $bla_{NDM\text{-}1}$. The genes that impart resistance to aminoglycosides in ESKAPE pathogens are *ant(3″)-IIa*, *aph(3″)-Ib*, *armA*, *aph(6)-Id* and *aph(3′)-Ia*. Some of the genes like *mphE, msrE* and *abaF* showed resistance to macrolides and lincosamides in bacteria. In *Enterococcus, tetM* and *tetL* genes are responsible for tetracycline resistance; *ermB* encode for erythromycin resistance; *vanA, vanB* and *vanM* genes encode for vancomycin resistance (Priyamvada et al. 2022). Genes implicated in drug resistance in *S. aureus* are *blaZ* which encode for penicillin resistance, *mecA* against methicillin, *msrA* for erythromycin, *ermA, ermB, ermC and ermT* towards erythromycin and clindamycin, *tetK, tetL* and *tetM* towards tetracycline resistance, *vanA* for vancomycin, *fusB* towards fusidic acid, *dfrA* and *dfrG* against trimethoprim and *aacA-aphD* towards gentamicin (Gordon et al. 2014).

1.6 Current Strategies to Compete Against ESKAPE Pathogens

In the above sections, we extensively discussed and came to know about the alarming prevalence of MDR ESKAPE pathogens in the environment. As per the last update of WHO, by 2050, drug-resistant infections are envisioned to result in ten million deaths/year and by 2030, it is anticipated that AMR will cause extreme poverty to 24 million people (WHO 2019). Successful development of new antibiotics will take many years. However, the situation is demanding urgent efficient antimicrobial strategies. Researchers and scientists from all around the world are looking for alternative, safer, novel and cost-effective antimicrobial modalities to fight against the current situation. Bacteriophage therapy, combination therapies, photodynamic therapy and therapy using antimicrobial peptides and nanomaterial are various current strategies introduced. An elaborate discussion of each technology is beyond the scope of this chapter, so we can concisely describe each technique. A few examples of various strategies introduced during the last 10 years are concluded in Table 1.1.

Table 1.1 Examples of current strategies to fight against drug-resistant ESKAPE pathogens

Sl. no.	Strategy	Brief description	Strategy-sensitive ESKAPE pathogen	Reference
1	Nanotherapy	Chemically synthesized silver nanoparticles (AgNPs)	MDR *K. pneumoniae* MDR *A. baumannii*	Siddique et al. (2020) and Hetta et al. (2021)
		Nanoemulsion of *Thymus daenensis* oil	MDR *A. baumannii*	Moghimi et al. (2018)
		Green-synthesized silver nanoparticles (AgNPs)	MDR *P. aeruginosa*, MDR *S. aureus*	Kasithevar et al. (2017)
		Amphiphilic silver nanoclusters (AgNCs)	MDR *P. aeruginosa*	Chen et al. (2020)
		Green synthesized copper oxide nanoparticles (CuO NPs)	MDR *K. pneumoniae*	Naseer et al. (2021)
		Copper sulphide nanozymes anchored to graphene oxide nanosheets (CuS/GO NC)	Methicillin-resistant *S. aureus* (MRSA)	Wang et al. (2020)
		Chitosan nano-carrier systems loaded with imipenem	*MDR A. baumannii*	Mufti et al. (2023)

(continued)

Table 1.1 (continued)

Sl. no.	Strategy	Brief description	Strategy-sensitive ESKAPE pathogen	Reference
2	Antimicrobial photodynamic therapy (aPDT)	Treatment using phenothiazinium dyes (Toluidine Blue O, Azure A and New Methylene Blue) exposed to 630 nm of light	*E. faecalis*, *K. pneumoniae*	Misba et al. (2017)
		Sulphonated polystyrene nanoparticles encapsulated with tetraphenylporphyrin (TPP-NP) photosensitizers and LED-based light source (414 nm)	Methicillin-resistant *S. aureus* (MRSA), *Enterococcus faecalis*	Malá et al. (2021)
		Menadione as a photosensitizer with blue light (405 nm) irradiation	Methicillin-resistant *S. aureus* (MRSA), *P. aeruginosa*	Negri et al. (2023)
		Methylene blue as PS and red light	*A. baumannii*, *E. faecium*, *E. faecalis*, *S. aureus*, *K. aerogenes*, *K. pneumoniae*, *P. aeruginosa*	Sabino et al. (2020)
		Toluidine blue as PS and LED with 630 emission combination therapy with colistin	*A. baumannii*	Boluki et al. (2017)
		Red-carbon dots (R-CDs) with a 350–700 nm range of spectral absorption	*Multidrug-resistant A. baumannii*, multidrug-resistant *S. aureus*	Liu et al. (2022)
		Aloe emodin as PS Xenon lamp with 435 ± 10 nm light	*Multidrug-resistant A. baumannii*	Li et al. (2020)
		Using methylene blue and gallium arsenide aluminium laser with a wavelength of 660 nm	*E. faecium* *E. faecalis*	Prado et al. (2017)
		Methylene blue and 633 nm red light LED	*XDR P. aeruginosa* *MDR K. pneumoniae*	Songsantiphap et al. (2022)

(continued)

Table 1.1 (continued)

Sl. no.	Strategy	Brief description	Strategy-sensitive ESKAPE pathogen	Reference
3	Antimicrobial peptide (AMPs)	Synthetic antimicrobial and antibiofilm peptides (SAAPs) SAAP-148	*Methicillin-resistant S. aureus (MRSA), MDR A. baumannii*	de Breij et al. (2018)
		Synthetic peptides DJK-5, DJK-6 and 1018	*MDR Carbapenemase producing K. pneumoniae*	Ribeiro et al. (2015)
		Short novel peptides—WR12, D-IK8	Methicillin-resistant *S. aureus* (MRSA) vancomycin-resistant *S. aureus*	Mohamed et al. (2016)
		ZY4, a cyclic peptide	MDR *P. aeruginosa* and *A. baumannii*	Mwangi et al. (2019)
		Optimized analogue of AMP jelleine-1 (Analogue 15)	MDR *P. aeruginosa*	Zhou et al. (2021)
		Short bovine lactoferrin-derived antimicrobial peptides	MDR *E. faecium*	Mishra et al. (2022)
		Tridecaptin-inspired antimicrobial peptide- $TriA_2$ (2, 8-D-Orn, 7-Orn)	MDR *A. baumannii, K. pneumoniae, E. cloacae*	Ballantine et al. (2019)
4	Bacteriophage therapy	*Pseudomonas*-targeting phage PEV31 delivered by the pulmonary route in a mouse lung infection model	MDR *P. aeruginosa*	Chang et al. (2022)
		Use of newly isolated phage, MMI-Ps1, which possesses strong lytic activity confirmed by *P. aeruginosa* mouse lung infection model	MDR *P. aeruginosa*	Abd El-Aziz et al. (2019)
		Bacteriophage ZCKP1, isolated from freshwater	*K. pneumoniae*	Taha et al. (2018)
		Treatment with either one or a combination of two lytic phages—Pharr (P1), a 40.6-kb podophage, and ϕKpNIH-2 (P2), a 49.4-kb siphophage	MDR *K. pneumoniae*	Hesse et al. (2021)
		ϕkm18p phage therapy in BALB/c and C57BL/6 mice models of XDR-*A. baumannii* bacteremia	XDR *A. baumannii*	Wang et al. (2018)
		Therapy using magistral preparations of two *Enterococcus* phages EFgrKN and EFgrNG in a 1-year-old child following a third liver transplantation	Vancomycin-resistant *E. faecium*	Paul et al. (2021)

(continued)

Table 1.1 (continued)

Sl. no.	Strategy	Brief description	Strategy-sensitive ESKAPE pathogen	Reference
5	Antibiotic adjuvants	Use of bis-2-aminoimidazole adjuvants with macrolide antibiotics	MDR *P. aeruginosa*	Hubble et al. (2018)
		Use of brominated carbazoles with β-lactam antibiotics	Methicillin-resistant *S. aureus* (MRSA)	Berndsen et al. (2022)
		Use of two non-antimicrobial peptides as ampicillin and oxacillin adjuvants	Methicillin-resistant *S. aureus* (MRSA)	Rishi et al. (2018)
		Use of plant secondary metabolite –linalool with antibiotic meropenem	Carbapenemase-producing *K. pneumoniae*	Yang et al. (2021)
		Bacteriophage-derived depolymerase Dpo71 with colistin	MDR *A. baumannii*	Chen et al. (2022)
		Synthetic nitrophenol compound, 2-(4-chlorophenyl)-6-methyl-4-nitrophenol, with polymyxin B	MDR *A. baumannii*	Kim et al. (2023)
		Derivatives of polyaminoisoprene with erythromycin, doxycycline, nalidixic acid and chloramphenicol	MDR *E. aerogenes*	Lieutaud et al. (2020)
6	Drug repurposing	Apramycin, aminocyclitol aminoglycoside-veterinary drug	MDR *A. baumannii* MDR *P. aeruginosa*	Kang et al. (2017)
		Anti-viral drug zidovudine (AZT) repurposing in combination with meropenem	MDR *K. pneumoniae*	DeSarno et al. (2020)
		Repurposing anti-viral zidovudine in synergistic combination with fosfomycin	MDR Enterobacterales	Antonello et al. (2021)
		Repurposing of antiprotozoal drug pentamidine and anti-rheumatic drug auranofin	MDR *A. baumannii*, *K. pneumoniae*	Yu et al. (2022)
		Repurposing thiram and disulfiram	MDR *S. aureus*	Long (2017)
		Repurposing Eltrombopag—Drug used in the treatment of thrombocytopenia	MDR *S. aureus*	Lee et al. (2021)

Although nanotherapeutics were introduced a long time ago, advanced research in the field obviously contributed a revolutionary action to combat infections in this post-antibiotic era. Owing to their unique physicochemical properties, different nanoscale materials like nanoparticles, nanotubes, nanofibres and various other designed nanoplatforms proved their effectiveness in the treatment of infections especially by multidrug-resistant strains. Moreover, these nanostructured compounds are widely adopted in the field of drug delivery (Zaidi et al. 2017). Antimicrobial photodynamic therapy (aPDT) is another modality currently accepted as a therapeutic alternative to various infections. It uses the combination of three components – photosensitizer (PS), light and oxygen to disrupt metabolically active microbial cells. Excitation by light leads to the production of triplet-excited PS, followed by cytotoxic reactive oxygen species (ROS) generation. The advantage of aPDT is that because it has different cellular targets, it is not likely to develop resistance (Nakonieczna et al. 2019). Antibiotic adjuvants are compounds that augment the activity of existing drugs and can decrease or halt resistance mechanisms. It can also enhance antibiotic activity by boosting host immune responses. Because it is delivered together with antibiotics and is considered as a combination drug (Wright 2016). Sometimes the adjuvant itself may have little antimicrobial property, whereas some adjuvants do not have any antimicrobial property on their own. Drug repurposing, in which existing approved drugs for one use are repurposed for another use is a promising strategy to combat drug-resistant organisms. For example, an antidiabetic drug metformin can be repurposed to treat *P. aeruginosa* infections (Abbas et al. 2017). This novel cost-effective strategy has several advantageous features, which will accelerate the process of drug development as their safeties are already approved and the formulation studies have already been accomplished, which reduces the cost and the time required for drug development. Apart from using a repurposing drug as an exclusive drug against a certain type of infection, it can also be used as an antibiotic adjuvant for existing drugs of choice as shown by She et al. They confirmed the adjuvant potential of FDA-approved, 9-aminoacridine, in combination with antibiotic rifampin for the treatment of infection with extensively drug-resistant and pan-drug-resistant *K. pneumoniae* (She et al. 2023).

1.7 Conclusion

ESKAPE organisms are a group of pathogens that account for the majority of nosocomial infections worldwide. These organisms now are not responding to various classes of antibiotics, including last-resort antibiotics. Because of their prevalence and antibiotic resistance, these pathogens are included in World Health Organization's priority level pathogens which are urgently demanding new antimicrobial strategies or drugs. Currently, there are various efficient strategies to treat infectious diseases introduced by scientists and researchers across the world. It includes combination therapy, antimicrobial adjuvants, photodynamic therapy, nanotherapies, drug repurposing, etc. But there are now limitations in these technologies to reach in translation medicine. Hence, future research should focus on solving the limitations and work on bringing these novel modalities into real-life applications.

References

Abbas HA, Elsherbini AM, Shaldam MA (2017) Repurposing metformin as a quorum sensing inhibitor in *Pseudomonas aeruginosa*. Afr Health Sci 17(3):808–819

Abd El-Aziz AM, Elgaml A, Ali YM (2019) Bacteriophage therapy increases complement-mediated lysis of bacteria and enhances bacterial clearance after acute lung infection with multidrug-resistant *Pseudomonas aeruginosa*. J Infect Dis 219(9):1439–1447

Alfouzan W, Dhar R, Abdo NM, Alali WQ, Rabaan AA (2021) Epidemiology and microbiological profile of common healthcare associated infections among patients in the intensive care unit of a general hospital in Kuwait: a retrospective observational study. J Epidemiol Glob Health 11(3):302

Al-Gethamy MM, Faidah HS, Adetunji HA, Haseeb A, Ashgar SS, Mohanned TK, Mohammed A-H, Khurram M, Hassali MA (2017) Risk factors associated with multi-drug-resistant *Acinetobacter baumannii* nosocomial infections at a tertiary care hospital in Makkah, Saudi Arabia—a matched case-control study. J Int Med Res 45(3):1181–1189

Alsharedeh RH, Yehya A, Yonis OB, Alameri O, Alshraiedeh N (2023) Characterization of ESKAPE pathogens in urinary tract infections among Jordanian patients. J Infect Dev Ctries 17(03):374–380

AMRSN (2021) Annual Report 2021. https://main.icmr.nic.in/sites/default/files/upload_documents/AMR_Annual_Report_2021.pdf

Antonello RM, Di Bella S, Betts J, La Ragione R, Bressan R, Principe L, Morabito S, Gigliucci F, Tozzoli R, Busetti M, Knezevich A, Furlanis F, Fontana F, Luzzaro F, Luzzati R, Lagatolla C (2021) Zidovudine in synergistic combination with Fosfomycin: an in vitro and in vivo evaluation against multidrug-resistant Enterobacterales. Int J Antimicrob Agents 58(1):106362

Arbune M, Gurau G, Niculet E, Iancu AV, Lupasteanu G, Fotea S, Vasile MC, Tatu AL (2021) Prevalence of antibiotic resistance of ESKAPE pathogens over five years in an infectious diseases hospital from south-east of Romania. Infect Drug Resist 14:2369–2378

Asokan GV, Ramadhan T, Ahmed E, Sanad H (2019) WHO global priority pathogens list: A bibliometric analysis of medline-pubmed for knowledge mobilization to infection prevention and control practices in Bahrain. Oman Med J 34(3):184–193

Ayobami O, Willrich N, Harder T, Okeke IN, Eckmanns T, Markwart R (2019) The incidence and prevalence of hospital-acquired (carbapenem-resistant) *Acinetobacter baumannii* in Europe, eastern Mediterranean and Africa: a systematic review and meta-analysis. Emerg Microb Infect 8(1):1747–1759

Baker AW, Dicks KV, Durkin MJ, Weber DJ, Lewis SS, Moehring RW, Chen LF, Sexton DJ, Anderson DJ (2016) Epidemiology of surgical site infection in a community hospital network. Infect Control Hosp Epidemiol 37(5):519–526

Ballantine RD, McCallion CE, Nassour E, Tokajian S, Cochrane SA (2019) Tridecaptin-inspired antimicrobial peptides with activity against multidrug-resistant gram-negative bacteria. Med Chem Commun 10(3):484–487

Bardi T, Pintado V, Gomez-Rojo M, Escudero-Sanchez R, Azzam Lopez A, Diez-Remesal Y, Martinez Castro N, Ruiz-Garbajosa P, Pestaña D (2021) Nosocomial infections associated to COVID-19 in the intensive care unit: clinical characteristics and outcome. Eur J Clin Microbiol 40(3):495–502

Benkő R, Gajdács M, Matuz M, Bodó G, Lázár A, Hajdú E, Papfalvi E, Hannauer P, Erdélyi P, Pető Z (2020) Prevalence and antibiotic resistance of ESKAPE pathogens isolated in the emergency department of a tertiary care teaching hospital in Hungary: a 5-year retrospective survey. Antibiotics 9(9):624

Berndsen R, Cunningham T, Kaelin L, Callender M, Boldog WD, Viering B, King A, Labban N, Pallock JA, Miller HB, Blackledge MS (2022) Identification and evaluation of brominated carbazoles as a novel antibiotic adjuvant scaffold in MRSA. ACS Med Chem Lett 13(3):483–491

Boluki E, Kazemian H, Peeridogaheh H, Alikhani MY, Shahabi S, Beytollahi L, Ghorbanzadeh R (2017) Antimicrobial activity of photodynamic therapy in combination with colistin against a pan-drug resistant *Acinetobacter baumannii* isolated from burn patient. Photodiagnosis Photodyn Ther 18:1–5

Buhl M, Peter S, Willmann M (2015) Prevalence and risk factors associated with colonization and infection of extensively drug-resistant *Pseudomonas aeruginosa*: a systematic review. Expert Rev Anti Infect Ther 13(9):1159–1170

Cabrera R, Fernández-Barat L, Motos A, López-Aladid R, Vázquez N, Panigada M, Álvarez-Lerma F, López Y, Muñoz L, Castro P, Vila J, Torres A (2020) Molecular characterization of methicillin-resistant *Staphylococcus aureus* clinical strains from the endotracheal tubes of patients with nosocomial pneumonia. Antimicrob Resist Infect Control 9(1):43

CARSS (2022) Canadian antimicrobial resistance surveillance system (CARSS) report, 2022. Public Health Agency of Canada, Ottawa, ON

CDC (2014). Available from https://www.cdc.gov/hai/infectiontypes.html. Accessed 20 Jul 2023

CDC (2019) Antibiotic resistance threats in the United States, 2019. U.S. Department of Health and Human Services, CDC, Atlanta, GA

Chan JD, Graves JA, Dellit TH (2010) Antimicrobial treatment and clinical outcomes of carbapenem-resistant *Acinetobacter baumannii* ventilator-associated pneumonia. J Intensive Care Med 25(6):343–348

Chang RYK, Chow MY, Wang Y, Liu C, Hong Q, Morales S, McLachlan AJ, Kutter E, Li J, Chan HK (2022) The effects of different doses of inhaled bacteriophage therapy for *Pseudomonas aeruginosa* pulmonary infections in mice. Clin Microbiol Infect 28(7):983–989

Chen Y, Ren L, Sun L, Bai X, Zhuang G, Cao B, Hu G, Zheng N, Liu S (2020) Amphiphilic silver nanoclusters show active nano–bio interaction with compelling antibacterial activity against multidrug-resistant bacteria. NPG Asia Mater 12(1):56

Chen X, Liu M, Zhang P, Xu M, Yuan W, Bian L, Liu Y, Xia J, Leung SS (2022) Phage-derived depolymerase as an antibiotic adjuvant against multidrug-resistant *Acinetobacter baumannii*. Front Microbiol 13:845500

Chiang C-H, Pan S-C, Yang T-S, Matsuda K, Kim HB, Choi YH, Hori S, Wang J-T, Sheng W-H, Chen Y-C, Chang F-Y, Chang SC (2018) Healthcare-associated infections in intensive care units in Taiwan, South Korea, and Japan: recent trends based on national surveillance reports. Antimicrob Resist Infect Control 7(1):129

Choo EJ (2017) Community-associated methicillin-resistant *Staphylococcus aureus* in nosocomial infections. Infect Chemother 49(2):158

Coombs GW, Daley DA, Thin Lee Y, Pang S, Pearson JC, Robinson JO, Johnson PD, Kotsanas D, Bell JM, Turnidge JD (2014). Australian group on antimicrobial resistance Australian Enterococcal Sepsis outcome Programme annual report, 2014

DBT (2021) Indian Priority Pathogen List (IPPL) jointly developed by DBT and WHO India Office. https://dbtindia.gov.in/latest-announcement/indian-priority-pathogen-list-jointly-developed-dbt-and-who-india-office#. Accessed 25 Aug 2023

de Breij A, Riool M, Cordfunke RA, Malanovic N, de Boer L, Koning RI, Ravensbergen E, Franken M, van der Heijde T, Boekema BK, Kwakman PH (2018) The antimicrobial peptide SAAP-148 combats drug-resistant bacteria and biofilms. Sci Transl Med 10(423):eaan4044

De Oliveira DM, Forde BM, Kidd TJ, Harris PN, Schembri MA, Beatson SA, Paterson DL, Walker MJ (2020) Antimicrobial resistance in ESKAPE pathogens. Clin Microbiol Rev 33(3):10–1128

Denissen J, Reyneke B, Waso-Reyneke M, Havenga B, Barnard T, Khan S, Khan W (2022) Prevalence of ESKAPE pathogens in the environment: antibiotic resistance status, community-acquired infection and risk to human health. Int J Hyg Environ Health 244:114006

DeSarno AE, Parcell BJ, Coote PJ (2020) Repurposing the anti-viral drug zidovudine (AZT) in combination with meropenem as an effective treatment for infections with multi-drug resistant, carbapenemase-producing strains of *Klebsiella pneumoniae*. Pathog Dis 78(9):ftaa063

Deslouches B, Di YP (2017) Antimicrobial peptides: a potential therapeutic option for surgical site infections. Clin Surg 2:1740

Divatia JV, Abraham B (2018) Multicenter observational study to evaluate epidemiology and resistance patterns of common intensive care unit-infections. Indian J Crit Care Med 22(1):20–26

ECDC (2023) Antimicrobial resistance surveillance in Europe 2023–2021 data. European Centre for Disease Prevention and Control and World Health Organization, Stockholm

Erdem I, Yıldırım I, Safak B, Karaali R, Erdal B, Ardic E, Dogan M, Kardan ME, Kavak C, Sahin Karadil K, Yildiz E, Topcu B, Kiraz N, Arar C (2022) A 5-year surveillance of healthcare-associated infections in a university hospital: a retrospective analysis. SAGE Open Med 10:205031212210917

Feleke T, Eshetie S, Dagnew M, Endris M, Abebe W, Tiruneh M, Moges F (2018) Multidrug-resistant bacterial isolates from patients suspected of nosocomial infections at the University of Gondar Comprehensive Specialized Hospital, Northwest Ethiopia. BMC Res Notes 11(1):602

Foster-Nyarko E, Cottingham H, Wick RR, Judd LM, Lam MM, Wyres KL, Stanton TD, Tsang KK, David S, Aanensen DM, Brisse S (2023) Nanopore-only assemblies for genomic surveillance of the global priority drug-resistant pathogen, *Klebsiella pneumoniae*. Microb Genom 9(2)

Garcia-Clemente M, de la Rosa D, Máiz L, Girón R, Blanco M, Olveira C, Canton R, Martinez-García MA (2020) Impact of *Pseudomonas aeruginosa* infection on patients with chronic inflammatory airway diseases. J Clin Med 9(12):3800

Garnacho-Montero J, Timsit J-F (2019) Managing *Acinetobacter baumannii* infections. Curr Opin Infect Dis 32(1):69–76

Giacobbe DR, Battaglini D, Enrile EM, Dentone C, Vena A, Robba C, Ball L, Bartoletti M, Coloretti I, Di Bella S, Di Biagio A, Brunetti I, Mikulska M, Carannante N, De Maria A, Magnasco L, Maraolo AE, Mirabella M, Montrucchio G, Bassetti M et al (2021) Incidence and prognosis of ventilator-associated pneumonia in critically ill patients with COVID-19: a multicenter study. J Clin Med 10(4):555

Gordon NC, Price JR, Cole K, Everitt R, Morgan M, Finney J, Kearns AM, Pichon B, Young B, Wilson DJ, Llewelyn MJ (2014) Prediction of *Staphylococcus aureus* antimicrobial resistance by whole-genome sequencing. J Clin Microbiol 52(4):1182–1191

Gotkowska-Płachta A (2021) The prevalence of virulent and multidrug-resistant enterococci in river water and in treated and untreated municipal and hospital wastewater. Int J Environ Res Public Health 18(2):563

Helal S, El Anany M, Ghaith D, Rabeea S (2015) The role of MDR-*Acinetobacter baumannii* in orthopedic surgical site infections. Surg Infect (Larchmt) 16(5):518–522

Hesse S, Malachowa N, Porter AR, Freedman B, Kobayashi SD, Gardner DJ, Scott DP, Adhya S, DeLeo FR (2021) Bacteriophage treatment rescues mice infected with multidrug-resistant *Klebsiella pneumoniae* ST258. MBio 12(1):10–1128

Hetta HF, Al-Kadmy IM, Khazaal SS, Abbas S, Suhail A, El-Mokhtar MA, Ellah NHA, Ahmed EA, Abd-Ellatief RB, El-Masry EA, Batiha GES (2021) Antibiofilm and antivirulence potential of silver nanoparticles against multidrug-resistant *Acinetobacter baumannii*. Sci Rep 11(1):10751

Hoang S, Georget A, Asselineau J, Venier AG, Leroyer C, Rogues AM, Thiébaut R (2018) Risk factors for colonization and infection by *Pseudomonas aeruginosa* in patients hospitalized in intensive care units in France. PloS One 13(3):e0193300

Hrenovic J, Durn G, Music MS, Dekic S, Troskot-Corbic T, Skoric D (2017) Extensively and multi drug-resistant *Acinetobacter baumannii* recovered from technosol at a dump site in Croatia. Sci Total Environ 607:1049–1055

Hubble VB, Hubbard BA, Minrovic BM, Melander RJ, Melander C (2018) Using small-molecule adjuvants to repurpose azithromycin for use against *Pseudomonas aeruginosa*. ACS Infect Dis 5(1):141–151

Jadimurthy R, Mayegowda SB, Nayak SC, Mohan CD, Rangappa KS (2022) Escaping mechanisms of ESKAPE pathogens from antibiotics and their targeting by natural compounds. Biotechnol Rep 34:e00728

Jung Y, Lee MJ, Sin HY, Kim NH, Hwang JH, Park J, Choe PG, Park WB, Kim ES, Park SW, Park KU (2012) Differences in characteristics between healthcare-associated and community-acquired infection in community-onset *Klebsiella pneumoniae* bloodstream infection in Korea. BMC Infect Dis 12(1):1–9

Kang CI, Kim SH, Park WB, Lee KD, Kim HB, Oh MD, Kim EC, Choe KW (2004) Bloodstream infections caused by Enterobacter species: predictors of 30-day mortality rate and impact of broad-spectrum cephalosporin resistance on outcome. Clin Infect Dis 39(6):812–818
Kang AD, Smith KP, Eliopoulos GM, Berg AH, McCoy C, Kirby JE (2017) In vitro apramycin activity against multidrug-resistant *Acinetobacter baumannii* and *Pseudomonas aeruginosa.* Diagn Microbiol Infect Dis 88(2):188–191
Kasithevar M, Periakaruppan P, Muthupandian S, Mohan M (2017) Antibacterial efficacy of silver nanoparticles against multi-drug resistant clinical isolates from post-surgical wound infections. Microb Pathog 107:327–334
Kaza P, Britto XB, Mahindroo J, Baker S, Nguyen TNT, Mavuduru RS, Mohan B, Taneja N (2021) Hypervirulent extensively-drug resistant (XDR) *Klebsiella pneumoniae* associated with complicated urinary tract infection in northern India. medRxiv, 2021–05
Khan HA, Baig FK, Mehboob R (2017) Nosocomial infections: epidemiology, prevention, control and surveillance. Asian Pac J Trop Biomed 7(5):478–482
Kim DW, Cha CJ (2021) Antibiotic resistome from the one-health perspective: understanding and controlling antimicrobial resistance transmission. Exp Mol Med 53(3):301–309
Kim SY, Seo HW, Park MS, Park CM, Seo J, Rho J, Myung S, Ko KS, Kim JS, Ryu CM (2023) Exploitation of a novel adjuvant for polymyxin B against multidrug-resistant *Acinetobacter baumannii.* J Antimicrob Chemother 78(4):923–932
Koulenti D, Lisboa T, Brun-Buisson C, Krueger W, Macor A, Sole-Violan J, EU-VAP/CAP Study Group (2009) Spectrum of practice in the diagnosis of nosocomial pneumonia in patients requiring mechanical ventilation in European intensive care units. Crit Care Med 37(8):2360–2369
Kramer TS, Schröder C, Behnke M, Aghdassi SJ, Geffers C, Gastmeier P, Remschmidt C (2019) Decrease of methicillin resistance in *Staphylococcus aureus* in nosocomial infections in Germany—a prospective analysis over 10 years. J Infect 78(3):215–219
Kumar S, Shankar B, Arya S, Deb M, Chellani H (2018) Healthcare associated infections in neonatal intensive care unit and its correlation with environmental surveillance. J Infect Public Health 11(2):275–279
Labovská S (2021) *Pseudomonas aeruginosa* as a cause of nosocomial infections. In: *Pseudomonas aeruginosa*—biofilm formation, infections and treatments. IntechOpen, London
Langford BJ, So M, Raybardhan S, Leung V, Westwood D, MacFadden DR, Soucy J-PR, Daneman N (2020) Bacterial co-infection and secondary infection in patients with COVID-19: a living rapid review and meta-analysis. Clin Microbiol Infect 26(12):1622–1629
Larru B, Gong W, Vendetti N, Sullivan KV, Localio R, Zaoutis TE, Gerber JS (2016) Bloodstream infections in hospitalized children: epidemiology and antimicrobial susceptibilities. J Pediatr Infect Dis 35(5):507–510
Lee H, Lee J, Hwang J, Park S, Kim N, Kim K, Lee H, Shum D, Jang S (2021) Repurposing Eltrombopag for multidrug resistant *Staphylococcus aureus* infections. Antibiotics 10(11):1372
Leonard AF, Morris D, Schmitt H, Gaze WH (2022) Natural recreational waters and the risk that exposure to antibiotic resistant bacteria poses to human health. Curr Opin Microbiol 65:40–46
Li J, Qin M, Liu C, Ma W, Zeng X, Ji Y (2020) Antimicrobial photodynamic therapy against multidrug-resistant *Acinetobacter baumannii* clinical isolates mediated by aloe-emodin: an in vitro study. Photodiagnosis Photodyn Ther 29:101632
Lieutaud A, Pieri C, Bolla JM, Brunel JM (2020) New polyaminoisoprenyl antibiotics enhancers against two multidrug-resistant gram-negative bacteria from Enterobacter and salmonella species. J Med Chem 63(18):10496–10508
Liu W, Gu H, Ran B, Liu W, Sun W, Wang D, Du J, Fan J, Peng X (2022) Accelerated antibacterial red-carbon dots with photodynamic therapy against multidrug-resistant *Acinetobacter baumannii.* Sci China Mater 65(3):845
Long TE (2017) Repurposing thiram and disulfiram as antibacterial agents for multidrug-resistant *Staphylococcus aureus* infections. Antimicrob Agents Chemother 61(9):10–1128
Ma Y, Wang C, Li Y, Li J, Wan Q, Chen J, Tay FR, Niu L (2020) Considerations and caveats in combating ESKAPE pathogens against nosocomial infections. Adv Sci 7(1):1901872

Ma L, Yang H, Guan L, Liu X, Zhang T (2022) Risks of antibiotic resistance genes and antimicrobial resistance under chlorination disinfection with public health concerns. Environ Int 158:106978

Malá Z, Žárská L, Malina L, Langová K, Večeřová R, Kolář M, Henke P, Mosinger J, Kolářová H (2021) Photodynamic effect of TPP encapsulated in polystyrene nanoparticles toward multiresistant pathogenic bacterial strains: AFM evaluation. Sci Rep 11(1):6786

Marturano JE, Lowery TJ (2019) ESKAPE pathogens in bloodstream infections are associated with higher cost and mortality but can be predicted using diagnoses upon admission. Open Forum Infect Dis 6(12):ofz503

Marutescu LG, Popa M, Gheorghe-Barbu I, Barbu IC, Rodríguez-Molina D, Berglund F, Blaak H, Flach CF, Kemper MA, Spießberger B, Wengenroth L (2023) Wastewater treatment plants, an "escape gate" for ESCAPE pathogens. Front Microbiol 14:1193907

Mateescu MC, Grigorescu S, Socea B, Bloanca V, Grigorescu OD (2023) Contribution to the personalized management of the nosocomial infections: a new paradigm regarding the influence of the community microbial environment on the incidence of the healthcare-associated infections (HAI) in emergency hospital surgical departments. J Pers Med 13(2):210

Matta R, Hallit S, Hallit R, Bawab W, Rogues A-M, Salameh P (2018) Epidemiology and microbiological profile comparison between community and hospital acquired infections: a multicenter retrospective study in Lebanon. J Infect Public Health 11(3):405–411

Meatherall BL, Gregson D, Ross T, Pitout JD, Laupland KB (2009) Incidence, risk factors, and outcomes of *Klebsiella pneumoniae* bacteremia. Am J Med 122(9):866–873

Misba L, Zaidi S, Khan AU (2017) A comparison of antibacterial and antibiofilm efficacy of phenothiazinium dyes between gram positive and gram negative bacterial biofilm. Photodiagnosis Photodyn Ther 18:24–33

Mishra B, Felix L, Basu A, Kollala SS, Chhonker YS, Ganesan N, Mylonakis E et al (2022) Design and evaluation of short bovine lactoferrin-derived antimicrobial peptides against multidrug-resistant *Enterococcus faecium*. Antibiotics 11(8):1085

Moghimi R, Aliahmadi A, Rafati H, Abtahi HR, Amini S, Feizabadi MM (2018) Antibacterial and anti-biofilm activity of nanoemulsion of thymus daenensis oil against multi-drug resistant *Acinetobacter baumannii*. J Mol Liq 265:765–770

Mohamed MF, Abdelkhalek A, Seleem MN (2016) Evaluation of short synthetic antimicrobial peptides for treatment of drug-resistant and intracellular *Staphylococcus aureus*. Sci Rep 6(1):29707

Mohd Sazlly Lim S, Zainal Abidin A, Liew SM, Roberts JA, Sime FB (2019) The global prevalence of multidrug-resistance among *Acinetobacter baumannii* causing hospital-acquired and ventilator-associated pneumonia and its associated mortality: a systematic review and meta-analysis. J Infect 79(6):593–600

Mouanga-Ndzime Y, Onanga R, Longo-Pendy NM, Bignoumba M, Bisseye C (2023) Epidemiology of community origin of major multidrug-resistant ESKAPE uropathogens in a paediatric population in south-East Gabon. Antimicrob Resist Infect Control 12(1):47

Mufti IU, Gondal A, Kiyani KM, Mufti SM, Shahid R, Ihsan A, Imran M (2023) Microstructural, physico-chemical, antibacterial and antibiofilm efficacy of imipenem loaded chitosan nanocarrier systems to eradicate multidrug resistant *Acinetobacter baumannii*. Mater Today Commun 35:105874

Mukherjee S, Mitra S, Dutta S, Basu S (2021) Neonatal sepsis: the impact of carbapenem-resistant and hypervirulent *Klebsiella pneumoniae*. Front Med 8:634349

Mulani MS, Kamble EE, Kumkar SN, Tawre MS, Pardesi KR (2019) Emerging strategies to combat ESKAPE pathogens in the era of antimicrobial resistance: a review. Front Microbiol 10:539

Munier A-L, Biard L, Legrand M, Rousseau C, Lafaurie M, Donay J-L, Flicoteaux R, Mebazaa A, Mimoun M, Molina J-M (2019) Incidence, risk factors and outcome of multi-drug resistant *Acinetobacter baumannii* nosocomial infections during an outbreak in a burn unit. Int J Infect Dis 79:179–184

Murray CJ, Ikuta KS, Sharara F, Swetschinski L, Aguilar GR, Gray A, Han C, Bisignano C, Rao P, Wool E, Johnson SC (2022) Global burden of bacterial antimicrobial resistance in 2019: a systematic analysis. Lancet 399(10325):629–655

Mwangi J, Yin Y, Wang G, Yang M, Li Y, Zhang Z, Lai R (2019) The antimicrobial peptide ZY4 combats multidrug-resistant *Pseudomonas aeruginosa* and *Acinetobacter baumannii* infection. Proc Natl Acad Sci U S A 116(52):26516–26522

Nakonieczna J, Wozniak A, Pieranski M, Rapacka-Zdonczyk A, Ogonowska P, Grinholc M (2019) Photoinactivation of ESKAPE pathogens: overview of novel therapeutic strategy. Future Med Chem 11(5):443–461

Naseer M, Ramadan R, Xing J, Samak NA (2021) Facile green synthesis of copper oxide nanoparticles for the eradication of multidrug resistant *Klebsiella pneumonia* and *Helicobacter pylori* biofilms. Int Biodeter Biodegr 159:105201

Navidinia M (2016) The clinical importance of emerging ESKAPE pathogens in nosocomial infections, pp 43–57

Negri LB, Mannaa Y, Korupolu S, Farinelli WA, Anderson RR, Gelfand JA (2023) Vitamin K3 (Menadione) is a multifunctional microbicide acting as a photosensitizer and synergizing with blue light to kill drug-resistant bacteria in biofilms. J Photochem Photobiol B Biol 244:112720

Ng QX, Ong NY, Lee DYX, Yau CE, Lim YL, Kwa ALH, Tan BH (2023) Trends in *Pseudomonas aeruginosa* (*P. aeruginosa*) bacteremia during the COVID-19 pandemic: a systematic review. Antibiotics 12(2):409

Nishiyama M, Praise S, Tsurumaki K, Baba H, Kanamori H, Watanabe T (2021) Prevalence of antibiotic-resistant bacteria ESKAPE among healthy people estimated by monitoring of municipal wastewater. Antibiotics 10(5):495

O'Toole RF (2021) The interface between COVID-19 and bacterial healthcare-associated infections. Clin Microbiol Infect 27(12):1772–1776

Onifade EO, Ogbonna IO, Ikwebe J, Aremu SO (2019) Profiling of the bacterial pathogens associated with hospital acquired infections in hospitals within Makurdi metropolis, Middle Belt, Nigeria. Bangladesh J Med Sci 18(2):368–378

Paul K, Merabishvili M, Hazan R, Christner M, Herden U, Gelman D, Khalifa L, Yerushalmy O, Coppenhagen-Glazer S, Harbauer T, Schulz-Jürgensen S (2021) Bacteriophage rescue therapy of a vancomycin-resistant *Enterococcus faecium* infection in a one-year-old child following a third liver transplantation. Viruses 13(9):1785

Pendleton JN, Gorman SP, Gilmore BF (2013) Clinical relevance of the ESKAPE pathogens. Expert Rev Anti Infect Ther 11(3):297–308

Peng X, Zhou W, Zhu Y, Wan C (2021) Epidemiology, risk factors and outcomes of bloodstream infection caused by ESKAPEEc pathogens among hospitalized children. BMC Pediatr 21:1–10

Prado ACC, De Barros PP, Santos JDD, De Oliveira LD, Carvalho CAT, Valera MC, Jorge AOC, Junqueira JC (2017) *Enterococcus faecium* and *Enterococcus faecalis* in endodontic infections: antibiotic resistance profile and susceptibility to photodynamic therapy. Lasers Dent Sci 1:91–99

Priyamvada P, Debroy R, Anbarasu A, Ramaiah S (2022) A comprehensive review on genomics, systems biology and structural biology approaches for combating antimicrobial resistance in ESKAPE pathogens: computational tools and recent advancements. World J Microbiol Biotechnol 38(9):153

Rangel K, Chagas TPG, De-Simone SG (2021) *Acinetobacter baumannii* infections in times of COVID-19 pandemic. Pathogens 10(8):1006

Raouf MR, Sayed M, Rizk HA, Hassuna NA (2018) High incidence of MBL-mediated imipenem resistance among *Pseudomonas aeruginosa* from surgical site infections in Egypt. J Infect Dev Ctries 12(07):520–525

Rawson TM, Moore LSP, Zhu N, Ranganathan N, Skolimowska K, Gilchrist M, Satta G, Cooke G, Holmes A (2020) Bacterial and fungal coinfection in individuals with coronavirus: a rapid review to support COVID-19 antimicrobial prescribing. Clin Infect Dis 71(9):2459–2468

Reynolds D, Kollef M (2021) The epidemiology and pathogenesis and treatment of *Pseudomonas aeruginosa* infections: an update. Drugs 81(18):2117–2131

Rezai MS, Ahangarkani F, Rafiei A, Hajalibeig A, Bagheri-Nesami M (2018) Extended-spectrum beta-lactamases producing *Pseudomonas aeruginosa* isolated from patients with ventilator associated nosocomial infection. Arch Clin Infect Dis 13(4):e13974

Ribeiro SM, De La Fuente-Núñez C, Baquir B, Faria-Junior C, Franco OL, Hancock RE (2015) Antibiofilm peptides increase the susceptibility of carbapenemase-producing *Klebsiella pneumoniae* clinical isolates to β-lactam antibiotics. Antimicrob Agents Chemother 59(7):3906–3912

Rishi P, Vij S, Maurya IK, Kaur UJ, Bharati S, Tewari R (2018) Peptides as adjuvants for ampicillin and oxacillin against methicillin-resistant *Staphylococcus aureus* (MRSA). Microb Pathog 124:11–20

Rout B, Behera B, Bahinipati S, Sahu KK, Otta S (2021) Prevalence and susceptibility of Carbapenem resistance gram-negative bacteria in an ICU setup. Indian J Med Paediatr Oncol 43:S1–S19

Rubin C, Myers T, Stokes W, Dunham B, Harris S, Lautner B, Annelli J (2013) Review of institute of medicine and national research council recommendations for one health initiative. Emerg Infect Dis 19(12):1913

Russo TP, Minichino A, Gargiulo A, Varriale L, Borrelli L, Pace A, Santaniello A, Pompameo M, Fioretti A, Dipineto L (2022) Prevalence and phenotypic antimicrobial resistance among ESKAPE bacteria and Enterobacterales strains in wild birds. Antibiotics 11(12):1825

Sabino CP, Wainwright M, Ribeiro MS, Sellera FP, Dos Anjos C, da Silva Baptista M, Lincopan N (2020) Global priority multidrug-resistant pathogens do not resist photodynamic therapy. J Photochem Photobiol B Biol 208:111893

Sandiumenge A, Rello J (2012) Ventilator-associated pneumonia caused by ESKAPE organisms: cause, clinical features, and management. Curr Opin Pulm Med 18(3):187–193

Santajit S, Indrawattana N (2016) Mechanisms of antimicrobial resistance in ESKAPE pathogens. Biomed Res Int 2016:2475067

She P, Li Y, Li Z, Liu S, Yang Y, Li L, Zhou L, Wu Y (2023) Repurposing 9-Aminoacridine as an adjuvant enhances the antimicrobial effects of rifampin against multidrug-resistant *Klebsiella pneumoniae*. Microbiol Spectr 11:e0447422

Shorr AF, Combes A, Kollef MH, Chastre J (2006) Methicillin-resistant *Staphylococcus aureus* prolongs intensive care unit stay in ventilator-associated pneumonia, despite initially appropriate antibiotic therapy. Crit Care Med 34(3):700–706

Siddique MH, Aslam B, Imran M, Ashraf A, Nadeem H, Hayat S, Khurshid M, Afzal M, Malik IR, Shahzad M, Qureshi U (2020) Effect of silver nanoparticles on biofilm formation and EPS production of multidrug-resistant *Klebsiella pneumoniae*. Biomed Res Int 2020:6398165

Songsantiphap C, Vanichanan J, Chatsuwan T, Asawanonda P, Boontaveeyuwat E (2022) Methylene blue–mediated antimicrobial photodynamic therapy against clinical isolates of extensively drug resistant gram-negative bacteria causing nosocomial infections in Thailand. An in vitro study. Front Cell Infect Microbiol 12:929242

Sultan AM, Seliem WA (2018) Identifying risk factors for healthcare-associated infections caused by Carbapenem-resistant *Acinetobacter baumannii* in a neonatal intensive care unit. Sultan Qaboos Univ Med J 18(1):75

Taha OA, Connerton PL, Connerton IF, El-Shibiny A (2018) Bacteriophage ZCKP1: a potential treatment for *Klebsiella pneumoniae* isolated from diabetic foot patients. Front Microbiol 9:2127

Tolera M, Abate D, Dheresa M, Marami D (2018) Bacterial nosocomial infections and antimicrobial susceptibility pattern among patients admitted at Hiwot Fana specialized university hospital, Eastern Ethiopia. Adv Med 2018:2127814

Torre-Cisneros J, Natera C, Mesa F, Trikic M, Rodríguez-Baño J (2018) Clinical predictors of methicillin-resistant *Staphylococcus aureus* in nosocomial and healthcare-associated pneumonia: a multicenter, matched case–control study. Eur J Clin Microbiol Infect Dis 37(1):51–56

Tsao LH, Hsin CY, Liu HY, Chuang HC, Chen LY, Lee YJ (2018) Risk factors for healthcare-associated infection caused by carbapenem-resistant *Pseudomonas aeruginosa*. J Microbiol Immunol Infect 51(3):359–366

Venkat Ramanan P, Sharma S, Krishna V (2014) Changing profile of pediatric community-acquired UTI in a hospital in South India. J Trop Pediatr 60(6):483–483

Voidazan S, Albu S, Toth R, Grigorescu B, Rachita A, Moldovan I (2020) Healthcare associated infections—a new pathology in medical practice? Int J Environ Res Public Health 17(3):760

Walia K, Madhumathi J, Veeraraghavan B, Chakrabarti A, Kapil A, Ray P, Singh H, Sistla S, Ohri VC (2019) Establishing antimicrobial resistance surveillance & research network in India: journey so far. Indian J Med Res 149(2):164

Wang JL, Kuo CF, Yeh CM, Chen JR, Cheng MF, Hung CH (2018) Efficacy of φkm18p phage therapy in a murine model of extensively drug-resistant *Acinetobacter baumannii* infection. Infect Drug Resist 11:2301–2310

Wang M, Wei H, Zhao Y, Shang L, Di L, Lyu C, Liu J (2019) Analysis of multidrug-resistant bacteria in 3223 patients with hospital-acquired infections (HAI) from a tertiary general hospital in China. Bosn J Basic Med Sci 19(1):86–93

Wang W, Li B, Yang H, Lin Z, Chen L, Li Z, Ge J, Zhang T, Xia H, Li L, Lu Y (2020) Efficient elimination of multidrug-resistant bacteria using copper sulfide nanozymes anchored to graphene oxide nanosheets. Nano Res 13:2156–2164

WHO (2017). Available from https://www.who.int/news/item/27-02-2017-who-publishes-list-of-bacteria-for-which-new-antibiotics-are-urgently-needed. Accessed 23 Aug 2023

WHO (2019). Available from https://www.who.int/news/item/29-04-2019-new-report-calls-for-urgent-action-to-avert-antimicrobial-resistance-crisis . Accessed on 23 Aug 2023

Wright GD (2016) Antibiotic adjuvants: rescuing antibiotics from resistance. Trends Microbiol 24(11):862–871

Xuan H, Yao X, Pan R, Gao Y, Wei J, Shao D, Liu K, Li Z, Qiu Y, Ma Z, Li B (2021) Antimicrobial resistance in *Enterococcus faecium* and *Enterococcus faecalis* isolates of swine origin from eighteen provinces in China. J Vet Med Sci 83(12):1952–1958

Yang SK, Yusoff K, Ajat M, Wee CY, Yap PSX, Lim SHE, Lai KS (2021) Combinatorial antimicrobial efficacy and mechanism of linalool against clinically relevant *Klebsiella pneumoniae*. Front Microbiol 12:635016

Yu Y, Zhao H, Lin J, Li Z, Tian G, Yang YY, Yuan P, Ding X (2022) Repurposing non-antibiotic drugs Auranofin and Pentamidine in combination to combat multidrug-resistant gram-negative bacteria. Int J Antimicrob Agents 59(5):106582

Zaidi S, Misba L, Khan AU (2017) Nano-therapeutics: a revolution in infection control in post antibiotic era. Nanomedicine 13(7):2281–2301

Zhang YJ, Hu HW, Chen QL, Singh BK, Yan H, Chen D, He JZ (2019) Transfer of antibiotic resistance from manure-amended soils to vegetable microbiomes. Environ Int 130:104912

Zhang G, Hu C, Luo L, Fang F, Chen Y, Li J, Peng Z, Pan H (2020) Clinical features and short-term outcomes of 221 patients with COVID-19 in Wuhan, China. J Clin Virol 127:104364

Zhou ZC, Feng WQ, Han Y, Zheng J, Chen T, Wei YY, Gillings M, Zhu YG, Chen H (2018) Prevalence and transmission of antibiotic resistance and microbiota between humans and water environments. Environ Int 121:1155–1161

Zhou X, Willems RJ, Friedrich AW, Rossen JW, Bathoorn E (2020) Enterococcus faecium: from microbiological insights to practical recommendations for infection control and diagnostics. Antimicrob Resist Infect Control 9:1–3

Zhou J, Zhang L, He Y, Liu K, Zhang F, Zhang H, Lu Y, Yang C, Wang Z, Fareed MS, Liang X (2021) An optimized analog of antimicrobial peptide Jelleine-1 shows enhanced antimicrobial activity against multidrug resistant *P. aeruginosa* and negligible toxicity in vitro and in vivo. Eur J Med Chem 219:113433

Zhu Y, Huang WE, Yang Q (2022) Clinical perspective of antimicrobial resistance in bacteria. Infect Drug Resist 15:735–746

Antibiotic Resistance Profile and Detection in ESKAPE Pathogens

2

Ankita Agrawal and Amiya Kumar Patel

Abstract

Antibiotic resistance indicates the presence of a genetically determined resistance mechanism in bacteria to withstand or combat the lethal or inhibitory antibiotic concentration. Increased antibiotic resistance of bacterial pathogens and their spread in the clinical sector, industry and environment is a grave threat to therapeutic applications and healthcare globally. Owing to uncontrolled spread, antibiotic resistance loomed as a global "imminent pandemic" with higher healthcare costs, longer hospitalization and illness duration, and greater morbidity and mortality. WHO predicted an increase in mortality rate to 7 lakhs annually due to antibiotic resistance and forecasting ten million deaths by 2050. This chapter deals with the antibiotic resistance profile of "ESKAPE" pathogens comprising Gram-positive and Gram-negative bacteria with multidrug resistance (MDR) causing public health crisis and hence designated as highest "priority status" by WHO. Although genetically versatile, the bacterial pathogens share similar resistance strategies including the decreased uptake of antibiotics, alteration in the target site, enzymatic inactivation of antibiotics and activation of multiple efflux pumps. Understanding the antibiotic resistance profile and mechanism in ESKAPE pathogens is a prerequisite not only to limit their spread or to develop newer antibiotics with novel mechanisms but also for the prediction of unknown resistance mechanisms to develop innovative diagnostic strategies. Numerous techniques (phenotypic, genotypic and emerging) have been developed for rapid detection that is crucial in the fight against pathogenic infections but still lacking somewhere. The study provides the background knowledge of various detection

A. Agrawal · A. K. Patel (✉)
Department of Biotechnology and Bioinformatics, Sambalpur University, Jyoti Vihar, Sambalpur, Odisha, India

S. Busi, R. Prasad (eds.), *ESKAPE Pathogens*,
https://doi.org/10.1007/978-981-99-8799-3_2

methods involved in antibiotic resistance profile, highlighting advantages and drawbacks, which necessitate newer and robust diagnostic strategies for prompt and accurate detection to tackle hard-to-treat fatal infections resulting from ESKAPE pathogens.

Keywords

Antibiotic resistance · β-Lactamase · Efflux · ESKAPE · Resistance mechanism

2.1 Introduction

Antibiotics are drugs that act as nemesis for bacterial infections (Yang et al. 2023). Antimicrobial resistance (AMR) is described as the resistance exhibited by microbes against lethal or inhibitory antimicrobial concentration (Boolchandani et al. 2019), whereas antibiotic resistance is a subset of AMR defined by the presence of structural and/or genetically determined mechanisms of resistance in bacteria to encounter the inhibitory effect of antimicrobial agents (Khan et al. 2019). In today's world, antibiotic resistance is the most significant challenge confronting mankind associated with higher healthcare costs, longer hospitalization, and higher mortality and morbidity rates resulting in clinical and economic complications (Chinemerem Nwobodo et al. 2022). The emergence of antibiotic resistance is a natural evolutionary phenomenon, but misuse and overuse of antibiotics, deficient sanitation and hygiene, lack of safe drinking water, inadequate infection control practices and lack of knowledge are the main drivers that accelerated the process to many folds causing severe damage to healthcare and economic sectors (Ramay et al. 2020). Presently, the increased rate of resistance against antibiotics has appeared as an "imminent pandemic" risking millions of lives. In addition, the spread of resistant strains within a population or from environmental sources significantly contributes to an upsurge in antibiotic resistance in bacterial isolates that synchronize with the increasing number of nosocomial outbreaks. The resistance to antibiotics is a major blockage in disease control strategies, since increased resistance and pathogenicity result in decreased treatment effectiveness, which prolongs infection time and expensive treatment for patients (Benkő et al. 2020). Antibiotic resistance also has a negative impact on cancer therapy, surgical events and transplantation (Perez and Van Duin 2013; Bharadwaj et al. 2022). Inadequate knowledge about the resistance mechanism involved in bacterial tolerance and persistence also contributes to antibiotic resistance (Fernández-García et al. 2018). Therefore, a greater understanding of the resistance pattern, in-depth research on different resistance mechanisms, and advancements in the diagnosis of resistant bacteria is the need of the hour for better therapeutic applications.

2.1.1 ESKAPE Pathogens and Their Significance

The World Health Organization ranked antibiotic resistance as one of the foremost global public healthcare and development ultimatum of the twenty-first century, which accounts for 0.7 million deaths per annum and predicts ten million deaths per annum by 2050 (Brogan and Mossialos 2016; WHO Antimicrobial Resistance 2021; Kalpana et al. 2023). Accordingly, WHO has developed a list of bacterial pathogens belonging to 12 different families named as "global priority list of antibiotic resistant bacteria" to counter antibiotic resistance and ranked them into 3 priority tiers based on the urgency to develop new, novel and efficient antibiotic agents to tackle these bacterial pathogens (De Oliveira et al. 2020). The bacterial list was designed by considering the mortality rate, prevalence of resistance, healthcare load, and the ability for treatment and transmission. Currently, WHO has prioritized six bacterial pathogens designated as "ESKAPE" with amplified multidrug resistance and virulence (Mancuso et al. 2021). The acronym "ESKAPE" stands for ***E**nterococcus faecium*, ***S**taphylococcus aureus*, ***K**lebsiella pneumoniae*, ***A**cinetobacter baumannii*, ***P**seudomonas aeruginosa* and ***E**nterobacter species,* which effectively escape the bactericidal action of antibiotics by sophisticated resistance mechanism, significantly enhancing the burden of AMR (Rice 2008; Paul et al. 2022; Kalpana et al. 2023). ESKAPE pathogens dubbed as "superbugs" cause various nosocomial infections and serve as models for pathogenesis, transmission and resistance. The rapid spreading of ESKAPE pathogens and antibiotic resistance has declined the number of effective antibiotics to a greater extent and thereby the global awareness and monitoring of antibiotic resistance need to be increased especially for ESKAPE pathogens. Furthermore, the basic understanding of antibiotic resistance profile and mechanisms in ESKAPE is crucial for novel antibiotic discovery, to guide antibiotic selection and prediction of unknown resistance mechanisms. Subsequently, the innovation and development of novel and quick diagnostic procedures is a prerequisite to substantiate effective treatment of difficult-to-treat infections attributed to ESKAPE bacteria and to combat the phenomenon of AMR (Santajit and Indrawattana 2016; Aslam et al. 2018).

2.1.2 Antibiotic Resistance in ESKAPE Pathogens

An enormous spike in resistance by ESKAPE bacteria results in more complicated and difficult-to-cure infections that raise the death numbers. Considering the rising menace of resistance to antibiotics, understanding the basis of bacterial resistance is crucial. The bacterial resistance may be natural (intrinsic) or acquired. The term intrinsic resistance is defined as the inbuilt structural or functional characteristic of bacteria to tolerate antibiotic effects and is independent of previous exposure (Blair et al. 2015; Mancuso et al. 2021). Intrinsic resistance is chromosomally mediated that includes reduced outer membrane permeability, enzyme production or the presence of non-specific efflux pumps (Reygaert 2018; Morrison and Zembower 2020).

For example, *K. pneumoniae* is intrinsically resistant to ampicillin, Gram-positive cocci to aztreonam and Gram-negative bacilli resistance to vancomycin (Morrison and Zembower 2020).

In contrast, the acquired resistance results when the previously susceptible bacteria become resistant reflecting the sign of danger in clinical sectors making the treatment much more complicated. It arises in one of two ways: either by vertical transmission (mutation in chromosomal gene) or by horizontal gene transfer (HGT), where resistance determinants are acquired exogenously (Wright 2011; Munita and Arias 2016). The development of mutation in susceptible bacterial populations alters the action of antibiotics through target modification, decreased drug uptake, efflux pump activation or alteration in the metabolic pathways, which make them resistant to various antibiotics. Besides, the mutational changes resulting in resistance are highly diverse and vary in complexity (Munita and Arias 2016). However, exogenous resistant genes are acquired by HGT via transformation, conjugation or transduction. Conjugation involves the sharing of genetic material directly by mating between plasmids and/or integrative and conjugative elements that are capable of incorporating into bacterial chromosomes. Transformation involves the direct uptake of genetic material from typically lyzed bacterial cells, whereas transduction involves the sharing of genetic material mediated by the bacteriophages (Fig. 2.1). The spreading of resistant genes from resistant bacteria to susceptible strains makes them resilient to antibiotics and thereby the antibiotics become worthless (Sun et al. 2019). Intrinsically, resistant bacteria are not the primary focus for an obstacle in antibiotic resistance, but the mechanism of resistance acquired by the resistance bacteria that are previously susceptible to the recommended dose of antibiotics is

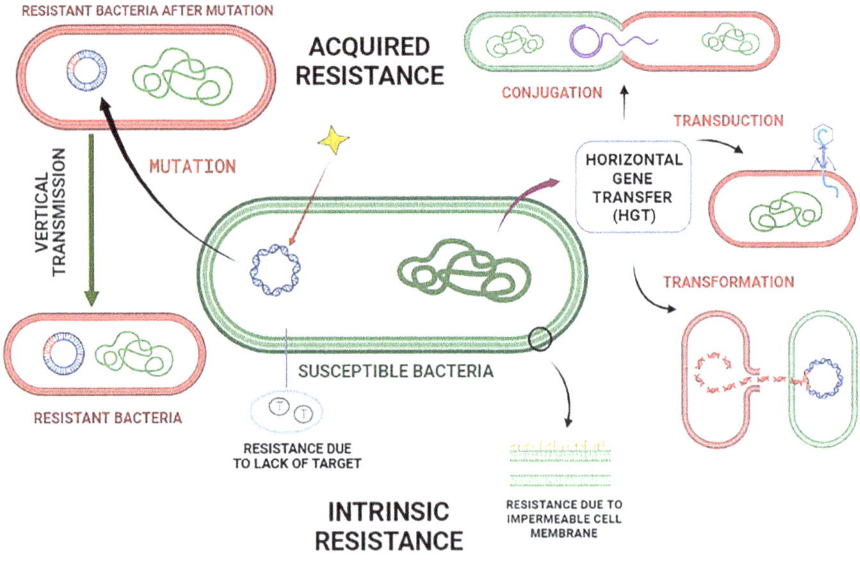

Fig. 2.1 Emergence and spreading of antibiotic resistance in ESKAPE pathogens

important, which is considered as the leading cause of antibiotic resistance and spreading (Munita and Arias 2016). Moreover, several studies revealed that each bacterial species prefers a specific resistance pattern against antibiotics over others that vary greatly among the bacterial populations (Munita and Arias 2016).

Before comprehending diverse resistance mechanisms in ESKAPE bacteria against antibiotics, the mode of action of different antibiotics needs to be briefly examined. Antibiotics affect the bacterial cells either by blocking the processes involved in nucleic acid (DNA and RNA) biosynthesis, protein synthesis, cell wall synthesis or by inhibiting various metabolic pathways (such as folic acid metabolism). Moreover, antibiotics also act by interfering with the cell membrane functioning or by competing with enzymes essential for their survival (Mancuso et al. 2021). Furthermore, the mechanism of different antibiotic actions depends on their structure and preference for distinct bacteria, which aids in understanding the antibiotic resistance.

2.2 Antibiotic Resistance Mechanism in ESKAPE Pathogens

ESKAPE bacteria serve as a role model for antibiotic resistance. Although ESKAPE pathogens have different genetic makeups, the overall mechanism of antibiotic resistance vital for the emergence and persistence of ESKAPE pathogens is universal and remains the same (Kalpana et al. 2023). Antibiotic resistance mechanisms by the ESKAPE include four broad categories such as antibiotic inactivation or alteration (enzyme-catalyzed irreversible cleavage), modification of antibiotic-binding target sites, decreased antibiotic intake (change in cellular permeability), efflux pump activation and/or biofilm production (Fig. 2.2) (Santajit and Indrawattana 2016). The resistance mechanism is complicated, involving different mechanisms that provide resistance against several antibiotics or a single resistance mechanism targeting different antibiotics gradually decreasing the number of effective drugs. For example, resistance against carbapenems in *A. baumannii* is achieved by β-lactamases, loss of CarO porin and modification of penicillin-binding protein (PBP) (Kyriakidis et al. 2021). Patterns of resistance mechanism differ greatly in Gram-negative and Gram-positive ESKAPE pathogens in connection to structural variations. For example, Gram-negative bacteria produce β-lactamases, whereas Gram-positive bacteria alter the penicillin-binding site as the major resistance strategy against β-lactams. Understanding the peculiarities of each resistance mechanism is fundamental to identifying the targets for successful medication.

2.2.1 Antibiotic Inactivation by Enzyme Production

Antibiotic inactivation or alteration is a prevalent resistance mechanism exhibited by ESKAPE bacterial pathogens achieved either by the irreversible inactivation of antibiotics (β-lactamases) or modification of antibiotic molecules (aminoglycoside-modifying enzymes [AMEs]). Inactivation of the β-lactam class of antibiotics that

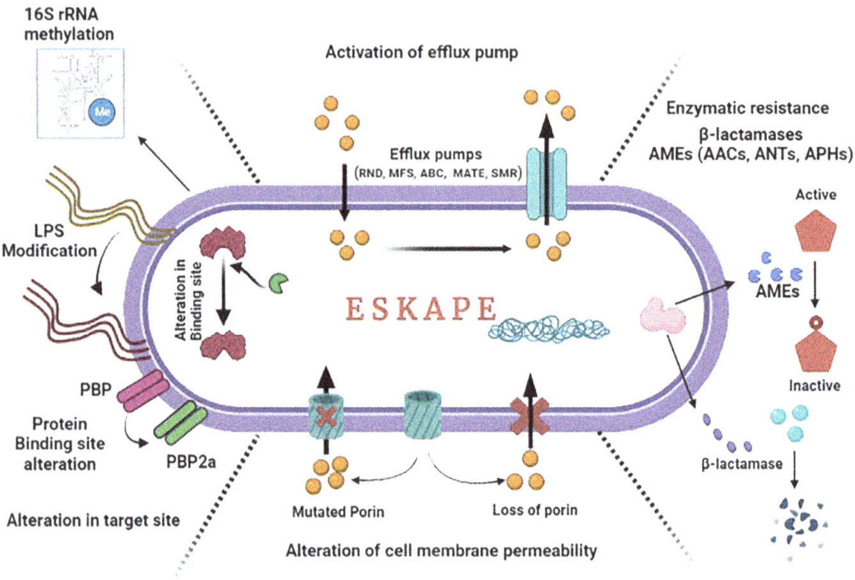

Fig. 2.2 Mechanism of antibiotic resistance exhibited by ESKAPE pathogens such as (**a**) enzymatic resistance (destruction or modification of antibiotics); (**b**) alterations in target site (LPS modification, expression of PBP2a with low affinity, 16S rRNA modification); (**c**) alteration in cell membrane permeability (loss of porins) and (**d**) activation of efflux pump to expel antibiotics preventing its accumulation. Five families of efflux pump: RND, MFS, ABC, MATE and SMR. *AACs* aminoglycoside acetyltransferases, *AMEs* aminoglycoside-modifying enzymes, *ANTs* aminoglycoside nucleotidyl transferases, *ABC* ATP-binding cassette, *APHs* aminoglycoside phosphotransferases, *LPS* lipopolysaccharide, *MATE* multidrug and toxic compound extrusion, *MFS* major facilitator superfamily, *PBP* penicillin-binding protein, *RND* resistance-nodulation division, *SMR* small multidrug resistance

comprise carbapenem, cephalosporin, penicillin, monobactam, and β-lactamase inhibitor occurs by β-lactamases production. The β-lactamase disintegrates the four-sided β-lactam loop before reaching the target site rendering antibiotics ineffective, especially in Gram-negative ESKAPE pathogens. β-Lactamases are grouped into four different classes (A-D) based on the sequence motifs that differ in hydrolytic mechanism (Kyriakidis et al. 2021).

Class A serine β-lactamases comprise important enzymes such as penicillinase, carbapenemases (KPCs), cephalosporinase and extended-spectrum β-lactamases (ESBLs) such as CTX-M, SHV and TEM, that are susceptible to compounds like clavulanate, tazobactam, ticarcillin and sulbactam (Jeon et al. 2015). Class B β-lactamases represent broad-spectrum metallo-β-lactamases (MBLs) with zinc at their active sites that hydrolyzes most β-lactam antibiotics (except monobactams) efficiently (Nordmann et al. 2012). Most prominent MBL enzymes include New Delhi metallo-β-lactamases-1 (NDM-1), Imipenemase (IMP) and Verona integron encoded metallo-β-lactamases (VIM) and were reported exclusively in Gram-negative ESKAPE pathogens (Rice 2010). Studies suggested that *P. aeruginosa*,

K. pneumonia, A. baumannii and *Enterobacter cloacae* have the majority of IMP-type MBLs; *A. baumannii* has VIM genes, whereas the NDM-1 gene was reported in *K. pneumonia* and *E. cloacae.* Unlike other β-lactamases, their activity is only inhibited by a chelating agent (EDTA, ethylenediaminetetraacetic acid) due to the existence of metal in their active site but not by the commercial inhibitors (clavulanic acid, tazobactam or sulbactam) (De Oliveira et al. 2020). The class C family are chromosomally encoded enzymes providing resistance against cephamycin, cephalosporins and penicillin and are non-susceptible to commercial β-lactamases inhibitors except avibactam leading to therapeutic issues (Poirel and Nordmann 2006; Jeon et al. 2015). Moreover, AmpC β-lactamase, a cephalosporinase, is reported in ESKAPE pathogens such as *P. aeruginosa, Acinetobacter* sp. and *Enterobacter* sp. Lastly, class D β-lactamase mainly consists of oxacillinases (OXAs), which can inactivate all β-lactam groups and exhibit varied resistance to β-lactam inhibitors (De Oliveira et al. 2020).

Furthermore, the ESKAPE pathogen produces aminoglycoside-modifying enzymes (AMEs), which alter and inactivate aminoglycoside molecules contributing to greater aminoglycoside resistance. The AMEs are divided into three different classes: acetyltransferase (AAC), adenyltransferase (ANT) and phosphotransferase (APH). The specific chemical group (hydroxyl or amino) modifications in aminoglycosides are catalyzed by these enzymes, which declines their binding affinity to target sites in bacterial pathogens (De Oliveira et al. 2020).

2.2.2 Alterations of Membrane Permeability

Another prevalent resistance strategy adopted by ESKAPE pathogens is reduced antibiotic uptake owing to alternations in cell membrane permeability. In particular, Gram-negative ESKAPE pathogens possess non-specific porins in the outer membrane that permit intracellular passage of hydrophilic antibiotic agents. However, loss, impairment or reduction in the gene expressing porin protein results in low-antibiotic intake making them antibiotic-resistant. For example, the loss of porin (OprD) in *P. aeruginosa* resulted in enhanced carbapenem resistance (Muntean et al. 2022).

2.2.3 Alterations in Antibiotic-Target Sites

ESKAPE pathogens also resist the action of various antibiotic molecules by altering or modifying the target sites, preventing the antibiotic from binding. For instance, resistance to macrolide and linezolid in ESKAPE pathogens is mediated by ribosomal RNA methylation catalyzed by rRNA methyltransferase encoded by the *emr* gene, which decreases antibiotic-binding affinity (Terreni et al. 2021). Subsequently, resistance against methicillin in *S. aureus* (MRSA) has been enhanced by the modification of the protein-binding site with poor affinity.

2.2.4 Efflux Pump Activation

Activation of efflux pump in ESKAPE pathogens expels antibiotics at a higher rate out of the cell making drug concentration insufficient for antibacterial activities. These resistance tactics provide resistance to a number of antibiotics (fluoroquinolones, β-lactam, carbapenem, polymyxins and protein synthesis inhibitors). There are five different families of efflux pumps that bacteria deploy as a self-defence strategy. The major facilitator super (MFS) family, multidrug and toxic compound extrusion (MATE) family, ATP-binding cassette (ABC) family, small multidrug resistance (SMR) family and resistance-nodulation division (RND) family (Munita and Arias 2016; Santajit and Indrawattana 2016).

2.3 Antibiotic Resistance Profile in ESKAPE Pathogens

The antibiotic resistance profile of bacterial pathogens grouped under "ESKAPE" is reviewed in depth later. The "ESKAPE" pathogens encompass both critical and high-priority categories as per WHO guideline—2017, which require urgent attention in research and development. The "Priority list 1: Critical" involves multidrug resistance Gram-negative ESKAPE pathogens such as *A. baumannii*, *P. aeruginosa*, *K. pneumonia* and *Enterobacter* sp., which cause a variety of fatal infections.

2.3.1 Antibiotic Resistance Profile of *Acinetobacter baumannii*

A. baumannii is a non-motile, aerobic and Gram-negative bacillus. An opportunistic ESKAPE pathogen is predominantly linked with acquired infections in hospital settings such as bacteraemia, urinary tract infections, pneumonia, bloodstream infections and nosocomial meningitis with a mortality rate of nearly 35% globally (Denissen et al. 2022; Cavallo et al. 2023). *A. baumannii* is a highly troublesome pathogen with impressive genetic plasticity, extended resistome and virolome, biofilm formation, ability to survive in extreme conditions and minimal metabolic rate with limited treatment options causing severe infections (Dahdouh et al. 2017; Vrancianu et al. 2020). The resistance profile of *A. baumannii* isolates from different regions revealed high resistance against carbapenem, aminoglycoside and colistin (Lin and Lan 2014; Rodríguez et al. 2016). According to scientific studies, *A. baumannii* exhibited relatively higher intrinsic resistance against antimicrobial agents (glycopeptides, macrolides, streptogramins and lincosamides) and has higher competency in incorporating foreign DNA determinants facilitating rapid genetic modification for acquired resistance leading to multidrug and pandrug resistance (Peleg et al. 2007; Perović et al. 2018). The strategies for resistance range from the production of β-lactamases and AMEs, permeability defects and activation of efflux pumps to antibiotic-target site alterations in *A. baumannii* strains (Lee et al. 2017; Vrancianu et al. 2020; Abdi et al. 2020). Antibiotic resistance profile along with phenotypic resistance (percentage), and resistance genes (genotypic resistance) from different sources have been depicted (Table 2.1).

The β-lactamases enzyme production in *A. baumannii* serves better survival benefits by conferring resistance to various antibiotics. All four β-lactamases have been identified that are incorporated through acquired resistance mechanisms resulting in multidrug resistance phenomenon (Kyriakidis et al. 2021). Moreover, class A β-lactamases include both broad spectrum β-lactamases (ESBL) such as CARB-10, CTX-M (2, 15), VEB-1, PER (1, 2, 7), GES-14, SHV-5, TEM-92 and narrow spectrum β-lactamases such as TEM-1, CARB-4 and SCO-1 (Vrancianu et al. 2020; Roy et al. 2022). Diverse class B β-lactamases (MBLs) include families of IMP (1, 2, 4–6, 8, 11, 19, 24), NDM (1–3), VIM (1–4), SIM-1, SPM-1, GIM-1 and FIM-1 (Potron et al. 2015). Recently, *A. baumannii* was reported to have an intrinsically chromosomally mediated *AmpC* gene expressing cephalosporins, a class C β-lactamases that confers resistance against the third-generation cephalosporins (Breijyeh et al. 2020; Roy et al. 2022). Furthermore, class D oxacillinases (OXAs) with more than 400 variants are predominantly present in carbapenem-resistant *A. baumannii* (CRAB) strains. Specific OXAs genes such as OXA (23, 24, 51, 58, 143, 235) were encoded chromosomally or mediated through plasmid in *A. baumannii,* conferring resistance to carbapenems and cephalosporins (Evans and Amyes 2014; Poirel and Nordmann 2006; Lee et al. 2017). Moreover, all three classes of AMEs in *A. baumannii* were reported to supplement the main resistance mechanism against aminoglycosides (Ramirez and Tolmasky 2010; Zhou et al. 2010). Besides, *ant(2″)-Ia* and *aac(3)-I* genes are recognized as key determinants of resistance to gentamicin and tobramycin. Genes that express AMEs are usually present in the plasmids and/or transposable elements (Lin and Lan 2014).

Subsequently, the presence of chromosomally encoded multiple efflux pumps in *A. baumannii* provides resistance to commercially available different antibiotic classes such as imipenem (Hu et al. 2007), tetracycline, chloramphenicol (Fournier et al. 2006), tigecycline (Xu et al. 2019), carbapenem and cephalosporin (Huang et al. 2008), erythromycin, trimethoprim, fluoroquinolone and various β-lactams (Chen et al. 2017). Studies revealed the existence of no fewer than four efflux pump classes in *A. baumannii* such as the RND superfamily (AdeABC, AdeFGH, AdeIJK), MATE family (AbeM), MFS family (AmvA, AbaF, CmlA, CraA, TetA, TetB) and SMR family (AbeS) (Pérez-Varela et al. 2019; Cavallo et al. 2023). AbeABC pump of the RND family is broad spectrum conferring multidrug resistance against aminoglycosides, tigecycline and non-fluoroquinolones (Vrancianu et al. 2020; Mancuso et al. 2021). Furthermore, efflux pumps were found to have a synergistic effect with biofilm development, substantiating the resistance mechanism against different antibiotics resulting in the rapid evolution of MDR bacterial pathogens (Yoon et al. 2015).

Furthermore, the existence of porins and their modification bestows *A. baumannii* with greater antibiotic resistance and pathogenicity (Asif et al. 2018; Roy et al. 2022). As a scenario, reduced expression of porins such as 29-kDa outer membrane protein (OmpA, Omp22–23, Omp33–36, Omp37, Omp43–44, Omp47) and loss of CarO porin provide resistance against carbapenem (imipenem and meropenem)

Table 2.1 Antibiotic resistance profile of ESKAPE pathogen (*Acinetobacter baumannii*)

ESKAPE pathogens	Isolates number	Antibiotic resistance profile (%) (phenotypic resistance)	Resistance gene (genotypic resistance)	Sources of isolation	References
Acinetobacter baumannii	*N* = 9	Resistance to aztreonam (53.8%); chloramphenicol, amoxicillin, clavulanate, cefpodoxime and cefoxitin (15.3%); cefuroxime, gentamicin and florphenicol (7.7%). All susceptible to penicillin, sulphamethoxazole/trimethoprim, ciprofloxacin, levofloxacin, nalidixic acid, tetracycline, colistin, norfloxacin, ampicillin, imipenem and ceftazidime	–	Tank milk samples	Güneri (2023)
	N = 1	Resistance to erythromycin, colistin. Intermediate resistance to tetracycline, cephalosporin, meropenem. Sensitive to amikacin, gentamicin, imipenem, tobramycin and tigecycline	–	Commercial strain	Glover et al. (2022)
	N = 8	All resistance to piperacillin/tazobactam, cefotaxime, ceftriaxone, ciprofloxacin, ceftazidime, vancomycin, cefepime, colistin, tetracycline, linezolid, gentamicin, meropenem and imipenem (100%)	*armA*, *aph(6)-Id*, *aph(3′)-Vla*, *aac(3)-Ia*, *Oxa-23*, *Oxa-66*, *Adc-25*, *tet(B)*, *sul1*, *aadA1*,	Clinical sample	D'Onofrio et al. (2020)
	N = 41	Resistance to ceftazidime, imipenem, meropenem and piperacillin/tazobactam, quinolone, ciprofloxacin (95%), gentamicin, levofloxacin and amikacin (92.8%); cefepime (82.9%). All susceptible to colistin (100%)	*Oxa-23* and *Oxa-73*	Hospitalized patients	Makke et al. (2020)
	N = 9	Resistance to ampicillin and piperacillin (100%), cefotaxime (88.9%), amoxicillin-clavulanic acid (88.9%), ceftriaxone (88.9%), ceftazidime (77.8%), ciprofloxacin (44.5%), gentamicin (88.9%), meropenem (33.3%), tetracycline (77.8%) and sulphamethoxazole-trimethoprim (66.7%)	–	Hospital settings	Motbainor et al. (2020)
	N = 4	All resistance to amikacin, piperacillin-tazobactam, ampicillin, aztreonam, cefepime, ceftazidime, ceftazidime, ciprofloxacin, cefotaxime, gentamicin, imipenem, meropenem and levofloxacin (100%)	–	Clinical sample	Havenga et al. (2019)

ESKAPE pathogens	Isolates number	Antibiotic resistance profile (%) (phenotypic resistance)	Resistance gene (genotypic resistance)	Sources of isolation	References
Acinetobacter baumannii	*N* = 154	Resistance to cefepime (96.2%), carbenicillin (88.4%), sulphamethoxazole-trimethoprim (75.6%), ticarcillin (74.2%), piperacillin (69.8%), ceftazidime (69.7%), ciprofloxacin (65.8%), imipenem (65.7%), gentamicin (60.8%), tigecycline (57.6%); amikacin and streptomycin (56.2%)	–	Clinical sample	Yang et al. (2019)
	N = 294	Resistant to amikacin, amoxicillin-clavulanate, cefotaxime, imipenem, ampicillin, nitrofurantoin, nalidixic acid, ciprofloxacin, aztreonam. Susceptible to fosfomycin, polymyxin-B and cefoperazone-sulbactam	–	Urinary tract infections	Rasool et al. (2019)
	N = 41	Resistance to amikacin, ampicillin-sulbactam, ceftazidime, cefepime, gentamicin, ciprofloxacin, piperacillin/tazobactam, imipenem, meropenem and tigecycline Susceptible to colistin (100%)	*Oxa-23* and *Oxa-51*	Clinical samples in ICUs	da Silva et al. (2018)
	N = 40	Resistance to ceftazidime (92.5%), imipenem (85%) and gentamicin (80%)	*Oxa-23* (77.5%) and *Oxa-24* (5%)	Hospital samples	Shamsizadeh et al. (2017)
	N = 3	Resistance to amikacin, meropenem, piperacillin/tazobactam, imipenem, ciprofloxacin, levofloxacin and ticarcillin/clavulanate	*Oxa-72* and *Oxa-23*	Technosol at dumpsite	Hrenovic et al. (2017)
	N = 1	Resistance to quinolones, carbapenems and fluoroquinolones, chloramphenicol, tetracycline, and tigecycline. Susceptible to amikacin, netilmicin, kanamycin, tobramycin and gentamicin	*Oxa-23*	River water	Girlich et al. (2010)
	N = 20	Resistance to imipenem, meropenem and ticarcillin (100%) and ceftazidime (95%)	*Oxa-23*	Intensive care units	Mugnier et al. (2010)
	N = 42	Resistance to imipenem and meropenem (100%); ampicillin-sulbactam (98%), ceftazidime (50%), tetracycline (24%) and tigecycline (9.5%). Susceptible to colistin	*Oxa-40*	Hospitals facilities	Lolans et al. (2006)
	N = 1	Resistance to amikacin, aztreonam, cefoperazone-sulbactam, piperacillin, piperacillin-tazobactam, ceftazidime, sulbactam, ciprofloxacin, imipenem, meropenem. Sensitive to tigecycline	*Oxa-51*	Intensive care units, Izmir	Vahaboglu et al. (2006)
	N = 121	Resistance to piperacillin (97.5%), ticarcillin-clavulanate (95.9%), ceftazidime (95.9%), aztreonam (93.4%), ciprofloxacin (92.6%), gentamicin (87.6%), ampicillin sulbactam (58.7%), netilmicin (56.2%), amikacin (29.8%). All sensitive to imipenem (100%)	–	Hospital settings	Maniatis et al. (2003)

(Hood et al. 2010; Jin et al. 2011; Roy et al. 2022). Furthermore, *A. baumannii* has acquired resistance against colistin by the depletion of lipopolysaccharides (LPS) owing to a mutation in genes (*lpxA*, *lpxC*, *lpxD*) involved in the biosynthesis of the LPS layer altering their permeability (Cavallo et al. 2023).

Furthermore, *A. baumannii* has acquired resistance against different antibiotics by the mechanism of target site alterations, for example carbapenem resistance through mutation in penicillin-binding protein (PBP-2), quinolone resistance by a spontaneous mutation of DNA gyrase and topoisomerase IV leading to decline in drug affinity, rifampin resistance due to mutation in gene encoding RNA polymerase (rpoB) and tetracycline resistance by mutation in ribosomal proteins (Kyriakidis et al. 2021; Roy et al. 2022). Moreover, mutations in 16S ribosomal RNA by the *armA* gene reduce their propensity for binding to the target site, enhancing resistance to aminoglycosides (Hasani et al. 2016; Cavallo et al. 2023). Moreover, *A. baumannii* capable of producing biofilms also increases antimicrobial resistance (Kyriakidis et al. 2021).

MDR *A. baumannii* utilizes various resistance mechanisms resulting in bioburden for healthcare with limited antibiotics for effective treatment. Carbapenem (imipenem and meropenem) is considered an effective antibiotic against *A. baumannii*-mediated infections but their effectiveness declined with the emergence of CRAB. Minocycline combined with colistin is highly effective against *A. baumannii* resistant to minocycline, whereas colistin plus rifampin is utilized for colistin-resistant *A. baumannii*. Currently, colistin combined with trimethoprim/sulphamethoxazole is highly effective against infections mediated by CRAB (Breijyeh et al. 2020). Thus, increasing antibiotic-resistant isolates demands novel strategies and effective screening-based approaches for the discovery of potent antibiotics against multidrug resistance *A. baumannii*.

2.3.2 Antibiotic Resistance Profile of *Pseudomonas aeruginosa*

P. aeruginosa is a Gram-negative bacillus that triggers various fatal nosocomial infections such as cystic fibrosis, endocarditis, wound infections, pneumonia, and urinary and respiratory tract infections in immunocompromised patients (Shortridge et al. 2019; Ibrahim et al. 2020; De Oliveira et al. 2020; Jurado-Martín et al. 2021). Studies indicated that increased prevalence of multidrug-resistant *P. aeruginosa* resulted in high-mortality rates in hospitalized patients and also substantial economic losses globally (Rezaloo et al. 2022). Eight different antibiotics group effective against *P. aeruginosa* includes cephalosporins, aminoglycosides (amikacin, gentamicin, tobramycin), fluoroquinolones (levofloxacin, ciprofloxacin), carbapenems, penicillin with β-lactamase-mediated inhibitor (ticarcillin-clavulanic acid, piperacillin-tazobactam), polymyxin (polymyxin B, colistin), monobactams (aztreonam) and phosphonic acid (fosfomycin) (Botelho et al. 2019; Ibrahim et al. 2020). The *P. aeruginosa* infection is problematic because this pathogen has an integration of intrinsic, acquired and adaptive resistance mechanisms (Table 2.2), which limit the choice of antimicrobial therapy (Oliver et al. 2015; Nguyen et al. 2018; Kunz Coyne et al. 2022).

The inherent structure of *P. aeruginosa* entails low penetrance of the outer envelope, a variety of efflux pumps and a chromosomal *AmpC* β-lactamase gene that not

Table 2.2 Antibiotic resistance profile of ESKAPE pathogen (*Pseudomonas aeruginosa*)

ESKAPE pathogens	Isolates number	Antibiotic resistance profile (%) (phenotypic resistance)	Resistance gene (genotypic resistance)	Sources of isolation	References
Pseudomonas aeruginosa	$N = 54$	Resistance to ampicillin (100%), tetracycline (100%), sulphamethoxazole-trimethoprim (100%), ceftriaxone (92.6), cefotaxime (92.6%), streptomycin (90.7%), amikacin (90.7%), amoxicillin-clavulanate (88.8%), erythromycin (77.7%), norfloxacin (18.5%) and colistin (18.5%)	bla_{TEM} (100%), *tetA* (100%), *sul*1 (100%), *aad*A1 (90.7%), $bla_{CTX\text{-}M}$ (88.8%), $bla_{OXA\text{-}1}$ (81.5%)	Non-clinical (boiler chicken)	Algammal et al. (2023)
	$N = 1$	Resistance to amikacin, gentamycin, ceftazidime, ticarcillin, trimethoprim/sulphamethoxazole, ticarcillin/clavulanic acid, piperacillin, pefloxacin, minocycline and ciprofloxacin Sensitive to cefepime and colistin		Hospitalized patients with wounds	Hassan et al. (2022)
	$N = 29$	Resistance towards ampicillin (89.7%), penicillin (86.2%), tetracycline (82.8%), gentamicin (51.7%), cefoxitin (37.9%), ciprofloxacin (34.48%), clindamycin (31.03%), sulphamethoxazole (27.8%), imipenem (20.7%) and chloramphenicol (17.24%)	bla_{DHA} (93.10%), $bla_{CTX\text{-}M}$ (83.65%), bla_{SHV} (48.27%); bla_{OXA} and bla_{VEB} (34.5%)	Non-clinical (meat and meat products)	Rezaloo et al. (2022)
	$N = 36$	Resistance to aztreonam (86.1%), ceftazidime (63.9%), piperacillin (58.3%), cefepime (55.6%), imipenem (50%), piperacillin/tazobactam (47.2%), meropenem (41.7%) and levofloxacin (30.6%)	bla_{SHV} (38%), bla_{TEM} (16.6%), $bla_{CTM\text{-}M}$ (8.3%)	Non-clinical (wastewater)	Hosu et al. (2021)
	$N = 11$	Resistance to amoxicillin-clavulanate, ampicillin, piperacillin, cefotaxime, ceftriaxone and sulphamethoxazole-trimethoprim (100%); tetracycline (90.9%), ceftazidime (63.6%), gentamicin (54.5%), meropenem (45.5%) and ciprofloxacin (36.4%)		Hospital settings	Motbainor et al. (2020)
	$N = 88$	Resistance to ceftazidime (98%), tigecycline (94%), aztreonam (37%), amikacin (37%), imipenem (30%), colistin (24%) and ciprofloxacin (17%)	*amp*C (100%), bla_{VIM} (12.5%), bla_{NDM} (18.7%)	Hospitalized patients	Saleem and Bokhari (2020)

(continued)

Table 2.2 (continued)

ESKAPE pathogens	Isolates number	Antibiotic resistance profile (%) (phenotypic resistance)	Resistance gene (genotypic resistance)	Sources of isolation	References
Pseudomonas aeruginosa	$N = 224$	Resistance to gentamycin (57.9%), cefepime (56.7%), fluoroquinolones (55.1%) and carbapenems (55%), piperacillin/tazobactam (49%) and colistin (2.6%)		Intensive care unit samples	Feretzakis et al. (2019)
	$N = 2$	Resistance to ceftazidime and meropenem (100%); amikacin (50%), aztreonam (50%), cefepime (50%), ciprofloxacin (50%), levofloxacin (50%), gentamicin (50%), imipenem (50%) and piperacillin-tazobactam (50%)		Clinical sample	Havenga et al. (2019)
	$N = 22$	Resistance to ceftazidime (100%), ciprofloxacin (18.2%), carbenicillin (18.2%), amikacin (9.1%), imipenem (4.5%), piperacillin-tazobactam (4.5%) and cefepime (4.5%)	*amp*C, *Oxa*	Clinical samples	Odumosu et al. (2016)
	$N = 27$	Resistance to cefepime (18.5%), ceftazidime (18.5%), ticarcillin (29.6%), ticarcillin-clavulanic acid (22.2%), and aztreonam (70.4%) and all isolates are sensitive to amikacin, gentamicin, tobramycin, imipenem, meropenem, piperacillin and colistin		Wastewater, Northern Poland	Luczkiewicz et al. (2015)
	$N = 32$	Resistance to cefepime and tetracycline families, moderately sensitive to penicillin and all isolates are highly sensitive to quinolone, polymyxin and aminoglycoside (100%)		Seawater, Indian peninsula	Nair et al. (2015)
	$N = 120$	Resistance to ciprofloxacin (93.3%), tobramycin (84.1%), ceftriaxone (81.6%), ceftazidime (80%), gatifloxacin (79.1%), cefoperazone (76.6%), gentamicin (70.8%), amikacin (52.5%), meropenem (54.1%), imipenem (45.8%) and polymyxin B (0%)	*Vim* 2 (12.5%), bla_{IMP} (1.7%)	Clinical samples	Qureshi and Bhatnagar (2015)
	$N = 47$	Resistance to cefotaxime, aztreonam, ceftriaxone (83%); imipenem (78.7%), meropenem (75%), ceftazidime and gentamicin (72.7%); ciprofloxacin (69.2%) and cefepime (67.57%)	$bla_{CTX\text{-}M\text{-}1}$ (8.5%) and bla_{IMP} (2.1%)	Burn patients	Vala et al. (2014)
	$N = 51$	Resistance to tobramycin, piperacillin, carbenicillin and cefoxitin (100%); imipenem, aztreonam, ticarcillin and cotrimoxazole (98%); amikacin, cefepime and ciprofloxacin (96%); meropenem (94%), ceftazidime (92%) and piperacillin/tazobactam (88%)	*Amp*C (60.8%), bla_{TEM} (39.2%), bla_{PER1}(21.6%)	Hospitalized patients with burns	Rafiee et al. (2014)
	$N = 25$	Resistance to ceftriaxone (48%), imipenem (24%), cefotaxime (12%), piperacillin (8%), ceftazidime (4%). Sensitive to cephalosporin (100%), cefepime (100%), ofloxacin (100%) and cefoperazone-sulbactam (100%)		Composts	Kaszab et al. (2011)

only provides relatively higher intrinsic resistance but also increases the efficiency of both acquired and adaptive resistance mechanisms (Breidenstein et al. 2011). For instance, intrinsic resistance supplementing acquired resistance in *P. aeruginosa* includes mutational greater expression of endogenous AmpC β-lactamase gene and Mex efflux pump systems (Saleem and Bokhari 2020). In addition, *P. aeruginosa* has various other mechanisms of acquired resistance achieved either by HGT or chromosomal mutation, which includes enzymatic resistance, upregulation of efflux pumps, alteration in target sites and porins (Mulcahy et al. 2010; Munita and Arias 2016) as well as adaptive resistance by induced stimuli followed by biofilm-mediated resistance and persister cell generation (Taylor et al. 2014).

P. aeruginosa is highly resistant to β-lactam antibiotics, aminoglycosides and fluoroquinolones by virtue of limited membrane permeability achieved by manipulating the embedded four different classes of porins (Rezaloo et al. 2022). Reduction in OprD protein production also leads to the decline in antibiotic uptake making them resistant to imipenem.

Antibiotic-inactivating enzymes such as β-lactamase and AMEs improve tolerance in *P. aeruginosa* (Pang et al. 2019). Furthermore, *P. aeruginosa* has been identified with all four β-lactamases (Dehbashi et al. 2020). Most commonly, it produces class A β-lactamases such as PSE (pseudomonas-specific enzymes), CARB (carbenicillinases) and TEM families. In addition, it possesses an inducible *AmpC* gene-expressing class C β-lactamase enzyme induced by various β-lactam antibiotics (benzylpenicillin, imipenem) that confer resistance against cefepime, ceftazidime and tazobactam (Breijyeh et al. 2020). Furthermore, *P. aeruginosa* also has many resistance genes such as *bla*SHV, blaVEB, blaCTX-M, *bla*TEM and *bla*OXA. *bla*TEM provides resistance against penicillin, *bla*CTX-M against cephalosporins and *bla*OXA against carbapenems. Moreover, *P. aeruginosa* has all three AMEs that provide resistance against aminoglycosides by inactivating the antibiotic molecules by the addition of a phosphate radical, acetyl or adenyl moiety and thereby decline the binding affinity to target sites. Furthermore, *P. aeruginosa* has been identified with four AMEs genes such as aac(6′)-Ib, aac(3)-IV, ant(2″)-Ia and aph(3)-Ia, out of which aac(6′)-Ib linked with higher multidrug resistance mechanism.

Out of all five efflux pump families in *P. aeruginosa,* proteins of the RND family confer major resistance via MexAB-OprM, MexCD-OprJ, MexEF-OprN and MexXY-OprM (Atzori et al. 2019; Pang et al. 2019). Specifically, MexAB-OprM is the most effective pump providing resistance against all β-lactam antibiotics (except imipenem) and quinolones (Dupont et al. 2005; Li et al. 2015). Subsequently, the remaining three efflux pumps expel out β-lactams, quinolones and aminoglycosides (Masuda et al. 2000; Alcalde-Rico et al. 2018). In addition, quinolone resistance is enhanced by point mutation in chromosomal-encoded *gyrA* and *parC* genes. Previous research indicated that efflux pump inhibitors are a potent antipseudomonal therapy for *P. aeruginosa*-mediated illness (Pang et al. 2019). Furthermore, colistin combined with anti-pseudomonas agents (aztreonam, ceftazidime, ciprofloxacin, imipenem, piperacillin or norfloxacin) is considered as a potent therapy for *P. aeruginosa* infections (Ontong et al. 2021; Algammal et al. 2023). In addition, fosfomycin therapy conjugated with other antibiotics (aminoglycosides, cephalosporins and penicillin) is regarded as an effective therapeutic strategy for controlling MDR *P. aeruginosa* (Pachori et al. 2019; Mancuso et al. 2021).

2.3.3 Antibiotic Resistance Profile of *Klebsiella pneumoniae*

K. pneumoniae is a non-fastidious, Gram-negative bacillus that promotes hospital and community-acquired health issues. *K. pneumoniae* causes pneumonia, liver abscess, respiratory tract and blood infections particularly in patients with immune deficiencies (Eghbalpoor et al. 2019; Ghamari et al. 2022). The rapid acquisition of resistance genes in *K. pneumoniae* either through mutations or HGT accounts for a major portion of its greater antibiotic resistance. Acquired genes encode various β-lactamase enzymes including ESBLs and carbapenemases (KPC, IMP, NDM and VIM), hydrolysing antibiotics such as carbapenem, cephalosporin and penicillin. Furthermore, *K. pneumoniae* contains OXA-type β-lactamases (OXA-48) encoded by plasmid making them resistant to almost all β-lactam antibiotics (Table 2.3). A higher prevalence of carbapenem-resistant *K. pneumoniae* (CRKP) harbouring the *bla*KPC-3 gene poses a significant danger to public healthcare since carbapenems are considered as the last-line treatment option against MDR bacterial pathogens (Gualtero et al. 2020). Furthermore, the emergence of *K. pneumoniae* with the *bla*NDM gene has subsequently added to higher carbapenem-resistant strains, encoding enzymes conferring resistance to β-lactams, fluoroquinolones and aminoglycosides. In addition to enzymatic resistance, *K. pneumoniae* has various resistance strategies involving efflux pump activation, decline in antibiotic uptake and target site modifications to encounter the bacteriocidal effects of different antibiotics such as aminoglycosides, fluoroquinolones and tetracyclines. Decreased antibiotic uptake caused by either porin loss (OmpK35 and OmpK36) or porin replacement (OmpK35 → OmpK36) has a synergistic effect with enzymatic resistance conferred by β-lactamases and carbapenemases A in *K. pneumoniae*, resulting in higher resistance to carbapenems and cephalosporins (Santajit and Indrawattana 2016). Moreover, target site modifications in *K. pneumonia* through point mutations in two genes (*par*C and *gyr*A) are considered potent-resistant mechanisms against fluoroquinolones (Denissen et al. 2022). Excluding the acquired resistance, several studies substantiated that *K. pneumoniae* has an inbuilt resistance mechanism against penicillin owing to the existence of chromosomally expressed SHV-1 penicillinase (Wand et al. 2015; Holt et al. 2015; Denissen et al. 2022). Although several practices have been used to control the pathogenesis and transmission of *K. pneumonia*-mediated infections, the spreading and prevalence of the carbapenem-resistant strains are continued, which necessitates effective treatment strategies to overcome the infections caused by these ESKAPE pathogens.

Table 2.3 Antibiotic resistance profile of ESKAPE pathogen (*Klebsiella pneumoniae*)

ESKAPE pathogens	Isolates number	Antibiotic resistance profile (%) (phenotypic resistance)	Resistance gene (genotypic resistance)	Sources of isolation	References
Klebsiella pneumoniae	*N* = 94	Resistance to ampicillin/sulbactam, cefazolin, cefuroxime, piperacillin /tazobactam, cefoperazone/sulbactam, ceftriaxone, ceftazidime, imipenem and meropenem (100%); cefepime (98%), ceftazidime/avibactam (76.6%), cefotetan (75.5%), aztreonam (71.3%), trimethoprim/sulphamethoxazole (24.5%), ciprofloxacin (23.4%), gentamicin (22.3%), tigecycline (21.3%), levofloxacin (17%), tobramycin (16%) and amikacin (14.9%)	*bla*$_{NDM}$ (72.3%), *bla*$_{KPC}$ (24.5%) and *bla*$_{VIM}$ (3.2%)	Hospitalized patients	Shen et al. (2023)
	N = 82	Resistance to ampicillin-sulbactam (18.3%), cefazolin (13.4%), trimethoprim/sulphamethoxazole (13.4%), ceftazidime (10.9%), ceftriaxone (8.5%), aztreonam (4.9%), cefepime (1.2%), imipenem (1.2%), tobramycin (1.2%) and amikacin (0%)	–	Hospitalized children with pneumonia	Mai et al. (2023)
	N = 54	Resistant to ampicillin (100%), amoxicillin/clavulanate (100%), ceftazidime (83.3%), cefotaxime (75.9%), cefepime (70.4%), aztreonam (70.37%), piperacillin/tazobactam (64.8%), ciprofloxacin (63%), gentamicin (55.5%), trimethoprim-sulphamethoxazole (55.5%), amikacin (48%), cefoxitin (48%), imipenem (37%) and meropenem (37%)	*bla*$_{SHV}$ (48%) and *bla*$_{TEM}$ (27.8%)	Clinical sample	Mohammed and Anwar (2022)
	N = 60	Resistance to ceftazidime, ciprofloxacin, cefoxitin, cefotaxime, levofloxacin, imipenem, meropenem, (100%); gentamicin (75%), kanamycin (72%) and tetracycline (53%)	OXA-48 (28.5%), NDM (22%), VIM (10%), *aac6′ Ib* (57%), *aac(3)-IVa* (28%), *aac3-Ia* (22%)	Clinical sample	Ghamari et al. (2022)
	N = 89	Resistance to ampicillin (97%), cefotaxime (78%), aztreonam, cotrimoxazole, gentamicin, cefuroxime, cephalothin and (65%), imipenem (38%), meropenem (31%) and colistin (1%)	*bla*$_{CTXM-1}$ (92%), *bla*$_{SHV}$ (43%), *bla*$_{TEM}$ (36%), *bla*$_{NDM}$ (26%), *bla*$_{GES}$ (20%), and *bla*$_{IMP-1}$ (8%)	Clinical sample (blood, pus, urine)	Indrajith et al. (2021)
	N = 22	Resistant to ampicillin (100%), trimethoprim/sulphamethoxazole, cefuroxime and ciprofloxacin (81%); piperacillin/tazobactam and cefotaxime (76.2%), amoxicillin/clavulanic acid (75.0%), gentamicin (19%), amikacin (19%), imipenem (13.6%), meropenem (13.6%) and ertapenem (13.6%)	–	Urine sample of patients	Kot et al. (2021)

(continued)

Table 2.3 (continued)

ESKAPE pathogens	Isolates number	Antibiotic resistance profile (%) (phenotypic resistance)	Resistance gene (genotypic resistance)	Sources of isolation	References
Klebsiella pneumoniae	*N* = 83	Resistance to cefotaxime (92%), piperacillin-tazobactam (91%), ampicillin-sulbactam (87%), levofloxacin and tobramycin (85%), ceftriaxone (84%), ciprofloxacin (80%), ceftazidime (78%), trimethoprim-sulphamethoxazole (73%), meropenem (70%), amikacin (67%), gentamicin (66%), imipenem (62%), nitrofurantoin (50%). Sensitive to colistin (100%)	–	Clinical sample (urine, wounds, blood, sputum)	Karimi et al. (2021)
	N = 335	Resistance to ticarcillin (100%), ampicillin (95.6%), piperacillin (89.5%), pefloxacin (71.4%); cefazolin and cefazolin (65.8%); aztreonam (60%), piperacillin-tazobactam (54.5%), ampicillin- sulbactam (47.3%), nitrofurantoin (46.3%), cefoxitin (40%), minocycline (38.1%), trimethoprim-sulphamethoxazole (28.3%), ciprofloxacin (26.5%), levofloxacin (26.3%), ceftriaxone (25%), gentamicin (22%), ceftazidime (21.7%), tobramycin (21.5%), cefepime (21%), meropenem (14.7%), ticarcillin-clavulanate (11.1%) and amikacin (6%)	–	Clinical sample (blood, urine, pus, sputum)	Giubelan et al. (2021)
	N = 150	Resistance to cefuroxime (76%), cefotaxime (73.5%), ciprofloxacin (66.9%), ceftazidime (72%), gentamicin (60.3%), piperacillin-tazobactam (48.3%), meropenem (44.6%), aztreonam (70%), netilmicin (46%) and colistin (0.6%)	NDM (23.34%), OXA-48 (8.0%) and KPC (7.30%)	Clinical sample	Aminul et al. (2021)
	N = 165	Resistance to tetracycline (71.5%), ciprofloxacin (56.9%), ceftriaxone (41.8%), cefotaxime (35.1%), cefazolin (23.6%), amikacin (19.7%), ceftazidime (18.2%), cefepime (12.7%), gentamicin (6%) and imipenem (6%)	–	Children with urinary tract infections	Jafari-Sales et al. (2020)
	N = 4	All resistance to ampicillin, amikacin, amoxicillin-clavulanate, piperacillin-tazobactam, all cephalosporins, all carbapenems, trimethoprim-sulphamethoxazole, ciprofloxacin and gentamicin (100%). Sensitive to colistin (100%)	*bla*$_{SHV-1}$(100%), *aac(3′)-II* (100%), *aac(6′)-Ib* (100%) and *aph(3′)-Ia* (100%)	River water	Jelić et al. (2019)
	N = 41	Resistance to ampicillin, cefazolin, ceftazidime, imipenem, piperacillin/tazobactam, ceftriaxone, ampicillin/sulbactam, aztreonam and cefotetan (100%); nitrofurantoin (58.5%), gentamicin (43.9%), trimethoprim-sulphamethoxazole (36.6%), tobramycin (29.3%), ciprofloxacin (26.8%), amikacin and levofloxacin (24.4%)	NDM-1 (56.1%), IMP (26.8%), KPC-2 (22.0%) and SHV (92.7%), TEM-1(68.3%), CTX-M-14 (43.9%), CTX-M-15 (43.9%), OXA-1(14.6%)	Clinical sample	Zhang et al. (2018)
	N = 1261	Resistance to ceftazidime (5.6%), levofloxacin (5.3%), ceftriaxone (4.9%), piperacillin-tazobactam (3.6%), ertapenem (2.5%), cefepime and imipenem (1.7%); amikacin (0.3%)	–	Hospital patients	Karlowsky et al. (2017)
	N = 99	Resistance to ampicillin (92.3%), tetracycline (31.3%), trimethoprim-sulphamethoxazole (18.2%), chloramphenicol (10.1%), gentamicin (5.1%), piperacillin-tazobactam (0%), ceftazidime (0%), aztreonam (0%), imipenem (0%), meropenem (0%), colistin (0%)	*bla*$_{SHV}$, *bla*$_{CTX-M-1}$, *and bla*$_{CTX-M-10}$, *aacA4, aacC2, aadA1*	Food sample	Guo et al. (2016)
	N = 180	Resistance to trimethoprim-sulphamethoxazole (86.7%), cefotaxime (78.8%), ceftazidime (76.6%), cefepime (75%), ciprofloxacin (73.3%), aztreonam (62.2%), kanamycin (43.3%), meropenem (23.3%), amikacin (22.2%), ertapenem (16.1%) and imipenem (7.7%)	*bla*$_{NDM}$, *bla*$_{VIM}$, *bla*$_{CTX-M}$, *bla*$_{SHV}$, *bla*$_{TEM}$, and *bla*$_{KPC}$	Clinical sample (urine, blood, faces)	Nobari et al. (2014)
	N = 153	Resistance to cefazolin (48.4%), cefuroxime (43.8%), ceftazidime (41.2%), cefotaxime (41.8%), cefepime (40.5%), gentamicin (27.4%), piperacillin-tazobactam (18.3%), levofloxacin (15%), amikacin (13.8%), imipenem (3.3%) and meropenem (2.6%)	*bla*$_{CTX-M}$ (41%), *bla*$_{DHA}$ (9.1%), *bla*$_{KPC}$ (3.2%)	Hospital settings	Du et al. (2014)

2.3.4 Antibiotic Resistance Profile of *Enterobacter* Species

Enterobacter species are mostly motile, Gram-negative, non-fastidious, rod-shaped bacterium of the Enterobacteriaceae family. More often, these are opportunistic pathogens exhibiting several antibiotic resistance mechanisms, which cause severe infections in immunocompromised patients. Several workers reported the existence of two clinically significant *Enterobacter* species such as *Enterobacter aerogenes* and *Enterobacter cloacae* and their resistance pattern is depicted in Table 2.4. The MDR *Enterobacter* bacterial pathogens have proven as tolerant to almost all antibiotics except tigecycline and colistin (Di Franco et al. 2021). Several isolates of *Enterobacter species* produce enzymes that include ESBLs, carbapenemases (VIM, OXA and KPC) and metallo-β-lactamse-1, conferring them intrinsic resistance against ampicillin, cefoxitin, amoxicillin and cephalosporins (Yuan et al. 2019; Halat and Moubareck 2022). Nevertheless, *E. aerogenes* with acquired ESBLs were observed to be resistant against last-line carbapenems and colistin, whereas *E. cloacae* strains producing β-lactamases exhibit high resistance against multiple antibiotics including cephalosporins, ampicillin and cefoxitin (Davin-Regli and Pagès 2015). Moreover, the permanent shutdown of chromosomally encoded AmpC β-lactamases, readily expressed at high levels via mutation, serves a vital role in antibiotic-resistant mechanisms in *Enterobacter* species. Earlier findings suggested the existence of several efflux pumps including AcrAB-TolC in *E. cloacae* and *E. aerogenes* expel out several commercially available antibiotics thereby conferring a high-resistance mechanism (Rosa et al. 2017). In addition, the efflux pumps including EefABC and OqxAB along with the several members belonging to RND and MATE families also reported to resist various antibiotics (Davin-Regli and Pagès 2015). Besides, the alternations in the outer membrane also enhance the development of multidrug resistance in *Enterobacter* species (Davin-Regli and Pagès 2015). Moreover, several scientific studies substantiated the fact that the emergences of pandrug-resistant *Enterobacter* species were found to resist even the last-line colistin (De Oliveira et al. 2020). Furthermore, previous studies suggested that *E. aerogenes* can harbour a subpopulation of colistin-resistant bacteria that are undetectable by existing diagnostic methods, which makes the treatment strategies more complicated leading to severe infections (Mancuso et al. 2021).

The "Priority list 2: High" contains the multidrug resistance Gram-positive ESKAPE pathogens such as *E. faecium* and *S. aureus* associated with severe life-threatening infections.

2.3.5 Antibiotic Resistance Profile of *Staphylococcus aureus*

S. aureus is a Gram-positive cocci, commensal bacterium that colonizes soft tissues and skin causing diverse infections ranging from nosocomial infections in the eye, intestine, skin, urogenital tract, nasopharynx to life-threatening pneumonia, bacteraemia, toxic shock syndrome resulting in increased death rates in hospital regions (Stephen et al. 2023). MDR *S. aureus* is highly prevalent in clinical and hospital

Table 2.4 Antibiotic resistance profile of ESKAPE pathogens (*Enterobacter cloacae* and *Enterobacter aerogenes*)

ESKAPE pathogens	Isolates number	Antibiotic resistance profile (%) (phenotypic resistance)	Resistance gene (genotypic resistance)	Sources of isolation	References
Enterobacter cloacae	$N = 2$	Resistance to ampicillin (100%), aztreonam (100%), cefotaxime (100%), ceftriaxone (100%), amoxicillin/clavulanate (100%), sulphamethoxazole/ trimethoprim (100%), cefoxitin (100%), ciprofloxacin (100%), meropenem (50%), gentamycin (50%), amikacin (0%)	–	Hospital environment	Sebre et al. (2020)
Enterobacter cloacae and *Enterobacter aerogenes*	$N = 9$	Resistance to ampicillin (100%), cephalothin (100%), cefoxitin (88.8%), ampicillin/clavulanate (55.5%), tetracycline (44.4%), ciprofloxacin (33.3%), sulphamethoxazole/ trimethoprim (22.2%), gentamicin (11.1%), trimethoprim (11.1%), ceftriaxone (33.3%), sulphisoxazole (33.3%)	–	Waste water samples from healthcare sector	Tesfaye et al. (2019)
Enterobacter species	$N = 77$	Resistance to cefepime (68.8%), cefotaxime (63.6%), ceftazidime (50.6%), nalidixic acid (46.7%), ciprofloxacin (20.7%), ofloxacin (11.6%), aztreonam (41.6%), tobramycin (87%), gentamicin (54.5%), colistin (0%)	bla_{TEM} (44.2%), $bla_{CTX\text{-}M}$ (36.4%), $bla_{CTX\text{-}M\text{-}2}$ (5.2%), bla_{IMP} (3.9%), $bla_{OXA\text{-}23}$ (2.6%), $bla_{VIM\text{-}2}$ and $bla_{NDM\text{-}1}$ (1.3%)	Hospital origin	Boutarfi et al. (2019)

ESKAPE pathogens	Isolates number	Antibiotic resistance profile (%) (phenotypic resistance)	Resistance gene (genotypic resistance)	Sources of isolation	References
Enterobacter cloacae	$N = 5$	Resistance to ampicillin (100%), cefotaxime and ceftriaxone (80%), amoxicillin/clavulanate (60%), ceftazidime (40%), amikacin (20%), tobramycin (20%), meropenem (0%)	–	Hospital area	Bitew (2018)
Enterobacter species	$N = 661$	Resistance to ceftriaxone (44%), ceftazidime (36.6%), piperacillin-tazobactam (9.5%), ertapenem (8.2%), cefepime (7.1%), levofloxacin (6.2%), imipenem (2.3%), amikacin (0.2%)	–	Hospital patients with intra-abdominal infections	Karlowsky et al. (2017)
Enterobacter cloacae and *Enterobacter aerogenes*	$N = 130$	Resistance to carbapenems (41.9%) in *E. aerogenes* and (32%) in *E. cloacae*. All sensitive to fosfomycin, polymyxin B and tigecycline	bla_{KPC}, bla_{TEM} and /or bla_{CTX} and AcrAB-TolC gene	Hospital settings	Rosa et al. (2017)

(continued)

Table 2.4 (continued)

ESKAPE pathogens	Isolates number	Antibiotic resistance profile (%) (phenotypic resistance)	Resistance gene (genotypic resistance)	Sources of isolation	References
Enterobacter species	$N = 29$	Resistance to ampicillin (97%), cefalotin (97%), cefotaxime (62%), nitrofurantoin (59%), netilmicin (59%), gentamicin (55%), sulphamethoxazole/trimethoprim (55%), chloramphenicol (52%), ceftriaxone (48%), cefepime (41%), levofloxacin (24%) and amikacin (17%)	–	Clinical isolates of diabetic foot ulcers	Sánchez-Sánchez et al. (2017)
Enterobacter species	$N = 7$	Resistance to ampicillin (100%), cotrimoxazole (43%), tetracycline (57%), cephalothin (43%), nalidixic acid (29%), streptomycin (29%), chloramphenicol (14%), ceftriaxone (14%), ciprofloxacin (14%), gentamycin (14%), cefotaxime (14%) and kanamycin (14%)	–	Hospital and non-hospital environment	Moges et al. (2014)
Enterobacter species	$N = 108$	Resistance to meropenem (100%), gentamicin (94.4%), ciprofloxacin (92.5%), cotrimoxazole (92.5%), tobramycin (81.4%), piperacillin-tazobactam (80.5%), ceftazidime (76.5%) and ceftriaxone (61%),	–	Hospital setting (ICU)	Ong et al. (2011)

ESKAPE pathogens	Isolates number	Antibiotic resistance profile (%) (phenotypic resistance)	Resistance gene (genotypic resistance)	Sources of isolation	References
Enterobacter species	$N = 57$	Resistance to cefepime (27.85%), cefoxitin (53.6%), aztreonam (34.4%), chloramphenicol (32.7%), varying percent of resistance to cefpodoxime, clavulanic acid, ceftadoxime and cefuroxime Sensitive to imipenem (100%)	bla_{TEM} (42%) and bla_{SHV} (44%)	River water sample	Sharma et al. (2008)

regions posing a significant threat to human healthcare. Furthermore, *S. aureus* has an extraordinary ability to resist commonly available antibiotics (penicillin, aminoglycosides, macrolides and tetracycline) and is capable of developing resistance to newer antibiotics leading to therapeutic difficulties (Table 2.5).

S. aureus has intrinsically meditated resistance mechanisms including efflux pump, enzymatic resistance and outer membrane permeability conferring resistance against β-lactam and aminoglycosides. In addition, *S. aureus* has acquired several resistance mechanisms through HGT including target site modifications, mutation in target genes, acquiring resistance gene and overexpression of efflux pump (*nor*A pump) against methicillin, vancomycin, daptomycin and linezolid (Guo et al. 2020; Denissen et al. 2022). Subsequently, *S. aureus has been recognized with various AMEs such as aac(3), aac(6′), ant(4′), aph (3′) that collectively confer resistance against amikacin, gentamicin and tobramycin (*De Oliveira et al. 2020*).* Resistance acquired by *S. aureus* involves mutation in target genes such as *gyr*A and *gyr*B making them resistant to fluoroquinolones or reduced synthesis of the outer membrane proteins.Remodelling of target sites by mecA and mecC genes expressing PBP2a protein with low affinity for almost all β-lactam, serve as main mechanism of resistance against antibiotics especially methicillin. This leads to the outbreak of methicillin resistance S. aureus (MRSA) (D'Costa et al., 2011; Mancuso et al., 2021). *S. aureus* acquires resistance against vancomycin (VRSA) by acquiring *van*A gene from vancomycin-resistant *Enterococcus* sp. *Staphylococcus aureus* was also reported to acquire resistance against trimethoprim-sulphamethoxazole and tetracyclines. The alarming rise of multidrug-resistant *S. aureus* is hazardous for public health. Moreover, *S. aureus* resistance to methicillin (MRSA) and vancomycin (VRSA) causes hindrances in treatment procedures making infection more critical and deadly. Thus, intensive and continuous research on resistance strategies is a prerequisite to dealing with MRSA- and VRSA-mediated infections through the development of novel antibiotics useful for public healthcare.

2.3.6 Antibiotic Resistance Profile of *Enterococcus faecium*

E. faecium is a facultative anaerobe, Gram-positive cocci, which is the primary cause of bacteraemia, neonatal sepsis, endocarditis and meningitis (Abamecha et al. 2015). Besides, *E. faecium* has been linked to a number of hospital-acquired infections, such as pelvic, intra-abdominal and urinary tract infections. *E. faecium* bacterial pathogens exhibited intrinsic and acquired resistance against different antibiotics, particularly aminoglycosides, ampicillin and glycopeptides specifically vancomycin (Gawryszewska et al. 2016). *E. faecium*-mediated infections are challenging because they have a combination of intrinsic, acquired and adaptive resistance mechanisms. The details of the resistance profile of *E. faecium* with resistance genes involved and sources of isolation have been depicted in Table 2.6.

The over-expression of β-lactamase enzymes as well as the alterations in target sites (penicillin-binding site) in *E. faecium* was reported to be the major resistance mechanism against the β-lactam group of antibiotics. For example,

Table 2.5 Antibiotic resistance profile of ESKAPE pathogen (*Staphylococcus aureus*)

ESKAPE pathogens	Isolates number	Antibiotic resistance profile (%) (phenotypic resistance)	Resistance gene (genotypic resistance)	Sources of isolation	References
S. aureus	$N = 28$	Resistance to ampicillin (82.1%), amoxicillin-clavulanic acid (75%), sulphamethoxazole/trimethoprim (64.3%), erythromycin (60.7%), gentamicin (57.1%), cefepime (53.6%), tetracycline (46.4%), ciprofloxacin (35.7%), ceftriaxone (32.1%), cefoxitin (28.6%) and imipenem (0%)	*nuc* (46.4%), *mecA* (32.1%), *tst* (14.3%), *pvl* (10.7%)	Clinical sample	Fazal et al. (2023)
	$N = 20$	Resistance to methicillin (100%), penicillin G (100%), colistin (70%), cefoxitin (55%), imipenem (5%) and levofloxacin (10%)	–	Milk sample	Zedan et al. (2023)
	$N = 11$	Resistance to amikacin (9.8%), gentamicin (9.8%), tetracycline (9.8%), meropenem (9.8%), ceftriaxone (9.8%), ceftazidime (9.8%), cefixime (9.8%), cephalexin (90.1%), cephradine (9.8%), cefuroxime (9.8%), levofloxacin (9.8%), ciprofloxacin (9.8%), clindamycin (9.8%), penicillin-G (9.8%), amoxicillin (9.8%), sulphamethoxazole/trimethoprim (9.8%), azithromycin (9.8%), nitrofurantoin (9.8%), fluconazole (9.8%) and colistin (9.8%)	–	Hospitalized patients	Nobel et al. (2022)
	$N = 13$	Resistance to ampicillin (76.9%), erythromycin (76.9%), chloramphenicol (61.5%), ciprofloxacin (53.8%), vancomycin (30.7%), rifampicin (7.6%) and teicoplanin (0%)	–	Hospital waste water sample	Alarjani and Skalicky (2021)
	$N = 24$	Resistance to amoxicillin and cephalexin (100%), ciprofloxacin (91.7%), cefixime (83.3%), gentamicin (79.2%), ceftriaxone (58.3%), netilmicin (45.8%), imipenem (0%), meropenem (0%), vancomycin (0%) and linezolid (0%)	–	ICU	Setu and Anijejob (2021)
	$N = 84$	Resistance to penicillin (64.3%), tetracycline (59.5%), cefoxitin (35.7%), erythromycin (30.9%), amoxicillin-clavulanic acid (21.4%), cefotaxime (15.5%) and ampicillin-sulbactam (13.1%)	*coa* (35.7%), *mecA* (35.7%), *spa* (30.9%) and *pvl* (10%)	Subclinical bovine mastitis	Algammal et al. (2020)

(continued)

Table 2.5 (continued)

ESKAPE pathogens	Isolates number	Antibiotic resistance profile (%) (phenotypic resistance)	Resistance gene (genotypic resistance)	Sources of isolation	References
S. aureus	*N* = 50	Resistant to erythromycin (82%), vancomycin (80%), tetracycline and amoxicillin (76%); oxacillin (74%), cefoxitin (68%), sulphamethoxazole (20%) and neomycin (8%)	–	Milk and beef sample	Bissong et al. (2020)
	N = 63	Resistance to penicillin (93.7%), cefoxitin (85.7%), erythromycin (49.2%), sulphamethoxazole/ trimethoprim (47.6%), doxycycline (38.1%), chloramphenicol (23.8%), ciprofloxacin (19%), gentamicin (15.9%) and clindamycin (11.3%)	–	Hospital environment	Sebre et al. (2020)
	N = 80	Resistance to lincomycin (100%), oxacillin (98.6%), cefoxitin and penicillin (97.5%), ampicillin (96.3%), cefozolin (72.5%), azithromycin (66.3%), amoxicillin/ clavulanic acid (52.5%), erythromycin (40%) and vancomycin (33.8%), clindamycin and rifampicin (22.5%), sulphamethoxazole/trimethoprim (17.5%), tetracycline (11.3%), gentamicin, imipenem and chloramphenicol (5%) and amikacin (2.5%)	*aac(6′)/aph(2″)* (56.25%), *blaZ* (70.00%), *erm*C (62.50%), *msr*A (22.50%) and *tet*K gene (70.00%)	Waste water effluent	Ramessar and Olaniran (2019)
	N = 14	Resistance to benzylpenicillin (85.7%), erythromycin (50%), clindamycin (42.8%); oxacillin, moxifloxacin ceftriaxone and ciprofloxacin (36%); amoxicillin-clavulanic acid (30.8%), gentamicin and tetracycline (21.4%); sulphamethoxazole/trimethoprim (7.1%). Sensitive to linezolid, teicoplanin, vancomycin, tigecycline and fusidic acid	*mec*A (35.7%), *tsst*-1 (7.1%)	Human derived sample	Vitale et al. (2019)
	N = 79	Resistant to ampicillin (100%), oxacillin and cefoxitin (68.4%), clindamycin (63.3%), cephalothin (59.5%), tetracycline (57%), sulphamethoxazole + trimethoprim (53.2%), bacitracin (53.2%) and erythromycin (51.9%)	–	Hospitalized patient	Tadesse et al. (2018)
	N = 15	Resistant to penicillin and ampicillin (86.7%), ceftriaxone (46.7%), cefoxitin and gentamicin (40%), trimethoprim-sulphamethoxazole (33.3%), erythromycin (26.7%) and clindamycin (20%)	–	Hospital area	Bitew (2018)

ESKAPE pathogens	Isolates number	Antibiotic resistance profile (%) (phenotypic resistance)	Resistance gene (genotypic resistance)	Sources of isolation	References
S. aureus	*N* = 30	Resistance to ampicillin and penicillin G (96.7%); clindamycin and rifampicin (80%); erythromycin (70%); oxacillin (73.3%). Susceptibility to imipenem (96.7%), levofloxacin (86.7%), chloramphenicol (83.3%), cefoxitin (76.7%), ciprofloxacin (66.7%), gentamycin (63.3%), tetracycline and sulphamethoxazole-trimethoprim (56.7%), vancomycin and doxycycline (50%)	*tet*M (72.7%), *erm*B (71.4%), *blaZ* (55.2%), *fem*A (53.3%), *rpoB* (45.8%), *mecA* (22.7%)	Beach water and sand samples	Akanbi et al. (2017)
	N = 10	Resistance to ampicillin (100%), nalidixic acid (60%), cefotaxime (20%), clindamycin (20%), erythromycin (20%), chloramphenicol (20%), methicillin (10%), ciprofloxacin (10%), gentamycin (10%), tetracycline (10%), streptomycin (10%), ceftriaxone (10%), cephalothin (10%). Sensitive to vancomycin and streptomycin (100%)	–	Hospital and non-hospital environment	Moges et al. (2014)
	N = 82	Resistance to tetracycline (97.6%), methicillin (75.6%); sulphamethoxazole, trimethoprim, and streptomycin (31.7%), gentamicin (29.3%), enrofloxacin (28%), chloramphenicol (20.73%) and cephalothin (17%)	*mecA* (82.92%), *msrB* (47.56%), *msrA* (34.14%), *tetK* (52.4%), *tet*M (46.34%) and *aac*A-D (39%)	Chicken meat	Momtaz et al. (2013)
	300	Resistance to penicillin (98.3%), tetracycline (56.3%), ciprofloxacin and levofloxacin (55.3%), methicillin (55%), gentamicin (54%), tobramicin, (53%), moxifloxacin (50.3%), clindamycin (20%), linezolid, vancomycin, teicoplanin and nitrofurantoin (0%)	–	Clinical sample	Yucel et al. (2011)
	N = 11	Resistance to methicillin and cefixime (100%), tetracycline (54.5%), ciprofloxacine (18.2%), sulphamethoxazole-trimethoprim (18.2%) and norfloxacine (0%)	–	Clinical specimen	Motamedi et al. (2010)
	N = 10	Resistance to methicillin (60%), erythromycin (40%), kanamycin (30%), sulphamethoxazole (30%), tetracycline (30%) and chloramphenicol (10%)	*ccr*B (50%), *erm*A (50%), *tet*K (20%) and *tet*M (20%)	Public marine beaches	Soge et al. (2009)

Table 2.6 Antibiotic resistance profile of ESKAPE pathogen (*Enterococcus faecium*)

ESKAPE pathogens	Isolates number	Antibiotic resistance profile (%) (phenotypic resistance)	Resistance gene (genotypic resistance)	Sources of isolation	References
E. faecium	*N* = 54	Resistance to levofloxacin (96.3%): penicillin, ampicillin, ampicillin-sulbactam and imipenem (92.6%), minocycline (3.7%), erythromycin (88.7%), rifampicin (44.4%), gentamicin (18.5%). Sensitive to vancomycin, daptomycin and teicoplanin (100%)	*msrC* (55.6%), *ermB* (38.9%), *aac(6′)-le-aph(2″)-la* (16.7%) and *aph(3′)-IIIa* (3.7%)	Clinical sample	Aung et al. (2023)
	N = 16	Resistance to erythromycin (100%), ampicillin (87.5%), tetracycline (87.5%), ciprofloxacin (81.3%), penicillin (50%), nitrofurantoin (50%), vancomycin (43.7%) and chloramphenicol (37.5%)	*aac(6′)-Ie-aph(2″)-Ia* (16.2%), *tet*M (13.5%), *erm*B (13.5%), *tet*K (10.8%), *tet*L (10.8%), *erm*A (0.01%), *van*A (0%)	Poultry sample	Mwikuma et al. (2023)
	N = 44	Resistant to erythromycin (100%). Susceptible to tetracycline (27.5%), levofloxacin (56.8%), gentamicin (63.6%), ampicillin (84.1%), nitrofurantoin (90.0%), teicoplanin (97.5%); vancomycin, tigecycline and linezolid (100%)	–	Clinical sample	Georges et al. (2022)
	N = 40	Resistance to fluoroquinolones (ciprofloxacin, levofloxacin and norfloxacin) (100%), erythromycin (92.5%), rifampin (80%), nitrofurantoin (60%), linezolid (35%), fosfomycin (27.5%), tetracycline (25%), doxycycline (25%), vancomycin (17.5%), minocycline (15%) and chloramphenicol (10%)	*erm*B (25%), *tet*L (22.5%), *tet*M (22.5%), *ant(6′)-la* (10%), *mef*AE (7.5%), *erm*C (5%), *van*C1 (5%), aac (6′)-Ie-*aph(2″)-la* (2.5%), *ant(4′) Ia* (2.5%), and *aph(3′)-IIIa* (2.5%)	Soil and water sample	Dos Santos et al. (2021)
	N = 12	Resistant to oxacillin (100%), erythromycin, vancomycin and linezolid (66.7%), penicillin and nitrofurantoin (58.3%) and tetracycline (0%)	*bla*Z (66.7%), *erm*B (41.7%), *van*A (33.3%), *optrA* (8.3%) and *tetM* (0%)	Sheep and goat milk sample	El-Zamkan and Mohamed (2021)

ESKAPE pathogens	Isolates number	Antibiotic resistance profile (%) (phenotypic resistance)	Resistance gene (genotypic resistance)	Sources of isolation	References
E. faecium	$N = 10$	Resistant to ampicillin (100%), imipenem (100%), ciprofloxacin (100%), trimethoprim/sulphamethoxazole (100%), gentamicin (40.0%) and vancomycin (40%)	–	Urine samples from patients	Kot et al. (2021)
	$N = 98$	Resistance to linezolid (41%), tetracycline (40%), ciprofloxacin (34%), ampicillin (16%), chloramphenicol (6%), vancomycin (6%) and gentamicin (1%)	*van*A (2%) and *van*B (2%)	Environmental water sample	Monteiro and Santos (2020)
	$N = 144$	Resistant to ciprofloxacin (95.1%), rifampin (93%), erythromycin (89.5%), tetracycline (71.5%), linezolid (70.8%), chloramphenicol (50%), penicillin (39.5%), ampicillin (36.8%), vancomycin (33.3%) and fosfomycin (24.3%)	*erm*A (20.8%), *erm*B (25.6%), *tet*A (9%), *tet*B (6.2%), *tetM* (54.1%), *tet*L (46.5%), *optrA* (3.4%)	Faecal sample from yaks	Cui et al. (2020)
	$N = 30$	Resistance to erythromycin (76.7%), ampicillin (70%), ciprofloxacin (70%), quinupristin/dalfopristin (56.7%), gentamicin (56.7%), tigecycline (30%), tetracycline (13.3%), chloramphenicol and vancomycin (0%)	–	Human clinical sample	Golob et al. (2019)
	$N = 108$	Resistant to erythromycin (100%), ciprofloxacin and clindamycin (96%); ampicillin (92.5%) and gentamicin (75%)	*Van*A and *aac (6′)-le-aph(2″)-la,*	Infected children	Sattari-Maraji et al. (2019)
	$N = 166$	Resistant to rifampin (62.1%), tetracycline (27.7%), erythromycin (27.1%), ciprofloxacin (17.5%), penicillin (4.8%), ampicillin (1.2%), chloramphenicol (3%) and vancomycin (0.6%)	*tet*M (21.6%), *erm*B (19.2%), *tet*L (10.2%), *tet*O (3.6%), *optr*A (0.6%)	Layer parent stock	Kim et al. (2019a, b)
	$N = 142$	Resistant to tetracycline (88.7%), erythromycin (82.3%), rifampin (40.1%), ciprofloxacin (26%), chloramphenicol (21.8%), quinupristin/dalfopristin (20.4%), gentamicin (8.4%), ampicillin (2%), teicoplanin (1.4%), nitrofurantoin (1.4%) and vancomycin (0.7%)	–	Broiler cloacal sample	Ünal et al. (2017)
	$N = 112$	Resistance to ciprofloxacin (99.1%), rifampin (92.9%), penicillin (91.7%), ampicillin (89.3%), gentamicin (86.6%), tetracycline (52.7%), vancomycin (7.1%), chloramphenicol (6.25%) and quinupristin/dalfopristin (5.4%)	–	Clinical sample	Gawryszewska et al. (2016)
	$N = 40$	Resistance to penicillin (100%), erythromycin (97.5%), tetracycline (92.5%), streptomycin (90%), ciprofloxacin (87.5%), ampicillin (87.5%), norfloxacin (85%), nitrofurantoin (45%) and chloramphenicol (27.5%)	–	Intestinal tracts of hospitalized patients	Abamecha et al. (2015)

E. faecium was found to express mutated penicillin-binding protein, i.e. PBP5 with a lower affinity that confers resistance to β-lactam enzymes (penicillin). Besides, it is evident from the earlier studies that *E. faecium* produces all three AMEs thereby conferring intrinsic resistance to aminoglycosides (Kim et al. 2021). Moreover, *E. faecium* has been identified with three different resistant mechanisms against cephalosporins (Zaheer et al. 2020). In addition, *E. faecium* has been reported to exhibit relatively higher resistance against vancomycin, once considered an effective antibiotic against them, by acquiring the resistance gene (*van*A) through mobile genetic elements. The *van*A complex mediates target alterations resistance mechanism through the alternations in the peptidoglycan crosslink targets from D-Ala-D-Ala to D-Ala-D-Lac thereby increasing resistance against the currently available commercial glycopeptides, including vancomycin and teicoplanin. Moreover, the *E. faecium* has also acquired relatively higher resistance against fluoroquinolones by two distinct mechanisms, by a point mutation in *gyrA* and *parC* genes and by the mechanism of activation of efflux pump (*Nor*A) (Jubeh et al. 2020). Keeping in view, intensive research on antibiotic resistance patterns and mechanisms is necessitated to deal with *E. faecium*-mediated infections to combat their pathogenesis for better public healthcare.

2.4 Detection Methods for Antibiotic Resistance in ESKAPE Pathogens

As mentioned earlier, antibiotic resistance represents an alarming danger to public health by turning routine infections into potentially lethal diseases (Kaprou et al. 2021). The pattern of antibiotic resistance varies globally due to the existence of pathogenic variants as well as variances in pathogen-drug interactions. Apart from drug development, rapid and early detection of antibiotic resistance contributes significantly to antibiotic stewardship, prolonging antibiotic effectiveness and treatment optimization (Vasala et al. 2020; Hay 2021; Kalpana et al. 2023). Clinicians are restricted to using broad-spectrum antibiotics irrespective of diagnosis due to longer turnover time following identification and analysis of bacterial pathogens. Therefore, rapid and effective diagnosis is considered an essential tool for the development of any strategies against MDR pathogens (Kaprou et al. 2021). With the tremendous spike in the percentage of resistant ESKAPE pathogens, there is an unmet need for fast, robust, highly sensitive, affordable and potable detection methods to elucidate the antibiotic resistance patterns and mechanisms for better and accurate treatment options using specific antibiotics to combat the difficulties and complications in disease management and thereby controlling infections caused by the ESKAPE pathogens.

2.4.1 Different Detection Tools Used for Antibiotic Resistance Profiling

Detection of antibiotic profiling and resistance mechanisms exhibited by ESKAPE pathogens has come a long way. Previously, conventional techniques have been used for addressing resistance patterns, antibiotic susceptibility tests and surveillance. However, with the development of technologies, non-conventional methods and microfluid techniques have been employed for detection purposes. The limitations and advantages of current detection methods are summarized. Besides, the development of innovative and emerging detection methods targeting faster detection is also discussed.

2.4.1.1 Conventional Detection Methods

Convention methods used for antibiotic resistance profiling of ESKAPE pathogens have been broadly classified into phenotypic methods and genotypic methods. Phenotypic strategies are culture-dependent techniques frequently utilized in the clinical and diagnostic sectors. These are standardized methods to identify phenotypic resistance and are critical for detecting new resistance profiles. They are further divided into manual and automated methods. Manual phenotypic approaches include disk diffusion, broth microdilution techniques, agar dilution and gradient test method, whereas automated method includes commercial platforms such as Vitek 2 system, Phoenix system, Sensititre™, MicroScan AutoScan 4, MicroScan WalkAway system and Aris™ 2X that run in a simplified workflow and time efficiency process (Khan et al. 2021). Each automated platform has inherent cons and pros, and the results vary significantly based on the antibiotics used, cards used for detection and software versions. Disk diffusion and broth dilution are widely used methods for evaluating antibiotic susceptibility against bacterial pathogens. Conventional phenotypic methods can detect both resistance and susceptibility patterns. In addition, minimum inhibitory concentration (MIC) through broth dilution and Epsilometer test have been performed using standard antibiotics following CLSI and EUCAST guidelines (Kaprou et al. 2021). Several phenotypic methods include the interpretation of antibiogram in Gram-positive ESKAPE pathogens through disk diffusion, which is considered as an accurate and economically feasible method for elucidating the β-lactam resistance pattern. The resistance pattern in vancomycin-resistant *Enterococcus* sp. is diagnosed via disk diffusion, broth dilution and breakpoint agar methods (Muntean et al. 2022). Furthermore, the susceptibility patterns of carbapenem-resistant *K. pneumonia* against polymyxins were detected by the automated Sensititre™ panel method offering both quantitative and qualitative attributes (Richter et al. 2018). However, the prolonged turn-around time (12–72 h) including prior bacterial identification, maintenance of bacterial culture, requirement of specific media and growth conditions are considered as the major limitations of these methods (Kalpana et al. 2023). In addition, they are limited to culture-dependent bacteria only. Therefore, there is a need for sensitive and specific molecular-based methods for analysing the antibiotic resistance profile in ESKAPE pathogens.

In contrast, the genotypic methods are molecular-based approaches that determine the resistance profile through amplification and hybridization of genetic determinants encoding resistance. It includes fast, highly specific and sensitive methods using polymerase chain reaction, isothermal-based amplification methods and DNA microarrays. They have several advantages over the phenotypic methods, which include shorter turn-around time, multiplex targeting, precise characterization and elimination of isolate purification step with low-biohazard risk (Kalpana et al. 2023). However, pitfalls associated with the molecular approaches include high cost and false-positive or false-negative results due to amplification of silent gene or pseudogene or alteration in primer-binding sites and detection. Moreover, the molecular assay is unable to determine the MIC value. The hypervariable regions in bacterial pathogens with rapidly changing mechanisms and wide diversity in resistance genes in ESKAPE pathogens especially in Gram-negative ESBLs strains and carbapenem resistance pathogens pose major challenges to be diagnosed by molecular approaches.

2.4.1.2 Non-Conventional Detection Methods

With the advancement in sequence technology, non-conventional methods for detecting antibiotic resistance in ESKAPE pathogens have emerged. These include genome sequencing, MALDI-TOF MS (matrix-assisted laser desorption/ionization-time of flight mass spectrometry) and FTIR (Fourier transform infrared) spectroscopy. Sequencing-based exploration of antibiotic profiling is an emerging technique, which provides huge databases useful for pinpointing antibiotic resistance and their determinates alongside correlate the molecular-based finding with phenotypic results more rapidly and precisely (Pereckaite et al. 2018). Besides, the utilization of next-generation sequencing (NGS) is beneficial for outlining the molecular basis of resistance mechanisms in pandrug-resistant *A. baumannii.* Moreover, whole genome sequencing is proving as one of the powerful tools that identifies all the prevalent genes and mutations conferring phenotypic resistance with higher accuracy, consistency and affordability. The ongoing epidemiological studies based on whole genome sequencing, especially in MRSA and carbapenem-resistant *K. pneumonia*, were found to be significant in controlling their transmission (Madigan et al. 2018; Fasciana et al. 2019). Besides, MALDI-TOF MS has accelerated antibiotic resistance detection, involving the monitoring of antibiotic modification by pathogenic bacteria, detection of the specific proteins responsible for resistance mechanism and thereby used for the detection of β-lactamase production and carbapenemases in ESKAPE pathogens (Gajdács et al. 2017). Subsequently, whole cell MALDI-TOF MS is utilized for the efficient detection of MRSA strain typing contributing significantly towards control strategies in hospital settings (Kalpana et al. 2023). Nevertheless, MALDI-TOF MS appears as a powerful tool with a precise, reliable, robust and eco-friendly methodology to profile the antibiotic resistance in ESKAPE pathogens (Vrioni et al. 2018), still not yet validated for all antibiotics and pathogenic species (Gajic et al. 2022). Furthermore, FTIR spectroscopy is also a robust and economical method that differentiates molecular changes linked to antibiotic resistance development and thereby predicts antibiotic resistance patterns.

2.4.1.3 Emerging Detection Methods

Various emerging detection techniques based on microfluidics used for the diagnosis of antibiotic resistance patterns acquired by the ESKAPE pathogens include spectroscopic and colorimetric-based techniques, point-of-care (POC), quartz-crystal microbalance (QCM), pH-based, multiplexing with single cell or single molecule with their limitations and advantages. Fast and high-throughput analysis, accurate results, minimal sample volume, cheaper, low-power consumption, automation, integration, compactness and portability are advantages of emerging detection methods over others. Even though potent and inventive detection tools have been developed, extensive research is still required to circumvent the drawbacks and limitations. Hence, more sustainable advancement is a prerequisite for the development of innovative detection platforms with superior performance and efficiency for speeding up the detection of antibiotic resistance patterns exhibited by ESKAPE pathogens. Improvement of detection techniques with an objective for appropriate antimicrobial therapy and designing potent antimicrobial agents is mandatory to battle against the ESKAPE pathogens in the future.

2.5 Conclusion

ESKAPE pathogens with different antibiotic resistance mechanisms and resistance profile reflect paradigms of tolerance, pathogenesis, disease severity and propagation. ESKAPE pathogens responsible for causing infectious diseases escaping the bacteriocidal therapeutic effects of antibiotics make the treatment strategies more complicated. Based on the overall healthcare and economic burden, the ESKAPE pathogens draw the uttermost attention from clinical as well as research and development perspectives because of their high tendency to acquire multidrug resistance and rapid spread of resistance mechanisms. In this regard, current research efforts based on the exploration of known and underlying antibiotic resistance mechanisms contribute significantly to the antibiotic stewardship programme and the development and advancement of potent antibiotics. Furthermore, rapid and early detection of antibiotic resistance exhibited by ESKAPE pathogens by innovative diagnostic techniques will be imperative to develop advanced therapeutic schemes for the prevention, mitigation and treatment of infectious diseases attributed by ESKAPE pathogens. Several factors may play a crucial role in overcoming the detrimental consequences of MDR pathogens ("superbugs") including focusing on their prevention by detecting the antibiotic profiling, development of potent antimicrobial agents, innovations in treatment strategies with clinical involvement and extensive research based on their epidemic nature of proliferation. Therefore, collective global action is required for the effective management of antibiotic resistance acquired by ESKAPE pathogens through the balancing of innovation access and stewardship.

Acknowledgements I would like to express my gratitude to the Department of Biotechnology and Bioinformatics, Sambalpur University for providing a scientific environment. I express my hearty thanks to all my seniors, especially Dr. Subhaswaraj Pattnaik and my dear friends for supporting me.

Funding The funding of the Department of Science and Technology (DST), Govt. of India, New Delhi through DST/INSPIRE FELLOWSHIP/2020/IF200208 is hereby acknowledged.

Conflict of Interest None.

References

Abamecha A, Wondafrash B, Abdissa A (2015) Antimicrobial resistance profile of *Enterococcus* species isolated from intestinal tracts of hospitalized patients in Jimma, Ethiopia. BMC Res Notes 8:1–7. https://doi.org/10.1186/s13104-015-1200-2

Abdi SN, Ghotaslou R, Ganbarov K, Mobed A, Tanomand A, Yousefi M et al (2020) *Acinetobacter baumannii* efflux pumps and antibiotic resistance. Infect Drug Resist 13:423–434. https://doi.org/10.2147/IDR.S22

Akanbi OE, Njom HA, Fri J, Otigbu AC, Clarke AM (2017) Antimicrobial susceptibility of *Staphylococcus aureus* isolated from recreational waters and beach sand in eastern Cape Province of South Africa. Int J Environ Res Public Health 14(9):1001. https://doi.org/10.3390/ijerph14091001

Alarjani KM, Skalicky M (2021) Antimicrobial resistance profile of Staphylococcus aureus and its in-vitro potential inhibition efficiency. J Infect Public Health 14(12):1796–1801. https://doi.org/10.1016/j.jiph.2021.10.018

Alcalde-Rico M, Olivares-Pacheco J, Alvarez-Ortega C, Cámara M, Martínez JL (2018) Role of the multidrug resistance efflux pump MexCD-OprJ in the *Pseudomonas aeruginosa* quorum sensing response. Front Microbiol 9:2752. https://doi.org/10.3389/fmicb.2018.02752

Algammal AM, Enany ME, El-Tarabili RM, Ghobashy MO, Helmy YA (2020) Prevalence, antimicrobial resistance profiles, virulence and enterotoxins-determinant genes of MRSA isolated from subclinical bovine mastitis in Egypt. Pathogens 9(5):362. https://doi.org/10.3390/pathogens9050362

Algammal AM, Eidaroos NH, Alfifi KJ, Alatawy M, Al-Harbi AI, Alanazi YF et al (2023) *opr*L gene sequencing, resistance patterns, virulence genes, quorum sensing and antibiotic resistance genes of XDR *Pseudomonas aeruginosa* isolated from broiler chickens. Infect Drug Resist 16:853–867. https://doi.org/10.2147/IDR.S401473

Aminul P, Anwar S, Molla MMA, Miah MRA (2021) Evaluation of antibiotic resistance patterns in clinical isolates of *Klebsiella pneumoniae* in Bangladesh. Biosaf Health 3(6):301–306. https://doi.org/10.1016/j.bsheal.2021.11.001

Asif M, Alvi IA, Rehman SU (2018) Insight into *Acinetobacter baumannii*: pathogenesis, global resistance, mechanisms of resistance, treatment options, and alternative modalities. Infect drug resist 8:1249–1260. https://doi.org/10.2147/IDR.S166750

Aslam B, Wang W, Arshad MI, Khurshid M, Muzammil S, Rasool MH et al (2018) Antibiotic resistance: a rundown of a global crisis. Infect Drug Resist 11:1645. https://doi.org/10.2147/IDR.S173867

Atzori A, Malviya VN, Malloci G, Dreier J, Pos KM, Vargiu AV, Ruggerone P (2019) Identification and characterization of carbapenem binding sites within the RND-transporter *Acr*B. Biochim Biophys Acta Biomembr 1861(1):62–74. https://doi.org/10.1016/j.bbamem.2018.10.012

Aung MS, Urushibara N, Kawaguchiya M, Ohashi N, Hirose M, Kudo K et al (2023) Antimicrobial resistance, virulence factors, and genotypes of *Enterococcus faecalis* and *Enterococcus faecium* clinical isolates in northern Japan: identification of optrA in ST480 *E. faecalis*. Antibiotics 12(1):108. https://doi.org/10.3390/antibiotics12010108

Benkő R, Gajdács M, Matuz M, Bodó G, Lázár A, Hajdú E et al (2020) Prevalence and antibiotic resistance of ESKAPE pathogens isolated in the emergency department of a tertiary care teaching hospital in Hungary: a 5-year retrospective survey. Antibiotics 9(9):624. https://doi.org/10.3390/antibiotics9090624

Bharadwaj A, Rastogi A, Pandey S, Gupta S, Sohal JS (2022) Multidrug-resistant bacteria: their mechanism of action and prophylaxis. Biomed Res Int 2022:5419874. https://doi.org/10.1155/2022/5419874

Bissong MEA, Tahnteng BF, Ateba CN, Akoachere JFTK (2020) Pathogenic potential and antimicrobial resistance profile of *Staphylococcus aureus* in milk and beef from the northwest and southwest regions of Cameroon. Biomed Res Int 2020:6015283. https://doi.org/10.1155/2020/6015283

Bitew A (2018) Multi-drug resistance profile of bacteria isolated from blood stream infection at Tikur anbessa specialized hospital, Addis Ababa, Ethiopia. EC Microbiol 14:119–126

Blair JM, Webber MA, Baylay AJ, Ogbolu DO, Piddock LJ (2015) Molecular mechanisms of antibiotic resistance. Nat Rev Microbiol 13(1):42–51. https://doi.org/10.1038/nrmicro3380

Boolchandani M, D'Souza AW, Dantas G (2019) Sequencing-based methods and resources to study antimicrobial resistance. Nat Rev Genet 20(6):356–370. https://doi.org/10.1038/s41576-019-0108-4

Botelho J, Grosso F, Peixe L (2019) Antibiotic resistance in *Pseudomonas aeruginosa*—mechanisms, epidemiology and evolution. Drug Resist Updat 44:100640. https://doi.org/10.1016/j.drup.2019.07.002

Boutarfi Z, Rebiahi SA, Morghad T, Pulido RP, Burgos MJG, Mahdi F et al (2019) Biocide tolerance and antibiotic resistance of *Enterobacter* spp. isolated from an Algerian hospital environment. J Glob Antimicrob Resist 18:291–297. https://doi.org/10.1016/j.jgar.2019.04.005

Breidenstein EB, de la Fuente-Núñez C, Hancock RE (2011) *Pseudomonas aeruginosa*: all roads lead to resistance. Trends Microbiol 19(8):419–426. https://doi.org/10.1016/j.tim.2011.04.005

Breijyeh Z, Jubeh B, Karaman R (2020) Resistance of gram-negative bacteria to current antibacterial agents and approaches to resolve it. Molecules 25(6):1340. https://doi.org/10.3390/molecules25061340

Brogan DM, Mossialos E (2016) A critical analysis of the review on antimicrobial resistance report and the infectious disease financing facility. Glob Health 12(1):1–7. https://doi.org/10.1186/s12992-016-0147-y

Cavallo I, Oliva A, Sivori F, Truglio M, Fabrizio G, Pasqua M et al (2023) *Acinetobacter baumannii* in the critically ill: complex infections get complicated. Front Microbiol 14:1196774. https://doi.org/10.3389/fmicb.2023.1196774

Chen H, Cao J, Zhou C, Liu H, Zhang X, Zhou T (2017) Biofilm formation restrained by subinhibitory concentrations of Tigecyclin in *Acinetobacter baumannii* is associated with downregulation of efflux pumps. Chemotherapy 62(2):128–133. https://doi.org/10.1159/000450537

Chinemerem Nwobodo D, Ugwu MC, Oliseloke Anie C, Al-Ouqaili MT, Chinedu Ikem J, Victor Chigozie U, Saki M (2022) Antibiotic resistance: the challenges and some emerging strategies for tackling a global menace. J Clin Lab Anal 36(9):e24655. https://doi.org/10.1002/jcla.24655

Cui P, Feng L, Zhang L, He J, An T, Fu X, Li C, Zhao X, Zhai Y, Li H, Yan W (2020) Antimicrobial resistance, virulence genes, and biofilm formation capacity among *Enterococcus* species from yaks in aba Tibetan autonomous prefecture, China. Front Microbiol 11:1250. https://doi.org/10.3389/fmicb.2020.01250

D'Costa VM, King CE, Kalan L, Morar M, Sung WWL, Schwarz C, Froese D, Zazula G, Calmels F, Debruyne R et al (2011) Antibiotic resistance is ancient. Nature 477:457–461. https://doi.org/10.1038/nature10388

D'Onofrio V, Conzemius R, Varda Brkić D, Bogdan M, Grisold A, Gyssens IC et al (2020) Epidemiology of colistin-resistant, carbapenemase-producing Enterobacteriaceae and *Acinetobacter baumannii* in Croatia. Infect Genet Evol 81:104263. https://doi.org/10.1016/j.meegid.2020.104263

da Silva KE, Maciel WG, Croda J, Cayô R, Ramos AC, Ro d S et al (2018) A high mortality rate associated with multidrug-resistant *Acinetobacter baumannii* ST79 and ST25 carrying OXA-23 in a Brazilian intensive care unit. PLoS One 13(12):e0209367. https://doi.org/10.1371/journal.pone.0209367

Dahdouh E, Gómez-Gil R, Pacho S, Mingorance J, Daoud Z, Suárez M (2017) Clonality, virulence determinants, and profiles of resistance of clinical *Acinetobacter baumannii* isolates obtained from a Spanish hospital. PLoS One 12(4):e0176824. https://doi.org/10.1371/journal.pone.0176824

Davin-Regli A, Pagès JM (2015) *Enterobacter aerogenes* and *Enterobacter cloacae*; versatile bacterial pathogens confronting antibiotic treatment. Front Microbiol 6:392. https://doi.org/10.3389/fmicb.2015.00392

De Oliveira DMP, Forde BM, Kidd TJ, Harris PNA, Schembri MA, Beatson SA, Paterson DL, Walker MJ (2020) Antimicrobial resistance in ESKAPE pathogens. Clin Microbiol Rev 33(3):e00181–e00119. https://doi.org/10.1086/533452

Dehbashi S, Tahmasebi H, Alikhani MY, Keramat F, Arabestani MR (2020) Distribution of class B and class A β-lactamases in clinical strains of *Pseudomonas aeruginosa*: comparison of phenotypic methods and high-resolution melting analysis (HRMA) assay. Infect Drug Resist 13(2020):2037–2052. https://doi.org/10.2147/IDR.S25

Denissen J, Reyneke B, Waso-Reyneke M, Havenga B, Barnard T, Khan S, Khan W (2022) Prevalence of ESKAPE pathogens in the environment: antibiotic resistance status, community-acquired infection and risk to human health. Int J Hyg Environ Health 244:114006. https://doi.org/10.1016/j.ijheh.2022.114006

Di Franco S, Alfieri A, Pace MC, Sansone P, Pota V, Fittipaldi C et al (2021) Blood stream infections from MDR bacteria. Life 11(6):575. https://doi.org/10.3390/life11060575

Dos Santos LDR, Furlan JPR, Gallo IFL, Ramos MS, Savazzi EA, Stehling EG (2021) Occurrence of multidrug-resistant *Enterococcus faecium* isolated from environmental samples. Lett Appl Microbiol 73(2):237–246. https://doi.org/10.1111/lam.13508

Du J, Li P, Liu H, Lü D, Liang H, Dou Y (2014) Phenotypic and molecular characterization of multidrug resistant *Klebsiella pneumoniae* isolated from a university teaching hospital, China. PLoS One 9(4):e95181. https://doi.org/10.1371/journal.pone.0095181

Dupont P, Hocquet D, Jeannot K, Chavanet P, Plésiat P (2005) Bacteriostatic and bactericidal activities of eight fluoroquinolones against MexAB-OprM-overproducing clinical strains of *Pseudomonas aeruginosa*. J Antimicrob Chemother 55(4):518–522. https://doi.org/10.1093/jac/dki030

Eghbalpoor F, Habibi M, Azizi O, Asadi Karam MR, Bouzari S (2019) Antibiotic resistance, virulence and genetic diversity of *Klebsiella pneumoniae* in community-and hospital-acquired urinary tract infections in Iran. Acta Microbiol Immunol Hung 66(3):349–366. https://doi.org/10.1556/030.66.2019.006

El-Zamkan MA, Mohamed HM (2021) Antimicrobial resistance, virulence genes and biofilm formation in *enterococcus* species isolated from milk of sheep and goat with subclinical mastitis. PLoS One 16(11):e0259584. https://doi.org/10.1371/journal.pone.0259584

Evans BA, Amyes SG (2014) OXA β-lactamases. Clin Microbiol Rev 27(2):241–263. https://doi.org/10.1128/cmr.00117-13

Fasciana T, Gentile B, Aquilina M, Ciammaruconi A, Mascarella C, Anselmo A et al (2019) Co-existence of virulence factors and antibiotic resistance in new *Klebsiella pneumoniae* clones emerging in south of Italy. BMC Infect Dis 19:1–10. https://doi.org/10.1186/s12879-019-4565-3

Fazal MA, Rana EA, Akter S, Alim MA, Barua H, Ahad A (2023) Molecular identification, antimicrobial resistance and virulence gene profiling of diverse *Staphylococcus* spp. involved in bovine sub-clinical mastitis, Bangladesh. Vet Anim Sci 21:100297. https://doi.org/10.1016/j.vas.2023.100297

Fernández-García L, Fernandez-Cuenca F, Blasco L, López-Rojas R, Ambroa A, Lopez M, Pascual Á, Bou G, Tomás M (2018) Relationship between tolerance and persistence mechanisms in *Acinetobacter baumannii* strains with AbkAB toxin-antitoxin system. Antimicrob Agents Chemother 62(5):10–128. https://doi.org/10.1128/aac.00250-18

Feretzakis G, Loupelis E, Sakagianni A, Skarmoutsou N, Michelidou S, Velentza A et al (2019) A 2-year single-centre audit on antibiotic resistance of *Pseudomonas aeruginosa*, *Acinetobacter baumannii* and *Klebsiella pneumoniae* strains from an intensive care unit and other wards in a general public hospital in Greece. Antibiotics 8(2):62. https://doi.org/10.3390/antibiotics8020062

Fournier PE, Vallenet D, Barbe V, Audic S, Ogata H, Poirel L et al (2006) Comparative genomics of multidrug resistance in *Acinetobacter baumannii*. PLoS Genet 2(1):e7. https://doi.org/10.1371/journal.pgen.0020007

Gajdács M, Spengler G, Urbán E (2017) Identification and antimicrobial susceptibility testing of anaerobic bacteria: Rubik's cube of clinical microbiology? Antibiotics 6(4):25. https://doi.org/10.3390/antibiotics6040025

Gajic I, Kabic J, Kekic D, Jovicevic M, Milenkovic M, Mitic Culafic D et al (2022) Antimicrobial susceptibility testing: a comprehensive review of currently used methods. Antibiotics 11(4):427. https://doi.org/10.3390/antibiotics11040427

Gawryszewska I, Żabicka D, Bojarska K, Malinowska K, Hryniewicz W, Sadowy E (2016) Invasive enterococcal infections in Poland: the current epidemiological situation. Eur J Clin Microbiol Infect Dis 35:847–856. https://doi.org/10.1007/s10096-016-2607-y

Georges M, Odoyo E, Matano D, Tiria F, Kyanya C, Mbwika D et al (2022) Determination of *Enterococcus faecalis* and *Enterococcus faecium* antimicrobial resistance and virulence factors and their association with clinical and demographic factors in Kenya. J Pathog 2022:3129439. https://doi.org/10.1155/2022/3129439

Ghamari M, Beigverdi R, Jabalameli F, Emaneini M (2022) Antimicrobial resistance pattern, virulence determinants and molecular analysis of carbapenem-resistant *Klebsiella pneumoniae* isolated from clinical samples in Iran. FEMS Microbiol Lett 369(1):fnac100. https://doi.org/10.1093/femsle/fnac100

Girlich D, Poirel L, Nordmann P (2010) First isolation of the blaOXA-23 carbapenemase gene from an environmental *Acinetobacter baumannii* isolate. Antimicrob Agents Chemother 54(1):578. https://doi.org/10.1128/AAC.00861-09

Giubelan L, Dragonu L, Stoian AC, Dumitrescu F (2021) Antibiotic resistance profile of *Klebsiella pneumoniae* strains isolated in an infectious diseases clinic. Romanian J Infect Dis 24(1):32–37. https://doi.org/10.37897/RJID.2021.1.4

Glover JS, Ticer TD, Engevik MA (2022) Profiling antibiotic resistance in *Acinetobacter calcoaceticus*. Antibiotics 11(7):978. https://doi.org/10.3390/antibiotics11070978

Golob M, Pate M, Kušar D, Dermota U, Avberšek J, Papić B, Zdovc I (2019) Antimicrobial resistance and virulence genes in *Enterococcus faecium* and *Enterococcus faecalis* from humans and retail red meat. Biomed Res Int 2019:1–12. https://doi.org/10.1155/2019/2815279

Gualtero S, Valderrama S, Valencia M, Rueda D, Muñoz-Velandia O, Ariza B et al (2020) Factors associated with mortality in infections caused by Carbapenem-resistant Enterobacteriaceae. J Infect Dev Ctries 14(06):654–659. https://doi.org/10.3855/jidc.12267

Güneri CÖ (2023) Determination of antibiotic resistance profiles in *Acinetobacter* species isolated from tank Milk. J Appl Biol Sci 17(1):39–45. https://jabsonline.org/index.php/jabs/article/view/1109

Guo Y, Zhou H, Qin L, Pang Z, Qin T, Ren H et al (2016) Frequency, antimicrobial resistance and genetic diversity of *Klebsiella pneumoniae* in food samples. PLoS One 11(4):e0153561. https://doi.org/10.1371/journal.pone.0153561

Guo Y, Song G, Sun M, Wang J, Wang Y (2020) Prevalence and therapies of antibiotic-resistance in *Staphylococcus aureus*. Front Cell Infect Microbiol 10:107. https://doi.org/10.3389/fcimb.2020.00107

Halat DH, Moubareck CA (2022) The intriguing Carbapenemases of *Pseudomonas aeruginosa*: current status, genetic profile, and global epidemiology. Yale J Biol Med 95(4):507–515

Hasani A, Sheikhalizadeh V, Ahangarzadeh Rezaee M, Rahmati-Yamchi M, Hasani A, Ghotaslou R, Goli HR (2016) Frequency of aminoglycoside-modifying enzymes and ArmA among different sequence groups of *Acinetobacter baumannii* in Iran. Microb Drug Resist 22(5):347–353. https://doi.org/10.1089/mdr.2015.0254

Hassan MA, Abd El-Aziz S, Elbadry HM, Samy A, Tamer TM (2022) Prevalence, antimicrobial resistance profile, and characterization of multi-drug resistant bacteria from various infected wounds in North Egypt. Saudi J Biol Sci 29(4):2978–2988. https://doi.org/10.1016/j.sjbs.2022.01.015

Havenga B, Ndlovu T, Clements T, Reyneke B, Waso M, Khan W (2019) Exploring the antimicrobial resistance profiles of WHO critical priority list bacterial strains. BMC Microbiol 19:1–16. https://doi.org/10.1186/s12866-019-1687-0

Hay AD (2021) Point-of-care tests to inform antibiotic prescribing. BMJ 374:n2253. https://doi.org/10.1136/bmj.n2253

Holt KE, Wertheim H, Zadoks RN, Baker S, Whitehouse CA, Dance D et al (2015) Genomic analysis of diversity, population structure, virulence, and antimicrobial resistance in *Klebsiella pneumoniae*, an urgent threat to public health. Proc Natl Acad Sci 112(27):E3574–E3581. https://doi.org/10.1073/pnas.1501049112

Hood MI, Jacobs AC, Sayood K, Dunman PM, Skaar EP (2010) Acinetobacter baumannii increases tolerance to antibiotics in response to monovalent cations. Antimicrob Agents Chemother 54(3):1029–1041. https://doi.org/10.1128/aac.00963-09

Hosu MC, Vasaikar S, Okuthe GE, Apalata T (2021) Molecular detection of antibiotic-resistant genes in *Pseudomonas aeruginosa* from nonclinical environment: public health implications in Mthatha, eastern cape province, South Africa. Int J Microbiol 2021:1–9. https://doi.org/10.1155/2021/8861074

Hrenovic J, Durn G, Music MS, Dekic S, Troskot-Corbic T, Skoric D (2017) Extensively and multi drug-resistant *Acinetobacter baumannii* recovered from technosol at a dump site in Croatia. Sci Total Environ 607:1049–1055. https://doi.org/10.1016/j.scitotenv.2017.07.108

Hu WS, Yao SM, Fung CP, Hsieh YP, Liu CP, Lin JF (2007) An OXA-66/OXA-51-like carbapenemase and possibly an efflux pump are associated with resistance to imipenem in *Acinetobacter baumannii*. Antimicrob Agents Chemother 51(11):3844–3852. https://doi.org/10.1128/aac.01512-06

Huang L, Sun L, Xu G, Xia T (2008) Differential susceptibility to carbapenems due to the AdeABC efflux pump among nosocomial outbreak isolates of *Acinetobacter baumannii* in a Chinese hospital. Diagn Microbiol Infect Dis 62(3):326–332. https://doi.org/10.1016/j.diagmicrobio.2008.06.008

Ibrahim D, Jabbour JF, Kanj SS (2020) Current choices of antibiotic treatment for *Pseudomonas aeruginosa* infections. Curr Opin Infect Dis 33(6):464–473. https://doi.org/10.1097/QCO.0000000000000677

Indrajith S, Mukhopadhyay AK, Chowdhury G, Al Farraj DA, Alkufeidy RM, Natesan S et al (2021) Molecular insights of Carbapenem resistance *Klebsiella pneumoniae* isolates with focus on multidrug resistance from clinical samples. J Infect Public Health 14(1):131–138. https://doi.org/10.1016/j.jiph.2020.09.018

Jafari-Sales A, Soleimani H, Moradi L (2020) Antibiotic resistance pattern in *Klebsiella pneumoniae* strains isolated from children with urinary tract infections from Tabriz hospitals. Health Biotechnol Biopharma 4(1):38–45. https://doi.org/10.22034/HBB.2020.03

Jelić M, Hrenović J, Dekić S, Goić-Barišić I, Andrašević AT (2019) First evidence of KPC-producing ST258 *Klebsiella pneumoniae* in river water. J Hosp Infect 103(2):147–150. https://doi.org/10.1016/j.jhin.2019.04.001

Jeon JH, Lee JH, Lee JJ, Park KS, Karim AM, Lee CR et al (2015) Structural basis for carbapenem-hydrolyzing mechanisms of carbapenemases conferring antibiotic resistance. Int J Mol Sci 16(5):9654–9692. https://doi.org/10.3390/ijms16059654

Jin JS, Kwon SO, Moon DC, Gurung M, Lee JH, Kim SI, Lee JC (2011) Acinetobacter baumannii secretes cytotoxic outer membrane protein A via outer membrane vesicles. PLoS One 6(2):e17027. https://doi.org/10.1371/journal.pone.0017027

Jubeh B, Breijyeh Z, Karaman R (2020) Resistance of gram-positive bacteria to current antibacterial agents and overcoming approaches. Molecules 25(12):2888. https://doi.org/10.3390/molecules25122888

Jurado-Martín I, Sainz-Mejías M, McClean S (2021) *Pseudomonas aeruginosa*: an audacious pathogen with an adaptable arsenal of virulence factors. Int J Mol Sci 22(6):3128. https://doi.org/10.3390/ijms22063128

Kalpana S, Lin WY, Wang YC, Fu Y, Lakshmi A, Wang HY (2023) Antibiotic resistance diagnosis in ESKAPE pathogens—a review on proteomic perspective. Diagnostics 13(6):1014. https://doi.org/10.3390/diagnostics13061014

Kaprou GD, Bergšpica I, Alexa EA, Alvarez-Ordóñez A, Prieto M (2021) Rapid methods for antimicrobial resistance diagnostics. Antibiotics 10(2):209. https://doi.org/10.3390/antibiotics10020209

Karimi K, Zarei O, Sedighi P, Taheri M, Doosti-Irani A, Shokoohizadeh L (2021) Investigation of antibiotic resistance and biofilm formation in clinical isolates of *Klebsiella pneumoniae*. Int J Microbiol 2021:1–6. https://doi.org/10.1155/2021/5573388

Karlowsky JA, Hoban DJ, Hackel MA, Lob SH, Sahm DF (2017) Antimicrobial susceptibility of Gram-negative ESKAPE pathogens isolated from hospitalized patients with intra-abdominal and urinary tract infections in Asia–Pacific countries: SMART 2013–2015. J Med Microbiol 66(1):61–69. https://doi.org/10.1099/jmm.0.000421

Kaszab E, Szoboszlay S, Dobolyi C, Háhn J, Pék N, Kriszt B (2011) Antibiotic resistance profiles and virulence markers of *Pseudomonas aeruginosa* strains isolated from composts. Bioresour Technol 102(2):1543–1548. https://doi.org/10.1016/j.biortech.2010.08.027

Khan ZA, Siddiqui MF, Park S (2019) Current and emerging methods of antibiotic susceptibility testing. Diagnostics 9(2):49. https://doi.org/10.3390/diagnostics9020049

Khan A, Arias CA, Abbott A, Dien Bard J, Bhatti MM, Humphries RM (2021) Evaluation of the Vitek 2, Phoenix, and MicroScan for antimicrobial susceptibility testing of *Stenotrophomonas maltophilia*. J Clin Microbiol 59(9):e00654–e00621. https://doi.org/10.1128/jcm.00654-21

Kim YB, Seo KW, Shim JB, Hyun Son S, Noh EB, Lee YJ (2019a) Molecular characterization of antimicrobial-resistant *Enterococcus faecalis* and *Enterococcus faecium* isolated from layer parent stock. Poult Sci 98(11):5892–5899. https://doi.org/10.3382/ps/pez288

Kim JM, Kim I, Chung SH, Chung Y, Han M, Kim JS (2019b) Rapid discrimination of methicillin-resistant *Staphylococcus aureus* by MALDI-TOF MS. Pathogens 8(4):214. https://doi.org/10.3390/pathogens8040214

Kim MH, Moon DC, Kim SJ, Mechesso AF, Song HJ, Kang HY, Choi JH, Yoon SS, Lim SK (2021) Nationwide surveillance on antimicrobial resistance profiles of Enterococcus faecium and Enterococcus faecalis isolated from healthy food animals in South Korea, 2010 to 2019. Microorganisms 9(5):925. https://doi.org/10.3390/microorganisms9050925

Kot B, Grużewska A, Szweda P, Wicha J, Parulska U (2021) Antibiotic resistance of uropathogens isolated from patients hospitalized in district hospital in Central Poland in 2020. Antibiotics 10(4):447. https://doi.org/10.3390/antibiotics10040447

Kunz Coyne AJ, El Ghali A, Holger D, Rebold N, Rybak MJ (2022) Therapeutic strategies for emerging multidrug-resistant *Pseudomonas aeruginosa*. Infect Dis Ther 11(2):661–682. https://doi.org/10.1007/s40121-022-00591-2

Kyriakidis I, Vasileiou E, Pana ZD, Tragiannidis A (2021) *Acinetobacter baumannii* antibiotic resistance mechanisms. Pathogens 10(3):373. https://doi.org/10.3390/pathogens10030373

Lee CR, Lee JH, Park M, Park KS, Bae IK, Kim YB et al (2017) Biology of *Acinetobacter baumannii:* pathogenesis, antibiotic resistance mechanisms, and prospective treatment options. Front Cell Infect Microbiol 7:55. https://doi.org/10.3389/fcimb.2017.00055

Li XZ, Plésiat P, Nikaido H (2015) The challenge of efflux-mediated antibiotic resistance in gram-negative bacteria. Clin Microbiol Rev 28(2):337–418. https://doi.org/10.1128/cmr.00117-14

Lin MF, Lan CY (2014) Antimicrobial resistance in *Acinetobacter baumannii*: from bench to bedside. World J Clin Cases 2(12):787–814. https://doi.org/10.12998/wjcc.v2.i12.787

Lolans K, Rice TW, Munoz-Price LS, Quinn JP (2006) Multicity outbreak of carbapenem-resistant *Acinetobacter baumannii* isolates producing the carbapenemase OXA-40. Antimicrob Agents Chemother 50(9):2941–2945. https://doi.org/10.1128/aac.00116-06

Luczkiewicz A, Kotlarska E, Artichowicz W, Tarasewicz K, Fudala-Ksiazek S (2015) Antimicrobial resistance of *Pseudomonas* spp. isolated from wastewater and wastewater-impacted marine coastal zone. Environ Sci Pollut Res 22:19823–19834. https://doi.org/10.1007/s11356-015-5098-y

Madigan T, Cunningham SA, Patel R, Greenwood-Quaintance KE, Barth JE, Sampathkumar P et al (2018) Whole-genome sequencing for methicillin-resistant *Staphylococcus aureus* (MRSA) outbreak investigation in a neonatal intensive care unit. Infect Control Hosp Epidemiol 39(12):1412–1418. https://doi.org/10.1017/ice.2018.239

Mai W, Liu Y, Meng Q, Xu J, Wu J (2023) Bacterial epidemiology and antimicrobial resistance profiles of respiratory specimens of children with pneumonia in Hainan, China. Infect Drug Resist 16:249–261. https://doi.org/10.2147/IDR.S397513

Makke G, Bitar I, Salloum T, Panossian B, Alousi S, Arabaghian H et al (2020) Whole-genome-sequence-based characterization of extensively drug-resistant *Acinetobacter baumannii* hospital outbreak. MSphere 5(1):e00934–e00919. https://doi.org/10.1128/msphere.00934-19

Mancuso G, Midiri A, Gerace E, Biondo C (2021) Bacterial antibiotic resistance: the most critical pathogens. Pathogens 10(10):1310. https://doi.org/10.3390/pathogens10101310

Maniatis AN, Pournaras S, Orkopoulou S, Tassios PT, Legakis NJ, Bacterial Resistance Study Group (2003) Multi resistant *Acinetobacter baumannii* isolates in intensive care units in Greece. Clin Microbiol Infect 9(6):547–553. https://doi.org/10.1046/j.1469-0691.2003.00558.x

Masuda N, Sakagawa E, Ohya S, Gotoh N, Tsujimoto H, Nishino T (2000) Substrate specificities of MexAB-OprM, MexCD-OprJ, and MexXY-oprM efflux pumps in *Pseudomonas aeruginosa*. Antimicrob Agents Chemother 44(12):3322–3327. https://doi.org/10.1128/aac.44.12.3322-3327.2000

Moges F, Endris M, Belyhun Y, Worku W (2014) Isolation and characterization of multiple drug resistance bacterial pathogens from waste water in hospital and non-hospital environments, Northwest Ethiopia. BMC Res Notes 7:1–6. https://doi.org/10.1186/1756-0500-7-215

Mohammed AB, Anwar KA (2022) Phenotypic and genotypic detection of extended spectrum beta lactamase enzyme in *Klebsiella pneumoniae*. PLoS One 17(9):e0267221. https://doi.org/10.1371/journal.pone.0267221

Momtaz H, Dehkordi FS, Rahimi E, Asgarifar A, Momeni M (2013) Virulence genes and antimicrobial resistance profiles of *Staphylococcus aureus* isolated from chicken meat in Isfahan province. Iran J Appl Poult Res 22(4):913–921. https://doi.org/10.3382/japr.2012-00673

Monteiro S, Santos R (2020) Incidence of enterococci resistant to clinically relevant antibiotics in environmental waters and in reclaimed waters used for irrigation. J Water Health 18(6):911–924. https://doi.org/10.2166/wh.2020.020

Morrison L, Zembower TR (2020) Antimicrobial resistance. Gastrointest Endosc Clin 30(4):619–635. https://doi.org/10.1016/j.giec.2020.06.004

Motamedi H, Mirzabeigi H, Shirali T (2010) Determining of antibiotic resistance profile in *Staphylococcus aureus* isolates. Asian Pac J Trop Med 3(9):734–737. https://doi.org/10.1016/S1995-7645(10)60176-9

Motbainor H, Bereded F, Mulu W (2020) Multi-drug resistance of blood stream, urinary tract and surgical site nosocomial infections of *Acinetobacter baumannii* and *Pseudomonas aeruginosa* among patients hospitalized at Felegehiwot referral hospital, Northwest Ethiopia: a cross-sectional study. BMC Infect Dis 20:1–11. https://doi.org/10.1186/s12879-020-4811-8

Mugnier PD, Poirel L, Naas T, Nordmann P (2010) Worldwide dissemination of the blaOXA-23 Carbapenemase gene of *Acinetobacter baumannii*1. Emerg Infect Dis 16(1):35. https://doi.org/10.3201/eid1601.090852

Mulcahy LR, Burns JL, Lory S, Lewis K (2010) Emergence of *Pseudomonas aeruginosa* strains producing high levels of persister cells in patients with cystic fibrosis. J Bacteriol 192(23):6191–6199. https://doi.org/10.1128/jb.01651-09

Munita JM, Arias CA (2016) Mechanisms of antibiotic resistance. In: Virulence mechanisms of bacterial pathogens, pp 481–511. https://doi.org/10.1128/9781555819286.ch17

Muntean MM, Muntean AA, Preda M, Manolescu LSC, Dragomirescu C, Popa MI, Popa GL (2022) Phenotypic and genotypic detection methods for antimicrobial resistance in ESKAPE pathogens. Exp Ther Med 24(2):1–12. https://doi.org/10.3892/etm.2022.11435

Mwikuma G, Kainga H, Kallu SA, Nakajima C, Suzuki Y, Hangombe BM (2023) Determination of the prevalence and antimicrobial resistance of *Enterococcus faecalis* and *Enterococcus fae-*

cium associated with poultry in four districts in Zambia. Antibiotics 12(4):657. https://doi.org/10.3390/antibiotics12040657

Nair AV, Joseph N, Krishna K, Sneha KG, Tom N, Jangid K, Nair S (2015) A comparative study of coastal and clinical isolates of *Pseudomonas aeruginosa*. Braz J Microbiol 46:725–734. https://doi.org/10.1590/S1517-838246320140502

Nguyen L, Garcia J, Gruenberg K, MacDougall C (2018) Multidrug-resistant *Pseudomonas infections:* hard to treat, but hope on the horizon? Curr Infect Dis Rep 20:1–10. https://doi.org/10.1007/s11908-018-0629-6

Nobari S, Shahcheraghi F, Rahmati Ghezelgeh F, Valizadeh B (2014) Molecular characterization of carbapenem-resistant strains of *Klebsiella pneumoniae* isolated from Iranian patients: first identification of blaKPC gene in Iran. Microb Drug Resist 20(4):285–293. https://doi.org/10.1089/mdr.2013.0074

Nobel FA, Islam S, Babu G, Akter S, Jebin RA, Sarker TC et al (2022) Isolation of multidrug resistance bacteria from the patients with wound infection and their antibiotics susceptibility patterns: a cross-sectional study. Ann Med Surg 84:104895. https://doi.org/10.1016/j.amsu.2022.104895

Nordmann P, Dortet L, Poirel L (2012) Carbapenem resistance in Enterobacteriaceae: here is the storm! Trends Mol Med 18(5):263–272. https://doi.org/10.1016/j.molmed.2012.03.003

Odumosu BT, Ajetunmobi O, Dada-Adegbola H, Odutayo I (2016) Antibiotic susceptibility pattern and analysis of plasmid profiles of *Pseudomonas aeruginosa* from human, animal and plant sources. Springerplus 5(1):1–7. https://doi.org/10.1186/s40064-016-3073-9

Oliver A, Mulet X, López-Causapé C, Juan C (2015) The increasing threat of *Pseudomonas aeruginosa* high-risk clones. Drug Resist Updat 21:41–59. https://doi.org/10.1016/j.drup.2015.08.002

Ong DS, Jongerden IP, Buiting AG, Leverstein-van Hall MA, Speelberg B, Kesecioglu J, Bonten MJ (2011) Antibiotic exposure and resistance development in *Pseudomonas aeruginosa* and *Enterobacter* species in intensive care units. Crit Care Med 39(11):2458–2463. https://doi.org/10.1097/CCM.0b013e318225756d

Ontong JC, Ozioma NF, Voravuthikunchai SP, Chusri S (2021) Synergistic antibacterial effects of colistin in combination with aminoglycoside, carbapenems, cephalosporins, fluoroquinolones, tetracyclines, fosfomycin, and piperacillin on multidrug resistant *Klebsiella pneumoniae* isolates. PLoS One 16(1):e0244673. https://doi.org/10.1371/journal.pone.0244673

Pachori P, Gothalwal R, Gandhi P (2019) Emergence of antibiotic resistance *Pseudomonas aeruginosa* in intensive care unit; a critical review. Genes Dis 6(2):109–119. https://doi.org/10.1016/j.gendis.2019.04.001

Pang Z, Raudonis R, Glick BR, Lin TJ, Cheng Z (2019) Antibiotic resistance in *Pseudomonas aeruginosa*: mechanisms and alternative therapeutic strategies. Biotechnol Adv 37(1):177–192. https://doi.org/10.1016/j.biotechadv.2018.11.013

Paul D, Verma J, Banerjee A, Konar D, Das B (2022) Antimicrobial resistance traits and resistance mechanisms in bacterial pathogens. In: Antimicrobial resistance. Springer, Berlin, pp 1–27. https://doi.org/10.1007/978-981-16-3120-7_1

Peleg AY, Adams J, Paterson DL (2007) Tigecycline efflux as a mechanism for nonsusceptibility in *Acinetobacter baumannii*. Antimicrob Agents Chemother 51(6):2065–2069. https://doi.org/10.1128/aac.01198-06

Pereckaite L, Tatarunas V, Giedraitiene A (2018) Current antimicrobial susceptibility testing for beta-lactamase-producing Enterobacteriaceae in clinical settings. J Microbiol Methods 152:154–164. https://doi.org/10.1016/j.mimet.2018.07.014

Perez F, Van Duin D (2013) Carbapenem-resistant Enterobacteriaceae: a menace to our most vul nerable patients. Cleve Clin J Med 80(4):225. https://doi.org/10.3949/ccjm.80a.12182

Pérez-Varela M, Corral J, Aranda J, Barbé J (2019) Roles of efflux pumps from different superfamilies in the surface-associated motility and virulence of *Acinetobacter baumannii* ATCC 17978. Antimicrob Agents Chemother 63(3):e02190–e02118. https://doi.org/10.1128/aac.02190-18

Perović S, Veinović G, Antic Stanković J (2018) A review on antibiotic resistance: origin and mechanisms of bacterial resistance as biological phenomenon. Genetika 50(3):1123–1135. https://doi.org/10.2298/GENSR1803123P

Poirel L, Nordmann P (2006) Carbapenem resistance in *Acinetobacter baumannii*: mechanisms and epidemiology. Clin Microbiol Infect 12(9):826–836. https://doi.org/10.1111/j.1469-0691.2006.01456.x

Potron A, Poirel L, Nordmann P (2015) Emerging broad-spectrum resistance in *Pseudomonas aeruginosa* and *Acinetobacter baumannii*: mechanisms and epidemiology. Int J Antimicrob Agents 45(6):568–585. https://doi.org/10.1016/j.ijantimicag.2015.03.001

Qureshi MA, Bhatnagar RK (2015) Molecular analysis of metallo beta lactamase in multi drug resistant *Pseudomonas aeruginosa* among the clinical isolates. J Infect Dis 1(1):1004

Rafiee R, Eftekhar F, Tabatabaei SA, Tehrani DM (2014) Prevalence of extended-spectrum and metallo β-lactamase production in AmpC β-lactamase producing *Pseudomonas aeruginosa* isolates from burns. Jundishapur J Microbiol 7(9):e16436. https://doi.org/10.5812/jjm.16436

Ramay BM, Caudell MA, Cordón-Rosales C, Archila LD, Palmer GH, Jarquin C et al (2020) Antibiotic use and hygiene interact to influence the distribution of antimicrobial-resistant bacteria in low-income communities in Guatemala. Sci Rep 10(1):13767. https://doi.org/10.1038/s41598-020-70741-4

Ramessar K, Olaniran AO (2019) Antibiogram and molecular characterization of methicillin-resistant *Staphylococcus aureus* recovered from treated wastewater effluent and receiving surface water in Durban, South Africa. World J Microbiol Biotechnol 35(142):1–11. https://doi.org/10.1007/s11274-019-2715-9

Ramirez MS, Tolmasky ME (2010) Aminoglycoside modifying enzymes. Drug Resist Updat 13(6):151–171. https://doi.org/10.1016/j.drup.2010.08.003

Rasool MS, Siddiqui F, Ajaz M, Rasool SA (2019) Prevalence and antibiotic resistance profiles of gram-negative bacilli associated with urinary tract infections (UTIs) in Karachi, Pakistan. Pak J Pharm Sci 32(6):2617–2623. https://doi.org/10.36721/PJPS.2019.32.6.REG.2617-2623.1

Reygaert WC (2018) An overview of the antimicrobial resistance mechanisms of bacteria. AIMS Microbiol 4(3):482. https://doi.org/10.3934/microbiol.2018.3.482

Rezaloo M, Motalebi A, Mashak Z, Anvar A (2022) Prevalence, antimicrobial resistance, and molecular description of *Pseudomonas aeruginosa* isolated from meat and meat products. J Food Qual 2022:1–11. https://doi.org/10.1155/2022/9899338

Rice LB (2008) Federal funding for the study of antimicrobial resistance in nosocomial pathogens: no ESKAPE. J Infect Dis 197(8):1079–1081. https://doi.org/10.1086/533452

Rice LB (2010) Progress and challenges in implementing the research on ESKAPE pathogens. Infect Control Hosp Epidemiol 31(1):S7–S10. https://doi.org/10.1086/655995

Richter SS, Karichu J, Otiso J, Van Heule H, Keller G, Cober E et al (2018) Evaluation of Sensititre broth microdilution plate for determining the susceptibility of carbapenem-resistant *Klebsiella pneumoniae* to polymyxins. Diagn Microbiol Infect Dis 91(1):89–92. https://doi.org/10.1016/j.diagmicrobio.2018.01.005

Rodríguez CH, Nastro M, Fiorilli G, Dabos L, Calvo JL, Fariña ME et al (2016) Trends in the resistance profiles of *Acinetobacter baumannii* endemic clones in a university hospital of Argentina. J Chemother 28(1):25–27. https://doi.org/10.1179/1973947814Y.0000000213

Rosa JF, Rizek C, Marchi AP, Guimaraes T, Miranda L, Carrilho C et al (2017) Clonality, outer-membrane proteins profile and efflux pump in KPC-producing *Enterobacter* sp. in Brazil. BMC Microbiol 17:1–9. https://doi.org/10.1186/s12866-017-0970-1

Roy S, Chowdhury G, Mukhopadhyay AK, Dutta S, Basu S (2022) Convergence of biofilm formation and antibiotic resistance in *Acinetobacter baumannii* infection. Front Med 9:793615. https://doi.org/10.3389/fmed.2022.793615

Saleem S, Bokhari H (2020) Resistance profile of genetically distinct clinical *Pseudomonas aeruginosa* isolates from public hospitals in Central Pakistan. J Infect Public Health 13(4):598–605. https://doi.org/10.1016/j.jiph.2019.08.019

Sánchez-Sánchez M, Cruz-Pulido WL, Bladinieres-Cámara E, Alcalá-Durán R, Rivera-Sánchez G, Bocanegra-García V (2017) Bacterial prevalence and antibiotic resistance in clinical isolates of diabetic foot ulcers in the northeast of Tamaulipas, Mexico. Int J Low Extrem Wounds 16(2):129–134. https://doi.org/10.1177/1534734617705254

Santajit S, Indrawattana N (2016) Mechanisms of antimicrobial resistance in ESKAPE pathogens. Biomed Res Int 2016:2475067. https://doi.org/10.1155/2016/2475067

Sattari-Maraji A, Jabalameli F, Node Farahani N, Beigverdi R, Emaneini M (2019) Antimicrobial resistance pattern, virulence determinants and molecular analysis of *Enterococcus faecium* isolated from children infections in Iran. BMC Microbiol 19(1):1–8. https://doi.org/10.1186/s12866-019-1539-y

Sebre S, Abegaz WE, Seman A, Awoke T, Desalegn Z, Mihret W et al (2020) Bacterial profiles and antimicrobial susceptibility pattern of isolates from inanimate hospital environments at Tikur Anbessa specialized teaching hospital, Addis Ababa, Ethiopia. Infect Drug Resist 13:4439–4448. https://doi.org/10.2147/IDR.S286293

Setu SK, Anijejob S (2021) Antimicrobial resistant profile of bacterial isolates in the intensive care unit of a tertiary Care Hospital in Bangladesh. Eur J Biomed Pharm Sci 8(2):06–12. https://doi.org/10.17605/OSF.IO/9JY8R

Shamsizadeh Z, Nikaeen M, Nasr Esfahani B, Mirhoseini SH, Hatamzadeh M, Hassanzadeh A (2017) Detection of antibiotic resistant *Acinetobacter baumannii* in various hospital environments: potential sources for transmission of *Acinetobacter* infections. Environ Health Prev Med 22:1–7. https://doi.org/10.1186/s12199-017-0653-4

Sharma A, Dour P, Singh TN (2008) The prevalence of extended-spectrum β-lactamase in environmental isolates of *Enterobacter*. Indian J Pathol Microbiol 51(1):130–136

Shen M, Chen X, He J, Xiong L, Tian R, Yang G et al (2023) Antimicrobial resistance patterns, sequence types, virulence and carbapenemase genes of carbapenem-resistant Klebsiella pneumoniae clinical isolates from a tertiary care teaching hospital in Zunyi, China. Infect Drug Resist 16:637–649. https://doi.org/10.2147/IDR.S398304

Shortridge D, Gales AC, Streit JM, Huband MD, Tsakris A, Jones RN (2019) Geographic and temporal patterns of antimicrobial resistance in *Pseudomonas aeruginosa* over 20 years from the SENTRY antimicrobial surveillance program, 1997–2016. Open Forum Infect Dis 6(1):S63–S68. https://doi.org/10.1093/ofid/ofy343

Soge OO, Meschke JS, No DB, Roberts MC (2009) Characterization of methicillin-resistant Staphylococcus aureus and methicillin-resistant coagulase-negative *Staphylococcus* spp. isolated from US west coast public marine beaches. J Antimicrob Chemother 64(6):1148–1155. https://doi.org/10.1093/jac/dkp368

Stephen J, Salam F, Lekshmi M, Kumar SH, Varela MF (2023) The major facilitator superfamily and antimicrobial resistance efflux pumps of the ESKAPEE pathogen *Staphylococcus aureus*. Antibiotics 12(2):343. https://doi.org/10.3390/antibiotics12020343

Sun D, Jeannot K, Xiao Y, Knapp CW (2019) Horizontal gene transfer mediated bacterial antibiotic resistance. Front Microbiol 10:1933. https://doi.org/10.3389/fmicb.2019.01933

Tadesse S, Alemayehu H, Tenna A, Tadesse G, Tessema TS, Shibeshi W, Eguale T (2018) Antimicrobial resistance profile of *Staphylococcus aureus* isolated from patients with infection at Tikur Anbessa specialized hospital, Addis Ababa, Ethiopia. BMC Pharmacol Toxicol 19(1):1–8. https://doi.org/10.1186/s40360-018-0210-9

Taylor PK, Yeung AT, Hancock RE (2014) Antibiotic resistance in *Pseudomonas aeruginosa* biofilms: towards the development of novel anti-biofilm therapies. J Biotechnol 191:121–130. https://doi.org/10.1016/j.jbiotec.2014.09.003

Terreni M, Taccani M, Pregnolato M (2021) New antibiotics for multidrug-resistant bacterial strains: latest research developments and future perspectives. Molecules 26(9):2671. https://doi.org/10.3390/molecules26092671

Tesfaye H, Alemayehu H, Desta AF, Eguale T (2019) Antimicrobial susceptibility profile of selected Enterobacteriaceae in wastewater samples from health facilities, abattoir, downstream rivers and a WWTP in Addis Ababa, Ethiopia. Antimicrob Resist Infect Control 8(1):1–11. https://doi.org/10.1186/s13756-019-0588-1

Ünal N, Aşkar Ş, Yildirim M (2017) Antibiotic resistance profile of enterococcus faecium and enterococcus faecalis isolated from broiler cloacal samples. Turkish J Vet Anim Sci 41(2):199–203. https://doi.org/10.3906/vet-1607-26

Vahaboglu H, Budak F, Kasap M, Gacar G, Torol S, Karadenizli A et al (2006) High prevalence of OXA-51-type class D β-lactamases among ceftazidime-resistant clinical isolates of *Acinetobacter* spp.: co-existence with OXA-58 in multiple centres. J Antimicrob Chemother 58(3):537–542. https://doi.org/10.1093/jac/dkl273

Vala MH, Hallajzadeh M, Hashemi A, Goudarzi H, Tarhani M, Tabrizi MS, Bazmi F (2014) Detection of ambler class a, B and D ß-lactamases among *Pseudomonas aeruginosa* and *Acinetobacter baumannii* clinical isolates from burn patients. Ann Burns Fire Disasters 27(1):8–13

Vasala A, Hytönen VP, Laitinen OH (2020) Modern tools for rapid diagnostics of antimicrobial resistance. Front Cell Infect Microbiol 10:308. https://doi.org/10.3389/fcimb.2020.00308

Vitale M, Galluzzo P, Buffa PG, Carlino E, Spezia O, Alduina R (2019) Comparison of antibiotic resistance profile and biofilm production of *Staphylococcus aureus* isolates derived from human specimens and animal-derived samples. Antibiotics 8(3):97. https://doi.org/10.3390/antibiotics8030097

Vrancianu CO, Gheorghe I, Czobor IB, Chifiriuc MC (2020) Antibiotic resistance profiles, molecular mechanisms and innovative treatment strategies of *Acinetobacter baumannii*. Microorganisms 8(6):935. https://doi.org/10.3390/microorganisms8060935

Vrioni G, Tsiamis C, Oikonomidis G, Theodoridou K, Kapsimali V, Tsakris A (2018) MALDI-TOF mass spectrometry technology for detecting biomarkers of antimicrobial resistance: current achievements and future perspectives. Ann Transl Med 6(12):240. https://doi.org/10.21037/atm.2018.06.28

Wand ME, Bock LJ, Bonney LC, Sutton JM (2015) Retention of virulence following adaptation to colistin in *Acinetobacter baumannii* reflects the mechanism of resistance. J Antimicrob Chemother 70(8):2209–2216. https://doi.org/10.1093/jac/dkv097

WHO Antimicrobial Resistance (2021). Available from https://www.who.int/news-room/fact-sheets/detail/antimicrobial-resistance. Accessed 6 Feb 2023

Wright GD (2011) Molecular mechanisms of antibiotic resistance. Chem Commun 47(1):4055–4061. https://doi.org/10.1039/C0CC05111J

Xu CF, Bilya SR, Xu W (2019) adeABC efflux gene in *Acinetobacter baumannii*. New Microbes New Infect 30:100549. https://doi.org/10.1016/j.nmni.2019.100549

Yang CH, Su PW, Moi SH, Chuang LY (2019) Biofilm formation in *Acinetobacter baumannii*: genotype-phenotype correlation. Molecules 24(10):1849. https://doi.org/10.3390/molecules24101849

Yang X, Guo R, Zhang B, Xie B, Zhou S, Zhang B, Lai Q (2023) Retrospective analysis of drug resistance characteristics and infection related risk factors of multidrug-resistant organisms (MDROs) isolated from the orthopaedics department of a tertiary hospital. Sci Rep 13(1):2199. https://doi.org/10.1038/s41598-023-28270-3

Yoon EJ, Nait Chabane Y, Goussard S, Snesrud E, Courvalin P, Dé E, Grillot-Courvalin C (2015) Contribution of resistance-nodulation-cell division efflux systems to antibiotic resistance and biofilm formation in *Acinetobacter baumannii*. MBio 6(2):e00309–e00315. https://doi.org/10.1128/mbio.00309-15

Yuan S, Wu G, Zheng B (2019) Complete genome sequence of an IMP-8, CTX-M-14, CTX-M-3 and QnrS1 co-producing *Enterobacter asburiae* isolate from a patient with wound infection. J Glob Antimicrob Resist 18:52–54. https://doi.org/10.1016/j.jgar.2019.05.029

Yucel N, Citak S, Bayhün S (2011) Antimicrobial resistance profile of *Staphylococcus aureus* isolated from clinical samples and foods of animal origin. Foodborne Pathog Dis 8(3):427–431. https://doi.org/10.1089/fpd.2010.0707

Zaheer R, Cook SR, Barbieri R, Goji N, Cameron A, Petkau A et al (2020) Surveillance of *Enterococcus* spp. reveals distinct species and antimicrobial resistance diversity across a one-health continuum. Sci Rep 10(1):3937. https://doi.org/10.1038/s41598-020-61002-5

Zedan A, Alatfeehy N, Marouf S (2023) Isolation and Antibiogram profiles of *Staphylococcus aureus* isolates from cow milk and dog samples. J Appl Vet Sci 8(1):38–44. https://doi.org/10.21608/javs.2022.164610.1181

Zhang X, Chen D, Xu G, Huang W, Wang X (2018) Molecular epidemiology and drug resistant mechanism in carbapenem-resistant *Klebsiella pneumoniae* isolated from pediatric patients in Shanghai, China. PLoS One 13(3):e0194000. https://doi.org/10.1371/journal.pone.0194000
Zhou Y, Yu H, Guo Q, Xu X, Ye X, Wu S et al (2010) Distribution of 16S rRNA methylases among different species of gram-negative bacilli with high-level resistance to aminoglycosides. Eur J Clin Microbiol Infect Dis 29:1349–1353. https://doi.org/10.1007/s10096-010-1004-1

Mechanistic Understanding of Antibiotic Resistance in ESKAPE Pathogens

3

Sampathkumar Ranganathan, Hemavathy Nagarajan, Siddhardha Busi, Dinakara Rao Ampasala, and Jung-Kul Lee

Abstract

Antibiotic resistance of *Enterococcus faecium, Staphylococcus aureus, Klebsiella pneumoniae, Acinetobacter baumannii, Pseudomonas aeruginosa,* and *Enterobacter* spp. (ESKAPE) is an increasing global threat to public health. Most nosocomial infections in hospitals are caused by these pathogens that have acquired resistance to commonly accessible drugs. Antibiotic resistance is a complex process involving genetic and epigenetic factors, and horizontal gene transfer is the most prevalent resistance mechanism in these pathogens. The propagation of multiple antibiotic-resistance genes among bacterial populations can occur through plasmids, transposons, or integrons, enabling their transmission. In contrast, the mechanisms involved in modifying bacterial cell walls, such

S. Ranganathan
Department of Chemical Engineering, Konkuk University, Seoul, Republic of Korea

Department of Bioinformatics, School of Life Sciences, Pondicherry University, Puducherry, India

H. Nagarajan
Centre for Bioinformatics, Kamalnayan Bajaj Institute for Research in Vision and Ophthalmology, Vision Research Foundation, Sankara Nethralaya, Chennai, Tamil Nadu, India

S. Busi
Department of Microbiology, School of Life Sciences, Pondicherry University, Puducherry, India

D. R. Ampasala
Department of Bioinformatics, School of Life Sciences, Pondicherry University, Puducherry, India

J.-K. Lee (✉)
Department of Chemical Engineering, Konkuk University, Seoul, Republic of Korea
e-mail: jkrhee@konkuk.ac.kr

S. Busi, R. Prasad (eds.), *ESKAPE Pathogens*,
https://doi.org/10.1007/978-981-99-8799-3_3

as the development of efflux pumps or alteration of outer cell membrane porins, can prevent antibiotics from entering the cells or facilitate their expulsion. Additionally, mutations in antibiotic targets or the upregulation of antibiotic degradation enzymes can confer resistance to specific antibiotics. Furthermore, misuse and improper practice of antibiotics in healthcare and agriculture have contributed to increased antibiotic resistance among ESKAPE pathogens. Selective pressure due to antibiotic misuse can lead to the development and selection of antibiotic-resistant strains, and incomplete or inadequate antibiotic utilization can promote the survival and proliferation of resistant pathogens. Hence, understanding the resistance mechanisms in ESKAPE pathogens is a significant priority for combating antibiotic resistance. Furthermore, it will pave the way for identifying novel antimicrobial agents to combat ESKAPE pathogens.

Keywords

Antimicrobial resistance · Antibiotic resistance mechanism · ESKAPE pathogens · Target site modification · Public health

Abbreviations

AAC Aminoglycoside N-acetyltransferase
AME Aminoglycoside-modifying enzyme
APH Aminoglycoside O-phosphotransferase
ANT Aminoglycoside O-nucleotidyltransferase
AMP Antimicrobial peptides
CAT Chloramphenicol acetyltransferases
CTX-M Cefotaximase from Munch
ERE Erythromycin esterase
ESBL Extended-spectrum β-lactamase
ESKAPE *Enterococcus faecium, Staphylococcus aureus, Klebsiella pneumoniae, Acinetobacter baumannii, Pseudomonas aeruginosa,* and *Enterobacter* spp.
GI Genomic islands
GES Guiana extended-spectrum β-lactamase
HGT Horizontal gene transfer
IS Insertion sequences
ICE Integrative conjugative elements
KPC *Klebsiella pneumoniae* carbapenemase
MRSA Methicillin-resistant *S. aureus*
MGE Mobile genetic element
MDR Multidrug resistance
OXA Oxacillin-hydrolyzing enzymes
PBP Penicillin-binding protein

PDT Photodynamic therapy
PER Pseudomonas extended resistant
SHV Sulfhydral reagent variable
TEM Temoniera
Tn Transposons
VEB Vietnam extended-spectrum β-lactamase
VRE Vancomycin-resistant *Enterococcus*

3.1 Introduction

ESKAPE pathogens are a cluster of bacterial species that "escape" the effects of currently available antibiotics (Navidinia 2016). The six significant bacteria, *Enterococcus faecium, Staphylococcus aureus, Klebsiella pneumoniae, Acinetobacter baumannii, Pseudomonas aeruginosa*, and *Enterobacter* spp., are represented by the acronym ESKAPE. The ESKAPE pathogens account for many bacterial diseases and are associated with elevated mortality rates (Navidinia 2016; Rice 2008).

The onset and propagation of antibiotic resistance in these organisms pose a serious risk to public health. Antibiotic resistance occurs when bacteria acquire genetic alterations that enable them to grow and propagate in the presence of antibiotics. Bacteria develop antibiotic resistance when undergoing genetic changes that enable them to persist and thrive in the presence of antibiotics (Mancuso et al. 2021; Aslam et al. 2018)). This phenomenon is primarily driven by several factors, including inappropriate and overuse of antibiotics and inadequate infection control systems, which allow bacteria to transmit resistance genes to one another (Serwecińska 2020; Uddin et al. 2021).

Antibiotic resistance among ESKAPE pathogens is critical as it affects patient outcomes and healthcare facilities. The broad-spectrum resistance of ESKAPE bacteria to various antibiotics restricts treatment options and increases morbidity, mortality, and healthcare expenditure (Denissen et al. 2022; Serwecińska 2020). Compared to antibiotic-susceptible strains, antibiotic-resistant ESKAPE pathogens prolong hospital stays with increased treatment failure rates and mortality (Mancuso et al. 2021).

Understanding the mechanisms underlying ESKAPE pathogen resistance is critical for developing practical approaches to combat antibiotic resistance. The fundamental objective of this chapter is to gain a deeper understanding of how bacteria obtain, maintain, and transmit resistance genes, along with the biological mechanisms that confer resistance (Denissen et al. 2022; Santajit and Indrawattana 2016; Venkateswaran et al. 2023). By deciphering their mechanisms, researchers can identify potential targets for novel antimicrobial drugs, identify more effective therapies, and establish approaches to counteract the propagation of antibiotic resistance (Belete 2019; Murugaiyan et al. 2022; Rice 2008).

Additionally, research on ESKAPE pathogens has provided insight into comprehensive antibiotic resistance mechanisms. Many resistance mechanisms linked to these pathogens, such as the production of β-lactamases or efflux pumps, are shared among other bacteria. Thus, understanding the mechanisms of ESKAPE pathogens can improve efforts to combat resistance in other bacterial species (De Oliveira et al. 2020; Mancuso et al. 2021).

In conclusion, increasing antibiotic resistance in ESKAPE pathogens is a primary concern worldwide, as it jeopardizes our ability to treat infections efficiently, substantially challenging healthcare systems. Understanding the precise mechanism of antibiotic resistance in ESKAPE pathogens is crucial for developing novel treatments and preserving the efficacy of current treatment options.

Consequently, the primary goal of this chapter is to provide a complete overview of antibiotic resistance in ESKAPE pathogen infections, including its clinical implications and challenges in healthcare settings. Additionally, we explored common and specific resistance mechanisms, molecular characterizations, and advances in genomic and molecular approaches. This chapter highlights the importance of studying ESKAPE pathogens to understand broader antibiotic resistance mechanisms. This study elucidates the implications of antibiotic resistance on patient outcomes, healthcare costs, and public health. Current strategies include the development of novel antimicrobial agents, combination therapies, and infection control practices.

3.2 Overview of Antibiotic Resistance in ESKAPE Pathogens

ESKAPE pathogens are a specific group of well-defined bacterial species that "escape" the effects of antibiotics. Although each group of ESKAPE pathogens has unique antibiotic resistance mechanisms, they share common traits that contribute to their ability to evade treatment (Marturano and Lowery 2019; Mulani et al. 2019). Here, we outline the antibiotic resistance process observed among ESKAPE pathogens.

3.2.1 *E. faecium*

Enterococcus spp. are Gram-positive facultative anaerobic bacteria found primarily in the gastrointestinal (GI) tract of humans and other animals. Although more than 20 *Enterococcus* spp. have been discovered, only *E. faecalis* and *E. faecium* are clinically significant species found in hospitalized patients (Dubin and Pamer 2017; Ramos et al. 2020). Enterococcal infections commonly occur endogenously, and cross-infection is only observed in hospitals (Dubin and Pamer 2017; García-Solache and Rice 2019). They are notorious for resisting the antibiotic vancomycin (vancomycin-resistant Enterococcus (VRE)), a last-resort drug for many infections (Ahmed and Baptiste 2018). VRE strains of *E. faecium* initially appeared in North America between the end of the 1980s and 2002. Their prevalence in Europe is lower than in North America (Miller et al. 2020).

Six types of VREs were observed in the clinical isolates (Van-A–E and Van-G). They are responsible for resistance in *E. faecium* and encode enzymes that modify the vancomycin target, preventing its binding and rendering it ineffective. However, resistance is primarily mediated by the acquisition of *van-A* genes and possesses the greatest level of resistance to glycopeptide antibiotics (Smith 2007).

Galloway-Peña et al. (2011) reported two different clades of *E. faecium,* clades A and B, with distinct genetic makeup. Clade A is commonly associated with hospital settings, and clade B is associated with the community; however, they both express the penicillin-binding protein (PBP) PBP5 at low levels, which weakly bind with β-lactam antibiotics (Arias and Murray 2012; Belloso Daza et al. 2021).

3.2.2 *S. aureus*

S. aureus is a Gram-positive coccus organized in a distinctive cluster resembling grapes. It is generally found in the normal skin flora, specifically in the nasal cavity and abdomen of humans and animals (Thomer et al. 2016). Methicillin-resistant *S. aureus* (MRSA) is a well-known, clinically significant ESKAPE pathogenic strain of *S. aureus*. It resists an extensive array of β-lactams like penicillin and cephalosporins. MRSA strains produce an altered PBP, PBP2a, that has a decreased affinity for β-lactam antibiotics, allowing bacteria to thrive and multiply in the presence of these drugs (Aedo and Tomasz 2016; Fergestad et al. 2020).

3.2.3 *K. pneumoniae*

K. pneumoniae is a Gram-negative bacterium that commonly thrives in hospital settings. It can be endogenous or acquired through absolute interactions with a contaminated host (Ashurst and Dawson 2018; Ko et al. 2002). A wide range of β-lactamase enzymes have recently been acquired by *K. pneumoniae*, enabling them to metabolize β-lactam antibiotics, such as penicillin, carbapenems, and cephalosporins (Bush 2018). *K. pneumoniae* is predominantly recognized for its carbapenem resistance, facilitated by the synthesis of carbapenemases, enzymes that hydrolyze and render carbapenem activity (Halat and Moubareck 2020). Carbapenemases, such as the *K. pneumoniae* carbapenemase (KPC), inactivate these antibiotics and are ineffective against *K. pneumoniae* infections (Aurilio et al. 2022).

3.2.4 *A. baumannii*

A. baumannii is a Gram-negative bacterium that thrives under natural conditions, and this group of bacteria is clinically significant for nosocomial infections (Howard et al. 2012; Nocera et al. 2021). *A. baumannii* generally causes respiratory and urinary tract infections in hospital settings (Nocera et al. 2021). It is a highly drug-resistant pathogen that often causes healthcare-related diseases in surgical wards and intensive care units (Bush and Bradford 2020). It demonstrates resistance to an array of antibiotics across various processes, including the synthesis of β-lactamases namely imipenem metallo-β-lactamase and oxacillinase serine β-lactamase encoded by bla_{IMP} and bla_{OXA} genes, respectively (Bush and Bradford 2020; Nguyen and Joshi 2021). It also acquires an efflux pump mechanism that expels antibiotics from the bacterial cell and alterations in outer membrane porins that reduce drug entry (Abdi et al. 2020; Coyne et al. 2011).

3.2.5 *P. aeruginosa*

P. aeruginosa is a Gram-negative opportunistic pathogen associated with various diseases in hospital facilities(Stover et al. 2000; Whiteley et al. 2017). This condition predominantly affects immunocompromised individuals. It possesses intrinsic resistance mechanisms and can acquire additional resistance to genetic mutations and horizontal gene transfer (HGT) (Pang et al. 2019). They exhibit resistance via the synthesis of β-lactamases, particularly efflux pumps, and alterations in the target sites of antibiotics (Glen and Lamont 2021; Qin et al. 2022). It mainly acquires carbapenem (imipenem) resistance in combination with the *AmpC* gene and porin changes in the outer membrane to overcome antibiotic stress (Glen and Lamont 2021).

3.2.6 *Enterobacter* spp.

Enterobacter is a Gram-negative bacteria, including *E. cloacae* and *E. aerogenes*, which are recognized for their capability for the emergency and propagate resistance genes (Intra et al. 2023; Liu et al. 2021). They primarily affect immunocompromised patients in healthcare settings (Intra et al. 2023). These pathogens possess a wide array of antibiotic resistance mechanisms, including the production of carbapenemases and extended-spectrum β-lactamases (ESBLs) (Davin-Regli and Pagès 2015). They can acquire resistance through alterations in porins and efflux pumps. *Enterobacter* spp. multidrug resistance (MDR) strains are now being managed with tigecycline and colistin (Davin-Regli and Pagès 2015).

Elucidating the detailed mechanisms underlying the development of antibiotic resistance in ESKAPE pathogens is essential for discovering novel strategies to combat resistance, designing novel antibiotics, and applying appropriate disease regulatory measures to prevent their spread.

3.3 Impact of Antibiotic Resistance on Treatment Options

Antibiotic resistance substantially influences the treatments available for ESKAPE pathogen infections, resulting in fewer alternatives, higher healthcare expenditures, treatment failure, and poor outcomes. Because ESKAPE pathogens are incredibly resistant to the broad spectrum of antibiotics currently in use, it is difficult to treat the diseases caused by them, which ultimately increases morbidity and mortality (Denissen et al. 2022). The emergence of alternative treatment options, such as the identification of new drug targets and novel antibiotics, and non-antibiotic techniques, such as phage or immune-based therapies, could be driven by an understanding of the process of antibiotic resistance in ESKAPE pathogens (Idris and Nadzir 2023).

A mechanistic understanding of antibiotic resistance would also aid in developing effective prevention strategies and infection control, such as minimizing the transmission of resistant strains and enhancing antimicrobial stewardship in hospital facilities. Hence, it is essential to investigate the antibiotic resistance of ESKAPE pathogens to ensure the development of effective therapies to combat them.

3.4 Mechanisms of Antibiotic Resistance in ESKAPE Pathogens

Antibiotic resistance in ESKAPE pathogens is heterogeneous and involves both genetic and molecular processes. The following are some of the significant mechanisms of resistance found in ESKAPE pathogens.

3.4.1 Production of Enzymes That Inactivate or Alter Antibiotics

Bacteria use drug alterations and inactivation strategies to develop antibiotic resistance. They produce enzymes that modify the structures of antibiotics and render them ineffective. These alterations enable bacteria to evade the harmful effects of antibiotics, thereby reducing their effectiveness in treating infections (Reygaert 2018).

For instance, bacteria may enzymatically modify the antibiotic by adding chemical groups, such as acetyl, phosphate, or adenyl groups, which may inhibit the capacity of the antibiotic to bind to its target site (Breijyeh et al. 2020). Enzymes, such as β-lactamases, chloramphenicol acetyltransferase, and aminoglycoside-modifying enzymes (AMEs), are examples that break down antibiotics and degrade their activity (Bush 2018).

3.4.1.1 Beta-Lactamases

One of the most studied enzymes produced by ESKAPE pathogens is β-lactamase, which alters and deactivates β-lactam antibiotics. Generally, β-lactam antibiotics hamper cell membrane synthesis by binding to PBPs, which are crucial for forming cross-links between peptidoglycans in bacteria (Bush and Bradford 2016). The β-lactamases hydrolyze the 4-membered β-lactam ring in the β-lactam antibiotics, such as penicillin, early cephalosporins, carbapenems, and monobactams. ESKAPE pathogens, including *K. pneumoniae*, *Enterobacter species*, and *P. aeruginosa*, produce β-lactamases that can hydrolyze the four-membered β-lactam ring of these antibiotics, rendering them active. This enzymatic inactivation occurs through the cleavage of the amide bond in the 4-membered β-lactam ring, annihilating the antimicrobial properties of the antibiotic (Tooke et al. 2019). Some β-lactamases have a limited array of activities and can hydrolyze certain types of β-lactam antibiotics. In contrast, other antibiotics have broad-spectrum activities and can inactivate a wide array of β-lactam antibiotics. For instance, Penicillinase, cephalosporinases, broad spectrum- and Metallo-β-lactamases, ESBL, and Carbapenemases remain as well-know lactamases, which are highly accountable for the antibiotic resistance in the ESKAPE pathogens. Based on the Ambler structural categorization of β-lactamases, they can be classified into four classes: A, B, C, and D (Bush and Jacoby 2010). The large family of clusters, the class A member of β-lactamases, has serine residue in its catalytic site. This introduces the hydroxyl (-OH) group into the 4-membered ring, which is located in the β-lactams, such as benzylpenicillin and many penicillin derivatives. The modified antibiotic loses its binding affinity for PBPs, leading to the survival and proliferation of pathogens (Table 3.1) (Bush and Bradford 2016, 2019; Bush and Jacoby 2010).

Table 3.1 Beta-lactam enzyme in ESKAPE pathogens with their classification

Pathogen	Class	Subclass	Enzyme names
Enterococcus faecium	Class A	TEM-type	TEM-1, TEM-2, TEM-3
	Class B		IMP, VIM, NDM
	Class C	AmpC-type	AmpC
	Class D	OXA-type	OXA-23, OXA-40, OXA-51, OXA-58
Staphylococcus aureus	Class A	TEM-type	TEM-1, TEM-2, TEM-3
	Class B		
	Class C	AmpC-type	AmpC
	Class D	OXA-type	OXA-1, OXA-2, OXA-10, OXA-48
Klebsiella pneumoniae	Class A	TEM-type	TEM-1, TEM-2, TEM-3
		SHV-type	SHV-1, SHV-2, SHV-3
		CTX-M	CTX-M-1, CTX-M-2, CTX-M-3
	Class B		IMP-1, IMP-2, VIM-1, NDM-1
	Class C	AmpC-type	AmpC
	Class D	OXA-type	OXA-1, OXA-2, OXA-48
Acinetobacter baumannii	Class A	TEM-type	TEM-1, TEM-2, TEM-3
	Class B		IMP-1, IMP-2, VIM-1, NDM-1
	Class C	AmpC-type	AmpC
	Class D	OXA-type	OXA-23, OXA-24, OXA-51, OXA-58
Pseudomonas aeruginosa	Class A	TEM-type	TEM-1, TEM-2, TEM-3
	Class B		IMP-1, IMP-2, VIM-1, NDM-1
	Class C	AmpC-type	AmpC
	Class D	OXA-type	OXA-1, OXA-2, OXA-10, OXA-48
Enterobacter species	Class A	TEM-type	TEM-1, TEM-2, TEM-3
		SHV-type	SHV-1, SHV-2, SHV-3
		CTX-M	CTX-M-1, CTX-M-2, CTX-M-3
	Class B		IMP-1, IMP-2, VIM-1, NDM-1
	Class C	AmpC-type	AmpC
	Class D	OXA-type	OXA-1, OXA-2, OXA-48

The majority of class A β-lactamase enzymes present in Gram-negative ESKAPE pathogens are ESBLs, TEM (Temoniera), SHV (sulfhydral reagent variable), CTX-M (Cefotaximase from Munch), KPC, gram-positive bacteria penicillinase, cephalosporinases, and broad-spectrum β-lactamases (Bush and Bradford 2020).

Most of the *blaZ* gene encoding penicillinase in *S. aureus* is responsible for resistance, which has also been observed in some *Enterococcus* spp. (De Oliveira et al. 2020). CTX-M ESBL type is predominantly found in *K. pneumoniae* among the significant clinical isolates. ESBL is a subtype of class A β-lactamases that hydrolyze a comprehensive range of β-lactam antibiotics, including third-generation cephalosporins. ESBLs, such as Guiana extended-spectrum β-lactamase (GES), Pseudomonas extended resistant (PER), and Vietnam extended-spectrum β-lactamase (VEB), primarily exist in the Gram-negative species of ESKAPE pathogens (Reygaert 2018; De Oliveira et al. 2020).

Most Class A enzymes were susceptible to clavulanate, except TEM-30, TEM-50, and SVH-10 (Bush and Jacoby 2010). In contrast, the KPC strain of *K. pneumoniae* is resistant to all types of β-lactamases, including carbapenems (Bush and Jacoby 2010). However, these KPC strains are currently treated with a blend of imipenem-cilastatin-relebactam and meropenem-vaborbactam (Nordmann et al. 2009).

Ambler classified group B β-lactamases are Metallo-β-lactamases (MBLs), which contain Zn^{+} at their catalytic site. These MBLs were predominantly detected in the Gram-negative strains of ESKAPE pathogens. MBLs, such as Verona integron-encoded MBL (VIM) (Hishinuma et al. 2019; Matsumura et al. 2017) and imipenemase (IMP), have been identified in *P. aeruginosa*, *K. pneumoniae*, *E. cloacae* isolate, and *Acinetobacter* spp. NDM (New Delhi Metallo-β-lactamases) is the most alarming strain since its resistance gene is encoded in the mobile genetic element of the Grams-negative strains of ESKAPE pathogens namely, *K. pneumoniae* and *E. cloacae,* and it easily transferability to the other antibiotic-resistant bacterial classes (Jovcic et al. 2011; Peirano et al. 2018).

Ambler class C β-lactamases, namely penicillinases and cephalosporinases, generally resist narrow-intermediate spectrum cephalosporin antibiotics encoded by the chromosomal *AmpC* gene. Class C is also composed of serine at the active site. This is commonly found in *P. aeruginosa*, *E. cloacae,* and some strains of *Enterobacter* spp., with low levels of expression without clinical significance. However, resistance can emerge during the course of antibiotic therapy (Bush and Bradford 2020; Craig and Andes 2014). *AmpC* resistance becomes a concern when transferred through the plasmid to other *AmpC* lacking strains such as *K. pneumonia.* This *AmpC* encoded β-lactamases, which inactivate all penicillin, wide varieties of cephalosporins, and aztreonam, which were not usually inactivated through other β-lactamase enzymes classified in the Ambler scheme (Philippon et al. 2002; Tamma et al. 2019).

Other class C enzymes (ESAC as extended-spectrum *AmpC*), including GC1 in *E. cloacae*, plasmid-mediated CMY (10, 19, and 37) (Wachino et al. 2006), and imipenem-resistant *P. aeruginosa AmpC* variants, have also been reported. These enzymes confer high resistance to ceftazidime and some oxyimino-β-lactams

(Nordmann and Mammeri 2007; Nukaga et al. 1995; Rodríguez-Martínez et al. 2009).

Ambler class D also catalyzes serine-mediated hydrolysis of carbenicillin and cloxacillin/oxacillin. It consists of several enzymes, namely oxacillin-hydrolyzing enzymes (OXA). The ESKAPE pathogen, *P. aeruginosa*, was identified based on the ESBL properties of enzymes, such as OXA-10, OXA-11, OXA-14, OXA-15, and OXA-16 (Bradford 2001; Pitout et al. 2020). It was also found that chromosome-encoded OXA enzymes in *A. baumannii* most frequently have higher resistance to carbapenem-hydrolyzing activities, combined with a decreased level of porin formation in the pathogen cell membrane (Thomson and Bonomo 2005; Walther-Rasmussen and Høiby 2006).

ESBLs are often plasmid-mediated, meaning they are encoded on a mobile genetic element (MGE) plasmid. This facilitates transfer among diverse classes of bacteria and contributes to the propagation of antibiotic resistance in other bacterial cells (Gekenidis et al. 2020).

However, some ESKAPE pathogens can produce enzymes that introduce chemical modifications, such as acetylation, phosphorylation, and adenylation, to overcome the effects of antibiotics (Kakoullis et al. 2021). They are broadly categorized as aminoglycoside-modifying enzymes (AMEs), rifamycin-modifying enzymes, macrolide phosphotransferases, chloramphenicol acetyltransferases (CAT), phosphomycin-modifying enzymes, and flavin-dependent monooxygenases. The most clinically significant enzymes are discussed below (Imran et al. 2022; Urban-Chmiel et al. 2022).

3.4.1.2 AMEs

AMEs play a significant role in the antibiotic resistance of ESKAPE pathogens. AMEs can introduce chemical modifications to aminoglycosides via acetylation, adenylation, or phosphorylation at specific positions (Ahmadian et al. 2021; Costello et al. 2019). This modification of aminoglycosides causes them to lose their binding affinity to the target site of bacterial ribosomes, making them ineffective. For example, streptomycin, gentamicin, and amikacin are aminoglycosides that can be inactivated by ESKAPE pathogens and are mostly isolated from *Actinomycetes* spp. These three types of aminoglycoside-modifying enzymes are described below (Jadimurthy et al. 2022; Ramirez and Tolmasky 2017; Zárate et al. 2018).

3.4.1.3 Aminoglycoside N-Acetyltransferases (AACs)

AACs are the largest class of AMEs, and this enzyme adds acetyl groups to the amino group present in the aminoglycoside antibiotics sites (1, 3, 2′ and 6′) by acetyl coenzyme A-dependent catalysis mechanism. This alteration changes the charge of the molecules and hinders their interaction with the bacterial target ribosomes. AACs are further categorized into four subgroups depending on the specific site of alterations (Ban et al. 2021; Costello et al. 2019). AAC(1) and AAC(3) can acetylate the amino group of the 2-deoxystreptamine central ring of the aminoglycosides at sites 1 and 3. In contrast, AAC(6′) and AAC(2″) act on the amine groups in the 2,6-dideoxy-2,6-diamino-glucose ring at the 2′ and 6′ sites of aminoglycosides (Krause et al. 2016). Contemporary epidemiological and statistical analyses in the United States, Asia, and Europe have shown that the most clinically significant AAC(3) and AAC(6) are commonly detected in *K. pneumonia, A. baumannii, P. aeruginosa,* and *Enterobacter* spp. They are primarily resistant to gentamicin, amikacin, neomycin, and tobramycin; *acc(3)-Ia* and *acc(6′)-Ia* encode these enzymes (Castanheira et al. 2019; Costello et al. 2019; Ramirez and Tolmasky 2010).

3.4.1.4 Aminoglycoside O-Phosphotransferases (APHs)

APHs are the second-highest observed class of AMEs. APHs modify the hydroxyl groups of aminoglycosides via ATP-mediated phosphorylation. This phosphorylation reduces antibiotic binding affinity to the target site in ESKAPE pathogens. APHs are categorized into seven subgroups based on the position at which they phosphorylate. They are APH(4), APH(6), APH(9), APH(3″), APH(3′), APH(2″), and APH(7″). The clinically significant APH enzymes observed among ESKAPE pathogens, namely *S. aureus* and *Enterococcus* spp., are *aph(3′)-IIIa* encoded APH(3′). This confers resistance to amikacin, primarily mediated by the plasmid (Ramirez and Tolmasky 2010).

3.4.1.5 Aminoglycoside O-Nucleotidyltransferase (ANTs)

ANTs are the third largest class of AMEs, capable of substituting nucleotide monophosphates with a hydroxyl group (-OH) of aminoglycosides. This transfer is mediated by Mg^{2+} ions and a catalytic base residue (Asp) in the catalytic core. This substitution results in the reduced binding of antibiotics to ribosomes. The substitutions occurred at the 2″, 3″, 4′, 6, and 9 positions of aminoglycosides, and based on these positional substitutions, the enzymes were categorized (Cox et al. 2015; Latorre et al. 2016). ANT(2″ and 4″) are the most clinically significant enzymes. ANT(4′) is encoded by the *ant(4′)-Ia* gene present in *S. aureus, Enterococcus* spp., *K. pneumoniae,* and *P. aeruginosa* are generally resistant to tobramycin, neomycin B (or C), kanamycin (A, B, and C), and amikacin. ANT(2″) is encoded by the *ant(2″)-Ia*, which confers resistance to gentamicin and tobramycin in Gram-negative ESKAPE pathogens (Table 3.2) (Benveniste and Davies 1971; Hirsch et al. 2014; Thacharodi and Lamont 2022).

3.4.1.6 Chloramphenicol Acetyltransferases

Chloramphenicol acetyltransferases (CAT) add an acetyl group to a chloramphenicol antibiotic and reduce its activity. In general, chloramphenicol binds to the 50S subunit of the ribosome and hinders protein synthesis by interrupting peptide bond formation. Acetylation of the 3-hydroxy group in chloramphenicol by CAT is mediated by acetyl-S-CoA-dependent acetylation. *Enterobacter* spp. genes *catA7*, *catA8*, and *catA9*; *P. aeruginosa* gene *catB7*; and *A. baumannii* genes *catB3*, *catA1*, and *catA2* encode CAT enzymes, which are responsible for chloramphenicol resistance (Fig. 3.1) (Pavlova et al. 2023; Rohana et al. 2023; Tevyashova 2021).

Table 3.2 Aminoglycosides resistance mechanisms in ESKAPE pathogens and its associated genes

ESKAPE pathogen	Aminoglycoside resistance mechanisms	Aminoglycoside-resistant genes	Examples of aminoglycosides
Enterococcus faecium	AMEs—Reduced permeability due to changes in membrane proteins	- *aac(6')-Ie -aph(2")-Ia*	Gentamicin streptomycin
Staphylococcus aureus	AMEs—Ribosomal target site mutations—Efflux pumps	- *aac(6')-Ie-aph(2")-Ia*	Gentamycin Tobramycin
Klebsiella pneumoniae	AMEs—Reduced uptake due to porin loss	- *aac(6')-Ib-cr*	Amikacin Gentamycin
Acinetobacter baumannii	AMEs—Efflux pumps—Ribosomal target site mutations	- *aac(6')-Ib*	Amikacin Gentamycin
Pseudomonas aeruginosa	AMEs—Efflux pumps—Ribosomal target site mutations	- *aac(6')-Ib-cr*	Amikacin Tobramycin
Enterobacter species	AMEs—Efflux pumps—Reduced uptake due to porin loss	- *aac(6')-Ib-cr*	Amikacin Streptomycin

3.4.1.7 Macrolide Esterases and Phosphotransferase

Macrolide antibiotics (erythromycin and clarithromycin) hinder bacterial protein synthesis by binding to the peptide exit tunnel of the 50S ribosomal subunit. This lumen blockage prohibits the extension of the extended polypeptides (Zieliński et al. 2021). However, the ESKAPE pathogen, *K. pneumoniae*, can hydrolyze erythromycin antibiotics by producing the macrolide esterase enzyme erythromycin esterase (ERE) encoded by the *EreC* gene. These enzymes break down the macrolide rings of various macrolide antibiotics, leading to their inactivation and protection against the effects of antibiotics. *K. pneumoniae* and *S. aureus* also carry the macrolide phosphotransferase (*mph*) and erythromycin methyltransferase (*erm*) genes to overcome the effects of macrolide antibiotics (Fig. 3.2) (Miklasińska-Majdanik 2021; Myers and Clark 2021).

Resistance genes encode AMEs on plasmids and integrons, promoting horizontal transmission among bacteria and contributing to the prevalence of aminoglycoside resistance. Some ESKAPE pathogens are highly resistant to various antibiotics and possess multiple AME mechanisms. This phenomenon is often observed in Gram-negative ESKAPE pathogens (Almaghrabi et al. 2014; Robicsek et al. 2005). The bifunctional AAC(6′)-APH(2″) enzymes are also seen in the MRSA and VRE. Globally, this confers high resistance to gentamicin (Lauretti et al. 1999). Another variant enzyme, *AAC(6′)-Ib-cr*, confers resistance against ciprofloxacin and has been observed in Gram-negative strains of ESKAPE pathogens (Libisch et al. 2008; Holbrook and Garneau-Tsodikova 2018).

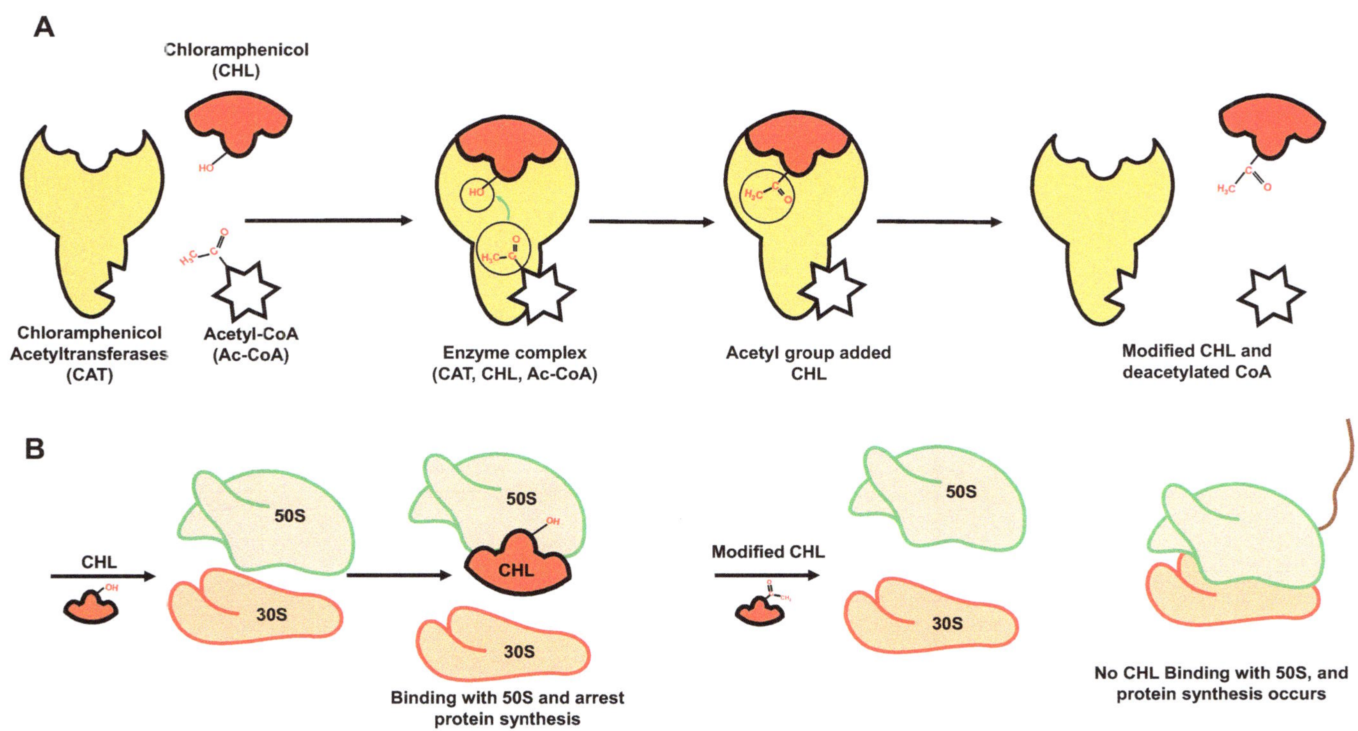

Fig. 3.1 Chloramphenicol acetyltransferases (CAT) mechanism of action against chloramphenicol antibiotics. (**a**) Modification of the acetyl group to the hydroxyl group of the chloramphenicol (CHL) with the involvement of Acetyl-CoA and (**b**) schematic representation of CHL modification, which lost its affinity to bind with 50S subunit of bacterial ribosome so there is no loss of protein synthesis process

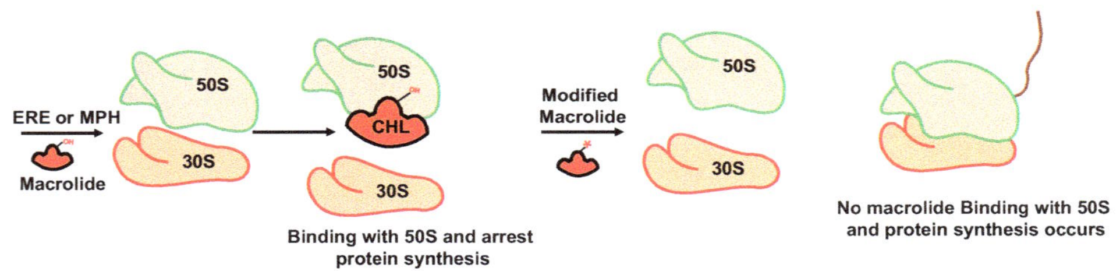

Fig. 3.2 Schematic representation of modified macrolide, which lost its ability to bind with 50S subunit of bacterial ribosome so there is no loss of protein synthesis process

3.4.2 Modification of Antibiotic Target Site

ESKAPE pathogens can protect themselves by modifying the targets of antibiotics, making them less susceptible to drug action. Three main modifications were observed in the ESKAPE pathogens: (I) replacement of the original target enzyme, (II) ribosomal binding site modification, and (III) modification of cell membrane glycoproteins (Jadimurthy et al. 2022).

3.4.3 Replacement of Original Target Enzymes

β-lactam antibiotics mainly target the PBP enzyme, which is crucial for cell wall synthesis in Gram-positive bacteria. Significantly, the MRSA strain of *S. aureus* predominantly expresses low-affinity PBP2a encoded by the *mecA* gene. PBP2a binds to broad-lactam antibiotics, including methicillin, with low affinity and prevents it from binding to traditional PBPs, which involve cell wall assembly (Lakhundi and Zhang 2018). Thus, PBPs help *S. aureus* survive extreme concentrations of β-lactam antibiotics. *mecA* is located within the staphylococcal cassette chromosome mec (SCCmec), which encodes a two-component regulatory system (TCRS*)*, site-specific *ccr* recombinase genes, and three joining regions that encompass resistance determinants, mobile genetic elements (MGEs), and regulators (Lakhundi and Zhang 2018). Thirteen SCCmec genes have been identified in *S. aureus*, and mecB and mecC have recently been determined to be more clinically significant in MRSA strains (Baig et al. 2018). In contrast, *Enterobacter* spp. ESKAPE pathogens, such as *E. faecalis* and *E. faecium*, were also shown to demonstrate resistance to β-lactam by expressing low-affinity chromosome-associated PBP5 at a significant level. In addition, specific uncommon alterations confer resistance to carbapenems, which have also been identified in *A. baumannii* (Arias and Murray 2012; Zapun et al. 2008).

Another significant target modification that confers resistance to ESKAPE pathogens is the use of fluoroquinolones. Fluoroquinolone antibiotics, namely gemifloxacin, ciprofloxacin, and norfloxacin, mainly target DNA gyrase and topoisomerase IV enzymes, which are crucial for bacterial replication and DNA repair (Wendorff and Berger 2018). The instinctive mutations in the *gyrA* and *parC* genes encoding these enzymes imply changes in the amino acid sequence, which specifically binds to the quinolone-binding site and confers resistance to fluoroquinolone antibiotics (Hooper 2001; Laponogov et al. 2009; Wendorff and Berger 2018). Plasmid-mediated quinolone resistance (PMQR) is observed in *K. pneumoniae* and *Enterobacter* spp. in which Qnr-family proteins, such as QnrA, QnrB, and QnrS, bind directly to antibiotics that target DNA gyrases (Horinouchi and Weisblum 1980; Robicsek et al. 2006).

3.4.4 Binding Site Modification of Antibiotics

ESKAPE pathogens, such as *S. aureus* and *Enterococcus* spp., are resistant to macrolide-lincosamide-streptogramin B antibiotics (MLS_B). Erm-encoded rRNA methyltransferases exhibit conserved or acquired resistance. Conserved expression of the *erm* gene encoded macrolide esterase is widely resistant to all MLS_B, whereas acquired gene expression is limited to 14- and 15-macrolides. Most of the identified erms were located in MGEs(De Oliveira et al. 2020). *ermA* is predominantly found in the HA-MRSA strain and is located on transposon Tn554 as a part of the SCCmec II cassette. The genes *erm(A)* and *erm(B)* are primarily found in *S. aureus* and *Enterococci* spp., respectively (Ali Alghamdi et al. 2023; Serwecińska 2020).

Another ribosomal-level alteration occurs in the gene encoding the *Cfr*-mediated methylation of 23S rRNA, which confers resistance against linezolid and aminoglycosides. The *Cfr* gene is present in MGEs and is most frequently seen in *S. aureus,* along with other resistance mechanisms for the MLS_B antibiotic (Doi et al. 2016; Locke et al. 2014). 16S rRNA methylation mediated by 16S rRNA methyltransferase confers resistance to a wide array of aminoglycosides (Costello et al. 2019; Mendes et al. 2008). To date, nearly ten different classes have been reported, including ARnA, RmFA-H, and NmpA, which are located on plasmids and are considered clinically significant (Doi et al. 2016; Long et al. 2006).

3.4.5 Chemical Modification of Cell Wall Composition

Gram-negative ESKAPE pathogens have advanced mechanisms of resistance to glycopeptide inhibitors. Glycopeptide inhibitors directly target the outer cell membrane peptidoglycan D-Ala-D-Ala. This type of resistance in MRSA and *Enterococcus* strains is mediated by the acquisition of the *van* gene cluster (*Van-A-D* and *Van-G*) (Li et al. 2023; Zeng et al. 2016). This gene cluster confers resistance through (1) the synthesis of altered peptidoglycan with D-Ala-D-lactate or D-Ala-D-serine in place of D-Ala-D-Ala and (2) the production of carboxypeptidase enzyme, which clears the host target D-Ala-D-Ala (Li et al. 2023). The most significant *Van* gene clusters are *VanA* and *VanB*, which have been identified in VRE-resistant *E. faecalis, S. aureus,* and *E. faecium* in humans. Generally, the Van gene cluster is present in either the transposon or plasmid and, in some cases, in both sites (e.g., *VanA*) of the pathogens(Kim et al. 2005; Li et al. 2023).

Resistance to antimicrobial peptides (AMP) in Gram-positive ESKAPE pathogens is mediated by changes in surface charge, phospholipid composition, metabolism, and membrane stress responses. For instance, AMP daptomycin resistance in *S. aureus* develops through this mechanism (Miller et al. 2016). Polymyxin (cationic AMP) resistance in *K. pneumonia, P. aeruginosa,* and *A. baumannii* develops via alterations in the structure of the outer membrane lipopolysaccharide lipid A. This modification reduces the net negative charge of the membrane and prevents the tight binding of polymyxin to the bacterial cell membrane, which ultimately hinders the uptake of AMP into cells (Liu et al. 2016).

3.4.6 Decreased Antibiotic Invasion and Subcellular Accumulation

The uptake and accumulation of subcellular concentrations are directly related to bacterial susceptibility to antibiotics. However, pathogens have devised techniques to reduce antibiotic uptake by modulating proteins involved in antibiotic influx (Prajapati et al. 2021). Enhancing the efflux mechanism of outer membrane proteins, known as efflux pumps, is another method of lowering their cytoplasmic concentrations. These two mechanisms are used by bacteria to achieve antibiotic resistance (Huang et al. 2022).

3.4.7 Reduced Antibiotic Uptake

Porins are outer membrane channel proteins observed in Gram-negative bacteria that allow hydrophilic compounds, such as β-lactam antibiotics, to enter the cell. Mutations in the pathogens confer reduced expression, function, or loss of porin, leading to antibiotic resistance in Gram-negative ESKAPE pathogens (Prajapati et al. 2021). For instance, imipenem and carbapenem resistance in *P. aeruginosa* arises due to the loss or alteration of OprD porins. The resistance of another clinically significant strain of *A. baumannii* to imipenem is caused by the loss or inactivation of the CarO porin (Masi et al. 2019). In the ESKAPE pathogens *K. pneumoniae* and a few *Enterococci* spp., the equilibrium between various porins is adjusted during antibiotic therapy (Li et al. 2015; Davin-Regli and Pagès 2015). For example, the smaller pore size of the Omp36 porin is replaced by that of Omp35; in this case, Omp36 restricts the entry of antibiotics and makes them resistant to sorbitol. However, they were sensitive to intermediate carbapenems. A strain lacking the Omp35 and Omp36 porins confers carbapenem resistance. The LamB porin loss or reduced porin expression in certain stains of ESKPE pathogens creates resistance against β-lactam antibiotics (Li et al. 2015).

3.4.8 Increased Efflux of Antibiotics

In general, the internal level of drug accumulation significantly determines the antibiotic susceptibility of bacteria. However, ESKAPE pathogens can effectively expel certain intracellularly accumulated antibiotics through membrane proteins, termed efflux pumps, which serve as exporters. Efflux pumps are protein complexes that dynamically pump antibiotics from bacterial cells. These pumps recognize and expel a broad spectrum of antibiotics, reducing their intracellular concentration and rendering them ineffective. Genes encoding efflux pumps primarily reside in the MEGs or chromosomes. To date, six major efflux pumps have been identified in ESKAPE. These include the resistance-nodulation-division (RND), major facilitator superfamilies (MSF), MATE (multidrug and toxic compound extrusion), ATPinding cassettes (ABC), and proteobacterial antimicrobial compound efflux

(PACE) families (Davin-Regli and Pagès 2015; Hassan et al. 2015). The most widely observed efflux pump in Gram-negative bacteria is of the RND type. For instance, *P. aeruginosa* primarily expresses chromosomally expressed multidrug resistance efflux pumps (MEX). MexAB-OprM and MexCD-OprJ efflux pumps confer resistance against a wide range of antibiotics, such as β-lactams, carbapenems, and fluoroquinolones. MexXY-OprM from *P. aeruginosa* confers resistance to specific cephalosporins,-lactams, and fluoroquinolones. MexCD-OprJ and MexEF-OprN were also expressed in these pathogens (Masuda et al. 2000).

Overproduction of the AcrAB efflux pump AcrAB-TolC with downregulated porin expression has been observed in *E. aerogenes* and *P. aeruginosa* confers imipenem resistance as well as resistance to the β-lactam, tetracycline, macrolides, chloramphenicol, and aminoglycosides. In *A. baumannii*, they exhibit multidrug resistance by expressing AdeABC, AdeFGH, and AdeIJK efflux pumps (Leus et al. 2018; Pagdepanichkit et al. 2016; Su et al. 2019). Recent studies have shown that the OqxAB efflux pump in *K. pneumoniae* confers resistance to chloramphenicol and quinolones (Wong et al. 2015).

3.4.9 Altered Cell Wall or Membrane Composition (Biofilm Formation)

Some ESKAPE pathogens can produce biofilms and thrive in complex microbial colonies. Biofilms comprise extracellular polymeric substances, predominantly lipopolysaccharides, peptidoglycans, proteins, lipids, and extracellular DNA from pathogenic bacteria inside the matrix. Microorganisms communicate within the biofilm via chemical signaling known as quorum sensing (Tuson and Weibel 2013). The biofilms of these microbial communities confer resistance to antibiotics and protect against other environmental stress factors. Typically, biofilms function as a physical barrier that prevents the penetration of antimicrobial agents into bacterial cells and upregulates specific genes associated with biofilms, contributing to antibiotic resistance (Lepape et al. 2020).

Biofilm formation comprises three significant steps. (1) Cell adhesion: At this stage, the planktonic cells reach the target site and anchor themselves to the surface. (2) Formation and growth of microcolonies: Bacterial cells adhering to surfaces multiply and form microcolonies in a polysaccharide, protein, lipid, and nucleic acid biofilm matrix. Extracellular polysaccharides influence cell architecture, stability, and nutrient availability, resulting in the microenvironmental conditions inside the biofilm (Laverty et al. 2014). (3) Biofilm maturation and dispersal: EPS accumulation, eDNA production, waste disposal channels, nutrient exchange channels, ionic concentrations, and quorum-sensing (QS) signals are all involved in the biofilm maturation stage (Petrova et al. 2012). Additionally, flagellar development genes were downregulated at this stage for stable biofilm development among the microbes harbored inside the matrix (Yu et al. 2019). Finally, the dispersal of pathogens inside microcolonies occurs in two ways: active and passive. Actively, detachment is triggered by the microbes inside the matrix via QS and enzymatic degradation of the biofilm. However, passive dispersion is caused by external factors, such as scraping, sharing, and human intervention (Laverty et al. 2014; Sharma et al. 2014).

It is possible to claim that the primary drivers of antibiotic resistance are not conventional antibiotic resistance mechanisms, namely target site alteration, porin or efflux pumps, or enzymatic degradation. The biofilm matrix most likely provides a physical and biochemical barrier to attenuate drug action (e.g., hydrophobicity, reduced water, O_2 and pH levels, and increased CO_2 levels) (Funari and Shen 2022). Under these conditions, it is challenging to eradicate pathogens using commonly available antibiotics. Furthermore, nutrient deprivation causes pathogens to attain an extreme level of resistance against antibiotics. This is evident from studies of bacteria isolated from biofilms that showed complete sensitivity to antibiotics (Høiby et al. 2010). This showed that resistance occurred at the phenotypic level rather than the genotypic level. *P. aeruginosa* is the most clinically significant strain among patients with cystic fibrosis, and other pathogens, including *S. aureus*, *A. baumannii,* and *K. pneumonia,* present in medical devices in hospital settings (del Pozo and Patel 2007; Percival et al. 2015).

3.4.10 Persister Cells and Antibiotic Tolerance

Apart from the antibiotic resistance conferred by the existence of inheritable resistance-encoding genes or mutations that result in a higher MIC, there is significant evidence that some ESKAPE pathogens can evade treatment through tolerance. Antibiotic tolerance allows the entire bacterial community to survive transitory exposure to increased concentrations (β-lactam and quinolones) without increasing the MIC (Balaban et al. 2019). This tolerance can be inherited or acquired under severe external circumstances, such as nutritional constraints, host factors, temperature, and antibiotic treatment. Under these circumstances, a subpopulation of pathogens undergoes a dormant state (persisters) and revives after clearance of antibiotic exposure (Brauner et al. 2016; Yan and Bassler 2019; Cohen et al. 2013). The emergence of antibiotic persisters has been observed in MRSA strains in recent investigations, and it was evidenced that this is through mutations and is likely associated with biofilm-mediated infections. The minimum duration required to kill 99% of the bacterial population (MDK_{99}) was used to quantitatively determine bacterial tolerance to antibiotics. Persisters are generally less susceptible to antibiotics and do not proliferate in their presence (Brauner et al. 2017). After antibiotic treatment, persister cells resume growth and contribute to chronic infection (Cohen et al. 2013; Yan and Bassler 2019).

3.5 Intracellular Survival Mechanism of Antibiotic Resistant Pathogens

Some ESKAPE pathogens can internalize and persist within host cells, contributing to the enhancement of antibiotic resistance. Recent experimental studies have demonstrated that *K. pneumoniae* and *E. faecalis* can endure and survive via the formation of unique intracellular vacuolar compartments after engulfment by alveolar macrophages, whereas *S. aureus* can attach to and remain in diverse phagocytes (Fraunholz and Sinha 2012; Zou and Shankar 2016). This implies that microorganisms can elude the host immune system and remain resistant to cell-impermeable antibiotics, thereby acting as a basin for widespread or latent infection. An antibiotic-treated mouse infection model demonstrated a 100-fold increase in the vancomycin MIC for MRSA isolates and greater proclivity for systemic dissemination (Lehar et al. 2015).

3.6 Genetic Determinants of Resistance

Genetic determinants of resistance in ESKAPE pathogens are crucial for identifying mechanisms underlying antibiotic resistance. These pathogens may be innately resistant to specific antibiotics, and genes that confer resistance may collectively accrue on MGEs (De Oliveira et al. 2020). MGEs generally play a particular role in that they capture genes and facilitate their intercellular mobility, that is, within or between DNA molecules, such as transposons (Tn), insertion sequences (IS), and gene cassettes/integrons (Partridge et al. 2018). Plasmids, integrative conjugative elements (ICE), and other genomic islands (GI), MGEs play roles in intracellular mobility (Partridge et al. 2018). Overall, MGEs play a substantial role in enabling and mediating HGT and promoting the development and propagation of resistance genes among microbial communities.

3.6.1 IS and Tns

ISs are the shortest DNA segments present in the bacterial genome and are flanked by repeated inverted sequences. Generally, an IS is less than 3 kb in size. These IS are capable of self-transposition and can be identified by the genes encoding transposases (Jones and Howe 2014; Qi et al. 2023). These transposase enzymes catalyze the mobility of the IS from one position in the genome to another, frequently resulting in modest target-site duplication at the insertion site, known as transposition (Ivics and Izsvák 2010). Excision and insertion can occur within similar DNA molecules (intermolecular transposition) or between other DNA molecules (intermolecular transposition), facilitating the acquisition and propagation of resistance genes (Lipszyc et al. 2022).

Tns are more complex than IS sequences. They comprise one or more genes flanked by an insertion sequence or other repetitive DNA segments. Transposases also mediate this process (Munoz-Lopez and Garcia-Perez 2010). Three types of mobility are possible for transposons. First, simple transposition entails entirely removing transposons from the donor site and inserting them in a new place. The second type of replicative transposition occurs by duplicating the transposon, in which one copy is preserved at the donor site and the other is incorporated into a target site. Finally, composite transposition involves the mobilization of the Tn and the adjacent DNA segment, which may contain additional transposons or resistance genes (Lipszyc et al. 2022). Classic examples are Tn9, Tn10, and Tn5, which possess specific insertion sequences for chloramphenicol, tetracycline, and aminoglycoside resistance, respectively (Haniford and Ellis 2015; Vandecraen et al. 2017). In ESKAPE pathogens, transposon-mediated resistance is commonly linked to the Tn3 family, Tn7-like units, and Tn55-like elements (Nicolas et al. 2015; Partridge et al. 2018).

In some ESKAPE pathogens, the transposition of IS genes may modulate the expression of adjacent genes by deactivation or acquiring promoter or terminator sequences. This deactivation of genes confers antibiotic resistance in pathogens. For instance, the IS-associated deactivation of the OmpK36 porin in *K. pneumonia* restricts the entry of hydrophobic antibiotics, resulting in a high level of resistance (Lee et al. 2007). The IS promoter inserted gene sequence upstream of some genes also confers antibiotic resistance to ESKAPE pathogens. For example, the incorporation of the *ISAba1* gene upstream of bla_{OXA-51} is related to carbapenem resistance in *A. baumannii* (Chen et al. 2008). The same type of resistance has been discovered in *K. pneumoniae* and *P. aeruginosa* (Aubert et al. 2003, 2006).

3.6.2 Plasmids

Plasmids are crucial for the propagation of antibiotic resistance genes among ESKAPE pathogens. Plasmids are self-replicating, circular, double-stranded, extrachromosomal genetic elements in bacteria that can carry and transfer resistance genes between bacterial cells, thereby facilitating the spread of antibiotic resistance (Aminov 2011). Some critical aspects of plasmids are genetic determinants of resistance to ESKAPE pathogens.

Resistance gene carriage: Plasmids commonly harbor resistance genes through IS and other MGEs that encode antibiotic-resistance enzymes or proteins (Munoz-Lopez and Garcia-Perez 2010; Shintani et al. 2015). These genes can confer resistance to certain classes of antibiotics, for instance, β-lactams, quinolones, and aminoglycosides. Plasmids encoded by resistance genes can be transferred horizontally across bacteria, including different species or genera, leading to resistance and propagation (Carattoli 2009; Guglielmini et al. 2011; Thomas and Nielsen 2005).

3.6.2.1 Transferability

Plasmids are MGEs capable of HGT. They can be transferred from one bacterium to another through conjugation, transduction, or transformation. Conjugative plasmids contain specific transfer genes that can be transferred via direct cell-to-cell contact (Botelho et al. 2023). Nevertheless, nonconjugative plasmids, such as transposons or integrons, can be mobilized through other MGEs (Ragupathi et al. 2019).

3.6.2.2 Co-Resistance

Plasmids frequently carry multiple antibiotic resistance genes, allowing bacteria to concurrently acquire resistance to various antibiotics. This phenomenon is termed co-resistance or multidrug resistance. Co-resistance is a significant concern in the ESKAPE treatment. Co-resistant plasmids may limit the treatment choices and make infections caused by these pathogens more challenging to manage (Santajit and Indrawattana 2016).

3.6.2.3 Plasmid Size and Replicon Types

Plasmids ranged from a few kilobases to hundreds. The plasmid size determines its stability and transferability. Plasmids also contain replicon types and DNA sequences that control plasmid replication. Different families and compatibility groups are associated with different types of replicons, determining their compatibility and ability to coexist within a bacterial cell (Al-Trad et al. 2023; Juraschek et al. 2022).

3.6.2.4 Evolution and Adaptation

Plasmids can evolve through the insertion, deletion, or modification of resistance genes. They can also acquire additional genetic elements such as IS or integrons, which enrich their stability, mobility, and resistance gene propagation (Souque et al. 2021). Plasmids can undergo rapid evolution in response to selective antibiotic pressure, influencing the advancement of novel antibiotic resistance mechanisms and the spread of resistance in ESKAPE pathogens(Bennett 2008).

For example, MDR *K. pneumoniae* and *Enterobacter* spp. harbor versatile varieties of incompatibility (Inc) plasmids, which are usually group F (multi-replicon groups), C, I, H, L, C, and N, are commonly associated with multidrug resistance in these pathogens (Rozwandowicz et al. 2018; Petty et al. 2014; Pitout 2010). Enterobacterales, which is an alarming clinically significant pathogen, confers the rise and propagation of ESBLs, specifically qnrB-specific and $bla_{CTX\text{-}M}$ categories, AmpC-type cephalosporinases ($bls_{CMY\text{-}2}$ and $bla_{DHA\text{-}1}$), colistin resistance is conferred by *mcr* gene, and carbapenem resistance is conferred by genes, such as bla_{NDM}, bla_{KPC}, and $bla_{OXA\text{-}48}$ (Juraschek et al. 2022). The Gram-negative pathogen *A. baumannii* confers plasmid (bla_{OXA})-mediated antibiotic resistance to carbapenems. Plasmid-mediated resistance was observed in *P. aeruginosa* (IncP-2 plasmid), the size of which ranged from approximately 300 to 500 kb. This plasmid confers carbapenem resistance. The genes involved in this process were bla_{IMP} and bla_{VIM} (Philippon et al. 2002; Walsh et al. 2011; Bauernfeind et al. 1989; Deshpande et al. 2006). The multi-resistant plasmid pSK1 is frequently identified in clinical isolates of *S. aureus* in hospital settings, and the pLW10143 conjugative plasmid confers multidrug resistance to aminoglycosides, vancomycin, and β-lactam (Morton et al. 1993; Partridge 2015; Liu et al. 2013). Inc18 and repA_N plasmids in *Enterococci* confer resistance, particularly to vancomycin, in the MRSA strain (Diep et al. 2006; Morton et al. 1995; Partridge 2015; Zhu et al. 2010).

3.6.3 GIs and ICEs

GIs and ICEs are MGEs in the genomes of ESKAPE pathogens that play a central role in the acquisition and spread of multidrug resistance (Botelho et al. 2023). Usually, the GI and ICE are transferred through HGT. GIs are large DNA segments frequently transferred horizontally during cell division. Multiple genes, including those that impart antibiotic resistance, virulence factors, metabolic capacity, and other adaptive features, are typically found in GIs (Huddleston 2014). They are typically flanked by IS or integrases during integration and excision. A common GIs found in clinical isolates such as *S. aureus* is the SCCmec (*Staphylococcal* Cassette Chromosome), which induces resistance to methicillin, penicillin, and lactams by expressing low-affinity PBPa2 enzymes (Wegener et al. 2020). Another pathogen, *E. faecalis* harbors a pathogenic island (*EfaPI*) that is responsible for antibiotic resistance, specifically to vancomycin(Shankar et al. 2002).

ICEs are self-transmissible MEGs that can be transferred from cell to cell through bacterial conjugation via an HGT mechanism. ICES have genes that encode integrative and conjugative elements within DNA fragments, which enable them to integrate into the genome and transfer to recipient cells (Johnson and Grossman 2015). These primarily comprise genes that encode antibiotic resistance and virulence factors. For example, in the ESKAPE pathogen *A. baumannii*, AbaR-type resistant islands are well-known ICE. This gene encodes multiple resistance genes, namely AMEs, β-lactamases, and efflux pumps. In *K. pneumonia*, the ICEKp element confers resistance through genes encoded by Carbapenemases and ESBL, which enables them to resist β-lactam antibiotics (Kalpana et al. 2023). *P. aeruginosa* possesses ICE, such as PAPI-1 and PAGI-2/3-like (Van-B Tn1549), acquired resistance genes responsible for vancomycin resistance (Cochetti et al. 2008).

3.7 HGT and Resistance Spread

HGT is a critical mechanism involved in the evolution and propagation of antibiotic resistance in ESKAPE pathogens. This allows the transfer of resistance genes among similar or diverse species of bacteria. HGT involves three mechanisms: transduction, transformation, and conjugation (Lerminiaux and Cameron 2019; Tao et al. 2022). Nonconjugative plasmids were mobilized using conjugative plasmids. ESKAPE pathogens, such as *Klebsiella*, *Pseudomonas*, *Acinetobacter*, and *Staphylococcus*, can acquire resistance genes through HGT. In particular, *K. pneumonia* genome analysis revealed that HGT-mediated carbapenem resistance was observed in *Enterobacteriaceae* in hospital settings. This resistance plasmid was also observed in *E. cloacae* and *E. asburiae* (De Oliveira et al. 2020; Wyres and Holt 2018).

3.8 Role of Antibiotic Use and Misuse

The advancement and propagation of antibiotic resistance among ESKAPE pathogens are highly dependent on antibiotic usage. Understanding the detailed mechanisms underlying the proper usage and misuse of antibiotics is crucial and may help develop ways to overcome antibiotic resistance(Kalpana et al. 2023). Selective pressure, overuse, improper usage, inadequate drug concentrations, antibiotic combination therapy, and antibiotic use in agriculture and livestock are crucial aspects of antibiotic misuse (Murugaiyan et al. 2022).

As susceptible bacteria are killed or suppressed, selective pressure promotes the emergence of resistance among the smaller proportion of bacteria that survive due to pre-existing mechanisms or acquired resistance naturally, such as mutations and HGT (Beceiro et al. 2013). These groups of bacteria that survive have a selection advantage and can reproduce, resulting in the progression and spread of resistance to susceptible bacterial strains. Antibiotic overuse and improper use contribute considerably to the development of resistance, with overuse being an inappropriate prescription and excessive usage also leading to resistance development (Murray et al. 2022; Uddin et al. 2021).

Suboptimal drug concentrations during therapy can potentially contribute to the development of resistance by exposing bacteria to low antibiotic concentrations and facilitating the selection of resistant subpopulations (Gullberg et al. 2011). Combination therapy employing various antibiotics can occasionally result in resistance development; nevertheless, it should be carefully organized and guided by susceptibility rests to reduce the likelihood of resistance development. Antibiotic resistance also results in cross- and co-resistance (Tängdén 2014). These phenomena can limit treatment options and make infection management more difficult. Hence, understanding and proper usage, formulating appropriate antibiotic prescription practices, antibiotic stewardship programs, enhancing antibiotic usage and resistance surveillance, and educating healthcare providers and the public are necessary steps to tackle antibiotic resistance problems (Doron and Davidson 2011; Sartelli et al. 2020).

3.9 Novel Therapeutic Targets

Currently, researchers are exploring novel therapeutic targets to overcome antibiotic resistance in ESKAPE pathogens. This targeted focus is on bacterial susceptibility and other essential processes in bacteria that are less likely to be related to existing resistance mechanisms. These include bacterial enzymes, virulence factors, quorum sensing, efflux pump inhibitors, two-component systems, bacterial toxin-antitoxin systems, type III secretion systems, biofilm formation, essential transporters, and host-directed therapeutics (Santajit et al. 2022; Yang et al. 2009). These targets indicate critical bacterial activity, impeding bacterial growth and survival. Further exploration and improvement should be performed to estimate these targets' effectiveness and safety and to transform them into clinically viable medicines.

3.10 Alternative Therapies

In the fight against antibiotic resistance, replacement treatments are being explored as potential strategies to combat ESKAPE pathogens caused by ESKAPE. These therapies are substitutes for standard antibiotics and can target bacterial pathogens via diverse mechanisms. Researchers have explored the following alternative therapies to overcome antibiotic resistance.

3.10.1 Bacteriophage Therapy

Bacteriophages are viruses that precisely target, kill, and eliminate bacteria. These bacteriophages are either engineered or specifically selected to effectively target ESKAPE pathogens, providing a precise and tailored approach to bacterial infections (Liu et al. 2022; Panwar et al. 2020). Preclinical and clinical investigations on bacteriophage treatment have demonstrated encouraging results, especially on multidrug-resistance bacterial species. Bacteriophage therapy has shown promising results in preclinical and clinical studies, particularly for multidrug-resistant bacteria (Gill et al. 2015). For instance (clincalTials.gov registry), NTC02116010 is a clinical single-blind trial to treat "PhagoBurn" infections caused by *P. aeruginosa* and *E. coli* (Servick 2016), and NCT03140085 (clincalTials.gov registry) is for testing "Pyophages" for the treatment of UTI associated with ESKAPE pathogens Enterococcus spp., *P. aeruginosa*, Staphylococcus spp., *E. coli,* and *P. mirabilis* (Leitner et al. 2017). Another remarkable clinical trial was a double-blind placebo and double-blind trial against *S. aureus* for diabetic food ulcer NCT02664740 (ClinicalTials.gov registry) (Leitner et al. 2017).

3.10.2 Antimicrobial Peptides (AMPs)

AMPs are naturally occurring small proteins (5–50 amino acids) that have the potential to act against pathogens because of their antimicrobial properties. They disrupt the bacterial cell membranes, inhibit essential bacterial enzymes, or modulate bacterial gene expression (Huan et al. 2020). AMPs exhibit broad-spectrum antibacterial activity against various bacterial genera, including ESKAPE pathogens. They can be used as alternative therapies or adjuvants to antibiotics. For instance, Maximin H5, Andropin, melittin, and N-acetylmuramoyl-L-alanine disrupt cell walls and interfere with nucleic acid and protein synthesis (Mukhopadhyay et al. 2020).

3.10.3 Probiotics

Probiotics contain adequate quantities of live microorganisms that provide health benefits. Certain probiotic strains have been demonstrated to reduce bacterial pathogenicity and virulence factors (Das et al. 2022). Probiotics can help restore microbiota homeostasis and prevent or cure infections caused by ESKAPE pathogens by colonizing and competing with them (Kuwelker et al. 2021). Recent studies on the probiotic species of breastfed infants, such as lactic acid bacteria LBM220, showed a potential antimicrobial effect against ESKAPE pathogens(Rastogi et al. 2021). Similarly, other studies have demonstrated that Lactobacillus (*L. sakei* and *L. gallinarum*) and Bacillus species (*B. celezensis* and *B. thuringiensis*) confer antimicrobial properties against ESKAPE pathogens (Neidhöfer et al. 2023).

3.10.4 Immunotherapies

Immunotherapies boost the host immune system to combat bacterial diseases. This involves using monoclonal antibodies (mAbs) that target specific bacterial antigens or virulence factors to improve the immune response to infection (Ramamurthy et al. 2021). Immunomodulatory therapies, such as cytokines or immune checkpoint inhibitors, can facilitate the enrichment of the immune system's capacity to eliminate bacterial infections. Currently, mAb KB001-A is a PEGylated fragment that targets the *P. aeruginosa* Type II secretion system (Jain et al. 2018). Another mAb, MAB1, targets and retards the growth of *E. coli* by hindering the-barrel assembly machinery associated with the proper assembly of OMPs of Gram-negative bacterial species (Kharga et al. 2023).

3.10.5 Photodynamic Therapy (PDT)

PDT involves the use of photosensitizing agents and light to produce reactive oxygen species that can cause lethality in bacteria. Antibiotic-resistant bacteria (Sai et al. 2021); PDT can be used separately or in combination with standard antibiotics to target ESKAPE pathogens, thereby providing a non-antibiotic treatment option. For instance, porphyrin (5-ALA/MAL), phenothiazines (EtNBS derivatives), and xanthene (RB) are the most commonly used photosensitizers against ESKAEP pathogens, namely *S. aureus*, *E. faecalis*, *K. pneumonia*, and *A. baumannii* (Almeida 2020; Nakonieczna et al. 2019).

3.10.6 Essential Oils and Plant Extracts

Plant-derived essential oils have antibacterial characteristics and have been studied for their potential to combat bacterial infections (Zhang et al. 2022). Certain plant extracts and active chemicals are antibacterial against ESKAPE pathogens and may serve as chemical sources for developing new therapies or antibiotic adjuvants. Recent studies on medicinal plant extracts, such as *Lavandula angustifolia Mill, Rosmarinus officinalis*, *Melaleuca alternifolia,* and *Eucalyptus obliqua* essential oils rich in limonene, α-pinene, γ-terpinene, and carbitol, have shown antimicrobial activity against *S. aureus* (Puvača et al. 2021). Another investigation of essential oils from *Cinnamomum verum*, *Syzygium aromaticum,* and *Lavandula angustifolia,* rich in carvacrol, cinnamaldehyde, citral, and eugenol, showed that they confer antimicrobial activity by modulating biofilm-associated protein (*bap,*), *sarA luxS,* and *agr* QS in *S. aureus* and *P. aeruginosa* (Panda et al. 2022).

3.10.7 Nanoparticles

Antimicrobial properties can be established by engineering NPs to interact directly with bacterial cells or by delivering antimicrobial agents (Sharmin et al. 2021). Nanoparticles can disrupt bacterial membranes, impair significant cellular processes, and function as carriers of antimicrobial drugs. Preclinical investigations have demonstrated that nanoparticles represent a novel method for tackling antibiotic resistance. For example, nanoparticles, such as SeNPs, ZnO NPs, CeO_2, and AgNPs, have shown antimicrobial activity against ESKAPE pathogens targeting the QS system (Dar et al. 2022; Mukherjee et al. 2023).

3.11 Future Directions for Research and Interventions

Further research and interventions should explore novel drug targets, develop alternative treatment strategies, and combine therapies to overcome antibiotic resistance in ESKAPE pathogens. Understanding and targeting HGT mechanisms is crucial for preventing the propagation of resistance genes and limiting the development of multidrug-resistant strains. Improved point-of-care diagnostics can guide treatment decisions and enable the development of targeted therapies. Enhanced infection control measures, including stringent hygiene practices, improved surveillance systems, and the promotion of antimicrobial stewardship programs, are fundamental for controlling the propagation of resistant ESKAPE pathogens. Investigating the influence of the environment on antibiotic resistance in ESKAPE pathogens is critical, as it involves identifying potential reservoirs and transmission pathways, and developing strategies to mitigate environmental contributions to resistance development. Therefore, researchers should develop more effective treatments and interventions against ESKAPE pathogens by focusing on these key directions.

3.12 Conclusion

In conclusion, a mechanistic understanding of antibiotic resistance in ESKAPE pathogens is necessary to address the global challenge of multidrug-resistant infections. The diverse mechanisms by which these pathogens evade the effects of antibiotics include enzymatic inactivation, target modification, reduced uptake, increased efflux, and acquisition of resistance genes, all contributing to their ability to persist and cause severe infections.

Studying the resistance mechanisms of ESKAPE pathogens will help identify vulnerabilities and develop targeted strategies to combat them. This has enabled the discovery of new drug targets, antibiotics, and alternative treatments. A mechanistic understanding also guides infection control measures, antimicrobial stewardship programs, and surveillance systems. Collaboration between researchers, healthcare professionals, policymakers, and the pharmaceutical industry is crucial for developing innovative solutions, supporting appropriate antibiotic use, and promoting the development of new antimicrobial agents. Enhanced public awareness and education are essential to avoid the propagation of resistance. A better understanding of the complex mechanisms driving antibiotic resistance in ESKAPE pathogens could lead to enhanced treatment options, effective infection control strategies, and a more sustainable future in the battle against multidrug-resistant infections.

References

Abdi SN, Ghotaslou R, Ganbarov K, Mobed A, Tanomand A, Yousefi M, Asgharzadeh M, Kafil HS (2020) *Acinetobacter baumannii* efflux pumps and antibiotic resistance. Infect Drug Resist 13:423–434. https://doi.org/10.2147/IDR.S228089

Aedo S, Tomasz A (2016) Role of the stringent stress response in the antibiotic resistance phenotype of methicillin-resistant *Staphylococcus aureus*. Antimicrob Agents Chemother 60(4):2311–2317. https://doi.org/10.1128/AAC.02697-15

Ahmadian L, Norouzi Bazgir Z, Ahanjan M, Valadan R, Goli HR (2021) Role of aminoglycoside-modifying enzymes (AMEs) in resistance to aminoglycosides among clinical isolates of *Pseudomonas aeruginosa* in the North of Iran. Biomed Res Int 2021:7077344. https://doi.org/10.1155/2021/7077344

Ahmed MO, Baptiste KE (2018) Vancomycin-resistant Enterococci: a review of antibiotic resistancemechanisms and perspectives of human and animal health. Microb Drug Resist 24(5):590–606. https://doi.org/10.1089/mdr.2017.0147

Ali Alghamdi B, Al-Johani I, Al-Shantibiotic Resistanceani JM, Musamed Alshantibiotic Resistanceani H, Al-Otaibi BG, Almazmomi K, Yusnoraini Yusof N (2023) Antibiotic resistance in methicillin-resistant *Staphylococcus aureus*. Saudi J Biol Sci 30(4):103604. https://doi.org/10.1016/j.sjbs.2023.103604

Almaghrabi R, Clancy CJ, Doi Y, Hao B, Chen L, Shields RK, Press EG, Iovine NM, Townsend BM, Wagener MM, Kreiswirth B, Nguyen MH (2014) Carbapenem-resistant *Klebsiella pneumoniae* strains exhibit diversity in aminoglycoside-modifying enzymes, which exert differing effects on plazomicin and other agents. Antimicrob Agents Chemother 58(8):4443–4451. https://doi.org/10.1128/AAC.00099-14

Almeida A (2020) Photodynamic therapy in the inactivation of microorganisms. Antibiotics 9(4):138. https://doi.org/10.3390/antibiotics9040138

Al-Trad EI, Chew CH, Che Hamzah AM, Suhaili Z, Rahman NIA, Ismail S, Puah SM, Chua KH, Kwong SM, Yeo CC (2023) The plasmidomic landscape of clinical methicillin-resistant *Staphylococcus aureus* isolates from Malaysia. Antibiotics 12(4):733. https://doi.org/10.3390/antibiotics12040733

Aminov RI (2011) Horizontal gene exchange in environmental microbiota. Front Microbiol 2(JULY):11234. https://doi.org/10.3389/fmicb.2011.00158

Arias CA, Murray BE (2012) The rise of the Enterococcus: beyond vancomycin resistance. Nat Rev Microbiol 10(4):266–278. https://doi.org/10.1038/nrmicro2761

Ashurst JV, Dawson A (2018) Pneumonia, Klebsiella. In: StatPearls. StatPearls, Treasure Island, FL. http://www.ncbi.nlm.nih.gov/pubmed/30085546

Aslam B, Wang W, Arshad MI, Khurshid M, Muzammil S, Rasool MH, Nisar MA, Alvi RF, Aslam MA, Qamar MU, Salamat MKF, Baloch Z (2018) Antibiotic resistance: a rundown of a global crisis. Infect Drug Resist 11:1645–1658. https://doi.org/10.2147/IDR.S173867

Aubert D, Naas T, Nordmann P (2003) IS1999 increases expression of the extended-spectrum β-lactamase VEB-1 in *Pseudomonas aeruginosa*. J Bacteriol 185(17):5314–5319. https://doi.org/10.1128/JB.185.17.5314-5319.2003

Aubert D, Naas T, Héritier C, Poirel L, Nordmann P (2006) Functional characterization of IS1999, an IS4 family element involved in mobilization and expression of β-lactam resistance genes. J Bacteriol 188(18):6506–6514. https://doi.org/10.1128/JB.00375-06

Aurilio C, Sansone P, Barbarisi M, Pota V, Giaccari LG, Coppolino F, Barbarisi A, Passavanti MB, Pace MC (2022) Mechanisms of action of Carbapenem resistance. Antibiotics 11(3):421. https://doi.org/10.3390/antibiotics11030421

Baig S, Johannesen TB, Overballe-Petersen S, Larsen J, Larsen AR, Stegger M (2018) Novel SCCmec type XIII (9A) identified in an ST152 methicillin-resistant Staphylococcus aureus. Infect Genet Evol 61:74–76. https://doi.org/10.1016/j.meegid.2018.03.013

Balaban NQ, Helaine S, Lewis K, Ackermann M, Aldridge B, Andersson DI, Brynildsen MP, Bumann D, Camilli A, Collins JJ, Dehio C, Fortune S, Ghigo JM, Hardt WD, Harms A, Heinemann M, Hung DT, Jenal U, Levin BR, Zinkernagel A et al (2019) Definitions and guidelines for research on antibiotic persistence. Nat Rev Microbiol 17(7):441–448. https://doi.org/10.1038/s41579-019-0196-3

Ban YH, Song MC, Jeong JH, Kwun MS, Kim CR, Ryu HS, Kim E, Park JW, Lee DG, Yoon YJ (2021) Microbial enzymatic synthesis of amikacin analogs with antibacterial activity against multidrug-resistant pathogens. Front Microbiol 12:725916. https://doi.org/10.3389/fmicb.2021.725916

Bauernfeind A, Schweighart S, Chong Y (1989) Extended broad spectrum β-lactamase in Klebsiella pneumoniae including resistance to cephamycins. Infection 17(5):316–321. https://doi.org/10.1007/BF01650718

Beceiro A, Tomás M, Bou G (2013) Antibiotic resistance and virulence: a successful or deleterious association in the bacterial world? Clin Microbiol Rev 26(2):185–230. https://doi.org/10.1128/CMR.00059-12

Belete TM (2019) Novel targets to develop new antibacterial agents and novel alternatives to antibacterial agents. Hum Microbiome J 11:100052. https://doi.org/10.1016/j.humic.2019.01.001

Belloso Daza MV, Cortimiglia C, Bassi D, Cocconcelli PS (2021) Genome-based studies indicate that the *Enterococcus faecium* clade B strains belong to *Enterococcus lactis* species and lack of the hospital infection associated markers. Int J Syst Evol Microbiol 71(8). https://doi.org/10.1099/ijsem.0.004948

Bennett PM (2008) Plasmid encoded antibiotic resistance: acquisition and transfer of antibiotic resistance genes in bacteria. Br J Pharmacol 153(SUPPL. 1):S347. https://doi.org/10.1038/sj.bjp.0707607

Benveniste R, Davies J (1971) R-factor mediated gentamicin resistance: a new enzyme which modifies aminoglycoside antibiotics. FEBS Lett 14(5):293–296. https://doi.org/10.1016/0014-5793(71)80282-X

Botelho J, Cazares A, Schulenburg H (2023) The ESKAPE mobilome contributes to the spread of antibiotic resistance and CRISPR-mediated conflict between mobile genetic elements. Nucleic Acids Res 51(1):236–252. https://doi.org/10.1093/nar/gkac1220

Bradford PA (2001) Extended-spectrum β-lactamases in the 21st century: characterization, epidemiology, and detection of this important resistance threat. Clin Microbiol Rev 14(4):933–951. https://doi.org/10.1128/CMR.14.4.933-951.2001

Brauner A, Fridman O, Gefen O, Balaban NQ (2016) Distinguishing between resistance, tolerance and persistence to antibiotic treatment. Nat Rev Microbiol 14(5):320–330. https://doi.org/10.1038/nrmicro.2016.34

Brauner A, Shoresh N, Fridman O, Balaban NQ (2017) An experimental framework for quantifying bacterial tolerance. Biophys J 112(12):2664–2671. https://doi.org/10.1016/j.bpj.2017.05.014

Breijyeh Z, Jubeh B, Karaman R (2020) Resistance of gram-negative bacteria to current antibacterial agents and approaches to resolve it. Molecules 25(6):1340. https://doi.org/10.3390/molecules25061340

Bush K (2018) Past and present perspectives on β-lactamases. Antimicrob Agents Chemother 62(10):e01076. https://doi.org/10.1128/AAC.01076-18

Bush K, Bradford PA (2016) β-lactams and β-lactamase inhibitors: an overview. Cold Spring Harb Perspect Med 6(8):a025247. https://doi.org/10.1101/cshperspect.a025247

Bush K, Bradford PA (2019) Interplay between β-lactamases and new β-lactamase inhibitors. Nat Rev Microbiol 17(5):295–306. https://doi.org/10.1038/s41579-019-0159-8

Bush K, Bradford PA (2020) Epidemiology of β-lactamase-producing pathogens. Clin Microbiol Rev 33(2):e00047. https://doi.org/10.1128/CMR.00047-19

Bush K, Jacoby GA (2010) Updated functional classification of β-lactamases. Antimicrob Agents Chemother 54(3):969–976. https://doi.org/10.1128/AAC.01009-09

Carattoli A (2009) Resistance plasmid families in Enterobacteriaceae. Antimicrob Agents Chemother 53(6):2227–2238. https://doi.org/10.1128/AAC.01707-08

Castanheira M, Davis AP, Serio AW, Krause KM, Mendes RE (2019) In vitro activity of Plazomicin against Enterobacteriaceae isolates carrying genes encoding aminoglycoside-modifying enzymes most common in US Census divisions. Diagn Microbiol Infect Dis 94(1):73–77. https://doi.org/10.1016/j.diagmicrobio.2018.10.023

Chen TL, Wu RCC, Shaio MF, Fung CP, Cho WL (2008) Acquisition of a plasmid-borne blaOXA-58 gene with an upstream IS1008 insertion conferring a high level of carbapenem resistance to *Acinetobacter baumannii*. Antimicrob Agents Chemother 52(7):2573–2580. https://doi.org/10.1128/AAC.00393-08

Cochetti I, Tili E, Mingoia M, Varaldo PE, Montanari MP (2008) erm(B)-carrying elements in tetracycline-resistant Pneumococci and correspondence between Tn1545 and Tn6003. Antimicrob Agents Chemother 52(4):1285–1290. https://doi.org/10.1128/AAC.01457-07

Cohen NR, Lobritz MA, Collins JJ (2013) Microbial persistence and the road to drug resistance. Cell Host Microbe 13(6):632–642. https://doi.org/10.1016/j.chom.2013.05.009

Costello SE, Deshpande LM, Davis AP, Mendes RE, Castanheira M (2019) Aminoglycoside-modifying enzyme and 16S ribosomal RNA methyltransferase genes among a global collection of Gram-negative isolates. J Glob Antibiot Resist 16:278–285. https://doi.org/10.1016/j.jgar.2018.10.020

Cox G, Stogios PJ, Savchenko A, Wright GD (2015) Structural and molecular basis for resistance to aminoglycoside antibiotics by the adenylyltransferase ANT(2″)-Ia. MBio 6(1):e02180. https://doi.org/10.1128/mBio.02180-14

Coyne S, Courvalin P, Périchon B (2011) Efflux-mediated antibiotic resistance in *Acinetobacter* spp. Antimicrob Agents Chemother 55(3):947–953. https://doi.org/10.1128/AAC.01388-10

Craig WA, Andes DR (2014) Cephalosporins. In: Mandell, Douglas, and Bennett's Principles and Practice of Infectious Diseases, vol 1. Elsevier, Amsterdam, pp 278–292. https://doi.org/10.1016/B978-1-4557-4801-3.00021-7

Dar M, Gul R, Karuppiah P, Al-Dhabi N, Alfadda A (2022) Antibacterial activity of cerium oxide nanoparticles against ESKAPE pathogens. Crystals 12(2):179. https://doi.org/10.3390/cryst12020179

Das TK, Pradhan S, Chakrabarti S, Mondal KC, Ghosh K (2022) Current status of probiotic and related health benefits. Appl Food Res 2(2):100185. https://doi.org/10.1016/j.afres.2022.100185

Davin-Regli A, Pagès JM (2015) Enterobacter aerogenes and Enterobacter cloacae; Versatile bacterial pathogens confronting antibiotic treatment. Front Microbiol 6(MAY):392. https://doi.org/10.3389/fmicb.2015.00392

De Oliveira DMP, Forde BM, Kidd TJ, Harris PNA, Schembri MA, Beatson SA, Paterson DL, Walker MJ (2020) Antibiotic resistance in ESKAPE pathogens. Clin Microbiol Rev 33(3):e00181. https://doi.org/10.1128/CMR.00181-19

del Pozo JL, Patel R (2007) The challenge of treating biofilm-associated bacterial infections. Clin Pharmacol Therapeut 82(2):204–209. https://doi.org/10.1038/sj.clpt.6100247

Denissen J, Reyneke B, Waso-Reyneke M, Havenga B, Barnard T, Khan S, Khan W (2022) Prevalence of ESKAPE pathogens in the environment: antibiotic resistance status, community-acquired infection and risk to human health. Int J Hyg Environ Health 244:114006. https://doi.org/10.1016/j.ijheh.2022.114006

Deshpande LM, Jones RN, Fritsche TR, Sader HS (2006) Occurrence and characterization of carbapenemase-producing enterobacteriaceae: report from the SENTRY Antimicrobial Surveillance Program (2000-2004). Microb Drug Resist 12(4):223–230. https://doi.org/10.1089/mdr.2006.12.223

Diep BA, Gill SR, Chang RF, Van Phan TH, Chen JH, Davidson MG, Lin F, Lin J, Carleton HA, Mongodin EF, Sensabaugh GF, Perdreau-Remington F (2006) Complete genome sequence of USA300, an epidemic clone of community-acquired meticillin-resistant *Staphylococcus aureus*. Lancet 367(9512):731–739. https://doi.org/10.1016/S0140-6736(06)68231-7

Doi Y, Wachino JI, Arakawa Y (2016) Aminoglycoside resistance: the emergence of acquired 16S ribosomal RNA methyltransferases. Infect Dis Clin North Am 30(2):523–537. https://doi.org/10.1016/j.idc.2016.02.011

Doron S, Davidson LE (2011) Antimicrobial stewardship. Mayo Clin Proc 86(11):1113–1123. https://doi.org/10.4065/mcp.2011.0358

Dubin K, Pamer EG (2017) Enterococci and their interactions with the intestinal microbiome. Microbiol Spectr 5(6). https://doi.org/10.1128/microbiolspec.bad-0014-2016

Fergestad ME, Stamsås GA, Morales Angeles D, Salehian Z, Wasteson Y, Kjos M (2020) Penicillin-binding protein PBP2a provides variable levels of protection toward different β-lactams in *Staphylococcus aureus* RN4220. MicrobiologyOpen 9(8):e1057. https://doi.org/10.1002/mbo3.1057

Fraunholz M, Sinha B (2012) Intracellular *Staphylococcus aureus*: live-in and let die. Front Cell Infect Microbiol 2:43. https://doi.org/10.3389/fcimb.2012.00043

Funari R, Shen AQ (2022) Detection and characterization of bacterial biofilms and biofilm-based sensors. ACS Sensors 7(2):347–357. https://doi.org/10.1021/acssensors.1c02722

Galloway-Peña J, Rice LB, Murray BE (2011) Analysis of pbp5 of early u.s. isolates of enterococcus faecium: sequence variation alone does not explain increasing ampicillin resistance over time. Antimicrob Agents Chemother 55(7):3272–3277. https://doi.org/10.1128/aac.00099-11

García-Solache M, Rice LB (2019) The enterococcus: a model of adaptability to its environment. Clin Microbiol Rev 32(2):e00058. https://doi.org/10.1128/CMR.00058-18

Gekenidis MT, Kläui A, Smalla K, Drissner D (2020) Transferable extended-spectrum β-lactamase (Esbl) plasmids in enterobacteriaceae from irrigation water. Microorganisms 8(7):1–15. https://doi.org/10.3390/microorganisms8070978

Gill EE, Franco OL, Hancock R, E. W. (2015) Antibiotic adjuvants: diverse strategies for controlling drug-resistant pathogens. Chem Biol Drug Des 85(1):56–78. https://doi.org/10.1111/cbdd.12478

Glen KA, Lamont IL (2021) β-lactam resistance in *Pseudomonas aeruginosa*: current status, future prospects. Pathogens 10(12):1638. https://doi.org/10.3390/pathogens10121638

Guglielmini J, Quintais L, Garcillán-Barcia MP, de la Cruz F, Rocha EPC (2011) The repertoire of ICE in prokaryotes underscores the unity, diversity, and ubiquity of conjugation. PLoS Genet 7(8):e1002222. https://doi.org/10.1371/journal.pgen.1002222

Gullberg E, Cao S, Berg OG, Ilbäck C, Sandegren L, Hughes D, Andersson DI (2011) Selection of resistant bacteria at very low antibiotic concentrations. PLoS Pathog 7(7):e1002158. https://doi.org/10.1371/journal.ppat.1002158

Halat DH, Moubareck CA (2020) The current burden of carbapenemases: review of significant properties and dissemination among gram-negative bacteria. Antibiotics 9(4):186. https://doi.org/10.3390/antibiotics9040186

Haniford DB, Ellis MJ (2015) Transposons Tn 10 and Tn 5. Microbiol Spectr 3(1):MDNA3-0002-2014. https://doi.org/10.1128/microbiolspec.mdna3-0002-2014

Hassan KA, Liu Q, Henderson PJF, Paulsen IT (2015) Homologs of the *Acinetobacter baumannii* AceI transporter represent a new family of bacterial multidrug efflux systems. MBio 6(1):e01982. https://doi.org/10.1128/mBio.01982-14

Hirsch DR, Cox G, D'Erasmo MP, Shakya T, Meck C, Mohd N, Wright GD, Murelli RP (2014) Inhibition of the ANT(2″)-Ia resistance enzyme and rescue of aminoglycoside antibiotic activity by synthetic α-hydroxytropolones. Bioorg Med Chem Lett 24(21):4943–4947. https://doi.org/10.1016/j.bmcl.2014.09.037

Hishinuma T, Tada T, Uchida H, Shimojima M, Kirikae T (2019) A Novel VIM-type metallo-lactamase variant, VIM-60, with increased hydrolyzing activity against fourth-generation cephalosporins in pseudomonas aeruginosa clinical isolates in Japan. Antimicrob Agents Chemother 63(6):e00124. https://doi.org/10.1128/AAC.00124-19

Høiby N, Bjarnsholt T, Givskov M, Molin S, Ciofu O (2010) Antibiotic resistance of bacterial biofilms. Int J Antimicrob Agents 35(4):322–332. https://doi.org/10.1016/j.ijantimicag.2009.12.011

Holbrook SYL, Garneau-Tsodikova S (2018) Evaluation of aminoglycoside and carbapenem resistance in a collection of drug-resistant pseudomonas aeruginosa clinical isolates. Microb Drug Resist 24(7):1020–1030. https://doi.org/10.1089/mdr.2017.0101

Hooper DC (2001) Emerging mechanisms of fluoroquinolone resistance. Emerg Infect Dis 7(2):337–341. https://doi.org/10.3201/eid0702.010239

Horinouchi S, Weisblum B (1980) Posttranscriptional modification of mRNA conformation: mechanism that regulates erythromycin-induced resistance. Proc Natl Acad Sci U S A 77(12 II):7079–7083. https://doi.org/10.1073/pnas.77.12.7079

Howard A, O'Donoghue M, Feeney A, Sleator RD (2012) *Acinetobacter baumannii*: an emerging opportunistic pathogen. Virulence 3(3):5. https://doi.org/10.4161/viru.19700

Huan Y, Kong Q, Mou H, Yi H (2020) Antimicrobial peptides: classification, design, application and research progress in multiple fields. Front Microbiol 11:582779. https://doi.org/10.3389/fmicb.2020.582779

Huang L, Wu C, Gao H, Xu C, Dai M, Huang L, Hao H, Wang X, Cheng G (2022) Bacterial multidrug efflux pumps at the frontline of antibiotic resistance: an overview. Antibiotics 11(4):520. https://doi.org/10.3390/antibiotics11040520

Huddleston JR (2014) Horizontal gene transfer in the human gastrointestinal tract: potential spread of antibiotic resistance genes. Infect Drug Resist 7:167–176. https://doi.org/10.2147/IDR.S48820

Idris FN, Nadzir MM (2023) Multi-drug resistant ESKAPE pathogens and the uses of plants as their antimicrobial agents. Arch Microbiol 205(4):115. https://doi.org/10.1007/s00203-023-03455-6

Imran M, Jha SK, Hasan N, Insaf A, Shrestha J, Shrestha J, Devkota HP, Khan S, Panth N, Warkiani ME, Dua K, Hansbro PM, Paudel KR, Mohammed Y (2022) Overcoming multidrug resistance of antibiotics via nanodelivery systems. Pharmaceutics 14(3):586. https://doi.org/10.3390/pharmaceutics14030586

Intra J, Carcione D, Sala RM, Siracusa C, Brambilla P, Leoni V (2023) Antibiotic resistance patterns of *Enterobacter cloacae* and *Klebsiella aerogenes* strains isolated from clinical specimens: a twenty-year surveillance study. Antibiotics 12(4):775. https://doi.org/10.3390/antibiotics12040775

Ivics Z, Izsvák Z (2010) The expanding universe of transposon technologies for gene and cell engineering. Mobile DNA 1(1):1–15. https://doi.org/10.1186/1759-8753-1-25

Jadimurthy R, Mayegowda SB, Nayak SC, Mohan CD, Rangappa KS (2022) Escaping mechanisms of ESKAPE pathogens from antibiotics and their targeting by natural compounds. Biotechnol Rep 34:e00728. https://doi.org/10.1016/j.btre.2022.e00728

Jain R, Beckett VV, Konstan MW, Accurso FJ, Burns JL, Mayer-Hamblett N, Milla C, VanDevanter DR, Chmiel JF (2018) KB001-A, a novel anti-inflammatory, found to be safe and well-tolerated in cystic fibrosis patients infected with *Pseudomonas aeruginosa*. J Cyst Fibros 17(4):484–491. https://doi.org/10.1016/j.jcf.2017.12.006

Johnson CM, Grossman AD (2015) Integrative and conjugative elements (ICEs): what they do and how they work. Annu Rev Genet 49:577–601. https://doi.org/10.1146/annurev-genet-112414-055018

Jones LS, Howe RA (2014) Microbial resistance and superbugs. In: Biofilms in infection prevention and control: a healthcare handbook. Elsevier, Amsterdam, pp 257–285. https://doi.org/10.1016/B978-0-12-397043-5.00015-3

Jovcic B, Lepsanovic Z, Suljagic V, Rackov G, Begovic J, Topisirovic L, Kojic M (2011) Emergence of NDM-1 metallo-β-lactamase in *Pseudomonas aeruginosa* clinical isolates from Serbia. Antimicrob Agents Chemother 55(8):3929–3931. https://doi.org/10.1128/AAC.00226-11

Juraschek K, Malekzadah J, Malorny B, Käsbohrer A, Schwarz S, Meemken D, Hammerl JA (2022) Characterization of qnrB-carrying plasmids from ESBL- and non-ESBL-producing *Escherichia coli*. BMC Genomics 23(1):365. https://doi.org/10.1186/s12864-022-08564-y

Kakoullis L, Papachristodoulou E, Chra P, Panos G (2021) Mechanisms of antibiotic resistance in important gram-positive and gram-negative pathogens and novel antibiotic solutions. Antibiotics 10(4):415. https://doi.org/10.3390/antibiotics10040415

Kalpana S, Lin W-Y, Wang Y-C, Fu Y, Lakshmi A, Wang H-Y (2023) Antibiotic resistance diagnosis in ESKAPE pathogens—a review on proteomic perspective. Diagnostics 13(6):1014. https://doi.org/10.3390/diagnostics13061014

Kharga K, Kumar L, Patel SKS (2023) Recent advances in monoclonal antibody-based approaches in the management of bacterial sepsis. Biomedicine 11(3):765. https://doi.org/10.3390/biomedicines11030765

Kim W-J, Song JY, Hwang IS, Eom JS, Cheong HJ, Bae K, Ho Park Y, Kim WJ (2005) Prevalence and molecular epidemiology of vancomycin-resistant Enterococci (VRE) strains isolated from animals and humans in Korea. Korean J Intern Med 20:55–62

Ko WC, Paterson DL, Sagnimeni AJ, Hansen DS, Von Gottberg A, Mohapatra S, Casellas JM, Goossens H, Mulazimoglu L, Trenholme G, Klugman KP, McCormack JG, Yu VL (2002) Community-acquired *Klebsiella pneumoniae* bacteremia: global differences in clinical patterns. Emerg Infect Dis 8(2):160–166. https://doi.org/10.3201/eid0802.010025

Krause KM, Serio AW, Kane TR, Connolly LE (2016) Aminoglycosides: an overview. Cold Spring Harb Perspect Med 6(6):a027029. https://doi.org/10.1101/cshperspect.a027029

Kuwelker K, Langeland N, Löhr IH, Gidion J, Manyahi J, Moyo SJ, Blomberg B, Klingenberg C (2021) Use of probiotics to reduce infections and death and prevent colonization with extended-spectrum beta-lactamase (ESBL)-producing bacteria among newborn infants in Tanzania (ProRIDE Trial): study protocol for a randomized controlled clinical trial. Trials 22(1):312. https://doi.org/10.1186/s13063-021-05251-3

Lakhundi S, Zhang K (2018) Methicillin-resistant staphylococcus aureus: molecular characterization, evolution, and epidemiology. Clin Microbiol Rev 31(4):e00020. https://doi.org/10.1128/CMR.00020-18

Laponogov I, Sohi MK, Veselkov DA, Pan XS, Sawhney R, Thompson AW, McAuley KE, Fisher LM, Sanderson MR (2009) Structural insight into the quinolone-DNA cleavage complex of type IIA topoisomerases. Nat Struct Mol Biol 16(6):667–669. https://doi.org/10.1038/nsmb.1604

Latorre M, Revuelta J, García-Junceda E, Bastida A (2016) 6-O-Nucleotidyltransferase: an aminoglycoside-modifying enzyme specific for streptomycin/streptidine. Med Chem Commun 7(1):177–183. https://doi.org/10.1039/c5md00496a

Lauretti L, Riccio ML, Mazzariol A, Cornaglia G, Amicosante G, Fontana R, Rossolini GM (1999) Cloning and characterization of bla(VIM), a new integron-borne metallo-β-lactamase gene from a *Pseudomonas aeruginosa* clinical isolate. Antimicrob Agents Chemother 43(7):1584–1590. https://doi.org/10.1128/aac.43.7.1584

Laverty G, Gorman SP, Gilmore BF (2014) Biomolecular mechanisms of *Pseudomonas aeruginosa* and *Escherichia coli* biofilm formation. Pathogens 3(3):596–632. https://doi.org/10.3390/pathogens3030596

Lee CH, Chu C, Liu JW, Chen YS, Chiu CJ, Su LH (2007) Collateral damage of flomoxef therapy: in vivo development of porin deficiency and acquisition of bla DHA-1 leading to ertapenem resistance in a clinical isolate of *Klebsiella pneumoniae* producing CTX-M-3 and SHV-5 β-lactamases. J Antimicrob Chemother 60(2):410–413. https://doi.org/10.1093/jac/dkm215

Lehar SM, Pillow T, Xu M, Staben L, Kajihara KK, Vandlen R, DePalatis L, Raab H, Hazenbos WL, Hiroshi Morisaki J, Kim J, Park S, Darwish M, Lee BC, Hernandez H, Loyet KM, Lupardus

P, Fong R, Yan D, Mariathasan S et al (2015) Novel antibody-antibiotic conjugate eliminates intracellular *S. aureus*. Nature 527(7578):323–328. https://doi.org/10.1038/nature16057

Leitner L, Sybesma W, Chanishvili N, Goderdzishvili M, Chkhotua A, Ujmajuridze A, Schneider MP, Sartori A, Mehnert U, Bachmann LM, Kessler TM (2017) Bacteriophages for treating urinary tract infections in patients undergoing transurethral resection of the prostate: a randomized, placebo-controlled, double-blind clinical trial. BMC Urol 17(1):90. https://doi.org/10.1186/s12894-017-0283-6

Lepape A, Jean A, De Waele J, Friggeri A, Savey A, Vanhems P, Gustin MP, Monnet DL, Garnacho-Montero J, Kohlenberg A (2020) European intensive care physicians' experience of infections due to antibiotic-resistant bacteria. Antibiot Resist Infect Control 9(1):1. https://doi.org/10.1186/s13756-019-0662-8

Lerminiaux NA, Cameron ADS (2019) Horizontal transfer of antibiotic resistance genes in clinical environments. Can J Microbiol 65(1):34–44. https://doi.org/10.1139/cjm-2018-0275

Leus IV, Weeks JW, Bonifay V, Smith L, Richardson S, Zgurskaya HI (2018) Substrate specificities and efflux efficiencies of RND efflux pumps of *Acinetobacter baumannii*. J Bacteriol 200(13):49–67. https://doi.org/10.1128/JB.00049-18

Li XZ, Plésiat P, Nikaido H (2015) The challenge of efflux-mediated antibiotic resistance in Gram-negative bacteria. Clin Microbiol Rev 28(2):337–418. https://doi.org/10.1128/CMR.00117-14

Li G, Walker MJ, De Oliveira DMP (2023) Vancomycin resistance in *Enterococcus* and *Staphylococcus aureus*. Microorganisms 11(1):24. https://doi.org/10.3390/microorganisms11010024

Libisch B, Poirel L, Lepsanovic Z, Mirovic V, Balogh B, Pászti J, Hunyadi Z, Dobák A, Füzi M, Nordmann P (2008) Identification of PER-1 extended-spectrum β-lactamase producing *Pseudomonas aeruginosa* clinical isolates of the international clonal complex CC11 from Hungary and Serbia. FEMS Immunol Med Microbiol 54(3):330–338. https://doi.org/10.1111/j.1574-695X.2008.00483.x

Lipszyc A, Szuplewska M, Bartosik D (2022) How do transposable elements activate expression of transcriptionally silent antibiotic resistance genes? Int J Mol Sci 23(15):8063. https://doi.org/10.3390/ijms23158063

Liu MA, Kwong SM, Jensen SO, Brzoska AJ, Firth N (2013) Biology of the staphylococcal conjugative multiresistance plasmid pSK41. Plasmid 70(1):42–51. https://doi.org/10.1016/j.plasmid.2013.02.001

Liu YY, Wang Y, Walsh TR, Yi LX, Zhang R, Spencer J, Doi Y, Tian G, Dong B, Huang X, Yu LF, Gu D, Ren H, Chen X, Lv L, He D, Zhou H, Liang Z, Liu JH, Shen J (2016) Emergence of plasmid-mediated colistin resistance mechanism MCR-1 in animals and human beings in China: a microbiological and molecular biological study. Lancet Infect Dis 16(2):161–168. https://doi.org/10.1016/S1473-3099(15)00424-7

Liu S, Fang R, Zhang Y, Chen L, Huang N, Yu K, Zhou C, Cao J, Zhou T (2021) Characterization of resistance mechanisms of *Enterobacter cloacae* complex co-resistant to carbapenem and colistin. BMC Microbiol 21(1):208. https://doi.org/10.1186/s12866-021-02250-x

Liu S, Lu H, Zhang S, Shi Y, Chen Q (2022) Phages against pathogenic bacterial biofilms and biofilm-based infections: a review. Pharmaceutics 14(2):427. https://doi.org/10.3390/pharmaceutics14020427

Locke JB, Zuill DE, Scharn CR, Deane J, Sahm DF, Denys GA, Goering RV, Shawa KJ (2014) Linezolid-resistant *Staphylococcus aureus* strain 1128105, the first known clinical isolate possessing the cfr multidrug resistance gene. Antimicrob Agents Chemother 58(11):6592–6598. https://doi.org/10.1128/AAC.03493-14

Long KS, Poehlsgaard J, Kehrenberg C, Schwarz S, Vester B (2006) The Cfr rRNA methyltransferase confers resistance to phenicols, lincosamides, oxazolidinones, pleuromutilins, and streptogramin A antibiotics. Antimicrob Agents Chemother 50(7):2500–2505. https://doi.org/10.1128/AAC.00131-06

Mancuso G, Midiri A, Gerace E, Biondo C (2021) Bacterial antibiotic resistance: the most critical pathogens. Pathogens 10(10):1310. https://doi.org/10.3390/pathogens10101310

Marturano JE, Lowery TJ (2019) ESKAPE pathogens in bloodstream infections are associated with higher cost and mortality but can be predicted using diagnoses upon admission. Open Forum Infect Dis 6(12):ofz503. https://doi.org/10.1093/ofid/ofz503

Masi M, Winterhalter M, Pagès JM (2019) Outer membrane porins. In: Subcellular biochemistry, vol 92. Springer, New York, pp 79–123. https://doi.org/10.1007/978-3-030-18768-2_4

Masuda N, Sakagawa E, Ohya S, Gotoh N, Tsujimoto H, Nishino T (2000) Substrate specificities of MexAB-OprM, MexCD-OprJ, and MexXY-OprM efflux pumps in *Pseudomonas aeruginosa*. Antimicrob Agents Chemother 44(12):3322–3327. https://doi.org/10.1128/AAC.44.12.3322-3327.2000

Matsumura Y, Peirano G, Devinney R, Bradford PA, Motyl MR, Adams MD, Chen L, Kreiswirth B, Pitout JDD (2017) Genomic epidemiology of global VIM-producing Enterobacteriaceae. J Antimicrob Chemother 72(8):2249–2258. https://doi.org/10.1093/jac/dkx148

Mendes RE, Deshpande LM, Castanheira M, DiPersio J, Saubolle MA, Jones RN (2008) First report of cfr-mediated resistance to linezolid in human staphylococcal clinical isolates recovered in the United States. Antimicrob Agents Chemother 52(6):2244–2246. https://doi.org/10.1128/AAC.00231-08

Miklasińska-Majdanik M (2021) Mechanisms of resistance to macrolide antibiotics among *Staphylococcus aureus*. Antibiotics 10(11):1406. https://doi.org/10.3390/antibiotics10111406

Miller WR, Bayer AS, Arias CA (2016) Mechanism of action and resistance to daptomycin in *Staphylococcus aureus* and *Enterococci*. Cold Spring Harb Perspect Med 6(11):a026997. https://doi.org/10.1101/cshperspect.a026997

Miller WR, Murray BE, Rice LB, Arias CA (2020) Resistance in Vancomycin-resistant Enterococci. Infect Dis Clin North Am 34(4):751–771. https://doi.org/10.1016/j.idc.2020.08.004

Morton TM, Eaton DM, Johnston JL, Archer GL (1993) DNA sequence and units of transcription of the conjugative transfer gene complex (trs) of *Staphylococcus aureus* plasmid pGO1. J Bacteriol 175(14):4436–4447. https://doi.org/10.1128/jb.175.14.4436-4447.1993

Morton TM, Johnston JL, Patterson J, Archer GL (1995) Characterization of a conjugative staphylococcal mupirocin resistance plasmid. Antimicrob Agents Chemother 39(6):1272–1280. https://doi.org/10.1128/AAC.39.6.1272

Mukherjee A, Bose S, Shaoo A, Das SK (2023) Nanotechnology based therapeutic approaches: an advanced strategy to target the biofilm of ESKAPE pathogens. Mater Adv 4(12):2544. https://doi.org/10.1039/d2ma00846g

Mukhopadhyay S, Bharath Prasad AS, Mehta CH, Nayak UY (2020) Antimicrobial peptide polymers: no escape to ESKAPE pathogens—a review. World J Microbiol Biotechnol 36(9):131–132. https://doi.org/10.1007/s11274-020-02907-1

Mulani MS, Kamble EE, Kumkar SN, Tawre MS, Pardesi KR (2019) Emerging strategies to combat ESKAPE pathogens in the era of antibiotic resistance: a review. Front Microbiol 10(APR):539. https://doi.org/10.3389/fmicb.2019.00539

Munoz-Lopez M, Garcia-Perez J (2010) DNA transposons: nature and applications in genomics. Curr Genomics 11(2):115–128. https://doi.org/10.2174/138920210790886871

Murray CJ, Ikuta KS, Sharara F, Swetschinski L, Robles Aguilar G, Gray A, Han C, Bisignano C, Rao P, Wool E, Johnson SC, Browne AJ, Chipeta MG, Fell F, Hackett S, Haines-Woodhouse G, Kashef Hamadani BH, Kumaran EAP, McManigal B, Naghavi M et al (2022) Global burden of bacterial antibiotic resistance in 2019: a systematic analysis. Lancet 399(10325):629–655. https://doi.org/10.1016/S0140-6736(21)02724-0

Murugaiyan J, Anand Kumar P, Rao GS, Iskandar K, Hawser S, Hays JP, Mohsen Y, Adukkadukkam S, Awuah WA, Jose RAM, Sylvia N, Nansubuga EP, Tilocca B, Roncada P, Roson Calero N, Moreno-Morales J, Amin R, Krishna Kumar B, Kumar A, van Dongen MBM et al (2022) Progress in alternative strategies to combat antibiotic resistance: focus on antibiotics. Antibiotics 11(2):200. https://doi.org/10.3390/antibiotics11020200

Myers AG, Clark RB (2021) Discovery of macrolide antibiotics effective against multi-drug resistant gram-negative pathogens. Acc Chem Res 54(7):1635–1645. https://doi.org/10.1021/acs.accounts.1c00020

Nakonieczna J, Wozniak A, Pieranski M, Rapacka-Zdonczyk A, Ogonowska P, Grinholc M (2019) Photoinactivation of ESKAPE pathogens: overview of novel therapeutic strategy. Future Med Chem 11(5):443–461. https://doi.org/10.4155/fmc-2018-0329

Navidinia M (2016) The clinical importance of emerging ESKAPE pathogens in nosocomial infections. J Paramed Sci 7(3):43–57. https://doi.org/10.22037/JPS.V7I3.12584

Neidhöfer C, Rathore K, Parčina M, Sieber MA (2023) ESKAPEE pathogen biofilm control on surfaces with probiotic *Lactobacillaceae* and *Bacillus* species. Antibiotics 12(5):871. https://doi.org/10.3390/antibiotics12050871

Nguyen M, Joshi SG (2021) Carbapenem resistance in *Acinetobacter baumannii*, and their importance in hospital-acquired infections: a scientific review. J Appl Microbiol 131(6):2715–2738. https://doi.org/10.1111/jam.15130

Nicolas E, Lambin M, Dandoy D, Galloy C, Nguyen N, Oger CA, Hallet B (2015) The Tn 3-family of replicative transposons. In: Mobile DNA III. ASM Press, Washington, DC, pp 693–726. https://doi.org/10.1128/9781555819217.ch32

Nocera FP, Attili AR, De Martino L (2021) *Acinetobacter baumannii*: its clinical significance in human and veterinary medicine. Pathogens 10(2):1–13. https://doi.org/10.3390/pathogens10020127

Nordmann P, Mammeri H (2007) Extended-spectrum cephalosporinases: structure, detection and epidemiology. Future Microbiol 2(3):297–307. https://doi.org/10.2217/17460913.2.3.297

Nordmann P, Cuzon G, Naas T (2009) The real threat of *Klebsiella pneumoniae* carbapenemase-producing bacteria. Lancet Infect Dis 9(4):228–236. https://doi.org/10.1016/S1473-3099(09)70054-4

Nukaga M, Haruta S, Tanimoto K, Kogure K, Taniguchi K, Tamaki M, Sawai T (1995) Molecular evolution of a class C β-lactamase extending its substrate specificity. J Biol Chem 270(11):5729–5735. https://doi.org/10.1074/jbc.270.11.5729

Pagdepanichkit S, Tribuddharat C, Chuanchuen R (2016) Distribution and expression of the ade multidrug efflux systems in *Acinetobacter baumannii* clinical isolates. Can J Microbiol 62(9):794–801. https://doi.org/10.1139/cjm-2015-0730

Panda SK, Buroni S, Swain SS, Bonacorsi A, da Fonseca Amorim EA, Kulshrestha M, da Silva LCN, Tiwari V (2022) Recent advances to combat ESKAPE pathogens with special reference to essential oils. Front Microbiol 13:1029098. https://doi.org/10.3389/fmicb.2022.1029098

Pang Z, Raudonis R, Glick BR, Lin TJ, Cheng Z (2019) Antibiotic resistance in *Pseudomonas aeruginosa*: mechanisms and alternative therapeutic strategies. Biotechnol Adv 37(1):177–192. https://doi.org/10.1016/j.biotechadv.2018.11.013

Panwar U, Aarthy M, Singh SK (2020) Bacteriophage as a therapeutic agent to combat bacterial infection: a journey from history to application. In: Biocommunication of phages. Springer International, Berlin, pp 347–370. https://doi.org/10.1007/978-3-030-45885-0_17

Partridge SR (2015) What's in a name? ISSwi1 corresponds to transposons related to Tn2 and Tn3. MBio 6(5):e01344. https://doi.org/10.1128/mBio.01344-15

Partridge SR, Kwong SM, Firth N, Jensen SO (2018) Mobile genetic elements associated with antibiotic resistance. Clin Microbiol Rev 31(4):e00088. https://doi.org/10.1128/CMR.00088-17

Pavlova JA, Tereshchenkov AG, Nazarov PA, Lukianov DA, Skvortsov DA, Polshakov VI, Vasilieva BF, Efremenkova OV, Kaiumov MY, Paleskava A, Konevega AL, Dontsova OA, Osterman IA, Bogdanov AA, Sumbatyan NV (2023) Conjugates of Chloramphenicol Amine and Berberine as antimicrobial agents. Antibiotics 12(1):370. https://doi.org/10.3390/antibiotics12010015

Peirano G, Matsumura Y, Adams MD, Bradford P, Motyl M, Chen L, Kreiswirth BN, Pitout JDD (2018) Genomic epidemiology of global carbapenemase-producing enterobacter spp., 2008–2014. Emerg Infect Dis 24(6):1010–1019. https://doi.org/10.3201/eid2406.171648

Percival SL, Suleman L, Vuotto C, Donelli G (2015) Healthcare-associated infections, medical devices and biofilms: risk, tolerance and control. J Med Microbiol 64(4):323–334. https://doi.org/10.1099/jmm.0.000032

Petrova OE, Schurr JR, Schurr MJ, Sauer K (2012) Microcolony formation by the opportunistic pathogen *Pseudomonas aeruginosa* requires pyruvate and pyruvate fermentation. Mol Microbiol 86(4):819–835. https://doi.org/10.1111/mmi.12018

Petty NK, Zakour NLB, Stanton-Cook M, Skippington E, Totsika M, Forde BM, Phan MD, Moriel DG, Peters KM, Davies M, Rogers BA, Dougan G, Rodriguez-Baño J, Pascual A, Pitout JDD, Upton M, Paterson DL, Walsh TR, Schembri MA, Beatson SA (2014) Global dissemination of a multidrug resistant *Escherichia coli* clone. Proc Natl Acad Sci U S A 111(15):5694–5699. https://doi.org/10.1073/pnas.1322678111

Philippon A, Arlet G, Jacoby GA (2002) Plasmid-determined AmpC-type β-lactamases. Antimicrob Agents Chemother 46(1):1–11. https://doi.org/10.1128/AAC.46.1.1-11.2002

Pitout JDD (2010) Infections with extended-spectrum β-lactamase-producing enterobacteriaceae: changing epidemiology and drug treatment choices. Drugs 70(3):313–333. https://doi.org/10.2165/11533040-000000000-00000

Pitout JDD, Peirano G, Kock MM, Strydom KA, Matsumura Y (2020) The global ascendency of OXA-48-type carbapenemases. Clin Microbiol Rev 33(1):e00102. https://doi.org/10.1128/CMR.00102-19

Prajapati JD, Kleinekathöfer U, Winterhalter M (2021) How to enter a bacterium: bacterial porins and the permeation of antibiotics. Chem Rev 121(9):5158–5192. https://doi.org/10.1021/acs.chemrev.0c01213

Puvača N, Milenković J, Galonja Coghill T, Bursić V, Petrović A, Tanasković S, Pelić M, Ljubojević Pelić D, Miljković T (2021) Antimicrobial activity of selected essential oils against selected pathogenic bacteria: in vitro study. Antibiotics 10(5):546. https://doi.org/10.3390/antibiotics10050546

Qi Q, Rajabal V, Ghaly TM, Tetu SG, Gillings MR (2023) Identification of integrons and gene cassette-associated recombination sites in bacteriophage genomes. Front Microbiol 14:1091391. https://doi.org/10.3389/fmicb.2023.1091391

Qin S, Xiao W, Zhou C, Pu Q, Deng X, Lan L, Liang H, Song X, Wu M (2022) *Pseudomonas aeruginosa*: pathogenesis, virulence factors, antibiotic resistance, interaction with host, technology advances and emerging therapeutics. Signal Transduct Target Ther 7(1):199. https://doi.org/10.1038/s41392-022-01056-1

Ragupathi NKD, Bakthavatchalam YD, Mathur P, Pragasam AK, Walia K, Ohri VC, Veeraraghavan B (2019) Plasmid profiles among some ESKAPE pathogens in a tertiary care centre in south India. Indian J Med Res 149(2):222–231. https://doi.org/10.4103/ijmr.IJMR_2098_17

Ramamurthy D, Nundalall T, Cingo S, Mungra N, Karaan M, Naran K, Barth S (2021) Recent advances in immunotherapies against infectious diseases. Immunother Adv 1(1):1–16. https://doi.org/10.1093/immadv/ltaa007

Ramirez MS, Tolmasky ME (2010) Aminoglycoside modifying enzymes. Drug Resist Updat 13(6):151–171. https://doi.org/10.1016/j.drup.2010.08.003

Ramirez MS, Tolmasky ME (2017) Amikacin: uses, resistance, and prospects for inhibition. Molecules 22(12):2267. https://doi.org/10.3390/molecules22122267

Ramos S, Silva V, de Dapkevicius M, Igrejas G, Poeta P (2020) Enterococci, from harmless bacteria to a pathogen. Microorganisms 8(8):1–12. https://doi.org/10.3390/microorganisms8081118

Rastogi S, Mittal V, Singh A (2021) Selection of potential probiotic bacteria from exclusively breastfed infant faeces with antagonistic activity against multidrug-resistant ESKAPE pathogens. Probiot Antimicrob Prot 13(3):739–750. https://doi.org/10.1007/s12602-020-09724-w

Reygaert CW (2018) An overview of the antibiotic resistance mechanisms of bacteria. AIMS Microbiol 4(3):482–501. https://doi.org/10.3934/microbiol.2018.3.482

Rice LB (2008) Federal funding for the study of antibiotic resistance in nosocomial pathogens: no ESKAPE. J Infect Dis 197(8):1079–1081. https://doi.org/10.1086/533452

Robicsek A, Strahilevitz J, Jacoby GA, Macielag MJ, Abbanat D, Park CH et al (2005) Fluoroquinolone-modifying enzyme: a new adaptation of a common aminoglycoside acetyltransferase. Nat Med 12(1):83–88. https://doi.org/10.1038/nm1347

Robicsek A, Jacoby GA, Hooper DC (2006) The worldwide emergence of plasmid-mediated quinolone resistance. Lancet Infect Dis 6(10):629–640. https://doi.org/10.1016/S1473-3099(06)70599-0

Rodríguez-Martínez JM, Poirel L, Nordmann P (2009) Extended-spectrum cephalosporinases in *Pseudomonas aeruginosa*. Antimicrob Agents Chemother 53(5):1766–1771. https://doi.org/10.1128/AAC.01410-08

Rohana H, Hager-Cohen A, Azrad M, Peretz A (2023) Trend of changes in chloramphenicol resistance during the years 2017–2020: a retrospective report from Israel. Antibiotics 12(2):196. https://doi.org/10.3390/antibiotics12020196

Rozwandowicz M, Brouwer MSM, Fischer J, Wagenaar JA, Gonzalez-Zorn B, Guerra B, Mevius DJ, Hordijk J (2018) Plasmids carrying antibiotic resistance genes in Enterobacteriaceae. J Antimicrob Chemother 73(5):1121–1137. https://doi.org/10.1093/jac/dkx488

Sai DL, Lee J, Nguyen DL, Kim YP (2021) Tailoring photosensitive ROS for advanced photodynamic therapy. Exp Mol Med 53(4):495–504. https://doi.org/10.1038/s12276-021-00599-7

Santajit S, Indrawattana N (2016) Mechanisms of antibiotic resistance in ESKAPE pathogens. Biomed Res Int 2016:2475067. https://doi.org/10.1155/2016/2475067

Santajit S, Sookrung N, Indrawattana N (2022) Quorum sensing in ESKAPE bugs: a target for combating antibiotic resistance and bacterial virulence. Biology 11(10):1466. https://doi.org/10.3390/biology11101466

Sartelli M, Hardcastle TC, Catena F, Chichom-Mefire A, Coccolini F, Dhingra S, Haque M, Hodonou A, Iskandar K, Labricciosa FM, Marmorale C, Sall I, Pagani L (2020) Antibiotic use in low and middle-income countries and the challenges of antibiotic resistance in surgery. Antibiotics 9(8):1–12. https://doi.org/10.3390/antibiotics9080497

Servick K (2016) Beleaguered phage therapy trial presses on: setbacks suggest difficult road for much-needed antibiotic alternatives. Science 352(6293):1506. https://doi.org/10.1126/science.352.6293.1506

Serwecińska L (2020) Antimicrobials and antibiotic-resistant bacteria: a risk to the environment and to public health. Water 12(12):3313. https://doi.org/10.3390/w12123313

Shankar N, Baghdayan AS, Gilmore MS (2002) Modulation of virulence within a pathogenicity island in vancomycin-resistant *Enterococcus faecalis*. Nature 417(6890):746–750. https://doi.org/10.1038/nature00802

Sharma G, Rao S, Bansal A, Dang S, Gupta S, Gabrani R (2014) *Pseudomonas aeruginosa* biofilm: potential therapeutic targets. Biologicals 42(1):1–7. https://doi.org/10.1016/j.biologicals.2013.11.001

Sharmin S, Rahaman MM, Sarkar C, Atolani O, Islam MT, Adeyemi OS (2021) Nanoparticles as antimicrobial and antiviral agents: a literature-based perspective study. Heliyon 7(3):e06456. https://doi.org/10.1016/j.heliyon.2021.e06456

Shintani M, Sanchez ZK, Kimbara K (2015) Genomics of microbial plasmids: classification and identification based on replication and transfer systems and host taxonomy. Front Microbiol 6:242. https://doi.org/10.3389/fmicb.2015.00242

Smith A (2007) Bacterial resistance to antibiotics. In: Hugo and Russell's pharmaceutical microbiology. Blackwell Science, London, pp 220–232. https://doi.org/10.1002/9780470988329.ch13

Souque C, Escudero JA, Maclean RC (2021) Integron activity accelerates the evolution of antibiotic resistance. Elife 10:1–47. https://doi.org/10.7554/eLife.62474

Stover CK, Pham XQ, Erwin AL, Mizoguchi SD, Warrener P, Hickey MJ, Brinkman FSL, Hufnagle WO, Kowallk DJ, Lagrou M, Garber RL, Goltry L, Tolentino E, Westbrock-Wadman S, Yuan Y, Brody LL, Coulter SN, Folger KR, Kas A, Olson MV et al (2000) Complete genome sequence of *Pseudomonas aeruginosa* PAO1, an opportunistic pathogen. Nature 406(6799):959–964. https://doi.org/10.1038/35023079

Su CC, Morgan CE, Kambakam S, Rajavel M, Scott H, Huang W, Emerson CC, Taylor DJ, Stewart PL, Bonomo RA, Yu EW (2019) Cryo-electron microscopy structure of an *Acinetobacter baumannii* multidrug efflux pump. MBio 10(4):e01295. https://doi.org/10.1128/mBio.01295-19

Tamma PD, Doi Y, Bonomo RA, Johnson JK, Simner PJ (2019) A primer on AmpC β-Lactamases: necessary knowledge for an increasingly multidrug-resistant world. Clin Infect Dis 69(8):1446–1455. https://doi.org/10.1093/cid/ciz173

Tängdén T (2014) Combination antibiotic therapy for multidrug-resistant Gram-negative bacteria. Ups J Med Sci 119(2):149–153. https://doi.org/10.3109/03009734.2014.899279

Tao S, Chen H, Li N, Wang T, Liang W (2022) The spread of antibiotic resistance genes in vivo model. Can J Infect Dis Med Microbiol 2022:3348695. https://doi.org/10.1155/2022/3348695

Tevyashova AN (2021) Recent trends in synthesis of chloramphenicol new derivatives. Antibiotics 10(4):370. https://doi.org/10.3390/antibiotics10040370

Thacharodi A, Lamont IL (2022) Aminoglycoside-modifying enzymes are sufficient to make pseudomonas aeruginosa clinically resistant to key antibiotics. Antibiotics 11(7):884. https://doi.org/10.3390/antibiotics11070884

Thomas CM, Nielsen KM (2005) Mechanisms of, and barriers to, horizontal gene transfer between bacteria. Nat Rev Microbiol 3(9):711–721. https://doi.org/10.1038/nrmicro1234

Thomer L, Schneewind O, Missiakas D (2016) Pathogenesis of *Staphylococcus aureus* bloodstream infections. Annu Rev Pathol Mech Dis 11(1):343–364. https://doi.org/10.1146/annurev-pathol-012615-044351

Thomson JM, Bonomo RA (2005) The threat of antibiotic resistance in Gram-negative pathogenic bacteria: β-lactams in peril! Curr Opin Microbiol 8(5):518–524. https://doi.org/10.1016/j.mib.2005.08.014

Tooke CL, Hinchliffe P, Bragginton EC, Colenso CK, Hirvonen VHA, Takebayashi Y, Spencer J (2019) β-Lactamases and β-Lactamase inhibitors in the 21st century. J Mol Biol 431(18):3472–3500. https://doi.org/10.1016/j.jmb.2019.04.002

Tuson HH, Weibel DB (2013) Bacteria-surface interactions. Soft Matter 9(17):4368–4380. https://doi.org/10.1039/c3sm27705d

Uddin TM, Chakraborty AJ, Khusro A, Zidan BRM, Mitra S, Emran TB, Dhama K, Ripon MKH, Gajdács M, Sahibzada MUK, Hossain MJ, Koirala N (2021) Antibiotic resistance in microbes: history, mechanisms, therapeutic strategies and future prospects. J Infect Public Health 14(12):1750–1766. https://doi.org/10.1016/j.jiph.2021.10.020

Urban-Chmiel R, Marek A, Stępień-Pyśniak D, Wieczorek K, Dec M, Nowaczek A, Osek J (2022) Antibiotic resistance in bacteria—a review. Antibiotics 11(8):1079. https://doi.org/10.3390/antibiotics11081079

Vandecraen J, Chandler M, Aertsen A, Van Houdt R (2017) The impact of insertion sequences on bacterial genome plasticity and adaptability. Crit Rev Microbiol 43(6):709–730. https://doi.org/10.1080/1040841X.2017.1303661

Venkateswaran P, Vasudevan S, David H, Shaktivel A, Shanmugam K, Neelakantan P, Solomon AP (2023) Revisiting ESKAPE pathogens: virulence, resistance, and combating strategies focusing on quorum sensing. Front Cell Infect Microbiol 13:1159798. https://doi.org/10.3389/fcimb.2023.1159798

Wachino JI, Kurokawa H, Suzuki S, Yamane K, Shibata N, Kimura K, Ike Y, Arakawa Y (2006) Horizontal transfer of blaCMY-bearing plasmids among clinical *Escherichia coli* and *Klebsiella pneumoniae* isolates and emergence of cefepime-hydrolyzing CMY-19. Antimicrob Agents Chemother 50(2):534–541. https://doi.org/10.1128/AAC.50.2.534-541.2006

Walsh TR, Weeks J, Livermore DM, Toleman MA (2011) Dissemination of NDM-1 positive bacteria in the New Delhi environment and its implications for human health: an environmental point prevalence study. Lancet Infect Dis 11(5):355–362. https://doi.org/10.1016/S1473-3099(11)70059-7

Walther-Rasmussen J, Høiby N (2006) OXA-type carbapenemases. J Antimicrob Chemother 57(3):373–383. https://doi.org/10.1093/jac/dki482

Wegener A, Damborg P, Guardabassi L, Moodley A, Mughini-Gras L, Duim B, Wagenaar JA, Broens EM (2020) Specific staphylococcal cassette chromosome mec (SCCmec) types and clonal complexes are associated with low-level amoxicillin/clavulanic acid and cefalotin resistance in methicillin-resistant *Staphylococcus pseudintermedius*. J Antimicrob Chemother 75(3):508–511. https://doi.org/10.1093/jac/dkz509

Wendorff TJ, Berger JM (2018) Topoisomerase VI senses and exploits both DNA crossings and bends to facilitate strand passage. Elife 7:e31724. https://doi.org/10.7554/eLife.31724

Whiteley M, Diggle SP, Greenberg EP (2017) Progress in and promise of bacterial quorum sensing research. Nature 551(7680):313–320. https://doi.org/10.1038/nature24624

Wong MHY, Chan EWC, Chen S (2015) Evolution and dissemination of OqxAB-like efflux pumps, an emerging quinolone resistance determinant among members of Enterobacteriaceae. Antimicrob Agents Chemother 59(6):3290–3297. https://doi.org/10.1128/AAC.00310-15

Wyres KL, Holt KE (2018) *Klebsiella pneumoniae* as a key trafficker of drug resistance genes from environmental to clinically important bacteria. Curr Opin Microbiol 45:131–139. https://doi.org/10.1016/j.mib.2018.04.004

Yan J, Bassler BL (2019) Surviving as a community: antibiotic tolerance and persistence in bacterial biofilms. Cell Host Microbe 26(1):15–21. https://doi.org/10.1016/j.chom.2019.06.002

Yang L, Rybtke MT, Jakobsen TH, Hentzer M, Bjarnsholt T, Givskov M, Tolker-Nielsen T (2009) Computer-aided identification of recognized drugs as *Pseudomonas aeruginosa* quorum-sensing inhibitors. Antimicrob Agents Chemother 53(6):2432–2443. https://doi.org/10.1128/AAC.01283-08

Yu MK, Kim MA, Rosa V, Hwang YC, Del Fabbro M, Sohn WJ, Min KS (2019) Role of extracellular dna in *Enterococcus faecalis* biofilm formation and its susceptibility to sodium hypochlorite. J Appl Oral Sci 27:e20180699. https://doi.org/10.1590/1678-7757-2018-0699

Zapun A, Contreras-Martel C, Vernet T (2008) Penicillin-binding proteins and β-lactam resistance. FEMS Microbiol Rev 32(2):361–385. https://doi.org/10.1111/j.1574-6976.2007.00095.x

Zárate SG, De La Cruz Claure ML, Benito-Arenas R, Revuelta J, Santana AG, Bastida A (2018) Overcoming aminoglycoside enzymatic resistance: design of novel antibiotics and inhibitors. Molecules 23(2):284. https://doi.org/10.3390/molecules23020284

Zeng D, Debabov D, Hartsell TL, Cano RJ, Adams S, Schuyler JA, McMillan R, Pace JL (2016) Approved glycopeptide antibacterial drugs: mechanism of action and resistance. Cold Spring Harb Perspect Med 6(12):a026989. https://doi.org/10.1101/cshperspect.a026989

Zhang L, Gao F, Ge J, Li H, Xia F, Bai H, Piao X, Shi L (2022) Potential of aromatic plant-derived essential oils for the control of foodborne bacteria and antibiotic resistance in animal production: a review. Antibiotics 11(11):1673. https://doi.org/10.3390/antibiotics11111673

Zhu W, Murray PR, Huskins WC, Jernigan JA, McDonald LC, Clark NC, Anderson KF, McDougal LK, Hageman JC, Olsen-Rasmussen M, Frace M, Alangaden GJ, Chenoweth C, Zervos MJ, Robinson-Dunn B, Schreckenberger PC, Reller LB, Rudrik JT, Patel JB (2010) Dissemination of an Enterococcus Inc18-like vanA plasmid associated with vancomycin-resistant *Staphylococcus aureus*. Antimicrob Agents Chemother 54(10):4314–4320. https://doi.org/10.1128/AAC.00185-10

Zieliński M, Park J, Sleno B, Berghuis AM (2021) Structural and functional insights into esterase-mediated macrolide resistance. Nat Commun 12(1):1–9. https://doi.org/10.1038/s41467-021-22016-3

Zou J, Shankar N (2016) The opportunistic pathogen *Enterococcus faecalis* resists phagosome acidification and autophagy to promote intracellular survival in macrophages. Cell Microbiol 18(6):831–843. https://doi.org/10.1111/cmi.12556

4 Standard Microbiological Techniques (Staining, Morphological and Cultural Characteristics, Biochemical Properties, and Serotyping) in the Detection of ESKAPE Pathogens

Paramanantham Parasuraman, Siddhardha Busi, and Jung-Kul Lee

Abstract

ESKAPE pathogens are multidrug-resistant bacteria that pose a significant threat to public health. The early and accurate detection of these pathogens is critical for effective treatment and infection control. Standard microbiological techniques, including staining, morphological and cultural characterization, biochemical analysis, and serotyping, are widely used to detect and identify ESKAPE pathogens. Staining techniques such as Gram staining provide valuable information on the cell wall structure of bacteria and are used to differentiate between gram-positive and gram-negative bacteria. Morphological and cultural characteristics such as colony morphology and growth requirements can provide clues regarding the identity of bacteria. Biochemical properties such as the ability to ferment specific sugars and produce certain enzymes can further help identify ESKAPE pathogens. Serotyping involves the identification of specific surface antigens on bacterial cells and is a highly sensitive and specific method for identifying bacterial strains. This technique is particularly useful for tracking the spread of bacterial infections and can help develop targeted vaccines. Therefore, standard microbiological techniques are critical for detecting and identifying ESKAPE pathogens. These techniques provide valuable information regarding the structure, growth requirements, and metabolic properties of these important bacterial species. Applying these techniques, in combination

P. Parasuraman · J.-K. Lee (✉)
Department of Chemical Engineering, Konkuk University, Seoul, Republic of Korea
e-mail: jkrhee@konkuk.ac.kr

S. Busi
Department of Microbiology, School of Life Sciences, Pondicherry University, Puducherry, India

S. Busi, R. Prasad (eds.), *ESKAPE Pathogens*,
https://doi.org/10.1007/978-981-99-8799-3_4

with advanced molecular methods, is essential for the accurate and timely diagnosis of ESKAPE pathogen infections.

Keywords

Multidrug-resistant bacteria · Staining · Morphological and cultural characteristics · Biochemical properties · Serotyping

4.1 Introduction

ESKAPE pathogens are highly virulent, drug-resistant bacteria that pose a significant threat to public health (Santajit and Indrawattana 2016). The acronym "ESKAPE" is from the initial letters of six prominent pathogens: *Enterococcus faecium, Staphylococcus aureus, Klebsiella pneumoniae, Acinetobacter baumannii, Pseudomonas aeruginosa*, and *Enterobacter* species. These pathogens can "escape" the effects of commonly used antimicrobial agents, leading to limited treatment options and increased morbidity and mortality rates (Santaniello et al. 2020). Comprehensive global surveillance systems for antimicrobial resistance (AMR) are currently lacking. However, various reports indicate that the United States alone experiences over two million cases of AMR infections annually, resulting in approximately 29,000 deaths. The associated healthcare costs directly linked to these infections amount to more than $4.7 billion (De Oliveira et al. 2020). In Europe, the impact of AMR is substantial, leading to over 33,000 deaths and causing 874,000 cases of overall disease burden annually, as a result of both hospital-acquired and community-acquired infections. The direct and indirect costs associated with these infections amount to approximately $1.5 billion (Cassini et al. 2019). In developing nations, where comprehensive economic evaluations are limited, communicable diseases continue to exert a substantial toll on public health. These diseases persist as primary contributors to the mortality rates and overshadow other health challenges faced by these nations (Hendrix et al. 2022). Furthermore, the emergence and re-emergence of bacterial infections further worsen the existing burden, presenting an intensified risk to healthcare systems and underserved populations in these regions (Spernovasilis et al. 2022).

In healthcare settings, ESKAPE pathogens remain a significant burden owing to their ability to cause various infectious diseases, including bloodstream infections, pneumonia, surgical site infections, and urinary tract complications (Loyola-Cruz et al. 2023). They are often associated with high rates of hospital-acquired infectious diseases and pose a significant challenge to healthcare professionals in maintaining patient safety. The significance of the ESKAPE pathogens extends beyond their intrinsic resistance to antimicrobial agents. These microorganisms also possess several virulence factors that contribute to their ability to cause infections and evade the immune system. Factors such as biofilm formation, toxin production, and genetic adaptability render ESKAPE pathogens a severe threat in healthcare settings (Santajit et al. 2022). The lack of quick diagnostic tools to precisely identify bacterial pathogens and detect AMR genes in clinical environments is a key reason for the overuse and inappropriate utilization of broad-spectrum antibiotics (Avershina et al. 2023). In the absence of timely and precise diagnostic information, healthcare providers often prescribe broad-spectrum antibiotics as a precautionary measure, leading to the misuse of these powerful medications. This practice not only poses risks to individual patients, such as adverse reactions and an increased likelihood of developing drug-resistant infections, but also contributes to the broader global issue of AMR (Uddin et al. 2021). Moreover, the rapid transmission of ESKAPE pathogens within healthcare facilities poses a significant risk to vulnerable patient populations, including those with compromised immune systems or who are undergoing invasive medical procedures. Their ability to colonize medical devices such as catheters or ventilators further complicates infection control measures and increases the likelihood of healthcare-associated outbreaks (Laxminarayan et al. 2013).

Given the limited treatment options and the potentially devastating consequences associated with ESKAPE infections, early and precise identification of these pathogens is crucial for effective infection control measures. Early detection of ESKAPE pathogens can promptly initiate appropriate treatment strategies. Timely identification allows health care professionals to administer the most suitable antibiotics and potentially prevent infections from worsening or spreading (Pogue et al. 2015). Early detection enables the implementation of appropriate infection control measures. This includes isolating infected patients, enforcing strict hygiene practices, and taking other necessary precautionary measures to prevent the transmission of ESKAPE pathogens to other patients. Strict implementation of these measures would effectively contain the spread of these pathogens and reduce the risk of outbreaks (Majumder et al. 2020). Lastly, early and accurate identification of ESKAPE pathogens is essential for antimicrobial stewardship. Identification of the specific pathogens causing the infection helps provide appropriate antibiotic treatments to target specific bacteria rather than relying on broad-spectrum antibiotics. This targeted approach helps minimize the unnecessary and excessive use of antibiotics, which, in turn, helps combat the growth and spread of AMR (Yang et al. 2021).

Therefore, the development and implementation of sensitive diagnostic methods are important to ensure timely and targeted treatments that can effectively control pathogen transmission. However, most traditional diagnostic methods are time-consuming and require several hours or days to yield results (Dong et al. 2022). It is necessary to develop appropriate diagnostic techniques that can rapidly identify ESKAPE pathogens within a short timeframe. Other concerns regarding traditional diagnostic methods include the accuracy, specificity, and identification of AMR. Some existing methods, such as culture-based techniques, have limitations that can lead to false-positive or false-negative results. This raises concerns about the sensitivity and specificity of the diagnostic procedure outcomes. Similarly, traditional methods to determine antibiotic susceptibility can be time-consuming and may not detect the full range of resistance mechanisms (Gajic et al. 2022). Advancements in diagnostic technologies are needed to enable the rapid, sensitive, specific, and comprehensive detection of AMR in ESKAPE pathogens. Addressing these challenges in the diagnostic methods for ESKAPE pathogens would greatly enhance our ability to rapidly and accurately identify these pathogens, leading to improved patient outcomes, better infection control measures, and more targeted antibiotic therapies.

In this chapter, we explore the use of standard microbiological techniques, including staining, morphological and cultural characteristics, biochemical analysis, serotyping, and recent advanced molecular methods, for the detection of ESKAPE pathogens. Understanding these techniques and their applications will enable the timely identification of targeted treatment strategies.

4.2 Microbial Staining Techniques

Microbial staining techniques are essential tools in microbiology that enable visualization and identification of microorganisms (Siguenza et al. 2019). Gram staining, one of the most widely used techniques, classifies bacteria into gram-positive or -negative based on the nature of their cell wall (Delfiner et al. 2016). Acid-fast staining is employed to detect bacteria such as *Mycobacterium tuberculosis*, whereas endospore staining allows the visualization of endospores produced by certain bacterial species (Wu et al. 2016). Additionally, fluorescent staining methods, such as fluorescent antibody staining and fluorescent in situ hybridization, provide specific and rapid identification of microbes in test samples (Gu et al. 2022). These microbial staining techniques aid in the characterization, classification, and diagnosis of infectious diseases, contributing to our understanding of microbial diversity and pathogenesis.

4.2.1 Gram Staining

Gram staining is the most commonly used staining technique. It involves a series of steps in which bacteria are first stained with crystal violet, followed by iodine treatment, alcohol decolorization, and counterstaining with safranin (Budin et al. 2012). The technique derives its name from the Danish bacteriologist, Hans Christian Gram, who reported this method in 1882, primarily for the identification of microorganisms that cause pneumonia (Wu and Yang 2020). This method aids in the classification of bacteria into two principal groups, gram-positive and gram-negative, by distinguishing the disparities in their cell wall properties (Liu et al. 2021). The cell walls of gram-positive bacteria feature a substantial peptidoglycan layer, making up a significant portion of their overall structure. This layer provides structural support and retains the crystal violet stain during Gram staining. In contrast, gram-negative bacteria exhibit a thinner peptidoglycan layer between their inner and outer membranes. Gram-positive bacteria appear purple, whereas gram-negative bacteria release the stain and appear pink after counterstaining (Silhavy et al. 2010). This differentiation is crucial, as it provides crucial information regarding the structural characteristics of the bacteria and helps guide appropriate treatment strategies.

Gram staining can be performed on various clinical samples. It is essential to collect these samples in sterile containers to maintain their integrity and prevent contamination (Woo et al. 2008).

4.2.1.1 Golden Standard Procedure for Gram Staining

A step-by-step description of the gold standard procedure for Gram staining is shown in Fig. 4.1 (Claus 1992).

Step 1: Prepare a thin, heat-fixed smear of the bacterial sample on a clean microscope slide. Heat fixation involves briefly passing the slide through a flame several times to kill the bacteria and then fixing them to the slide.
Step 2: Flood the smear with a crystal violet stain, covering the entire area of the glass slide. Permit the stain to absorb by sample for approximately 1 min.
Step 3: Gently wash the glass slide with running tap water to remove the unbounded crystal violet stain.
Step 4: Apply iodine solution (Gram's iodine) to the smear and cover it completely. Leave the iodine solution on the slide for 1 min. This forms a complex with the crystal violet, aiding in the retention of the stain.
Step 5: Rinse the slide again with water to remove excess iodine solution.
Step 6: Decolorize the slide using a decolorizing agent, typically 95% ethanol or a mixture of ethanol and acetone. Decolorization is crucial and requires careful timing control. Slowly pour the decolorizing agent onto the slide until the runoff becomes clear. Immediately wash the glass slide with water to stop the excessive removal of the stain.

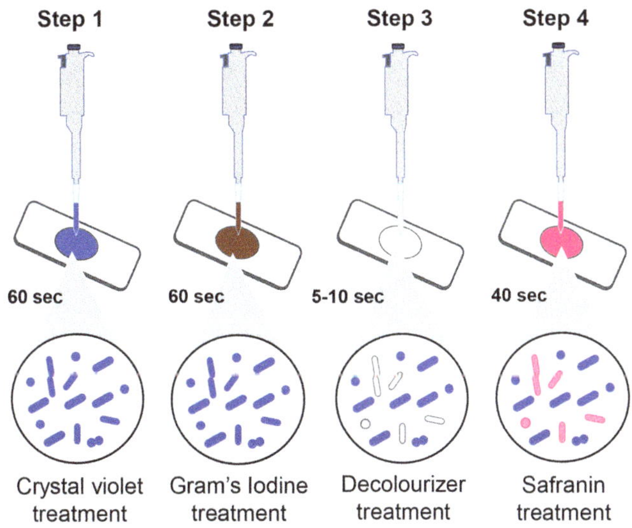

Fig. 4.1 Diagrammatic representation of Gram staining procedures

Table 4.1 Morphological and gram nature characteristics of ESKAPE pathogens

Pathogen	Gram nature	Morphology
E. faecium	Gram-positive	Coccus-shaped
S. aureus	Gram-positive	Coccus-shaped
K. pneumoniae	Gram-negative	Rod-shaped
A. baumannii	Gram-negative	Short-rod-shaped
P. aeruginosa	Gram-negative	Rod-shaped
Enterobacter spp.	Gram-negative	Rod-shaped

Step 7: Counterstain the slide with safranin solution and cover the entire smear. Allow the safranin to act for approximately 1 min.
Step 8: Gently wash the glass slide once again with water to remove excess safranin and gently blot the slide with bibulous paper or allow it to air dry.
Step 9: Examine the slide under a light microscope by oil immersion. Gram-positive bacteria exhibit a purple color, retaining the crystal violet stain, while gram-negative bacteria exhibit a pink or red color, taking up the counterstain safranin.

The morphological and Gram characteristics of the ESKAPE pathogens are presented in Table 4.1. This table highlights the following characteristics (Table 4.1).

E. faecium is a gram-positive coccus-shaped bacterium that typically exists in pairs or chains (Pankratova et al. 2018). *S. aureus* is a gram-positive coccus-shaped bacterium that commonly occurs in clusters (Jubeh et al. 2020). *K. pneumoniae* is a gram-negative, rod-shaped bacterium, also referred to as bacillus (Chang et al. 2021). *A. baumannii* is a gram-negative, short, rod-shaped bacterium often known as coccobacillus (Micelli et al. 2023). *P. aeruginosa* is a gram-negative, rod-shaped bacterium (Diggle and Whiteley 2020). *Enterobacter* is a family of rod-shaped bacteria exhibiting gram-negative characteristics (Davin-Regli et al. 2019).

4.2.2 Limitations and Troubleshooting Staining Techniques of ESKAPE Pathogens

Gram staining is a vital laboratory procedure for bacterial identification and for distinguishing between gram-positive and -negative bacteria. This method is simple, cost-effective, and rapid for identifying bacterial strains. However, this technique has some limitations. First, Gram staining provides preliminary information about the bacteria present in the sample and requires further confirmation through additional procedures, such as culture techniques, to identify the bacterial pathogen. False-negative results can occur because of procedural errors or low bacterial counts in the sample (Claus 1992). Moreover, certain bacteria, such as *Mycobacterium tuberculosis*, may not be easily stained by Gram staining, leading to potential inaccuracies (Smith 2003; Yang et al. 2023). Interpretation can be affected by various conditions, including the age of the bacterial isolate (sample), the presence of background materials, samples from patients on antibiotics, or cultures transferred from antibiotic-containing media (Pauter et al. 2020). These factors can interfere with the staining process and affect the accuracy of results.

Here, some troubleshooting guidelines are provided to avoid false results in Gram staining:

(a) Inadequate heat fixation can result in the poor adherence of bacteria to the slide, leading to improper staining. Ensure that the smear is sufficiently heat-fixed by briefly moving the slide through the flame but not excessively.
(b) If gram-positive bacteria appeared to be gram-negative, the decolorization time was adjusted. Decrease the decolorization time or dilute the decolorizing agent to prevent excessive decolorization.
(c) Ensure the quality and freshness of staining reagents, including crystal violet, iodine solution, decolorizing agent, and counterstain. Expired or degraded reagents can affect the staining results.
(d) The recommended staining times were used for each step of the Gram staining protocol. Over-staining or under-staining may lead to inaccurate results.
(e) Ensure that the bacterial culture used for Gram staining is in the logarithmic growth phase (active culture). Old cultures, samples, cultures, or samples subjected to antibiotic treatment can affect the staining characteristics of bacteria.

Therefore, while Gram staining is valuable, it is important to consider its limitations and further confirm the results with other diagnostic techniques to ensure the accurate identification and characterization of bacterial infections.

4.3 Morphological and Cultural Characteristics

The morphological and cultural characteristics of the ESKAPE pathogens differed among individual species within each group. These characteristics provide important insights into the initial identification and differentiation of ESKAPE pathogens. It is crucial to acknowledge that these traits are not absolute and can vary within each pathogen group owing to strain variation and other factors. Morphologically, ESKAPE pathogens can be classified based on their shape. As mentioned in Sect. 4.2.2, Gram-positive pathogens such as *E. faecium* and *S. aureus* appear as cocci, which are round bacteria. Gram-negative pathogens, such as *K. pneumoniae*, *A. baumannii*, *P. aeruginosa*, and *Enterobacter* spp., are rod-shaped bacteria commonly referred to as bacilli. *A. baumannii* can also exhibit coccobacillus morphology, which is characterized by short rods. The culture characteristics included colony morphology, growth rate, and pigmentation. Colony morphology refers to the appearance of bacterial colonies on agar plates. The size, shape, surface texture, and color of colonies can provide valuable information for identification. The growth rate of ESKAPE pathogens can vary, with some exhibiting rapid growth and others at a slower pace. Pigmentation is another characteristic that varies among ESKAPE pathogens. While some strains may not be pigmented, others may exhibit pigmentation of various colors. Some of the cultural characteristics of the ESKAPE pathogens are listed in Table 4.2.

Table 4.2 Morphological and cultural characteristics of ESKAPE pathogens on nutrient agar or blood agar

	Colony character/growth rate on nutrient agar or blood agar						
ESKAPE pathogens	Size	Color	Surface	Margin	Hemolysis	Pigmentation	Growth rate
E. faecium	Medium (~ 1 mm)	Creamy or white	Smooth	Entire	Hemolytic	Non-pigmented	Slow-growing
S. aureus	Medium (~ 1 mm)	Golden or yellowish	Shiny	Entire	Hemolytic	Golden-yellow	Moderate to fast-growing
K. pneumoniae	Large (>1 mm)	White to creamy or pale yellow	Smooth	Undulate	Non-hemolytic	Non-pigmented	Fast-growing
A. baumannii	Medium (~ 1 mm)	White to pale yellow or light brown	Smooth or slightly rough surface	Entire	Non-hemolytic	Non-pigmented	Moderate to fast-growing
P. aeruginosa	Large (>1 mm)	Greenish-blue	Smooth	Entire	Hemolytic	Pyocyanin pigment	Fast-growing
Enterobacter spp.	Medium (~ 1 mm)	White to creamy or pale yellow	Smooth	Entire	Non-hemolytic	Non-pigmented	Moderate to fast-growing

4.3.1 Key Morphological and Cultural Characteristics of ESKAPE Pathogens

Enterococcus faecium belongs to the ESKAPE pathogens. It has certain morphological and cultural characteristics. *E. faecium* appears on agar plates as small- to medium-sized colonies that are smooth and circular. These colonies typically have a creamy or white appearance (Egido et al. 2016; Mao et al. 2020; Risqiyah et al. 2022). In terms of growth rate, *E. faecium* is generally considered to be slow-growing and typically non-pigmented; however, some strains of *E. faecium* may exhibit pale yellow or pink pigmentation (Čermák et al. 2009; Devriese et al. 1987). *Staphylococcus aureus*, another ESKAPE pathogen, exhibits distinctive morphological and cultural characteristics. *S. aureus* appears as medium-sized, round, opaque colonies, often exhibiting a golden or yellowish color. This characteristic color is attributed to the production of a golden-yellow pigment known as staphyloxanthin (Arjyal et al. 2020; Yang et al. 2020). *S. aureus* is classified as a moderate-to-fast-growing bacterium. These morphological and cultural characteristics, including clustered arrangement, round and opaque colonies with a golden or yellowish hue, and the production of staphyloxanthin pigment, serve as important diagnostic features for the preliminary identification of *S. aureus* (von Eiff et al. 2000). *Klebsiella pneumoniae* has distinct morphological and cultural characteristics. The colony morphology on the agar plates was large colonies with mucoid consistency and a smooth or glistening appearance (Mitrea and Vodnar 2019). The color of these colonies can vary, ranging from white to creamy or pale yellow (Munoz et al. 2006). *K. pneumoniae* is a fast-growing bacterium that is commonly nonpigmented (Bhardwaj et al. 2017); however, certain strains of *K. pneumoniae* may exhibit pale yellow or pink pigmentation (Sajjan et al. 2010). These morphological and cultural features, including bacillus-shaped rods, large mucoid colonies with smooth or glistening textures, and variable colony colors, provide valuable initial information for identification.

Acinetobacter baumannii are short, plump, gram-negative rods, often described as coccobacilli. Colonies of *A. baumannii* on agar plates form small- to medium-sized colonies with smooth or slightly rough surfaces. The color of these colonies can vary, ranging from white to pale yellow or light brown (Moubareck and Halat 2020; Peleg et al. 2008). Usually, *A. baumannii* is typically classified as a moderate-to-fast-growing non-pigmented bacterium (Ramirez et al. 2020). However, some strains of *A. baumannii* exhibit pale yellow or brown pigmentation (Doughari et al. 2011; Park et al. 2019). *Pseudomonas aeruginosa* is characterized by gram-negative rods that are often described as bacilli (Chevalier et al. 2017). Upon culturing on agar plates, it develops large, flat colonies that tend to spread. These colonies typically exhibit a distinctive greenish-blue or blue-green coloration (Abdelaziz et al. 2023; Wu and Li 2015). In some cases, certain strains of *P. aeruginosa* display a metallic sheen on the colony surface (Roulová et al. 2022). The growth rate of *P. aeruginosa* is generally high, allowing for rapid proliferation (LaBauve and Wargo 2012). One of the remarkable attributes of this bacterium is its ability to produce a pigment called pyocyanin, which contributes to the blue-green color observed in colonies (DeBritto et al. 2020). This pigmentation served as a marker for identifying *P. aeruginosa*. The combination of bacilli-shaped rods, large, flat, spreading colonies with a greenish-blue or blue-green color, and the generation of pyocyanin distinguishes *P. aeruginosa* from other bacteria. *Enterobacter* spp., a group of bacteria within the ESKAPE pathogens, share common morphological and cultural characteristics. Morphologically, *Enterobacter* spp. is characterized by gram-negative rods, commonly described as bacilli (Davin-Regli and Pagès 2015). Colonies grown on agar plates were medium-sized with a smooth or slightly rough surface (Kus 2014). The colors of these colonies can vary, ranging from white to creamy or pale yellow. *Enterobacter* spp. are moderate-to-fast-growing bacteria that are generally non-pigmented (Rogers et al. 2016). However, certain strains of *Enterobacter* spp. exhibit pale yellow or pink pigmentation (Sass and Fisher 2009). Further confirmatory tests and analyses are required for accurate species-level identification and differentiation from other bacteria. These characteristics are in accordance with the general guidelines and may vary among different strains and species within each ESKAPE pathogens. Therefore, additional techniques and tests are necessary for an accurate and definitive identification.

4.3.2 Limitations and Troubleshooting in Cultural Characteristics of ESKAPE Pathogens

Cultural characteristics play crucial roles in the identification and classification of ESKAPE pathogens. However, certain limitations and potential troubleshooting considerations are required for the accurate identification of pathogens.

(a) The cultural characteristics of ESKAPE pathogens can vary based on growth conditions such as temperature, pH, and nutrient availability. Suboptimal growth conditions may affect the colony morphology, pigmentation, and growth rate, leading to inconsistent results.
(b) Strain variations within the same ESKAPE pathogen species may exhibit slightly different cultural characteristics. These variations make it challenging to precisely identify pathogens based on their colony appearance.
(c) Ensuring the quality of the culture media to provide optimal growth conditions for ESKAPE pathogens. Contaminated or improperly prepared media can lead to aberrant colony morphology and growth.
(d) Proper aseptic technique is crucial during inoculation to prevent contamination and ensure the growth of pure cultures. Contamination can interfere with the accurate assessment of cultural characteristics.
(e) Maintain appropriate incubation conditions, including temperature, humidity, and duration, to facilitate optimal growth of ESKAPE pathogens. Deviations from the recommended conditions may affect colony morphology, pigmentation, and growth rate.

4.4 Biochemical Properties

Microorganisms exhibit a wide range of biochemical properties that serve as important indicators of their metabolic capabilities and physiological characteristics (Martínez-Espinosa 2020). These properties are fundamental for identifying and classifying microorganisms. One of the key aspects of biochemical properties is carbohydrate metabolism, which includes the ability to ferment specific sugars and produce acids and gases as metabolic by-products (Wang et al. 2021). Additionally, the utilization of various sugars as carbon sources provides further insights into the metabolic potential of microorganisms (Thomas 2015). Protein metabolism is another crucial aspect encompassing the generation of indole from tryptophan, hydrolysis of gelatin and casein, and urease activity, which breaks down urea into ammonia (Mamarasulov et al. 2022; Mitzscherling et al. 2022). Enzymatic activities, such as those of catalase, oxidase, and coagulase, play essential roles in microbial physiology and can be assessed to determine specific enzyme production (Ogodo et al. 2022).

Furthermore, the metabolism of compounds, such as citrate and nitrate, along with the generation of enzymes, such as lipases, amylases, and proteases, contributes to the overall biochemical profile of microorganisms (Rathakrishnan and Gopalan 2022; Shaik et al. 2017). In addition to specific metabolic activities, the generation of gases, such as hydrogen or carbon dioxide, during metabolic reactions is also considered. These biochemical properties are evaluated using a series of tests including agar-based, broth, and enzymatic assays (Khushboo et al. 2023). Combining the results of these tests with other characteristics such as colony morphology and growth patterns enables microbiologists to accurately identify and differentiate between microorganisms. It is important to recognize that the biochemical properties can vary among different species and strains of microorganisms (Janda and Abbott 2002). Therefore, a comprehensive approach incorporating multiple tests is necessary for precise identification and classification. Advancements in automated systems and molecular techniques have significantly enhanced the efficiency and accuracy of microbial identification in recent years, thereby expanding our understanding of the biochemical properties of microorganisms.

4.4.1 Major Biochemical Tests for the Detection of ESKAPE Pathogens

Numerous biochemical tests are available for bacterial detection, and specific tests may vary depending on the bacterial species. However, there is a list of commonly used biochemical tests to identify ESKAPE pathogens.

4.4.1.1 Catalase Test

This is a biochemical test used to detect the presence of catalase in bacteria. Catalase helps break down hydrogen peroxide into water and oxygen. This test is particularly useful for differentiating catalase-positive and catalase-negative bacteria (Bascomb

Table 4.3 Biochemical test results for ESKAPE pathogens

S. no.	Biochemical tests	*E. faecium*	*S. aureus*	*K. pneumoniae*	*A. baumannii*	*P. aeruginosa*	*Enterobacter* spp.
1	Catalase	−ve	+ve	+ve	+ve	+ve	+ve
2	Oxidase	−ve	−ve	−ve	−ve	+ve	−ve
3	Urease	−ve	+ve	+ve	−ve	−ve	−ve
4	Gelatin hydrolysis	−ve	+ve	−ve	−ve	+ve	−ve
5	Nitrate reduction	−ve	+ve	+ve	−ve	+ve	+ve
6	Methyl red	−ve	+ve	−ve	−ve	−ve	−ve
7	Voges-Proskauer	+ve	+ve	+ve	−ve	−ve	+ve
8	Citrate	−ve	+ve	−ve	+ve	+ve	+ve

+ve = 90% or greater positive within 48 h; −ve = 90% or greater negative within 48 h

and Manafi 1998). The principle of the catalase test is that some bacteria produce the enzyme catalase, which rapidly decomposes hydrogen peroxide into water and oxygen. The production of bubbles or effervescence upon the addition of hydrogen peroxide indicates a positive response, suggesting the presence of catalase in bacterial cells. In contrast, catalase-negative bacteria do not produce catalase; therefore, no bubbles or effervescence are observed (Shen and Zhang 2022).

This test is relatively simple to perform. The bacterial culture is taken and placed on clean glass slides. A few drops of hydrogen peroxide solution (usually 3% hydrogen peroxide) are mixed directly into the bacterial culture. If bacteria produce catalase, hydrogen peroxide undergoes rapid decomposition, leading to the release of oxygen in the form of bubbles or effervescence (Wanger et al. 2017). It is crucial to emphasize that the catalase test is a preliminary test and should be interpreted in comparison with other tests for the accurate detection of bacteria. Catalase activities of ESKAPE pathogens are listed in Table 4.3. The catalase test results for ESKAPE pathogens were as follows. *S. aureus* showed a positive result, as indicated by the presence of bubbles or effervescence when hydrogen peroxide was added to the bacterial colony (Pumipuntu et al. 2017). However, methicillin-resistant *S. aureus* isolated from clinical samples tested negative (Bertrand et al. 2002). *E. faecium* exhibited a negative result with no bubbles or effervescence (Day et al. 2001), *K. pneumoniae* (Kang et al. 2020), *A. baumannii* (Ahmad and Mohammad 2020), *P. aeruginosa* (Banerjee et al. 2017), and *Enterobacter* spp. (Chi et al. 2018). All patients displayed positive results, characterized by the presence of bubbles or effervescence. These results offer valuable insights into the catalase activity of each pathogen and can be utilized as a preliminary step in their identification.

4.4.1.2 Oxidase Test

This biochemical test was used to determine the presence of cytochrome c oxidase, an enzyme involved in the electron transport chain in certain microorganisms. This test is based on the principle that the enzyme oxidase can catalyze the transfer of electrons to a specific substrate, resulting in a color change (Kuss et al. 2017). Tetramethyl-*p*-phenylenediamine dihydrochloride (TMPD) is used for the oxidase test. This reagent is colorless in its reduced form but turns purple when oxidized. The oxidase enzyme in the microorganism being tested causes the oxidation of TMPD and a subsequent color change (Mitra et al. 2020). To perform the test, a small sample of the microorganism is taken, and two to three drops of the oxidase reagent are added. The reaction develops a purple color within a specified time frame, usually approximately 10–30 s. A positive result is indicated by the presence of a purple color, indicating the presence of the oxidase enzyme. A negative result is indicated by the absence of color change (Dawodu and Akanbi 2021).

The oxidase test characteristics of ESKAPE pathogens are listed in Table 4.3. In the oxidase test, *P. aeruginosa* is the only ESKAPE pathogen that yields a positive result, indicating the presence of an oxidase (Navidinia et al. 2017). Other pathogens, including *E. faecium*, *S. aureus*, *K. pneumoniae*, *A. baumannii*, and *Enterobacter* spp., did not produce oxidase (Bhatia et al. 2021; Imran et al. 2021). The oxidase test results serve as additional characteristics for the detection and classification of these pathogens. Overall, the oxidase test serves as a valuable tool for microbial identification and classification, aiding in the differentiation of ESKAPE pathogens and assisting in the determination of bacterial characteristics and taxonomic classification.

4.4.1.3 Urease Test

This biochemical test was used to evaluate the ability of the microorganisms to produce urease (Mekonnen et al. 2021). It is based on the hydrolysis of urea by urease, which results in the production of ammonia and carbon dioxide (Estiu and Merz 2004). The ammonia produced increases the pH of the medium, leading to a color change or the development of turbidity. For the urease test, a specific urea-containing medium is inoculated with the microorganisms under investigation. Over time, if an organism possesses urease activity, urea is broken down into ammonia and carbon dioxide (Sigurdarson et al. 2020). Ammonia production increases the pH of the medium, inducing the pH indicator (phenol red) to change color. The test results are indicated by a pink or deep red color change, indicating alkaline conditions due to ammonia production. In contrast, the negative results showed no color change, indicating the absence of urease activity (Graham and Miftahussurur 2018).

S. aureus and *K. pneumoniae* were found to be urease-positive, indicating the generation of urease enzymes and their capability to hydrolyze urea into ammonia and CO_2. This was evident from the formation of a pink to deep pink coloration in the medium, indicating a positive result (Celik et al. 2023; Duran Ramirez et al. 2022). *P. aeruginosa, E. faecium,* and *Enterobacter* spp. were negative for the urease test, indicating the absence of urease activity (Dahlén et al. 2018; Suman and Tanuja 2021; Vancanneyt et al. 2002). No color change was observed in the media containing these pathogens. The urease test provides valuable information regarding the metabolic capabilities of microorganisms and assists in their identification and differentiation.

4.4.1.4 Gelatin Hydrolysis Test

A gelatin hydrolysis test was performed to determine the ability of microorganisms to produce gelatinase. Gelatinases hydrolyze gelatin, a collagen-derived protein

(Nursyam et al. 2018). The principle of the gelatin hydrolysis test is based on the enzymatic degradation of gelatin by microorganisms. This test is typically performed by inoculating microorganisms onto a solid agar medium containing gelatin. Gelatin is added to the medium as a solidifying agent. If the microorganism produces gelatinase, it hydrolyzes gelatin, resulting in the liquefaction of the gelatin-containing medium. This is observed as a clear zone or liquefied area around the bacterial growth (Medina and Baresi 2007).

E. faecium is typically negative for gelatin hydrolysis, indicating an absence of gelatinase activity (da Silva Fernandes et al. 2015). *S. aureus*, on the contrary, is positive for gelatin hydrolysis (Karmakar et al. 2016). *K. pneumoniae*, *A. baumannii*, and *Enterobacter* spp. are negative for gelatin hydrolysis and lack the ability to produce gelatinases (Abbas et al. 2014; Sapkota et al. 2020; Seleem et al. 2020). Finally, *P. aeruginosa* is positive for gelatin hydrolysis (Kathiravan and Krishnani 2014). These results help distinguish ESKAPE pathogens based on their gelatinase activity. However, the gelatin hydrolysis test should be interpreted alongside other tests and clinical observations to accurately identify microorganisms and assess their pathogenic potential. Additionally, variations in the composition and preparation of the gelatin medium can influence test results and should be considered during interpretation.

4.4.1.5 Nitrate Reduction Test

This is a biochemical test applied to determine the capability of microorganisms to reduce nitrate (NO_3^-) to nitrite (NO_2^-) or to other nitrogenous compounds. The principle behind this test was based on nitrate reductase, which catalyzes the reduction of nitrate (Tiso and Schechter 2015). In this test, a culture of microorganisms is inoculated into a nitrate-containing medium. After incubation, sulfanilic acid and alpha-naphthylamine are added to the cultures. If nitrite is present, it reacts with the reagent to form a red diazo compound. This indicates a positive result for nitrate reduction (Tassadaq et al. 2013). However, if no color change occurs after the addition of the reagent, this suggests that nitrate was not reduced to nitrite. In such cases, further testing is required to differentiate between complete nitrate reduction to nitrogen gas or the presence of other nitrogenous compounds. Nitrate-reduction tests are valuable for differentiating bacteria based on their ability to reduce nitrate, which is an important step in the nitrogen cycle (Baskaran et al. 2020). This helps identify organisms that can utilize nitrate and play the role of a final electron acceptor in anaerobic respiration.

This test was conducted to assess the ability of ESKAPE pathogens to reduce nitrate to nitrite or to further metabolize it. Among the tested pathogens, *E. faecium* showed negative results, indicating the absence of nitrate reduction (Ren Loi et al. 2023). *S. aureus, P. aeruginosa,* and *K. pneumoniae* demonstrated positive results, indicating their ability to convert nitrate into nitrite (Das et al. 2019; Jia et al. 2017; Li et al. 2023). *A. baumannii* showed negative results, suggesting the absence of nitrate reduction (Nkem et al. 2016). *Enterobacter* spp. also displayed positive results, suggesting that they can convert nitrate to nitrite (Panigrahi et al. 2020).

4.4.1.6 Methyl Red Test

The methyl red (MR) test has been used to detect the production of mixed acid fermentation by microorganisms. Certain bacteria convert glucose into different organic acids, such as formic, lactic, acetic, and succinic acids through fermentation. The MR test helps to differentiate between bacteria that predominantly produce stable acidic end products and those that produce other fermentation products (Bhutia et al. 2021). The principle behind the MR test involves inoculating the microorganism in a medium in which glucose is the exclusive carbon source, such as MR-VP broth. During the incubation period, bacteria that produce large amounts of stable acidic end products from glucose metabolism will reduce the pH of the medium. After incubation, the pH indicator, methyl red, is added to the medium. Methyl red turns red at a low pH (acidic conditions) and remains yellow at a higher pH (alkaline conditions). If the bacteria produce enough stable acidic end products, the pH of the medium drops and exhibits a red color change when methyl red is added. This showed a positive result on the MR test. A negative result was indicated by the absence of a significant drop in pH and the resulting yellow color of the medium (Cowan 1953).

This test was performed to assess the ability of ESKAPE pathogens to develop stable acidic end-products from glucose fermentation. The results revealed the varying metabolic capabilities of the pathogens. *E. faecium* showed a negative result, indicating the absence of significant acid production, resulting in a yellow color in the medium. In contrast, *S. aureus* exhibited a positive MR test result, indicating sufficient acid generation, which led to a significant pH drop and a red color change (Banik et al. 2018). *K. pneumoniae, A. baumannii, P. aeruginosa*, and *Enterobacter* spp. showed negative MR test results, indicating a lack of significant acid production, resulting in yellow color in the medium (Ahmad and Mohammad 2020; Chi et al. 2018; Mukherjee et al. 2017; Permatasari et al. 2020; Rahim et al. 2017). The MR test provides valuable information regarding the metabolic capabilities of these pathogens and aids in their differentiation based on their ability to ferment glucose and produce acids.

4.4.1.7 Voges–Proskauer Test

This test is a biochemical assay used to detect the production of acetoin, a metabolic end-product, by microorganisms. This test involved several key steps. First, the microorganism of interest is cultured in a suitable medium containing glucose. The culture is then incubated under specific conditions to facilitate metabolic reactions. After incubation, two reagents are added to the culture: Reagent A, which contains alpha-naphthol, and Reagent B, which contains potassium hydroxide (KOH). The alpha-naphthol in Reagent A reacts with the acetyl methyl carbinol (acetoin) produced by microorganisms. This reaction leads to the formation of pink or red complexes. The formation of a pink or red color in the culture indicated a positive result for acetoin production, whereas the absence of a color change indicated a negative result (Barry and Feeney 1967; Bryn et al. 1973). It serves as a valuable tool for bacterial identification and classification and provides insights into the metabolic pathways of microorganisms in clinical and laboratory settings.

The Voges–Proskauer test has been performed to determine acetoin production by ESKAPE pathogens. Among the pathogens tested, *E. faecium, S. aureus*, and *K. pneumoniae* showed positive results, indicating their ability to produce acetoin (Mao et al. 2020; Sharma et al. 2015; Taylor et al. 1979). Acetoin production was observed by the formation of a pink to red color after mixing with VP reagents (Son and Taylor

2011). This metabolic characteristic can be useful for differentiating pathogens. Conversely, *A. baumannii* and *P. aeruginosa* yielded negative VP test results, indicating the absence of significant acetoin production (Ahmad and Mohammad 2020; Owlia et al. 2014). However, *Enterobacter* spp. showed positive results for acetoin production (De Graef et al. 2003). These VP test results provide insights into the metabolic profiles and potential identification markers of ESKAPE pathogens, aiding their accurate and reliable identification in clinical and laboratory settings.

4.4.1.8 Citrate Test

The citrate test is a biochemical test used to determine the ability of microorganisms to utilize citrate as the sole carbon source (Brocker et al. 2009). Bacteria produce citrate permease, which allows them to transport and utilize citrate from the medium. The test is performed in a medium containing sodium citrate as the sole carbon source and bromothymol blue as a pH indicator (Mendes Ferreira and Mendes-Faia 2020). The principle behind this test is that if an organism can utilize citrate, it converts it into pyruvate and releases alkaline byproducts, causing the pH of the medium to increase. This increase in pH led to a color change in the pH indicator from green to blue. The citrate test is particularly useful for differentiating the ESKAPE pathogens. While some organisms can utilize citrate, others cannot, and this difference in metabolic capabilities aids in their identification and differentiation. By observing the color change in the medium, we can determine whether the organism is citrate-positive (able to utilize citrate) or citrate-negative (unable to utilize citrate). The citrate test provides valuable information on the metabolic characteristics of microorganisms and can aid in their identification and classification in clinical and laboratory settings.

E. faecium was detected as negative, indicating that it cannot use citrate as a carbon source. *S. aureus* tested positive, indicating its ability to utilize citrate (Mahmudul Islam et al. 2015). *K. pneumoniae* was also negative, while *A. baumannii* and *P. aeruginosa* showed positive results (Banerjee et al. 2017; Lal et al. 2019; Sivakumar et al. 2016). Similarly, *Enterobacter* spp. exhibited a positive result (Iimura and Hosono 1996). The citrate test serves as a valuable tool for differentiating these organisms based on their ability to utilize citrate, thus aiding in their characterization and potential clinical implications.

4.4.1.9 Carbohydrate Utilization Test

Carbohydrate fermentation tests were performed to determine the ability of the microorganisms to ferment specific carbohydrates. These tests play a crucial role in differentiating bacterial groups or species based on their distinct fermentation patterns (Bronpenbre and Schlesinger 1918). Valuable information can be obtained for the classification and identification of bacteria by observing the metabolic reactions of microorganisms exposed to different carbohydrates (Slifkin and Pouchet 1977). The ability to ferment specific carbohydrates varies among bacterial species, leading to variations in acid or gas production, pH, and other observable indicators (Sánchez et al. 2000). These fermentation patterns serve as important diagnostic markers and aid in understanding the metabolic diversity and characteristics of ESKAPE.

During fermentation, microorganisms use organic substrates as the final electron acceptors (Spang et al. 2019). As a result, the breakdown of carbohydrates during fermentation contributes to the production of acids and the generation of gas (Putatunda et al. 2023). These end products of carbohydrate fermentation play a significant role in differentiating between bacterial species or groups based on their specific metabolic profiles (Yu et al. 2019). Carbohydrate fermentation can result in the production of diverse end products, which are influenced by the specific microorganisms involved in the fermentation process, the type of substrate being fermented, the enzymes present, and environmental conditions such as pH and temperature (Lu et al. 2022). The potential end-products include lactic acid, formic acid, acetic acid, butyric acid, butyl alcohol, acetone, ethyl alcohol, carbon dioxide, and hydrogen (Jones and Clifton 1953). Detection of fermentation reactions relies on observing the color change of a pH indicator when acidic byproducts are generated (Prabhu et al. 2020). To perform this test, a specific carbohydrate is added to a basal medium containing a pH indicator. Bacteria can utilize peptones in the medium to produce alkaline by-products. Therefore, a significant pH change occurs only when excess acid is produced by carbohydrate fermentation (Goresline 1933). Phenol red is a widely used pH indicator in carbohydrate fermentation tests because of its ability to detect the generation of organic acids, which are the predominant end products of carbohydrate utilization (Yoon and Mekalanos 2006).

Fermentation tubes, also known as Durham tubes, are used to detect gas production during carbohydrate fermentation (Oberg et al. 2021). These tubes consist of small, slender test tubes (6 mm × 50 mm) that are inverted and placed inside larger test tubes (13 mm × 100 mm). The tube is added to the fermentation medium before sterilization. During gas production, gas accumulates inside the Durham tube, resulting in the formation of visible air bubbles. This serves as an indicator of gas production during fermentation (Sitepu et al. 2019). The fermentation of a specific carbohydrate can exhibit three characteristic reactions that are used to classify bacteria as follows:

1. Culture tubes with acid production only: Some bacteria can ferment carbohydrates and produce organic acids as end products. These bacteria demonstrated a color shift in the pH indicator owing to acid production.
2. Culture tubes with acid and gas production: Certain bacteria not only produce organic acids during carbohydrate fermentation but also generate gas as a by-product. Gas is detected by the formation of a visible air bubble in the Durham tube.
3. Nonfermenters: Some bacteria do not possess the enzymes necessary to ferment carbohydrates, resulting in no significant color change or gas production in the fermentation medium.

By observing these characteristic reactions, bacteria can be classified into one of three groups according to their capacity for carbohydrate fermentation and the associated acid and gas production. Carbohydrate utilization tests are performed to assess the ability of bacteria to metabolize various carbohydrates as carbon sources. ESKAPE pathogens were subjected to a series of carbohydrate utilization tests, and the expected results are given in Table 4.4, based on Bergey's Manual (Table 4.4) (Grimont and Grimont 2015a, b; Juni 2015; Palleroni 2015; Schleifer and Bell 2015; Švec and Devriese 2015).

Table 4.4 Carbohydrate utilization profiles of ESKAPE pathogens

S. no.	Carbohydrates	*E. faecium*	*S. aureus*	*K. pneumoniae*	*A. baumannii*	*P. aeruginosa*	*Enterobacter* spp.
1	Adonitol	NA	–	d	NA	–	–
2	Arabinose	+/–	–	+	+	–	+
3	Arabitol	–	–	–	NA	NA	–
4	Cellobiose	+	–	NA	NA	–	+
5	Dextrin	NA	–	NA	NA	NA	NA
6	Dulcitol	–	–	d	NA	NA	+/–
7	Erythritol	–	–	–	NA	–	–
8	Erythrose	NA	–	NA	NA	NA	NA
9	Fructose	+	+	NA	NA	+	NA
10	Fucose	–	–	NA	NA	NA	–
11	Galactose	+	+/–	NA	+	–	NA
12	Gentiobiose	+	–	NA	NA	NA	NA
13	Glucose	+	+	+	NA	+	+
14	Glycerol	d	+	NA	NA	NA	+/–
15	Inositol	–	–	NA	NA	–	+/–
16	Lactose	+	+/–	+	+/–	–	+
17	Lyxose	–	–	NA	NA	NA	+/–
18	Maltose	+	+	+	–	–	+
19	Mannitol	d	+/–	+	NA	+	+
20	Mannose	+	+	+	+	–	+
21	Melezitose	NA	–	–	NA	NA	NA
22	Melibiose	d	–	NA	NA	NA	+
23	Raffinose	d	–	NA	NA	NA	+

24	Rhamnose	d	–	+	+	–	+
25	Ribose	NA	+	NA	+	+	NA
26	Salicin	NA	–	NA	NA	NA	+
27	Sorbitol	d	–	+	NA	–	+/–
28	Sorbose	NA	–	d	NA	NA	+
29	Sucrose	d	+	+	–	–	+
30	Trehalose	+	+	NA	NA	–	+
31	Turanose	–	+/–	+/–	NA	NA	d
32	Xylitol	–	–	d	NA	NA	NA
33	Xylose	–	–	+	+	–	+

+ve = 90% or more of strains are positive; –ve = 90% or more of strains are negative; d = 11–89% of strains are positive; +/– = strain instability (not the same as "d"); *NA* no information available

4.4.2 Limitations of Biochemical Tests

Biochemical tests are widely used in microbiology to detect and characterize microorganisms. However, it is important to be aware of these test limitations (Salmonová and Bunešová 2017). Some limitations of biochemical tests include:

1. Interpreting the results of biochemical tests can sometimes be subjective because the interpretation depends on visual observations or color changes. This subjectivity can introduce potential errors or inconsistencies in the interpretation of results.
2. Different strains or species of microorganisms exhibit variations in their biochemical properties, leading to inconsistent or inconclusive results. This makes it challenging to identify or differentiate closely related organisms accurately.
3. Some biochemical tests may yield false-positive or false-negative results. False positives can be observed when a test indicates the presence of a specific characteristic or enzymatic activity that is not actually present. However, false negatives can occur when a test fails to identify the presence of a special characteristic or enzyme activity.
4. Performing and interpreting biochemical tests can be time-consuming, particularly when multiple tests are required for comprehensive identification. This can delay the reporting of results and affect timely decision-making in clinical or research settings.
5. Biochemical tests require viable, well-grown cultures to obtain accurate results. Growth conditions, such as medium composition, temperature, and incubation time, can influence the performance of biochemical tests. However, suboptimal culture conditions can lead to inconclusive or inaccurate results.

4.4.3 Recent Advancements in Biochemical Analyses

Despite these limitations, biochemical tests remain valuable tools in microbiology for the preliminary identification and characterization of microorganisms. An automated biochemical test analyzer is a sophisticated instrument used in clinical laboratories to perform tests on various biological samples. These analyzers are designed to streamline and automate the process of conducting multiple biochemical tests simultaneously, offering several advantages over traditional manual methods (Carroll and Patel 2015). One of the key benefits of automated biochemical test analyzers is their ability to improve laboratory efficiency and productivity. These analyzers can handle a large volume of samples, allowing for rapid and simultaneous testing, which significantly reduces turnaround time and increases throughput. This is particularly advantageous in situations where quick and accurate test results are crucial, such as emergency settings (Zhang et al. 2022). Automated biochemical test analyzers offer enhanced accuracy and precision. They minimize the risk of human error by eliminating manual pipetting and handling, thereby reducing the potential for sample contamination and mishandling. Analyzers are equipped with advanced robotics and precision instruments that ensure precise and consistent test execution, leading to reliable and reproducible results (Buchan et al. 2012).

Another advantage of automated analyzers is their ability to perform a broad range of biochemical tests. These analyzers are programmed with an extensive menu of tests, including enzymatic assays, metabolic panels, electrolyte measurements, and other specialized tests. The flexibility to perform multiple tests on a single instrument optimizes laboratory space and reduces costs (Franco-Duarte et al. 2019). Automated biochemical test analyzers often feature user-friendly interfaces and software systems. This simplifies its operation, allowing laboratory personnel to easily set up and monitor tests, input patient information, and retrieve results. Analyzers also generate comprehensive reports that can be integrated with laboratory information systems to facilitate efficient data management and traceability. It is important to note that although automated biochemical test analyzers offer numerous advantages, they are not without limitations. These analyzers require regular maintenance, calibration, and quality control to ensure accurate and reliable performance. Additionally, some specialized tests may still require manual or specialized methods that are not available on automated platforms.

4.5 Serotyping

Serotyping is a valuable method for detecting and classifying bacteria, including ESKAPE pathogens. Serotyping involves the characterization of microorganisms based on specific antigens present on cell surfaces. These antigens include surface proteins, lipopolysaccharides, capsules, and other surface structures (Henriksen 1978). Several serotyping methods can be employed for ESKAPE pathogens, depending on the specific pathogen and antigens of interest. Some commonly used serotyping methods for ESKAPE pathogens include antibodies that bind to surface antigens of the pathogen. These antibodies cause the pathogen to clump or agglutinate, which can be visually observed (Koch et al. 2004). Many bacteria, including *K. pneumoniae,* possess capsules that play a role in virulence and immune evasion. Capsular typing involves the identification and characterization of specific capsule types using serological or molecular methods (Huang et al. 2022). Phages, or viruses that infect bacteria, can be used for serotyping certain bacterial pathogens, which is also known as phage typing. Phage typing involves testing the ability of different phages to infect and lyse the bacterial strains. The susceptibility pattern to different phages can be used to classify and identify bacterial strains, such as *S. aureus* (Kali 2013). With advancements in molecular techniques, serotyping can be performed using genetic methods. These methods involve the detection and analysis of specific genes or gene clusters associated with surface antigen expression. Molecular serotyping is often used to detect *P. aeruginosa* (Wang et al. 2022). It is crucial to emphasize that the choice of serotyping method depends on the specific pathogen and antigens relevant to its identification and classification. Different serotyping methods can provide valuable information for understanding the epidemiology, virulence, and diversity of ESKAPE strains.

4.5.1 Serological Agglutination Test

The serological agglutination test, also known as agglutination assay, is a serotyping method commonly employed for the detection and characterization of bacteria, including ESKAPE pathogens (Slotved et al. 2004). This test relies on the specific reaction between antibodies and surface antigens present in bacterial cells (Fig. 4.2). For ESKAPE, the serological agglutination test involves the use of specific antibodies that can bind to the surface antigens of the bacterium (Rai and Mitchell 2020). These antibodies are raised against the respective antigens of each pathogen in the ESKAPE pathogen group and are typically produced in animals or using monoclonal antibody production techniques (Zurawski and McLendon 2020). In general, a serological agglutination test is conducted as follows: A bacterial suspension containing the target strain is mixed with specific antibodies. Agglutination occurs when antibodies recognize and bind to the surface antigens of the pathogen. Agglutination is the visible clumping of bacterial cells due to the cross-linking of antibodies that bind to antigens on the bacteria. Agglutination reactions are typically observed either macroscopically or microscopically. Positive agglutination indicates the existence of particular antigens on the surface of the pathogenic strain, whereas lack of agglutination suggests the absence of these target antigens (Kruse et al. 2021). Serological agglutination tests of ESKAPE pathogens can be used to determine the serotype or serogroup of each bacterium. Different strains may possess distinct surface antigens that allow their classification into specific serotypes. This information is valuable for epidemiological studies, monitoring the spread of specific strains, and understanding the diversity and virulence of pathogens. Crucially, serological agglutination tests for specific pathogens require specific antibodies raised against the relevant surface antigens. The availability and specificity of these antibodies can vary, and the choice of target antigens in the test may depend on the research or diagnostic objectives. Serological agglutination tests play a significant role in the serotyping and identification of ESKAPE pathogenic strains by detecting the presence of specific surface antigens and providing valuable insights into the epidemiology and diversity of these pathogens.

To evaluate the presence of *Enterococcus* species in drinking water in South Africa, serological agglutination tests were performed using two specific antigens, Lancefield groups D and G, to detect contamination. The experimental procedure involved subjecting the isolates to a test in which a cell wall-specific carbohydrate

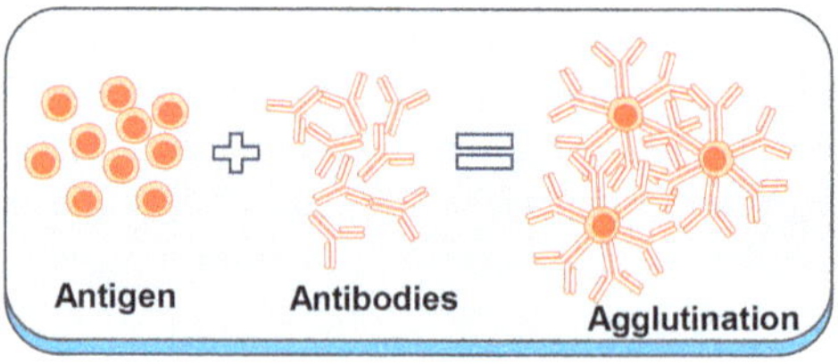

Fig. 4.2 Schematic representation of serological agglutination test

extract was reacted with nitrous acid reagents, specifically the D and G antisera. Positive reactions were observed when the extracts were agglutinated with the corresponding reagents, indicating the presence of Lancefield groups. Antiserum A was used as a negative control to control for potential false-positive results. This serological agglutination test provides valuable insights into the contamination levels and distribution of *Enterococcus* species, enabling a better understanding of the water quality and potential health risks related to *Enterococcus* contamination in the studied region (Collins and Merapelo 2011). In order to investigate that the existence of vancomycin-resistant enterococci in water used by the people of North West Province, South Africa, was investigated. The isolates obtained from the water samples were subjected to serological detection of *Enterococcus* species derived from the Lancefield grouping. The Lancefield grouping system categorizes Streptococci into groups A, B, C, D, F, and G. In this study, *Enterococcus* species were identified using a latex agglutination test, in which specific latex particles covered with antibodies against Lancefield group antigens were used. Based on the agglutination reactions observed, *Enterococcus* isolates were grouped according to their Lancefield group(Matlou et al. 2019).

For quick and authentic routine detection of *S. aureus*, a latex agglutination test that detects the clumping factor and protein A is highly recommended. This test allows for the efficient detection of *S. aureus* by relying on the presence of these specific antigens. In a recent investigation, the focus was on identifying two different capsular polysaccharides, type 5 and type 8, which have clinical relevance owing to their predominance among clinical infection isolates from various geographic regions. These capsular polysaccharides are present in *S. aureus* (clinical isolates) and play a significant role in its pathogenicity and virulence. This study aimed to determine the distribution of types 5 and 8 capsular polysaccharides in *S. aureus* isolates obtained from patients with clinical infections. By utilizing the latex slide agglutination test, researchers have been able to rapidly and accurately identify the presence of clinically relevant capsular polysaccharides in clinical samples (Verdier et al. 2007).

Previously, researchers have aimed to purify the capsular polysaccharide antigen of *K. pneumoniae* and generate hyperimmune sera against it. This step is crucial for developing specific antibodies that target capsular polysaccharides and aid in accurate diagnosis. They developed a rapid and simple antigen detection immunoassay using capsular polysaccharide antisera. This immunoassay is a quick and efficient method for diagnosing *K. pneumoniae* infections. Additionally, researchers have attempted to compare the effectiveness of monoclonal and capsular polysaccharide polyclonal antibodies in the detection of *K. pneumoniae* (Sikarwar and Batra 2011). A previous study used 20 different serovars in agglutination tests to identify *A. baumannii* (Traub 1989). In the case of *P. aeruginosa*, 16 O-antigens were used in the agglutination test to identify *P. aeruginosa* from the test samples (Brokopp et al. 1977). Similarly, for the identification of *Enterobacter* spp., 28 types of O-antigen were used (Gaston et al. 1983).

4.5.2 Molecular Serotyping

Molecular serotyping is a modern approach that is used to detect and classify bacteria based on specific genetic markers associated with surface antigens. It involves the detection and analysis of genes or gene regions that encode serotype-specific antigens, such as capsule polysaccharides or surface proteins (Deng et al. 2019). Molecular serotyping utilizes methods such as polymerase chain reaction (PCR), DNA sequencing, or microarray-based assays to detect and characterize these genetic markers. Molecular serotyping offers several advantages over traditional serotyping methods. It provides a high level of resolution, allowing for the differentiation of closely related serotypes or subtypes. It is also more rapid and efficient than other methods as it eliminates the need for time-consuming serological testing (Fratamico et al. 2016; Ghita et al. 2020). Molecular serotyping can be used for a broad range of microorganisms, including bacteria and viruses, and it enables the identification of emerging serotypes or variants. Molecular serotyping plays a crucial role in epidemiological surveillance, outbreak investigations, and monitoring of the spread of specific serotypes or clones within populations (Deng et al. 2014). This aids in understanding the genetic diversity, evolution, and pathogenicity of microorganisms. In addition, molecular serotyping data can contribute to vaccine development by identifying key serotype-specific targets for immunization strategies (Jagtap et al. 2023).

Molecular serotyping is an advanced technique that is used to detect and characterize ESKAPE pathogens. This involves the use of molecular techniques to detect and analyze specific surface antigens or genes associated with serotypes. Molecular serotyping of *E. faecium* often focuses on determining serotypes based on the antigenic determinants present on the cell surface. This can be achieved through the detection and analysis of specific genes or gene clusters that encode surface antigens such as capsule polysaccharides or surface proteins (Palmer et al. 2012). Molecular serotyping of *S. aureus* typically involves the detection and characterization of specific genes or gene regions that encode surface antigens, such as those for capsule polysaccharides or protein A. These molecular methods provide information on the serotypes or clonal complexes of *S. aureus* isolates (Dendani Chadi et al. 2022). In the case of *K. pneumoniae*, the detection and analysis of specific genes or genetic markers associated with capsule polysaccharides (*magA* and *rmpA*) are the major determinants of serotypes in this pathogen. Molecular methods, such as PCR-based assays or sequencing, can be employed to identify the existence of specific capsule genes and determine their serotypes (Siu et al. 2011).

Molecular serotyping of *A. baumannii* focuses on the detection and characterization of specific genes or genetic markers associated with surface antigens such as the outer membrane protein OmpA or capsule polysaccharides. Molecular methods can provide information on the serotypes of *A. baumannii* isolates (Hu et al. 2013). *P. aeruginosa* is used for the detection and analysis of specific genes or gene regions related to the expression of surface antigens, such as the B-band O-antigen or flagellar antigens (Nasrin et al. 2022). Molecular serotyping of *Enterobacter* spp. involves the use of specific genetic markers, such as the O-antigen, which is known for its high variability as a component of the cell surface and has been utilized to identify and classify different serotypes (Li et al. 2020). Molecular serotyping is a potential tool for the detection and characterization of different ESKAPE pathogen serotypes. This enhances our understanding of the epidemiology and pathogenicity of these bacteria and provides important information for clinical management and public health interventions.

4.6 Future Perspectives

In clinical microbiology, the detection of ESKAPE pathogens using standard microbiological techniques is crucial for understanding their characteristics, behavior, and impact on human health. However, as technology advances and our understanding of microbial diversity and pathogenicity increases, several future perspectives can further improve the detection and characterization of ESKAPE pathogens. Whole genome sequencing (WGS) has revolutionized microbial detection and classification. In the future, WGS could provide comprehensive genomic data for ESKAPE pathogens, enabling rapid and accurate identification, detection of AMR genes, and tracking of outbreaks. This could significantly improve our understanding of the genomic diversity and evolution of these pathogens. The development and implementation of high-throughput screening methods could help streamline the detection and identification of ESKAPE pathogens. Automated platforms, such as mass spectrometry-based proteomics and next-generation sequencing, enable rapid and simultaneous analysis of multiple pathogens, saving time and resources. The incorporation of bioinformatics tools and databases can enhance the analysis and interpretation of microbiological data. This includes the development of comprehensive databases of ESKAPE pathogens, allowing for an efficient and accurate comparison of genomic and phenotypic data. Gaining insights into the impact of the microbiome on health and disease is an emerging field. Future research should focus on studying the interactions between ESKAPE pathogens and the human microbiome to elucidate the factors that contribute to their colonization and pathogenicity. Metagenomic approaches can provide insights into complex microbial communities and their impact on ESKAPE pathogen colonization and infection. The development of rapid and user-friendly diagnostic tests for ESKAPE pathogens are crucial for timely and effective clinical management. Advances in technologies, such as microfluidics, biosensors, and miniaturized molecular assays, can enable point-of-care testing, facilitating early detection and targeted treatment. These advancements will enhance our ability to detect, characterize, and combat challenging pathogens, ultimately improving patient outcomes and public health.

4.7 Conclusion

Using standard microbiological techniques to detect and identify ESKAPE pathogens has been fundamental for understanding their characteristics and their impact on public health. These techniques, including staining, morphological and cultural characteristics, biochemical analysis, and serotyping, have provided valuable insights into the diversity and behavior of these pathogens. However, with technological advances, future perspectives can further enhance our understanding and detection of ESKAPE pathogens. However, the future of ESKAPE pathogens detection lies in integrating advanced technologies with genomics. Genomic analysis, including WGS, offers rapid and accurate identification and detection of AMR. High-throughput screening methods and bioinformatic tools can streamline data analysis. Studying the microbiome and metagenomics provides insights into pathogen-host interactions. It is essential to expand rapid point-of-care diagnostics using microfluidic and molecular assays. By combining advanced technologies with a deeper understanding of microbial ecology, we can improve the detection, characterization, and management of pathogenic ESKAPE infections, leading to better patient outcomes and public health.

References

Abbas SZ, Rafatullah M, Ismail N, Lalung J (2014) Isolation, identification, characterization, and evaluation of cadmium removal capacity of *Enterobacter* species. J Basic Microbiol 54(12):1279–1287. https://doi.org/10.1002/jobm.201400157

Abdelaziz AA, Kamer AMA, Al-Monofy KB, Al-Madboly LA (2023) *Pseudomonas aeruginosa's* greenish-blue pigment pyocyanin: its production and biological activities. Microb Cell Fact 22:110. https://doi.org/10.1186/s12934-023-02122-1

Ahmad NH, Mohammad GA (2020) Identification of *Acinetobacter baumannii* and determination of MDR and XDR strains. Baghdad Sci J 17(3):0726. https://doi.org/10.21123/bsj.2020.17.3.0726

Arjyal C, Kc J, Neupane S (2020) Prevalence of methicillin-resistant *Staphylococcus aureus* in shrines. Int J Microbiol 2020:7981648. https://doi.org/10.1155/2020/7981648

Avershina E, Khezri A, Ahmad R (2023) Clinical diagnostics of bacterial infections and their resistance to antibiotics—current state and whole genome sequencing implementation perspectives. Antibiotics 12(4):781. https://doi.org/10.3390/antibiotics12040781

Banerjee S, Batabyal K, Joardar SN, Isore DP, Dey S, Samanta I, Samanta TK, Murmu S (2017) Detection and characterization of pathogenic *Pseudomonas aeruginosa* from bovine subclinical mastitis in West Bengal, India. Vet World 10(7):738–742. https://doi.org/10.14202/vetworld.2017.738-742

Banik A, Mohammad N, Akter T, Fatema K, Abony M (2018) Prevalence, identification and antibiotic susceptibility of *Enterococcus* species isolated from chicken and pigeon meat in Gazipur Area of Bangladesh. Open J Med Microbiol 08(03):74–83. https://doi.org/10.4236/ojmm.2018.83007

Barry AL, Feeney KL (1967) Two quick methods for Voges-Proskauer test. Appl Microbiol 15(5):1138–1141. https://doi.org/10.1128/am.15.5.1138-1141.1967

Bascomb S, Manafi M (1998) Use of enzyme tests in characterization and identification of aerobic and facultatively anaerobic gram-positive cocci. Clin Microbiol Rev 11(2):318–340

Baskaran V, Patil PK, Antony ML, Avunje S, Nagaraju VT, Ghate SD, Nathamuni S, Dineshkumar N, Alavandi SV, Vijayan KK (2020) Microbial community profiling of ammonia and nitrite oxidizing bacterial enrichments from brackishwater ecosystems for mitigating nitrogen species. Sci Rep 10(1):5201. https://doi.org/10.1038/s41598-020-62183-9

Bertrand X, Huguenin Y, Talon D (2002) First report of a catalase-negative methicillin-resistant *Staphalococcus aureus*. Diagn Microbiol Infect Dis 43(3):245–246. https://doi.org/10.1016/S0732-8893(02)00398-X

Bhardwaj G, Shah R, Joshi B, Patel P (2017) *Klebsiella pneumoniae* VRE36 as a PGPR isolated from *Saccharum officinarum* cultivar Co99004. J Appl Biol Biotechnol 5:047–052. https://doi.org/10.7324/jabb.2017.50108

Bhatia P, Sharma A, George AJ, Anvitha D, Kumar P, Dwivedi VP, Chandra NS (2021) Antibacterial activity of medicinal plants against ESKAPE: an update. Heliyon 7(2):e06310. https://doi.org/10.1016/j.heliyon.2021.e06310

Bhutia MO, Thapa N, Tamang JP (2021) Molecular characterization of bacteria, detection of enterotoxin genes, and screening of antibiotic susceptibility patterns in traditionally processed meat products of Sikkim. India Front Microbiol 11:599606. https://doi.org/10.3389/fmicb.2020.599606

Brocker M, Schaffer S, Mack C, Bott M (2009) Citrate utilization by *Corynebacterium glutamicum* is controlled by the CitAB two-component system through positive regulation of the citrate transport genes citH and tctCBA. J Bacteriol 191(12):3869–3880. https://doi.org/10.1128/JB.00113-09

Brokopp CD, Gomez-Lus R, Farmer JJ (1977) Serological typing of *Pseudomonas aeruginosa*: use of commerical antisera and live antigens. J Clin Microbiol 5(6):640–649. https://doi.org/10.1128/jcm.5.6.640-649.1977

Bronpenbre J, Schlesinger MJ (1918) Carbohydrate fermentation by bacteria as influenced by the composition of the medium. Exp Biol Med 16(3):44–46. https://doi.org/10.3181/00379727-16-27

Bryn K, Ulstrup JC, Størmer FC (1973) Effect of acetate upon the formation of acetoin in *Klebsiella* and *Enterobacter* and its possible practical application in a rapid Voges-Proskauer test. Appl Microbiol 25(3):511–512. https://doi.org/10.1128/am.25.3.511-512.1973

Buchan BW, Riebe KM, Ledeboer NA (2012) Comparison of the MALDI Biotyper system using Sepsityper specimen processing to routine microbiological methods for identification of bacteria from positive blood culture bottles. J Clin Microbiol 50(2):346–352. https://doi.org/10.1128/JCM.05021-11

Budin G, Chung HJ, Lee H, Weissleder R (2012) A magnetic gram stain for bacterial detection. Angew Chem Int Ed 51(31):7752–7755. https://doi.org/10.1002/anie.201202982

Carroll KC, Patel R (2015) Systems for identification of bacteria and fungi. In: Manual of clinical microbiology. ASM Press, Washington, DC, pp 29–43. https://doi.org/10.1128/9781555817381.ch4

Cassini A, Högberg LD, Plachouras D, Quattrocchi A, Hoxha A, Simonsen GS, Colomb-Cotinat M, Kretzschmar ME, Devleesschauwer B, Cecchini M, Ouakrim DA, Oliveira TC, Struelens MJ, Suetens C, Monnet DL, Strauss R, Mertens K, Struyf T, Catry B, Hopkins S et al (2019) Attributable deaths and disability-adjusted life-years caused by infections with antibiotic-resistant bacteria in the EU and the European economic area in 2015: a population-level modelling analysis. Lancet Infect Dis 19(1):56–66. https://doi.org/10.1016/S1473-3099(18)30605-4

Celik C, Demir NY, Duman M, Ildiz N, Ocsoy I (2023) Red cabbage extract-mediated colourimetric sensor for swift, sensitive and economic detection of urease-positive bacteria by naked eye and smartphone platform. Sci Rep 13(1):2056. https://doi.org/10.1038/s41598-023-28604-1

Čermák P, Landfeld A, Měřička P, Houška M (2009) *Enterococcus faecium* growth model. Czech J Food Sci 27(5):361–371

Chang D, Sharma L, Dela Cruz CS, Zhang D (2021) Clinical epidemiology, risk factors, and control strategies of *Klebsiella pneumoniae* infection. Front Microbiol 12:750662. https://doi.org/10.3389/fmicb.2021.750662

Chevalier S, Bouffartigues E, Bodilis J, Maillot O, Lesouhaitier O, Feuilloley MGJ, Orange N, Dufour A, Cornelis P (2017) Structure, function and regulation of *Pseudomonas aeruginosa* porins. FEMS Microbiol Rev 41(5):698–722. https://doi.org/10.1093/femsre/fux020

Chi Q, Tang W, Liu L, Meng J, Dong X, Chen W, Li X (2018) Isolation and properties of *Enterobacter* sp. LX3 capable of producing indoleacetic acid. Appl Sci 8(11):2108. https://doi.org/10.3390/app8112108

Claus D (1992) A standardized gram staining procedure. World J Microbiol Biotechnol 8(4):451–452. https://doi.org/10.1007/BF01198764

Collins NA, Merapelo DM (2011) Detection of enterococcus species in groundwater from some rural communities in the Mmabatho area, South Africa: a risk analysis. Afr J Microbiol Res 5(23):3930–3935. https://doi.org/10.5897/AJMR11.645

Cowan ST (1953) Micromethod for the methyl red test. J Gen Microbiol 9(1):101–109. https://doi.org/10.1099/00221287-9-1-101

da Silva Fernandes M, Fujimoto G, de Souza LP, Kabuki DY, da Silva MJ, Kuaye AY (2015) Dissemination of enterococcus faecalis and enterococcus faecium in a ricotta processing plant and evaluation of pathogenic and antibiotic resistance profiles. J Food Sci 80(4):M765–M775. https://doi.org/10.1111/1750-3841.12824

Dahlén G, Hassan H, Blomqvist S, Carlén A (2018) Rapid urease test (RUT) for evaluation of urease activity in oral bacteria in vitro and in supragingival dental plaque ex vivo. BMC Oral Health 18(1):89. https://doi.org/10.1186/s12903-018-0541-3

Das A, Behera BK, Acharya S, Paria P, Chakraborty HJ, Parida PK, Das BK (2019) Genetic diversity and multiple antibiotic resistance index study of bacterial pathogen, *Klebsiella pneumoniae* strains isolated from diseased Indian major carps. Folia Microbiol (Praha) 64(6):875–887. https://doi.org/10.1007/s12223-019-00701-7

Davin-Regli A, Pagès JM (2015) *Enterobacter aerogenes* and *Enterobacter cloacae*; versatile bacterial pathogens confronting antibiotic treatment. Front Microbiol 6(MAY):392. https://doi.org/10.3389/fmicb.2015.00392

Davin-Regli A, Lavigne JP, Pagès JM (2019) *Enterobacter* spp.: update on taxonomy, clinical aspects, and emerging antimicrobial resistance. Clin Microbiol Rev 32(4):e00002. https://doi.org/10.1128/CMR.00002-19

Dawodu OG, Akanbi RB (2021) Isolation and identification of microorganisms associated with automated teller machines on Federal Polytechnic Ede campus. PLoS One 16(8):e0254658. https://doi.org/10.1371/journal.pone.0254658

Day AM, Sandoe JAT, Cove JH, Phillips-Jones MK (2001) Evaluation of a biochemical test scheme for identifying clinical isolates of *Enterococcus faecalis* and *Enterococcus faecium*. Lett Appl Microbiol 33(5):392–396. https://doi.org/10.1046/j.1472-765X.2001.01017.x

De Graef EM, Devriese LA, Vancanneyt M, Baele M, Collins MD, Lefebvre K, Swings J, Haesebrouck F (2003) Description of *Enterococcus canis* sp. nov. from dogs and reclassification of *Enterococcus porcinus* Teixeira et al. 2001 as a junior synonym of enterococcus villorum Vancanneyt et al. 2001. Int J Syst Evol Microbiol 53(4):1069–1074. https://doi.org/10.1099/ijs.0.02549-0

De Oliveira DMP, Forde BM, Kidd TJ, Harris PNA, Schembri MA, Beatson SA, Paterson DL, Walker MJ (2020) Antimicrobial resistance in ESKAPE pathogens. Clin Microbiol Rev 33:e00181. https://doi.org/10.1128/CMR

DeBritto S, Gajbar TD, Satapute P, Sundaram L, Lakshmikantha RY, Jogaiah S, Ito SI (2020) Isolation and characterization of nutrient dependent pyocyanin from *Pseudomonas aeruginosa* and its dye and agrochemical properties. Sci Rep 10(1):1542. https://doi.org/10.1038/s41598-020-58335-6

Delfiner MS, Martinez LR, Pavia CS (2016) A gram stain hands-on workshop enhances first year medical students' technique competency in comprehension and memorization. PLoS One 11(10):e0163658. https://doi.org/10.1371/journal.pone.0163658

Dendani Chadi Z, Dib L, Zeroual F, Benakhla A (2022) Usefulness of molecular typing methods for epidemiological and evolutionary studies of *Staphylococcus aureus* isolated from bovine intramammary infections. Saudi J Biol Sci 29(8):103338. https://doi.org/10.1016/j.sjbs.2022.103338

Deng X, Desai PT, den Bakker HC, Mikoleit M, Tolar B, Trees E, Hendriksen RS, Frye JG, Porwollik S, Weimer BC, Wiedmann M, Weinstock GM, Fields PI, McClelland M (2014) Genomic epidemiology of *Salmonella enterica* serotype enteritidis based on population structure of prevalent lineages. Emerg Infect Dis 20(9):1481–1489. https://doi.org/10.3201/eid2009.131095

Deng L, Li Y, Geng Y, Zheng L, Rehman T, Zhao R, Wang K, OuYang P, Chen D, Huang X, He C, Yang Z, Lai W (2019) Molecular serotyping and antimicrobial susceptibility of *Streptococcus agalactiae* isolated from fish in China. Aquaculture 510:84–89. https://doi.org/10.1016/j.aquaculture.2019.05.046

Devriese LA, Van De Kerckhove A, Kilpper-Balz R, Schleifer KH (1987) Characterization and identification of *Enterococcus* species isolated from the intestines of animals. Int J Syst Bacteriol 37(3):257

Diggle SP, Whiteley M (2020) Microbe profile: *Pseudomonas aeruginosa*: opportunistic pathogen and lab rat. Microbiology 166(1):30–33. https://doi.org/10.1099/mic.0.000860

Dong B, He Z, Li Y, Xu X, Wang C, Zeng J (2022) Improved conventional and new approaches in the diagnosis of tuberculosis. Front Microbiol 13:924410. https://doi.org/10.3389/fmicb.2022.924410

Doughari HJ, Ndakidemi PA, Human IS, Benade S (2011) The ecology, biology and pathogenesis of *Acinetobacter* spp.: an overview. Microbes Environ 26(2):101–112. https://doi.org/10.1264/jsme2.ME10179

Duran Ramirez JM, Gomez J, Obernuefemann CLP, Gualberto NC, Walker JN (2022) Semi-quantitative assay to measure urease activity by urinary catheter-associated uropathogens. Front Cell Infect Microbiol 12:859093. https://doi.org/10.3389/fcimb.2022.859093

Egido SH, Ruiz MS, Inés Revuelta S, García IG, Bellido JLM (2016) *Enterococcus faecium* small colony variant endocarditis in an immunocompetent patient. New Microb New Infect 9:47–49. https://doi.org/10.1016/j.nmni.2015.11.006

Estiu G, Merz KM (2004) The hydrolysis of urea and the proficiency of urease. J Am Chem Soc 126(22):6932–6944. https://doi.org/10.1021/ja049327g

Franco-Duarte R, Černáková L, Kadam S, Kaushik S, Salehi B, Bevilacqua A, Corbo MR, Antolak H, Dybka-Stępień K, Leszczewicz M, Relison Tintino S, Alexandrino de Souza VC, Sharifi-Rad J, Melo Coutinho HD, Martins N, Rodrigues CF (2019) Advances in chemical and biological methods to identify microorganisms—from past to present. Microorganisms 7(5):130. https://doi.org/10.3390/microorganisms7050130

Fratamico PM, DebRoy C, Liu Y, Needleman DS, Baranzoni GM, Feng P (2016) Advances in molecular serotyping and subtyping of *Escherichia coli*. Front Microbiol 7:644. https://doi.org/10.3389/fmicb.2016.00644

Gajic I, Kabic J, Kekic D, Jovicevic M, Milenkovic M, Mitic Culafic D, Trudic A, Ranin L, Opavski N (2022) Antimicrobial susceptibility testing: a comprehensive review of currently used methods. Antibiotics 11(4):427. https://doi.org/10.3390/antibiotics11040427

Gaston MA, Bucher C, Pitt TL (1983) O serotyping scheme for *Enterobacter cloacae*. J Clin Microbiol 18(5):1079–1083. https://doi.org/10.1128/jcm.18.5.1079-1083.1983

Ghita BT, Bennani L, Berrada S, Benboubker M, Bennani B (2020) Molecular serotyping and antibiotic resistance patterns of *Escherichia coli* isolated in Hospital Catering Service in Morocco. Int J Microbiol 2020:1–8. https://doi.org/10.1155/2020/5961521

Goresline HE (1933) Studies of agar-digesting bacteria. J Bacteriol 26(5):435–457. https://doi.org/10.1128/jb.26.5.435-457.1933

Graham DY, Miftahussurur M (2018) *Helicobacter pylori* urease for diagnosis of *Helicobacter pylori* infection: a mini review. J Adv Res 13:51–57. https://doi.org/10.1016/j.jare.2018.01.006

Grimont PAD, Grimont F (2015a) Enterobacter. In: Bergey's manual of systematics of archaea and bacteria. Wiley, New York, pp 1–17. https://doi.org/10.1002/9781118960608.gbm01145

Grimont PAD, Grimont F (2015b) Klebsiella. In: Bergey's manual of systematics of archaea and bacteria. Wiley, New York, pp 1–26. https://doi.org/10.1002/9781118960608.gbm01150

Gu J, Wang H, Zhang M, Xiong Y, Yang L, Ren B, Huang R (2022) Application of fluorescence in situ hybridization (FISH) in oral microbial detection. Pathogens 11(12):1450. https://doi.org/10.3390/pathogens11121450

Hendrix N, Kwete X, Bolongaita S, Megiddo I, Memirie ST, Mirkuzie AH, Nonvignon J, Verguet S (2022) Economic evaluations of health system strengthening activities in low-income and middle-income country settings: a methodological systematic review. BMJ Glob Health 7(3):e007392. https://doi.org/10.1136/bmjgh-2021-007392

Henriksen SD (1978) Chapter I. Serotyping of bacteria, pp 1–13. https://doi.org/10.1016/S0580-9517(08)70355-6

Hu D, Liu B, Dijkshoorn L, Wang L, Reeves PR (2013) Diversity in the major polysaccharide antigen of *Acinetobacter baumannii* assessed by DNA sequencing, and development of a molecular serotyping scheme. PLoS One 8(7):e70329. https://doi.org/10.1371/journal.pone.0070329

Huang X, Li X, An H, Wang J, Ding M, Wang L, Li L, Ji Q, Qu F, Wang H, Xu Y, Lu X, He Y, Zhang J-R (2022) Capsule type defines the capability of *Klebsiella pneumoniae* in evading Kupffer cell capture in the liver. PLoS Pathog 18(8):e1010693. https://doi.org/10.1371/journal.ppat.1010693

Iimura K, Hosono A (1996) Biochemical characteristics of *Enterobacter agglomerans* and related strains found in buckwheat seeds. Int J Food Microbiol 30(3):243–253. https://doi.org/10.1016/0168-1605(96)00949-X

Imran F, Malik M, Asif A, Akhtar G, Zaman S, Ajmal AN (2021) Changing antimicrobial susceptibility and resistance pattern of *Acinetobacter* species over the last eight years in a tertiary Care Hospital in Lahore, Pakistan. J Islamabad Med Dent Coll 10(2):83–88. https://doi.org/10.35787/jimdc.v10i2.615

Jagtap S, Pattabiraman C, Sankaradoss A, Krishna S, Roy R (2023) Evolutionary dynamics of dengue virus in India. PLoS Pathog 19(4):e1010862. https://doi.org/10.1371/journal.ppat.1010862

Janda JM, Abbott SL (2002) Bacterial identification for publication: when is enough enough? J Clin Microbiol 40(6):1887–1891. https://doi.org/10.1128/JCM.40.6.1887-1891.2002

Jia R, Yang D, Xu D, Gu T (2017) Mitigation of a nitrate reducing *Pseudomonas aeruginosa* biofilm and anaerobic biocorrosion using ciprofloxacin enhanced by D-tyrosine. Sci Rep 7(1):6946. https://doi.org/10.1038/s41598-017-07312-7

Jones LW, Clifton CE (1953) Metabolism and nutrition of clostridium FESERI. J Bacteriol 65(5):560–564. https://doi.org/10.1128/jb.65.5.560-564.1953

Jubeh B, Breijyeh Z, Karaman R (2020) Resistance of gram-positive bacteria to current antibacterial agents and overcoming approaches. Molecules 25(12):2888. https://doi.org/10.3390/molecules25122888

Juni E (2015) Acinetobacter. In: Bergey's manual of systematics of archaea and bacteria. Wiley, New York, pp 1–26. https://doi.org/10.1002/9781118960608.gbm01203

Kali A (2013) Bacteriophage types of methicillin-resistant *Staphylococcus aureus* in a Tertiary Care Hospital. Aust Med J 6(10):496–503. https://doi.org/10.4066/AMJ.2013.1742

Kang S-M, Bilal S, Shahzad R, Kim Y-N, Park C-W, Lee K-E, Lee J-R, Lee I-J (2020) Effect of ammonia and Indole-3-acetic acid producing endophytic *Klebsiella pneumoniae* YNA12 as a bio-herbicide for weed inhibition: special reference with evening primroses. Plan Theory 9(6):761. https://doi.org/10.3390/plants9060761

Karmakar A, Dua P, Ghosh C (2016) Biochemical and molecular analysis of *Staphylococcus aureus* clinical isolates from hospitalized patients. Can J Infect Dis Med Microbiol 2016:9041636. https://doi.org/10.1155/2016/9041636

Kathiravan V, Krishnani KK (2014) *Pseudomonas aeruginosa* and *Achromobacter* sp.: nitrifying aerobic denitrifiers have a plasmid encoding for denitrifying functional genes. World J Microbiol Biotechnol 30(4):1187–1198. https://doi.org/10.1007/s11274-013-1543-6

Khushboo, Karnwal A, Malik T (2023) Characterization and selection of probiotic lactic acid bacteria from different dietary sources for development of functional foods. Front Microbiol 14:1170725. https://doi.org/10.3389/fmicb.2023.1170725

Koch S, Hufnagel M, Theilacker C, Huebner J (2004) Enterococcal infections: host response, therapeutic, and prophylactic possibilities. Vaccine 22(7):822–830. https://doi.org/10.1016/j.vaccine.2003.11.027

Kruse RL, Huang Y, Smetana H, Gehrie EA, Amukele TK, Tobian AAR, Mostafa HH, Wang ZZ (2021) A rapid, point-of-care red blood cell agglutination assay detecting antibodies against SARS-CoV-2. Biochem Biophys Res Commun 553:165–171. https://doi.org/10.1016/j.bbrc.2021.03.016

Kus JV (2014) Infections due to Citrobacter and Enterobacter☆. In: Reference module in biomedical sciences. Elsevier, Amsterdam. https://doi.org/10.1016/B978-0-12-801238-3.05089-3

Kuss S, Tanner EEL, Ordovas-Montanes M, Compton RG (2017) Electrochemical recognition and quantification of cytochrome c expression in: *Bacillus subtilis* and aerobe/anaerobe *Escherichia coli* using N, N, N′, N′-tetramethyl- Para -phenylene-diamine (TMPD). Chem Sci 8(11):7682–7688. https://doi.org/10.1039/c7sc03498a

LaBauve AE, Wargo MJ (2012) Growth and laboratory maintenance of *Pseudomonas aeruginosa*. Curr Protoc Microbiol 25. https://doi.org/10.1002/9780471729259.mc06e01s25

Lal B, Vijayakumar S, Anandan S, Veeraraghavan B (2019) Specimen collection, processing, culture, and biochemical identification of *Acinetobacter* spp. Methods Mol Biol 1946:1–15. https://doi.org/10.1007/978-1-4939-9118-1_1

Laxminarayan R, Duse A, Wattal C, Zaidi AKM, Wertheim HFL, Sumpradit N, Vlieghe E, Hara GL, Gould IM, Goossens H, Greko C, So AD, Bigdeli M, Tomson G, Woodhouse W, Ombaka E, Peralta AQ, Qamar FN, Mir F, Cars O et al (2013) Antibiotic resistance-the need for global solutions. Lancet Infect Dis 13(12):1057–1098. https://doi.org/10.1016/S1473-3099(13)70318-9

Li Y, Huang J, Wang X, Xu C, Han T, Guo X (2020) Genetic characterization of the O-antigen and development of a molecular serotyping scheme for *Enterobacter cloacae*. Front Microbiol 11:727. https://doi.org/10.3389/fmicb.2020.00727

Li Y, Pan T, Cao R, Li W, He Z, Sun B (2023) Nitrate reductase NarGHJI modulates virulence via regulation of agr expression in methicillin-resistant *Staphylococcus aureus* strain USA300 LAC. Microbiol Spectr 11(3):e0359622. https://doi.org/10.1128/spectrum.03596-22

Liu K, Ke Z, Chen P, Zhu S, Yin H, Li Z, Chen Z (2021) Classification of two species of gram-positive bacteria through hyperspectral microscopy coupled with machine learning. Biomed Opt Express 12(12):7906. https://doi.org/10.1364/boe.445041

Loyola-Cruz MÁ, Gonzalez-Avila LU, Martínez-Trejo A, Saldaña-Padilla A, Hernández-Cortez C, Bello-López JM, Castro-Escarpulli G (2023) ESKAPE and beyond: the burden of coinfections in the COVID-19 pandemic. Pathogens 12(5):743. https://doi.org/10.3390/pathogens12050743

Lu H, Yadav V, Zhong M, Bilal M, Taherzadeh MJ, Iqbal HMN (2022) Bioengineered microbial platforms for biomass-derived biofuel production—a review. Chemosphere 288:132528. https://doi.org/10.1016/j.chemosphere.2021.132528

Mahmudul Islam AFM, Khalik MF, Uddin N, Hossain MS, Hossain MM, Hasan MM, Fahad SM, Saha P (2015) Evaluation of microbiological and physicochemical profile of some herbal preparations manufactured by pharmaceutical and herbal manufacturers in Bangladesh. J Pharm Investig 45(2):223–235. https://doi.org/10.1007/s40005-014-0168-6

Majumder MAA, Rahman S, Cohall D, Bharatha A, Singh K, Haque M, Gittens-St Hilaire M (2020) Antimicrobial stewardship: fighting antimicrobial resistance and protecting global public health. Infect Drug Resist 13:4713–4738. https://doi.org/10.2147/IDR.S290835

Mamarasulov B, Davranov K, Jahan MS, Jabborova D, Nasif O, Ansari MJ, Danish S, Datta R (2022) Characterization, enzymatic and biochemical properties of endophytic bacterial strains of the medicinal plant *Ajuga turkestanica* (Rgl.) brig (Lamiaceae). J King Saud Univ Sci 34(6):102183. https://doi.org/10.1016/j.jksus.2022.102183

Mao Q, Sun X, Sun J, Zhang F, Lv A, Hu X, Guo Y (2020) A candidate probiotic strain of *Enterococcus faecium* from the intestine of the crucian carp *Carassius auratus*. AMB Express 10(1):40. https://doi.org/10.1186/s13568-020-00973-0

Martínez-Espinosa RM (2020) Microorganisms and their metabolic capabilities in the context of the biogeochemical nitrogen cycle at extreme environments. Int J Mol Sci 21(12):1–19. https://doi.org/10.3390/ijms21124228

Matlou DP, Bissong MEA, Tchatchouang C-DK, Adem MR, Foka FET, Kumar A, Ateba CN (2019) Virulence profiles of vancomycin-resistant Enterococci isolated from surface and ground water utilized by humans in the north West Province, South Africa: a public health perspective. Environ Sci Pollut Res 26(15):15105–15114. https://doi.org/10.1007/s11356-019-04836-5

Medina P, Baresi L (2007) Rapid identification of gelatin and casein hydrolysis using TCA. J Microbiol Methods 69(2):391–393. https://doi.org/10.1016/j.mimet.2007.01.005

Mekonnen E, Kebede A, Nigussie A, Kebede G, Tafesse M (2021) Isolation and characterization of urease-producing soil bacteria. Int J Microbiol 2021:1–11. https://doi.org/10.1155/2021/8888641

Mendes Ferreira A, Mendes-Faia A (2020) The role of yeasts and lactic acid bacteria on the metabolism of organic acids during winemaking. Foods 9(9):1231. https://doi.org/10.3390/foods9091231

Micelli C, Dai Y, Raustad N, Isberg RR, Dowson CG, Lloyd AJ, Geisinger E, Crow A, Roper DI (2023) A conserved zinc-binding site in *Acinetobacter baumannii* PBP2 required for elongasome-directed bacterial cell shape. Proc Natl Acad Sci U S A 120(8):e2215237120. https://doi.org/10.1073/pnas.2215237120

Mitra M, Nguyen KM-A-K, Box TW, Gilpin JS, Hamby SR, Berry TL, Duckett EH (2020) Isolation and characterization of a novel bacterial strain from a tris-acetate-phosphate agar medium plate of the green micro-alga *Chlamydomonas reinhardtii* that can utilize common environmental pollutants as a carbon source. F1000Res 9:656. https://doi.org/10.12688/f1000research.24680.1

Mitrea L, Vodnar DC (2019) *Klebsiella pneumoniae*—a useful pathogenic strain for biotechnological purposes: diols biosynthesis under controlled and uncontrolled pH levels. Pathogens 8(4):293. https://doi.org/10.3390/pathogens8040293

Mitzscherling J, Maclean J, Lipus D, Bartholomäus A, Mangelsdorf K, Lipski A, Roddatis V, Liebner S, Wagner D (2022) *Nocardioides alcanivorans* sp. nov., a novel hexadecane-degrading species isolated from plastic waste. Int J Syst Evol Microbiol 72(4). https://doi.org/10.1099/ijsem.0.005319

Moubareck CA, Halat DH (2020) Insights into *Acinetobacter baumannii*: a review of microbiological, virulence, and resistance traits in a threatening nosocomial pathogen. Antibiotics 9(3):119. https://doi.org/10.3390/antibiotics9030119

Mukherjee S, Yadav V, Mondal M, Banerjee S, Halder G (2017) Characterization of a fluoride-resistant bacterium *Acinetobacter* sp. RH5 towards assessment of its water defluoridation capability. Appl Water Sci 7(4):1923–1930. https://doi.org/10.1007/s13201-015-0370-3

Munoz MA, Ahlström C, Rauch BJ, Zadoks RN (2006) Fecal shedding of *Klebsiella pneumoniae* by dairy cows. J Dairy Sci 89(9):3425–3430. https://doi.org/10.3168/jds.S0022-0302(06)72379-7

Nasrin S, Hegerle N, Sen S, Nkeze J, Sen S, Permala-Booth J, Choi M, Sinclair J, Tapia MD, Johnson JK, Sow SO, Thaden JT, Fowler VG, Krogfelt KA, Brauner A, Protonotariou E, Christaki E, Shindo Y, Kwa AL, Tennant SM et al (2022) Distribution of serotypes and antibiotic resistance of invasive *Pseudomonas aeruginosa* in a multi-country collection. BMC Microbiol 22(1):13. https://doi.org/10.1186/s12866-021-02427-4

Navidinia M, Goudarzi M, Rameshe SM, Farajollahi Z, Ebadi Asl P, Khosravi SZ, Mounesi MR (2017) Molecular characterization of resistance genes in MDR-ESKAPE pathogens. J Pure Appl Microbiol 11(2):779–792. https://doi.org/10.22207/JPAM.11.2.17

Nkem BM, Halimoon N, Yusoff FM, Johari WLW, Zakaria MP, Medipally SR, Kannan N (2016) Isolation, identification and diesel-oil biodegradation capacities of indigenous hydrocarbon-degrading strains of *Cellulosimicrobium cellulans* and *Acinetobacter baumannii* from tarball at Terengganu beach, Malaysia. Mar Pollut Bull 107(1):261–268. https://doi.org/10.1016/j.marpolbul.2016.03.060

Nursyam H, Prihanto AA, Warasari NI, Saadah M, Masrifa RE, Nabila NA, Istiqfarin N, Siddiq IJ (2018) The isolation and identification of endophytic bacteria from mangrove (*Sonneratia alba*) that produces gelatinase. IOP Conf Ser Earth Environ Sci 137(1):012056. https://doi.org/10.1088/1755-1315/137/1/012056

Oberg C, Sorensen K, Oberg T, Young S, Domek M, Culumber M, McMahon D (2021) Gluconate metabolism and gas production by *Paucilactobacillus wasatchensis* WDC04. J Dairy Sci 104(10):10586–10593. https://doi.org/10.3168/jds.2021-20232

Ogodo AC, Agwaranze DI, Daji M, Aso RE (2022) Microbial techniques and methods: basic techniques and microscopy. In: Analytical techniques in biosciences. Elsevier, Amsterdam, pp 201–220. https://doi.org/10.1016/B978-0-12-822654-4.00003-8

Owlia P, Nosrati R, Alaghehbandan R, Lari AR (2014) Antimicrobial susceptibility differences among mucoid and non-mucoid *Pseudomonas aeruginosa* isolates. GMS Hyg Infect Control 9(2):Doc13. https://doi.org/10.3205/dgkh000233

Palleroni NJ (2015) Pseudomonas. In: Bergey's manual of systematics of archaea and bacteria. Wiley, New York, pp 1–1. https://doi.org/10.1002/9781118960608.gbm01210

Palmer KL, Godfrey P, Griggs A, Kos VN, Zucker J, Desjardins C, Cerqueira G, Gevers D, Walker S, Wortman J, Feldgarden M, Haas B, Birren B, Gilmore MS (2012) Comparative genomics of Enterococci: variation in *Enterococcus faecalis*, clade structure in *E. faecium*, and defining characteristics of *E. gallinarum* and *E. casseliflavus*. MBio 3(1):e00318. https://doi.org/10.1128/mBio.00318-11

Panigrahi S, Mohanty S, Rath CC (2020) Characterization of endophytic bacteria *Enterobacter cloacae* MG00145 isolated from *Ocimum sanctum* with indole acetic acid (IAA) production and plant growth promoting capabilities against selected crops. S Afr J Bot 134:17–26. https://doi.org/10.1016/j.sajb.2019.09.017

Pankratova G, Leech D, Gorton L, Hederstedt L (2018) Extracellular electron transfer by the gram-positive bacterium *Enterococcus faecalis*. Biochemistry 57(30):4597–4603. https://doi.org/10.1021/acs.biochem.8b00600

Park G-N, Kang H-S, Kim H-R, Jung B-K, Kim D-H, Chang K-S (2019) A comparison of genospecies of clinical isolates in the *Acinetobacter* spp. complex obtained from hospitalized patients in Busan, Korea. Biomed Sci Lett 25(1):40–53. https://doi.org/10.15616/bsl.2019.25.1.40

Pauter K, Szultka-Młýnska M, Buszewski B (2020) Determination and identification of antibiotic drugs and bacterial strains in biological samples. Molecules 25(11):2556. https://doi.org/10.3390/molecules25112556

Peleg AY, Seifert H, Paterson DL (2008) *Acinetobacter baumannii*: emergence of a successful pathogen. Clin Microbiol Rev 21(3):538–582. https://doi.org/10.1128/CMR.00058-07

Permatasari DA, Witaningrum AM, Wibisono FJ, Effendi MH (2020) Detection and prevalence of multidrug-resistant *Klebsiella pneumoniae* strains isolated from poultry farms in Blitar, Indonesia. Biodiversitas 21(10):4642–4647. https://doi.org/10.13057/biodiv/d211024

Pogue JM, Kaye KS, Cohen DA, Marchaim D (2015) Appropriate antimicrobial therapy in the era of multidrug-resistant human pathogens. Clin Microbiol Infect 21(4):302–312. https://doi.org/10.1016/j.cmi.2014.12.025

Prabhu A, Nandagopal MS, Peralam Yegneswaran P, Prabhu V, Verma U, Mani NK (2020) Thread integrated smart-phone imaging facilitates early turning point colourimetric assay for microbes. RSC Adv 10(45):26853–26861. https://doi.org/10.1039/D0RA05190J

Pumipuntu N, Kulpeanprasit S, Santajit S, Tunyong W, Kong-ngoen T, Hinthong W, Indrawattana N (2017) Screening method for *Staphylococcus aureus* identification in subclinical bovine mastitis from dairy farms. Vet World 10(7):721–726. https://doi.org/10.14202/vetworld.2017.721-726

Putatunda C, Behl M, Solanki P, Sharma S, Bhatia SK, Walia A, Bhatia RK (2023) Current challenges and future technology in photofermentation-driven biohydrogen production by utilizing algae and bacteria. Int J Hydrogen Energy 48(55):21088–21109. https://doi.org/10.1016/j.ijhydene.2022.10.042

Rahim K, Saleha S, Basit A, Zhu X, Ahmed I, Huo L, Zhang P, Usman B, Munir S, Luiz Franco O (2017) Pseudomonas aeruginosa as a powerful biofilm producer and positive action of amikacin against isolates from chronic wounds. Jundishapur J Microbiol 10(10):e57564. https://doi.org/10.5812/jjm.57564

Rai AK, Mitchell AM (2020) Enterobacterial common antigen: synthesis and function of an enigmatic molecule. MBio 11(4):e01914. https://doi.org/10.1128/mBio.01914-20

Ramirez MS, Bonomo RA, Tolmasky ME (2020) Carbapenemases: transforming *Acinetobacter baumannii* into a yet more dangerous menace. Biomolecules 10(5):720. https://doi.org/10.3390/biom10050720

Rathakrishnan D, Gopalan AK (2022) Isolation and characterization of halophilic isolates from Indian salterns and their screening for production of hydrolytic enzymes. Environ Chall 6:100426. https://doi.org/10.1016/j.envc.2021.100426

Ren Loi H, Abbasiliasi S, Bothi Raja P, Shamzi Mohamed M, Tan WN, Suan Ng H, Chi-Wei Lan J, Shun Tan J (2023) Biosynthesis of silver nanoparticles using nitrate reductase produced by lactobacillus plantarum CAM 4: characterization and in vitro evaluation of its antimicrobial efficiency. J Mol Liq 376:121476. https://doi.org/10.1016/j.molliq.2023.121476

Risqiyah W, Narulita E, Rofiqoh A, Ludfi AS, Iqbal M (2022) Morphological and molecular identification of multi-antibiotic resistant bacteria in the wound site of diabetic ulcers. Biodiversitas 23(2):663–670. https://doi.org/10.13057/biodiv/d230207

Rogers L, Power K, Gaora PÓ, Fanning S (2016) *Escherichia coli* and other Enterobacteriaceae: occurrence and detection. In: Encyclopedia of food and health. Elsevier, Amsterdam, pp 545–551. https://doi.org/10.1016/B978-0-12-384947-2.00259-2

Roulová N, Motková P, Brožková I, Pejchalová M (2022) Antibiotic resistance of *Pseudomonas aeruginosa* isolated from hospital wastewater in The Czech Republic. J Water Health 20(4):692–701. https://doi.org/10.2166/wh.2022.101

Sajjan S, Kulkarni G, Yaligara V, Kyoung L, Karegoudar TB (2010) Purification and physiochemical characterization of melanin pigment from *Klebsiella* sp. GSK J Microbiol Biotechnol 20(11):1513–1520. https://doi.org/10.4014/jmb.1002.02006

Salmonová H, Bunešová V (2017) Methods of studying diversity of bacterial communities: a review. Sci Agric Bohem 48(3):154–165. https://doi.org/10.1515/sab-2017-0022

Sánchez I, Palop L, Ballesteros C (2000) Biochemical characterization of lactic acid bacteria isolated from spontaneous fermentation of 'Almagro' eggplants. Int J Food Microbiol 59(1–2):9–17. https://doi.org/10.1016/S0168-1605(00)00256-7

Santajit S, Indrawattana N (2016) Mechanisms of antimicrobial resistance in ESKAPE pathogens. Biomed Res Int 2016:1–8. https://doi.org/10.1155/2016/2475067

Santajit S, Sookrung N, Indrawattana N (2022) Quorum sensing in ESKAPE bugs: a target for combating antimicrobial resistance and bacterial virulence. Biology 11(10):1466. https://doi.org/10.3390/biology11101466

Santaniello A, Sansone M, Fioretti A, Menna LF (2020) Systematic review and meta-analysis of the occurrence of ESKAPE bacteria group in dogs, and the related zoonotic risk in animal-assisted therapy, and in animal-assisted activity in the health context. Int J Environ Res Public Health 17(9):3278. https://doi.org/10.3390/ijerph17093278

Sapkota A, Thapa A, Budhathoki A, Sainju M, Shrestha P, Aryal S (2020) Isolation, characterization, and screening of antimicrobial-producing actinomycetes from soil samples. Int J Microbiol 2020:2716584. https://doi.org/10.1155/2020/2716584

Sass LA, Fisher RG (2009) Enterobacter. In: Feigin and Cherry's textbook of pediatric infectious diseases. Elsevier, Amsterdam, pp 1519–1525. https://doi.org/10.1016/B978-1-4160-4044-6.50118-7

Schleifer K, Bell JA (2015) Staphylococcus. In: Bergey's manual of systematics of archaea and bacteria. Wiley, New York, pp 1–43. https://doi.org/10.1002/9781118960608.gbm00569

Seleem NM, Abd El Latif HK, Shaldam MA, El-Ganiny A (2020) Drugs with new lease of life as quorum sensing inhibitors: for combating MDR *Acinetobacter baumannii* infections. Eur J Clin Microbiol Infect Dis 39(9):1687–1702. https://doi.org/10.1007/s10096-020-03882-z

Shaik M, Girija Sankar G, Iswarya M, Rajitha P (2017) Isolation and characterization of bioactive metabolites producing marine *Streptomyces parvulus* strain sankarensis-A10. J Genet Eng Biotechnol 15(1):87–94. https://doi.org/10.1016/j.jgeb.2017.02.004

Sharma L, Verma AK, Kumar A, Rahat AN, Nigam R (2015) Incidence and pattern of antibiotic resistance of *Staphylococcus aureus* isolated from clinical and subclinical mastitis in cattle and buffaloes. Asian J Anim Sci 9(3):100–109. https://doi.org/10.3923/ajas.2015.100.109

Shen C, Zhang Y (2022) Biochemistry test of bacteria-1 (urease test, carbohydrate fermentation, catalase test, oxidase test). In: Introductory microbiology lab skills and techniques in food science. Elsevier, Amsterdam, pp 67–76. https://doi.org/10.1016/B978-0-12-821678-1.00005-8

Siguenza N, Jangid A, Strong EB, Merriam J, Martinez AW, Martinez NW (2019) Micro-staining microbes: an alternative to traditional staining of microbiological specimens using microliter volumes of reagents. J Microbiol Methods 164:105654. https://doi.org/10.1016/j.mimet.2019.105654

Sigurdarson JJ, Svane S, Karring H (2020) Development of a M9-based urea medium (M9U) for sensitive and real-time monitoring of ureolytic activity of bacteria and cell-free urease. Microbiology 9(3):e976. https://doi.org/10.1002/mbo3.976

Sikarwar AS, Batra HV (2011) Identification of *Klebsiella pneumoniae* by capsular polysaccharide polyclonal antibodies. Int J Chem Eng Applications 2:130–134. https://doi.org/10.7763/IJCEA.2011.V2.89

Silhavy TJ, Kahne D, Walker S (2010) The bacterial cell envelope. Cold Spring Harb Perspect Biol 2(5):a000414. https://doi.org/10.1101/cshperspect.a000414

Sitepu IR, Enriquez LL, Nguyen V, Doyle C, Simmons BA, Singer SW, Fry R, Simmons CW, Boundy-Mills K (2019) Ethanol production in switchgrass hydrolysate by ionic liquid-tolerant yeasts. Bioresour Technol Rep 7:100275. https://doi.org/10.1016/j.biteb.2019.100275

Siu LK, Fung C-P, Chang F-Y, Lee N, Yeh K-M, Koh TH, Ip M (2011) Molecular typing and virulence analysis of serotype K1 *Klebsiella pneumoniae* strains isolated from liver abscess patients and stool samples from noninfectious subjects in Hong Kong, Singapore, and Taiwan. J Clin Microbiol 49(11):3761–3765. https://doi.org/10.1128/JCM.00977-11

Sivakumar T, Shankar T, Thangapandian V, Mahendran S (2016) Media optimization for exopolysaccharide producing *Klebsiella pneumoniae* KU215681 under varying cultural conditions. Int J Biochem Biophys 4(2):16–23. https://doi.org/10.13189/ijbb.2016.040202

Slifkin M, Pouchet GR (1977) Rapid carbohydrate fermentation test for confirmation of the pathogenic Neisseria using a Ba(OH)2 indicator. J Clin Microbiol 5(1):15–19. https://doi.org/10.1128/jcm.5.1.15-19.1977

Slotved H-C, Kaltoft M, Skovsted IC, Kerrn MB, Espersen F (2004) Simple, rapid latex agglutination test for serotyping of pneumococci (Pneumotest-latex). J Clin Microbiol 42(6):2518–2522. https://doi.org/10.1128/JCM.42.6.2518-2522.2004

Smith I (2003) Mycobacterium tuberculosis pathogenesis and molecular determinants of virulence. Clin Microbiol Rev 16(3):463–496. https://doi.org/10.1128/CMR.16.3.463-496.2003

Son MS, Taylor RK (2011) Genetic screens and biochemical assays to characterize vibrio cholerae O1 biotypes: classical and El tor. Curr Protoc Microbiol 22(1):6A.2.1. https://doi.org/10.1002/9780471729259.mc06a02s22

Spang A, Stairs CW, Dombrowski N, Eme L, Lombard J, Caceres EF, Greening C, Baker BJ, Ettema TJG (2019) Proposal of the reverse flow model for the origin of the eukaryotic cell based on comparative analyses of Asgard archaeal metabolism. Nat Microbiol 4(7):1138–1148. https://doi.org/10.1038/s41564-019-0406-9

Spernovasilis N, Tsiodras S, Poulakou G (2022) Emerging and re-emerging infectious diseases: humankind's companions and competitors. Microorganisms 10(1):98. https://doi.org/10.3390/microorganisms10010098

Suman S, Tanuja (2021) Isolation and characterization of a bacterial strain *Enterobacter cloacae* (accession no. KX438060.1) capable of degrading DDTs under aerobic conditions and its use in bioremediation of contaminated soil. Microbiol Insights 14:117863612110242. https://doi.org/10.1177/11786361211024289

Švec P, Devriese LA (2015) Enterococcus. In: Bergey's manual of systematics of archaea and bacteria. Wiley, New York, pp 1–25. https://doi.org/10.1002/9781118960608.gbm00600

Tassadaq H, Aneela R, Shehzad M, Iftikhar A, Jafar K, Veronique EH, Kil YK, Muhammad A (2013) Biochemical characterization and identification of bacterial strains isolated from drinking water sources of Kohat, Pakistan. Afr J Microbiol Res 7(16):1579–1590. https://doi.org/10.5897/AJMR12.2204

Taylor SL, Guthertz LS, Leatherwood M, Lieber ER (1979) Histamine production by *Klebsiella pneumoniae* and an incident of scombroid fish poisoning. Appl Environ Microbiol 37(2):274–278. https://doi.org/10.1128/aem.37.2.274-278.1979

Thomas E (2015) Microbial growth and physiology: a call for better craftsmanship. Front Microbiol 6(MAR):287. https://doi.org/10.3389/fmicb.2015.00287

Tiso M, Schechter AN (2015) Nitrate reduction to nitrite, nitric oxide and ammonia by gut bacteria under physiological conditions. PLoS One 10(3):e0119712. https://doi.org/10.1371/journal.pone.0119712

Traub WH (1989) *Acinetobacter baumannii* serotyping for delineation of outbreaks of nosocomial cross-infection. J Clin Microbiol 27(12):2713–2716. https://doi.org/10.1128/jcm.27.12.2713-2716.1989

Uddin TM, Chakraborty AJ, Khusro A, Zidan BRM, Mitra S, Emran TB, Dhama K, Ripon MKH, Gajdács M, Sahibzada MUK, Hossain MJ, Koirala N (2021) Antibiotic resistance in microbes: history, mechanisms, therapeutic strategies and future prospects. J Infect Public Health 14(12):1750–1766. https://doi.org/10.1016/j.jiph.2021.10.020

Vancanneyt M, Lombardi A, Andrighetto C, Knijff E, Torriani S, Björkroth KJ, Franz CMAP, Foulquié Moreno MR, Revets H, De Vuyst L, Swings J, Kersters K, Dellaglio F, Holzapfel WH (2002) Intraspecies genomic groups in enterococcus faecium and their correlation with origin and pathogenicity. Appl Environ Microbiol 68(3):1381–1391. https://doi.org/10.1128/AEM.68.3.1381-1391.2002

Verdier I, Durand G, Bes M, Taylor KL, Lina G, Vandenesch F, Fattom AI, Etienne J (2007) Identification of the capsular polysaccharides in *Staphylococcus aureus* clinical isolates by PCR and agglutination tests. J Clin Microbiol 45(3):725–729. https://doi.org/10.1128/JCM.01572-06

von Eiff C, Proctor RA, Peters G (2000) *Staphylococcus aureus* small colony variants: formation and clinical impact. Int J Clin Pract Suppl 115:44–49

Wang Y, Wu J, Lv M, Shao Z, Hungwe M, Wang J, Bai X, Xie J, Wang Y, Geng W (2021) Metabolism characteristics of lactic acid bacteria and the expanding applications in food industry. Front Bioeng Biotechnol 9:612285. https://doi.org/10.3389/fbioe.2021.612285

Wang C, Ye Q, Ding Y, Zhang J, Gu Q, Pang R, Zhao H, Wang J, Wu Q (2022) Detection of *Pseudomonas aeruginosa* serogroup G using real-time PCR for novel target genes identified through comparative genomics. Front Microbiol 13:928154. https://doi.org/10.3389/fmicb.2022.928154

Wanger A, Chavez V, Huang RSP, Wahed A, Actor JK, Dasgupta A (2017) Biochemical tests and staining techniques for microbial identification. In: Microbiology and molecular diagnosis in pathology. Elsevier, Amsterdam, pp 61–73. https://doi.org/10.1016/B978-0-12-805351-5.00005-3

Woo PCY, Lau SKP, Teng JLL, Tse H, Yuen KY (2008) Then and now: use of 16S rDNA gene sequencing for bacterial identification and discovery of novel bacteria in clinical microbiology laboratories. Clin Microbiol Infect 14(10):908–934. https://doi.org/10.1111/j.1469-0691.2008.02070.x

Wu M, Li X (2015) *Klebsiella pneumoniae* and *Pseudomonas aeruginosa*. In: Molecular medical microbiology. Elsevier, Amsterdam, pp 1547–1564. https://doi.org/10.1016/B978-0-12-397169-2.00087-1

Wu KY, Yang TX (2020) A novel improved gram staining method based on the capillary tube. Pol J Microbiol 69(4):503–508. https://doi.org/10.33073/PJM-2020-043

Wu ML, Gengenbacher M, Dick T (2016) Mild nutrient starvation triggers the development of a small-cell survival morphotype in mycobacteria. Front Microbiol 7(JUN):947. https://doi.org/10.3389/fmicb.2016.00947

Yang Y, Wang H, Zhou H, Hu Z, Shang W, Rao Y, Peng H, Zheng Y, Hu Q, Zhang R, Luo H, Rao X, Yang CY (2020) Protective effect of the Golden Staphyloxanthin biosynthesis pathway on *Staphylococcus aureus* under cold atmospheric plasma treatment. Appl Environ Microbiol 86:e01998. https://doi.org/10.1128/AEM

Yang B, Fang D, Lv Q, Wang Z, Liu Y (2021) Targeted therapeutic strategies in the battle against pathogenic bacteria. Front Pharmacol 12:673239. https://doi.org/10.3389/fphar.2021.673239

Yang JL, Li D, Zhan XY (2023) Concept about the virulence factor of legionella. Microorganisms 11(1):74. https://doi.org/10.3390/microorganisms11010074

Yoon SS, Mekalanos JJ (2006) 2,3-butanediol synthesis and the emergence of the vibrio cholerae El tor biotype. Infect Immun 74(12):6547–6556. https://doi.org/10.1128/IAI.00695-06

Yu H, Yu Z, Huang H, Li P, Tang Q, Wang X, Shen S (2019) Gut microbiota signatures and lipids metabolism profiles by exposure to polyene phosphatidylcholine. Biofactors 45(3):439–449. https://doi.org/10.1002/biof.1495

Zhang Y, Zhou Y, Hou M, Zhang S (2022) Analysis of cerebrospinal fluid routine biochemical level, pathogenic bacteria distribution, and risk factors in patients with secondary intracranial infection after brain tumor surgery. Evid Based Complement Alternat Med 2022:1–8. https://doi.org/10.1155/2022/7716205

Zurawski DV, McLendon MK (2020) Monoclonal antibodies as an antibacterial approach against bacterial pathogens. Antibiotics 9(4):155. https://doi.org/10.3390/antibiotics9040155

Nucleic Acid Amplification and Molecular Diagnostic Techniques in the Detection of ESKAPE Bacterial Pathogens

5

Santhilatha Pandrangi, G. Kishore,
Gantala Sarva Sai Nikhilesh, and Suseela Lanka

Abstract

The word ESKAPE denotes a particular group of microbial species viz., *Enterococcus faecium*, *Staphylococcus aureus*, *Klebsiella pneumonia*, *Acinetobacter baumannii*, *Pseudomonas aeruginosa,* and *Enterobacter* species, all of which were found to play role in causing nosocomial infections and also have the potential to develop drug resistance owing to their ability to "escape" the outcome action of antibiotics. The current chapter gives an overview of the ESKAPE pathogens, diseases associated with these pathogens, available methods to diagnose these pathogens with a special emphasis on various nucleic acid-based diagnostic methods.

Keywords

ESKAPE pathogens · Nosocomial infections · Drug resistance · Nucleic acid-based diagnostic methods

S. Pandrangi
Department of Biochemistry and Bioinformatics, GITAM School of Sciences, GITAM (Deemed to be) University, Visakhapatnam, India

G. Kishore · S. Lanka (✉)
Department of Biosciences and Biotechnology, Krishna University, Machilipatnam, Andhra Pradesh, India

G. S. S. Nikhilesh
Andhra University, Visakhapatnam, Andhra Pradesh, India

S. Busi, R. Prasad (eds.), *ESKAPE Pathogens*,
https://doi.org/10.1007/978-981-99-8799-3_5

5.1 Bacterial Pathogens of ESKAPE

Each letter in the word ESKAPE denotes a particular microbial species, all of which were found to play role in causing nosocomial infections especially in seriously ill patients and also in patients with impaired immune system. In the word ESKAPE, "E" stands for *Enterococcus faecium*; "S" for *Staphylococcus aureus*; "K" for *Klebsiella pneumonia*; "A" for *Acinetobacter baumannii*; "P" for *Pseudomonas aeruginosa*; and "E" for *Enterobacter* species. These organisms also have potential to develop drug resistance because they have the ability to "escape" the outcome action of antibiotics (Rice 2008). ESKAPE microbes, which are categorized as HAI's abbreviated for Healthcare-Associated Infections are hospital-derived diseases affecting hospitalized persons in a period of less than 48 h after admission, 3 days (72 h) after patients discharge, or 30-day period following surgery (Inweregbu et al. 2005). Greater than 40% infections associated with hospital ICU's (intensive care units) are the result of these ESKAPE pathogens, which impose huge load on nations with low- and middle-income (Rice 2010). In due course, many microbial species found to become antibiotic resistant due to widespread drug abuse and overuse (Pacios et al. 2020).

Drug resistance in ESKAPE diseases manifests by the way of varied mechanisms viz., inactivating the drug candidates with enzymes, changing the binding location of the drug, by lowering the permeability capacity of the drug, by promoting absorption of the drug to reduce its accumulation, and biofilm formation (Mulani et al. 2019). Antibiotics that were once used to get rid of microbial infections turned out to be redundant and unsuccessful due to development of infections that are resistant to antibiotics, allowing microorganisms to flourish when high drug concentrations are present. It is highly crucial to unravel new targets in order to stop the progression and disease spread. One such target is caseinolytic proteins, which are widely distributed in many species and are essential for maintaining cellular protein homeostasis (AhYoung et al. 2015).

5.1.1 *Enterococcus faecium*

Previously, *Enterococcus* species were categorized under genus *Streptococcus*. They are facultative Gram+ve anaerobes and congregate either in chain or pairs. The human stomachs and other animals are their natural habitats. According to reports, more than 20 different species of *Enterococcus* exist; however, *E. faecalis* and *E. faecium* are dominant types that are most clinically significant. Around 20 years ago, a typing method based on fingerprint, amplified fragment length polymorphism (AFLP) was used to identify the presence of these distinct subpopulations (Willems et al. 2000). Even though endogenous acquisition accounts for the bulk of *Enterococcus* infections, hospitalized patients are nevertheless susceptible to cross-infection (Elsner et al. 2000). The subgroup of *E. faecalis* isolates aroused from the hospital-associated (HA) Ampicillin resistance and pathogenicity islands is usually linked to hospital outbreaks (Willems et al. 2005). Additionally, whole-genome analyses revealed acquisition of varied traits by these HA isolates that helped these isolates thrive in hospital-based settings, due to increased resistance to antibiotics, and production of virulence genes that facilitate the biofilm formation and as well colonization. To stop the spread of third-generation cephalosporin-resistant enterococci, vancomycin was first made available in the 1970s. Vancomycin-resistant enterococci (VRE), which became increasingly prevalent when vancomycin was used more often in the 1990s, are now the second most common nosocomial infection (Ramos et al. 2020; Zaheer et al. 2020). Mobile genetic elements like transposons and plasmids can help *E. faecium* acquire new genes (Jubeh et al. 2020). *Enterococcus faecium* is a multidrug-resistant bacterium because it is naturally resistant to aminoglycosides like kanamycin, tobramycin, and gentamicin owing to its capability to generate AMEs (aminoglycoside-modifying enzymes), AACs (aminoglycoside acetyltransferases), APHs (aminoglycoside phosphotransferases), and ANTs (aminoglycoside nucleotidyl transferases) (Kim et al. 2021).

5.1.2 *Staphylococcus aureus*

It is a Gram+ve, facultative anaerobe having catalase and coagulase activity, and forms clusters of unusual grapes (Tigabu and Getaneh 2021). *S. aureus* generates diseases that range from moderate to fatal, including infections of soft tissues and skin, pleuropulmonary of bacterial origin, infections like endocarditis, and those caused by devices, etc. (Tong et al. 2015). It is a common dangerous member of the genus staphylococci and is the causative agent with a broad disease range, including skin infections (superficial abscesses), food poisoning, as well as potentially fatal disorders such as in bacteremia, necrotic pneumonia in infants, and endocarditis (Shaw et al. 2004). It causes mastitis in infected cows, dermatitis in dogs, botryomycosis in horses, septicemia and arthritis in chickens (Zunita et al. 2008; Luzzago et al. 2014). Minor skin and Impetigo is a type of soft tissue infection caused by *S. aureus*, also includes folliculitis, and cutaneous abscesses. Pyomyositis (Tong et al. 2015) is a rare but serious infection in the neighborhood. *Staphylococcus aureus* can cause diseases in nosocomial settings by infecting surgical sites or medical devices that are implanted viz., catheters, artificial heart valves, orthopedic implants, and prosthetic joints (Richards et al. 1999; Hogan et al. 2015). *S. aureus,* widespread in the environment infects the blood during bacteremia and can seed important organs, as a result of widespread Endocarditis, osteomyelitis, and descending urinary tract infections (Wertheim et al. 2005). This pathogen's capacity to endure in a variety of situations, with wide host niches, such as skin (Montgomery et al. 2015) and abiotic devices (Scherr et al. 2014), makes things tough to eliminate it, resulting in recurring infections. Antibiotic resistance in this particular species was originally documented in 1940s, the period in which a strain gained resistance to penicillin by producing a hydrolytic enzyme, penicillinase (Basset et al. 2011).

5.1.3 *Klebsiella pneumonia*

K. pneumoniae are frequently encapsulated Gram−ve bacilli that belong to Enterobacteriaceae family (Santajit and Indrawattana 2016). It is a nonmotile with wide habitat range that includes soil and all sorts of surface waters, as well as on medical-related equipment (Bagley 1985). *K. pneumoniae* quickly enter mucosal surfaces of humans such as the GI (gastrointestinal) tract and the oropharynx, causing ill effects (Dao et al. 2014). These strains have the ability to enter different tissues and produce severe illnesses in humans. Many components of the immune system are evaded and survived by bacteria, rather than aggressively suppressed, and bacteria proliferate at multiple places in hosts. The strain is utmost prevalent cause of hospital-induced pneumonia, representing 3–8% of all nosocomial infections caused by bacteria (Jondle et al. 2018).

5.1.4 *Acinetobacter baumannii*

It is a nonfermenter, Gram−ve coccobacillus, positive for catalase activity, negative for oxidase, nonmotile, and was formerly thought to be a low-level pathogen but currently emerged as a prominent infectious agent with origin from hospitals and community-acquired diseases. In immunocompromised people, it is a frequent contributor of septicemia and pneumonia. Especially those who have spent more than 90 days in the hospital (Montefour et al. 2008), will be infected by this agent. Also commonly found in aquatic environments. *Acinetobacters* that can be recognized to the genus level are the creatures that are frequently tough to clean and, as a result, are frequently misidentified as Gram+ve bacteria. Till date not a single conclusive metabolic examination is known that identifies *Acinetobacters* different from other Gram−ve bacteria that do not ferment (Peleg et al. 2008). Organisms of *Acinetobacter* genus are frequently thought to be found everywhere in nature because they can be collected from nearly all samples of soil and surface waters (Baumann et al. 1968). *A. baumannii* is discovered extremely infrequently as a part of natural skin microbiota, according to one research indicating that the bacterium colonizes only 3% of the population (Seifert et al. 1997). *A. baumannii* mainly infects wet tissues like mucous membranes or exposed skin patches created by a mishap or an injury. Infected skin and soft tissues exhibit an orange coloration, followed by a sandpaper-like surface, and finally clear vesicles appear on the skin (Sebeny et al. 2008).

5.1.5 *Pseudomonas aeruginosa*

The organism is an aerobic, Gram−ve bacterium with rod shape and can be obtained from varied sources viz., soil, plants, and tissues of mammals (Stover et al. 2000). Making use of its influential binding elements, this organism can thrive on water, various surfaces of diverse nature, and medical equipment with the aid of their pili, flagella, and biofilms. Hence is common in natural, as well in man-made habitats, such as lakes, residential sink drains, and hospitals (Remold et al. 2011). It is a symbiotic microorganism involved in a variety of human illnesses. It is now a significant source of nosocomial infections and even exhibits resistance to antimicrobial agents (Shariati et al. 2018). It is also an opportunistic organism causing healthcare infections such as infections connected with ventilators, infections in intensive care units, bloodstream infections, infections at the surgical location, infections of urinary tract, and burn wounds, otitis media, and keratitis (Ramos et al. 2013). It can infect patients with immune deficiencies because of its capacity to resist through inherent and learned immune defenses, adhesion, colonization, and formation of biofilm, as well as to create factors that cause virulence resulting in considerable damage of tissue. It even results in high mortality in people afflicted with cystic fibrosis, infections of newborn, serious burns, cancer, etc. (Nathwani et al. 2014).

5.1.6 Enterobacter Species

Enterobacter, a Gram−ve facultative anaerobe bacilli with ability to grow with or without oxygen, do not generate spores, belongs to Enterobacteriaceae family. *E. cloacae* and *Enterobacter aerogenes*, two of its well-known species cause nosocomial infections in patients of intensive cares, mainly those with mechanical ventilation (Mezzatesta et al. 2012). After being placed in the genus Enterobacter in 1960, *Aerobacter aerogenes* became *Enterobacter aerogenes*. As a result of its peritrichous flagellum, mobility, and genetic similarity to the *Klebsiella* genus, this was recommended to called as *Klebsiella mobilis* in the year 1971. It is worth noting that the phenotypic differences between *E. aerogenes* and *Klebsiella* share not only with respect to motility but also the existence of ODC (ornithine decarboxylase) activity and whereas *E. aerogenes* has no urease activity (Farmer III et al. 1985).

Enterobacter cloacae species are common in nature, and act as pathogenic microbes. Both *E. cloacae* and *E. hormaechei* can be commonly separated from patient clinical samples. *E. cloacae,* one of frequently encountered *Enterobacter* sp. causes nosocomial infections in hospital setups and much have been written about these germs' antibiotic resistance. Regardless of importance of pathogenic mechanisms of *E. cloacae* as nosocomial microbe, much information is not available with regard to its involvement in other diseases. Its pathogenicity is dependent on its capability to build biofilms and to produce various cytotoxins (pore-forming toxins, enterotoxins, and hemolysins) (Mezzatesta et al. 2012). Some of its members were found to be associated with clinical specimens, particularly sputum and urine samples (Izdebski et al. 2014). Interestingly, *E. cloacae* has recently become the third *Enterobacteriaceae* sp. with a broad span engaged with nosocomial infections following *E. coli* and *K. pneumoniae*, owing to its wide spectrum carbapenemases and β-lactamases (ESBLs) (Mancuso et al. 2021).

5.2 Diseases Associated with ESKAPE Bacterial Pathogens

An abundant Gram−ve aerobic bacterium in the environment is *Pseudomonas aeruginosa*. As a most common pathogen, it causes varied chronic and acute nosocomial diseases, includes life-threatening respiratory problems in patients with compromised immune defenses (Jurado-Martín et al. 2021; Moradali et al. 2017) and was found to be the third most prevalent Gram−ve bacteria causing bloodstream infections of nosocomial type (Recio et al. 2020). The major causes for this resistance include increased expression of efflux pumps, a reduction in the outer membranes permeability, and the attainment of mutations in resistance genes that code for proteins that regulate the antibiotics passive diffusion through the outer membrane (Hwang and Yoon 2019; Pang et al. 2019; Henrichfreise et al. 2007; Langendonk et al. 2021). The cephalosporin antibiotics such as cefepime and Ceftazidime, which belong to third and fourth generations, belong to broad-spectrum and are effective toward *P. aeruginosa* (Sader et al. 2017). *P. aeruginosa* has been discovered to exhibit resistance to major 4 β-lactamases families, A, B, C, and D (Dehbashi et al. 2020). Several β-lactams, such as imipenem and benzylpenicillin, can result in endogenous β-lactamases like AmpC-β lactamase (Pachori et al. 2019). Additionally, a gene mutation that causes the overexpression of AmpC β-lactamases in this species might lead to the development of resistance (Berrazeg et al. 2015). Colistin in concert with antipseudomonas drugs like imipenem, aztreonam, piperacillin, ceftazidime/ciprofloxacin are used in general to treat MDR *P. aeruginosa*. Fosfomycin has been used successfully in association with aminoglycosides, cephalosporins, and penicillins to control *P. aeruginosa* that has developed drug resistance (Ontong et al. 2021).

Even though there are more than 50 different enterococci species, the majority of infections in human beings are mainly due to two of the species—*E. faecium* and *E. faecalis* (García-Solache and Rice 2019). Among these two, *E. faecalis* is the dangerous species, and also exhibits greater resistant to many of the existing antimicrobial drugs, and shows its effects mainly in hosts with weak immune systems, resulting in substantial mortality and morbidity. In case of healthy individuals, these microbes are not that much hazardous, but in immunocompromised patients, they have a role in hospital-acquired disorders like urinary tract infections that are associated with catheter, bacteremia, endocarditis, etc. (Shiadeh et al. 2019). *Enterococci* are gaining resistant to several of antimicrobial agents, mainly because of (1) the increased and over use of antibiotics of broad-spectrum type (cephalosporins and penicillins) in hospitals, which potentially increase the normal Gram−ve intestinal microbes and as well promote intestinal colonization of *E. faecium* (the associated reasons could be mutations in PBP and increased expression of β-lactamase which in turn lead to increased resistance to antibiotics of beta-lactam type) (Pöntinen et al. 2021); (2) the intrinsic resistance of the species to several of the routinely used antibiotics; (3) the capacity of these organisms to pick up and propagate antibiotic resistance determinants. In these particular enterococci, there are at least three varied resistance pathways to cephalosporin that have been identified (Ramos et al. 2020). In order to stop the spread of resistant enterococci sp. to third-generation cephalosporins, vancomycin was introduced in the 1970s. After that, VRE (vancomycin-resistant enterococci) became nosocomial pathogen with second most frequency owing to wide use of vancomycin in the 1990s (Zaheer et al. 2020). Clusters of Van genes like *vanS*, *vanR*, *vanH*, *vanX*, and *vanZ*, were thought to mediate this resistance to vancomycin by switching out D-Alanyl-D-Alanine termini for D-alanyl-D-lactate termini. Vancomycin shows a poor affinity binding for D-Ala-D-Lac because it binds to it much less strongly than it does to the typical dipeptide product. Gene clusters were found on a 10,581-bp *E. faecium* jumping gene (transposon, Tn1546) and they are of most common type (Jubeh et al. 2020; Cetinkaya et al. 2000). Due to its inherent resistance to aminoglycosides like tobramycin, kanamycin, and gentamicin as well as its capacity to produce AMEs (aminoglycoside-modifying enzymes) like ANTs, AACs (aminoglycoside nucleotidyl transferases and aminoglycoside acetyltransferases respectively), *E. faecium* is categorized as an MDR bacteria. Furthermore, mutations of point nature in the genes *parC* and *gyrA*, which respectively code subunit A of topoisomerase IV and DNA gyrase, as well as the efflux transporter NorA, which secretes these drugs, are frequently linked to high levels of fluoroquinolone resistance in *E. faecium* (Kim et al. 2021).

Human clinical specimens such as the respiratory system, urinary tract, blood, or gastrointestinal tract are used to isolate *Enterobacter aerogenes*. The organism's epidemiology has drawn attention in Europe, owing to its link to nosocomial disease epidemics since 1993 (Chevalier et al. 2008). Several of these strains produce carbapenemases and ESBLs, including OXA, VIM, KPC, and metallo-lactamase-1 (Halat and Moubareck 2020). This microbial group contributes appreciably to the evolution of drug resistance due to the complete suppression of ampC β-lactamases, that are expressed in large amounts (Uzunović et al. 2018).

MDR bacteria are resistant to nearly all the commercial antimicrobial drug candidates, with the exception of tigecycline and colistin (De Oliveira et al. 2020; Di Franco et al. 2021). Additionally, a recent study found that *K. aerogenes* has become resistant to every type of antibiotic, even the last resort drug colistin (De Oliveira et al. 2020). The management of bacterial infections is made more difficult by *K. aerogenes'* capacity to harbor sub populations of bacteria resistant to colistin that are untraceable by present diagnostic test methods (Band et al. 2016).

Coming to *K. pneumonia*, particularly in patients with compromised immune systems, it causes a variety of community-acquired and nosocomial diseases, viz., bloodstream infections, liver abscesses, pneumonia, and urinary tract infections (Eghbalpoor et al. 2019). Because *Klebsiella* cannot spread over the air, direct contact between people is required to become infected (Young et al. 2020). Klebsiella has increased levels of antibiotic resistance as a result of the extensive emergence of genes producing ESBLs and carbapenemases (Effah et al. 2020). The most common Enterobacteriaceae (CRE) strains that are carbapenem resistant are those from *K. pneumonia* (Lasko and Nicolau 2020). They pose a greater public health risk since KPC strains (carbapenemase-producing *K. pneumonia)* with the said enzyme expressing *blaKPC-3* gene typically serve as the final line of defense against Gram−ve infections of persistent nature (Gualtero et al. 2020; Sheu et al. 2019).

Soft tissue and skin infections, endocarditis of bacterial origin, pleuropulmonary, and infections brought on by medical equipment are only a few of the minor to fatal illnesses that *S. aureus* can cause (Tong et al. 2015). This bacterium is a significant human pathogen due to its high potential to acquire resistance to both traditional and new therapies, as well as the fact that it is exceedingly contagious and has the ability to cause chronic infections that last for a long period (Gimza and Cassat 2021). Three years after the discovery of the antibiotic, the first cases of *S. aureus* resistant to penicillin with plasmid-encoded β-lactamases that can hydrolyze the β-lactam ring of said antibiotic were reported. Beta-lactamase located on jumping genes (transposable elements) placed into plasmids frequently contained genes that were resistant to antibiotics, like gentamicin and erythromycin (Guo et al. 2020). Although the first strain of *S. aureus* resistant to methicillin was originally recognized in 1961 (Vestergaard et al. 2019), a semi-synthetic penicillin, methicillin was initially created in 1959 to treat bacterial infections that are penicillin resistant. PBPs (Penicillin-binding proteins), bound by methicillin and other antibiotics of β-lactam type, stop *S. aureus* from multiplying. *S. aureus* evolved methicillin resistance (MRSA) after acquiring *mecA* and *mecC* genes, which render methicillin inactive by creating a different PBP, known as PBP2a, which showed a decreased affinity for practically all antibiotics of β-lactam category (Lee et al. 2018; Turner et al. 2019). The antibiotic vancomycin is frequently used as a final option to treat serious Gram+ve infections like MRSA (Mühlberg et al. 2020). But by the end of 1980s, enterococci (VRE) and *S. aureus* (VRSA) had both developed vancomycin resistance (Cong et al. 2020). The VanA operon, which mediates the resistance mechanism of VRSA, is carried by the Tn1546, a jumping gene that was isolated from VRE (vancomycin-resistant Enterococcus) (Peérichon and Courvalin 2009; Kest and Kaushik 2019). The pathogenicity of *S. aureus* and its pattern of antibiotic resistance, according to the WHO (World Health Organization), pose a severe threat to human health on a global scale. MRSA, VISA (Vancomycin intermediate *S. aureus*), and VRSA are widely classified as priority agents owing to their ability to cause hospital-acquired infections and also cause fatal infections that are not possible to manage globally in the absence of effective containment and therapeutic solutions (Shariati et al. 2020).

Acinetobacter baumannii is pathogen of an opportunistic type responsible for infections in hospitals all over globe and acquired resistance to antibiotics through varied methods, the creation of beta-lactam antibiotic-degrading enzymes. The quick progression of this organism toward multiresistance would be caused by its genome incorporating exogenous DNA that produces all four classes of A, B, C, and D β-lactamases (Harding et al. 2018; Kyriakidis et al. 2021). Additionally, the genes responsible for production of narrow-spectrum β-lactamases (SCO-1, TEM-1, and CARB-4) as well as ESBL (GES-11 and CTX-M) have been discovered in *Acinetobacter* spp. (Vrancianu et al. 2020). All β-lactam antibiotics can be inhibited by class B lactamases (metallo-β lactamases; MBLs), with the exception of monobactams (Mojica et al. 2016). A class of widely distributed enzymes known as class C-lactamases are typically resistant to cephalosporins, penicillins, and cephamycins (cefoxitin and cefotetan) (Shaikh et al. 2015). Additionally, extended-spectrum cephalosporins and carbapenems can be hydrolyzed by Class DorOXAs β-lactamases found in *A. baumannii* (Halat and Moubareck 2020; Poirel and Nordmann 2006). *A. baumannii* also possesses an inherent ampC cephalosporinase (Beceiro et al. 2004). Enzymatic alteration is the main common cause of resistance to aminoglycosides (Ramirez et al. 2013). Three groups of enzymes acetyltransferases, adenylyl transferases, and phosphotransferases took part a noteworthy part in this organism's resistance to aminoglycosides. Plasmids and transposons can transmit the genes for the enzymes that alter aminoglycosides (Chen et al. 2021). The resistance of *A. baumannii* to-varied antibiotics from diverse chemical classes, viz., aminoglycosides, erythromycin, tetracyclines, trimethoprim, chloramphenicol, fluoroquinolones, and other beta-lactams, is mediated by efflux pumps (Abdi et al. 2020; Basatian-Tashkan et al. 2020). Numerous studies indicated that the antimicrobial resistance of the said species is linked to the efflux pumps, which include MFS (major facilitator superfamily), RND superfamily (resistance nodulation division), MATE family (multidrug and toxic compound extrusion), and SMR family (small multidrug resistance) transporters (Pérez-Varela et al. 2019). The increased expression of the Ade ABC efflux pump (RND member) was found to be connected with tigecycline resistance in this species (Xu et al. 2019).

The most successful medications for treating infections of *A. baumannii* were carbapenems like imipenem and meropenem (Vázquez-López et al. 2020). The mentioned two medications swapped out for minocycline/tigecycline before this bacterium developed appreciable resistance to them (Shankar et al. 2017). For MDR *A. baumannii* bacteremia, ampicillin, sulbactam, and carbapenem combination therapy is the most effective (Karaiskos et al. 2019). Despite notable rates of resistance being noted, minocycline treatment is still effective. The most successful treatment for *A. baumannii* infections that are colistin-resistant is colistin/rifampin, while minocycline- and colistin-resistant infections are treated separately (Bagińska et al. 2021). Furthermore, *A. baumannii* that is resistant to carbapenem is quickly killed by the combination of colistin and trimethoprim-sulfamethoxazole (Konca et al. 2021; Nepka et al. 2016). However, isolates of strains that are resistant to these medicines are also common. It is clear from the above that every effort must be made to discover new medicines that can kill MDR *A. baumannii*.

5.3 Currently Available Diagnostic Techniques to Identify the ESKAPE Bacterial Pathogens

Bacterial antibiotic resistance is caused by a variety of processes (Munita and Arias 2016). Genetic or mechanical factors can contribute to antibiotic resistance. A genetic foundation for resistance includes mutational changes to antibacterial targets like reduced drug absorption, Drug efflux has increased, or regulatory networks have been altered. Gene variations result from the horizontal gene transfer antibiotic resistance genes via a number of methods, including conjugation, which mobilizes mobile genetic elements (MGEs), particularly transposons and plasmids. MGEs play an important part in the rise and dissemination of resistance to antibiotics in biologically important organisms.

Bacteria develop resistance via changing the antibiotic (chemically altering or destroying it), decreasing the antibiotic's penetration and efflux, changing the target site (modifying, mutating, or bypassing it), and resistance as a result of broad-scale cell changes. Virulence factors like the development of biofilms linked to desiccation can also cause resistance (Peleg et al. 2008; Roca et al. 2012).

Microbial resistance profiling has advanced significantly. While biochemical tests are required for identifying bacterial strains and species, microbial-culture-based approaches primarily determine the bacterial phenotypic (Tsalik et al. 2018). The Clinical and Laboratory Standards Institute mandated protocols, Antibiotic resistance testing using diffusion of discs or dilution of broth methods objectively MIC-based resistance analysis (Reller et al. 2009). Many growth-based tests have limitations due to their prolonged turnaround periods of 12–72 h, (Van Belkum et al. 2019; Yang and Rothman 2004). Additional 18–24 h are required for biochemical characterization due to the bacterial cultivation (Tenover 2018). Collection circumstances and requirements for particular growing media, which produce errors and reduce sensitivity, are drawbacks of culturing techniques. Additionally, disease-causing microorganisms that cannot be cultured are not effectively studied using standard bacterial culture techniques (Chen et al. 2020; Lecuit and Eloit 2015). Because of this, the extremely sensitive and a molecule-specific method have also been used to identify resistance of bacteria (Li et al. 2017).

When compared to culture-based methods, molecular-based approaches have higher sensitivity, specificity, and turnaround times when amplifying or hybridizing genetic sequences particular resistance characteristics can be identified using traditional PCRs (polymerase chain reactions), RT-PCRs (quantitative real-time PCRs), or DNA microarrays (Maugeri et al. 2019; Hicks et al. 2019). However, in the presence of heteroresistance and low-gene abundance, these approaches require for detectable quantities of DNA. Methods that are not culturally reliant have varying clinical sensitivity (Young et al. 2019). Without the need for earlier culture enrichment, digital PCR methods quickly detect low-abundance targets and heteroresistance analyses (Boolchandani et al. 2019). Nucleic acid testing (NAT) approaches, which are culture-independent, provide faster diagnosis with higher sensitivity (Nilsson et al. 2008), but they also need advance knowledge of the pathogen being tested as well as its DNA sequences. The techniques incorporate molecular ones like amplification technologies include the Polymerase Chain Reaction, reverse transcriptase-PCR, nucleic acid sequence-based amplification (NASNBA), loop-mediated isothermal amplification (LAMP), strand displacement amplification (SDA), and transcription-mediated amplification (TMA) (Niemz et al. 2011). Extremely PCR with multiplexing panels can find bacterial infections that frequently produce clinically distinct symptoms at the same time. The additional unintended consequences techniques include next-generation sequencing (NGS) technologies (Boers et al. 2019; Khodakov et al. 2016). This, when paired with bioinformatics, results in allow for reliable pathogen detection and characterization as well as the prediction of strains that will escape vaccines (Maljkovic Berry et al. 2019).

Accurate information on the makeup of microbial communities that are not cultivable is provided by metagenomics NGS (mNGS) (Goldberg et al. 2015).

Clinical microbiology has also concentrated on using Aspartate aminotransferase (AST) with genomics. The main genomic method, known as whole-genome sequencing (WGS), accurately and consistently predicts strains of all common resistance traits. By screening numerous loci, it ensures the concurrent detection phenotypes of antibiotic resistance throughout the entire genome. False-negative results are reduced by the digital storage of genome sequence data, which is independent of primer specificity (Su et al. 2019). Antimicrobial resistance determinants are easily recognized by whole-genome and NGS technologies due to the abundance of publicly available data (Pereckaite et al. 2018). It quickly detects antibiotic resistance as a primer-independent technique, but it can only identify known pathways (McDermott et al. 2016; Kalpana et al. 2023).

5.3.1 Molecular Diagnostic Techniques Used for Detection

The identification of bacterial microorganisms and their associated antimicrobial resistance (AMR) genes has been difficulty, due to the lack of quick diagnostic methods. In clinical scenarios, this challenge has often been addressed by resorting to broad-spectrum antibiotics (Gerace et al. 2022). Rapid detection of bacterial microorganisms is essential to promptly initiate timely and robust antibiotic treatment, preventing the progression of persistent infections that could result in complications like wound infections, pneumonia, and bloodstream infections from catheters, ultimately culminating in sepsis. As a result, there is an urgent requirement for speedy and highly sensitive molecular techniques to swiftly identify virulent drug-resistant bacteria in clinical samples (Gerace et al. 2022).

Historically, microbial identification has relied on phenotypic techniques, involving the utilization of specialized culture media to isolate pathogens, automatic biochemical testing, and antimicrobial susceptibility assessments to develop treatment (Liu et al. 2023). However, this conventional approach has limitations such as the requirement for media tailored to specific microbes, extended incubation periods for cultivation, and delayed results acquisition (Gerace et al. 2022). In contrast, molecular diagnostics enable the direct detection of microbes in clinical samples, reducing the potential exposure of laboratory personnel to infectious agents. The evolution of technology coupled with advancements in bioinformatics tools has driven the creation of novel methodologies that have revolutionized the landscape of diagnosing bacterial infections (Gerace et al. 2022). Table 5.1 shows various nucleic acid-based methods for ESKAPE pathogen detection (Table 5.1).

5.3.2 Polymerase Chain Reaction (PCR)

Molecular diagnostic techniques like nucleic acid amplification and PCR are progressively replacing culture-based bacterial identification methods (Renner et al.

Table 5.1 Various nucleic acid methods used for detection of ESKAPE pathogens

Technique	Description	Advantages	Limitations	References
Polymerase chain reaction (PCR)	Amplifies DNA segments for pathogen identification	High sensitivity and specificity	Requires thermal cycling equipment	Gerace et al. (2022); Renner et al. (2017)
Recombinase polymerase amplification (RPA)	Isothermal amplification method for DNA	Rapid, sensitive, and low-temperature	Reagent costs and assay complexity	Gerace et al. (2022); Renner et al. (2017)
Real-time PCR	Sensitive and specific method for quick diagnosis	Quick results with fluorescent probes	Limited target options	Gerace et al. (2022)
BioFire FilmArray panels	In vitro diagnostic tool for nucleic acid analysis	Comprehensive panels for various infections	Cost of the equipment and reagents	Gerace et al. (2022)
DNA microarrays	High multiplexing for simultaneous gene detection	Valuable for investigating genetic AMR	High costs and method intricacies	Gerace et al. (2022); Renner et al. (2017)
Pulsed field gel electrophoresis	Separates DNA fragments for epidemiological studies	Replaced by more advanced methods	Labor-intensive and costly	Renner et al. (2017)
Multilocus sequence typing (MLST)	Uses conserved genes for bacterial isolate differentiation	High cost and limited discrimination	Suitable for epidemiological use	Gerace et al. (2022)
Pyrosequencing	Measures pyrophosphate release during nucleotide incorporation	Limited sequence length	Requires skilled technicians	Renner et al. (2017)
Whole-genome sequencing	Determines the complete nucleotide sequence of a microorganism	Useful for various applications	Clinical uptake is gradual	Renner et al. (2017)
Next-generation sequencing (NGS)	Swift, economical, and parallel sequencing technique	Proficient for clinical microbiology	Variability in NGS platforms	Renner et al. (2017)
Microfluidics	Lab-on-a-chip technique for detecting antibiotic-resistant bacteria	Affordable and automated	Reliance on costly microscopes	Gao et al. (2022)
Proteomic analysis	Measures protein levels for functional insights	Reflects cellular functions accurately	Technique complexity and cost	Leung et al. (2017)

(continued)

Table 5.1 (continued)

Technique	Description	Advantages	Limitations	References
Isothermal nucleic acid amplification	Rapidly multiplies nucleic acids at a constant temperature	Eliminates the need for thermal cycling	Primer design challenges for LAMP	Renner et al. (2017)
Nucleic acid sequence-based amplification (NASBA)	Isothermal RNA amplification technique	High specificity and sensitivity	Costly and in research stage	Renner et al. (2017); Gao et al. (2022)
Loop-mediated amplification (LAMP)	Isothermal DNA amplification method with high sensitivity	Exceptional sensitivity and cost-effective	Aerosol contamination risk	Gao et al. (2022)
Helicase-dependent amplification (HDA)	Mimics natural DNA replication process isothermally	Useful for detecting specific pathogens	Limited to short-length DNA fragments	Gao et al. (2022)
Recombinase polymerase amplification (RPA)	Isothermal amplification technique	Rapid and sensitive, low-temperature	Primer design challenges	Gerace et al. (2022)
Rolling circle amplification (RCA)	Catalyzed by DNA polymerase at a constant temperature	High specificity for pathogenic microorganisms	Requires specific reaction conditions	Liu et al. (2023)

2017). While PCR-based point-of-care diagnostic tools have been limited by thermocycling requirements, isothermal techniques for nucleic acid amplification and identification have addressed these challenges, making point-of-care assays feasible (Gerace et al. 2022). Recombinase polymerase amplification (RPA) is an isothermal method utilizing a recombinase enzyme to guide primer binding and amplify target DNA through strand displacement. RPA can detect even ~10 DNA target copies at a constant low temperature (~42 °C), minimizing power usage (Renner et al. 2017). RPA has been successful in detecting various pathogens, including *Francisella tularensis*, *Plasmodium falciparum, Mycobacterium tuberculosis*, yellow fever virus, and HIV. Challenges to implementing RPA at the point of care include reagent costs, simplifying assay steps, and the availability of components for quantification, result communication, and therapeutic recommendations (Gerace et al. 2022).

Real-time PCR, a powerful technique derived from PCR, is widely used for its sensitivity, specificity, and quick diagnosis of infectious diseases (Gerace et al. 2022). It enables quick capturing of target amplification by the use of either nonspecific fluorescent dyes or particular labeled probes (Liu et al. 2023). The fluorescent signal is developed upon probe-target hybridization, proportional to the PCR amplicon count (Gerace et al. 2022). Despite available commercial kits for direct pathogen and resistance gene detection from biological samples, inconsistent results may arise due to factors like gene rearrangement and horizontal gene transfer (Renner et al. 2017). Due to more number of β-lactamase-encoding genes, the diagnosis of resistance to β-lactam antibiotics in Gram "−ve" bacteria is difficult (Gerace et al. 2022).

BioFire FilmArray Panels serves as an in vitro diagnostic testing tool that combines isolation of nucleic acid, amplification, and transcription techniques, and subsequent melting curve analysis. It is purpose-built to be used alongside comprehensive panels targeting a range of infections, including respiratory viruses, pneumonia, bloodstream infections, gastrointestinal infections, and meningitis encephalitis, along with genes responsible for antimicrobial resistance. The FilmArray offering encompasses Gastrointestinal, Pneumonia, Meningitis Encephalitis, and Blood Culture Identification panels (Gerace et al. 2022). For instance, the FilmArray Gastrointestinal panel identifies 22 prevalent pathogens linked to gastrointestinal symptoms, the major cause for child mortality in less developed countries. The Pneumonia panel detects seven genetic markers indicating antibiotic resistance and the infections caused in lower respiratory tract, Pneumonia panel detects 27 commonly found pathogens. Aligning with sepsis management guidelines, prompt identification of pathogens enables targeted antibiotic therapy. The FilmArray Meningitis Encephalitis panel tests cerebral spinal fluid for 14 common pathogens linked to community-acquired meningitis, crucial for rapid diagnosis due to the condition's swift progression and severity (Gerace et al. 2022).

5.3.3 DNA Microarray

PCR techniques are widely accepted for identifying specific genes, though not suitable for broad gene detection, such as those tied to antimicrobial resistance (AMR) (Renner et al. 2017). Microarrays, in contrast, excel at high multiplexing, enabling simultaneous detection of numerous genes, making them valuable for investigating genetic AMR in bacteria. This technology employs oligonucleotide sequences to hybridize and multiply few molecules in the sample, facilitating the detection of AMR genes like ESBLs and carbapenemases (Gerace et al. 2022). ESKAPE pathogens may harbor multiple resistance genes, necessitating swift identification to curb multiresistant strain propagation. Plasmid-borne ESBL genes can transfer readily between enteric Gram−ve bacteria, common causes of urinary tract infections (Renner et al. 2017). Another imperative category is vancomycin resistance genes that are carried by, implicated in vancomycin-resistant S. aureus (VRSA), often resistant to diverse antibiotics (Gerace et al. 2022). Microarrays' capability to identify numerous multidrug resistance genes in one strain helps doctors in targeting antimicrobial therapy[1]. Nonetheless, limited clinical implementation of microarrays results from high costs and method intricacies (Renner et al. 2017).

Numerous methods exist for analyzing ESBL and carbapenemase genes as well as for epidemiological investigations (Renner et al. 2017). These include

Pulsed field gel electrophoresis separates DNA fragments after enzyme digestion using an electric field on a gel matrix. Yet, PFGE has been replaced by superior methods like NGS and MALDI due to their enhanced performance. MLST is an allele-based system using conserved housekeeping genes to distinguish bacterial isolates, yielding sequence types (ST) without whole-genome sequencing (Renner et al. 2017). MLST's high cost and limited discrimination make it suitable for epidemiological use (Gerace et al. 2022). Pyrosequencing employs bioluminescence to measure inorganic pyrophosphate release upon nucleotide incorporation. However, its constraint is short sequence length. Generally, these methods suffer from high costs and the need for skilled technicians to perform them (Renner et al. 2017).

Whole-Genome Sequencing involves determining a microorganism's complete nucleotide sequence in a single test. WGS has found success in diverse areas such as pinpointing pathogen virulence factors, tracing disease transmission during outbreaks, antimicrobial resistance profiling, and uncovering origin of repeating infections and transmission between patients. Even though with its significant potential, clinical uptake of WGS remains gradual. However, the introduction of cost-effective benchtop sequencing platforms might accelerate the integration of bacterial WGS into clinical practice (Renner et al. 2017).

Next-Generation Sequencing (NGS) stands out as a swift, economical, and extensively parallel sequencing technique. With a focus on the exome, it emerges as a pragmatic alternative to complete genome sequencing. Particularly in clinical microbiology, second-generation NGS is preferred due to its proficient short-read technique and cost friendly. The NGS workflow for second-generation platforms entails library and template preparation, succeeded by sequencing. Disparities in NGS platforms arise from these distinct methods and associated technologies like sequencing by hybridization and synthesis. These two methods have demonstrated success in scrutinizing outbreaks and delineating resistance genes in ESKAPE pathogens (Renner et al. 2017).

Microfluidics is the "lab-on-a-chip" technique that holds significant promise for detecting antibiotic-resistant bacteria (Liu et al. 2023). In contrast to conventional larger-scale methods, microfluidics boasts advantages like affordability, swifts high-range analysis, sample volume is less, and automated. Microfluidic-based detection approaches fall into two categories: genotypic and phenotypic assays (Gao et al. 2022). PCR, LAMP are the examples of genotypic on-chip methods that target genetic marker like 16S rRNA genes, enabling swift bacterial identification without the need for bacterial growth (Renner et al. 2017). However, these assays are not suitable for determining bacterial antibiotic susceptibility profiles (Renner et al. 2017). Conversely, phenotypic on-chip assays monitor bacterial growth when exposed to antibiotics, yielding precise antibiotic susceptibility testing (AST) outcomes (Gao et al. 2022). Bacterial cells are often confined within chambers, droplets, or agarose-containing compartments, either monitored by antibodies on membranes or magnetic beads. Drawbacks include reliance on costly microscopes and limited antibodies targeting diverse bacterial strains. Overcoming these drawbacks is crucial for making these microfluidic platforms commercially viable (Gao et al. 2022).

Proteomic analysis provides insights that hold greater functional and clinical relevance compared to genetic/genomic testing, as protein levels accurately mirror cellular functions. Bacterial proteomics embraces methods with gel and without gel

(Leung et al. 2017). Initial steps involve employing 2D gel electrophoresis, followed by differential in gel electrophoresis (DIGE) utilizing fluorescent dyes for further examination (Renner et al. 2017). Enhanced technologies permit direct quantification of proteins and assessment of their functional status through mass spectrometry analysis (Leung et al. 2017).

Peptides undergo labeling as tandem mass tags (TMTs) with the help of isobaric mass tags to enable quantification. The quantification of relative protein levels across diverse sources is done by the method called isobaric tags for relative and absolute quantitation (iTRAQ) with a single experiment (Leung et al. 2017). After protein digestion, peptides are covalently attached to isobaric tags, samples are combined, followed by fractionation using LC, and subjected to analysis through tandem mass spectrometry. The combination of iTRAQ with MALDI or ESI-MS/MS accurately furnishes the data regarding relative protein concentrations. This comprehensive strategy illuminates functional facets of bacterial proteomics and broadens our understanding of cellular behavior (Leung et al. 2017).

5.4 Nucleic Acid Amplification and Importance in Diagnosis of ESKAPE Pathogens

Isothermal nucleic acid amplification is a technique in nucleic acid amplification technologies that rapidly multiplies nucleic acids by using specific enzymes and primers at a constant temperature (Renner et al. 2017). Developed since the late twentieth century, different isothermal methods offer simplicity, speed, and efficiency, eliminating the need for complex thermal cycling equipment (Gao et al. 2022). This lowers the demands on the environment and equipment, pointing toward isothermal amplification as a viable substitute for PCR. Nucleic acid sequence-based amplification (NASBA), Loop-mediated amplification (LAMP), recombinase polymerase amplification (RPA), rolling circle amplification (RCA), and helicase-dependent amplification (HDA) are the isothermal technologies that are used widely (Gao et al. 2022).

Notomi et al. in 2000 has introduced the LAMP, necessitating the design of four to six primers for the amplification process (Gao et al. 2022). Operating within a temperature range of 60–65 °C, Bst DNA polymerase facilitates the elongation of primers with the template while catalyzing strand displacement for approximately 1 h, resulting in target repeat fragments of varying lengths (Gao et al. 2022). ELISA, gel electrophoresis, real-time turbidimetry, lateral flow test strips, and fluorescent probes can be used for the identification of amplified products (Renner et al. 2017). Over the past couple of decades, LAMP has gained considerable popularity in the development of point-of-care testing due to its exceptional sensitivity minimal instrument prerequisites, and the convenience of interpreting results through visual cues such as turbidity or fluorescence (Gao et al. 2022). Yet, LAMP has drawbacks. Aerosol contamination during amplification can lead to false-positive results. Challenging primer design for certain pathogen gene sequences and varied fragment lengths in the end product restrict downstream application possibilities

(Liu et al. 2023). As a result, these limitations hinder the widespread adoption and practical use of LAMP (Liu et al. 2023).

In 1991, Compton has introduced an in vitro nucleic acid amplification technique known as NASBA (Gao et al. 2022). In this process, At 65 °C for 5 min, the template is heated to eliminate RNA secondary structures, followed by a constant 42 °C amplification reaction in a standard setup (Gao et al. 2022). The initial primer conducts reverse transcription and attaches a T7 promoter to generate cDNA (Liu et al. 2023). RNase H degradation removes RNA from the newly formed heteroduplex, permitting the second DNA primer to hybridize (Liu et al. 2023). Reverse transcriptase (RT) extension forms double-stranded DNA (dsDNA), further transcribed by T7 RNA polymerase, leading to exponential amplification (Renner et al. 2017). Detection methods involve electrophoresis, ELISA, or the molecular beacon approach. NASBA offers high specificity and sensitivity (Gao et al. 2022). It minimizes contamination during amplification and RNA virus detection remains unaffected by template DNA interference (Renner et al. 2017). It is valuable for identifying RNA viruses like respiratory viruses and hepatitis A viruses (Renner et al. 2017). Yet, NASBA is relatively costly and mostly remains in the research stage. Achieving true constant amplification temperature is challenging due to preheating requirements before testing (Gao et al. 2022).

Drawbacks of RT-PCR encompass limited target options because of dye limitations, decreased sensitivity and specificity with increased multiplexing, and the costly maintenance of instruments. Progress in RT-PCR has prompted the emergence of isothermal amplification methods, such as LAMP, HDA, NASBA, and TMA, as alternatives (Gao et al. 2022).

LAMP, a cost-effective diagnostic test, stands out (Renner et al. 2017). It is equipped with high-strand displacement activity DNA polymerase and distinct regions which can be recognized by four primers, forming a stem loop DNA structure for immediate amplification (Renner et al. 2017). Detection involves a DNA-binding dye. Compared with conventional PCR, LAMP can produce nearly 109 copies within 1 h which makes it more sensitive (Gao et al. 2022). LAMP kits that are available in the market target resistance genes such as blaKPC and blaNDM-1 in *K. pneumonia* and *A. baumannii (*Gao et al. 2022*)*.

The isothermal amplification reactions that use mRNA as target sequences are NASBA and TMA. These are gold standards for diagnosing gonorrhea and chlamydial infection (Gao et al. 2022). HDA employs helicase's unwinding activity to divide dsDNA into single strands that helps as templates for new DNA synthesis (Liu et al. 2023). SSBs maintain single-strand templates, preventing double helix reconstitution. HDA successfully detects *Clostridium difficile*, *S. aureus*, and *N. gonorrhoeae* (Gao et al. 2022).

In essence, NAATs, including PCR variants and isothermal methods like LAMP, NASBA, TMA, and HDA, have revolutionized pathogen detection (Renner et al. 2017). These techniques address challenges in rapid identification of bacterial pathogens and their resistance mechanisms, significantly impacting the management of infectious diseases (Gao et al. 2022).

In 2006, Piepenburget et al. introduced the RPA technique (Gao et al. 2022). RPA stands for an isothermal amplification technique which makes use of constituents like recombinase T4 uvsX, polymerase Bsu, single-strand-binding protein gp32, and a pair of primers. While recombinase binds to the primer via similar sequences in ds DNA, it triggers DNA strand exchange (Gao et al. 2022). Subsequently, polymerase initiates primer-directed DNA synthesis, leading to accelerated duplication of the specific sites on the template. The displaced DNA strand binds to ssDNA-binding protein, preventing another replacement (Renner et al. 2017). Notably, RPA eliminates the need for thermal template denaturation and generates amplified products within 30 min at temperatures of 37–42 °C. Methods like real-time detection using fluorescent probes, lateral flow chromatography, and like agarose gel electrophoresis are used to obtained analytical results (Renner et al. 2017).

RPA's rapidity, high sensitivity, and mild constant-temperature conditions contribute to its rapid growth as a leading isothermal amplification technique (Gerace et al. 2022). Despite limitations such as longer primer probes, primer design challenges, and enzyme storage restrictions, RPA is viewed as a promising alternative to PCR in the realm of isothermal nucleic acid amplification techniques (Gerace et al. 2022).

At a constant temperature, RCA is an amplification process catalyzed by a DNA polymerase along with strand displacement activity. It involves hybridizing a padlock probe with the target sequence via DNA ligase, followed by ligation into a circular template (Renner et al. 2017). DNA polymerase helps in extending the primer aligned with the circular template along the loop, continuously replacing previous extension chains to generate recurring long ssDNA products (Renner et al. 2017). RCA's advantages include single reaction primers and high specificity for pathogenic microorganism detection. However, RCA's reliance on DNA ligase demands a specific reaction system and its high cost and time-consuming process hinder cost-effective and rapid detection (Liu et al. 2023). The detectable threshold gets disturbed by the huge background signals generated from the padlock probe and DNA template present in the reaction system. Given its status as an isothermal amplification approach, RCA necessitates enhancements in different points like probe configuration, mitigation of background interference, cost-effectiveness, and the duration of the reaction process (Liu et al. 2023).

HDA is a nucleic acid amplification technique which mimics the natural DNA replication process. It has isothermal property at 65 °C and employs DNA helicase to convert double-stranded DNAs into single-stranded DNAs (Liu et al. 2023). Target primers (P1 and P2) attach to target sequences, forming partial DNA duplexes that are extended by DNA polymerase. These newly formed DNA duplexes are then unraveled by a heat-resistant helicase, initiating the subsequent amplification cycle and achieving isothermal DNA amplification. HDA is limited to short-length DNA fragment amplification and presents drawbacks like slow amplification (60–120 min), restricting its point-of-care testing applicability (Gao et al. 2022).

In the last 20 years, there has been remarkable progress in the advancement of different techniques in isothermal nucleic acid amplification. These developments aim to enable the sensitive, uncomplicated, and swift to recognize the pathogenic nucleic acids (Liu et al. 2023).

References

Abdi SN, Ghotaslou R, Ganbarov K, Mobed A, Tanomand A, Yousefi M, Asgharzadeh M, Kafil HS (2020) Acinetobacter baumannii efflux pumps and antibiotic resistance. Infect Drug Resist 13:423–434

AhYoung AP, Koehl A, Cascio D, Egea PF (2015) Structural mapping of the ClpB ATPases of Plasmodium falciparum: targeting protein folding and secretion for antimalarial drug design. Protein Sci 24:1508–1520

Bagińska N, Cieślik M, Górski A, Jończyk-Matysiak E (2021) The role of antibiotic resistant A. baumannii in the pathogenesis of urinary tract infection and the potential of its treatment with the use of bacteriophage therapy. Antibiotics 10:281

Bagley ST (1985) Habitat association of Klebsiella species. Infect Control 6:52–58

Band VI, Crispell EK, Napier BA, Herrera CM, Tharp GK, Vavikolanu K, Pohl J, Read TD, Bosinger SE, Trent MS (2016) Antibiotic failure mediated by a resistant subpopulation in Enterobacter cloacae. Nat Microbiol 1:16053

Basatian-Tashkan B, Niakan M, Khaledi M, Afkhami H, Sameni F, Bakhti S, Mirnejad R (2020) Antibiotic resistance assessment of Acinetobacter baumannii isolates from Tehran hospitals due to the presence of efflux pumps encoding genes (adeA and adeS genes) by molecular method. BMC Res Notes 13:543

Basset P, Feil EJ, Zanetti G, Blanc DS (2011) Chapter 25—The evolution and dynamics of methicillin-resistant Staphylococcus aureus. In: Tibayrenc M (ed) Genetics and evolution of infectious disease. Elsevier, London, pp 669–688. https://doi.org/10.1016/B978-0-12-384890-1.00025-X

Baumann P, Doudoroff M, Stanier RY (1968) A study of the Moraxella group. II. Oxidative-negative species (genus Acinetobacter). J Bacteriol 95:1520–1541

Beceiro A, Dominguez L, Ribera A, Vila J, Molina F, Villanueva R, Eiros JM, Bou G (2004) Molecular characterization of the gene encoding a new AmpC β-lactamase in a clinical strain of Acinetobacter genomic species 3. Antimicrob Agents Chemother 48:1374–1378

Berrazeg M, Jeannot K, Enguéné VYN, Broutin I, Loeffert S, Fournier D, Plésiat P (2015) Mutations in β-Lactamase AmpC increase resistance of Pseudomonas aeruginosa isolates to antipseudomonal cephalosporins. Antimicrob Agents Chemother 59:6248–6255

Boers SA, Jansen R, Hays J (2019) Understanding and overcoming the pitfalls and biases of next-generation sequencing (NGS) methods for use in the routine clinical microbiological diagnostic laboratory. Eur J Clin Microbiol Infect Dis 38:1059–1070

Boolchandani M, D'Souza AW, Dantas G (2019) Sequencing-based methods and resources to study antimicrobial resistance. Nat Rev Genet 20:356–370

Cetinkaya Y, Falk P, Mayhall CG (2000) Vancomycin-resistant Enterococci. Clin Microbiol Rev 13:686–707

Chen C-Y, Clark CG, Langner S, Boyd DA, Bharat A, McCorrister SJ, McArthur AG, Graham MR, Westmacott GR, Van Domselaar G (2020) Detection of antimicrobial resistance using proteomics and the comprehensive antibiotic resistance database: a case study. Proteomics Clin Appl 14:1800182

Chen L, Tan P, Zeng J, Yu X, Cai Y, Liao K, Guo P, Chen Y, Wu Z, Qu P et al (2021) Impact of an intervention to control imipenem-resistant Acinetobacter baumannii and its resistance mechanisms: an 8-year survey. Front Microbiol 11:610109

Chevalier J, Mulfinger C, Garnotel E, Nicolas P, Davin-Régli A, Pagès JM (2008) Identification and evolution of drug efflux pump in clinical Enterobacter aerogenes strains isolated in 1995 and 2003. PLoS One 3:e3203. https://doi.org/10.1371/journal.pone.0003203

Cong Y, Yang S, Rao X (2020) Vancomycin resistant Staphylococcus aureus infections: a review of case updating and clinical features. J Adv Res 21:169–176
Dao TT, Liebenthal D, Tran TK, Ngoc Thi VB, Ngoc Thi Nguyen D, Thi Tran HK, Thi Nguyen CK, Thi Vu HL, Fox A, Horby P, Van Nguyen K, Wertheim HFL (2014) Klebsiella pneumoniae oropharyngeal carriage in rural and urban Vietnam and the effect of alcohol consumption. PLoS One 9:e91999. https://doi.org/10.1371/journal.pone.0091999
De Oliveira DMP, Forde BM, Kidd TJ, Harris PNA, Schembri MA, Beatson SA, Paterson DL, Walker MJ (2020) Antimicrobial resistance in ESKAPE pathogens. Clin Microbiol Rev 33:181
Dehbashi S, Tahmasebi H, Alikhani MY, Keramat F, Arabestani MR (2020) Distribution of class B and class A β-lactamases in clinical strains of Pseudomonas aeruginosa: comparison of phenotypic methods and high-resolution melting analysis (HRMA) assay. Infect Drug Resist 13:2037–2052
Di Franco S, Alfieri A, Pace M, Sansone P, Pota V, Fittipaldi C, Fiore M, Passavanti M (2021) Blood stream infections from MDR bacteria. Life 11:575
Effah CY, Sun T, Liu S, Wu Y (2020) Klebsiella pneumoniae: an increasing threat to public health. Ann Clin Microbiol Antimicrob 19:1–9
Eghbalpoor F, Habibi M, Azizi O, Karam MRA, Bouzari S (2019) Antibiotic resistance, virulence and genetic diversity of Klebsiella pneumoniae in community- and hospital-acquired urinary tract infections in Iran. Acta Microbiol Immunol Hung 66:349–366
Elsner H-A, Sobottka I, Mack D, Claussen M, Laufs R, Wirth R (2000) Virulence factors of *Enterococcus faecalis* and *Enterococcus faecium* blood culture isolates. Eur J Clin Microbiol Infect Dis 19(1):39–42. https://doi.org/10.1007/s100960050007
Farmer JJ III, Davis BR, Hickman-Brenner FW, McWorther A, HuntleyCarter GP, Asbury MA et al (1985) Biochemical identification of new species and biogroups of Enterobacteriaceae isolated from clinical specimens. J Clin Microbiol 21:46–76
Gao D, Guo X, Yang Y et al (2022) Microfluidic chip and isothermal amplification technologies for the detection of pathogenic nucleic acid. J Biol Eng 16:33. https://doi.org/10.1186/s13036-022-00312-w
García-Solache M, Rice LB (2019) The Enterococcus: a model of adaptability to its environment. Clin Microbiol Rev 32:e00058–e00018
Gerace E, Mancuso G, Midiri A, Poidomani S, Zummo S, Biondo C (2022) Recent advances in the use of molecular methods for the diagnosis of bacterial infections. Pathogens 11(6):663. https://doi.org/10.3390/pathogens11060663
Gimza BD, Cassat JE (2021) Mechanisms of antibiotic failure during Staphylococcus aureus osteomyelitis. Front Immunol 12:638085
Goldberg B, Sichtig H, Geyer C, Ledeboer N, Weinstock GM (2015) Making the leap from research laboratory to clinic: challenges and opportunities for next-generation sequencing in infectious disease diagnostics. MBio 6:e01888–e01815
Gualtero S, Valderrama S, Valencia M, Rueda D, Muñoz-Velandia O, Ariza B, Cortes G, Salgado D, Porras Y, Niño A (2020) Factors associated with mortality in infections caused by carbapenem-resistant Enterobacteriaceae. J Infect Dev Ctries 14:654–659
Guo Y, Song G, Sun M, Wang J, Wang Y (2020) Prevalence and therapies of antibiotic-resistance in Staphylococcus aureus. Front Cell Infect Microbiol 10:107
Halat DH, Moubareck CA (2020) The current burden of carbapenemases: review of significant properties and dissemination among gram-negative bacteria. Antibiotics 9:186
Harding CM, Hennon SW, Feldman MF (2018) Uncovering the mechanisms of Acinetobacter baumannii virulence. Nat Rev Genet 16:91–102
Henrichfreise B, Wiegand I, Pfister W, Wiedemann B (2007) Resistance mechanisms of multiresistant Pseudomonas aeruginosa strains from Germany and correlation with hypermutation. Antimicrob Agents Chemother 51:4062–4070
Hicks AL, Wheeler N, Sánchez-Busó L, Rakeman JL, Harris SR, Grad YH (2019) Evaluation of parameters affecting performance and reliability of machine learning-based antibiotic susceptibility testing from whole genome sequencing data. PLoS Comput Biol 15:e1007349

Hogan S, Stevens NT, Humphreys H et al (2015) Current and future approaches to the prevention and treatment of staphylococcal medical device-related infections. Curr Pharm Des 21:100–113
Hwang W, Yoon SS (2019) Virulence characteristics and an action mode of antibiotic resistance in multidrug-resistant Pseudomonas aeruginosa. Sci Rep 9:1–15
Inweregbu K, Dave J, Pittard A (2005) Nosocomial infections. Contin Educ Anaesth Crit Care Pain 5:14–17
Izdebski R, Baraniak A, Herda M, Fiett J, Bonten MJ, Carmeli Y et al (2014) MLST reveals potentially high-risk international clones of Enterobacter cloacae. J Antimicrob Chemother 70:48–56. https://doi.org/10.1093/jac/dku359
Jondle CN, Gupta K, Mishra BB, Sharma J (2018) Klebsiella pneumoniae infection of murine neutrophils impairs their efferocytic clearance by modulating cell death machinery. PLoS Pathog 14(10):e1007338
Jubeh B, Breijyeh Z, Karaman R (2020) Resistance of gram-positive bacteria to current antibacterial agents and overcoming approaches. Molecules 25:2888
Jurado-Martín I, Sainz-Mejías M, McClean S (2021) Pseudomonas aeruginosa: an audacious pathogen with an adaptable arsenal of virulence factors. Int J Mol Sci 22:3128
Kalpana S, Lin W-Y, Wang Y-C, Fu Y, Lakshmi A, Wang H-Y (2023) Antibiotic resistance diagnosis in ESKAPE pathogens—a review on proteomic perspective. Diagnostics 13:1014. https://doi.org/10.3390/diagnostics13061014
Karaiskos I, Lagou S, Pontikis K, Rapti V, Poulakou G (2019) The "old" and the "new" antibiotics for MDR gram-negative pathogens: for whom, when, and how. Front Public Health 7:151
Kest H, Kaushik A (2019) Vancomycin-resistant Staphylococcus aureus: formidable threat or silence before the storm? Infect Dis Epidemiol 5:93
Khodakov D, Wang C, Zhang DY (2016) Diagnostics based on nucleic acid sequence variant profiling: PCR, hybridization, and NGS approaches. Adv Drug Deliv Rev 105:3–19
Kim M, Moon D, Kim S-J, Mechesso A, Song H-J, Kang H, Choi J-H, Yoon S-S, Lim S-K (2021) Nationwide surveillance on antimicrobial resistance profiles of Enterococcus faecium and Enterococcus faecalis isolated from healthy food animals in South Korea, 2010 to 2019. Microorganisms 9:925
Konca C, Tekin M, Geyik M (2021) Susceptibility patterns of multidrug-resistant Acinetobacter baumannii. Indian J Pediatr 88:120–126
Kyriakidis I, Vasileiou E, Pana Z, Tragiannidis A (2021) Acinetobacter baumannii antibiotic resistance mechanisms. Pathogens 10:373
Langendonk RF, Neill DR, Fothergill JL (2021) The building blocks of antimicrobial resistance in Pseudomonas aeruginosa: implications for current resistance-breaking therapies. Front Cell Infect Microbiol 11:665759
Lasko MJ, Nicolau DP (2020) Carbapenem-resistant Enterobacterales: considerations for treatment in the era of new antimicrobials and evolving enzymology. Curr Infect Dis Rep 22:1–12
Lecuit M, Eloit M (2015) The potential of whole genome NGS for infectious disease diagnosis. Expert Rev Mol Diagn 15:1517–1519
Lee AS, De Lencastre H, Garau J, Kluytmans J, Malhotra-Kumar S, Peschel A, Harbarth S (2018) Methicillin-resistant Staphylococcus aureus. Nat Rev Dis Primers 4:18033
Leung LM, Fondrie WE, Doi Y et al (2017) Identification of the ESKAPE pathogens by mass spectrometric analysis of microbial membrane glycolipids. Sci Rep 7:6403. https://doi.org/10.1038/s41598-017-04793-4
Li Y, Yang X, Zhao W (2017) Emerging microtechnologies and automated systems for rapid bacterial identification and antibiotic susceptibility testing. SLAS Technol 22:585–608
Liu Q, Jin X, Cheng J, Zhou H, Zhang Y, Dai Y (2023) Advances in the application of molecular diagnostic techniques for the detection of infectious disease pathogens (Review). Mol Med Rep 27(5):104. https://doi.org/10.3892/mmr.2023.12991
Luzzago C, Locatelli C, Franco A, Scaccabarozzi L, Gualdi V, Vigano R, Sironi G, Besozzi M, Castiglioni B, Lanfranchi P, Cremonesi P (2014) Clonal diversity, virulence-associated genes and antimicrobial resistance profile of Staphylococcus aureus isolates from nasal cavities and

soft tissue infections in wild ruminants in Italian Alps. Vet Microbiol 170(1):157–161. https://doi.org/10.1016/j.vetmic.2014.01.016
Maljkovic Berry I, Melendrez MC, Bishop-Lilly KA, Rutvisuttinunt W, Pollett S, Talundzic E, Morton L, Jarman RG (2019) Next-generation sequencing and bioinformatics methodologies for infectious disease research and public health: approaches, applications, and considerations for development of laboratory capacity. J Infect Dis 221:S292–S307
Mancuso G, Midiri A, Gerace E, Biondo C (2021) Bacterial antibiotic resistance: the most critical pathogens. Pathogens 10:1310. https://doi.org/10.3390/pathogens10101310
Maugeri G, Lychko I, Sobral R, Roque ACA (2019) Identification and antibiotic-susceptibility profiling of infectious bacterial agents: a review of current and future trends. Biotechnol J 14:1700750
McDermott PF, Tyson GH, Kabera C, Chen Y, Li C, Folster JP, Ayers SL, Lam C, Tate HP, Zhao S (2016) Whole-genome sequencing for detecting antimicrobial resistance in nontyphoidal salmonella. Antimicrob Agents Chemother 60:5515–5520
Mezzatesta ML, Gona F, Stefani S (2012) Enterobacter cloacae complex: clinical impact and emerging antibiotic resistance. Future Microbiol 7:887–902. https://doi.org/10.2217/fmb.12.61
Mojica MF, Bonomo RA, Fast W (2016) B1-Metallo-β-lactamases: where do we stand? Curr Drug Targets 17:1029–1050
Montefour K, Frieden J, Hurst S, Helmich C, Headley D, Martin M et al (2008) Acinetobacter baumannii: an emerging multidrug-resistant pathogen in critical care. Crit Care Nurse 28:15–25; quiz 26; PMID: 18238934
Montgomery CP, David MZ, Daum RS (2015) Host factors that contribute to recurrent staphylococcal skin infection. Curr Opin Infect Dis 28:253–258
Moradali MF, Ghods S, Rehm BHA (2017) Pseudomonas aeruginosa lifestyle: a paradigm for adaptation, survival, and persistence. Front Cell Infect Microbiol 7:39
Mühlberg E, Umstätter F, Kleist C, Domhan C, Mier W, Uhl P (2020) Renaissance of vancomycin: approaches for breaking antibiotic resistance in multidrug-resistant bacteria. Can J Microbiol 66:11–16
Mulani MS, Kamble E, Kumkar SN, Tawre MS, Pardesi KR (2019) Emerging strategies to combat ESKAPE pathogens in the era of antimicrobial resistance: a review. Front Microbiol 10:539
Munita JM, Arias C (2016) Mechanisms of antibiotic resistance. Microbiol Spectr 4:481–511
Nathwani D, Raman G, Sulham K, Gavaghan M, Menon V (2014) Clinical and economic consequences of hospital-acquired resistant and multidrug-resistant Pseudomonas aeruginosa infections: a systematic review and meta-analysis. Antimicrob Resist Infect Control 3:32
Nepka M, Perivolioti E, Kraniotaki E, Politi L, Tsakris A, Pournaras S (2016) In vitro bactericidal activity of trimethoprim-sulfamethoxazole alone and in combination with colistin against carbapenem-resistant Acinetobacter baumannii clinical isolates. Antimicrob Agents Chemother 60:6903–6906
Niemz A, Ferguson T, Boyle D (2011) Point-of-care nucleic acid testing for infectious diseases. Trends Biotechnol 29:240–250
Nilsson AC, Björkman P, Persson K (2008) Polymerase chain reaction is superior to serology for the diagnosis of acute Mycoplasma pneumoniae infection and reveals a high rate of persistent infection. BMC Microbiol 8:93
Ontong JC, Ozioma NF, Voravuthikunchai SP, Chusri S (2021) Synergistic antibacterial effects of colistin in combination with aminoglycoside, carbapenems, cephalosporins, fluoroquinolones, tetracyclines, fosfomycin, and piperacillin on multidrug resistant Klebsiella pneumoniae isolates. PLoS One 16:e0244673
Pachori P, Gothalwal R, Gandhi P (2019) Emergence of antibiotic resistance Pseudomonas aeruginosa in intensive care unit; a critical review. Genes Dis 6:109–119
Pacios O, Blasco L, Bleriot I, Fernandez-Garcia L, Bardanca MG, Ambroa A, López M, Bou G, Tomás M (2020) Strategies to combat multidrug-resistant and persistent infectious diseases. Antibiotics 9:65
Pang Z, Raudonis R, Glick BR, Lin T-J, Cheng Z (2019) Antibiotic resistance in Pseudomonas aeruginosa: mechanisms and alternative therapeutic strategies. Biotechnol Adv 37:177–192

Peérichon B, Courvalin P (2009) VanA-Type vancomycin-resistant Staphylococcus aureus. Antimicrob Agents Chemother 53:4580–4587

Peleg AY, Seifert H, Paterson DL (2008) *Acinetobacter baumannii:* emergence of a successful pathogen. Clin Microbiol Rev 21:538–582. https://doi.org/10.1128/CMR.00058-07

Pereckaite L, Tatarunas V, Giedraitiene A (2018) Current antimicrobial susceptibility testing for beta-lactamase-producing Enterobacteriaceae in clinical settings. J Microbiol Methods 152:154–164

Pérez-Varela M, Corral J, Aranda J, Barbé J (2019) Roles of efflux pumps from different superfamilies in the surface-associated motility and virulence of Acinetobacter baumannii ATCC 17978. Antimicrob Agents Chemother 63:e02190–e02118

Poirel L, Nordmann P (2006) Carbapenem resistance in Acinetobacter baumannii: mechanisms and epidemiology. Clin Microbiol Infect 12:826–836

Pöntinen AK, Top J, Arredondo-Alonso S, Tonkin-Hill G, Freitas AR, Novais C, Gladstone RA, Pesonen M, Meneses R, Pesonen H et al (2021) Apparent nosocomial adaptation of Enterococcus faecalis predates the modern hospital era. Nat Commun 12:1523

Ramirez MS, Nikolaidis N, Tolmasky ME (2013) Rise and dissemination of aminoglycoside resistance: the aac(6′)-Ib paradigm. Front Microbiol 4:121

Ramos GP, Rocha JL, Tuon FF (2013) Seasonal humidity may influence Pseudomonas aeruginosa hospital-acquired infection rates. Int J Infect Dis 17:e757–e761

Ramos S, Silva V, Dapkevicius M, Igrejas G, Poeta P (2020) Enterococci, from harmless bacteria to a pathogen. Microorganisms 8:1118

Recio R, Mancheño M, Viedma E, Villa J, Orellana MÁ, Lora-Tamayo J, Chaves F (2020) Predictors of mortality in bloodstream infections caused by Pseudomonas aeruginosa and impact of antimicrobial resistance and bacterial virulence. Antimicrob Agents Chemother 64:e01759–e01719

Reller LB, Weinstein M, Jorgensen JH, Ferraro MJ (2009) Antimicrobial susceptibility testing: a review of general principles and contemporary practices. Clin Infect Dis 49:1749–1755

Remold SK, Brown CK, Farris JE, Hundley TC, Perpich JA, Purdy ME (2011) Differential habitat use and niche partitioning by Pseudomonas species in human homes. Microb Ecol 62:505–517

Renner LD, Zan J, Hu LI, Martinez M, Resto PJ, Siegel AC, Torres C, Hall SB, Slezak TR, Nguyen TH, Weibel DB (2017) Detection of ESKAPE bacterial pathogens at the point of care using isothermal DNA-based assays in a portable degas-actuated microfluidic diagnostic assay platform. Appl Environ Microbiol 83(4):e02449–e02416. https://doi.org/10.1128/AEM.02449-16

Rice LB (2008) Federal funding for the study of antimicrobial resistance in nosocomial pathogens: no ESKAPE. J Infect Dis 197(8):1079–1081. https://doi.org/10.1086/533452

Rice LB (2010) Progress and challenges in implementing the research on ESKAPE pathogens. Infect Control Hosp Epidemiol 31(Suppl. 1):S7–S10

Richards MJ, Edwards JR, Culver DH et al (1999) Nosocomial infections in medical intensive care units in the United States. National Nosocomial Infections Surveillance System. Crit Care Med 27:887–892

Roca I, Espinal P, Vila-Farrés X, Vila J (2012) The Acinetobacter baumannii oxymoron: commensal hospital dweller turned pan-drug-resistant menace. Front Microbiol 3:148

Sader HS, Huband MD, Castanheira M, Flamm RK (2017) Pseudomonas aeruginosa antimicrobial susceptibility results from four years (2012 to 2015) of the international network for optimal resistance monitoring program in the United States. Antimicrob Agents Chemother 61:e02252–e02216

Santajit S, Indrawattana N (2016) Mechanisms of antimicrobial resistance in ESKAPE pathogens. Biomed Res Int 2016:2475067

Scherr TD, Heim CE, Morrison JM et al (2014) Hiding in plain sight: interplay between staphylococcal biofilms and host immunity. Front Immunol 5:37

Sebeny PJ, Riddle MS, Petersen K (2008) Acinetobacter baumannii skin and soft-tissue infection associated with war trauma. Clin Infect Dis 47:444–449. https://doi.org/10.1086/590568. PMID: 18611157

Seifert H, Dijkshoorn L, Gerner-Smidt P, Pelzer N, Tjernberg I, Vaneechoutte M (1997) Distribution of Acinetobacter species on human skin: comparison of phenotypic and genotypic identification methods. J Clin Microbiol 35:2819–2825
Shaikh S, Fatima J, Shakil S, Rizvi SMD, Kamal MA (2015) Antibiotic resistance and extended spectrum beta-lactamases: types, epidemiology and treatment. Saudi J Biol Sci 22:90–101
Shankar C, Nabarro LEB, Anandan S, Veeraraghavan B (2017) Minocycline and tigecycline: what is their role in the treatment of carbapenem-resistant gram–negative organisms? Microb Drug Resist 23:437–446
Shariati A, Azimi T, Ardebili A, Chirani AS, Bahramian A, Pormohammad A, Sadredinamin M, Erfanimanesh S, Bostanghadiri N, Shams S et al (2018) Insertional inactivation of oprD in carbapenem-resistant Pseudomonas aeruginosa strains isolated from burn patients in Tehran, Iran. New Microbes New Infect 21:75–80
Shariati A, Dadashi M, Moghadam MT, Van Belkum A, Yaslianifard S, Darban-Sarokhalil D (2020) Global prevalence and distribution of vancomycin resistant, vancomycin intermediate and heterogeneously vancomycin intermediate Staphylococcus aureus clinical isolates: a systematic review and meta-analysis. Sci Rep 10:1–16
Shaw L, Golonka E, Potempa J, Foster SJ (2004) The role and regulation of the extracellular proteases of Staphylococcus aureus. Microbiology 150(1):217–228. https://doi.org/10.1099/mic.0.26634-0
Sheu C-C, Chang Y-T, Lin S-Y, Chen Y-H, Hsueh P-R (2019) Infections caused by carbapenem-resistant Enterobacteriaceae: an update on therapeutic options. Front Microbiol 10:80
Shiadeh SMJ, Pormohammad A, Hashemi A, Lak P (2019) Global prevalence of antibiotic resistance in blood-isolated Enterococcus faecalis and Enterococcus faecium: a systematic review and meta-analysis. Infect Drug Resist 12:2713–2725
Stover CK, Pham XQ, Erwin AL, Mizoguchi SD, Warrener P, Hickey MJ, Brinkman FS, Hufnagle WO, Kowalik DJ, Lagrou M et al (2000) Complete genome sequence of Pseudomonas aeruginosa PAO1, an opportunistic pathogen. Nature 406:959–964
Su M, Satola SW, Read TD (2019) Genome-based prediction of bacterial antibiotic resistance. J Clin Microbiol 57:e01405–e01418
Tenover FC (2018) The role for rapid molecular diagnostic tests for infectious diseases in precision medicine. Expert Rev Precis Med Drug Dev 3:69–77
Tigabu A, Getaneh A (2021) Staphylococcus aureus, ESKAPE bacteria challenging current health care and community settings: a literature review. Clin Lab 67:930
Tong SYC, Davis JS, Eichenberger E, Holland TL, Fowler VG (2015) Staphylococcus aureus infections: epidemiology, pathophysiology, clinical manifestations, and management. Clin Microbiol Rev 28:603–661
Tsalik EL, Bonomo RA, Fowler VG Jr (2018) New molecular diagnostic approaches to bacterial infections and antibacterial resistance. Annu Rev Med 69:379–394
Turner NA, Sharma-Kuinkel BK, Maskarinec S, Eichenberger E, Shah PP, Carugati M, Holland TL, Fowler VG (2019) Methicillin-resistant Staphylococcus aureus: an overview of basic and clinical research. Nat Rev Genet 17:203–218
Uzunović S, Ibrahimagić A, Bedenić B (2018) Antibiotic resistance in Enterobacter cloacae strains with derepressed/partly derepressed/inducible AmpC and extendedspectrum beta-lactamases in Zenica-Doboj Canton, Bosnia and Herzegovina. Med Glas 15:37–45
Van Belkum A, Bachmann TT, Lüdke G, Lisby JG, Kahlmeter G, Mohess A, Becker K, Hays JP, Woodford N, Mitsakakis K (2019) Developmental roadmap for antimicrobial susceptibility testing systems. Nat Rev Microbiol 17:51–62
Vázquez-López R, Solano-Gálvez SG, Vignon-Whaley JJJ, Vaamonde JAA, Alonzo LAP, Reséndiz AR, Álvarez MM, López ENV, Franyuti-Kelly G, Álvarez-Hernández DA et al (2020) Acinetobacter baumannii resistance: a real challenge for clinicians. Antibiotics 9:205
Vestergaard M, Frees D, Ingmer H (2019) Antibiotic resistance and the MRSA problem. Microbiol Spectr 7:57

Vrancianu CO, Gheorghe I, Czobor IB, Chifiriuc MC (2020) Antibiotic resistance profiles, molecular mechanisms and innovative treatment strategies of Acinetobacter baumannii. Microorganisms 8:935
Wertheim HF, Melles DC, Vos MC et al (2005) The role of nasal carriage in Staphylococcus aureus infections. Lancet Infect Dis 5:751–762
Willems RJ, Top J, van Den Braak N, van Belkum A, Endtz H, Mevius D (2000) Host specificity of vancomycin-resistant Enterococcus faecium. J Infect Dis 182(3):816–823
Willems RJ, Top J, van Santen M, Robinson DA, Coque TM, Baquero F et al (2005) Global spread of vancomycin-resistant Enterococcus faecium from distinct nosocomial genetic complex. Emerg Infect Dis 11(6):821–828
Xu C, Bilya S, Xu W (2019) adeABC efflux gene in Acinetobacter baumannii. New Microbes New Infect 30:100549
Yang S, Rothman RE (2004) PCR-based diagnostics for infectious diseases: uses, limitations, and future applications in acute-care settings. Lancet Infect Dis 4:337–348
Young BA, Hanson KE, Gomez CA (2019) Molecular diagnostic advances in transplant infectious diseases. Curr Infect Dis Rep 21:52
Young TM, Bray AS, Nagpal RK, Caudell DL, Yadav H, Zafar MA (2020) Animal model to study Klebsiella pneumonia gastrointestinal colonization and host-to-host transmission. Infect Immun 88:e00071–e00020
Zaheer R, Cook SR, Barbieri R, Goji N, Cameron A, Petkau A, Polo RO, Tymensen L, Stamm C, Song J (2020) Surveillance of Enterococcus spp. reveals distinct species and antimicrobial resistance diversity across a One-Health continuum. Sci Rep 10:3937
Zunita Z, Bashir A, Hafizal A (2008) Occurrence of multidrug resistant Staphylococcus aureus in horses in Malaysia. Vet World 1(6):165–167

Biochemical, Molecular, and Computational Techniques for the Determination of Virulence Factors of ESKAPE Pathogens

6

Archana Priyadarshini Jena
and Vemuri Venkateswara Sarma

Abstract

Emerging antibiotic resistance and its spreading dynamics have reached an alarming level, a global challenge. Overuse and Misuse of effective antibiotics raised the emergence of multidrug-resistant (MDR) and extensively drug-resistant bacteria (XDR) that ultimately cause treatment failures. Among all the clinical pathogens the World Health Organization (WHO) has reported the most common critical pathogen in hospital settings. The most listed six pathogenic bacteria that possess multidrug resistance and virulence are *Enterococcus faecium, Staphylococcus aureus, Klebsiella pneumoniae, Acinetobacter baumannii, Pseudomonas aeruginosa,* and *Enterobacter* spp. that refer as ESKAPE pathogens. Acquisition of resistance in bacteria is mediated by the transcription of virulence genes that produce virulence proteins/factors. Virulence factors significantly influence pathogenesis and infections evading the host immune system. Therefore, the determination and quantification of these virulence factors via biochemical, molecular, and computational techniques are essential for discovering new therapies that can combat antibiotic resistance in the future. In this chapter, we have listed different techniques for the determination of virulence factors in ESKAPE pathogens.

Keywords

ESKAPE pathogen · Antibiotic resistance · Virulence factors · Biochemical techniques · Molecular techniques

A. P. Jena · V. V. Sarma (✉)
Department of Biotechnology, School of Life Sciences, Pondicherry University, Pondicherry, India

S. Busi, R. Prasad (eds.), *ESKAPE Pathogens*,
https://doi.org/10.1007/978-981-99-8799-3_6

6.1 Introduction

In the last century, antimicrobial resistance (AMR) has been one of the most significant challenges that have been achieved progressively. Bacterial AMR has been recognized as a serious global health hazard. By the year 2050, World Health Organization (WHO) estimates that ten million people will be killed yearly. Therefore, an urgent and coordinated strategy is required to tackle the imminent threat of AMR. According to the Centers for Disease Control and Prevention (CDC), United States, there are 18 identified threats related to AMR. Similarly, the European Union and the European economic area have reported 16 AMR threats caused by eight selective pathogens (Kalpana et al. 2023). The WHO approved a list of six virulent multidrug-resistant bacteria collectively known as ESKAPE (*Enterococcus faecium*, *Staphylococcus aureus*, *Klebsiella pneumoniae*, *Acinetobacter baumannii*, *Pseudomonas aeruginosa*, and *Enterobacter species*) that cause severe and potentially lethal infections in hospital settings (Zhu et al. 2022). Therefore, more attention is needed to understand the respective virulence factors that facilitate the AMR of the abovementioned pathogens as depicted in Fig. 6.1.

Pathogenic bacteria acquire a wide range of virulence proteins/factors that play a crucial role in their ability to survive in adverse conditions and cause serious illness. To facilitate the prompt implementation of infection control measures, precise and sensitive microbial detection methods need to be implemented (Gerace et al. 2022). So rapid and accurate diagnostic methods should take place to identify the

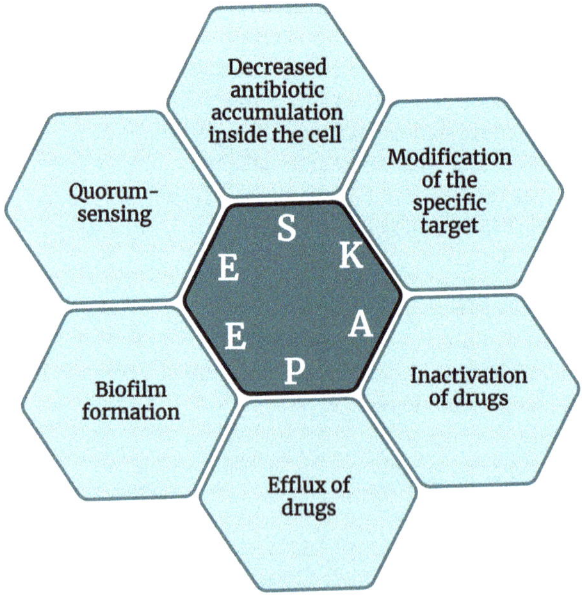

Fig. 6.1 Diagrammatic representation of various resistance mechanisms activated by virulence factors of pathogens

significant virulence factors of these pathogens to initiate timely and targeted antibiotic therapy which can also be useful in managing outbreaks and reducing the development of AMR in the near future. In this book chapter, our objective is to determine the virulence factors exhibited by ESKAPE pathogens that spread pathogenesis. Furthermore, we provide a comprehensive overview of methodologies employed to identify and analyze these virulence factors at different levels, including biochemical, molecular, and computational approaches.

6.2 Virulence Factors

6.2.1 *Enterococcus faecium* (*E. faecium*)

Enterococci are Gram-positive and can survive in both oxygen-rich and oxygen-deprived environments. They have the unique ability to grow in high concentrations of salt (6.5% NaCl) and under alkaline pH conditions. Additionally, they can break down bile-esculin and L-pyrrolidonyl-beta-naphthylamide (PYR). In the past, *E. faecium* was classified as one of the Lancefield group D *Streptococcus* members. However, further research based on DNA homology studies indicated that it belongs to a definite genus (Hollenbeck and Rice 2012).

In recent years, *Enterococcus* has become increasingly a significant source of nosocomial infections. While they are naturally present in the normal microbial population of humans and animals, they appeared as the second most frequent cause of urinary tract infections and the third most prevalent cause of bloodstream infections acquired in healthcare settings, particularly among individuals with weakened immune systems (Comerlato et al. 2013). Currently, the proliferation of Vancomycin-Resistant *Enterococci* (VRE) has raised challenges. Numerous strains of *Enterococci* that are acquired in healthcare settings and resistant to Vancomycin also exhibit resistance to Penicillin, along with high-level resistance (HLR) to aminoglycosides. This resistance pattern to multiple antibiotics has become a major cause for concern (Malani et al. 2014).

Enterococci typically inhabit the gastrointestinal tract of mammals but can also be found in water, soil, and food. Among the various species of *Enterococci*, *Enterococcus faecalis,* and *E. faecium* are important for the majority of infections. However, the increasing prevalence of these bacteria in healthcare settings and their association with healthcare-associated infections can be attributed to several factors. These include the acquisition of new mechanisms of resistance to antibiotics, the possession of specific virulence traits shown in Fig. 6.2, and the ability to develop biofilms (Fig. 6.2). These factors contribute to the spread and persistence of enterococcal infections in hospitals, veterinary, and, environments (Hollenbeck and Rice 2012).

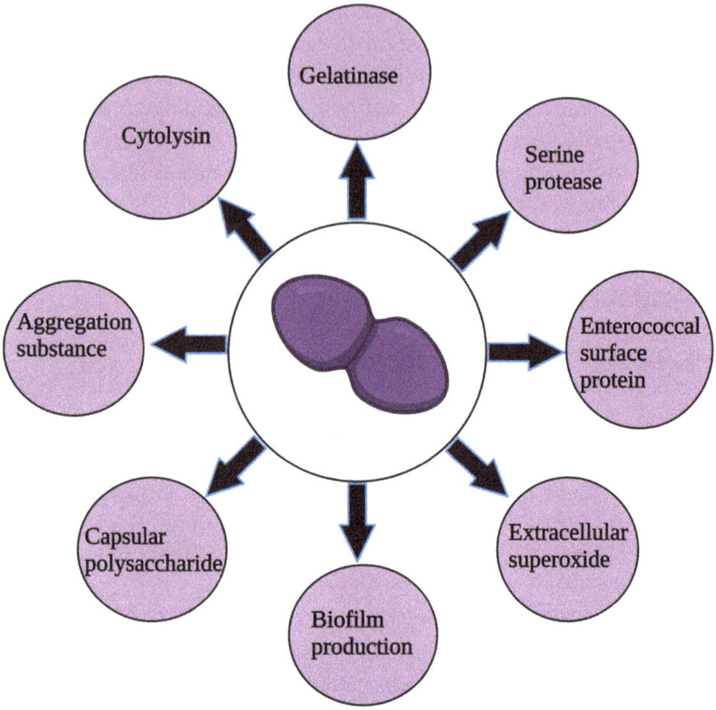

Fig. 6.2 An overview of *Enterococcus faecium-associated* numerous virulence factors that contribute toward pathogenesis

6.2.2 *Staphylococcus aureus* (*S. aureus*)

It is a Gram-positive bacterium, appears unremarkable, and lacks motility; however, it poses a significant threat as a human pathogen. It can cause a wide range of infections in both community settings and healthcare facilities including food poisoning, suppurative diseases, pneumoniae, and toxic shock syndrome. In today's scenario, the emerging methicillin-resistant *S. aureus* (MRSA) has further exacerbated the problem, as it is particularly challenging to treat and commonly leads to nosocomial infections (Lindsay and Holden 2004). One of the fundamental characteristics of this bacterium is its capacity to asymptomatically inhabit the bodies of healthy individuals. People who are carriers of *S. aureus* have an increased susceptibility to infections, and they can easily transmit the infections among individuals. Hence, carriers serve as a potential reservoir of multidrug-resistant *S. aureus* strains to spread and transmit diseases to other individuals (Chambers and DeLeo 2009).

S. aureus exhibits a wide range of factors that contribute to its pathogenicity, activating it to colonize tissues, cause tissue damage, and lead to infections at distant sites as well. It has the ability to survive inside host cells and can invade various nonspecialized phagocytes like osteoblasts, fibroblasts, endothelial cells, and epithelial cells. Once internalized, *S. aureus* can either persist by evading the host's

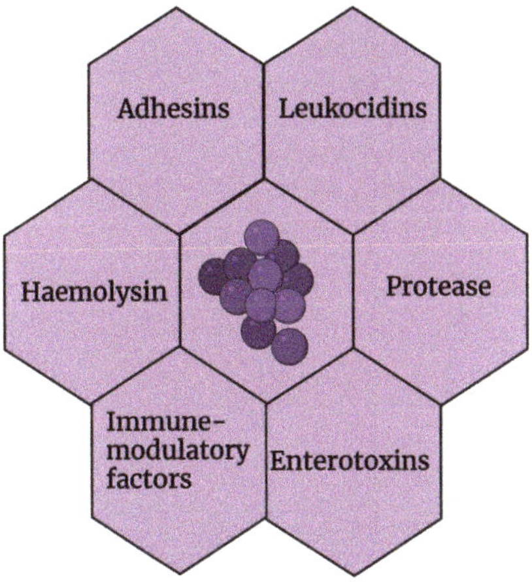

Fig. 6.3 Schematic representation of virulence factors produced by *Staphylococcus aureus* that causes staphylococcal infections

immune defenses and antibacterial treatments, or it can multiply and spread further within the body. This behavior is controlled by global regulatory systems that detect changes in the surrounding environment, including bacterial density. These regulators can trigger the release of virulence proteins that disrupt host cells, facilitating bacterial propagation. Consequently, the invasion of host cells not only offers a protective refuge for the bacteria but can also be part of a clever strategy resembling a game of hide-and-seek, similar to what has been observed with enteric bacteria (Bien et al. 2011).

S. aureus has been widely recognized for the production of numerous virulence factors as depicted in Fig. 6.3, that are crucial for causing severe infections in the human body (Fig. 6.3). However, the expression of these virulence factors is controlled and regulated throughout the growth of the bacterium. The respective regulatory system involved in this process is *agr* system, known as the quorum sensing (QS) system. It acts as important in governing the regulation of virulence factors while it operates by detecting and responding to the density of bacterial cells, allowing the coordination of gene expression and the production of virulence factors (Novick and Geisinger 2008).

6.2.3 *Klebsiella pneumoniae* (*K. pneumoniae*)

It is a Gram-negative, encapsulated, nonmotile, facultative anaerobic bacteria isolated from the airways of a pneumonia patient. It is commonly known as "Friedlander's bacillus" after being explained by Carl Friedlander in 1882. *K. pneumoniae* cause infections at multiple sites such as lung, urinary tract, bloodstream, wound or surgical site, and brain. Therefore, those having preexisting health conditions are more susceptible to *K. pneumoniae*-associated infections (Chang et al. 2021).

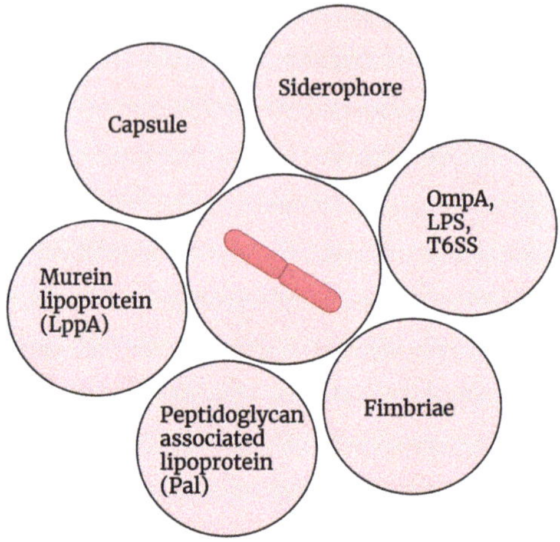

Fig. 6.4 Diagrammatic overview of numerous virulence factors released by *Klebsiella pneumoniae* that evade host immune system which leads to disease progression

Two distinct types of *K. pneumoniae* strains exist, i.e., classic *K. pneumoniae* (c*Kp*) and hypervirulent *K. pneumoniae* (hv*Kp*). The c*Kp* strain is considered non-virulent and typically produces extended-spectrum beta-lactamases (ESBLSs). This strain is commonly associated with infections acquired in healthcare settings (nosocomial infections). On the contrary, hv*Kp* strains are known for their hypervirulence and tend to cause invasive infections in healthy individuals, while most strains show susceptibility to antimicrobial agents. In recent decades, it has become a significant global threat due to the upsurge in both hypervirulent and carbapenem-resistant strains (Khaertynov et al. 2018).

K. pneumoniae can lead to infections that are acquired in the community or healthcare settings, and these infections are often associated with high rates of illness and death. While infections can occur in various parts of the body, the respiratory and urinary tracts are the most common sites to be infected by this bacterium. Although the infections can develop at any age, they are most frequently observed in the neonates and elderly populations, as well as those with weakened immune systems due to underlying conditions like alcoholism or diabetes, or as a result of immunosuppressive therapies. Virulence factors of *K. pneumoniae* as depicted in Fig. 6.4 play a critical role in invading host cells, contributing to disease development, and evading host defenses (Fig. 6.4). These factors facilitate the establishment and persistence of the bacteria within the host, leading to the progression of the infection (Highsmith and Jarvis 1985).

6.2.4 *Acinetobacter baumannii* (*A. baumannii*)

It is a Gram-negative coccobacillus. It is strictly aerobic and commonly found as a commensal organism. In recent decades, it has become a significant opportunistic pathogen, particularly in hospital environments for which it gained more attention and is listed among one of the ESKAPE pathogens. As depicted in Fig. 6.5, the

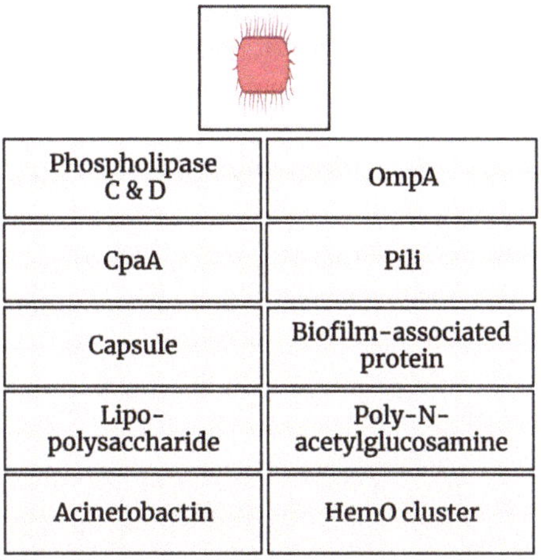

Fig. 6.5 Diagrammatic representation of Acinetobacter baumannii-associated virulence proteins that cause nosocomial infections in hospital settings

mentioned virulence traits are responsible for its persistence in healthcare environments, leading to its pathogenicity (Fig. 6.5). In addition to this, it is resistant to a wide range of antimicrobials, can easily spread from patient to patient, and most importantly can survive in dry conditions as well for a longer period. There are a total of 19 biotypes described within the *A. baumannii* species (Braun and Vidotto 2004).

Clinical isolates of *A. baumannii* exhibit significant phenotypic variations in virulence-associated strains which include biofilm development, adherence to human epithelial cells, invasiveness, motility, and cytotoxicity. Undoubtedly, it is difficult for the bacterium to adhere to the surface of the medical devices and spread within the hospital wards and among patients. In addition to surface contamination, it can able to cause various septic infections such as ventilator-associated pneumonia, bacteremia, soft tissue infections, and secondary meningitis. From the recent study, it has become a major concern in healthcare settings due to its rapid acquisition of resistance. Therefore, it has emerged as a global social and healthcare issue. Its highly adaptive traits and virulence factors have allowed it to evade the host immune defense and deactivate the current strategies that have been employed by the researchers to kill the organism. This poses ongoing challenges in combating *A. baumannii* infections and finding effective therapies (Eijkelkamp et al. 2014; Harding et al. 2018).

Over time, few strains of *A. baumannii* developed resistance to all conventional antibiotics, including carbapenems, that were previously considered as the primary treatment option for multidrug-resistant strain (MDR) *A. baumannii* infections. These are highly spreadable pathogens among hospitalized patients; hence it causes nosocomial infections as well which may lead to outbreaks. The emergence of these MDR strains has become a critical therapeutic challenge as there is no effective treatment available to combat these strains. Based on the global surveillance data, the rates of infections caused by this Gram-negative pathogen are almost four times higher as compared to other similar pathogens such as *Pseudomonas aeruginosa* and *Klebsiella pneumoniae* that are already in the list of ESKAPE pathogens (Amala Reena et al. 2017; Zhou et al. 2023).

6.2.5 *Pseudomonas aeruginosa* (*P. aeruginosa*)

It is a Gram-negative bacillus that commonly occurs in environments including soil, water, plants, and the human body as well. It possesses remarkable adaptability and can survive in nutrient-deprived environments and temperature range, from 4 to 42 °C. Notably, it is extremely resilient and can persist for up to 6 months on dry, abiotic surfaces in hospitals. Although major progress has been made to understand the pathogenesis of the opportunistic pathogen *P. aeruginosa* in humans, our perception regarding its pathogenesis in causing infections remains confined. It is responsible for several infections in humans, starting from acute localized infections (urinary tract infections, ulcerative keratitis, malignant otitis media, peritonitis, and ventilator-associated pneumonia in patients) to chronic localized infections like destructive lung infections among individuals with cystic fibrosis (CF). Furthermore, it is a nosocomial pathogen specifically affecting patients with severe underlying conditions that compromise their physical health (mechanically ventilated individuals, burn patients) or immune defenses (neutropenia, AIDS). So, patients suffering from the abovementioned diseases are at significant risk of developing systemic disease from localized infections, associated with increased mortality rates (van Delden 2004; Liao et al. 2022). WHO has prioritized this bacterium as a critical pathogen that exhibits a high level of antibiotic resistance, emphasizing the urgent need for the discovery of new antibiotics (French 2017).

Treatment of *P. aeruginosa*-associated infections can be a major challenge due to its inherent resistance to multiple classes of drugs, as well as its ability to acquire resistance to more conventional effective antibiotics. In Rrecent studies, it has reported that the MDR strain of *P. aeruginosa* acquired resistance to at least three main categories of antibiotics such as aminoglycosides, carbapenems, antipseudomonal penicillin, quinolones, and cephalosporins. It possesses numerous virulence factors, depicted in Fig. 6.6. These factors are strictly regulated by a cell to cell signaling systems. Among these virulence factors, *exotoxin* A is the primary factor produced by most of the *P. aeruginosa* isolates and plays a crucial role in the acquisition of infections (Fig. 6.6). Additionally, an extracellular enzyme, i.e., neuraminidase is believed to contribute to the attachment of *P. aeruginosa* on surfaces; however, the genetic mechanisms underlying this process remain unclear (Corehtash et al. 2015).

As compared to the strains that initially caused the infection, the MDR *P. aeruginosa* strains found in the airways of CF patients exhibit a greater variation in genotypic and phenotypic levels due to specific evolutionary changes. So, a comprehensive understanding of the mechanisms that regulate the adaptation and evolution of *P. aeruginosa* during chronic respiratory CF will be more attractive in developing targeted therapies to effectively eliminate this nosocomial pathogen (Jurado-Martín et al. 2021).

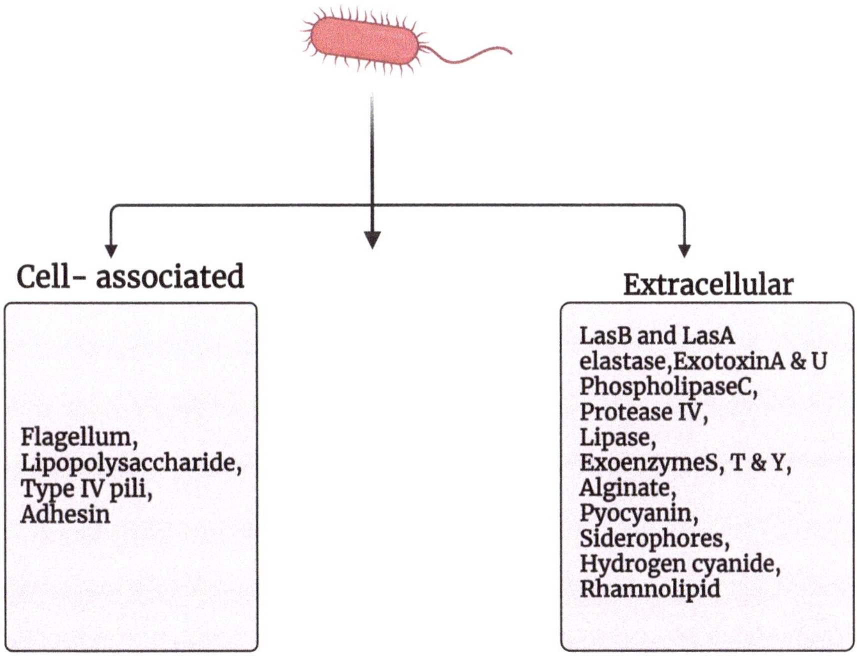

Fig. 6.6 Overview of Pseudomonas aeruginosa-associated virulence factors regulated by cell signaling systems that cause severe infections in human beings

6.2.6 *Enterobacter* Species

Enterobacter species are Gram-negative, rod-shaped bacteria that come under the *Enterobacteriaceae* family. They can thrive in both aerobic and anaerobic environments. Most of them are flagellated and exhibit class 1 fimbriae. Approximately, 80% of the *Enterobacter* are encapsulated. In the environment, they are considered saprophytic due to their common occurrence in sewage and soil. Moreover, being a part of the commensal enteric flora, they too reside in the gastrointestinal (GI) tract in humans (Mezzatesta et al. 2012).

As an opportunistic pathogen, several *Enterobacter* species such as *E. cloacae*, *E. sakazakii*, *E. gergoviae*, *E. agglomerans*, and *E. aerogenes* have been identified in humans. They are able to cause a diverse array of infections such as skin and eye infections, pneumonia, meningitis, bacteremia, urinary tract infections, and intestinal and surgical site infections. *Enterobacter* that are is associated with extraintestinal infections possesses specific virulence-associated features. These strains are capable of adhering to and invading eukaryotic cells (Hussain and Alammar 2013; Mishra et al. 2020).

Over recent years, *Enterobacter* species have also been declared as a nosocomial pathogen among patients in Intensive Care Units (ICUs), constituting only 5% of nosocomial bacteremia. Further advanced research is needed to elucidate the

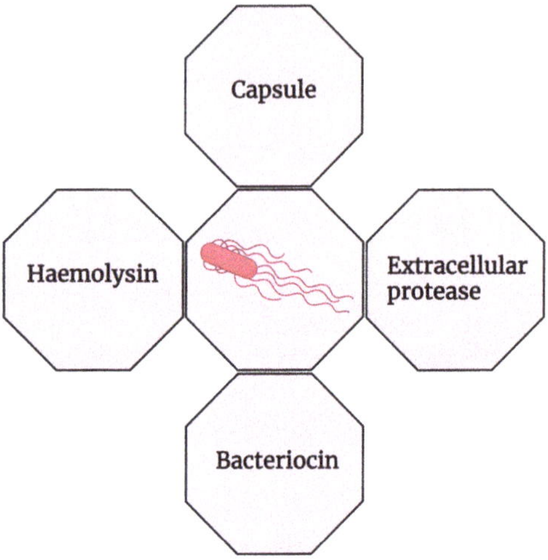

Fig. 6.7 Schematic overview of virulence factors produced by *Enterobacter species* that cause a wide range of infections including extraintestinal infections

underlying mechanisms and factors associated with *Enterobacter* infections (Mezzatesta et al. 2012).

The *Enterobacter cloacae* complex (ECC) consists of a set of pathogenic strains that commonly occur in hospitals, thus responsible for causing various infections. Surprisingly, before the discovery of antibiotics, ECC was not listed under nosocomial pathogen. In the 1970s for the first time, it has identified as nosocomial pathogens (Paauw et al. 2009). Currently, the ECC is comprised of six recognized species, i.e., *E. cloacae*, *E. kobei*, *E. hormaechei*, *E. asburiae*, *E. ludwigii*, and *E. nimipressuralis*. These species collectively contribute to the pathogenicity and infections that are associated with the *Enterobacter* species. Naturally, bacteria release virulence proteins/factors shown in Fig. 6.7 into the extracellular space via a specialized secretion system as part of their normal growth (Fig. 6.7). These proteins typically contribute to nutrient acquisition and the fitness of bacteria. However, in the case of pathogenic bacteria, there is a different scenario, i.e., the virulence proteins involved in causing disease by promoting host colonization and modulating immune responses. As a result, the release of these proteins can disintegrate the host cells or the extracellular release of these proteins can enhance the competition with neighbor microbes (Kumari et al. 2023).

6.3 Biochemical, Molecular, and Computational Techniques for the Identification of Virulence Factors of ESKAPE Pathogens

Identification and characterization of virulence factors in ESKAPE pathogens can be critical for understanding their pathogenesis and developing effective strategies to combat infections. Here are some methods shown in Table 6.1 routinely used for the identification of virulence factors (Table 6.1).

Table 6.1 Tabular representation of biochemical, molecular, and computational approaches designed for the identification and determination of numerous virulence factors produced by ESKAPE pathogens

Approaches	Different methods to determine the virulence factors
Biochemical	Traditional methods
	MALDI-TOF MS (matrix-assisted laser desorption ionization time flight mass spectrometry)
	Microfluidics
Molecular	PCR (polymerase chain reaction)
	RT-PCR (real-time PCR)
	BioFire FilmArrays
	Microarrays
	PFGE (Pulse field gel electrophoresis)
	WGS (whole-genome sequencing)
	NGS (next-generation sequencing)
	Biosensor
Computational	PathoFact
	Virulence factor databases and servers: MvirDB, VFDB, VirulentPred

6.3.1 Biochemical Methods

6.3.1.1 Traditional Methods

Traditional biochemical methods in microbiology have been used for the identification of microorganisms or pathogens, primarily depended on culturing techniques. These methods involve performing various tests such as the catalase test, oxidase test, triple sugar iron test, urease test, etc. However, these abovementioned methods are not in routine use due to the development of new molecular and computational techniques that can generate faster results. So, these advancements have led to a deeper understanding of biochemistry and improved instrumentation that offers a broad spectrum of applications described below.

6.3.1.2 Matrix-Assisted Laser Desorption Ionization Time Flight Mass Spectrometry (MALDI-TOF MS)

It is a recently developed tool that is utilized for the rapid identification and classification of microorganisms. Unlike traditional methods that are mostly time-consuming such as identification of bacterial species and antibiotic susceptibility testing, MALDI-TOF provides a rapid, convenient, and cost-effective approach. This tool relies upon spectral libraries to analyze the mass fingerprinting of peptides that are produced by each microorganism which enables accurate identification. Concisely, in MALDI-TOF MS, bacterial colonies are combined with a matrix solution and affixed on a stainless-steel plate well. After the solvent evaporates, a laser beam is supplied, which leads to the desorption (removal) and ionization of the matrix. Subsequently, the energy from the laser beam is transferred to sample molecules, accelerated them forward, and separating them based on their mass-to-charge ratio under the impact of an electric field. So, microorganisms can be identified by comparing the obtained spectrum with a database of preexisting spectra of known organisms (Dierig et al. 2015). The MALDI-TOF MS platforms which are in use are the Bruker Biotyper, developed by Bruker Daltonik in Bremen, Germany, and the VitekMS, created by bioMérieux in Marcy l'Etoile, France (Torres-Sangiao and Rodriguez 2021).

One of the major challenges in medical microbiology is using the MALDI-TOF MS for the identification of AMR in microorganisms including emerging ESKAPE pathogens. However, a recent study reported that it has been successfully utilized to identify the presence of carbapenemases, that enable active detection of AMR (Gato et al. 2022; Florio et al. 2020).

6.3.1.3 Microfluidics

The microfluidics, commonly named the "lab-on-a-chip" technique shows significant results in detecting antibiotic-resistant bacteria. As compared to conventional methods on a large scale, microfluidics offers greater advantages, i.e., inexpensive, rapid, and high-throughput analysis, minute quantity of sample, and automation capabilities. Generally, two primary microfluidic detection methods exist: genotypic assays and phenotypic assays. Genotypic on-chip assays target specific genetic markers like 16s rRNA genes. However, they are not suitable for determining AST profiles. On the contrary, phenotypic on-chip assays track bacterial growth in the existence of antibiotics, which provides precise results for AST. Overall microfluidic offers a promising approach to the detection of antibiotic-resistant bacteria (Zhang et al. 2018).

6.3.2 Molecular Methods

6.3.2.1 Polymerase Chain Reaction (PCR)

In 1985, Kary Mullis first introduced PCR, the pioneering DNA amplification technique to generate millions of copies of a DNA template that initiates from a minute quantity of DNA molecule. This tool is highly effective and can be used for the rapid and accurate identification of pathogen-specific nucleic acids. In general, the process depends upon thermal cycling, where the reaction mixture goes through repeated cycles to continue the DNA synthesis. This process utilizes primers which are short DNA fragments, which have sequences complementary to that of the template region. Primers bind to target DNA, which allows the thermostable DNA polymerase to initiate replication and amplify the target sequence continuously. As the current scenario demands rapid identification of pathogens and their resistance mechanisms, several types of PCR have been developed since its invention. These different types of PCR are designed to achieve fast and accurate pathogen detection and to specify specific virulence factors-encoded gene that contributes to antibiotic resistance. So, researchers can rapidly detect pathogens and the respective antibiotic-resistant genes by using nucleic acid amplification technologies (NAAT) like PCR and its variants (Vosberg 1989; Zhu et al. 2020).

6.3.2.2 Real-Time PCR (RT-PCR)

This technique is extensively used worldwide, especially in clinical settings to diagnose infectious diseases due to its remarkable sensitivity, accuracy, and rapidity. It has become one of the most commonly employed methods in clinical practice for the detection of pathogens as well as their virulence mechanisms (Zhu et al. 2020).

Commonly there are two primary methods used for detecting the amplified PCR products in RT PCR. The first method consists of the utilization of a fluorescent dye, such as SYBR® Green, that binds nonspecifically to double-stranded DNA. In this method, the DNA dye complex assimilates blue light and subsequently emits green light. The second common method involves the use of Fluorescent Resonance Energy Transfer (FRET) probes, that specifically bind to the amplified DNA (Maurin 2012).

Different types of commercial kits are available for pathogen detection and resistance gene identification from samples. Meanwhile, it is worth mentioning that there have been such cases where these kits have shown false-positive and false-negative outcomes when targeting different genes. In other words, it has been reported that in some instances kits have either incorrectly identified the presence of a pathogen or somehow failed in it, ultimately leading to inaccurate results. It is crucial to understand that the accuracy of these kits can also differ depending on several factors, such as the specific genes targeted for detection, the quality and quantity of the samples, and the presence of any interfering substances (Chen et al. 2018).

Though in today's clinical practices, it has been used tremendously, however, it does have certain limitations and drawbacks. A notable drawback of RT-PCR is the limitation in simultaneously detecting a limited number of targets primarily due to the restricted availability of fluorescent dyes. Furthermore, as the multiplexing level rises, there is a decrease in sensitivity and accuracy. In addition to this, the maintenance of the instrument requires regular servicing, calibration, and monitoring of its performance. So it can be a financial burden for laboratories as well as research institutes that are using RT-PCR technology (Elnifro et al. 2000).

Over the years, recent advancements and technologies have taken place in the field of RT PCR leading to the emergence of different isothermal techniques. One of these techniques is loop-mediated amplification (LAMP) which utilizes a set of designed primers and DNA polymerase with strand displacement activity to achieve rapid amplification. It has gained attention due to its efficiency and high sensitivity. Another technique is helicase-dependent amplification (HDA) which uses helicase enzymes to separate the double-stranded DNA strands and primers to initiate the amplification of the single-stranded DNA. Nucleic acid sequence-based amplification (NASBA) is another technique that employs reverse transcriptase and RNA polymerase to amplify RNA targets which is specifically used for detecting RNA viruses for disease diagnosis. These abovementioned isothermal techniques have a significant contribution toward the development of NAAT (Obande and Singh 2020).

6.3.2.3 BioFire FilmArrays

This is an in vitro rapid diagnostic test designed for nucleic acid analysis through various steps. It integrates several processes like DNA/RNA isolation, reverse transcription, nested multiplex PCR, and melting curve analysis to provide a comprehensive and significant approach for identifying and analyzing target nucleic acids. However, among all the steps, melting curve analysis is the key step as it is able to determine the melting temperature of the PCR products. Thereafter, this information can be employed to identify and differentiate target sequences based on their distinct melting profiles (Young et al. 2020).

It is exclusively designed to operate with broad panels that include a wide range of assays for pathogens. These panels cover several types of infections, including respiratory viral infections, encephalitis, bloodstream infections, gastrointestinal infections, meningitis, and pneumonia. In addition to this, the panels also incorporate assays that are meant to detect virulence factor-associated genes (Mitton et al. 2021).

The gastrointestinal panel comprised 22 common pathogens that contribute to gastrointestinal symptoms including diarrhea which is a major health concern worldwide, particularly in developing countries where they are a main cause of child death. It encompasses a range of infectious agents, such as viruses, bacteria, and protozoa which cause diarrheal symptoms (Torres-miranda et al. 2020). Therefore, this testing method aids in accurate diagnosis, appropriate treatment decisions, and effective management of patients who are suffering from diarrheal diseases.

6.3.2.4 DNA Microarrays

DNA microarray is an advanced technology that involves the immobilization of nucleic acids such as DNA or RNA sequences on a surface. They are used to determine the relative concentrations of different nucleic acid sequences within a sample through a process called hybridization. In the hybridization steps, the target nucleic acid from the sample is labeled with fluorescent or radioactive tags which are then applied to the microarray. Then this microarray is subjected to a detection step that depends on the type of labeling used for the target nucleic acids, i.e., fluorescent labeling or radioactive labeling (Bumgarner 2013).

The multiplexing capability of microarrays, which enables the simultaneous detection and identification of a number of genes, makes this technique highly efficient for understanding AMR in bacteria. It also offers the ability to detect genes associated with ESBLs and carbapenemases. Microarrays that are designed for AMR investigations contain specific oligonucleotide probes that are complementary to the target one such as ESBLs and carbapenemases (Fishbain et al. 2012).

To prevent the spreading of MDR strains, it is necessary to promptly identify multiple resistance genes in bacteria including ESKAPE pathogens. As earlier mentioned, microarrays can detect multiple MDR genes within a single strain at one time. This feature enables clinicians to quickly determine the various resistance genes that contribute toward the production of respective virulence factors (Dally et al. 2013). In recent studies, it has been mentioned that microarrays can also enable the detailed analysis of gene expression changes in response to drug treatments, providing important information for drug discovery, personalized medicine, and deeper understanding at the molecular level (Debouck and Goodfellow 1999). So, the ability to carry out high-throughput analysis and examine multiple target genes proves microarrays a powerful detection tool for AMR research.

6.3.2.5 Pulse Field Gel Electrophoresis (PFGE)

This method is utilized to segregate the DNA fragments that result from the restriction enzyme digestion. In this technique, an electric field is applied to a gel matrix, that changes the direction periodically. Different types of PFGE protocols have been created over time to identify various pathogens and investigate outbreaks. However, over time, this technique has been replaced by several advanced techniques which can yield better results. To date, it is widely accepted as the "gold standard" for bacterial typing due to its effectiveness and enduring popularity as compared to other methods that have been established around the same period (Neoh et al. 2019).

6.3.2.6 Whole-Genome Sequencing (WGS)

It is a developed method used to determine the entire nucleotide sequence of an organism in a single analysis. In recent decades, WGS has been found to get more attention due to its wide array of applications in various fields such as identifying the virulence proteins of certain pathogens, tracing the pathways of pathogen dissemination during outbreak investigations, and profiling AMR patterns. In brief, WGS has proven to be an efficient tool for analyzing complete genomes and has gained significant insights in multiple areas such as pathology, epidemiology, and clinical diagnostics for which it is most suited for use in laboratories specifically in public health and clinical sectors (Gilchrist et al. 2015; Kwong et al. 2015).

6.3.2.7 Next-Generation Sequencing (NGS)

This is a high-throughput and cost-effective sequencing method that allows sequencing for multiple DNA fragments at the same time. Hence it substituted the WGS by specifically targeting the exome. The protein-coding genes in the genome are referred to as exomes. Due to its cost-effectiveness and short-read technology, clinical microbiology should primarily adopt second-generation NGS technologies. It consists of three primary phases: library preparation, template generation, and sequencing. Major distinction among various NGS platforms lies in the DNA amplification technique, i.e., sequencing by synthesis and sequencing by hybridization. Both sequencing methods have effectively contributed to outbreak investigations and the mapping of resistance genes in ESKAPE pathogens. These advanced technologies have encouraged the identification and tracking of infectious disease outbreaks and enabled scientists to map the genetic mechanisms underlying antibiotic resistance in these pathogens (Hu et al. 2021).

NGS systems are commonly represented by several systems such as SOLiD/Ion Torrent PGM from Life Sciences, GS FLX Titanium/GS Junior from Roche, and Genome Analyzer/HiSeq 2000/MiSeq from Illumina. Beijing Genomics Institute (BGI) is well-known for having the world's largest sequencing capacity, and possesses numerous NGS systems including 137 HiSeq 2000, one Ion Torrent PGM, 27 SOLiD, one 454 sequencer, and one MiSeq (Liu et al. 2012).

6.3.2.8 Biosensor

A biosensor is a device that is designed to detect and measure alteration in physical quantities and convert them into signals that resemble the concentration of a specific analyte in a reaction. These are categorized based on the type of physical object or analyte that they are designed to measure. Generally, they are used to detect and quantify analytes such as proteins, enzymes, DNA, gases, chemicals, and other biological samples (Naresh and Lee 2021).

In a recent study, researchers employed an electrochemical biosensor which is based on molecularly imprinted polymers (MIP) to detect and analyze virulence factors, i.e., cagA and vacA in *Helicobacter pylori* that causes gastric cancer (Saxena et al. 2022, 2023).

6.3.3 Computational Methods

As earlier mentioned, pathogenic bacteria remain a major global health concern, it is required to identify their virulence factors at the computational level as well. Given the large amount of protein sequence data generated in the postgenomic era, there is a need to develop computational techniques that can rapidly and accurately identify virulence factors based on sequence information. A few computational methods and virulence factors' databases that have been extensively used for the prediction of virulence factors are described here.

6.3.3.1 PathoFact

It is a software tool designed for UNIX-based systems, that generally offer a command-like interface. It combines three different processes for the estimation of virulence factors, bacterial toxins, and AMR genes from metagenomic data. These workflows can be used independently or integrated with other workflows for a comprehensive analysis. The tool has been implemented in Python 3.6 and utilizes the snakemake workflow management software (version 5.5.4). This execution has brought up multiple advantages such as efficient workflow assembly, parallel processing capabilities, and the ability to resume analysis in case of any errors. In the near future, with the upgradation of PathoFact, researchers can efficiently analyze metagenomic data to determine virulence factors, toxins, and AMR bacteria without any data interruptions (De Nies et al. 2021).

6.3.3.2 Virulence Factor Databases and Servers

6.3.3.2.1 MvirDB

MvirDB is a complete dataset that contains a wide range of information that is related to toxins, virulence factors, and AMR-related genes. It serves as a valuable asset for researchers working in the field of biodefense studies. It combines DNA and protein sequence data from various sources, i.e., Tox-Prot, Islander, SCORPION, VFDB, PRINTS virulence factors, TVFac, ARGO, and a subset of VIDA. Moreover, the database is managed through the microbial annotation database (MannDB) system. One of the main features is its ability to aid quick identification and characterization of genes. By using BLAST search tools, researchers can query DNA or protein sequences of their interest within the databases. The following tool is able to extract descriptions, sequences, and categorization of virulence proteins. Furthermore, weekly it receives automatic updates to remain up to date (Zhou et al. 2007).

6.3.3.2.2 Virulence Factor Database (VFDB)

Like MvirDB, VFDB is also another comprehensive and integrated database that contains information regarding virulence factors in pathogenic bacteria. It includes extensive details about the characteristics, functions, and mechanisms. These factors have been selected on a detailed literature search on PubMed. Additionally, VFDB also provides different browsing options for users, including species-specific browsing, text search as well as BLAST and PSI-BLAST (Chen et al. 2005).

In an updated version of VFDB (2008), distinct features have been introduced to enhance its processing. These features included a tabular contrast of pathogenomic structure pointing to virulence factors, various alignments, and statistical assessment of homologous virulence genes that will provide data on their evolutionary relationship and functional similarities. Lastly, it involves the graphical comparison of the pathogenomic arrangement of virulence factors (Yang et al. 2008).

The latest version of VFDB (2012), imported several user interfaces to improve the user experience and promote intergenera comparative analysis of virulence factors. One major addition is the expanded trees that will allow for a more detailed representation of the hierarchical relations between virulence factors and their associated organism. In addition to this, collapsible menus are also introduced in order to navigate through the database more efficiently. This upgraded version of VFDB will allow researchers to acquire in-depth knowledge of the evolutionary patterns of virulence factors (Chen et al. 2012).

6.3.3.2.3 VirulentPred

It is also a predictive appliance for the determination of virulence factors. Unlike other tools, a dual-layer Support Vector Machine (SVM) is employed in VirulentPred, trained on recognized virulence factors that are compiled from SWISS-PROT and VFDB databases. The initial layer is trained and improved with several features such as amino acid composition, dipeptide, and high-order dipeptide composition that aids in capturing essential patterns and characteristics of virulence proteins. Similarly, the second layer of the SVM model is trained with the SVM score generated from the first layer along with PSI-BLAST outcomes. According to one study, VirulentPred can accomplish an accuracy rate of up to 82% which means it can effectively identify the factors associated with diseases. The major advantage of using this tool is, that it can determine potential virulence factors that may not have been added to the current databases. Therefore by using its predictive models and introducing new information, it can discover novel virulence factors that have not been characterized earlier (Garg and Gupta 2008).

Recently reported few techniques depicted in Table 6.2 have been successfully experimented for identification and detection of virulence factors as well as virulence factor-encoding genes in ESKAPE pathogens (Table 6.2).

6.4 Conclusion

The rapid and accurate detection of infectious diseases as well as disease-associated virulence factors is crucial for initiating novel antimicrobial therapy while minimizing the use of conventional antibiotics and their complications and healthcare costs. Specifically, the ESKAPE pathogens possess virulence factors that ultimately contribute toward pathogenesis and cause major threat due to AMR genes which cause severe infection and death around the world. Therefore, development in biochemical, molecular, and computational approaches in the determination of associated virulence factors is needed that may offer several advantages such as generating faster results, direct application of clinical samples for quick responses, reducing hospitalization hence decreasing the nosocomial infections, and minimizing risks associated with complications and mortality.

Table 6.2 Tabular depiction of reported techniques used for the detection of ESKAPE virulence proteins/virulence factors-encoded genes

Serial no.	Pathogens	Virulence factors	Techniques	References
1	*Enterococcus faecium*	Enterococcal surface protein (esp)	Multiplex PCR	Arshadi et al. (2018)
		Hyaluronidase (hyl)	Multiplex PCR	Heidari et al. (2017)
		Collagen-binding adhesin (acm)	Multiplex PCR	Shafeek et al. (2018); Trivedi et al. (2011)
		asa1 (aggregation substance), cylA (cytolysin), ace/acm (adhesion bind to collagen) esp (enterococcal surface protein), gelE (gelatinase), and hyl (hyaluronidase)	PCR and multiplex PCR	Shahraki and Rabi Nezhad Mousavi (2017)
2	*Staphylococcus aureus*	Serine hydrolases (FphB)	Activity-based protein profiling (ABPP)	Lentz et al. (2018)
		Staphylococcal enterotoxin A (sea), Staphylococcal enterotoxin B (seb), and panton-valentine leukocidin (pvl) genes	Cross-priming amplification (CPA) method	Jiang et al. (2021)
		Hemolysin, proteases, leukocidin, hyaluronidase, staphylokinase, lipase, nuclease	PCR, multiplex PCR	Lin et al. (2021)
		Identification of staphylococcal isolates	MALDI TOF-MS	Tegegne et al. (2021)
		Classical enterotoxin-related (sea, seb, sec, sed, and see), toxin from toxic shock syndrome (tsst-1), biofilm formation (icaA, icaB, and icaC), quorum-sensing agr system (agr), and methicillin resistance (mecA)	PCR, quantitative reverse transcription PCR	Castro et al. (2020)
3	*Klebsiella pneumonia*	Virulence genes—*rmpA*, *entB*, *ybtS*, *kfu*, *iutA*, *mrkD*, *allS*	Multiplex PCR	Compain et al. (2014)
		ybtS, *entB*, *mrkD*, *magA*, *kfu*, *iutA*, *rmpA*, and *K2* genes	PCR	Shakib et al. (2018)
		Capsules, adhesins, lipopolysaccharide, type 3 fimbriae	Signature-tagged mutagenesis (STM)	Lawlor et al. (2005)
		Virulence genes—magA, k2A, rmpA, wabG, uge, allS, entB, ycfM, kpn, wcaG, fimH, mrkD, iutA, iroN, hly ve, cnf-1	Multiplex PCR	Candan and Aksöz (2015)
		la_{KPC}, *rmpA*, *magA*, and serotype-specific genes	PCR	Xu et al. (2018)

(continued)

Table 6.2 (continued)

Serial no.	Pathogens	Virulence factors	Techniques	References
4	*Acinetobacter baumannii*	Colicin V, yersiniabactin, invasin, aerobactin, pathogenicity-associated island serum resistance	PCR	Mohajeri et al. (2016)
		Virulence genes (cnf1, csgA, cvaC, iutA)	PCR	AL-Kadmy et al. (2018)
		Virulence factor-encoded genes (FimH, afa/draBC, csgA, cnf1, cnf2, and *iutA)*	PCR	Tavakol et al. (2018)
		Phospholipase C (plc N), elastase (las B)	Conventional PCR	Kareem et al. (2017)
		Virulence genes	DNA microarray-based assay	Wolff et al. (2021)
		Biofilm-associated genes	PCR	Ali et al. (2017)
		Virulence genes (cnf1, csgA, cvaC, iutA)	PCR	Darvishi (2016)
5	*Pseudomonas aeruginosa*	Virulence genes (oprI, oprL, toxA, exoS, nan1, and lasB)	PCR	Mona et al. (2015); Aslani et al. (2012)
		Virulence genes	Multiplex PCR-based signature-tagged mutagenesis (STM)	Potvin et al. (2003)
		Virulence genes (*toxA*, *aprA*, *rhlAB*, *plcH*, *lasB*, and *fliC)*	PCR	Sabharwal et al. (2014)
		Exotoxin A (toxA), exoenzyme S (exoS), outer membrane protein (oprL), outer membrane lipoprotein I (oprI)	PCR	Neamah (2017)
		Virulence genes (tox A, exo U, exo S)	PCR	Bahador et al. (2019)
		Virulence genes (*OprI*, *OprL*, *LasB*, *PlcH*, *ExoS*, and *ToxA*)	PCR	Ullah et al. (2017)

(continued)

Table 6.2 (continued)

Serial no.	Pathogens	Virulence factors	Techniques	References
6	*Enterobacter* spp.	Siderophores, T6SSD, fimbriae	VFDB, BIGSdb	Manandhar et al. (2022)
		Virulence factors-encoding genes (siderophores, capsule, bacteriocin)	PCR	Hussain and Alammar (2013)
		Mannose-resistant, mannose-sensitive hemagglutination (MRHA, MSHA), hemolysin production, biofilm formation, serum resistance, and lipase, protease, and lecithinase	Phenotypic detection and PCR	Hassan et al. (2011)
		allS (allantoin metabolism), *entB*, *ybtS*, and *iutA* (siderophores), *kfu* (iron transport and phosphotransferase function), *mrkD* (adhesin type 3 fimbriae), *rmpA* (regulator of mucoid phenotype A), *ycfM* (outer membrane lipoprotein), and *fimH* (adhesive subunit of type 1 fimbriae)	PCR	Azevedo et al. (2018)

References

Ali HM, Salem MZM, El-Shikh MS, Megeed AA, Alogaibi YA, Talea IA (2017) Investigation of the virulence factors and molecular characterization of the clonal relations of multidrug-resistant Acinetobacter baumannii isolates. J AOAC Int 100:152–158. https://doi.org/10.5740/jaoacint.16-0139

Al-Kadmy IMS, Ali ANM, Salman IMA, Khazaal SS (2018) Molecular characterization of Acinetobacter baumannii isolated from Iraqi hospital environment. New Microbes New Infect 21:51–57. https://doi.org/10.1016/j.nmni.2017.10.010

Amala Reena AA, Subramaniyan A, Kanungo R (2017) Biofilm formation as a virulence factor of Acinetobacter baumannii: an emerging pathogen in critical care units. J Curr Res Sci Med 3:74. https://doi.org/10.4103/jcrsm.jcrsm_66_17

Arshadi M, Mahmoudi M, Motahar MS, Soltani S, Pourmand MR (2018) Virulence determinants and antimicrobial resistance patterns of vancomycin-resistant enterococcus faecium isolated from different sources in southwest Iran. Iran J Public Health 47:264–272

Aslani MM, Nikbin VS, Sharafi Z, Hashemipour M, Shahcheraghi F, Ebrahimipour GH (2012) Molecular identification and detection of virulence genes among Pseudomonas aeruginosa isolated from different infectious origins. Iran J Microbiol 4:118–123

Azevedo PAA, Furlan JPR, Oliveira-Silva M, Nakamura-Silva R, Gomes CN, Costa KRC, Stehling EG, Pitondo-Silva A (2018) Detection of virulence and β-lactamase encoding genes in Enterobacter aerogenes and Enterobacter cloacae clinical isolates from Brazil. Braz J Microbiol 49:224–228. https://doi.org/10.1016/j.bjm.2018.04.009

Bahador N, Shoja S, Faridi F, Dozandeh-Mobarrez B, Qeshmi FI, Javadpour S, Mokhtary S (2019) Molecular detection of virulence factors and biofilm formation in pseudomonas aeruginosa

obtained from different clinical specimens in bandar abbas. Iran J Microbiol 11:25–30. https://doi.org/10.18502/ijm.v11i1.701
Bien J, Sokolova O, Bozko P (2011) Characterization of virulence factors of Staphylococcus aureus: novel function of known virulence factors that are implicated in activation of airway epithelial proinflammatory response. J Pathog 2011:1–13. https://doi.org/10.4061/2011/601905
Braun G, Vidotto MC (2004) Evaluation of adherence, hemagglutination, and presence of genes codifying for virulence factors of Acinetobacter baumannii causing urinary tract infection. Mem Inst Oswaldo Cruz 99:839–844. https://doi.org/10.1590/S0074-02762004000800010
Bumgarner R (2013) Overview of DNA microarrays: types, applications, and their future. pp 1–11. https://doi.org/10.1002/0471142727.mb2201s101
Candan ED, Aksöz N (2015) Klebsiella pneumoniae: characteristics of carbapenem resistance and virulence factors*. Acta Biochim Pol 62:3–10
Castro RD, Pedroso SHSP, Sandes SHC, Silva GO, Luiz KCM, Dias RS, Filho RAT, Figueiredo HCP, Santos SG, Nunes AC, Souza MR (2020) Virulence factors and antimicrobial resistance of Staphylococcus aureus isolated from the production process of Minas artisanal cheese from the region of Campo das Vertentes, Brazil. J Dairy Sci 103:2098–2110. https://doi.org/10.3168/jds.2019-17138
Chambers HF, DeLeo FR (2009) Waves of resistance: Staphylococcus aureus in the antibiotic era. Nat Rev Microbiol 7:629–641. https://doi.org/10.1038/nrmicro2200
Chang D, Sharma L, Dela Cruz CS, Zhang D (2021) Clinical epidemiology, risk factors, and control strategies of Klebsiella pneumoniae infection. Front Microbiol 12:1–9. https://doi.org/10.3389/fmicb.2021.750662
Chen L, Yang J, Yu J, Yao Z, Sun L, Shen Y, Jin Q (2005) VFDB: a reference database for bacterial virulence factors. Nucleic Acids Res 33:325–328. https://doi.org/10.1093/nar/gki008
Chen L, Xiong Z, Sun L, Yang J, Jin Q (2012) VFDB 2012 update: toward the genetic diversity and molecular evolution of bacterial virulence factors. Nucleic Acids Res 40:641–645. https://doi.org/10.1093/nar/gkr989
Chen JW, Lau YY, Krishnan T, Chan KG, Chang CY (2018) Recent advances in molecular diagnosis of Pseudomonas aeruginosa infection by state-of-the-art genotyping techniques. Front Microbiol 9:1–8. https://doi.org/10.3389/fmicb.2018.01104
Comerlato CB, de Resende MCC, Caierão J, d'Azevedo PA (2013) Presence of virulence factors in Enterococcus faecalis and Enterococcus faecium susceptible and resistant to vancomycin. Mem Inst Oswaldo Cruz 108:590–595. https://doi.org/10.1590/S0074-02762013000500009
Compain F, Babosan A, Brisse S, Genel N, Audo J, Ailloud F (2014) Multiplex PCR for detection of seven virulence factors and K1/K2 capsular serotypes of Klebsiella pneumoniae. J Clin Microbiol 52:4377. https://doi.org/10.1128/JCM.02316-14
Corehtash ZG, Khorshidi A, Firoozeh F, Akbari H, Aznaveh AM (2015) Biofilm formation and virulence factors among pseudomonas aeruginosa isolated from burn patients. Jundishapur J Microbiol 8:e22345. https://doi.org/10.5812/jjm.22345
Dally S, Lemuth K, Kaase M, Rupp S, Knabbe C (2013) DNA microarray for genotyping antibiotic resistance determinants in Acinetobacter baumannii clinical isolates. Antimicrob Agents Chemother 57:4761–4768. https://doi.org/10.1128/AAC.00863-13
Darvishi M (2016) Virulence factors profile and antimicrobial resistance of Acinetobacter baumannii strains isolated from various infections recovered from immunosuppressive patients. Biomed Pharmacol J 9:1057–1062. https://doi.org/10.13005/bpj/1048
De Nies L, Lopes S, Busi SB, Galata V, Heintz-buschart A, Laczny CC, May P, Wilmes P (2021) PathoFact: a pipeline for the prediction of virulence factors and antimicrobial resistance genes in metagenomic data. Microbiome 9:1–14
Debouck C, Goodfellow PN (1999) DNA microarrays in drug discovery and development. Nat Genet 21:1–3
Dierig A, Frei R, Egli A (2015) The fast route to microbe identification matrix assisted laser desorption/ionization—time of flight mass spectrometry (MALDI-TOF MS). Pediatr Infect Dis J 34:97–99. https://doi.org/10.1097/INF.0000000000000601

Eijkelkamp BA, Stroeher UH, Hassan KA, Paulsen IT, Brown MH (2014) Comparative analysis of surface-exposed virulence factors of Acinetobacter baumannii. BMC Genomics 15:1–12. https://doi.org/10.1186/1471-2164-15-1020
Elnifro EM, Ashshi AM, Cooper RJ, Klapper PE (2000) Multiplex PCR: optimization and application in diagnostic virology. Clin Microbiol Rev 13:559–570. https://doi.org/10.1128/CMR.13.4.559-570.2000
Fishbain JT, Sinyavskiy O, Riederer K, Hujer AM, Bonomo RA (2012) Detection of extended-spectrum β-lactamase and Klebsiella pneumoniae carbapenemase genes directly from blood cultures by use of a nucleic acid microarray. J Clin Microbiol 50:2901. https://doi.org/10.1128/JCM.01023-12
Florio W, Baldeschi L, Rizzato C, Tavanti A, Ghelardi E (2020) Detection of antibiotic-resistance by MALDI-TOF mass spectrometry: an expanding area. Front Cell Infect Microbiol 10:1–7. https://doi.org/10.3389/fcimb.2020.572909
French N (2017) One Health and controlling zoonotic infections. Univ Otago
Garg A, Gupta D (2008) VirulentPred: a SVM based prediction method for virulent proteins in bacterial pathogens. BMC Bioinformatics 12:1–12. https://doi.org/10.1186/1471-2105-9-62
Gato E, Anantharajah A, Arroyo MJ, Artacho MJ (2022) Multicenter performance evaluation of MALDI-TOF MS for rapid detection of carbapenemase activity in Enterobacterales: the future of networking data analysis with online software. Front Microbiol 12:1–10. https://doi.org/10.3389/fmicb.2021.789731
Gerace E, Mancuso G, Midiri A, Poidomani S, Zummo S, Biondo C (2022) Recent advances in the use of molecular methods for the diagnosis of bacterial infections. Pathogens 11:1–13. https://doi.org/10.3390/pathogens11060663
Gilchrist CA, Turner SD, Riley MF, Petri WA, Hewlett L (2015) Whole-genome sequencing in outbreak analysis. Clin Microbiol Rev 28:541–563. https://doi.org/10.1128/CMR.00075-13
Harding CM, Hennon SW, Feldman MF (2018) Uncovering the mechanisms of Acinetobacter baumannii virulence. Nat Rev Microbiol 16:91–102. https://doi.org/10.1038/nrmicro.2017.148
Hassan R, El-Naggar W, El-Sawy E, El-Mahdy A (2011) Characterization of some virulence factors associated with Enterbacteriaceae isolated from urinary tract infections in Mansoura hospitals. Egypt J Med Microbiol 20:9–17
Heidari H, Hasanpour S, Ebrahim-Saraie HS, Motamedifar M (2017) High incidence of virulence factors among clinical Enterococcus faecalis isolates in Southwestern Iran. Infect Chemother 49:51–56. https://doi.org/10.3947/ic.2017.49.1.51
Highsmith AK, Jarvis WR (1985) Klebsiella pneumoniae: selected virulence factors that contribute to pathogenicity. Infect Control 6:75–77. https://doi.org/10.1017/S0195941700062640
Hollenbeck BL, Rice LB (2012) Intrinsic and acquired resistance mechanisms in enterococcus. Virulence 3:421–569. https://doi.org/10.4161/viru.21282
Hu T, Chitnis N, Monos D, Dinh A (2021) Next-generation sequencing technologies: an overview. Hum Immunol 82:801–811. https://doi.org/10.1016/j.humimm.2021.02.012
Hussain M, Alammar M (2013) Molecular study of some virulence factors encoding genes of Enterobacter spp. isolated from different clinical specimens. Al-Kufa Univ J Biol 5(2). https://journal.uokufa.edu.iq/index.php/ajb/article/view/7854
Jiang H, Wang K, Yan M, Ye Q, Lin X, Chen L, Ye Y (2021) Pathogenic and virulence factor detection on viable but non-culturable methicillin-resistant Staphylococcus aureus. Front Microbiol 12:1–8. https://doi.org/10.3389/fmicb.2021.630053
Jurado-Martín I, Sainz-Mejías M, McClean S (2021) Pseudomonas aeruginosa: an audacious pathogen with an adaptable arsenal of virulence factors. Int J Mol Sci 22:1–37. https://doi.org/10.3390/ijms22063128
Kalpana S, Lin W-Y, Wang Y-C, Fu Y, Lakshmi A, Wang H-Y (2023) Antibiotic resistance diagnosis in ESKAPE pathogens—a review on proteomic perspective. Diagnostics 13:1014. https://doi.org/10.3390/diagnostics13061014
Kareem SM, Al-Kadmy IMS, Al-Kaabi MH, Aziz SN, Ahmad M (2017) Acinetobacter baumannii virulence is enhanced by the combined presence of virulence factors genes phospho-

lipase C (plcN) and elastase (lasB). Microb Pathog 110:568–572. https://doi.org/10.1016/j.micpath.2017.08.001
Khaertynov KS, Anokhin VA, Rizvanov AA, Davidyuk YN, Semyenova DR, Lubin SA, Skvortsova NN (2018) Virulence factors and antibiotic resistance of Klebsiella pneumoniae strains isolated from neonates with sepsis. Front Med 5:1–9. https://doi.org/10.3389/fmed.2018.00225
Kumari K, Sharma PK, Singh RP (2023) Unraveling the virulence factors and secreted proteins of an environmental isolate Enterobacter sp. S-16. Curr Microbiol 80:1–11. https://doi.org/10.1007/s00284-023-03197-0
Kwong JC, McCallum N, Sintchenko V, Howden BP (2015) Whole genome sequencing in clinical and public health microbiology. Pathology 47:199–210. https://doi.org/10.1097/PAT.0000000000000235
Lawlor MS, Hsu J, Rick PD, Miller VL (2005) Identification of Klebsiella pneumoniae virulence determinants using an intranasal infection model. Mol Microbiol 58:1054–1073. https://doi.org/10.1111/j.1365-2958.2005.04918.x
Lentz CS, Sheldon JR, Crawford LA, Cooper R, Garland M, Amieva MR, Weerapana E, Skaar EP, Bogyo M (2018) Identification of a S. aureus virulence factor by activity-based protein profiling (ABPP). Nat Chem Biol 14:609. https://doi.org/10.1038/s41589-018-0060-1
Liao C, Huang X, Wang Q, Yao D, Lu W (2022) Virulence factors of Pseudomonas aeruginosa and antivirulence strategies to combat its drug resistance. Front Cell Infect Microbiol 12:1–17. https://doi.org/10.3389/fcimb.2022.926758
Lin T, Li Q, Jin D, Liu W, Tang C, Zhang X (2021) Investigation of virulence genes of Staphylococcus aureus isolated from sterile body fluid samples and their correlation with clinical symptoms and outcomes. Can J Infect Dis Med Microbiol 2021:5354747
Lindsay JA, Holden MTG (2004) Staphylococcus aureus: superbug, super genome? Trends Microbiol 12:378–385. https://doi.org/10.1016/j.tim.2004.06.004
Liu L, Li Y, Li S, Hu N, He Y, Pong R, Lin D, Lu L, Law M (2012) Comparison of next-generation sequencing systems. J Biomed Biotechnol 2012:251364. https://doi.org/10.1155/2012/251364
Malani PN, Kauffman CA, Zervos MJ (2014) Enterococcal disease, epidemiology, and treatment. In: The Enterococci. pp 385–408. https://doi.org/10.1128/9781555817923.ch10
Manandhar S, Nguyen Q, Nguyen Thi Nguyen T, Pham DT, Rabaa MA, Dongol S, Basnyat B, Dixit SM, Baker S, Karkey A (2022) Genomic epidemiology, antimicrobial resistance and virulence factors of Enterobacter cloacae complex causing potential community-onset bloodstream infections in a tertiary care hospital of Nepal. JAC Antimicrob Resist 4:1–8. https://doi.org/10.1093/jacamr/dlac050
Maurin M (2012) Real-time PCR as a diagnostic tool for bacterial diseases. Expert Rev Mol Diagn 12:731–754. https://doi.org/10.1586/erm.12.53
Mezzatesta ML, Gona F, Stefani S (2012) Enterobacter cloacae complex: clinical impact and emerging antibiotic resistance. Future Microbiol 7:887–902. https://doi.org/10.2217/fmb.12.61
Mishra M, Panda S, Barik S, Sarkar A, Singh DV, Mohapatra H (2020) Antibiotic resistance profile, outer membrane proteins, virulence factors and genome sequence analysis reveal clinical isolates of Enterobacter are potential pathogens compared to environmental isolates. Front Cell Infect Microbiol 10:1–13. https://doi.org/10.3389/fcimb.2020.00054
Mitton B, Rule R, Said M (2021) Laboratory evaluation of the BioFire FilmArray Pneumonia plus panel compared to conventional methods for the identification of bacteria in lower respiratory tract specimens: a prospective cross-sectional study from South Africa. Diagn Microbiol Infect Dis 99:115236. https://doi.org/10.1016/j.diagmicrobio.2020.115236
Mohajeri P, Sharbati S, Farahani A, Rezaei Z (2016) Evaluate the frequency distribution of nonadhesive virulence factors in carbapenemase-producing Acinetobacter baumannii isolated from clinical samples in Kermanshah. J Nat Sci Biol Med 7:58–61. https://doi.org/10.4103/0976-9668.175071
Mona S, Nour MAK, ElSheshtawy NM (2015) Genetic identification of Pseudomonas aeruginosa virulence genes among different isolates. J Microb Biochem Technol 07:274–277. https://doi.org/10.4172/1948-5948.1000224

Naresh V, Lee N (2021) A review on biosensors and recent development of nanostructured materials-enabled biosensors. Sensors (Basel) 21(4):1109

Neamah AA (2017) Molecular detection of virulence factor genes in pseudomonas aeruginosa isolated from human and animals in Diwaniya province. Kufa J Vet Med Sci 8:218–230

Neoh H, Tan X, Fazilla H, Leong T (2019) Pulsed-field gel electrophoresis (PFGE): a review of the "gold standard" for bacteria typing and current alternatives. Infect Genet Evol 74:103935. https://doi.org/10.1016/j.meegid.2019.103935

Novick RP, Geisinger E (2008) Quorum sensing in staphylococci. Annu Rev Genet 42:541–564. https://doi.org/10.1146/annurev.genet.42.110807.091640

Obande GA, Singh KKB (2020) Current and future perspectives on isothermal nucleic acid amplification technologies for diagnosing infections. Infect Drug Resist 13:455–483. https://doi.org/10.2147/IDR.S217571

Paauw A, Caspers MPM, Leverstein-van Hall MA, Schuren FHJ, Montijn RC, Verhoef J, Fluit AC (2009) Identification of resistance and virulence factors in an epidemic Enterobacter hormaechei outbreak strain. Microbiology 155:1478–1488. https://doi.org/10.1099/mic.0.024828-0

Potvin E, Lehoux DE, Kukavica-Ibrulj I, Richard KL, Sanschagrin F, Lau GW, Levesque RC (2003) In vivo functional genomics of Pseudomonas aeruginosa for high-throughput screening of new virulence factors and antibacterial targets. Environ Microbiol 5:1294–1308. https://doi.org/10.1046/j.1462-2920.2003.00542.x

Sabharwal N, Dhall S, Chhibber S, Harjai K (2014) Molecular detection of virulence genes as markers in Pseudomonas aeruginosa isolated from urinary tract infections. Int J Mol Epidemiol Genet 5:125–134

Saxena K, Murti BT, Yang P, Malhotra BD, Chauhan N, Jain U (2022) Fabrication of a molecularly imprinted nano-interface-based electrochemical biosensor for the detection of CagA virulence factors of *H. pylori*. Biosensors 12:1066. https://doi.org/10.3390/bios12121066

Saxena K, Chauhan N, Malhotra BD, Jain U (2023) A molecularly imprinted polymer-based electrochemical biosensor for detection of VacA virulence factor of H. pylori causing gastric cancer. Process Biochem 130:87–95. https://doi.org/10.1016/j.procbio.2023.03.024

Shafeek MY, El-malt LM, Hameed KGA, El-zamkan MA (2018) Some virulence genes of pathogenic enterococci isolated from raw milk and some milk products. SVU-Int J Vet Sci 1:102–113

Shahraki S, Rabi Nezhad Mousavi M (2017) Determination of virulence factors in clinical multidrug resistance enterococci isolates at Southeast of Iran. Jundishapur J Microbiol 10:e45514. https://doi.org/10.5812/jjm.45514

Shakib P, Kalani MT, Ramazanzadeh R, Rouhi S (2018) Molecular detection of virulence genes in Klebsiella Pneumoniae clinical isolates from Kurdistan Province. Iran Biomed Press 5:2581–2589

Tavakol M, Momtaz H, Mohajeri P, Shokoohizadeh L, Tajbakhsh E (2018) Genotyping and distribution of putative virulence factors and antibiotic resistance genes of Acinetobacter baumannii strains isolated from raw meat. Antimicrob Resist Infect Control 7:1–11. https://doi.org/10.1186/s13756-018-0405-2

Tegegne DT, Mamo G, Waktole H, Messele YE (2021) Molecular characterization of virulence factors in Staphylococcus aureus isolated from bovine subclinical mastitis in central. Ann Microbiol 71:28

Torres-miranda D, Akselrod H, Karsner R, Secco A, Silva-cantillo D, Siegel MO, Roberts AD, Simon GL (2020) Use of BioFire FilmArray gastrointestinal PCR panel associated with reductions in antibiotic use, time to optimal antibiotics, and length of stay. BMC Gastroenterol 20:1–7

Torres-Sangiao E, Rodriguez CL (2021) Application and perspectives of MALDI-TOF mass spectrometry in clinical microbiology laboratories. Microorganisms 9:1–19

Trivedi K, Cupakova S, Karpiskova R (2011) Virulence factors and antibiotic resistance in enterococci isolated from food-stuffs. Vet Med 56:352–357

Ullah W, Qasim M, Rahman H, Jie Y, Muhammad N (2017) Beta-lactamase-producing Pseudomonas aeruginosa: phenotypic characteristics and molecular identification of virulence genes. J Chin Med Assoc 80:173–177. https://doi.org/10.1016/j.jcma.2016.08.011

van Delden C (2004) Virulence factors In Pseudomonas aeruginosa. In: Granada S, Ramos J-L (eds) CSIC. Virulence and gene regulation. Springer, Boston, MA, pp 03–45

Vosberg HP (1989) The polymerase chain reaction: an improved method for the analysis of nucleic acids. Hum Genet 83:1–15. https://doi.org/10.1007/BF00274139

Wolff N, Hendling M, Schroeder F, Schönthaler S, Geiss AF, Bedenic B, Barišić I (2021) Full pathogen characterisation: species identification including the detection of virulence factors and antibiotic resistance genes via multiplex DNA-assays. Sci Rep 11:1–14. https://doi.org/10.1038/s41598-021-85438-5

Xu M, Fu Y, Kong H, Chen X, Chen Y, Li L, Yang Q (2018) Bloodstream infections caused by Klebsiella pneumoniae: prevalence of bla KPC, virulence factors and their impacts on clinical outcome. BMC Infect Dis 18:1–9

Yang J, Chen L, Sun L, Yu J, Jin Q (2008) VFDB 2008 release: an enhanced web-based resource for comparative pathogenomics. Nucleic Acids Res 36:539–542. https://doi.org/10.1093/nar/gkm951

Young I, Huh K, Jin H, Ae S, Na Y (2020) Evaluation of the BioFire FilmArray Pneumonia Panel for rapid detection of respiratory bacterial pathogens and antibiotic resistance genes in sputum and endotracheal aspirate specimens. Int J Infect Dis 95:326–331. https://doi.org/10.1016/j.ijid.2020.03.024

Zhang D, Bi H, Liu B, Qiao L (2018) Detection of pathogenic microorganisms by microfluidics based analytical methods. Anal Chem 90(9):5512–5520. https://doi.org/10.1021/acs.analchem.8b00399

Zhou CE, Smith J, Lam M, Zemla A, Dyer MD, Slezak T (2007) MvirDB—a microbial database of protein toxins, virulence factors and antibiotic resistance genes for bio-defence applications. Nucleic Acids Res 35:391–394. https://doi.org/10.1093/nar/gkl791

Zhou JX, Feng DY, Li X, Zhu JX, Wu WB, Zhang TT (2023) Advances in research on virulence factors of Acinetobacter baumannii and their potential as novel therapeutic targets. J Appl Microbiol 134:1–9. https://doi.org/10.1093/jambio/lxac089

Zhu H, Zhang H, Xu Y, Laššáková S, Korabečná M, Neužil P (2020) PCR past, present and future. Biotechniques 69:317–325. https://doi.org/10.2144/BTN-2020-0057

Zhu Y, Huang WE, Yang Q (2022) Clinical perspective of antimicrobial resistance in bacteria. Infect Drug Resist 15:735–746. https://doi.org/10.2147/IDR.S345574

Enterococcus faecium Virulence Factors and Biofilm Components: Synthesis, Structure, Function, and Inhibitors

7

Suseela Lanka, Anitha Katta, Mounika Kovvali, and Santhilatha Pandrangi

Abstract

Enterococcus faecium, a Gram-positive bacterium belonging to the family Enterococcaceae is known for its clinical significance due to its ability to cause infection, especially in healthcare. It exists in anaerobic conditions and is frequently found in the human intestinal tract. This strain is extremely resistant to aminoglycosides, penicillin, and vancomycin as it possesses innate and acquired antibiotic resistance. This chapter provides an overview of virulence factors and biofilm formation that enhance the organism's pathogenicity. The virulence factors are mainly categorized into secreted factors and cell surface factors. The secreted virulence factors and their genes concerned include cytolysin (clyA, clyB, clyM), secreted antigen A (sagA), gelatinase (efaAfs, efaAfm), and Glucosyl hydrolase (hyl_{Efm}). The cell surface virulence factors include Aggregation Substances (asp1, asc10 (prgB), asa1), MSCRAMMs (ace, acm, scm), Pili (piliA, piliB, piliF), Esp (espfs, espfm), which cause damage to the host cell and help in the formation of biofilm. Biofilm formation involves coordinated action of various components, enzymes, and genetic pathways. Polysaccharides such as exopolysaccharides are synthesized and secreted by the bacterium; lipids & proteins stabilize the structure of biofilm; while nucleic acid (eDNA) gives the structural integrity to the matrix and exchange of genetic material. Management and control of *E. faecium* infection includes the development of inhibitors targeting biofilm formation and virulence factors. Potential inhibitors include a quorum sensing system that interferes with bacterial communica-

S. Lanka (✉) · A. Katta · M. Kovvali
Department of Biosciences and Biotechnology, Krishna University, Machilipatnam, Andhra Pradesh, India

S. Pandrangi
Department of Biochemistry and Bioinformatics, GITAM School of Sciences, GITAM (Deemed to be) University, Visakhapatnam, India

S. Busi, R. Prasad (eds.), *ESKAPE Pathogens*,
https://doi.org/10.1007/978-981-99-8799-3_7

tion, an electrochemical method for the degradation of biofilm, degradation of the EPS matrix of biofilm, external membrane structure, and enzyme-mediated biofilm control that degrades the biofilm. Additionally, the discovery of new antibiotics and therapeutic strategies remains essential for managing *E. faecium* infection. Understanding the virulence factors, biofilm formation, and potential inhibitors of *E. faecium* is critical for developing effective treatments and preventing the spread of this pathogen in healthcare settings.

Keywords

Enterococcus faecium · Virulence factors · Biofilm components · Biofilm formation · Antibiotic resistance · Inhibitors

7.1 Introduction

The Gram-positive lactic acid-producing bacteria *Enterococcus faecium*, which has a spherical appearance, exists in pairs or chains. The developed colonies have a length of 1–2 mm and seem moist. It is an anaerobic bacterium that may exist with or without oxygen. It is a member of the family Enterrococcaceae, genus Enterococcus, species *Enterococcus faecium,* and belongs to the kingdom Bacteria. Both humans and animals' digestive tracts can include it as a commensal. However, it can also result in illnesses like endocarditis or newborn meningitis.

E. faecium, which is frequently found in the human intestinal tract and is resistant to vancomycin (VRE), was originally discovered in 1986 and has been linked to a higher death rate in patients with bacteremia. It is acknowledged to have the populations with the greatest physiological variability. It does not create spores or capsules like certain bacteria do. It is next to *E. faecalis* in causing hospital-acquired infections, accounting for roughly 5–16% of *Enterococci* detected in human illnesses (Gray et al. 1991; Livornese Jr et al. 1992; Sader et al. 1994). As per the studies of Facklam and Collins (1989), Ruoff et al. (1990), Gordon et al. (1992), and Stern et al. (1994), it is second most common *Enterococcus* species to be isolated from human illnesses.

E. faecium has also demonstrated a propensity to evolve resistance to a variety of antimicrobial drugs (Chenoweth and Schaberg 1996; Spera Jr and Farber 1992), making it one of the *Enterococcal* species with the greatest physical diversity (Teixeira et al. 1995). *E. faecium* strains that are extremely resistant to aminoglycosides, penicillin, and vancomycin have recently been the subject of an increasing number of studies (Gordon et al. 1992). Because it possesses both innate and acquired antibiotic resistance, *E. faecium* is among them and is especially worrisome (Klare et al. 2003). Because of its extraordinary development, *Enterococcus faecium* multidrug-resistant has been referred to as the "nosocomial pathogen of the 1990s" (Spera Jr and Farber 1992).

Today, *E. faecium* is regarded as an important infection acquired in hospitals. Depending on the species, there are differences in the frequency and severity of *Enterococcal* infections, as well as in mortality and drug resistance. A

PCR-mediated assay was developed to swiftly identify this bacterium because conventional biochemical methods had difficulty in identifying *E. faecium* apart from recently identified *Enterococcal* species (Cheng et al. 1997).

7.2 Virulence Factors

Pathogen invasion, infection persistence, and the potential for biofilm development in host tissue are all correlated with virulence factors. Along with virulence factors, cell wall components such as capsular polysaccharides, lipoteichoic acid, and wall teichoic acid (WTA) increase the pathogenicity of Gram-positive bacteria. They play a vital role in immunogenic processes viz., phagocytosis, complement activation, antimicrobial resistance, biofilm development, and host/surface attachment. Due to its dominant virulence characteristics, *E. faecalis* continues to play the predominant role in nosocomial-type infections. Though not a significant problem in the genus *Enterococcus*, virulence factors are helpful for environmental survival and can be linked to the development of biofilms and surface adhesion (Gao et al. 2018).

7.2.1 Virulence Factors (Secreted Nature)

Cytolysin, one of the first *Enterococci* virulence factors was discovered by Segarra et al. (1991). In addition to cytolytic activity, it also exhibits bactericidal activity (Brock and Davie 1963; Ike et al. 1984; Gilmore et al. 1990). Cytolysin has been linked to endophthalmitis and endocarditis, as well as some action against mouse white and red blood cells (Miyazaki et al. 1993; Jett et al. 1992; Chow et al. 1993). Recent studies have not found any connection between endophthalmitis and cytolysin-producing Enterococci (Todokoro et al. 2017; Chilambi et al. 2021), but studies of Jett et al. (1992) using a rabbit model suggested a potential link between the two conditions.

Only *E. faecium* isolates have SagA (secreted antigen A) (Solheim et al. 2009). It is a protein that is associated to stress (Muller et al. 2006) and may interfere with cell wall metabolism, which in turn may impact cell development (Teng et al. 2003). SagA can also bind to proteins in the extracellular matrix (Pedicord et al. 2016), improve the function of the intestinal barrier, and promote tolerance to enteric pathogenic microbes like *Clostridium difficile* and *Salmonella typhimurium,* possibly as a result of activating the host's innate immune mechanisms (Rangan et al. 2016; Kim et al. 2019; Paganelli et al. 2015). However, only Enterococci from clade A1 have been linked to formation of biofilm (Paganelli et al. 2015).

The majority of clinical isolates, especially *E. faecium* CC17, consists of the *hylEfm* genes, which encode a glucosyl hydrolase (Rice et al. 2003; Top et al. 2008). It has been proposed that it might propagate antibiotic resistance genes even if there are few studies that support its potential as a virulence factor (Arias et al. 2009).

7.2.2 Cell Surface Virulence Factors

The production of biofilms and defense against the host immune system are two bacterial defensive mechanisms that depend heavily on cell surface components (Fabretti et al. 2006; Geiss-Liebisch et al. 2012). According to Hendrickx et al. (2009a, b), the surface proteins on enterococci are often of the type LPxTG and contain leucine (L), proline (P), any amino acid (X), threonine (T), and glycine (G), as well as pili and MSCRAMMs (microbial surface components recognizing adhesive matrix molecules).

There are three adhesins—Asp1, Asc10 (known as PrgB), and Asa1—that form part of the cell surface proteins group that includes the substances resulting in aggregation (called AS or Agg). These have a similar amino acid sequence, and encoded by three different conjugative type plasmids, pPD1, pCF10, and pAD1, that play role in Enterococcal adherence to cells (Hendrickx et al. 2009a, b). It has been hypothesized that these adhesins are what cause Enterococci to adhere to renal tubular cells (Kreft et al. 1992) and the intestinal tract (Isenmann et al. 2000; Sartingen et al. 2000; Wells et al. 2000). AS has been linked to phagocytosis survival, vanA (an operon associated with resistance to vancomycin) Paoletti et al. (2007) co-transference in *E. faecalis*, and adherence to immune cells, particularly Asa1 and Asc10 aggregation factors (Vanek et al. 1999; Süßmuth et al. 2000). Asc10, produced by the *prgB* gene, has also been linked to increased virulence in endocarditis (infective endocarditis) and biofilm development (Schlievert et al. 1998; Bhatty et al. 2015).

There are three main MSCRAMMs, belong to a subfamily of bacterial adhesins which recognize and bind to ECM (extracellular matrix) components, that have been identified in enterococci: Ace (Rich et al. 1999), Acm (Nallapareddy et al. 2003) and Scm (Sillanpää et al. 2008), (E. faecium collagen adhesions).

Enterococci have been found to harbor the MSCRAMMs Ace, Acm, and Scm (Rich et al. 1999; Nallapareddy et al. 2003; Sillanpää et al. 2008). While Scm binds to fibrinogen, Ace can stick to various substances such laminin and dentin (Nallapareddy et al. 2000; Kowalski et al. 2006). Ace and Acm both contribute to the emergence of enterococcal-infected endocarditis (Nallapareddy et al. 2008a, b; Singh et al. 2010, 2018). The reason that this complex became so important as a nosocomial pathogen is that Acm is mostly disseminated among *E. faecium* clinical isolates, with a greater incidence among CC17 (Nallapareddy et al. 2008a, b).

Pili are virulence factors that stretch as lengthy filaments arising from cell surfaces, and are made up of multimeric fibers comprised by pilin subunits. These fibers are encoded by an operon of several genes in enterococci known as the pili gene clusters (PGC). According to Hendrickx et al. (2009a, b), Sillanpää et al. (2008), and Qin et al. (2012), *E. faecium* has four gene clusters, PGC1-4. *EmpA*, *EmpB*, and *EmpC* genes are found in PGC3 (EMP). In a model of urinary tract infection, EMP is associated with biofilm development, adhesion to EMCs, and colonization of the kidney and bladder (Sillanpää et al. 2010; Montealegre et al. 2016).

Esp is a different cell surface protein that is primarily present in clinical isolates of *E. faecium* and *E. faecalis* (Shankar et al. 1999; Eaton and Gasson 2002). Through an amyloid-based mechanism (Taglialegna et al. 2020), this protein is found to be crucial for the formation of biofilms, but it has also been linked to the development of infectious endocarditis and urinary tract infections (Heikens et al. 2011; Shankar et al. 2001). A more recent study (Weng et al. 2019) has confirmed Meredith et al. (2009)'s hypothesis that the existence of this protein may impact *E. faecium*'s vulnerability to β-lactams. PrpA (Prieto et al. 2015), which binds to fibronectin, fibrinogen, and platelets, as well SgrA and EcbA, the two adhesins that are typically found in clinical isolates of Enterococcus (Hendrickx et al. 2009a, b), are other cell surface proteins that have the potential to be virulence factors. Table 7.1 shows Enterococcus faecium virulence factors—types, genes concerned, and associated functions (Tendolkar et al. 2004; Toledo-Arana et al. 2001) (Table 7.1).

7.3 Biofilm Formation

Planktonic (free-floating) and sessile (adhering to a surface) are the two states in which bacterial cells can be found. The genes involved for the manufacture and maturation of extracellular material are altered as a result of the bacterial cells' adhesion to a surface, and this process results in the formation of a protective biofilm. A structured polymeric matrix that is adhering to a surface and encloses bacterial cells is the classic definition of a biofilm (Bjarnsholt 2011). Both Gram-positive and Gram-negative bacteria are able to form biofilms, including the ESKAPE pathogens which are the leading cause of nosocomial infections throughout the world (Mei et al. 2021; Chen et al. 2013; Santajit and Indrawattana 2016).

7.3.1 Biofilm Components

Exopolysaccharides, proteins, lipids, and extracellular DNA (eDNA) make up the majority of bacterial EPS (extracellular polymeric substances) (Flemming et al. 2007). Cells must expend a large amount of energy and resources to produce EPS. However, the biofilm chemicals seen in Fig. 7.1 provide a separate environment with particular capabilities for bacterial cells. Figure 7.1 shows the different components of a biofilm (Fig. 7.1).

7.3.1.1 Polysaccharides

Both Gram-positive and Gram-negative bacteria use polysaccharides as a large part of the EPS matrix (Ruhal and Kataria 2021). Although certain exopolysaccharides, such as glucans, fructans, and cellulose, are homopolysaccharides, the majority are heteropolysaccharides, which combine neutral and charged sugar moieties. Well-documented polyanionic exopolysaccharides include xanthan, alginate, and colanic acid, even as polycationic exopolysaccharides include PNAG (Poly-(1-6)-*N*-acetylglucosamine), and Pel. EPS polysaccharides play wide roles in the

Table 7.1 *Enterococcus faecium* virulence factors—types, genes concerned, and associated functions

S. no	Types of factors	Genes concerned	Virulence factors	Functions
1	Secreted factors	*clyA, clyB, clyM*	Cytolysin (Cyl)	Activation, transport, and posttranslational modification of cytolysin
		sagA	Secreted antigen A	Stress-related protein interferes with cell wall metabolism; it could potentially be involved in virulence and pathogenicity; interacts with the host's immune system or other cells during infection
		efaAfs, efaAfm	Gelatinase (gelE)	Matrix metalloproteinase; interferes with complement-mediated immunity; hydrolyses gelatin and collagen; participates in the development of infectious endocarditis; cell wall adhesion
		hyl_{Efm}	Glucosyl hydrolase(hyl)	Encode a glucosyl hydrolase; antibiotic resistance genes
2	Cell-surface factors	*asp1, asc10 (prgB), asa1*	Aggregation substances (As/Agg)	Adhesion to the surface of intestinal and renal tubular cells; **Asp1** associated with extracellular matrix protein adhesion; **Asa1**-promoting infection and shielding them from host's immune responses; **Asc10** includes promoting cell to cell interactions in the biofilm formation
		ace, acm, scm	MSCRAMMs	Bind to extracellular matrix proteins; **Ace**—*E. faecium* adheres to collagen in host tissue; **Acm**—adhesion of *E. faecium* to collagen; **Scm**—it enables *E. faecium* to adhere to fibrinogen
		piliA, piliB, piliF	Pili	Biofilm, formation; bind to extracellular matrix protein; colonizing both the kidneys and the bladder
		espfs, espfm	Esp	Adhesion of *E. faecium* to host tissues; biofilm formation; interferes with host immune response and helps the *E. faecium* to evade immune detection and clearance

development and maintenance of biofilms. They act as a physical framework for cell and surface attachment, a barrier between biofilm cells and the environment, a means of surface colonization, a catalyst for bacterial aggregation, a source of nutrients, a hydrated polymer network that stabilizes the biofilm, and a means of resistance to host defense (Flemming and Wingender 2010).

7.3.1.2 Lipids

Recent research suggests that the structural stability of the biofilm may depend on the lipids found on the plasma membrane and ECM (Rella et al. 2016; Pierce et al. 2017). It has been demonstrated that lipids are essential for controlling microbial biofilms. The distribution of phospholipids and their molecular species differs

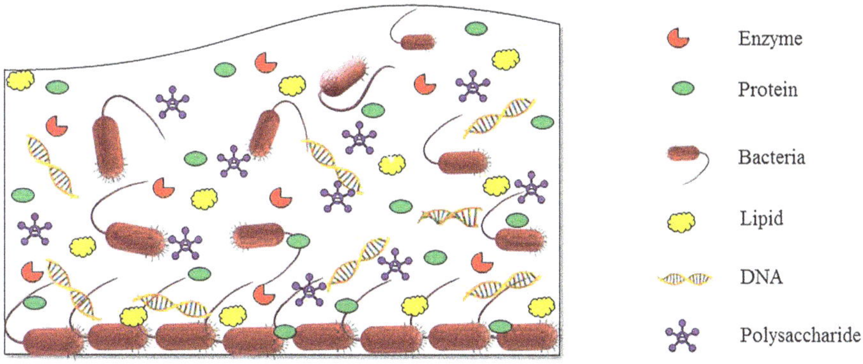

Fig. 7.1 Components of a biofilm

between planktonic and biofilm-forming cells (Mukherjee et al. 2003). A single inner phospholipid bilayer is present in the Gram-positive bacteria *E. faecium* (Auer and Weibel 2017).

7.3.1.3 Proteins

Proteins and polysaccharides found in EPS composition are crucial for morphogenesis, adhesion, pathogenicity, and the maintenance of biofilms. Additionally, they defend cells from phagocytes and toxic environments (Chaffin et al. 1998; Bridier et al. 2011). By joining cells to EPS, structural proteins maintain the structure of biofilms (Fong and Yildiz 2015). The Esp is a surface protein that plays a significant role in biofilm formation, an important aspect of infection pathogenesis, and is the first known determinant in *E. faecium* CC17 (Heikens et al. 2007).

7.3.1.4 Nucleic Acids

eDNA, another essential component of EPS, contributes to the stability, nutrition provision, genetic exchange, biofilm integrity, and resistance to drugs (Donlan 2002; Martins et al. 2010). The biological functions of eDNA in the biofilm matrix are confirmed (Montanaro et al. 2011; Yu et al. 2019) and differ with the varied stages of biological cycle (adhesion, early stages of biofilm development, biofilm matrix stabilization, horizontal gene transfer, phagocytosis prevention, inflammation inhibition). Table 7.2 shows different components of a biofilm and their role in its formation Thomas et al. (2009) (Table 7.2).

7.3.2 Synthesis of Biofilm

Numerous elements, including as hydrodynamic parameters, nutritional content, bacterial movement, and intercellular communication, have an impact on the development of biofilms. Cell mobility, electrostatic and hydrophobic interactions,

Table 7.2 Biofilm components and their role in the biofilm formation

S. no	Components	Roles in the biofilm formation	Reference
1.	Bacterial cells	The structural unit of biofilm embedded in the EPS matrix	Segev-Zarko et al. (2015)
2.	Nucleic acids	eDNA gives structural integrity to the matrix Exchange of genetic material	Guo et al. (2016)
3.	Polysaccharides	EPS matrix adheres to the cell surface for the formation of a network Absorption of organic and inorganic ions Protective barriers to the EPS matrix Water retention within the biofilm	Guo et al. (2016)
4.	Structural proteins	Stabilize the structure of biofilm by connecting cells to the EPS matrix Protective barriers to the EPS matrix	Segev-Zarko et al. (2015)
5.	Lipids	Nutrient source	Guo et al. (2016)
6.	Enzymes	Enzymatic activity nutrient source It promotes the restructuring of biofilm and dispersal	Guo et al. (2016)
7.	Water	Lubricates the cells within the matrix Distribution of nutrients to bacterial cells It acts as a simple circulatory system	Guo et al. (2016)

adhesin expression, and other variables all affect bacterial adherence to formation sites (Flemming and Wingender 2010; Trentin et al. 2013).

1. **Attachment/Reversible Attachment Phase:** During the early stage, weak interactions such as acid-base, hydrophobic, Vander Waals, and electrostatic forces are used to reversibly attach free-swimming planktonic cells to the biotic or abiotic surfaces.
2. **Colonization/Irreversible Attachment Phase**: The bacterial pathogens permanently affix to the surface through stronger contacts, which involve the adherence of the bacteria through adhesins like collagen-binding sticky proteins, lipopolysaccharides, flagella, and pili.
3. **Proliferation:** Extracellular Polymeric Substances (EPS) are created in large quantities when the multilayered bacterial cells are extensively accumulated.
4. **Maturation:** The creation of multilayered bacterial cells is developed into the matured biofilm with the typical 3D biofilm structure when bacterial cells synthesis and release signaling molecules to communicate with one another.
5. **Dispersion/Detachment Phase:** After biofilms have fully developed, they are dismantled or dispersed through mechanical and active procedures, where bacteria detach from them and return to their planktonic state.

Figure 7.2 shows the process of biofilm formation (Fig. 7.2).

7.3.2.1 Structure and Functions of Biofilm

A colony of prokaryotic organisms that are attached to a surface and covered with a polysaccharide layer makes up a microbial biofilm. The slime layer is made up of

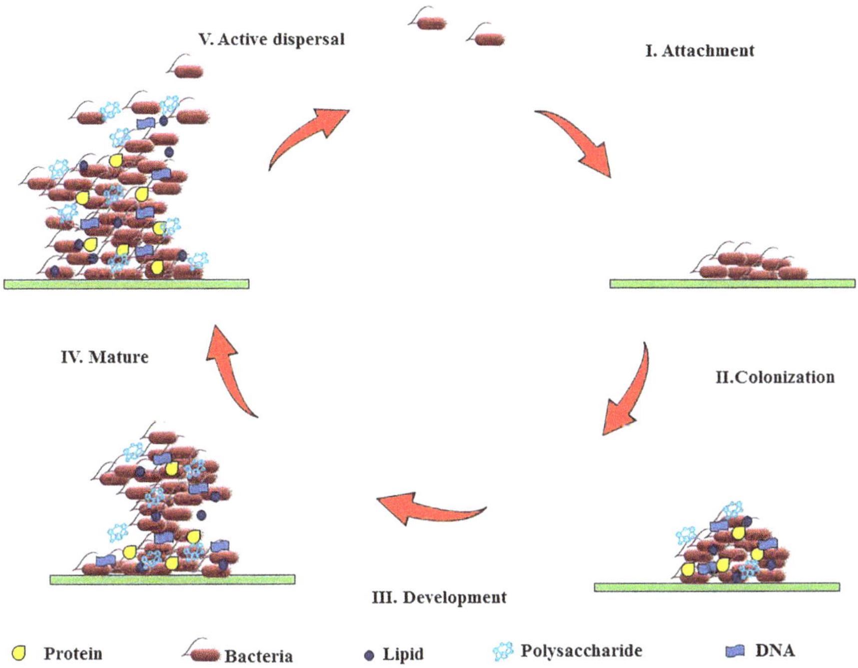

Fig. 7.2 Synthesis of biofilm

porous layers and channels that let the colony's central cells take in nutrients and flush out waste, establishing and sustaining the biofilm through cell-to-cell interaction. When a few cells adhere to a surface, the biofilm is created. Proteins made by the cells then serve as signals for neighboring cells, causing the colony to swell and the biofilm to spread. According to Dunny et al. (2014), the proteins also indicate the emergence of the polysaccharides that make up the slime layer and contain eDNA, which exchanges genetic material (Fig. 7.3).

Microorganisms gather and establish a colony in a biofilm as a means of cooperating metabolically. Through stronger defense, easier access to nutrients, and better chances for cellular communication and genetic material transfer, this cooperative growth technique improves cells' chances of surviving. Fighting off physical threats like immune system elimination or displacement by flowing fluids requires effective cellular defense. The polysaccharide coating on the biofilm serves as an adhesive, nutrient supply, cohesion of the structure, and source of cohesion, securing the colony to a surface and inhibiting physical removal of cells as well as immune system or antibiotic penetration of the biofilm. Particularly eDNA, nucleic acids exchange the genetic material in bacterial cells. Biofilms can be difficult to eradicate and present health hazards to people, like in cystic fibrosis, where the development of a biofilm in the lungs can cause uncomfortable symptoms. For bacteria, the biofilm provides the perfect environment. The attachment of the cells to a surface rich in nutrients enables them to flourish in a particular niche.

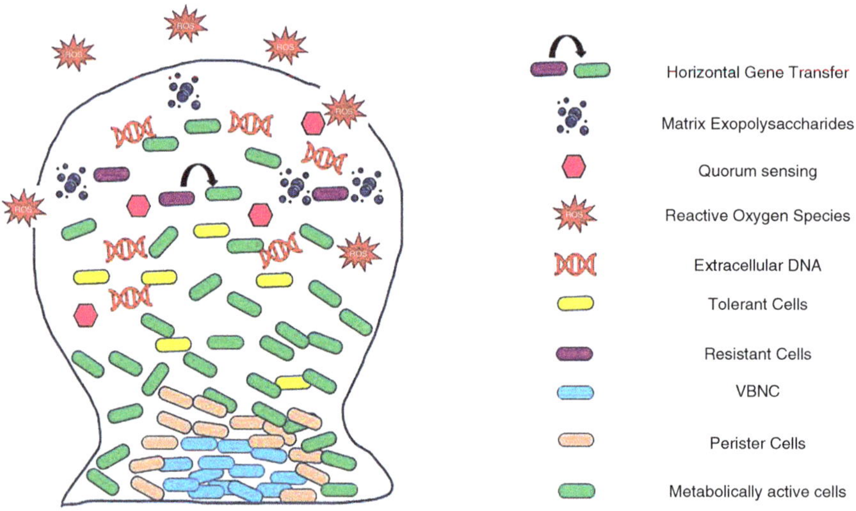

Fig. 7.3 Mature biofilm

7.4 Inhibitors

Bacterial biofilms are an increasing issue because they lead to nosocomial diseases like chronic tissue infections and infections of urinary catheters. They also offer immunity to the host's immune system and medicines. In addition to standard antibiotics, biofilm inhibitors are promising methods to lower virulence and prevent the development of microorganisms that are responsible for 80% of human illnesses (Shunmugaperumal 2010). Biofilms can withstand the human immune system because of the EPS matrix in them. Antimicrobial peptides that act on the matrix, defensins, and biofilm inhibitors, are very important (Lewis 2001). The usage of biofilm inhibitors, which include a variety of special substances such as phenols, imidazoles, furanone, indole, and bromopyrrole, is strongly emphasized in the field of biofilm remediation research. However, there is a lack of study on the substances that target and prevent bacterial growth in biofilms. Numerous cutting-edge strategies to attack biofilms have been found and extensively discussed.

7.4.1 Quorum Sensing (QS)

Numerous studies have demonstrated that inhibiting the QS system is a reliable method of stopping microbial biofilm development. Additionally, bacteria that cause disease can create virulence factors and biofilms in the host by activating QS signals. The pathogens may thus become more susceptible to the host's immune system and antibiotic treatment if this bacterial communication is inhibited by QS inhibitors (Li et al. 2020; Ravindran et al. 2018; Jiang et al. 2019). By down-regulating or silencing the QS system, quorum quenching can be used to control biofilm-mediated bacterial infections. Three crucial tactics may be used to disable the QS system: Acyl-homoserine lactone (AHL) and autoinducing peptide (AIP), respectively, are two distinct quorum sensing mechanisms used by Gram-negative and Gram-positive bacteria. However, according to Mangwani et al. (2012), both kinds of bacteria also employ the autoinducer-2 (AI-2) mechanism.

7.4.2 Electrochemical Method for Degradation of Biofilm

A possible way to prevent bacterial biofilm development is through electrochemical method. The "Bioelectric effect" breaks up developing or mature biofilms by combining a weak electric field with modest dosages of antibiotics. According to research, the electric potential kills the organisms that make up the biofilm and lowers the amount of antibiotics required to inactivate it (Srinivasan et al. 2021).

7.4.3 Degradation of the EPS Matrix of Biofilm

Proteins, nucleic acids (e-DNA), and polysaccharides make up the EPS matrix of biofilms. As it dissolves the biofilm, disrupting the matrix is a successful biofilm suppression method. The microbial strain, age, pH, oxygen tension, and nutrient availability are few examples of the variables that affect the composition of the biofilm EPS matrix (Flemming and Wingender 2010).

7.4.4 External Membrane Structure

Hydrophilic antimicrobials cannot enter the cell because of the lipopolysaccharide layer, while hydrophobic compounds are rejected by the membrane proteins on the outside of the cell. To modify the cellular membrane and target certain areas that regulate antibiotic resistance, antibacterial drugs must enter bacterial cells (Arciola et al. 2015; Bi et al. 2021).

7.4.5 Enzyme-Mediated Biofilm Control

Some enzymes damage the bacterial species' biofilms. However, endogenous matrix-degrading enzymes produced by bacteria, such as glycosidases, proteases, and DNase, cause biofilm dispersion. DNase prevents Gram-positive and Gram-negative bacteria from forming biofilms (Tetz et al. 2009). DNase dissolves the phosphodiester bonds in eDNA molecules that are necessary for EPS aggregation and initial adhesion in order to maintain an unbroken biofilm for a long time. eDNA is required for the development, maintenance, and control of biofilms (Das et al. 2010; Periasamy et al. 2012; Chagnot et al. 2013). Table 7.3 shows the mechanism of inhibition of biofilm formation (Table 7.3).

Table 7.3 Inhibitors for biofilm formation

Mechanism of biofilm inhibition	Characteristics	References
Quorum sensing (QS)	Inhibiting bacterial communication through QS inhibitors makes the pathogens susceptible to the host's immune system, it regulates the controlling of biofilm formation and pectolinarin is a secondary metabolite that inhibits the formation of biofilm. So, this QS system is called quorum quenching	Ravindran et al. (2018); Jiang et al. (2019); Li et al. (2020)
Electrochemical method for degradation of biofilm	It combines a weak electric field with lower doses of antibiotics to disintegrate biofilm formation or mature biofilms, known as the "Bioelectric effect." This phenomenon of electric potential decreases the antibiotic dosage needed to inactivate the biofilm and has a lethal effect on the biofilm organisms	Srinivasan et al. (2021)
Degradation of the EPS matrix of biofilm	The presence of divalent ions, such as Ca and Mg, affects the EPS matrix in various ways. OligoG is responsible for regulating the association of EPS and e-DNA in the matrix, while DspB protein affects the linkage of the glycosidic bond in the polysaccharide of the EPS matrix. As a result, different biofilm structure is detached	Flemming and Wingender (2010); Izano et al. (2007)
External membrane structure	The bacterial cell's outer layer, consisting of lipopolysaccharides and external membrane proteins, blocks hydrophilic and hydrophobic molecules, which makes it difficult for antibacterial agents to penetrate the cell and target specific sites responsible for antibiotic resistance	Arciola et al. (2015); Bi et al. (2021)
Enzyme-mediated biofilm control	Certain enzymes can disrupt the biofilm formation in some of the bacterial species. The enzymes such as glycosidases, proteases, and DNase, for biofilm dispersion. DNase inhibits biofilm formation in both Gram-positive and Gram-negative bacteria. While DNase breaks down phosphodiester linkages in e-DNA molecules, which are essential for initial attachment and aggregation of EPS to form an intact biofilm for extended periods	Tetz et al. (2009); Das et al. (2010); Periasamy et al. (2012); Chagnot et al. (2013)

7.5 Conclusion

The synthesis, structure, roles, and inhibitors of biofilm compounds and virulence factors specific to *Enterococcus faecium* are covered in detail in this chapter. It discusses the elements that lead to the development of biofilms in Enterococcus species, especially those pathogenic to humans as *E. faecalis* and *E. faecium*. The chapter highlights molecules influencing eDNA release by cell lysis (AtlA, gelatinase, cytolysin, serin protease) as well as surface proteins (Acm, Scm, Aggregation substance) and secreted virulence factors (sagA, gelatinase, cytolysin) as factors involved in biofilm development. Adhesion, aggregation, matrix formation, and stabilization are all caused by these elements. Proteins, polysaccharides, and extracellular DNA are among the constituents of biofilms. Various bacteria use a variety of polysaccharides to create the biofilm matrix, some of which are involved in the initial adhesion of cells to surfaces and others of which provide the biofilm structure resilience and stability. Infections linked to biofilms are currently a major health issue. For the purpose of eradicating persistent biofilms and discovering new biofilm inhibitors, researchers have looked into a variety of sources. Since many chronic illnesses are linked to the development of bacterial biofilms, conventional antibiotic therapy is frequently insufficient to eradicate biofilm-related infections. On the other hand, biofilm inhibitors may not result in antibiotic resistance and show great potential for the treatment of biofilm-based illnesses in healthcare settings in the future.

References

Arciola CR, Campoccia D, Ravaioli S, Montanaro L (2015) Polysaccharide intercellular adhesin in biofilm: structural and regulatory aspects. Front Cell Infect Microbiol 5:7

Arias CA, Panesso D, Singh K, Rice LB, Murray BE (2009) Co-transfer of antibiotic resistance genes and a hyl_{Efm}-containing virulence plasmid in *Enterococcus faecium*. Antimicrob Agents Chemother 53:4240–4246

Auer GK, Weibel DB (2017) Bacterial cell mechanics. Biochemistry 56(29):3710–3724

Bhatty M, Cruz MR, Frank KL, Laverde-Gomez JA, Andrade F, Garsin DA, Dunny GM, Kaplan HB, Christie PJ (2015) *Enterococcus faecalis* pCF10-encoded surface proteins PrgA, PrgB (Aggregation Substance), and PrgC contribute to plasmid transfer, biofilm formation, and virulence. Mol Microbiol 95:660–677

Bi Y, Xia G, Shi C, Wan J, Liu L, Chen Y et al (2021) Therapeutic strategies against bacterial biofilms. Fundam Res 1(2):193–212

Bjarnsholt T (2011) Introduction to biofilms. In: Biofilm infections. Springer, Berlin/Heidelberg, pp 1–9

Bridier A, Briandet R, Thomas V, Dubois-Brissonnet F (2011) Resistance of bacterial biofilms to disinfectants: a review. Biofouling 27:1017–1032

Brock TD, Davie JM (1963) Probable identity of a group D hemolysin with a bacteriocine. J Bacteriol 86:708–712

Chaffin WL, Lopez-Ribot JL, Casanova M, Gozalbo D, Martinez JP (1998) Cell wall and secreted proteins of Candida albicans: identification, function, and expression. Microbiol Mol Biol Rev 62(1):130–180

Chagnot C, Zorgani MA, Astruc T, Desvaux M (2013) Proteinaceous determinants of surface colonization in bacteria: bacterial adhesion and biofilm formation from a protein secretion perspective. Front Microbiol 4:303

Chen M, Yu Q, Sun H (2013) Novel strategies for the prevention and treatment of biofilm related infections. Int J Mol Sci 14:18488–18501

Cheng S, Mccleskey FK, Gress MJ, Petroziello JM, Liu R, Namdari H, DelVecchio VG (1997) A PCR assay for identification of Enterococcus faecium. J Clin Microbiol 35(5):1248–1250

Chenoweth CE, Schaberg DR (1996) Enterococcus species Hospital epidemiology and infection control. Williams and Wilkins, Baltimore, pp 334–345

Chilambi GS, Nordstrom HR, Evans DR, Kowalski RP, Dhaliwal DK, Jhanji V, Shanks RMQ, van Tyne D (2021) Genomic and phenotypic diversity of *Enterococcus faecalis* isolated from endophthalmitis. PLoS One 16:e0250084

Chow JW, Thal LA, Perri MB, Vazquez JA, Donabedian SM, Clewell DB, Zervos MJ (1993) Plasmid-associated hemolysin and aggregation substance production contribute to virulence in experimental enterococcal endocarditis. Antimicrob Agents Chemother 37:2474–2477

Das T, Sharma PK, Busscher HJ, Van Der Mei HC, Krom BP (2010) Role of extracellular DNA in initial bacterial adhesion and surface aggregation. Appl Environ Microbiol 76:3405–3408
Donlan RM (2002) Biofilms: microbial life on surfaces. Emerg Infect Dis 8(9):881
Dunny GM, Hancock LE, Shankar N (2014) Enterococcal biofilm structure and role in colonization and disease. https://www.ncbi.nlm.nih.gov/books/NBK190433/
Eaton TJ, Gasson MJ (2002) A variant enterococcal surface protein Esp_{fm} in *Enterococcus faecium*; distribution among food, commensal, medical, and environmental isolates. FEMS Microbiol Lett 216:269–275
Fabretti F, Theilacker C, Baldassarri L, Kaczynski Z, Kropec A, Holst O, Huebner J, Infettive M, Istituto I (2006) Alanine esters of enterococcal lipoteichoic acid play a role in biofilm formation and resistance to antimicrobial peptides. Infect Immun 74:4164–4171
Facklam RR, Collins MD (1989) Identification of Enterococcus species isolated from human infections by a conventional test scheme. J Clin Microbiol 27:731–734
Flemming HC, Wingender J (2010) The biofilm matrix. Nat Rev Microbiol 8:623–633
Flemming HC, Neu TR, Wozniak DJ (2007) The EPS matrix: the "house of biofilm cells". J Bacteriol 189:7945–7947
Fong JN, Yildiz FH (2015) Biofilm matrix proteins. In: Microbial biofilms. pp 201–222
Gao W, Howden BP, Stinear TP (2018) Evolution of virulence in Enterococcus faecium, a hospital-adapted opportunistic pathogen. Curr Opin Microbiol 41:76–82
Geiss-Liebisch S, Rooijakkers SHM, Beczala A, Sanchez-Carballo P, Kruszynska K, Repp C, Sakinc T, Vinogradov E, Holst O, Huebner J et al (2012) Secondary cell wall polymers of *Enterococcus faecalis* are critical for resistance to complement activation via mannose-binding lectin. J Biol Chem 287:37769–37777
Gilmore MS, Segarra RA, Booth MC (1990) An HlyB-type function is required for the expression of the *Enterococcus faecalis* hemolysin/bacteriocin. Infect Immun 58:3914–3923
Gordon S, Swenson JM, Hill BC, Pigott NE, Facklam RR, Cooksey RC, Thornsberry C, Jarvis WR, Tenover FC (1992) Antimicrobial susceptibility patterns of common and unusual species of enterococci causing infections in the United States. Enterococcal Study Group. J Clin Microbiol 30:2373–2378
Gray JW, Stewart D, Pedler SJ (1991) Species identification and antibiotic susceptibility testing of enterococci isolated from hospitalized patients. Antimicrob Agents Chemother 35:1943–1945
Guo X, Wang X, Liu J (2016) Composition analysis of fractions of extracellular polymeric substances from an activated sludge culture and identification of dominant forces affecting microbial aggregation. Sci Rep 6(1):28391
Heikens E, Bonten MJ, Willems RJ (2007) Enterococcal surface protein Esp is important for biofilm formation of Enterococcus faecium E1162. J Bacteriol 189(22):8233–8240
Heikens E, Singh K, Jacques-Palaz KD, van Luit-Asbroek M, Oostdijk EAN, Bonten MJM, Murray BE, Willems RJL (2011) Contribution of the enterococcal surface protein Esp to pathogenesis of *Enterococcus faecium* endocarditis. Microbes Infect 13:1185–1190
Hendrickx APA, van Luit-Asbroek M, Schapendonk CME, van Wamel WJB, Braat JC, Wijnands LM, Bonten MJM, Willems RJL (2009a) SgrA, a nidogen-binding LPXTG surface adhesin implicated in biofilm formation, and EcbA, a collagen binding MSCRAMM, are two novel adhesins of hospital-acquired *Enterococcus faecium*. Infect Immun 77:5097–5106
Hendrickx APA, Willems RJL, Bonten MJM, van Schaik W (2009b) LPxTG surface proteins of enterococci. Trends Microbiol 17:423–430
Ike Y, Hashimoto H, Clewell DB (1984) Hemolysin of *Streptococcus faecalis* subspecies *zymogenes* contributes to virulence in mice. Infect Immun 45:528–530
Isenmann R, Schwarz M, Rozdzinski E, Marre R, Beger HG (2000) Aggregation substance promotes colonic mucosal invasion of *Enterococcus faecalis* in an ex vivo model. J Surg Res 89:132–138
Izano E, Wang H, Ragunath C, Ramasubbu N, Kaplan J (2007) Detachment and killing of Aggregatibacter actinomycetemcomitans biofilms by dispersin B and SDS. J Dent Res 86:618–622

Jett BD, Jensen HG, Nordquist RE, Gilmore MS (1992) Contribution of the pAD1-encoded cytolysin to the severity of experimental *Enterococcus faecalis* endophthalmitis. Infect Immun 60:2445–2452

Jiang Q, Chen J, Yang C, Yin Y, Yao K (2019) Quorum sensing: a prospective therapeutic target for bacterial diseases. Biomed Res Int 2019:2015978

Kim B, Wang YC, Hespen CW, Espinosa J, Salje J, Rangan KJ, Oren DA, Kang JY, Pedicord VA, Hang HC (2019) *Enterococcus faecium* secreted antigen A generates muropeptides to enhance host immunity and limit bacterial pathogenesis. Elife 8:e45343

Klare I, Konstabel C, Badstübner D, Werner G, Witte W (2003) Occurrence and spread of antibiotic resistances in Enterococcus faecium. Int J Food Microbiol 88(2–3):269–290

Kowalski WJ, Kasper EL, Hatton JF, Murray BE, Nallapareddy SR, Gillespie MJ (2006) *Enterococcus faecalis* adhesin, Ace, mediates attachment to particulate dentin. J Endod 32:634–637

Kreft B, Marre R, Schramm U, Wirth R (1992) Aggregation substance of *Enterococcus faecalis* mediates adhesion to cultured renal tubular cells. Infect Immun 60:25–30

Lewis K (2001) Riddle of biofilm resistance. Antimicrob Agents Chemother 45:999–1007

Li D, Ramanathan S, Wang G, Wu Y, Tang Q, Li G (2020) Acetylation of lysine 7 of AhyI affects the biological function in Aeromonashydrophila. Microbial Pathog 140:103952

Livornese LL Jr, Dias S, Samel C, Romanowski B, Taylor S, May P, Pitsakis P, Woods G, Kaye D, Levinson ME, Johnson CC (1992) Hospital-acquired infection with vancomycin-resistant Enterococcus faecium transmitted by electronic thermometers. Ann Intern Med 117:112–116

Mangwani N, Dash HR, Chauhan A, Das S (2012) Bacterial quorum sensing: functional features and potential applications in biotechnology. J Mol Microbiol Biotechnol 22(4):215–227

Martins M, Uppuluri P, Thomas DP, Cleary IA, Henriques M, Lopez-Ribot JL et al (2010) Presence of extracellular DNA in the Candida albicans biofilm matrix and its contribution to biofilms. Mycopathologia 169:323–331

Mei L, Wang X, Yin Y, Tang G, Wang C (2021) Conservative production of galactosaminogalactan in Metarhizium is responsible for appressorium mucilage production and topical infection of insect hosts. PLoS Pathog 17:1–25

Meredith K, Bolhuis AO, Neill MAA (2009) Enterococcal surface protein Esp affects antibiotic sensitivity in *Enterococcus faecium*. Int J Antimicrob Agents 34:392–393

Miyazaki S, Ohno A, Kobayashi I, Uji T, Yamaguchi K, Goto S (1993) Cytotoxic effect of hemolytic culture supernatant from *Enterococcus faecalis* on mouse polymorphonu-neutrophils and macrophages. Microbiol Immunol 37:265–270

Montanaro L, Poggi A, Visai L, Ravaioli S, Campoccia D, Speziale P, Arciola CR (2011) Extracellular DNA in biofilms. Int J Artif Organs 34:824–831

Montealegre MC, Singh K, Somarajan SR, Yadav P, Chang C, Spencer R, Sillanpää J, Ton-That H, Murray BE (2016) Role of the Emppilus subunits of *Enterococcus faecium* in biofilm formation, adherence to host extracellular matrix components and experimental infection. Infect Immun 84:1491–1500

Mukherjee PK, Chandra J, Kuhn DM, Ghannoum MA (2003) Mechanism of fluconazole resistance in *Candida albicans* biofilms: phase-specific role of efflux pumps and membrane sterols. Infect Immun 71:4333–4340

Muller C, Breton Y, Morin T, Benachour A, Auffray Y, Rincé A, Auffray Y, Rincé A (2006) The response regulator CroR modulates expression of the secreted stress induced SalB protein in *Enterococcus faecalis*. J Bacteriol 188:2636–2645

Nallapareddy SR, Qin X, Weinstock GM, Höök M, Murray BE (2000) *Enterococcus faecalis* adhesin, Ace, mediates attachment to extracellular matrix proteins collagen type IV and laminin as well as collagen type I. Infect Immun 68:5218–5224

Nallapareddy SR, WeinstockG M, Murray BE (2003) Clinical isolates of *Enterococcus faecium* exhibit strain-specific collagen binding mediated by Acm, a new member of the MSCRAMM family. Mol Microbiol 47:1733–1747

Nallapareddy SR, Singh K, Murray BE (2008a) Contribution of the collagen adhesin Acm to pathogenesis of *Enterococcus faecium* in experimental endocarditis. Infect Immun 76:4120–4128

Nallapareddy SR, Singh K, Okhuysen PC, Murray BE (2008b) A functional collagen adhesin gene, *acm*, in clinical isolates of *Enterococcus faecium* correlates with the recent success of this emerging nosocomial pathogen. Infect Immun 76:4110–4119

Paganelli FL, de Been M, Braat JC, Hoogenboezem T, Vink C, Bayjanov J, Rogers MRC, Huebner J (2015) Distinct SagA from hospital-associated clade A1 *Enterococcus faecium* strains contributes to biofilm formation. Appl Environ Microbiol 81:6873–6882

Paoletti C, Foglia G, Princivalli MS, Magi G, Guaglianone E, Donelli G, Pruzzo C, Biavasco F, Facinelli B (2007) Co-transfer of *vanA* and aggregation substance genes from *Enterococcus faecalis* isolates in intra- and interspecies matings. J Antimicrob Chemother 59:1005–1009

Pedicord VA, Lockhart AAK, Rangan KJ, Craig JW, Loschko J, Rogoz A, Hang H, Mucida D (2016) Exploiting a host-commensal interaction to promote intestinal barrier function and enteric pathogen tolerance. Sci Immunol 1:eaai7732

Periasamy S, Joo HS, Duong AC, Bach THL, Tan VY, Chatterjee SS et al (2012) How Staphylococcus aureus biofilms develop their characteristic structure. Proc Natl Acad Sci U S A 109:1281–1286

Pierce CG, Vila T, Romo JA, Montelongo-Jauregui D, Wall G, Ramasubramanian A, Lopez-Ribot JL (2017) The *Candida albicans* biofilm matrix: composition, structure and function. J Fungi 3:14

Prieto AMG, Urbanus RT, Zhang X, Bierschenk D, Koekman CA, van Luit-Asbroek M, Ouwerkerk JP, Pape M, Paganelli FL, Wobser D et al (2015) The N-terminal domain of the thermo-regulated surface protein PrpA of *Enterococcus faecium* binds to fibrinogen, fibronectin and platelets. Nat Publ Group 5:18255

Qin X, Galloway-Peña JR, Sillanpää J, Roh JH, Nallapareddy SR, Chowdhury S, Bourgogne A, Choudhury T, Muzny DM, Buhay CJ et al (2012) Complete genome sequence of *Enterococcus faecium* strain TX16 and comparative genomic analysis of *Enterococcus faecium* genomes. BMC Microbiol 12:135

Rangan KJ, Pedicord VA, Wang YC, Kim B, Lu Y, Shaham S, Mucida D, Hang H (2016) A secreted bacterial peptidoglycan hydrolase enhances tolerance to enteric pathogens. Science 353:1434–1437

Ravindran D, Ramanathan S, Arunachalam K, Jeyaraj G, Shunmugiah K, Arumugam V (2018) Phytosynthesized silver nanoparticles as antiquorum sensing and antibiofilm agent against the nosocomial pathogen Serratia marcescens: an in vitro study. J Appl Microbiol 124:1425–1440

Rella A, Farnoud AM, Del Poeta M (2016) Plasma membrane lipids and their role in fungal virulence. Prog Lipid Res 61:63–72

Rice LB, Carias L, Rudin S, Vael C, Goossens H, Konstabel C, Klare I, Nallapareddy SR, Huang W, Murray BE (2003) A potential virulence gene, *hylEfm*, predominates in *Enterococcus faecium* of clinical origin. J Infect Dis 187:508–512

Rich RL, Kreikemeyer B, Owens RT, LaBrenz S, Narayana SVL, Weinstock GM, Murray BE, Höök M (1999) Ace is a collagen-binding MSCRAMM from *Enterococcus faecalis*. J Biol Chem 274:26939–26945

Ruhal R, Kataria R (2021) Biofilm patterns in gram-positive and gram-negative bacteria. Microbiol Res 251:126829

Ruoff KL, La Maza L, Murtagh MJ, Spargo JD, Ferraro MJ (1990) Species identities of enterococci isolated from clinical specimens. J Clin Microbiol 28:435–437

Sader HS, Pfaller MA, Tenover FC, Hollis RJ, Jones RN (1994) Evaluation and characterization of multiresistant Enterococcus faecium from 12 U.S. medical centers. J Clin Microbiol 32:2840–2842

Santajit S, Indrawattana N (2016) Mechanisms of antimicrobial resistance in ESKAPE pathogens. Biomed Res Int 2016:1–8

Sartingen S, Rozdzinski E, Muscholl-Silberhorn A, Marre R (2000) Aggregation substance increases adherence and internalization, but not translocation, of *Enterococcus faecalis* through different intestinal epithelial cells in vitro. Infect Immun 68:6044–6047

Schlievert PM, Gahr PJ, Assimacopoulos AP, Dinges MM, Stoehr JA, Harmala JW, Hirt H, Dunny GM (1998) Aggregation and binding substances enhance pathogenicity in rabbit models of *Enterococcus faecalis* endocarditis. Infect Immun 66:218–223

Segarra RA, Booth MC, Morales DA, Huycke MM, Gilmore MS (1991) Molecular characterization of the *Enterococcus faecalis* hemolysin/bacteriocin determinant. Infect Immun 59:1239–1246

Segev-Zarko LA, Saar-Dover R, Brumfeld V, Mangoni ML, Shai Y (2015) Mechanisms of biofilm inhibition and degradation by antimicrobial peptides. Biochem J 468(2):259–270

Shankar V, Baghdayan AS, Huycke MM, Lindahl G, Gilmore MS (1999) Infection-derived *Enterococcus faecalis* strains are enriched in *esp*, a gene encoding a novel surface protein. Infect Immun 67:193–200

Shankar N, Lockatell CV, Baghdayanm AS, Drachenberg C, Gilmore MS, Johnson DE (2001) Role of *Enterococcus faecalis* surface protein Esp in the pathogenesis of ascending urinary tract infection. Infect Immun 69:4366–4372

Shunmugaperumal T (2010) Biofilm eradication and prevention: a pharmaceutical approach to medical device infections. Wiley, Hoboken, NJ

Sillanpää J, Nallapareddy SR, Prakash VP, Qin X, Hook M, Weinstock GM, Murray BE (2008) Identification and phenotypic characterization of a second collagen adhesin, Scm, and genome-based identification and analysis of 13 other predicted MSCRAMMs, including four distinct pilus loci, in *Enterococcus faecium*. Microbiology 154:3199–3211

Sillanpää J, Nallapareddy SR, Singh K, Parakash VP, Fothergill T, Ton-That H, Murray BE (2010) Characterization of the ebp_{fm} pilus-encoding operon of *Enterococcus faecium* and its role in biofilm formation and virulence in a murine model of urinary tract infection. Virulence 1:236–246

Singh K, NallapareddyS R, Sillanpää J, Murray BE (2010) Importance of the collagen adhesin Ace in pathogenesis and protection against *Enterococcus faecalis* experimental endocarditis. PLoS Pathog 6:e1000716

Singh K, Pinkston KL, Gao P, Harvey BR, Murray BE (2018) Anti-Ace monoclonal antibody reduces *Enterococcus faecalis* aortic valve infection in a rat infective endocarditis model. Pathog Dis 76:fty084

Solheim M, Aakra Å, Snipen LG, Brede DA, Nes IF (2009) Comparative genomics of *Enterococcus faecalis* from healthy Norwegian infants. BMC Genomics 10:194

Spera RV Jr, Farber BF (1992) Multiple-resistant Enterococcus faecium. The nosocomial pathogen of the 1990s. JAMA 268:2563–2564

Srinivasan R, Santhakumari S, Poonguzhali P, Geetha M, Dyavaiah M, Xiangmin L (2021) Bacterial biofilm inhibition: a focused review on recent therapeutic strategies for combating the biofilm mediated infections. Front Microbiol 12:676458

Stern CS, Carvalho MGS, Teixeira LM (1994) Characterization of enterococci isolated from human and nonhuman sources in Brazil. Diagn Microbiol Infect Dis 20:61–67

Süßmuth SD, Muscholl-Silberhorn A, Wirth R, Susa M, Marre R, Rozdzinski E (2000) Aggregation substance promotes adherence, phagocytosis, and intracellular survival of *Enterococcus faecalis* within human macrophages and suppresses respiratory burst. Infect Immun 68:4900–4906

Taglialegna A, Matilla-Cuenca L, Dorado-Morales P, Navarro S, Ventura S, Garnett JA, Lasa I, Valle J (2020) The biofilm-associated surface protein Esp of *Enterococcus faecalis* forms amyloid-like fibers. Nat Partn J 6:15

Teixeira LM, Facklam RR, Steigerwalt AG, Pigott NE, Merquior VLC, Brenner DJ (1995) Correlation between phenotypic characteristics and DNA relatedness within Enterococcus faecium strains. J Clin Microbiol 33:1520–1523

Tendolkar PM, Baghdayan AS, Gilmore MS, Shankar N (2004) Enterococcal surface protein, Esp, enhances biofilm formation by *Enterococcus faecalis*. Infect Immun 72:6032–6039

Teng F, Kawalec M, Weinstock GM, Hryniewicz W, Murray BE (2003) An *Enterococcus faecium* secreted antigen, SagA, exhibits broad-spectrum binding to extracellular matrix proteins and appears essential for *E. faecium* growth. Infect Immun 71:5033–5041

Tetz GV, Artemenko NK, Tetz VV (2009) Effect of DNase and antibiotics on biofilm characteristics. Antimicrob Agents Chemother 53:1204–1209

Thomas VC, Hiromasa Y, Harms N, Thurlow L, Tomich J, Hancock LE (2009) A fratricidal mechanism is responsible for eDNA release and contributes to biofilm development of Enterococcus faecalis. Mol Microbiol 72:1022–1036

Todokoro D, Suzuki T, Kobayakawa S, Tomita H, Ohashi Y, Akiyama H (2017) Postoperative *Enterococcus faecalis* endophthalmitis: virulence factors leading to poor visual outcome. Jpn J Ophthalmol 61:408–414

Toledo-Arana A, Valle J, Solano C, Arrizubieta MJ, Cucarella C, Lamata M, Amorena B, Leiva J, Penadés JR, Lasa I (2001) The enterococcal surface protein, Esp, is involved in *Enterococcus faecalis* biofilm formation. Appl Environ Microbiol 67:4538–4545

Top J, Willems R, Bonten M (2008) Emergence of CC17 *Enterococcus faecium*: from commensal to hospital-adapted pathogen. FEMS Immunol Med Microbiol 52:297–308

Trentin DS, Giordani RB, Macedo AJ (2013) Biofilmes bacterianos patogênicos: Aspectos geais, importância clínica e estratégias de combate. Revista 14:113–238

Vanek NN, Simon SI, Jacques-Palaz K, Mariscalco MM, Dunny GM, Rakita RM (1999) *Enterococcus faecalis* aggregation substance promotes opsonin-independent binding to human neutrophils via a complement receptor type 3-mediated mechanism. FEMS Immunol Med Microbiol 26:49–60

Wells CL, Moore EA, Hoag JA, Hirt H, Dunny GM, Erlandsen SL (2000) Inducible expression of *Enterococcus faecalis* aggregation substance surface protein facilitates bacterial internalization by cultured enterocytes. Infect Immun 68:7190–7194

Weng PL, Ramli R, Hamat RA (2019) Antibiotic susceptibility patterns, biofilm formation and *esp* gene among clinical enterococci: is there any association? Int J Environ Res Public Health 16:3439

Yu MK, Kim MA, Rosa V, Hwang YC, Del Fabbro M, Sohn WJ, Min KS (2019) Role of extracellular DNA in Enterococcus faecalis biofilm formation and its susceptibility to sodium hypochlorite. J Appl Oral Sci 27:e20180699

Staphylococcus aureus Virulence Factors and Biofilm Components: Synthesis, Structure, Function and Inhibitors

8

Zarin Taj and Indranil Chattopadhyay

Abstract

Staphylococcus aureus is classified as an ESKAPE pathogen, possessing both multi-drug resistance and virulence factors. *S. aureus* exhibits opportunistic behaviour and is frequently encountered in healthcare settings. This bacterium is recognised for its ability to induce a wide array of infectious diseases, often linked to the formation of biofilms. These illnesses can range from minor infections affecting the skin and soft tissues to severe and potentially life-threatening conditions like septicaemia. *S. aureus* is a Gram-positive coccus that forms sporadic clusters resembling grapes and establishes a secure and functional environment for cell survival by means of a viscous extracellular polymer matrix known as Biofilm. Biofilms are capable of generating intricate bacterial communities through either surface adhesion or the formation of aggregates in the absence of surface adhesion. Enhanced adaptability and adherence are important characteristics of biofilms that serve as a means of survival in the face of various external challenges. These challenges include fluctuations in temperature, limited food availability, dehydration and most notably, the presence of antibacterial drugs and the host's immune responses. Through the processes of genome evolution and gene transfer, *Staphylococcus aureus* demonstrates genomic flexibility and adaptability. As a consequence, this bacterium plays an important part in the evolution of antimicrobial resistance (AMR). This phenomenon is of significant concern since, in the contemporary period, which is typified by the diminishing efficiency of antibiotics, it poses a substantial threat to the general population's health. The present study offers a comprehensive examination of several issues concerning *Staphylococcus aureus* virulence factors and biofilm development.

Z. Taj · I. Chattopadhyay (✉)
Department of Biotechnology, School of Life Sciences, Central University of Tamil Nadu, Thiruvarur, India
e-mail: indranil@cutn.ac.in

S. Busi, R. Prasad (eds.), *ESKAPE Pathogens*,
https://doi.org/10.1007/978-981-99-8799-3_8

These aspects encompass the constituents, overall structure, clinical significance, pathogenicity and resistance mechanisms associated with *S. aureus*. This review examines the various methodologies utilised to impede and eliminate *S. aureus* biofilms as a strategy to combat antimicrobial resistance.

Keywords

S. aureus · Biofilm · Genomic evolution · Antimicrobial resistance · Pathogenesis · Anti-Biofilm agents

8.1 Introduction

Staphylococcus aureus is a bacterium that is Gram-positive, non-motile and non-spore forming. It exhibits a various diameter size of 0.5–1.5 μm and displays a morphological appearance reminiscent of clustered grapes. This bacterium is able to live in conditions that are either aerobic or anaerobic at temperatures that range from 18 to 40 °C (Taylor and Unakal 2023) and can be found in the environment as well as in human flora such as the skin and mucous membranes. *S. aureus* is an opportunistic pathogen due to its inherent ability to penetrate the human immune system via virulence factor production and biofilm formation. Because of this ability, people with immune systems that are compromised are more likely to acquire either acute or chronic illnesses (Malani 2014; Struelens and Denis 2000). Moreover, *Staphylococcus aureus* is classified as one of the significant antibiotic-resistant pathogens known as 'ESKAPE' pathogens. These pathogens, which include *Acinetobacter baumannii*, *Enterobacter* sp. *Enterococcus faecium*, *Klebsiella pneumoniae*, *Pseudomonas aeruginosa* and *Staphylococcus aureus* are responsible for causing high-mortality rates in patients having nosocomial infections (De Oliveira et al. 2020; Santajit and Indrawattana 2016). *Staphylococcus aureus* possesses a remarkable capacity to develop resistance against various antimicrobial agents, including penicillin, tetracycline, macrolides, aminoglycosides and chloramphenicol (Weigel et al. 2007). The extensive methicillin and other semisynthetic penicillin use during the latter half of the 1960s was a significant factor in the emergence of methicillin-resistant *Staphylococcus aureus* (MRSA), a pathogen that continues to be widespread in both healthcare and community settings (Malani 2014).

Research has demonstrated that *S. aureus* plays a significant role in the development of various types of infections, encompassing bacteraemia, infective endocarditis, osteoarticular, skin and soft tissue as well as pleuropulmonary infections (Howden et al. 2010). In the past few decades, there have been two major changes in *S. aureus* infections epidemiology: first, a tremendous increase in infections have been caused by medical interventions, most notably infections caused by prosthetic devices and infective endocarditis; second, there has been an epidemic of skin and soft tissue infections brought on by beta-lactam antibiotics resistance strains with specific virulence factors are two interconnected trends (Laupland 2013). Bacteraemia is a widely studied manifestation of *Staphylococcus aureus* infection, with a well-documented prevalence, prognosis and outcome in industrially developed regions across the globe (Frimodt-Møller et al. 1997).

The following is a breakdown of the average patient population by main infection site: From there, the percentages rise to 26% for line-related infections, 9% for pleuropulmonary infections, 24% for cases with no defined focus or unknown origin and 11% for other infection foci. Infectious endocarditis accounts for 5%, SSTI for 19%, osteoarticular infections for 8% and so on. MRSA is often associated with infections that occur in healthcare settings, known as healthcare-associated (HCA) infections which are primarily manifest as skin and soft tissue infections (SSTIs) (Isobe et al. 2012; Kaasch et al. 2014; Pastagia et al. 2013). Prior to the extensive utilisation of antibiotics, *Staphylococcus aureus* bacteraemia (SAB) exhibited a case fatality rate (CFR) of approximately 80%. Despite the significant impact of penicillin's discovery in reducing the mortality rate associated with SAB, the case fatality rates (CFRs) for SAB have exhibited a consistent range of 15% to 50% over the course of several decades (Liu et al. 2011).

S. aureus acquires virulence factors encoding genes when the bacterium undergoes adaptation to novel ecological niches characterised by diverse environmental conditions, nutritional availability and stress factors. According to Malachowa and DeLeo (2010), *S. aureus* causes diverse array of infections through expressing a cluster of components involved pathogenesis. Bacterial adherence to surfaces and tissues, immune system evasion and attack and induction of detrimental toxic effects on the host system are performed by these components, which are together referred to as virulence factors. The destructive determinants known as virulence factors can be categorised into two types: cell-surface-associated factors, which are involved in adherence, and secreted factors, specifically exotoxins (Costa et al. 2013). *S. aureus* is involved in intercellular communication through its interaction with a diverse range of cell wall-anchored (CWA) proteins covalently attached to bacterial cell wall peptidoglycan layer. The demonstrated potential harmfulness of different CWA proteins notwithstanding, there exists a scarcity of data pertaining to their specific functions in the context of SSTIs (Lacey et al. 2016).

Furthermore, the biofilm development by the bacterial consortium is universally present in both normal and pathogenic bacteria, and is not limited to bacteria with virulence factors. The biofilm-forming phenomenon has been observed in various pathogens as a significant mechanism for bacterial survival in different host environments (Hammer et al. 2014). Biofilms are responsible for HCA infections as they exhibit a high propensity to colonise biomedical devices, posing a significant threat to patient health and well-being. As a commensal, Staphylococcal genera can be found in the skin, lower gastrointestinal tract, upper respiratory system and urogenital tract of humans. Therefore, it is highly probable that these entities possess the capability to inhabit biomedical devices utilised in pharmaceutical applications (Ebrey et al. 2004). Biofilm-producing pathogens are associated with chronic infections, including periodontitis, pneumonia with cystic fibrosis, urinary tract infections and other acute wound infections. The capacity of pathogens to persist within the human body for extended durations, facilitated by biofilm production conferring resistance, the immune response and antimicrobial agents (Davies 2003; Del Pozo and Patel 2007). The presence of polymers in biofilm structures renders contagious bacterial populations resistant to attack and destruction by antimicrobial agents (Dufour et al. 2010). The authors Guo et al. (2020) discuss the significant virulence factors and biofilm components of *Staphylococcus aureus*, highlighting their synergistic effects in evading current antibiotic treatments and developing antibiotic resistance. Therefore, it is of considerable significance to investigate virulence factors in order to effectively address *S. aureus* infection.

8.2 Virulence Factors

Bacterial organisms possess the capability to flourish within the inhospitable environment of the host's biological system due to their secretion of virulence factors, contribute to the facilitation of infection and the onset of disease. The virulence factors of *Staphylococcus aureus* include antigens (capsule, adhesins), proteins (coagulase, staphylokinase, nuclease, lipase, hyaluronidase) and toxins (Enterotoxin, Exfoliative Toxin, Toxic Shock Disorder Toxin, β-Toxin, δ-Toxin, P-V Leukocidins) (Table 8.1) (Chavakis et al. 2005; Ghasemian et al. 2015; Zecconi and Scali 2013).

8.2.1 Capsular Polysaccharides

Microorganisms frequently produce extracellular capsular polysaccharides, which can lead to invasive infections (Vann and Liu 1980). The enhancement of microbial pathogenicity is facilitated by the capsules' ability to confer immunity to phagocytosis on bacteria. The production of capsules by *S. aureus* was initially described by Gilbert in 1931. They increase the virulence of staphylococcal infections by inhibiting the process of phagocytosis, leading to bacterial persistence in the circulatory systems of affected persons. The capsules can be categorised into two types: 'macrocapsules', which give rise to mucoid colonies and are infrequently linked to disease, and 'microcapsules', which produce non-mucoid colonies with a thinner capsular layer and exhibit greater pathogenicity. The predominant *S. aureus* capsule types associated with clinical isolates are CP5 and CP8, while CP serotypes 1 and 2 have also been extensively studied. The prevalence of CP5 production among human isolates ranges from approximately 16–26%, whereas CP8 production is observed in approximately 55–65% of these isolates. The capsular polysaccharides are composed of *N*-acetyl mannosaminuronic acid (ManNAc), *N*-acetyl-D-fucosamine (D-FucNAc) and *N*-acetyl-L-fucosamine (L-FucNAc) repeating units. These polysaccharides contribute to the formation of abscesses, which can subsequently result in the development of chronic infections (Jones 2005; O'Riordan and Lee 2004; Tzianabos et al. 2001). Despite their structural similarities, CP5 and CP8 exhibit distinct immunological functions. The synthesis of CP5 and CP8 is significantly impacted by environmental factors, with the presence of iron limitation being found to enhance CP8 production. The production of CP5 is augmented under conditions of elevated oxygen levels, while it is restricted under alkaline growth

Table 8.1 Virulence factors of *Staphylococcus aureus* with their respective biological functions

Virulence factor	Related genes	Biological function	References
Autolysin	*atl*	Participates in biofilm formation and provides binding to host cell-extracellular factors Fn, Fg, He & Vn	Houston et al. (2011)
CWA fibronectin-binding protein	*ebh*	Contributes to bacterial cell growth and envelope assembly enabling complement resistance	Speziale and Pietrocola (2020)
Clumping factor A/B	*clfA/clfB*	Binds fibrinogen and contributes to platelet aggregation, evasion of the complement activation pathway	Herman-Bausier et al. (2018)
Collagen adhesion	*cna*	Mediates bacterial adherence to host tissue's collagen and inhibits the activation of the classical complement pathway	Kang et al. (2013)
Elastin-binding protein	*ebp*	Mediates bacterial cells to bind soluble elastin	Downer et al. (2002)
Extracellular adherence protein	*Eap/map*	DNA-binding protein capable to block neutrophil extracellular trap formation	Eisenbeis et al. (2018)
Fibrinogen-binding protein	*efb*	Facilitates bacterial survival and aggravation in blood and inhibits platelet aggregation	Zhang et al. (2011)
Fibronectin-binding protein	*fnbA/fnbB*	Confers adherence to fibrinogen and provides ability in the invasion of endothelial cells	Sinha et al. (2000)
Intercellular adhesion	*icaA-D, icaR*	Involves in biofilm formation	Fluckiger et al. (2005)
Ser-Asp-rich fibrinogen-binding protein	*sdrC-H*	Involves in biofilm formation and adherence to the surface	Foster et al. (2014)
Staphylococcal protein A	*spa*	Binds to the Fc region of Ig and inhibits opsonophagocytosis	Kobayashi and DeLeo (2013)
Capsular polysaccharide	*cap*	Promotes bacterial existence in the bloodstream and abscess formation through phagocytosis impedance	O'Riordan and Lee (2004)
Enzyme			
Cysteine protease	*sspB/sspC*	Blocks the phagocytosis by neutrophils and monocytes	Smagur et al. (2009a)
Serine v8 protease	*sspA*	Binds to the Fc region of IgG and evades the immune response.	Pietrocola et al. (2017)
Hyaluronate lyase	*hysA*	Degrades hyaluronic acid and establishes bacterial dissemination through disrupts ECM	Makris et al. (2004)
Lipase	*geh/lip*	Interacts with the host granulocyte and inactivates bactericidal lipids to provide bacterial survival inside the host	Hu et al. (2012)

(continued)

Table 8.1 (continued)

Virulence factor	Related genes	Biological function	References
Staphylocoagulase	*coa*	Actively binds with prothrombin to induce blood clots	Panizzi et al. (2010)
Staphylokinase	*sak*	Forms complex with plasmin and fibrin and prevents complementary pathway opsonisation	Rooijakkers et al. (2005)
Thermonuclease	*nuc*	Involves in evasion of neutrophil extracellular traps through nuclease activity	Berends et al. (2010)
Toxin			
Alpha haemolysin	*hla*	Binds to plasma membrane ADAM10 protein and invades epithelial cells, endothelial cells, T cells, monocytes and macrophages	Oliveira et al. (2018)
Beta haemolysin	*hlb*	Involves the hydrolysis of sphingomyelin in host tissue and provides an escape from phagosome	Vandenesch et al. (2012)
Delta haemolysin	*hld*	Involves in lysis of lipid bilayer and results in DNA fragmentation and apoptosis	Vandenesch et al. (2012)
Enterotoxin A–E, G–J	*Sea, seb, sec, sed, see, seg, seh*	Involves binding to class II MHC molecules on antigen-presenting cells and stimulating massive T-cell activation	Pinchuk et al. (2010)
Exfoliative toxin type A–D	*eta, etb, etc, etd*	Interacts with T-cell and exhibits mitogenic activity	Bukowski et al. (2010)
Gamma haemolysin	*hlg*	Involves in lysis of red blood cell	Dinges et al. (2000)
Leukocidin M, D, E	*lukM, lukD, lukE*	Promotes bacterial virulence by acquiring iron by erythrocyte lysis	Spaan et al. (2017)
Panton-valentine leukocidin	*lukF/S-PV*	A major contributor to the epidemic spread of CA-MRSA	Mohamadou et al. (2022)
Toxic shock syndrome toxin	*tsst*	Major contributes to clinical manifestation through stimulation of IL-1, IL-2 and TNF	Bohach et al. (1990)

conditions. Both CP5 and CP8 are subject to positive regulation by the *agr* regulatory system. The mucoid capsules played a crucial role as virulence factors that hindered phagocytosis by concealing C3b molecules attached to the surface of bacterial cells, thereby impeding recognition by phagocytic cells (Kuipers et al. 2016).

8.2.2 Cell Wall-Anchored (CWA) Proteins

Many different proteins that are covalently linked to the peptidoglycan layer of the bacterial cell wall are expressed by *S. aureus*. Sortase proteins are responsible for attaching around 25 different types of CWA to the cell wall in *S. aureus* (Hendrickx et al. 2011). In the early stages of abscess formation and SSTIs, the CWA proteins play a vital function in enhancing bacterial adhesion to the surface of the host.

8.2.2.1 Staphylococcal Protein A

Staphylococcal protein A (SpA), a protein that is anchored to the cell wall with the ability to interact with various host factors, including TNF receptor 1 (TNFR1), IgG and von Willebrand factor (VWF) suggesting that it is one of a pathogenic factor (Foster et al. 2014; Palmqvist et al. 2002). The SpA molecule plays a significant role in various immune evasion pathways due to its possession of immunosuppressive characteristics. In this case, it has been observed that SpA interacts with the Fcγ region of IgG, thereby inhibiting phagocytosis through its ability to disrupt the recognition process of neutrophils due to its incorrect binding orientation (Foster 2005). Furthermore, the bacterial attachment to surface of B cell VH3+ immunoglobulins leads to clonal proliferation. This, in turn, results in B cells depletion and the subsequent production of antibodies (Goodyear and Silverman 2003). In addition to its capacity to thwart the immune system, SpA possesses the capability to elicit a pro-inflammatory reaction through its interaction with TNFR1, leading to MAPK kinase pathway activation and succeeding synthesis of IL-8 and various other cytokines (Gómez et al. 2004). The study conducted by Gonzalez et al. (2019) demonstrates that *S. aureus* SSTIs are primarily influenced by the involvement of SpA, which triggers the activation of TNFR1 factors found in keratinocytes. This activation leads to a heightened affinity of *S. aureus* for the skin surface.

8.2.2.2 Fibronectin (Fn)-Binding Adhesins (Fn-BPA & Fn-BPB)

Two fibronectin (Fn)-binding adhesins also known as Fn-binding proteins A and B (FnBPA and FnBPB), are specifically binding to molecules like fibrinogen, histones, elastin and plasminogen and are facilitated by a domain present in these adhesins (Pietrocola et al. 2019; Speziale and Pietrocola 2020). FnBPs are of significant importance in the intricate process of biofilm formation. Multiple haemophilic low-affinity connections are formed between the A domains of FnBPs on nearby cells, facilitating this process. The functionality of this mechanism is contingent upon the existence of Zn^{2+} ions (Herman-Bausier et al. 2015). In the log growth phase, fnb genes are highly expressed, a phenomenon regulated by the accessory gene regulator (*Agr*) and the Staphylococcal accessory regulator (*Sar*) (Robinson et al. 2005; Wolz et al. 2000). The Leu-Pro-X-Thr-Gly (LPXTG) motif is found in the N-terminal signal sequence of FnBPs, which is followed by a hydrophobic membrane region and finally a C-terminal cytosolic tail. The invasion of *S. aureus* in host epithelial cells, fibroblasts, endothelial cells and osteoblasts is potentially facilitated by FnBPs adherence. These FnBPs bind to fibrinogen present in tissues' extracellular matrix through Dock, Lock and Latch (DLL) mechanism (Stemberk et al. 2014). The study conducted by Sinha et al. (1999) demonstrates that the interaction between FnBPs and α5β1 integrin on the host cell surface engages in recruiting the proteins necessary for effective bacterial invasion. FnBPs are also implicated in the processes of platelet activation and aggregation, as well as thrombus formation, during the pathogenesis of endocarditis.

8.2.2.3 Clumping Factors A and B (ClfA & ClfB)

Another fibrinogen-binding surface protein known as clumping factor A (ClfA) has significant influence in different infection pathogenesis with sequence homology to a protein produced by *S. aureus* called FnBPA. The bacteria are cross-linked to integrin GPIIB/IIIa by ClfA that also binds independently to the blood's fibrinogen and fibronectin. This causes platelets to aggregate, much like when FnBPs connect to platelets during infective endocarditis. The interaction between ClfA and fibrinogen, facilitated by a specific sequence of 12 residues, is subject to regulation by calcium ions present in the plasma of blood. The expression of ClfA has been observed throughout the growth cycles of the most of clinical isolates. According to Higgins et al. (2006), ClfA has the ability to avoid neutrophil phagocytosis by binding to fibrinogen on the surface and additionally, ClfA acts as a strong stimulator of T cells. Unlike ClfA, Clumping Factor B (ClfB) exists solely during the rapid proliferation phase. Subsequently, during the process of biofilm colonisation, ClfB is removed from the cell surface by proteolysis mediated by metalloprotease as a consequence of cell division. The increased binding affinity of clumping factor B (ClfB) is achieved through its interaction with the cornified envelope, leading to an enhanced ability to colonise the nasal cavity. The ClfB protein demonstrates a strong affinity for plasma fibrinogen, cytokeratin and loricrin, which are found in the squamous cells of nasal regions. This interaction plays a crucial role in the establishment of SSTIs (Lacey et al. 2019).

8.2.2.4 Serine-Aspartate Repeat Protein (SdrC, SdrD and SdrE)

Serine-aspartate repeat protein (Sdr) proteins are an assorted group of surface proteins characterised by the presence of R domain that contains tandem repeats of serine and aspartic acid (SD) dipeptides. SdrC, SdrD and SdrE proteins possess R domain that is accompanied by an additional B-repeat region situated between the distinct A and R regions. The B-repeat is comprised of 110–113 amino acid residues and encompasses a classical EF-hand motif that exhibits a strong affinity for calcium ions. This interaction with calcium ions is of considerable significance for maintaining the structural integrity of the B-repeat. The N-terminal signal sequences and the C-terminal wall- and membrane-spanning domains found in ClfA, ClfB, SdrC, SdrD and SdrE exhibit similarities. Every Sdr protein comprises the LPXTG cell wall-anchoring motif. The Sdr proteins are produced by a tandem array of three closely related genes, namely sdrC, sdrD and sdrE (Liu et al. 2015; Moormeier and Bayles 2017; Sabat et al. 2006; Schneewind and Missiakas 2019). The molecular mechanism underlying the pathogenesis of the Sdr protein remains unclear. However, Barbu et al. (2014) conducted a study that demonstrated the role of SdrC in facilitating biofilm formation. This process is facilitated by homophilic interactions between the N2 subdomains, which are likely to occur between adjacent bacteria. Consequently, this mechanism aids in the bacterial attachment and contributes to the overall pathogenicity.

8.2.2.5 Collagen Adhesion Protein (Cna)

The collagen adhesins are a subset of the microbial surface component recognising adhesive matrix molecules (MSCRAMMs) found in *S. aureus*. The Cna protein of *S. aureus* facilitates the attachment to collagen, a key component that provides structural integrity to various tissues. Cna proteins significantly involve in various staphylococcal infections, such as arthritis, endocarditis, osteomyelitis, mastitis and keratitis. During the initial stages of interaction, the MSCRAMM-binding region facilitates the conformational adaptation of the N1N2 domains to effectively enclose the triple helix structure and establish close contact with the collagen monomer. The collagen-binding A region of Cna is succeeded by the B region, which is characterised by its repetitive nature. The number of 23 kDa repeat units (B(1)–B(4)) in the B region vary for strain's origin. The affinity of collagen in the A area is not influenced by the B region (Ghasemian et al. 2015; Vazquez et al. 2011). The B repeat units have been hypothesised to serve as a structural element that extends the A region away from the bacterial surface, thereby facilitating bacterial attachment to collagen. Previous studies have noted that Cna demonstrates selectivity for collagen triple helix monomers, while exhibiting no discernible affinity for collagen fibres or fibrils. The binding ability to collagenous domain of C1q in the complement pathway has been observed to disrupt the interactions between C1r and C1q. Cna, also referred to as complement inhibitors, is involved in the adherence and colonisation strategies employed by *S. aureus* to escape the immune defence system (Arora et al. 2021; Kang et al. 2013).

8.2.2.6 *S. aureus* Surface Protein X (SasX)

SasX is a protein that involves in biofilm growth that facilitates cell aggregation, thereby decreasing neutrophil phagocytosis (Li et al. 2012; Otto 2012) and desquamation cell adhesion (Foster et al. 2014). These processes have been related with the severity of skin and lung infections. In addition to its role in encoding a colonisation factor, the *sasX* gene has been involved in pathogenicity by impeding the immune response, as demonstrated by Nakaminami et al. (2017). According to a study conducted by Li et al. (2012), it can be inferred that SasX played a substantial role in the prevalence of MRSA infections within Asian hospital settings.

8.2.2.7 Iron-Regulated Surface Proteins (Isd)

The *S. aureus* membrane surface proteins encompass a group known as the Iron-Regulated Surface Proteins (IRSPs). This family comprises four distinct surface proteins, namely IsdABCH, as well as a membrane ABC transporter referred to as IsdEF. Also, two intracellular enzymes responsible for haem degradation, IsdGI, are also included within this family. The proteins in question possess a NEAT motif, which plays a role in the internalisation of haem by interacting with haemoglobin and haem (Foster et al. 2014; Grigg et al. 2010). Iron limitation is a prevalent occurrence during the interaction between mammalian hosts and pathogens. The role of IsdA in adhering to squamous under conditions of limited iron and its essentiality for nasal carriage have been demonstrated (Clarke et al. 2007). The Isd protein contributes to the hydrophilic and negatively charged characteristics, which confer resistance against innate immune response, fatty acids and antimicrobial peptides. As a result, *S. aureus* is able to effectively colonise human skin (Clarke and Foster 2008). The IsdH protein plays a prominent role in the evasion of phagocytosis through its ability to inactivate the opsonin C3b (Mathelié-Guinlet et al. 2020). Additionally, it has been observed that IsdH binds to the integrin GPIIb/IIIa on platelets, leading to firm attachment and internalisation in non-phagocytic human cells (Zapotoczna et al. 2013).

8.2.3 Staphyloxanthins (STX)

Staphyloxanthin (STX) is a carotenoid pigment of a yellow hue of many *S. aureus* isolates. This pigment is recognised as a significant virulence factor and evading the host's innate immune defence mechanisms (Pelz et al. 2005). The extensive delocalisation of π-electrons in polyene systems found in carotenoids enables the absorption of visible light, resulting in the vibrant colours ranging from yellow to red. STX possesses the attribute of being an antioxidant due to its possession of multiple conjugated double bonds to detoxify reactive oxygen species. The studies conducted by Clauditz et al. (2006) investigate the mechanism by which neutrophils utilise defuse myeloperoxidase to evade oxidative stress imposed on *S. aureus* during the process of engulfment. This scavenging mechanism is believed to occur due to the interaction between the highly conjugated isoprenyl tails of STX and these oxidising agents. The regulation of STX biosynthesis is mediated by a series of crucial enzymes, including 4,4′-Diapophytoene synthase (CrtM), 4,4′-diapophytoene desaturase (CrtN), 4,4′-diaponeurosporene oxidase (CrtP), glycosyltransferase (CrtQ) and acyltransferase (CrtO) (Liu et al. 2008; Elmesseri et al. 2022).

8.2.4 Extracellular Enzymes

S. aureus also bestows production of enzymatic virulence factors known to be as exoenzymes or extracellular enzymes which functions to break down host or bacterial molecules for nutritional acquisition, survival and bacterial dissemination. Each exoenzymes have different substrates and pathogenesis mechanism to evade host immune system for effective invasion (Tam and Torres 2019).

8.2.4.1 Coagulase

S. aureus can activate host zymogens through the utilisation of three cofactors, namely von Willebrand factor-binding protein (vWbp), coagulase (Coa) and staphylokinase (Sak), despite their lack of inherent enzymatic activity. These three proteins impede with multiple sections of the host's coagulation system, deceiving the host's inherent defences into favouring bacterial survival and propagation. The existence of Coa and vWbp factor in a host system is a crucial condition for producing coagulation during acute infection (Bjerketorp et al. 2004). Coa and vWbp bind to prothrombin to form staphylothrombin resultant in zymogen activation thereby provoke fibrinogen to fibrin conversion leading to fibrinous clots formation contributes to infective endocarditis (Friedrich et al. 2003; Kroh et al. 2009). In abscess formation, Coa and vWbp involve in formation of pseudocapsule and fibrin formation around that pseudocapsule to evade immune response (Guggenberger et al. 2012).

8.2.4.2 Staphylokinase

Staphylokinase (Sak) is a monomeric protein that interacts with plasmin to induce the plasminogen activation, thereby promoting the degradation of fibrin clots and aiding in the dispersion of bacteria. The synthesis of Sak takes place during the lysogenic stage of staphylococci, while the prophage carries genetic instructions for additional virulence factors, including enterotoxin A and chemotaxis inhibitory proteins (Coleman et al. 1989; De Haas et al. 2004). The study conducted by Jin et al. (2004) investigates the involvement of Sak in bacterial infection, specifically focusing on its interaction with α-defensins found in polymorphonuclear cells. This interaction significantly involved in providing protection against infection by bacterial cell wall disruption. The activation of matrix metalloprotease 1 (MMP-1) involves in leukocyte migration and activation, which is facilitated by the formation of Sak-plasmin complexes (Santala et al. 1999). On the other hand, Sak-fibrin-bound complexes have been found to inhibit the opsonisation of bacteria through the complementary pathway (Rooijakkers et al. 2005). In addition to initiating primary skin infection, Sak also serves to restrict bacterial proliferation, thereby mitigating the severity of the infection (Kwiecinski et al. 2013). These limitations are facilitated by the activation of plasminogen, which provides biofilm detachment and reduction in biofilm formation (Kwiecinski et al. 2016).

8.2.4.3 Staphylococcal Nuclease

Staphylococcal nuclease (Nuc) is a monomeric Ca^{2+}-dependent enzyme that cleaves DNA or RNA substrates which are also known as 'Micrococcal Nuclease' identified by Cunnigham et al. in 1956. Nuc is a heat-stable thermonuclease that serves as both an endonuclease and an exonuclease by hydrolysing the 5′-phosphoryl ester bond (Cuatrecasas et al. 1967; Cunningham et al. 1956). Staphylococcal nuclease encoded by two different genes *nuc* and *nuc2* regulated by different promoter and their cellular localisation (Kiedrowski et al. 2011). *S. aureus* employs the mechanism of neutrophil extracellular trap (NET) formation facilitated by staphylococcal nuclease (Nuc) to evade macrophage efferocytosis. *S. aureus* that counteracts against host deployment of NET by executing the degradation of NETs DNA provides monophosphate nucleotides. These substrates are used by adenosine synthase A (AdsA) cell-anchored enzyme to convert deoxyadenosine which induces caspase-3-activation as an anti-inflammatory molecule and promotes bacterial survival in abscess (Howden et al. 2023; Thammavongsa et al. 2013). The increased risk of metastatic infections was achieved by these Nuc by promoting the dispersal of staphylococcal cells through degrading NETs DNA (Bhattacharya et al. 2020).

8.2.4.4 Proteases

Staphylococcus aureus is recognised for its ability to synthesise extracellular proteases, including serine, cysteine and metalloenzymes. The potential danger of these substances arises from the fact that most human plasma protease inhibitors tend to disregard them, and furthermore, some of these substances can be rendered inactive by the inhibitors (Dubin 2002). Tam and Torres (2019) have identified a total of 12 proteases produced by *S. aureus*, including one metalloprotease (aureolysin/Aur), two cysteine proteases (staphopain A (ScpA) and staphopain B (SspB)) and nine serine proteases. Proteases not only involve in acquiring nutrients but also in evading the host immune system through interactions with neutrophils, plasma proteins and antimicrobial peptides (Prokešová et al. 1992; Smagur et al. 2009b).

8.2.4.4.1 Metalloprotease: Aureolysin (Aur)

Aur is categorised as a constituent of the thermolysin family, exhibiting characteristics of a metalloprotease with the presence of Zn and Ca^{2+} ions for both its functional and structural integrity. The protein contains total of 301 amino acids, which undergo a process of folding to create β-pleated domain at the N-terminus and an α-helical C-terminal domain. The zymogen form of Aur is distinguished by the inclusion of its N-terminal fungalysin-thermolysin-propeptide (FTP) domain. Aur FTP undergoes autocatalytic processing, which triggers the staphylococcal proteolytic cascade by activating the serine and cysteine proteases SspA and SspB. The identified protease exhibits enzymatic activity by selectively hydrolysing peptide bonds located on the N-terminal side that possess significant hydrophobicity. As per the findings of Drapeau (1978), the metalloprotease involves in the activation of the V8 or serine protease proenzyme, and it may also be essential for the post-translational modification of other proteases. Aureolysin plays a role in the suppression of phagocytosis and neutrophil-mediated attacks by repressing the complement pathway system. The successful infection is facilitated by the role of Aur, which mimics the C3 convertase of the host system and performs the cleavage of the C3 molecule into C3a and C3b. The Aur hindered the C3b deposition, consequently impeding opsonisation and phagocytosis. This, in turn, inhibits the generation of C5a and the chemotaxis of neutrophils. Aur additionally collaborated with SspA, FnBPs and Protein A in order to facilitate the process of complement inhibition. The modulation of Ig degradation, Cathelicidin (a peptide with antimicrobial properties), serpin-type plasma protease inhibitors α1-proteinase and α1-antichymotrypsin by Aur has significant implications for the host immune response maturation. The proteolytic cleavage of antimicrobial peptides serves to safeguard the staphylococci residing within phagocytes. Aur additionally participates in the promotion of staphylocoagulation, as well as bacterial dissemination and invasion, as discussed by Joo and Otto in 2015.

8.2.4.4.2 Cysteine Protease

Staphylococcus aureus possesses a variety of extracellular cysteine (thiol) proteases, which are papain-like proteins exhibiting a wide range of substrate specificity. The entity's structural conformation is composed of L-domain and the R-domain, which are present in a three-dimensional arrangement. The L-domain is represented by the N-terminal region with the active site helix and a nucleophilic cysteine. The R domain is comprised of a pseudo barrel structure, characterised by size-stranded antiparallel arrangement, which accommodates the catalytic histidine and asparagine residues. The staphopain enzymes namely staphopain A (ScpA) and staphopain B (SspB) have the ability to disrupt the epithelium and the connective tissues of the host (Nickerson et al. 2010). SCP-A is initially produced as an inactive zymogen precursor within the cell. It subsequently undergoes self-activation outside the cell, results in 20 kDa protein formation. This protein is capable of degrading the host tissue, including collagen, elastin, fibrinogen, fibronectin and kininogen. SCP-A inhibits the host's innate immune response by impairing the migration and activation of neutrophils in response to CXCR2 chemokines. SCP-A has been observed to exhibit the capacity to impede the initiation of the complement pathway system, while concurrently neutralising α-1-protease and α 1-antichymotrypsin inhibitor to counteract the immune response. SspB, with a molecular weight of 20 kDa, exhibits structural similarities to ScpA. However, the activation of SspB is contingent upon the proteolytic activation mediated by aureolysin, a metalloproteinase, or the serine protease V8 or ScpA. SspB has the capability to disrupt the homeostasis of host immune cells through its interaction with neutrophils, resulting in the shedding of CD31, a protein involved in leukocyte transmigration and integrin activation. SspB has been documented to exhibit inhibitory effects on the opsonisation of bacteria, as well as its ability to form a dimer with SspB-monocyte, thereby attenuating the immune response (Singh and Phukan 2019; Smagur et al. 2009a).

8.2.4.4.3 Serine Protease (SspA)

The *S. aureus* serine (V8) protease is of narrow substrate-specific protease which preferentially cleaves glutamoyl peptide bonds of proteins. The activation and maturation of serine protease also require the involvement of aureolysin. SspA disrupts the effector function of immunoglobins by degrading Fc region thereby leads to loss of partial detection of antigenic determinants. The SspA plays an important role in the development of inflammatory skin infections. This SspA specifically induces damages in skin integrity by damaging stratum corneum and also induces keratinocyte damage (Rice et al. 2001; Frey et al. 2021).

8.2.4.4.4 Hyaluronidase

The production of hyaluronidase is a vital factor facilitating bacterial dissemination in infections, was first described in 1933 by Duran-Reynals as a 'spreading factor'. The staphylococcal hyaluronidase enzyme facilitates the hydrolysis of the β–1,4glycosidic bonds that connect the *N*-acetylglucosamine and glucuronic acid repeated units within the hyaluronic acid (HA) found in the extracellular matrices (ECM) of the host. It is generally accepted that the agr system does not involve in the regulation of the hyaluronidase that is generated by *Staphylococcus aureus* and that is encoded by the *hysA* gene. The larger presence of extracellular matrix in the skin and lung of the host facilitates the dissemination of bacteria. This is achieved through the involvement of bacterial hyaluronidase, which breaks down hyaluronic acid within the extracellular matrices and biofilms (Pecharki et al. 2008).

8.2.4.5 Lipase

Lipases are enzymes that function as glycerol ester hydrolases, participating in the process of breaking down long-chain triacylglycerols found in lipid surfaces. *Staphylococcus aureus* has the capability to synthesise lipases and esterases, enzymes that catalyse the hydrolysis of glycerol esters that are soluble in water. These enzymes are frequently observed in cases of septicaemia infections (Sargison et al. 2021). The extracellular matrix facilitates the secretion of pre-pro-lipase through the process of transcriptional expression of lipase encoding *geh* genes, by *agr* system regulation. The formation of active form of lipase catalase activity is also regulated through aureolysin activation (Nguyen et al. 2018). Both systemic and localised infections demonstrate elevated lipolytic activity which is important for bacterial metabolism by facilitating nutrient application and enhancing the emergence of biofilms. This process is further enhanced by the lipases' ability to confer resistance to polyamines, thereby ensuring the long-term survival and bacterial persistence. *S. aureus* could able to produce various enzymes, such as phospholipase C, β-toxin, phosphatidylinositol-specific phospholipase (PI-PLC) and fatty acid-modifying enzyme (FAME). These enzymes have been found to involve abscesses formation by facilitating the esterification of cholesterol (Lu et al. 2012; Vossen et al. 2010).

8.2.5 *Staphylococcus aureus* Toxins

The toxins produced by *S. aureus* are excreted substances that differ from other factors contributing to virulence due to their direct impact on the host system. These toxins also significantly involve in biofilm formation, the destruction of the host defence response, as well as the elicitation of cell death thus responsible for its proficiency. The toxins can be categorised into three primary groups based on their functional roles within the host system. According to Handler and Schwartz (2014), the three groups of toxins are as follows: (1) Pore-forming toxins (PFTs), (2) Exfoliative toxins (ETs) and (3) Superantigens (SAgs). The toxins possess the ability to induce harm to host cell membrane, either by intercellular connections degradation or by immune response modulation. The toxins generated by *S. aureus* correlated with a range of diseases, such as deep-seated skin infections, necrotising pneumonia, toxic shock syndrome (TSS) and staphylococcal scalded skin syndrome (SSSS) (Bukowski et al. 2010; Dinges et al. 2000; Holtfreter and Bröker 2005).

8.2.5.1 Pore-Forming Toxins (PFTs)

1. α-Haemolysin (Hla or α-toxin)

   The α-Haemolysin (Hla) is a polypeptide with a molecular weight of 33 kilodaltons (kDa) present in approximately 95% of *S. aureus* clinical isolates. HaemolysinA (Hla) is a protein that adopts a beta-barrel conformation and exhibits the ability to bind to various molecules. Furthermore, it has the capacity to oligomerise, forming a heptameric protein structure that subsequently attaches to host cell membrane. The Hla protein forms a hydrophilic transmembrane channel by protruding its barrel through the lipid layer when it binds to target cells (Divyakolu et al. 2019). The Hla binds to the transmembrane protein ADAM10 located on the host plasma membrane, which functions as a cellular receptor for α-toxin. Subsequently, the oligomerisation process of the toxin occurs to form a heptamer and pre-pore structure as a transmembrane channel. The phenomenon of pore formation and subsequent cell lysis induced by Hla toxin is widely acknowledged as a prominent feature of *S. aureus* infection, impacting T cells, epithelial cells, macrophages, endothelial cells and monocytes. The presence of pores facilitates the entry of extracellular calcium into the cellular environment, consequently initiating membrane phospholipids breakdown and arachidonic acid into leukotriene, prostanoids and thromboxane conversion through metabolic processes. The pivotal significance of HLA toxin is further underscored by the emergence of serum antibody responses within the host system (Wilke and Wardenburg 2010; Inoshima et al. 2011; Oliveira et al. 2018; Divyakolu et al. 2019).

2. β-Haemolysin (Sphingomyelinase C)

   β-toxin, a Mg^{2+}-dependent sphingomyelinase encoded by the *hlb* gene. β-toxin involves in the invaginations of host cell membrane which is less toxic than α-toxin. Sphingomyelinase is phosphoric diester hydrolase which cleaves the eukaryotic membranal sphingomyelins abundantly seen in human skin chronic infections. *S. aureus* deficient in β-toxin showed reduced pathogenicity in disease infection (Tajima et al. 2009; Giese et al. 2011). The activity of β-toxin enhanced by Mn^{2+} and Co^{2+}along with Mg^{2+} whereas Ca^{2+} and Zn^{2+} are inhibitory for sphingomyelin to phosphorylcholine and ceramide hydrolysis. The cytotoxicity of β-haemolysin extends to various human cell types, including keratinocytes, monocytes, polymorphonuclear leukocytes and T lymphocytes. Furthermore, it has been observed that β-haemolysin exerts an inhibitory effect on the expression of interleukin-8 (IL-8) in endothelial cells and escapes phagosomal ingestion and also facilitates biofilm formation (Kristensen et al. 1991; Tsuiji et al. 2019).

3. δ-Toxin

   The delta-toxin is a toxic protein of 26 amino acids residue cytolytic peptide secreted by *S. aureus* encoded by the *hld* gene. The delta-toxin articulated the structural similarity with bee venom melittin exhibiting cytolytic activity against erythrocytes, tissue culture cells and bacterial proteins. The presence of delta-toxin is most abundant in clinical *S. aureus* isolates. It lyses the lipid bilayer through forming a channels by oligomerise to form six δ-toxins aggregates (Dhople and Nagaraj 2005; Janzon and Arvidson 1990; Mellor et al. 1988; Otto 2014).

4. Leukotoxin

   It is well-established that *Staphylococcus aureus* has the ability to synthesise leukotoxins that possess pore-forming properties. The leukotoxins can be categorised into four distinct bi-component groups: gamma (γ)-Haemolysin (HlgA, HlgC, HlgB), Panton-Valentine Leukocidin (PVL), Leukotoxin AB/GH (LukAB/LukGH) and Leukotoxin ED (LukE, LukD). The PFTs consist of two discrete protein constituents, specifically denoted as 'S' and 'F'. The process of pore formation involves the release of monomers, with the S-component specifically binding to the cell surface receptor (Alonzo Iii et al. 2013; Seilie and Bubeck Wardenburg 2017; Spaan et al. 2017). The recruitment of F-components subsequently takes place through the process of dimerisation. Transmembrane channels are formed when these dimers undergo oligomerisation at the plasma membrane. According to Aslam et al. (2013), (γ)-Haemolysin and LukED are the only leukotoxins that specifically target the red blood cells. In contrast, other leukotoxins are responsible for lysing leukocytic lineage cells and causing neutrophil death.

The Panton and Valentine (PV) leukotoxin has two polypeptide chains LukS-PV of 32 kDa and LukF-PV of 38 kDa combined together to induce membrane damage on to macrophages and neutrophils encoded by *lukS-PV* and *lukF-PV* genes. The initial binding occurs with binding of LukF-PV two subunits component and then by LukS-PV. The PV leukotoxin specifically associated with severe skin infections like cutaneous abscesses, severe necrotic skin infections and furuncles rather than diseases such as endocarditis, urinary tract infections, mediastinitis, hospital-acquired pneumonia, etc. (Knudsen et al. 2016; Okolie et al. 2013).

Gamma (γ)-toxins express three proteins HlgAB, HlgCB and HlgACB encoded by *hlg* gene locus. HlgA and HlgC exhibit similar lytic activity related to LukS-PV, however HlgB effective in erythrocyte lysis activity (Cooney et al. 1993). This toxin helps in continued viability and growth of *S. aureus* during a bloodstream infection for its pathogenesis resulting from erythrocyte haem release and macrophage evasion (Cohen et al. 2016).

The cytolytic activity of the leukotoxin LukED is responsible for the pathogenicity of 88–99% of MRSA strains, and it is more common in epidemic strains. Panton-Valentine leucocidin (PVL), which aids in evasion of phagocytes and neutrophils, is a target of LukAB/GH as well as monocytes, dendritic cells and leukocytes. Systemic infection can be triggered by monomeric LukAB/GH that subsequently heterodimerises to bind cell-surface receptors (Aslam et al. 2013; Wang et al. 2007).

8.2.5.2 Phenol-Soluble Modulins

Phenol-soluble modulins (PSMs) express of three peptides PSMα, PSMβ and PSMγ which are extracted by hot phenol extraction process during pro-inflammatory complexes (Syed et al. 2015). The genes of PSMs localised in different operon systems were PSMα in *psmα*; PSMβ includes (PSMβ1 and PSMβ2) in *psmβ* operon and PSMγ in coding sequence of RNAIII. The PSMs attach through non-specific binding to the cytoplasmic membrane thereby leads to membrane disintegration and aggregates in oligomers to produce a pore. PSMs facilitate the formation of biofilm through these short-lived pores thereby involved in pathogenesis. *S.*

aureus-producing invasive infections with the high production of PSMs are correlated with their ability in stimulating inflammatory responses as well as lyse human neutrophils (Peschel and Otto 2013).

8.2.5.3 Exfoliative Toxins

S. aureus-produced exfoliative toxins are crucially implicated in staphylococcal scalded skin syndrome (SSSS) with epidermolytic activity. SSSS is most commonly observed staphylococcal illness in patients predominantly children with the characteristics of blisters and epidermal splitting. Two distinct forms of antigens are expressed by exfoliative toxin such ETA and ETB. The chromosomally located *eta* gene encodes ETA whereas plasmid originated *etb* gene translates as ETB toxin. Both the toxin found to be articulate the structural similarity with serine-proteinase with the lacking of proteolytic activity meanwhile with esterase activity (Aalfs et al. 2010; Bukowski et al. 2010).

8.2.5.4 Superantigens (Ags)

Staphylococcal enterotoxins (SEs) are another name for superantigens since they signpost the classic symptoms of *S. aureus* food poisoning, including nausea, vomiting and diarrhoea, when they are secreted into the host system (Balaban and Rasooly 2000; Johnson et al. 1991). Included in this group are 11 staphylococcal superantigen-like (SSL) toxins (SEIK-SEIQ, SEIU-SEIX) and the toxic shock syndrome toxin (TSST-1) (Grumann et al. 2011, 2014; Holtfreter and Bröker 2005; Lina et al. 2004). Direct cross-linking of the V domains of T-cell receptors with their conserved structures on MHC-II molecules activates T cells, causing them to produce an excess of pro-inflammatory cytokines (IL-2, IFN-γ and TNF). An overabundance of cytokines limits T-cell responses to antigens by inhibiting IL-2 release and resulting to cell death. High fever, skin rashes, desquamation, nausea, vomiting, low blood pressure and diarrhoea are only few of the symptoms that might develop, eventually leading to organ failure. SAgs are thought to be effective immunogens and neutralise the antibody response (Kwiecinski et al. 2013; Ono et al. 2008; Wilson et al. 2011) due to their function in disorders such as sepsis, skin infection and allergies.

8.3 *Staphylococcus aureus* Biofilm

The nasal carriage of *S. aureus* is a common occurrence that facilitates the dissemination of bacteria to the circulatory system. This is achieved through various mechanisms such as increased adherence factors, planktonic growth and the presence of epithelial fissures, all of which contribute to the successful establishment of infection (Wertheim et al. 2005). Biofilms possess a unique architectural structure that facilitates the formation of microbial communities, enabling staphylococci to invade host tissues. These biofilms are able to adhere to a wide variety of surfaces and eventually become encased within an extracellular polymeric matrix. Biofilm production is critically dependent on pathogenic bacteria's ability to avoid phagocytosis and also have a significant influence on the immune response of the host, leading to an anti-inflammatory state. The microorganism is presently receiving considerable scrutiny due to its escalating association with infections linked to foreign objects, its rapid acquisition of resistance to multiple antibiotics and its tendency to transition from a minor infection to a constant, persistent and recurring condition. The biofilms allow the bacterium to adhere and persist on the bones and heart valves like host tissues which develop to osteomyelitis and endocarditis. Additionally, *S. aureus* can colonise catheters, prosthetic joints and pacemakers, causing chronic infections (Møretrø et al. 2003; Speziale and Geoghegan 2015).

8.3.1 Components of Biofilm

The biofilms of *Staphylococcus aureus* are primarily composed of water (up to 97%), exopolysaccharides (EPS) and microcolonies. The exopolysaccharides consist of nucleic acids, specifically eDNA and eRNA, which make up less than 1% of the composition. Proteins account for less than 1–2%, lipids comprise less than 1% and other biomolecules are also present (Donlan 2002; Idrees et al. 2020; Guzmán-Soto et al. 2021). The primary constituent of EPS comprises polysaccharide intercellular adhesion (PIA), which is composed of poly-β(1-6)-*N*-acetylglucosamine (PNAG). Multiple biological activities rely on the cationic characteristics of PIA, including infection, colonisation, immunological evasion, phagocytosis and antimicrobial resistance. Accumulation-associated proteins (Aap), surface-binding protein A (Spa), extracellular matrix-binding protein (Embp), fibrinogen-binding proteins (FnBP) A and B, *S. aureus* surface-binding protein (SasG) and amyloid fibres, are just some of the proteins that make up the extracellular polymeric substance (EPS) in *S. aureus* (Idrees et al. 2021; Ko et al. 2013; Kuroda et al. 2008; Speziale and Pietrocola 2020). The MSCRAMM, three-helical bundle proteins, near iron transporters (NEAT) and G5-E repeat proteins all contribute significant roles in biofilm formation (Ghasemian et al. 2015). The CWA proteins are essential for adherence on the extracellular matrix (ECM) of host surfaces and to neighbouring cells. The proteins SasG and SbpA play a role in surface attachment and cell adhesion, whereas Aap interacts with PIA to facilitate biofilm maturation (Corrigan et al. 2007; Ilk et al. 2002; Wang et al. 2022). The maintenance of biofilm matrix stability is achieved through the utilisation of amyloid fibres as a scaffold. The EPS is composed of various charged groups, including carboxyl groups, glutamic acid, aspartic acid, sulphates, phosphates and amino sugars. The eDNA of *S. aureus*, in conjunction with the EPS proteins and PIA, is involved in antimicrobial resistance and deception of host immune system along with irreversible attachment and the preservation of biofilm stability (Idrees et al. 2021).

8.3.2 Biofilm Formation

Staphylococcus aureus must undergo a series of five essential stages in order to successfully form biofilms on various surfaces. (1) Attachment: In the beginning, planktonic cells that are capable of free-swimming can adhere to surfaces for an extended period of time as a result of weak interactions. (2) Colonisation: Bacterial pathogens establish irreversible attachment to surfaces through the utilisation of stronger connections such as adhesins, lipopolysaccharides and adhesive proteins that bind to collagen. (3) Formation: The process involves the significant aggregation of bacterial cells in multiple layers, accompanied by a substantial production of EPS. (4) Maturation: The adhered bacteria undergo a process of development, resulting in the formation of their characteristic three-dimensional biofilm structure. (5) Dispersal: It is the completion of biofilm formation, wherein the biofilm is fragmented or disseminated through the utilisation of mechanical and active methodologies (Moormeier and Bayles 2017).

The free-floating, planktonic cells of *S. aureus* colonise the host/accessible surfaces to initially form the biofilms. Interactions between hydrophobic and hydrophilic molecules on the host cell surface regulate bacterial adhesion. The development of microcolonies takes place after successful adhesion of cells. An extracellular polymeric substance (EPS) is produced following the establishment of microcolonies, and this EPS develops into a biofilm over time. When the biofilm is fully grown, the bacteria that reside within it release certain substances, such as D-amino acids and EPS-degrading enzymes lead to degradation of biofilm matrix (Lee et al. 2013). The bacterial dissemination takes place with the help of these dispersed planktonic cells from the matrix which either colonise the nearer surface for spatial spreading of infection or to other location for continuing the process of biofilm formation (Fig. 8.1).

8.3.3 Genetic Regulation in Biofilm Formation and Dispersal

S. aureus biofilms are characterised by the altered gene expressions of the resident cells. Genes essential to biofilm formation are differently expressed, giving rise to variations in biofilm's nature, development, functionality and structure. FnBPs (*fib*) gene, fibronectin-binding proteins (*fnbA* and *fnbB*) genes, intercellular adhesion (*icaA, B, C* and *D*) genes, clumping factor (*clfA* and *clfB*), elastin-binding protein (*ebps*), laminin-binding protein (*eno*) and collagen-binding protein (*cna*) gene and others are all included in *S. aureus* biofilm. Fibrinogen-binding proteins encoded by the *fib* gene facilitate initial adhesion to the surface, while collagen-binding proteins are encoded by the equivalent *cna* genes. Intercellular adhesion genes *icaABCD* are upregulated during biofilm formation (Kot et al. 2020), initiating the process of cell adhesion to one another. Co-expression of *fnbA* and *fnbB* genes contributes to biofilm production via invasin-mediated host cell penetration. CWA proteins are responsible for cell adhesion to surface fibrinogen (Deivanayagam et al. 1999). By binding to soluble fibrinogen, these clumping factors promote efficient

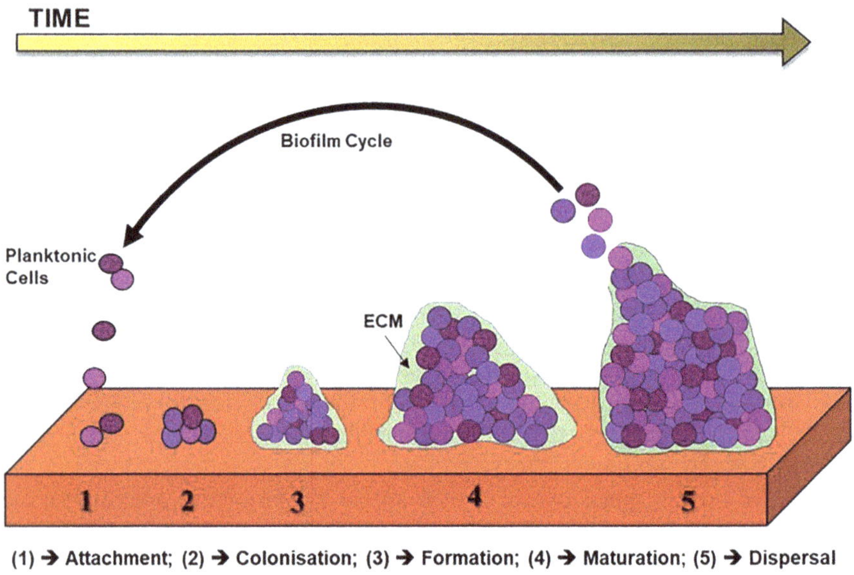

Fig. 8.1 The graphical illustration of biofilm formation in *Staphylococcus aureus* through five different steps. ECM denotes the extracellular matrix formed during biofilm

colonisation, biofilm development and immunological evasion. *SdrCD* are regulated by their respective *sdr* genes, allowed for desquamated epithelial cells and nasal colonisation (Liu et al. 2015). Elastin-binding proteins (encoded by the *ebps* gene) and laminin-binding proteins (encoded by the *eno* gene) are involved in colonisation and biofilm development (Kot et al. 2018).

The biofilm maturation process is influenced by the extracellular adherence protein (EAP) encoded by the *eap* genes and the beta toxin protein, as demonstrated by Haggar et al. (2010). Autolysis of biofilm cells by murein hydrolases encoded by the *atl* and *lytM* genes, generates eDNA (Komatsuzawa et al. 1997; Singh et al. 2010). The upregulation of these genes facilitates autolysis process that is controlled by two operons, namely *cidABC* and *lrgAB*. The *CidA* protein is essential because it helps the cell membrane to oligomerise, leading to the development of pores through which murein hydrolase may be transported. On the other hand, LrgAB functions as an antiholin, preventing CidA from doing its intended function. Micro-environmental niches near the bottom of the biofilm matrix are created when operons are regulated (Endres et al. 2022; Rice et al. 2005; Windham et al. 2016). When an autoinducing peptide (AIP) is detected, the *agr* operon is triggered to activate the *agr* system. The *agr* system regulates the protease secretion and phenol-soluble modulins which aid in biofilm dissemination and promote the transition from biofilm to planktonic growth (Johnson et al. 2015; Lyon et al. 2002).

8.4 Regulation of Virulence Factors

Staphylococcus aureus is capable of initiating and maintaining infection, leading to the development of both short- and long-term illnesses within the host's tissues through deploying varied virulence factors such as exoenzymes, haemolysins, super-antigens as well cell wall-associated factors depending on its invasion (Wardenburg and Schneewind 2008). The regulation of *S. aureus* is highly multifaceted and arbitrated by a hefty amount of regulatory systems. The virulence factors frequently associated with *Staphylococcus aureus* are usually represented by mobile genetic elements (MGEs) or the accessory genome, rather than the chromosomal genome. MGEs include plasmids, transposons, prophages, insertion sequences and antibiotic resistance determinants (Queck et al. 2009). *S. aureus* consists of staphylococcal pathogenicity islands (SaPIs) which contrast to MGEs and linked to pathogenicity along with characteristic for the entire species (Haag et al. 2021; Novick and Subedi 2007). SaPIs encoded for toxins and virulence determinants include enterotoxins and TSST. *S. aureus,* MSCRAMMs also have a key role in virulence factors that are not encoded by *agr* system. The virulence factors are influenced by various regulatory systems, including the Agr system, the Sar family of DNA-binding proteins, *Staphylococcus aureus* exoprotein expression (*sae*), alternative sigma factor SigB-dependent regulation and the staphylococcal respiratory regulator (*SrrAB*) (Cerca et al. 2008; Tiwari et al. 2020).

8.4.1 Accessory Gene Regulator (*Agr*) System

The agr system is an extensively studied regulatory system that governs virulence factors expression in the bacterium *S. aureus*. The phenomenon also acknowledged as a quorum-sensing system in the context of the growth cycle phase. This system consists of two promoters P2- and P3-regulated transcription units. AgrA and AgrC, two components of the P2 operon, interact with the regulatory RNA molecule RNAIII to control virulence factors. AgrB and AgrD are two entities that are responsible for encoding and secreting autoinducing peptide (AIP) (Huntzinger et al. 2005; Thoendel and Horswill 2009). Activation of P2 and P3 agr promoters is the primary function of the four gene units (*agrB, agrC, agrD* and *agrA*). AgrC and AgrA are two proteins that work together as a system to transmit signals. The AgrC protein is a transmembrane receptor and a histidine kinase sensor. The P2 and P3 promoters' transcriptional activation is mediated by the AgrA and the SarA protein. An AIP of cyclic octapeptide is synthesised by digesting the agrD protein generated by *agrD* gene. This AIP is crucial in facilitating the autoinduction process of the agr system. The specificity of AgrD processing, which results in the creation of a cyclic thiolactone bond between cysteine molecule and the C-terminal carboxyl group, is in part determined by the transmembrane protein AgrB (Lina et al. 1998; Sully et al. 2014; Zhang and Ji 2004).

Cell density serves as the initial trigger for activating the *agr* system. During exponential cell growth, it is probable that cells exhibit a minimal basal expression of the *agr* P2 operon. Consequently, the AgrD octapeptide pheromone undergoes expression and processing, subsequently culminating in its release (Zhang et al. 2004). According to Tan et al. (2018), it is postulated that once a specific concentration threshold is reached, the octapeptide binds to a specific region of the AgrC receptor. This region is supposed to be accessible from the cytoplasmic membrane outer surface. The activation of AgrC's protein kinase activity leads to an increase in P2 and P3 promoters' transcription, thereby stimulating the response regulator, AgrA (Srivastava et al. 2014). P2 operon's autoinduction leads to the agr P3 promoter transcript activation, which in turn produces the P3 transcript RNAIII. This RNAIII molecule acts as the effector for regulating the target genes specific to the agr system. The 5′ end of RNAIII plays a role in translational regulation while 3′ end appears to be involved in transcriptional regulation and repression functions. RNAIII also encodes for δ-toxin which does not involve in regulatory function of *agr* system. α-toxin, β-toxin, δ-toxin, TSST-1, SEB, SEC, FAME, serine protease, lipase, staphylokinase and CP5 are regulated by RNAIII. Some surface-associated proteins like protein A, FnBPA, FnBPB and coagulase are negatively regulated by RNAIII. However Cna, ClfB and EbpS are regulated independently of *agr* (Morfeldt et al. 1996; Huntzinger et al. 2005).

8.4.2 Staphylococcal Accessory Regulator (*sar*) System

The staphylococcal *sar* locus represents a comprehensive regulatory system encompassing quorum-sensing cascades, which effectively govern the virulence factors production and biofilm formation. The *sar* locus is comprised of three transcripts, namely *sarA* (0.56 kb), *sarB* (1.2 kb) and *sarC* (0.8 kb) that overlap with each other and share a common 3′ end. These transcripts are transcribed from three separate promoter regions and are responsible for encoding the DNA-binding SarA protein. The SarA protein enhances RNAII and the subsequent RNAIII transcription by preferentially binding to P2 rather than P3 promoter region. The major sigma factor, σA, activates the P2 and PI promoters that control the sarB and sarA transcripts. The alternative sigma factor, σB, activates the P3 promoter in stationary phase, which promotes the sarC transcript. Along with *agr* regulation, SarA directly impacts the expression of many exoprotein genes. Recently, an identical global regulator called SarHl (Sar Homologues1), which is similar to SarA, was discovered in *S. aureus*. Study on the effects of SarH1 on Hla (α-toxin) transcription, Spa (protein A) and Ssp (serine protease) has revealed that both RNAIII and SarA have an impact. It is conceivable for SarHl to bind directly to target gene promoters and control the transcription of those genes without involving SarA or Agr (Wesson et al. 1998; Wright et al. 2005).

8.4.3 Repressor of Toxins (Rot) System

The regulatory protein known as the 'repressor of toxins' (Rot) possesses an approximately 15.6 kilodaltons (kDa) protein that is comparable to the SarA protein family. The Rot protein was found to exert repression on Enterotoxin B (*seb*), alpha-toxin (*hla*), protease (*spl/ssp*) and lipase (*geh*). The repression of Rot production involves the participation of RNAIII, SarA and Sigma factor (SigB) as reported by Chatterjee et al. (2013) and Killikelly et al. (2015). The Rot proteins serve as an affirmative regulator of various virulence factors, including protein A (Spa), superantigen-like protein (Ssl) and the SarA-family protein SarH1.

8.4.4 Multiple Antibiotic Resistance Regulator (MgrA) System

The *S. aureus* 'MgrA' system is a type of regulatory system that aggregates virulence components and promotes biofilm formation. The activation of MgrA occurs through activation of ArlRS two-component system which provides unknown signalling for subsequent activation. The regulation of MgrA is of significant importance in the process of capsular production, specifically in the context of *cap5*, within the context of endocarditis infections. Alpha-toxin, Protein A, coagulase, extracellular serine proteases and nuclease are virulence factors that are controlled by this system. MgrA inhibits the generation of large surface proteins (Ebh, SasG and SraP), although promoting clumping via dimeric formation of Fg molecules by strengthening the interaction of ClfA/ClfB with fibrinogen. MgrA also known as 'Rat' regulator of autolysis as it regulates with *lytRS*, *lrgAB* and *arlRS* TCS in *S. aureus* autolysis (Luong et al. 2006; Sun et al. 2011, 2012).

8.4.5 *Staphylococcus aureus* Exoprotein (sae) System

The *sae* locus, also known as the *Staphylococcus aureus* exoprotein expression locus, is responsible for encoding a two-component signal transduction system. This system involves in initiating the transcriptional synthesis of various exoproteins, such as α-toxin, β-toxin, DNase and coagulase. Importantly, this pathway operates independently of agr or sarA. The *sae* locus is comprised of the *saeR* and *saeS* genes, which are responsible for encoding the SaeR and SaeS proteins. Co-transcription is a phenomenon that takes place between the *saeR* and *saeS* genes. The sensor histidine-kinase is commonly referred to as SaeS, whereas the corresponding response regulator is denoted as SaeR. An incomplete open reading frame (ORF) exhibiting similarity to the *B. Subtilis csbB* gene is located downstream of *saeS*. The open reading frame (ORF) may potentially have a role in the transcriptional termination of the sae locus, as suggested by previous studies (Clements and Foster 1999; Li and Cheung 2008; Miller 2015).

8.4.6 Sigma Factor-Dependent Regulation

S. aureus possesses a total of four sigma factors, namely Sigma A (σA), Sigma B (σB), Sigma H (σH) and Sigma S (σS). Sigma A is primarily responsible for regulating the promoter regions associated with housekeeping functions, while the remaining three sigma factors, σB, σH and σS, have been identified as alternative sigma factors. The protein SigB (σB) is upregulated during stress-responsive conditions. The primary sigma factor σA is essential for the proliferation of *S. aureus*, whereas an alternative sigma factor (σB) transcribes numerous virulence genes that exhibit significant homology to σA. Previous studies showed the role of σB as a pleiotropic regulator in virulence factors production, which is essential for the development of pathogenicity. The transcription of SigB has been found to be associated with the regulation of the *sar* locus and also involves in virulence genes expression. Moreover, the regulation of SigB has been associated with the synthesis of thermonuclease and lipase, which are crucial enzymes involved in the formation of abscesses. Hence, it can be inferred that SigB exerts an influence on the virulence genes expression through both direct and indirect mechanisms. The sigma factor A (SigA or σA) is implicated in the activation of the *rsbU* gene, which is known to be produced under conditions of heat shock. The activation of transcription at the SigB (σB) promoter is facilitated by RsbU, as demonstrated in studies conducted by Cerca et al. (2008), Giachino et al. (2001) and Tahmasebi et al. (2021).

8.4.7 Staphylococcal Respiratory Regulator (SrrAB) System

The SrrAB, alternatively known as SrhSR, functions as a universal regulator of virulence factors under hypoxic conditions. The genetic locus referred to as *srrAB* is accountable for the synthesis of two separate proteins, specifically SrrB, which serves as a histidine kinase, and SrrA, which assumes the role of a response regulator. In addition to positively regulating the expression of genes like *kat*, *ahpC* and *dps* that are involved in resistance to hydrogen peroxide, the SrrAB system confers resistance to nitrosative stress. The SrrAB proteins exert a negative regulatory effect on the agr system, as well as on Spa and TSST-1 virulence factors. SrrAB serves as a significant positive regulator for the expression of polysaccharide intercellular adhesin and phosphatidylinositol-specific phospholipase C enzyme. This regulatory function is essential bacterial survival in human blood, as it enables the evasion of neutrophil-mediated immune responses (Pragman et al. 2004; Tiwari et al. 2020; Yarwood et al. 2001). The Fig. 8.2 illustrates the regulatory system of *S. aureus*, specifically focusing on the interaction between its virulence factors (Fig. 8.2).

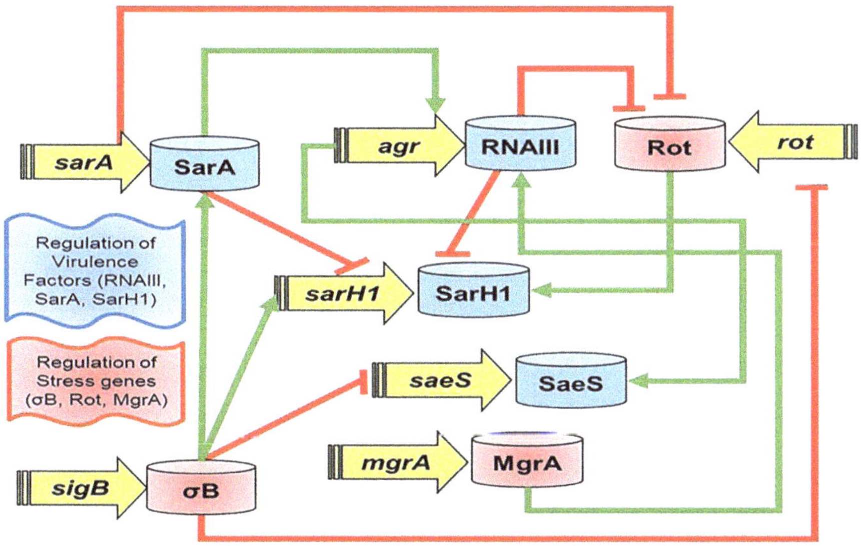

Fig. 8.2 Interaction between the regulatory systems of virulence factors in *S. aureus*. The striped arrow indicates the transcripts encoding proteins visualised as cylinders. The green arrows indicated the positive regulation of respective protein, while the red inhibit arrows represent the negative regulation of respective genes/protein

8.5 *S. aureus* in Antimicrobial Drug Resistance

Staphylococcus aureus serves as the primary causative agent for wide range of bacterial infections in the human population, displaying a range of severity levels. The facilitation of bacterial transmission to deeper tissue is observed in medical invasive procedures that involve devices like artificial implants and catheters. The pathogenicity is closely linked to bacterial potentiality to invade and persist within host tissues, and also its capacity to counteract the immune response and acquire antibiotic resistance. Previous studies have underscored the significance of advancing novel therapeutic approaches for the complete elimination of *S. aureus* infection (Rehm and Tice 2010; Smith and Jarvis 1999).

8.5.1 Evolutionary Origin of Multi-Drug-Resistant *Staphylococcus aureus* (MRSA, VRSA)

The crucial identification of penicillin through the astute observations made by Alexander Fleming regarding inhibition of *S. aureus* marked the commencement of the antibiotic era. Beta-lactam antibiotics have emerged as the pre-eminent class of antibiotics for *S. aureus* infections. The acquisition of resistance towards antibiotics is facilitated by beta-lactamases localised in plasmids (Novick and Richmond 1965). Consequently, the development and subsequent approval of penicillin, cephalosporins, Penicillin-resistant *S. aureus* have evolved due to the spread of the *blaZ* gene, coding beta-lactamase enzyme that hydrolyses the beta-lactam ring in antibiotics (Bondi and Dietz 1945). Penicillin resistance is prevalent among the isolates from hospitals and other healthcare settings, with an estimated 99% incidence (Vestergaard et al. 2019). The foremost clinical isolate of methicillin-resistant *Staphylococcus aureus* (MRSA) was discovered in the United Kingdom in the 1960s (Boyle-Vavra and Daum 2016), prompting the introduction of methicillin in 1961. The horizontal gene transfer (HGT) mechanism involves the transfer of the genetic element recognised as *staphylococcal cassette chromosome mec* (SCCmec), which encodes for PBP2a. PBP2a is an alternative transpeptidase that demonstrates reduced affinity towards the majority of beta-lactam rings. The aforementioned mechanism involves in the development of MRSA, a strain that poses considerable challenges. The increase in MRSA infections prompted the wider adoption of Vancomycin, an antibiotic that is typically reserved as a last-line treatment for MRSA infections. The development of Vancomycin-resistant *Staphylococcus aureus* (VRSA) occurs because of alteration in the assembly of D-ala-D-ala residues in cell wall peptidoglycan layer, a specific target of vancomycin. The VRSA isolates are characterised by the presence of the *vanA* genes with the enzymatic breakdown of the D-ala-D-ala precursors or the D-ala-D-lactate precursor production. The eventual consequence of these processes is that vancomycin only binds to a limited extent (Okano et al. 2017). *S. aureus* isolates susceptible to vancomycin are known as vancomycin-susceptible *S. aureus* (VSSA), and their minimum inhibitory concentrations (MICs) are lower than 2 μg/mL. On the other hand, strains that possess intermediate resistance (known as VISA) demonstrate minimum inhibitory concentrations (MICs) that span from 4 to 16 μg/mL. In contrast, strains that are fully resistant exhibit MICs that surpass 16 μg/mL (Arthur et al. 1996; Lakhundi and Zhang 2018).

8.5.2 Mechanism of Antibiotic Resistance *S. aureus*

The antibiotic resistance by *S. aureus* against many of antibiotics acquires through mobile genetic elements (MGEs) through HGT mechanism. The mutation in these determinants is crucially important for resistance development that can alter the drug target-binding sites, efflux pump activation, protein inactivation, etc. The cell envelope, Porins, protein and nucleic acid synthesis are common antibiotic targets. The induction of antibiotic resistance caused by primary resistance mechanisms includes (1) the enzymatic degradation of antimicrobial agents, (2) modification in potential antibiotic targets of pathogens and (3) alteration in membrane permeability for antibiotics influx (Pantosti et al. 2007; Schaenzer and Wright 2020).

The β-lactam antibiotics are predominantly used antibiotics targeting the *S. aureus* bifunctional transglycosylase-transpeptidase PBP2 which is responsible for development of peptidoglycan formation. *S. aureus* attains the resistance towards β-lactam antibiotics through activation of serine β-lactamase (BlaZ) which mimic the acyl enzyme intermediate as TP of PBP2. In case of methicillin and oxacillin, the resistance acquired through expression of gene-encoding PBP2a, a homolog to PBP2. Thus the inhibition of antibiotic binding to PBP2 occurs and induced resistance against methicillin and oxacillin. Mutations in the *mrpF* gene, coding for integral membrane protein MrpF and confer resistance to vancomycin and daptomycin, are responsible for this phenomenon. This protein increases the amount of lysyl-phosphatidyl glycerol (Lys-PG) in the membrane's outer face, which opposes the binding of the Ca-Dap (Calcium-bound daptomycin) complex and so prevents the onset of membrane damage by depolarisation, permeabilisation and ion leakage. The cell walls of VISA and VRSA are more robust because of an increased thickness of uncross-linked peptidoglycan. Subsequently, VISA and Dap resistance strains were found to have mutations in the *walKR, vraSR, rpoC* and *dtl* genes.

Many antibiotics block translation by interfering with cytoplasmic proteins that are necessary for ribosome binding or ribosomal translation. Several antibiotics have been developed to specifically target protein synthesis. These include tetracycline, erythromycin, streptogramin, lincomycin, linezolid and aminoglycosides such as neomycin and gentamicin (Ayliffe et al. 1977; Pantosti et al. 2007). Tetracycline exhibits binding affinity towards the encoding centre of the 30S ribosomal subunit, resulting in the dissociation of aminoacyl-tRNA (aa-tRNA) from the A site. Erythromycin and Streptogramin inhibit the function of the polypeptide exit tunnel located within the 50S subunit, while Lincomycin, Streptogramin, Pleuromutilin and Florphenicol disrupt the binding of aminoacyl tRNA at the peptidyl transferase centre. Aminoglycosides exert an influence on the process of mRNA misreading. During ribosome translocation, fusidic acid that binds with high affinity to elongation factor G disrupts the movement of transfer RNA from the P-site to the A-site. According to Engel et al. (1980), the functionality of isopentyl tRNA transferase is impeded by Muciprocin. *Staphylococcus aureus* has acquired its resistance mechanism to drugs that block protein synthesis through either horizontal gene transfer or chromosomal gene changes (Engel et al. 1980; Pantosti et al. 2007). Staphylococci exhibit resistance to tetracycline through the participation of two closely associated tetracycline efflux pumps, namely TetA(K) and TetA(L). The study conducted by Guay and Rothstein in 1993 involved the identification of a small plasmid called pT181. This plasmid contained multiple copies and encoded the *TetK* gene. Notably, the pT181 plasmid was found to be integrated within the chromosomal SCCmecIII cassette of MRSA strains. The tetracycline resistance in *S. aureus* has been observed through the presence of conjugative transposons Tn916 and Tn1545, which encode TetO/M determinants (Clewell et al. 1995). Similar to EF-G, the TetO/M molecule applies force to the ribosome, displacing Tet off its binding site. The protein FusB binds to EF-G, causing the displacement of the drug that is bound to EF-G and facilitating the process of translocation. The acquisition of neomycin resistance is mediated by either an adenyltransferase (aadD) or a phosphotransferase (aphA), as indicated by the presence of plasmid pUB110. On the other hand, the development of gentamicin resistance is attributed to a bifunctional acetyltransferase-phosphotransferase (aacA-aphD) carried by Tn4001, as reported by Rouch et al. (1987). The acquisition of 50S subunit-binding antibiotics resistance is mediated by genes such as *erm* and *cfr*, which are associated with the modification of the target site through rRNA methyltransferase activity. Additionally, other genes such as *Vga, Lsa, Sal, Msr* and *OptrA* are involved in ribosomal protection through the ABC-F mechanism. The activation of Fex efflux and Vgb, Vat enzymes is involved in drug modification, thereby providing a mechanism for resistance. The occurrence of chromosomal mutations in 23SrRNA, L3 and L4 has been linked with Linezolid resistance, as reported by Matsuoka et al. (2003). Similarly, resistance to Tigecycline has been linked to mutations in rRNA and MepA.

Successful antimicrobial medications are generally agreed upon to be those that interfere with the manufacture of nucleic acids. The presence of fluorine in the C6 position is the defining feature of the synthetic antibacterial drugs known as fluoroquinolones. They are derivatives of the prototypical nalidixic acid, and they are highly active and potent. At the molecular level, Topoisomerase IV and DNA gyrase are the bacteria that fluoroquinolones are designed to kill (Alt et al. 2011). Overexpression of efflux pumps like NorA, NorB and NorC, as well as a chromosomal mutation resulting in mono amino acid substitution in topoisomerase ParC subunit, constitute the resistance mechanism against drug. The MgrA regulatory system controls the expression of NorA, NorB and NorC. Truong-Bolduc et al. (2006) and Truong-Bolduc and Hooper (2010) showed that *S. aureus* has the remarkable capacity to build resistance against antibiotics that it has encountered.

Several studies suggested that biofilm-associated *S. aureus* strains show intense resistance/tolerance to various antibiotics. The extracellular matrix formed during biofilm hinders the penetration of antibiotic thereby limiting the exposure of antibiotics to reach the inner layers of cell. In spite of this, ECM also served as pool for enzyme accumulation for antibiotic inactivation providing frontline defence against antimicrobial agents. The heterogeneous cells population in the biofilm contributes for antibiotics susceptibility profile with different metabolic profiles. Biofilms also involve in antibiotic resistance prevalence through HGT mechanism (Ma et al. 2019). The role of virulence factors in host system is articulated in Fig. 8.3.

8.6 Inhibitors and Novel Therapeutics for *S. aureus* Infection

Antibiotics are commonly prescribed as initial treatments that inhibit bacterial growth by disrupting the synthesis of crucial components necessary for bacterial survival. The limited efficacy of conventional therapies in addressing infections is caused by antibiotic-resistant strains leading narrow spectrum of action exhibited by existing antibiotics, which predominantly focus on a restricted range of specific proteins. The utilisation of targeted anti-virulence agents to inhibit pathogenic virulence factors represents a significant advancement in combating resistance compared to conventional antibiotic treatment. This approach minimises the need for resistance development and reduces its impact, making it a highly valuable alternative. Multiple studies have documented the investigation of diverse anti-virulence strategies or inhibitors targeting the virulence factors and biofilm (Bal and Gould 2005; Guo et al. 2020). The utilisation of antimicrobial peptides (AMPs) derived from indigenous sources has demonstrated inhibitory activity against virulence

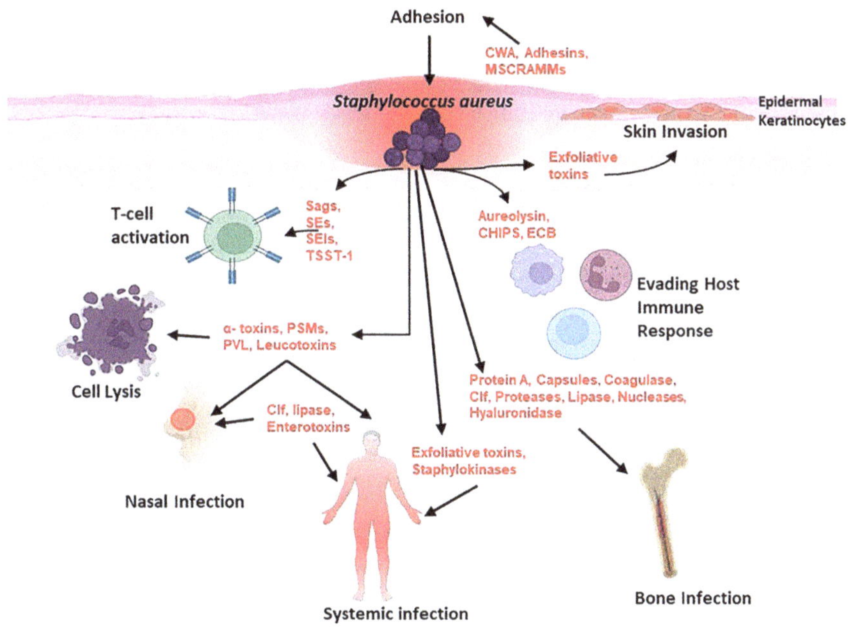

Fig. 8.3 The graphical illustration of role of virulence factors of *Staphylococcus aureus* in host system

factors and biofilm components and formation. The disruption of quorum sensing, a mechanism that can effectively impede the primary virulence factor of *S. aureus* and disrupt the Agr regulatory system, has been proposed as a novel therapeutic strategy that can slow the spread of the bacteria and prevent infection. The participation of lectin inhibition is implicated in the attenuation of the complement activation pathway by the extracellular adherence protein (Eap). Phage therapy offers advantages over other strategies due to its ability to exert host specificity, utilising phage particles to selectively restrain the growth and bacterial dissemination (Woo et al. 2017). The utilisation of nanotechnology in the field of bacterial treatment has demonstrated significant progress in terms of both diagnosis and therapeutics. The nanoparticles are very effective at penetrating the bacterial cell membrane and breaking up biofilms. Another potential therapeutic strategy involves the utilisation of CRISPR/Cas9 gene-editing technology to address the downregulation of virulence factors. Additionally, the development of vaccines targeting virulence proteins presents an alternative approach for combating staphylococcal infections. In addition to the aforementioned strategies aimed at reducing virulence, recent scientific research has been dedicated to investigating the potential of utilising plant extracts, essential oils and other naturally occurring bioactive compounds as anti-virulence therapies (Appelbaum 2007; Ohlsen 2009; Wu et al. 2019).

8.7 Conclusion

The extensive array and variability of virulence factors displayed by *S. aureus* play a significant role in its capacity to induce a broad spectrum of infections, with severity ranges from acute to chronic presentations. *Staphylococcus aureus* is capable of forming biofilms, which serve as a protective environment for bacterial survival and facilitate the chronic infections development. The primary aim of this review was to investigate various virulence factors and the regulation and functions of biofilms, as well as to analyse their implications in the immune responses of the host. The challenges encountered in the post-antibiotic era, characterised by the widespread presence of virulence and resistance mechanisms can be effectively addressed through continuous efforts aimed at developing specific anti-virulence strategies, vaccination approaches and novel therapeutics to combat detrimental *S. aureus* infections. Further investigation is warranted to explore the potential effectiveness of combining anti-virulence agents with antibiotic treatment in inhibiting the progression and manifestation of staphylococcal disease. The strategic use of these medications to target different virulence factors and biofilm-associated diseases caused by the widely recognised bacterium *Staphylococcus aureus* will ultimately contribute to the advancement of initiatives aimed at addressing antimicrobial resistance (AMR).

8.8 Future Prospective

The involvement of newer drugs and combinatorial therapy is feasible, cost-effective, long-lasting activity that ensures the constant ability to treat various Staphylococcal infections. Despite of potentiality of anti-virulence factors in treatment of illness, there exists numerous challenges in utilisation of this anti-virulence for clinical practices. The detailed understanding on to the pathogenesis of *S. aureus* virulence factors and biofilm provides an extent opportunity to successfully compete the acute or chronic staphylococcal infections.

References

Aalfs AS, Oktarina DAM, Diercks GFH, Jonkman MF, Pas HH (2010) Staphylococcal scalded skin syndrome: loss of desmoglein 1 in patient skin. Eur J Dermatol 20(4):451–456. https://doi.org/10.1684/ejd.2010.1007

Alonzo Iii F, Kozhaya L, Rawlings SA, Reyes-Robles T, DuMont AL, Myszka DG, Landau NR, Unutmaz D, Torres VJ (2013) CCR5 is a receptor for Staphylococcus aureus leukotoxin ED. Nature 493(7430):7430. https://doi.org/10.1038/nature11724

Alt S, Mitchenall LA, Maxwell A, Heide L (2011) Inhibition of DNA gyrase and DNA topoisomerase IV of Staphylococcus aureus and Escherichia coli by aminocoumarin antibiotics. J Antimicrob Chemother 66(9):2061–2069. https://doi.org/10.1093/jac/dkr247

Appelbaum PC (2007) Microbiology of antibiotic resistance in Staphylococcus aureus. Clin Infect Dis 45(Suppl. 3):S165–S170. https://doi.org/10.1086/519474

Arora S, Gordon J, Hook M (2021) Collagen binding proteins of gram-positive pathogens. Front Microbiol 12:628798. https://www.frontiersin.org/articles/10.3389/fmicb.2021.628798

Arthur M, Reynolds PE, Depardieu F, Evers S, Dutka-Malen S, Quintiliani R, Courvalin P (1996) Mechanisms of glycopeptide resistance in enterococci. J Infect 32(1):11–16. https://doi.org/10.1016/s0163-4453(96)80003-x

Aslam R, Laventie B-J, Marban C, Prévost G, Keller D, Strub J-M, Dorsselaer AV, Haikel Y, Taddei C, Metz-Boutigue M-H (2013) Activation of neutrophils by the two-component Leukotoxin LukE/D from Staphylococcus aureus: proteomic analysis of the secretions. J Proteome Res 12(8):3667–3678. https://doi.org/10.1021/pr400199x

Ayliffe GA, Green W, Livingston R, Lowbury EJ (1977) Antibiotic-resistant Staphylococcus aureus in dermatology and burn wards. J Clin Pathol 30(1):40–44. https://doi.org/10.1136/jcp.30.1.40

Bal AM, Gould IM (2005) Antibiotic resistance in Staphylococcus aureus and its relevance in therapy. Expert Opin Pharmacother 6(13):2257–2269. https://doi.org/10.1517/14656566.6.13.2257

Balaban N, Rasooly A (2000) Staphylococcal enterotoxins. Int J Food Microbiol 61(1):1–10. https://doi.org/10.1016/S0168-1605(00)00377-9

Barbu EM, Mackenzie C, Foster TJ, Höök M (2014) SdrC induces staphylococcal biofilm formation through a homophilic interaction. Mol Microbiol 94(1):172–185. https://doi.org/10.1111/mmi.12750

Berends ET, Horswill AR, Haste NM, Monestier M, Nizet V (2010) Nuclease expression by Staphylococcus aureus facilitates escape from neutrophil extracellular traps. J Innate Immun 2:576–586. https://doi.org/10.1159/000319909

Bhattacharya M, Berends ETM, Zheng X, Hill PJ, Chan R, Torres VJ, Wozniak DJ (2020) Leukocidins and the nuclease Nuc prevent neutrophil-mediated killing of Staphylococcus aureus biofilms. Infect Immun 88(10):e00372–e00320. https://doi.org/10.1128/IAI.00372-20

Bjerketorp J, Rosander A, Nilsson M, Jacobsson K, Frykberg L (2004) Sorting a Staphylococcus aureus phage display library against ex vivo biomaterial. J Med Microbiol 53(10):945–951. https://doi.org/10.1099/jmm.0.45638-0

Bohach GA, Fast DJ, Nelson RD, Schlievert PM (1990) Staphylococcal and streptococcal pyrogenic toxins involved in toxic shock syndrome and related illnesses. Crit Rev Microbiol 17(4):251–272. https://doi.org/10.3109/10408419009105728

Bondi A, Dietz CC (1945) Penicillin resistant staphylococci. Exp Biol Med 60(1):55–58. https://doi.org/10.3181/00379727-60-15089

Boyle-Vavra S, Daum RS (2016) Molecular strategies of Staphylococcus aureus for resisting antibiotics. In: Staphylococcus: genetics and physiology. Caister Academic Press, pp 249–300. https://doi.org/10.21775/9781910190494.11

Bukowski M, Wladyka B, Dubin G (2010) Exfoliative toxins of Staphylococcus aureus. Toxins 2(5):1148–1165. https://doi.org/10.3390/toxins2051148

Cerca N, Brooks JL, Jefferson KK (2008) Regulation of the intercellular adhesin locus regulator (icaR) by SarA, sigmaB, and IcaR in Staphylococcus aureus. J Bacteriol 190(19):6530–6533. https://doi.org/10.1128/JB.00482-08

Chatterjee SS, Joo H-S, Duong AC, Dieringer TD, Tan VY, Song Y, Fischer ER, Cheung GYC, Li M, Otto M (2013) Essential Staphylococcus aureus toxin export system. Nat Med 19(3):364–367. https://doi.org/10.1038/nm.3047

Chavakis T, Wiechmann K, Preissner KT, Herrmann M (2005) Staphylococcus aureus interactions with the endothelium: the role of bacterial "secretable expanded repertoire adhesive molecules" (SERAM) in disturbing host defense systems. Thromb Haemost 94(08):278–285. https://doi.org/10.1160/TH05-05-0306

Clarke SR, Foster SJ (2008) IsdA protects Staphylococcus aureus against the bactericidal pro tease activity of apolactoferrin. Infect Immun 76(4):1518–1526. https://doi.org/10.1128/IAI.01530-07

Clarke SR, Mohamed R, Bian L, Routh AF, Kokai-Kun JF, Mond JJ, Tarkowski A, Foster SJ (2007) The Staphylococcus aureus surface protein IsdA mediates resistance to innate defenses of human skin. Cell Host Microbe 1(3):199–212. https://doi.org/10.1016/j.chom.2007.04.005

Clauditz A, Resch A, Wieland K-P, Peschel A, Götz F (2006) Staphyloxanthin plays a role in the fitness of Staphylococcus aureus and its ability to cope with oxidative stress. Infect Immun 74(8):4950–4953. https://doi.org/10.1128/IAI.00204-06

Clements MO, Foster SJ (1999) Stress resistance in Staphylococcus aureus. Trends Microbiol 7(11):458–462. https://doi.org/10.1016/s0966-842x(99)01607-8

Clewell DB, Flannagan SE, Jaworski DD, Clewell DB (1995) Unconstrained bacterial promiscuity: the Tn916-Tn1545 family of conjugative transposons. Trends Microbiol 3(6):229–236. https://doi.org/10.1016/s0966-842x(00)88930-1

Cohen TS, Hilliard JJ, Jones-Nelson O, Keller AE, O'Day T, Tkaczyk C, DiGiandomenico A, Hamilton M, Pelletier M, Wang Q, Diep BA, Le VTM, Cheng L, Suzich J, Stover CK, Sellman BR (2016) Staphylococcus aureus α toxin potentiates opportunistic bacterial lung infections. Sci Transl Med 8(329):329ra31. https://doi.org/10.1126/scitranslmed.aad9922

Coleman DC, Sullivan DJ, Russel RJ, Arbuthnott JP, Carey BF, Pomeroy HM (1989) Staphylococcus aureus bacteriophages mediating the simultaneous lysogenic conversion of beta-lysin, staphylokinase and enterotoxin A: molecular mechanism of triple conversion. Microbiology 135(6):1679–1697. https://doi.org/10.1099/00221287-135-6-1679

Cooney J, Kienle Z, Foster TJ, O'Toole PW (1993) The gamma-hemolysin locus of Staphylococcus aureus comprises three linked genes, two of which are identical to the genes for the F and S components of leukocidin. Infect Immun 61(2):768–771. https://doi.org/10.1128/iai.61.2.768-771.1993

Corrigan RM, Rigby D, Handley P, Foster TJ (2007) The role of Staphylococcus aureus surface protein SasG in adherence and biofilm formation. Microbiology 153(8):2435–2446. https://doi.org/10.1099/mic.0.2007/006676-0

Costa SS, Viveiros M, Amaral L, Couto I (2013) Multidrug efflux pumps in Staphylococcus aureus: an update. Open Microbiol J 7:59–71. https://doi.org/10.2174/1874285801307010059

Cuatrecasas P, Fuchs S, Anfinsen CB (1967) Catalytic properties and specificity of the extracellular nuclease of Staphylococcus aureus. J Biol Chem 242(7):1541–1547

Cunningham L, Catlin BW, De Garilhe MP (1956) A deoxyribonuclease of micrococcus pyogenes1. J Am Chem Soc 78(18):4642–4645. https://doi.org/10.1021/ja01599a031

Davies D (2003) Understanding biofilm resistance to antibacterial agents. Nat Rev Drug Discov 2(2):114–122. https://doi.org/10.1038/nrd1008

De Haas CJC, Veldkamp KE, Peschel A, Weerkamp F, Van Wamel WJB, Heezius ECJM, Poppelier MJJG, Van Kessel KPM, Van Strijp JAG (2004) Chemotaxis inhibitory protein of Staphylococcus aureus, a bacterial antiinflammatory agent. J Exp Med 199(5):687–695. https://doi.org/10.1084/jem.20031636

De Oliveira DMP, Forde BM, Kidd TJ, Harris PNA, Schembri MA, Beatson SA, Paterson DL, Walker MJ (2020) Antimicrobial resistance in ESKAPE pathogens. Clin Microbiol Rev 33(3):e00181–e00119. https://doi.org/10.1128/cmr.00181-19

Deivanayagam CCS, Perkins S, Danthuluri S, Owens RT, Bice T, Nanavathy T, Foster TJ, Höök M, Narayana SVL (1999) Crystallization of ClfA and ClfB fragments: the fibrinogen-binding surface proteins of Staphylococcus aureus. Acta Crystallogr Sect D Biol Crystallogr 55(2):554–556. https://doi.org/10.1107/s0907444998012426

Del Pozo JL, Patel R (2007) The challenge of treating biofilm-associated bacterial infections. Clin Pharmacol Ther 82(2):204–209. https://doi.org/10.1038/sj.clpt.6100247

Dhople VM, Nagaraj R (2005) Conformation and activity of δ-lysin and its analogs. Peptides 26(2):217–225. https://doi.org/10.1016/j.peptides.2004.09.013

Dinges MM, Orwin PM, Schlievert PM (2000) Exotoxins of Staphylococcus aureus. Clin Microbiol Rev 13(1):16–34. https://doi.org/10.1128/cmr.13.1.16

Divyakolu S, Chikkala R, Ratnakar KS, Sritharan V (2019) Hemolysins of *Staphylococcus aureus*—an update on their biology, role in pathogenesis and as targets for anti-virulence therapy. Adv Infect Dis 09(02):80–104. https://doi.org/10.4236/aid.2019.92007

Donlan RM (2002) Biofilms: microbial life on surfaces. Emerg Infect Dis 8(9):881–890. https://doi.org/10.3201/eid0809.020063

Downer R, Roche F, Park PW, Mecham RP, Foster TJ (2002) The elastin-binding protein of Staphylococcus aureus (EbpS) is expressed at the cell surface as an integral membrane protein and not as a cell wall-associated protein. J Biol Chem 277(1):243–250. https://doi.org/10.1074/jbc.m107621200

Drapeau GR (1978) Role of metalloprotease in activation of the precursor of staphylococcal protease. J Bacteriol 136(2):607–613. https://doi.org/10.1128/jb.136.2.607-613.1978

Dubin G (2002) Extracellular proteases of Staphylococcus spp. Biol Chem 383(7–8):1075–1086. https://doi.org/10.1515/BC.2002.116

Dufour D, Leung V, Lévesque CM (2010) Bacterial biofilm: structure, function, and antimicrobial resistance: bacterial biofilm. Endod Top 22(1):2–16. https://doi.org/10.1111/j.1601-1546.2012.00277.x

Duran-Reynals F (1933) Studies on a certain spreading factor existing in bacteria and its significance for bacterial invasiveness. J Exp Med 58(2):161–181. https://doi.org/10.1084/jem.58.2.161

Ebrey R, Shea Hamilton M, Cairns G, Lappin-Scott HM (2004) Biofilms and hospital-acquired infections. In: Microbial biofilms. Wiley, pp 294–313. https://doi.org/10.1128/9781555817718.ch16

Eisenbeis J, Saffarzadeh M, Peisker H, Jung P, Thewes N, Preissner KT, Herrmann M, Molle V, Geisbrecht BV, Jacobs K, Bischoff M (2018) The Staphylococcus aureus extracellular adherence protein Eap is a DNA binding protein capable of blocking neutrophil extracellular trap formation. Front Cell Infect Microbiol 8:235. https://doi.org/10.3389/fcimb.2018.00235

Elmesseri RA, Saleh SE, Elsherif HM, Yahia IS, Aboshanab KM (2022) Staphyloxanthin as a potential novel target for deciphering promising anti-Staphylococcus aureus agents. Antibiotics 11(3):3. https://doi.org/10.3390/antibiotics11030298

Endres JL, Chaudhari SS, Zhang X, Prahlad J, Wang S-Q, Foley LA, Luca S, Bose JL, Thomas VC, Bayles KW (2022) The Staphylococcus aureus CidA and LrgA proteins are functional holins involved in the transport of by-products of carbohydrate metabolism. MBio 13(1):e02827–e02821. https://doi.org/10.1128/mbio.02827-21

Engel HW, Soedirman N, Rost JA, Van Leeuwen WJ, Van Embden JD (1980) Transferability of macrolide, lincomycin, and streptogramin resistances between group A, B, and D streptococci, Streptococcus pneumoniae, and Staphylococcus aureus. J Bacteriol 142(2):407–413. https://doi.org/10.1128/jb.142.2.407-413.1980

Fluckiger U, Ulrich M, Steinhuber A, Döring G, Mack D, Landmann R, Goerke C, Wolz C (2005) Biofilm formation, icaADBC transcription, and polysaccharide intercellular adhesin synthesis by staphylococci in a device-related infection model. Infect Immun 73(3):1811–1819. https://doi.org/10.1128/iai.73.3.1811-1819.2005

Foster TJ (2005) Immune evasion by staphylococci. Nat Rev Microbiol 3(12):948–958. https://doi.org/10.1038/nrmicro1289

Foster TJ, Geoghegan JA, Ganesh VK, Höök M (2014) Adhesion, invasion and evasion: the many functions of the surface proteins of Staphylococcus aureus. Nat Rev Microbiol 12(1):49–62. https://doi.org/10.1038/nrmicro3161

Frey AM, Chaput D, Shaw LN (2021) Insight into the human pathodegradome of the V8 protease from Staphylococcus aureus. Cell Rep 35(1):108930. https://doi.org/10.1016/j.celrep.2021.108930

Friedrich R, Panizzi P, Fuentes-Prior P, Richter K, Verhamme I, Anderson PJ, Kawabata S-I, Huber R, Bode W, Bock PE (2003) Staphylocoagulase is a prototype for the mechanism of cofactor-induced zymogen activation. Nature 425(6957):6957. https://doi.org/10.1038/nature01962

Frimodt-Møller N, Espersen F, Skinhøj P, Rosdahl VT (1997) Epidemiology of Staphylococcus aureus bacteremia in Denmark from 1957 to 1990. Clin Microbiol Infect 3(3):297–305. https://doi.org/10.1111/j.1469-0691.1997.tb00617.x

Ghasemian A, Najar Peerayeh S, Bakhshi B, Mirzaee M (2015) The microbial surface components recognizing adhesive matrix molecules (MSCRAMMs) genes among clinical isolates of Staphylococcus aureus from hospitalized children. Iran J Pathol 10(4):258–264

Giachino P, Engelmann S, Bischoff M (2001) Sigma(B) activity depends on RsbU in Staphylococcus aureus. J Bacteriol 183(6):1843–1852. https://doi.org/10.1128/JB.183.6.1843-1852.2001

Giese B, Glowinski F, Paprotka K, Dittmann S, Steiner T, Sinha B, Fraunholz MJ (2011) Expression of δ-toxin by Staphylococcus aureus mediates escape from phago-endosomes of human epithelial and endothelial cells in the presence of β-toxin. Cell Microbiol 13(2):316–329. https://doi.org/10.1111/j.1462-5822.2010.01538.x

Gilbert I (1931) Dissociation in an encapsulated Staphylococcus. J Bacteriol 21(3):157–160. https://doi.org/10.1128/jb.21.3.157-160.1931

Gómez MI, Lee A, Reddy B, Muir A, Soong G, Pitt A, Cheung A, Prince A (2004) Staphylococcus aureus protein A induces airway epithelial inflammatory responses by activating TNFR1. Nat Med 10(8):842–848. https://doi.org/10.1038/nm1079

Gonzalez CD, Ledo C, Cela E, Stella I, Xu C, Ojeda DS, Frenette PS, Gómez MI (2019) The good side of inflammation: Staphylococcus aureus proteins SpA and Sbi contribute to proper abscess formation and wound healing during skin and soft tissue infections. Biochim Biophys Acta (BBA) - Mol Basis Dis 1865(10):2657–2670. https://doi.org/10.1016/j.bbadis.2019.07.004

Goodyear CS, Silverman GJ (2003) Death by a B cell superantigen: in vivo VH-targeted apoptotic supraclonal B cell deletion by a Staphylococcal Toxin. J Exp Med 197(9):1125–1139. https://doi.org/10.1084/jem.20020552

Grigg JC, Ukpabi G, Gaudin CFM, Murphy MEP (2010) Structural biology of heme binding in the Staphylococcus aureus Isd system. J Inorg Biochem 104(3):341–348. https://doi.org/10.1016/j.jinorgbio.2009.09.012

Grumann D, Ruotsalainen E, Kolata J, Kuusela P, Järvinen A, Kontinen VP, Bröker BM, Holtfreter S (2011) Characterization of infecting strains and superantigen-neutralizing antibodies in Staphylococcus aureus bacteremia. Clin Vaccine Immunol 18(3):487–493. https://doi.org/10.1128/CVI.00329-10

Grumann D, Nübel U, Bröker BM (2014) Staphylococcus aureus toxins—their functions and genetics. Infect Genet Evol 21:583–592. https://doi.org/10.1016/j.meegid.2013.03.013

Guay GG, Rothstein DM (1993) Expression of the tetK gene from Staphylococcus aureus in Escherichia coli: comparison of substrate specificities of TetA(B), TetA(C), and TetK efflux proteins. Antimicrob Agents Chemother 37(2):191–198. https://doi.org/10.1128/AAC.37.2.191

Guggenberger C, Wolz C, Morrissey JA, Heesemann J (2012) Two distinct coagulase-dependent barriers protect Staphylococcus aureus from neutrophils in a three dimensional in vitro infection model. PLoS Pathog 8(1):e1002434. https://doi.org/10.1371/journal.ppat.1002434

Guo Y, Song G, Sun M, Wang J, Wang Y (2020) Prevalence and therapies of antibiotic-resistance in Staphylococcus aureus. Front Cell Infect Microbiol 10:107. https://doi.org/10.3389/fcimb.2020.00107

Guzmán-Soto I, McTiernan C, Gonzalez-Gomez M, Ross A, Gupta K, Suuronen EJ, Mah T-F, Griffith M, Alarcon EI (2021) Mimicking biofilm formation and development: recent progress in in vitro and in vivo biofilm models. IScience 24(5):102443. https://doi.org/10.1016/j.isci.2021.102443

Haag AF, Podkowik M, Ibarra-Chávez R, Gallego Del Sol F, Ram G, Chen J, Marina A, Novick RP, Penadés JR (2021) A regulatory cascade controls Staphylococcus aureus pathogenicity Island activation. Nat Microbiol 6(10):1300–1308. https://doi.org/10.1038/s41564-021-00956-2

Haggar A, Flock J-I, Norrby-Teglund A (2010) Extracellular adherence protein (Eap) from Staphylococcus aureus does not function as a superantigen. Clin Microbiol Infect 16(8):1155–1158. https://doi.org/10.1111/j.1469-0691.2009.03058.x

Hammer ND, Cassat JE, Noto MJ, Lojek LJ, Chadha AD, Schmitz JE, Creech CB, Skaar EP (2014) Inter- and intraspecies metabolite exchange promotes virulence of antibiotic-resistant Staphylococcus aureus. Cell Host Microbe 16(4):531–537. https://doi.org/10.1016/j.chom.2014.09.002

Handler MZ, Schwartz RA (2014) Staphylococcal scalded skin syndrome: diagnosis and management in children and adults. J Eur Acad Dermatol Venereol 28(11):1418–1423. https://doi.org/10.1111/jdv.12541

Hendrickx APA, Budzik JM, Oh S-Y, Schneewind O (2011) Architects at the bacterial surface—sortases and the assembly of pili with isopeptide bonds. Nat Rev Microbiol 9(3):166. https://doi.org/10.1038/nrmicro2520

Herman-Bausier P, El-Kirat-Chatel S, Foster TJ, Geoghegan JA, Dufrêne YF (2015) Staphylococcus aureus fibronectin-binding protein A mediates cell-cell adhesion through low-affinity homophilic bonds. MBio 6(3):e00413–e00415. https://doi.org/10.1128/mBio.00413-15

Herman-Bausier P, Labate C, Towell AM, Derclaye S, Geoghegan JA, Dufrêne YF (2018) Staphylococcus aureus clumping factor A is a force-sensitive molecular switch that activates bacterial adhesion. Proc Natl Acad Sci 115(21):5564–5569. https://doi.org/10.1073/pnas.1718104115

Higgins J, Loughman A, Van Kessel KPM, Van Strijp JAG, Foster TJ (2006) Clumping factor A of Staphylococcus aureus inhibits phagocytosis by human polymorphonuclear leucocytes. FEMS Microbiol Lett 258(2):290–296. https://doi.org/10.1111/j.1574-6968.2006.00229.x

Holtfreter S, Bröker BM (2005) Staphylococcal superantigens: do they play a role in sepsis? Arch Immunol Ther Exp 53(1):13–27

Houston P, Rowe SE, Pozzi C, Waters EM, O'Gara JP (2011) Essential role for the major autolysin in the fibronectin-binding protein-mediated Staphylococcus aureus biofilm phenotype. Infect Immun 79(3):1153–1165. https://doi.org/10.1128/IAI.00364-10

Howden BP, Davies JK, Johnson PDR, Stinear TP, Grayson ML (2010) Reduced vancomycin susceptibility in Staphylococcus aureus, including vancomycin-intermediate and heterogeneous vancomycin-intermediate strains: resistance mechanisms, laboratory detection, and clinical implications. Clin Microbiol Rev 23(1):99–139. https://doi.org/10.1128/cmr.00042-09

Howden BP, Giulieri SG, Wong Fok Lung T, Baines SL, Sharkey LK, Lee JYH, Hachani A, Monk IR, Stinear TP (2023) Staphylococcus aureus host interactions and adaptation. Nat Rev Microbiol 21(6):380–395. https://doi.org/10.1038/s41579-023-00852-y

Hu C, Xiong N, Zhang Y, Rayner S, Chen S (2012) Functional characterization of lipase in the pathogenesis of Staphylococcus aureus. Biochem Biophys Res Commun 419:617–620. https://doi.org/10.1016/j.bbrc.2012.02.057

Huntzinger E, Boisset S, Saveanu C, Benito Y, Geissmann T, Namane A, Lina G, Etienne J, Ehresmann B, Ehresmann C, Jacquier A, Vandenesch F, Romby P (2005) Staphylococcus aureus RNAIII and the endoribonuclease III coordinately regulate spa gene expression. EMBO J 24(4):824–835. https://doi.org/10.1038/sj.emboj.7600572

Idrees M, Mohammad AR, Karodia N, Rahman A (2020) Multimodal role of amino acids in microbial control and drug development. Antibiotics 9(6):6. https://doi.org/10.3390/antibiotics9060330

Idrees M, Sawant S, Karodia N, Rahman A (2021) Staphylococcus aureus biofilm: morphology, genetics, pathogenesis and treatment strategies. Int J Environ Res Public Health 18(14):7602. https://doi.org/10.3390/ijerph18147602

Ilk N, Völlenkle C, Egelseer EM, Breitwieser A, Sleytr UB, Sára M (2002) Molecular characterization of the S-layer gene, sbpA, of Bacillus sphaericus CCM 2177 and production of a functional S-layer fusion protein with the ability to recrystallize in a defined orientation while presenting the fused allergen. Appl Environ Microbiol 68(7):3251–3260. https://doi.org/10.1128/AEM.68.7.3251-3260.2002

Inoshima I, Inoshima N, Wilke GA, Powers ME, Frank KM, Wang Y, Wardenburg JB (2011) A Staphylococcus aureus pore-forming toxin subverts the activity of ADAM10 to cause lethal infection. Nat Med 17(10):1310–1314. https://doi.org/10.1038/nm.2451

Isobe M, Seki M, Yamagishi Y, Miyawaki K, Yabuno K, Hamaguchi S, Yoshioka N, Tomono K, Isobe M, Uejima E, Masaoka M (2012) Methicillin-resistant Staphylococcus aureus bacteremia at a university hospital in Japan. J Infect Chemother 18(6):841–847. https://doi.org/10.1007/s10156-012-0423-6

Janzon L, Arvidson S (1990) The role of the delta-lysin gene (hld) in the regulation of virulence genes by the accessory gene regulator (agr) in Staphylococcus aureus. EMBO J 9(5):1391–1399

Jin T, Bokarewa M, Foster T, Mitchell J, Higgins J, Tarkowski A (2004) Staphylococcus aureus resists human defensins by production of staphylokinase, a novel bacterial evasion mechanism. J Immunol 172(2):1169–1176. https://doi.org/10.4049/jimmunol.172.2.1169

Johnson HM, Russell JK, Pontzer CH (1991) Staphylococcal enterotoxin microbial superantigens. FASEB J 5(12):2706–2712. https://doi.org/10.1096/fasebj.5.12.1916093

Johnson JG, Wang B, Debelouchina GT, Novick RP, Muir TW (2015) Increasing AIP macrocycle size reveals key features of agr activation in Staphylococcus aureus. Chembiochem 16(7):1093–1100. https://doi.org/10.1002/cbic.201500006

Jones C (2005) Revised structures for the capsular polysaccharides from Staphylococcus aureus Types 5 and 8, components of novel glycoconjugate vaccines. Carbohydr Res 340(6):1097–1106. https://doi.org/10.1016/j.carres.2005.02.001

Joo H-S, Otto M (2015) Mechanisms of resistance to antimicrobial peptides in staphylococci. Biochim Biophys Acta 1848(11 Pt B):3055–3061. https://doi.org/10.1016/j.bbamem.2015.02.009

Kaasch AJ, Barlow G, Edgeworth JD, Fowler VG, Hellmich M, Hopkins S, Kern WV, Llewelyn MJ, Rieg S, Rodriguez-Baño J, Scarborough M, Seifert H, Soriano A, Tilley R, Tőrők ME, Weiß V, Wilson APR, Thwaites GE (2014) Staphylococcus aureus bloodstream infection: a pooled analysis of five prospective, observational studies. J Infect 68(3):242–251. https://doi.org/10.1016/j.jinf.2013.10.015

Kang M, Ko Y-P, Liang X, Ross CL, Liu Q, Murray BE, Höök M (2013) Collagen-binding microbial surface components recognizing adhesive matrix molecule (MSCRAMM) of gram-positive bacteria inhibit complement activation via the classical pathway. J Biol Chem 288(28):20520–20531. https://doi.org/10.1074/jbc.M113.454462

Kiedrowski MR, Kavanaugh JS, Malone CL, Mootz JM, Voyich JM, Smeltzer MS, Bayles KW, Horswill AR (2011) Nuclease modulates biofilm formation in community-associated methicillin-resistant Staphylococcus aureus. PLoS One 6(11):e26714. https://doi.org/10.1371/journal.pone.0026714

Killikelly A, Benson MA, Ohneck EA, Sampson JM, Jakoncic J, Spurrier B, Torres VJ, Kong X-P (2015) Structure-based functional characterization of repressor of toxin (Rot), a central regulator of Staphylococcus aureus virulence. J Bacteriol 197(1):188–200. https://doi.org/10.1128/JB.02317-14

Knudsen TA, Skov R, Petersen A, Larsen AR, Benfield T, for the Danish Staphylococcal Bacteremia Study Group (2016) Increased age-dependent risk of death associated with lukF-PV-positive Staphylococcus aureus bacteremia. Open Forum Infect Dis 3(4):ofw220. https://doi.org/10.1093/ofid/ofw220

Ko Y-P, Kuipers A, Freitag CM, Jongerius I, Medina E, Van Rooijen WJ, Spaan AN, Van Kessel KPM, Höök M, Rooijakkers SHM (2013) Phagocytosis escape by a Staphylococcus aureus protein that connects complement and coagulation proteins at the bacterial surface. PLoS Pathog 9(12):e1003816. https://doi.org/10.1371/journal.ppat.1003816

Kobayashi SD, DeLeo FR (2013) Staphylococcus aureus protein A promotes immune suppression. MBio 4(5):e00764–e00713. https://doi.org/10.1128/mBio.00764-13

Komatsuzawa H, Sugai M, Nakashima S, Yamada S, Matsumoto A, Oshida T, Suginaka H (1997) Subcellular localization of the major autolysin, ATL and its processed proteins in Staphylococcus aureus. Microbiol Immunol 41(6):469–479. https://doi.org/10.1111/j.1348-0421.1997.tb01880.x

Kot B, Sytykiewicz H, Sprawka I (2018) Expression of the biofilm-associated genes in methicillin-resistant Staphylococcus aureus in biofilm and planktonic conditions. Int J Mol Sci 19(11):3487. https://doi.org/10.3390/ijms19113487

Kot B, Sytykiewicz H, Sprawka I, Witeska M (2020) Effect of manuka honey on biofilm-associated genes expression during methicillin-resistant Staphylococcus aureus biofilm formation. Sci Rep 10(1):13552. https://doi.org/10.1038/s41598-020-70666-y

Kristensen MS, Paludan K, Larsen CG, Deleuran BW, Jensen PKA, Jørgensen P, Thestrup-Pedersen K, Zachariae COC (1991) Quantitative determination of IL-1 alpha-induced IL-8 mRNA levels in cultured human keratinocytes, dermal fibroblasts, endothelial cells, and monocytes. J Investig Dermatol 97(3):506–510. https://doi.org/10.1111/1523-1747.ep12481543

Kroh HK, Panizzi P, Bock PE (2009) Von Willebrand factor-binding protein is a hysteretic conformational activator of prothrombin. Proc Natl Acad Sci 106(19):7786–7791. https://doi.org/10.1073/pnas.0811750106

Kuipers A, Stapels DAC, Weerwind LT, Ko Y-P, Ruyken M, Lee JC, Van Kessel KPM, Rooijakkers SHM (2016) The Staphylococcus aureus polysaccharide capsule and Efb-dependent fibrinogen shield act in concert to protect against phagocytosis. Microbiology 162(7):1185–1194. https://doi.org/10.1099/mic.0.000293

Kuroda M, Ito R, Tanaka Y, Yao M, Matoba K, Saito S, Tanaka I, Ohta T (2008) Staphylococcus aureus surface protein SasG contributes to intercellular autoaggregation of Staphylococcus aureus. Biochem Biophys Res Commun 377(4):1102–1106. https://doi.org/10.1016/j.bbrc.2008.10.134

Kwiecinski J, Jacobsson G, Karlsson M, Zhu X, Wang W, Bremell T, Josefsson E, Jin T (2013) Staphylokinase promotes the establishment of Staphylococcus aureus skin infections while decreasing disease severity. J Infect Dis 208(6):990–999. https://doi.org/10.1093/infdis/jit288

Kwiecinski J, Peetermans M, Liesenborghs L, Na M, Björnsdottir H, Zhu X, Jacobsson G, Johansson BR, Geoghegan JA, Foster TJ, Josefsson E, Bylund J, Verhamme P, Jin T (2016) Staphylokinase control of Staphylococcus aureus biofilm formation and detachment through host plasminogen activation. J Infect Dis 213(1):139–148. https://doi.org/10.1093/infdis/jiv360

Lacey K, Geoghegan J, McLoughlin R (2016) The role of Staphylococcus aureus virulence factors in skin infection and their potential as vaccine antigens. Pathogens 5(1):22. https://doi.org/10.3390/pathogens5010022

Lacey KA, Mulcahy ME, Towell AM, Geoghegan JA, McLoughlin RM (2019) Clumping factor B is an important virulence factor during Staphylococcus aureus skin infection and a promising vaccine target. PLoS Pathog 15(4):e1007713. https://doi.org/10.1371/journal.ppat.1007713

Lakhundi S, Zhang K (2018) Methicillin-resistant Staphylococcus aureus: molecular characterization, evolution, and epidemiology. Clin Microbiol Rev 31(4):e00020–e00018. https://doi.org/10.1128/CMR.00020-18

Laupland KB (2013) Incidence of bloodstream infection: a review of population-based studies. Clin Microbiol Infect 19(6):492–500. https://doi.org/10.1111/1469-0691.12144

Lee J-H, Park J-H, Cho HS, Joo SW, Cho MH, Lee J (2013) Anti-biofilm activities of quercetin and tannic acid against Staphylococcus aureus. Biofouling 29(5):491–499. https://doi.org/10.1080/08927014.2013.788692

Li D, Cheung A (2008) Repression of hla by rot is dependent on sae in Staphylococcus aureus. Infect Immun 76(3):1068–1075. https://doi.org/10.1128/IAI.01069-07

Li M, Du X, Villaruz AE, Diep BA, Wang D, Song Y, Tian Y, Hu J, Yu F, Lu Y, Otto M (2012) MRSA epidemic linked to a quickly spreading colonization and virulence determinant. Nat Med 18(5):816–819. https://doi.org/10.1038/nm.2692

Lina G, Jarraud S, Ji G, Greenland T, Pedraza A, Etienne J, Novick RP, Vandenesch F (1998) Transmembrane topology and histidine protein kinase activity of AgrC, the agr signal receptor in Staphylococcus aureus. Mol Microbiol 28(3):655–662. https://doi.org/10.1046/j.1365-2958.1998.00830.x

Lina G, Bohach GA, Nair SP, Hiramatsu K, Jouvin-Marche E, Mariuzza R, International Nomenclature Committee for Staphylococcal Superantigens (2004) Standard nomenclature for the superantigens expressed by Staphylococcus. J Infect Dis 189(12):2334–2336. https://doi.org/10.1086/420852

Liu C-I, Liu GY, Song Y, Yin F, Hensler ME, Jeng W-Y, Nizet V, Wang AH-J, Oldfield E (2008) A cholesterol biosynthesis inhibitor blocks Staphylococcus aureus virulence. Science 319(5868):1391–1394. https://doi.org/10.1126/science.1153018

Liu C, Bayer A, Cosgrove SE, Daum RS, Fridkin SK, Gorwitz RJ, Kaplan SL, Karchmer AW, Levine DP, Murray BE, Rybak MJ, Talan DA, Chambers HF (2011) Clinical practice guidelines by the Infectious Diseases Society of America for the treatment of methicillin-resistant Staphylococcus aureus infections in adults and children. Clin Infect Dis 52(3):e18–e55. https://doi.org/10.1093/cid/ciq146

Liu H, Lv J, Qi X, Ding Y, Li D, Hu L, Wang L, Yu F (2015) The carriage of the serine-aspartate repeat protein-encoding sdr genes among Staphylococcus aureus lineages. Braz J Infect Dis 19(5):498–502. https://doi.org/10.1016/j.bjid.2015.07.003

Lu T, Park JY, Parnell K, Fox LK, McGuire MA (2012) Characterization of fatty acid modifying enzyme activity in staphylococcal mastitis isolates and other bacteria. BMC Res Notes 5(1):323. https://doi.org/10.1186/1756-0500-5-323

Luong TT, Dunman PM, Murphy E, Projan SJ, Lee CY (2006) Transcription profiling of the mgrA regulon in Staphylococcus aureus. J Bacteriol 188(5):1899–1910. https://doi.org/10.1128/JB.188.5.1899-1910.2006

Lyon GJ, Wright JS, Muir TW, Novick RP (2002) Key determinants of receptor activation in the agr autoinducing peptides of Staphylococcus aureus. Biochemistry 41(31):10095–10104. https://doi.org/10.1021/bi026049u

Ma D, Mandell JB, Donegan NP, Cheung AL, Ma W, Rothenberger S, Shanks RMQ, Richardson AR, Urish KL (2019) The toxin-antitoxin MazEF drives Staphylococcus aureus biofilm formation, antibiotic tolerance, and chronic infection. MBio 10(6):e01658–e01619. https://doi.org/10.1128/mBio.01658-19

Makris G, Wright JD, Ingham E, Holland KT (2004) The hyaluronate lyase of Staphylococcus aureus—a virulence factor? Microbiology 150(6):2005–2013. https://doi.org/10.1099/mic.0.26942-0

Malachowa N, DeLeo FR (2010) Mobile genetic elements of Staphylococcus aureus. Cell Mol Life Sci 67(18):3057–3071. https://doi.org/10.1007/s00018-010-0389-4

Malani PN (2014) National burden of invasive methicillin-resistant Staphylococcus aureus infection. Clin Rev Educ 311(14):1438. https://doi.org/10.1001/jama.2014.1666

Mathelié-Guinlet M, Viela F, Alfeo MJ, Pietrocola G, Speziale P, Dufrêne YF (2020) Single-molecule analysis demonstrates stress-enhanced binding between Staphylococcus aureus surface protein IsdB and host cell Integrins. Nano Lett 20(12):8919–8925. https://doi.org/10.1021/acs.nanolett.0c04015

Matsuoka M, Inoue M, Endo Y, Nakajima Y (2003) Characteristic expression of three genes, msr(A), mph(C) and erm(Y), that confer resistance to macrolide antibiotics on Staphylococcus aureus. FEMS Microbiol Lett 220(2):287–293. https://doi.org/10.1016/S0378-1097(03)00134-4

Mellor IR, Thomas DH, Sansom MSP (1988) Properties of ion channels formed by Staphylococcus aureus delta-toxin. Biochim Biophys Acta 942(2):280–294. https://doi.org/10.1016/0005-2736(88)90030-2

Miller LS (2015) Adipocytes armed against Staphylococcus aureus. N Engl J Med 372(14):1368–1370. https://doi.org/10.1056/NEJMcibr1500271

Mohamadou M, Essama SR, Ngonde Essome MC, Akwah L, Nadeem N, Gonsu Kamga H, Sattar S, Javed S (2022) High prevalence of Panton-Valentine leukocidin positive, multidrug resistant, Methicillin-resistant Staphylococcus aureus strains circulating among clinical setups in Adamawa and Far North regions of Cameroon. PLoS One 17(7):e0265118. https://doi.org/10.1371/journal.pone.0265118

Moormeier DE, Bayles KW (2017) Staphylococcus aureus biofilm: a complex developmental organism. Mol Microbiol 104(3):365–376. https://doi.org/10.1111/mmi.13634

Møretrø T, Hermansen L, Holck AL, Sidhu MS, Rudi K, Langsrud S (2003) Biofilm formation and the presence of the intercellular adhesion locus Ica among Staphylococci from food and food processing environments. Appl Environ Microbiol 69(9):5648–5655. https://doi.org/10.1128/AEM.69.9.5648-5655.2003

Morfeldt E, Tegmark K, Arvidson S (1996) Transcriptional control of the agr-dependent virulence gene regulator, RNAIII, in Staphylococcus aureus. Mol Microbiol 21(6):1227–1237. https://doi.org/10.1046/j.1365-2958.1996.751447.x

Nakaminami H, Ito T, Han X, Ito A, Matsuo M, Uehara Y, Baba T, Hiramatsu K, Noguchi N (2017) First report of sasX-positive methicillin-resistant Staphylococcus aureus in Japan. FEMS Microbiol Lett 364(16). https://doi.org/10.1093/femsle/fnx171

Nguyen M-T, Luqman A, Bitschar K, Hertlein T, Dick J, Ohlsen K, Bröker B, Schittek B, Götz F (2018) Staphylococcal (phospho)lipases promote biofilm formation and host cell invasion. Int J Med Microbiol 308(6):653–663. https://doi.org/10.1016/j.ijmm.2017.11.013

Nickerson N, Ip J, Passos DT, McGavin MJ (2010) Comparison of Staphopain A (ScpA) and B (SspB) precursor activation mechanisms reveals unique secretion kinetics of proSspB

(Staphopain B), and a different interaction with its cognate Staphostatin, SspC. Mol Microbiol 75(1):161–177. https://doi.org/10.1111/j.1365-2958.2009.06974.x

Novick RP, Richmond MH (1965) Nature and interactions of the genetic elements governing penicillinase synthesis in Staphylococcus aureus. J Bacteriol 90(2):467–480. https://doi.org/10.1128/jb.90.2.467-480.1965

Novick RP, Subedi A (2007) The SaPIs: mobile pathogenicity islands of Staphylococcus. Chem Immunol Allergy 93:42–57. https://doi.org/10.1159/0000100857

O'Riordan K, Lee JC (2004 Jan) Staphylococcus aureus capsular polysaccharides. Clin Microbiol Rev 17(1):218–234. https://doi.org/10.1128/CMR.17.1.218-234.2004

Ohlsen K (2009) Novel antibiotics for the treatment of Staphylococcus aureus. Expert Rev Clin Pharmacol 2(6):661–672. https://doi.org/10.1586/ecp.09.26

Okano A, Isley NA, Boger DL (2017) Total syntheses of vancomycin-related glycopeptide antibiotics and key analogues. Chem Rev 117(18):11952–11993. https://doi.org/10.1021/acs.chemrev.6b00820

Okolie CE, Cockayne A, Penfold C, James R (2013) Engineering of the LukS-PV and LukF-PV subunits of Staphylococcus aureus Panton-Valentine leukocidin for Diagnostic and Therapeutic Applications. BMC Biotechnol 13(1):103. https://doi.org/10.1186/1472-6750-13-103

Oliveira D, Borges A, Simões M (2018) Staphylococcus aureus toxins and their molecular activity in infectious diseases. Toxins 10(6):252. https://doi.org/10.3390/toxins10060252

Ono HK, Omoe K, Imanishi K, Iwakabe Y, Hu D-L, Kato H, Saito N, Nakane A, Uchiyama T, Shinagawa K (2008) Identification and characterization of two novel staphylococcal enterotoxins, types S and T. Infect Immun 76(11):4999–5005. https://doi.org/10.1128/iai.00045-08

Otto M (2012) MRSA virulence and spread. Cell Microbiol 14(10):1513–1521. https://doi.org/10.1111/j.1462-5822.2012.01832.x

Otto M (2014) Staphylococcus aureus toxins. Curr Opin Microbiol 17:32–37. https://doi.org/10.1016/j.mib.2013.11.004

Palmqvist N, Foster T, Tarkowski A, Josefsson E (2002) Protein A is a virulence factor in Staphylococcus aureus arthritis and septic death. Microb Pathog 33(5):239–249. https://doi.org/10.1006/mpat.2002.0533

Panizzi P, Kroh HK, Fuentes-Prior P, Bock PE (2010) Staphylocoagulase. In: Kini R, Clemetson K, Markland F, McLane M, Morita T (eds) Toxins and hemostasis. Springer. https://doi.org/10.1007/978-90-481-9295-3_32

Pantosti A, Sanchini A, Monaco M (2007) Mechanisms of antibiotic resistance in Staphylococcus aureus. Future Microbiol 2(3):323–334. https://doi.org/10.2217/17460913.2.3.323

Pastagia M, Schuch R, Fischetti VA, Huang DB (2013) Lysins: the arrival of pathogen-directed anti-infectives. J Med Microbiol 62(10):1506–1516. https://doi.org/10.1099/jmm.0.061028-0

Pecharki D, Petersen FC, Scheie AA (2008) Role of hyaluronidase in Streptococcus intermedius biofilm. Microbiology 154(3):932–938. https://doi.org/10.1099/mic.0.2007/012393-0

Pelz A, Wieland K-P, Putzbach K, Hentschel P, Albert K, Götz F (2005) Structure and biosynthesis of Staphyloxanthin from Staphylococcus aureus*. J Biol Chem 280(37):32493–32498. https://doi.org/10.1074/jbc.M505070200

Peschel A, Otto M (2013) Phenol-soluble modulins and staphylococcal infection. Nat Rev Microbiol 11(10):667–673. https://doi.org/10.1038/nrmicro3110

Pietrocola G, Nobile G, Rindi S, Speziale P (2017) Staphylococcus aureus manipulates innate immunity through own and host-expressed proteases. Front Cell Infect Microbiol 7:166. https://doi.org/10.3389/fcimb.2017.00166

Pietrocola G, Nobile G, Alfeo MJ, Foster TJ, Geoghegan JA, De Filippis V, Speziale P (2019) Fibronectin-binding protein B (FnBPB) from Staphylococcus aureus protects against the antimicrobial activity of histones. J Biol Chem 294(10):3588–3602. https://doi.org/10.1074/jbc.RA118.005707

Pinchuk IV, Beswick EJ, Reyes VE (2010) Staphylococcal enterotoxins. Toxins 2(8):2177–2197. https://doi.org/10.3390/toxins2082177

Pragman AA, Yarwood JM, Tripp TJ, Schlievert PM (2004) Characterization of virulence factor regulation by SrrAB, a two-component system in Staphylococcus aureus. J Bacteriol 186(8):2430–2438. https://doi.org/10.1128/JB.186.8.2430-2438.2004
Prokešová L, Potužníková B, Potempa J, Zikán J, Radl J, Hachová L, Baran K, Porwit-Bobr Z, John C (1992) Cleavage of human immunoglobulins by serine proteinase from Staphylococcus aureus. Immunol Lett 31(3):259–265. https://doi.org/10.1016/0165-2478(92)90124-7
Queck SY, Khan BA, Wang R, Bach T-HL, Kretschmer D, Chen L, Kreiswirth BN, Peschel A, DeLeo FR, Otto M (2009) Mobile genetic element-encoded cytolysin connects virulence to methicillin resistance in MRSA. PLoS Pathog 5(7):e1000533. https://doi.org/10.1371/journal.ppat.1000533
Rehm SJ, Tice A (2010) Staphylococcus aureus: methicillin-susceptible S. aureus to methicillin-resistant S. aureus and vancomycin-resistant S. aureus. Clin Infect Dis 51(S2):S176–S182. https://doi.org/10.1086/653518
Rice K, Peralta R, Bast D, De Azavedo J, McGavin MJ (2001) Description of Staphylococcus serine protease (ssp) operon in Staphylococcus aureus and nonpolar inactivation of sspA-encoded serine protease. Infect Immun 69(1):159–169. https://doi.org/10.1128/IAI.69.1.159-169.2001
Rice KC, Nelson JB, Patton TG, Yang S-J, Bayles KW (2005) Acetic acid induces expression of the Staphylococcus aureus cidABC and lrgAB murein hydrolase regulator operons. J Bacteriol 187(3):813–821. https://doi.org/10.1128/JB.187.3.813-821.2005
Robinson DA, Monk AB, Cooper JE, Feil EJ, Enright MC (2005) Evolutionary genetics of the accessory gene regulator (agr) locus in Staphylococcus aureus. J Bacteriol 187(24):8312–8321. https://doi.org/10.1128/jb.187.24.8312-8321.2005
Rooijakkers SHM, Ruyken M, Roos A, Daha MR, Presanis JS, Sim RB, Van Wamel WJB, Van Kessel KPM, Van Strijp JAG (2005) Immune evasion by a staphylococcal complement inhibitor that acts on C3 convertases. Nat Immunol 6(9):920–927. https://doi.org/10.1038/ni1235
Rouch DA, Byrne ME, Kong YC, Skurray RA (1987) The aacA-aphD gentamicin and kanamycin resistance determinant of Tn4001 from Staphylococcus aureus: expression and nucleotide sequence analysis. Microbiology 133(11):3039–3052. https://doi.org/10.1099/00221287-133-11-3039
Sabat A, Melles DC, Martirosian G, Grundmann H, Van Belkum A, Hryniewicz W (2006) Distribution of the serine-aspartate repeat protein-encoding sdr genes among nasal-carriage and invasive Staphylococcus aureus strains. J Clin Microbiol 44(3):1135–1138. https://doi.org/10.1128/JCM.44.3.1135-1138.2006
Santajit S, Indrawattana N (2016) Mechanisms of antimicrobial resistance in ESKAPE pathogens. Biomed Res Int 2016:1–8. https://doi.org/10.1155/2016/2475067
Santala A, Saarinen J, Kovanen P, Kuusela P (1999) Activation of interstitial collagenase, MMP-1, by Staphylococcus aureus cells having surface-bound plasmin: a novel role of plasminogen receptors of bacteria. FEBS Lett 461(3):153–156. https://doi.org/10.1016/S0014-5793(99)01440-4
Sargison F, Goncheva MI, Alves J, Pickering A, Fitzgerald JR (2021) Staphylococcus aureus secreted lipases do not inhibit innate immune killing mechanisms. Wellcome Open Res 5:286. https://doi.org/10.12688/wellcomeopenres.16194.2
Schaenzer AJ, Wright GD (2020) Antibiotic resistance by enzymatic modification of antibiotic targets. Trends Mol Med 26(8):768–782. https://doi.org/10.1016/j.molmed.2020.05.001
Schneewind O, Missiakas D (2019) Sortases, surface proteins, and their roles in Staphylococcus aureus disease and vaccine development. Microbiol Spectr 7(1):7.1.09. https://doi.org/10.1128/microbiolspec.psib-0004-2018
Seilie ES, Bubeck Wardenburg J (2017) Staphylococcus aureus pore-forming toxins: the interface of pathogen and host complexity. Semin Cell Dev Biol 72:101–116. https://doi.org/10.1016/j.semcdb.2017.04.003
Singh V, Phukan UJ (2019) Interaction of host and Staphylococcus aureus protease-system regulates virulence and pathogenicity. Med Microbiol Immunol 208(5):585–607. https://doi.org/10.1007/s00430-018-0573-y

Singh VK, Carlos MR, Singh K (2010) Physiological significance of the peptidoglycan hydrolase, LytM, in Staphylococcus aureus. FEMS Microbiol Lett 311(2):167–175. https://doi.org/10.1111/j.1574-6968.2010.02087.x

Sinha B, Francois PP, Nusse O, Foti M, Hartford OM, Vaudaux P, Foster TJ, Lew DP, Herrmann M, Krause K-H (1999) Fibronectin-binding protein acts as Staphylococcus aureus invasin via fibronectin bridging to integrin alpha5beta1. Cell Microbiol 1(2):101–117. https://doi.org/10.1046/j.1462-5822.1999.00011.x

Sinha B, Francois P, Que Y-A, Hussain M, Heilmann C, Moreillon P, Lew D, Krause K-H, Peters G, Herrmann M (2000) Heterologously expressed Staphylococcus aureus fibronectin-binding proteins are sufficient for invasion of host cells. Infect Immun 68(12):6871–6878. https://doi.org/10.1128/IAI.68.12.6871-6878.2000

Smagur J, Guzik K, Bzowska M, Kuzak M, Zarebski M, Kantyka T, Walski M, Gajkowska B, Potempa J (2009a) Staphylococcal cysteine protease staphopain B (SspB) induces rapid engulfment of human neutrophils and monocytes by macrophages. Biol Chem 390(4):361–371. https://doi.org/10.1515/BC.2009.042

Smagur J, Guzik K, Magiera L, Bzowska M, Gruca M, Thøgersen IB, Enghild JJ, Potempa J (2009b) A new pathway of staphylococcal pathogenesis: apoptosis-like death induced by Staphopain B in human neutrophils and monocytes. J Innate Immun 1(2):98–108. https://doi.org/10.1159/000181014

Smith TL, Jarvis WR (1999) Antimicrobial resistance in Staphylococcus aureus. Microbes Infect 1(10):795–805. https://doi.org/10.1016/s1286-4579(99)80082-3

Spaan AN, Van Strijp JAG, Torres VJ (2017) Leukocidins: staphylococcal bi-component pore-forming toxins find their receptors. Nat Rev Microbiol 15(7):435–447. https://doi.org/10.1038/nrmicro.2017.27

Speziale P, Geoghegan JA (2015) Biofilm formation by staphylococci and streptococci: structural, functional, and regulatory aspects and implications for pathogenesis. Front Cell Infect Microbiol 5:31. https://www.frontiersin.org/articles/10.3389/fcimb.2015.00031

Speziale P, Pietrocola G (2020) The multivalent role of fibronectin-binding proteins A and B (FnBPA and FnBPB) of Staphylococcus aureus in host infections. Front Microbiol 11:2054. https://doi.org/10.3389/fmicb.2020.02054

Srivastava SK, Rajasree K, Fasim A, Arakere G, Gopal B (2014) Influence of the AgrC-AgrA complex on the response time of Staphylococcus aureus quorum sensing. J Bacteriol 196(15):2876–2888. https://doi.org/10.1128/JB.01530-14

Stemberk V, Jones RPO, Moroz O, Atkin KE, Edwards AM, Turkenburg JP, Leech AP, Massey RC, Potts JR (2014) Evidence for steric regulation of fibrinogen binding to Staphylococcus aureus fibronectin-binding protein A (FnBPA). J Biol Chem 289(18):12842–12851. https://doi.org/10.1074/jbc.M113.543546

Struelens MJ, Denis O (2000) Methicillin resistant Staphylococcus aureus: toward a coordinated response to a continuing challenge. Eurosurveillance 5(3):25–26. https://doi.org/10.2807/esm.05.03.00030-en

Sully EK, Malachowa N, Elmore BO, Alexander SM, Femling JK, Gray BM, DeLeo FR, Otto M, Cheung AL, Edwards BS, Sklar LA, Horswill AR, Hall PR, Gresham HD (2014) Selective chemical inhibition of agr quorum sensing in Staphylococcus aureus promotes host defense with minimal impact on resistance. PLoS Pathog 10(6):e1004174. https://doi.org/10.1371/journal.ppat.1004174

Sun F, Zhou L, Zhao B-C, Deng X, Cho H, Yi C, Jian X, Song C-X, Luan C-H, Bae T, Li Z, He C (2011) Targeting MgrA-mediated virulence regulation in Staphylococcus aureus. Chem Biol 18(8):1032–1041. https://doi.org/10.1016/j.chembiol.2011.05.014

Sun F, Ding Y, Ji Q, Liang Z, Deng X, Wong CCL, Yi C, Zhang L, Xie S, Alvarez S, Hicks LM, Luo C, Jiang H, Lan L, He C (2012) Protein cysteine phosphorylation of SarA/MgrA family transcriptional regulators mediates bacterial virulence and antibiotic resistance. Proc Natl Acad Sci 109(38):15461–15466. https://doi.org/10.1073/pnas.1205952109

Syed AK, Reed TJ, Clark KL, Boles BR, Kahlenberg JM (2015) Staphlyococcus aureus phenol-soluble modulins stimulate the release of proinflammatory cytokines from keratinocytes and

are required for induction of skin inflammation. Infect Immun 83(9):3428–3437. https://doi.org/10.1128/iai.00401-15
Tahmasebi H, Dehbashi S, Arabestani MR (2021) Antibiotic resistance alters through iron-regulating Sigma factors during the interaction of Staphylococcus aureus and Pseudomonas aeruginosa. Sci Rep 11(1):18509. https://doi.org/10.1038/s41598-021-98017-5
Tajima A, Iwase T, Shinji H, Seki K, Mizunoe Y (2009) Inhibition of endothelial interleukin-8 production and neutrophil transmigration by Staphylococcus aureus beta-hemolysin. Infect Immun 77(1):327–334. https://doi.org/10.1128/iai.00748-08
Tam K, Torres VJ (2019) Staphylococcus aureus secreted toxins and extracellular enzymes. Microbiol Spectr 7(2):7.2.16. https://doi.org/10.1128/microbiolspec.GPP3-0039-2018
Tan L, Li SR, Jiang B, Hu XM, Li S (2018 Jan) Therapeutic Targeting of the Staphylococcus aureus Accessory Gene Regulator (agr) System. Front Microbiol 25(9):55. https://doi.org/10.3389/fmicb.2018.00055
Taylor TA, Unakal CG (2023) Staphylococcus aureus infection. In: StatPearls. StatPearls Publishing. http://www.ncbi.nlm.nih.gov/books/NBK441868/
Thammavongsa V, Missiakas DM, Schneewind O (2013) Staphylococcus aureus degrades neutrophil extracellular traps to promote immune cell death. Science 342(6160):863–866. https://doi.org/10.1126/science.1242255
Thoendel M, Horswill AR (2009) Identification of Staphylococcus aureus AgrD residues required for autoinducing peptide biosynthesis. J Biol Chem 284(33):21828–21838. https://doi.org/10.1074/jbc.M109.031757
Tiwari N, López-Redondo M, Miguel-Romero L, Kulhankova K, Cahill MP, Tran PM, Kinney KJ, Kilgore SH, Al-Tameemi H, Herfst CA, Tuffs SW, Kirby JR, Boyd JM, McCormick JK, Salgado-Pabón W, Marina A, Schlievert PM, Fuentes EJ (2020) The SrrAB two-component system regulates Staphylococcus aureus pathogenicity through redox sensitive cysteines. Proc Natl Acad Sci 117(20):10989–10999. https://doi.org/10.1073/pnas.1921307117
Truong-Bolduc QC, Hooper DC (2010) Phosphorylation of MgrA and its effect on expression of the NorA and NorB efflux pumps of Staphylococcus aureus. J Bacteriol 192(10):2525–2534. https://doi.org/10.1128/JB.00018-10
Truong-Bolduc QC, Strahilevitz J, Hooper DC (2006) NorC, a new efflux pump regulated by MgrA of Staphylococcus aureus. Antimicrob Agents Chemother 50(3):1104–1107. https://doi.org/10.1128/AAC.50.3.1104-1107.2006
Tsuiji M, Shiohara K, Takei Y, Shinohara Y, Nemoto S, Yamaguchi S, Kanto M, Itoh S, Oku T, Miyashita M, Seyama Y, Kurihara M, Tsuji T (2019) Selective cytotoxicity of staphylococcal α-hemolysin (α-toxin) against human leukocyte populations. Biol Pharm Bull 42(6):982–988. https://doi.org/10.1248/bpb.b18-01024
Tzianabos AO, Wang JY, Lee JC (2001) Structural rationale for the modulation of abscess formation by Staphylococcus aureus capsular polysaccharides. Proc Natl Acad Sci U S A 98(16):9365–9370. https://doi.org/10.1073/pnas.161175598
Vandenesch F, Lina G, Henry T (2012) Staphylococcus aureus hemolysins, bi-component leukocidins, and cytolytic peptides: a redundant arsenal of membrane-damaging virulence factors? Front Cell Infect Microbiol 2:12. https://doi.org/10.3389/fcimb.2012.00012
Vann RJBSRWBEW, Liu D (1980) Virulence properties of bacterial capsular polysaccharides-unanswered questions. In: Skehel HSJ, Turner M (eds) The molecular basis of microbial pathogenicity. Verlag Chemie GmbH, Weinheim, pp 115–132
Vazquez V, Liang X, Horndahl JK, Ganesh VK, Smeds E, Foster TJ, Hook M (2011) Fibrinogen is a ligand for the Staphylococcus aureus microbial surface components recognizing adhesive matrix molecules (MSCRAMM) bone sialoprotein-binding protein (Bbp). J Biol Chem 286(34):29797–29805. https://doi.org/10.1074/jbc.M110.214981
Vestergaard M, Frees D, Ingmer H (2019) Antibiotic resistance and the MRSA problem. Microbiol Spectr 7(2):7.2.18. https://doi.org/10.1128/microbiolspec.GPP3-0057-2018
Vossen JH, Abd-El-Haliem A, Fradin EF, Van Den Berg GCM, Ekengren SK, Meijer HJG, Seifi A, Bai Y, Ten Have A, Munnik T, Thomma BPHJ, Joosten MHAJ (2010) Identification of tomato phosphatidylinositol-specific phospholipase-C (PI-PLC) family members and the

role of PLC4 and PLC6 in HR and disease resistance. Plant J 62(2):224–239. https://doi.org/10.1111/j.1365-313X.2010.04136.x

Wang R, Braughton KR, Kretschmer D, Bach T-HL, Queck SY, Li M, Kennedy AD, Dorward DW, Klebanoff SJ, Peschel A, DeLeo FR, Otto M (2007) Identification of novel cytolytic peptides as key virulence determinants for community-associated MRSA. Nat Med 13(12):12. https://doi.org/10.1038/nm1656

Wang C, Chantraine C, Viljoen A, Herr AB, Fey PD, Horswill AR, Mathelié-Guinlet M, Dufrêne YF (2022) The staphylococcal biofilm protein Aap mediates cell–cell adhesion through mechanically distinct homophilic and lectin interactions. PNAS Nexus 1(5):pgac278. https://doi.org/10.1093/pnasnexus/pgac278

Wardenburg JB, Schneewind O (2008) Vaccine protection against Staphylococcus aureus pneumonia. J Exp Med 205(2):287–294. https://doi.org/10.1084/jem.20072208

Weigel LM, Donlan RM, Shin DH, Jensen B, Clark NC, McDougal LK, Zhu W, Musser KA, Thompson J, Kohlerschmidt D, Dumas N, Limberger RJ, Patel JB (2007) High-level vancomycin-resistant Staphylococcus aureus isolates associated with a polymicrobial biofilm. Antimicrob Agents Chemother 51(1):231–238. https://doi.org/10.1128/aac.00576-06

Wertheim HF, Melles DC, Vos MC, Van Leeuwen W, Van Belkum A, Verbrugh HA, Nouwen JL (2005) The role of nasal carriage in Staphylococcus aureus infections. Lancet Infect Dis 5(12):751–762. https://doi.org/10.1016/S1473-3099(05)70295-4

Wesson CA, Liou LE, Todd KM, Bohach GA, Trumble WR, Bayles KW (1998) Staphylococcus aureus Agr and Sar global regulators influence internalization and induction of apoptosis. Infect Immun 66(11):5238–5243. https://doi.org/10.1128/IAI.66.11.5238-5243.1998

Wilke GA, Wardenburg JB (2010) Role of a disintegrin and metalloprotease 10 in Staphylococcus aureus α-hemolysin–mediated cellular injury. Proc Natl Acad Sci 107(30):13473–13478. https://doi.org/10.1073/pnas.1001815107

Wilson GJ, Seo KS, Cartwright RA, Connelley T, Chuang-Smith ON, Merriman JA, Guinane CM, Park JY, Bohach GA, Schlievert PM, Morrison WI, Fitzgerald JR (2011) A novel core genome-encoded superantigen contributes to lethality of community-associated MRSA necrotizing pneumonia. PLoS Pathog 7(10):e1002271. https://doi.org/10.1371/journal.ppat.1002271

Windham IH, Chaudhari SS, Bose JL, Thomas VC, Bayles KW (2016) SrrAB modulates Staphylococcus aureus cell death through regulation of cidABC transcription. J Bacteriol 198(7):1114–1122. https://doi.org/10.1128/JB.00954-15

Wolz C, Pohlmann-Dietze P, Steinhuber A, Chien Y-T, Manna A, Van Wamel W, Cheung A (2000) Agr-independent regulation of fibronectin-binding protein(s) by the regulatory locus sar in Staphylococcus aureus. Mol Microbiol 36(1):230–243. https://doi.org/10.1046/j.1365-2958.2000.01853.x

Woo S-G, Lee S-Y, Lee S-M, Lim K-H, Ha E-J, Eom Y-B (2017) Activity of novel inhibitors of Staphylococcus aureus biofilms. Folia Microbiol 62(2):157–167. https://doi.org/10.1007/s12223-016-0485-4

Wright JS, Jin R, Novick RP (2005) Transient interference with staphylococcal quorum sensing blocks abscess formation. Proc Natl Acad Sci 102(5):1691–1696. https://doi.org/10.1073/pnas.0407661102

Wu S-C, Liu F, Zhu K, Shen J-Z (2019) Natural products that target virulence factors in antibiotic-resistant Staphylococcus aureus. J Agric Food Chem 67(48):13195–13211. https://doi.org/10.1021/acs.jafc.9b05595

Yarwood JM, McCormick JK, Schlievert PM (2001) Identification of a novel two-component regulatory system that acts in global regulation of virulence factors of Staphylococcus aureus. J Bacteriol 183(4):1113–1123. https://doi.org/10.1128/JB.183.4.1113-1123.2001

Zapotoczna M, Jevnikar Z, Miajlovic H, Kos J, Foster TJ (2013) Iron-regulated surface determinant B (IsdB) promotes Staphylococcus aureus adherence to and internalization by non-phagocytic human cells. Cell Microbiol 15(6):1026–1041. https://doi.org/10.1111/cmi.12097

Zecconi A, Scali F (2013) Staphylococcus aureus virulence factors in evasion from innate immune defenses in human and animal diseases. Immunol Lett 150(1–2):12–22. https://doi.org/10.1016/j.imlet.2013.01.004

Zhang L, Ji G (2004) Identification of a staphylococcal AgrB segment(s) responsible for group-specific processing of AgrD by gene swapping. J Bacteriol 186(20):6706–6713. https://doi.org/10.1128/JB.186.20.6706-6713.2004

Zhang L, Lin J, Ji G (2004) Membrane anchoring of the AgrD N-terminal amphipathic region is required for its processing to produce a quorum-sensing pheromone in Staphylococcus aureus. J Biol Chem 279(19):19448–19456. https://doi.org/10.1074/jbc.M311349200

Zhang X, Liu Y, Gao Y, Dong J, Mu C, Lu Q, Shao N, Yang G (2011) Inhibiting platelets aggregation could aggravate the acute infection caused by Staphylococcus aureus. Platelets 22(3):228–236. https://doi.org/10.3109/09537104.2010.543962

Klebsiella pneumoniae Virulence Factors and Biofilm Components: Synthesis, Structure, Function, and Inhibitors

9

Bicky Jerin Joseph, Maya Mathew, Riya Rachel, Jyothis Mathew, and E. K. Radhakrishnan

Abstract

Klebsiella pneumoniae is a Gram-negative pathogen capable of causing a diverse array of infections in humans. It belongs to the group of ESKAPE pathogens, which are responsible for causing severe infections in immunocompromised individuals. The virulence of *K. pneumoniae* is attributed to several factors, including the synthesis of capsular polysaccharides, lipopolysaccharides, fimbriae, and iron acquisition systems. Additionally, *K. pneumoniae* can form a biofilm, which protects it from both host immune defense and antimicrobial agents. Its ability to form biofilm thus adds to its virulence potential significantly as biofilm-associated infections are notoriously difficult to treat. Thus, the chapter has been designed to provide an overview on the synthesis, structure, and function of virulence factors and biofilm components of *K. pneumoniae,* which contribute to its pathogenicity and also its multidrug resistance (MDR). Furthermore, the chapter discusses the various inhibitors developed to target these virulence factors and biofilm components, which can potentially be used as therapeutic agents to combat infections caused by *K. pneumoniae.*

Keywords

ESKAPE pathogens · MDR · *Klebsiella pneumoniae* · Virulence factors · Biofilm · Inhibitors

---

B. J. Joseph · M. Mathew · J. Mathew · E. K. Radhakrishnan (✉)
School of Biosciences, Mahatma Gandhi University, Kottayam, Kerala, India
e-mail: radhakrishnanek@mgu.ac.in

R. Rachel
St. Berchmans College, Kottayam, Kerala, India

S. Busi, R. Prasad (eds.), *ESKAPE Pathogens*,
https://doi.org/10.1007/978-981-99-8799-3_9

9.1 Introduction

The rise of antibiotic-resistant microorganisms has posed a significant and alarming threat to human health. The emergence of drug-resistant bacteria has led to an escalation in the severity and frequency of infections, highlighting the urgent need for the development of more potent antibiotics. This escalating global health crisis prompted the World Health Organization (WHO) to take action to release a list of diseases that were particularly concerning due to the urgent requirement for new and effective treatments (Tacconelli 2017). Among the list of diseases identified as needing immediate attention were those caused by ESKAPE pathogens. The term "ESKAPE" is an acronym representing six specific drug-resistant bacteria, where "E" stands for *Enterococcus faecium,* "S" for *Staphylococcus aureus,* "K" for *Klebsiella pneumoniae,* "A" for *Acinetobacter baumannii,* "P" for *Pseudomonas aeruginosa*, and "E" for *Enterobacter* species. These pathogens possess the capability to "escape" the effects of commonly used antimicrobial agents, making them particularly problematic in hospital settings where vulnerable patients with compromised immune systems are at a higher risk of infection (Jadimurthy et al. 2022).

9.2 *Klebsiella pneumoniae:* A Potent ESKAPE Pathogen

Klebsiella are rod-shaped, non-motile, Gram-negative facultative aerobic gammaproteobacteria named after the microbiologist Edward Klebs (Chang et al. 2021). These encapsulated bacteria are ubiquitous in our environment, being found in soil, water, and plants. This bacterial group is rapidly evolving and various *Klebsiella* sp. have already been reported which include *Klebsiella pneumoniae (*subsp. *pneumoniae and* subsp. *ozaenae), Klebsiella oxytoca, Klebsiella ornithinolytica, Klebsiella planticola, K. terrigena, K. variicola (*subsp. *Tropicalensis), K. granulomatis, K. aerogenes, K. africanensis, K. grimontii,* and *K. quasipneumoniae* (Thorpe et al. 2022).

Klebsiella pneumoniae are part of common flora of humans and animals with opportunistic pathogenic behavior (Dong et al. 2022). The bacterium typically colonizes oropharynx and gastrointestinal (GI) tract mucosa of humans and animals (Wang et al. 2020). The respiratory system and feces of healthy people contain *K. pneumoniae.* They are classified as classical, hypervirulent, and MDR strains. Nosocomial infections are usually associated with classical strains, whereas hypervirulent strains cause community-acquired infections (Guerra et al. 2022).

In addition to being linked to chronic intestinal illnesses, *K. pneumoniae* is a prominent source of antimicrobial-resistant (AMR) acquired healthcare-associated infections, newborn sepsis, and also community-acquired liver abscess. Due to the rise in drug-resistant phenotypes and decline in antibiotic effectiveness, there are significant therapeutic concerns. These infections have the potential to raise the mortality rates of seriously ill and immunosuppressed patients receiving hospital care in intensive care units (ICUs) and to increase the financial burden of their hospitalization globally. Surprisingly, *K. pneumoniae*-related pneumonia patients had a death rate of nearly 50%. According to a recent meta-analysis, the incidence of carbapenem-resistant *K. pneumoniae* (CRKP) spectrum of colonization extends from 2 to 73% with a pooled incidence of 22.3%, while the frequency of CRKP colonization ranges from 0.13 to 22% with a pooled prevalence of 5.43% worldwide (Tesfa et al. 2022).

Thus, this chapter aims to provide a comprehensive exploration of the synthesis, structure, function, and inhibitors of virulence factors and biofilm components in *Klebsiella pneumoniae.* By delving into these aspects, we can gain a deeper understanding on the molecular mechanisms that underlie the pathogenicity of this bacterium and explore potential strategies for combating *Klebsiella pneumoniae* infections. Their virulence factors and biofilm components have been a new area of research emphasis because of their critical roles in pathogenesis.

9.3 Virulence Factors: Structure and Its Function

Klebsiella pneumoniae has undergone various evolutionary changes, resulting in multiple variations that possess various types of resistance and virulence characteristics. Virulence factors are molecules that provide the bacterium with the ability to colonize, get through the host defenses, and harm host tissues. The major virulence factors contributing to host immune evasion, role in pathogenesis, and colonization are its capsular structure (CPS), lipopolysaccharides (LPS), biofilm formation, siderophores (iron uptaking systems), and adhesins (fimbriae) (Riwu et al. 2022).

9.3.1 Capsular Polysaccharides (CPS)

Klebsiella pneumoniae capsule is a key component of its pathogenicity, and studies have shown a correlation between different capsular forms and specific diseases or infection severity.

It has the ability to mask the surface antigens and increase the tolerance to drug therapy. *Klebsiella pneumoniae* has repeating sugar units enveloping the bacterial surface, corresponding to the polysaccharide termed K-antigen. It consists of linear or branched oligosaccharides and provides a barrier against the complement system. There are currently more than 80 capsular (K-antigen) forms that have been identified, based on sugar composition, and linked to various *Klebsiella* species (Tsai et al. 2023).

The capsule, found in *K. pneumoniae* and many other harmful bacteria, is an outermost layer and an important factor contributing to their virulence, or their capacity to cause disease. This protective structure helps them to evade the body's natural defense mechanisms. Specifically, *K. pneumoniae* can create over 80 distinct kinds of polysaccharide capsules (Huang et al. 2022). Each variety of these capsules has a unique structural pattern based on the repeating units of capsular polysaccharide (CPS). Every known capsule variant in *K. pneumoniae* is produced using the Wzx/Wzy polysaccharide assembly system, which is controlled by a singular capsule polysaccharide (cps) biosynthesis locus. The major monosaccharide units are D-galactose, D-glucose, L-fucose, L-rhamnose, and D-mannose. Capsular (K-antigens) forms are negatively charged as they consist of pyruvate and or uronic acid (Yang 1974). CPS plays a major role in contributing virulence to *Klebsiella* species due to their variability in sugar composition.

The Wzx/Wzy-Dependent Secretion Pathway: Genes responsible for capsule biosynthesis are observed to be clustered in particular parts of the genome. The *cps* gene cluster, found in chromosomes, holds critical genes that play key roles in the synthesis of precursor sugar molecules, the assembly of repeating units, the translocation of these units to the periplasmic side, the polymerization of these repeating units, the transport of the emergent capsule polysaccharide (CPS), and the anchorage of the CPS onto the *Klebsiella* species' surface. (Whitfield and Paiment 2003; Sachdeva et al. 2017).

The pathway begins with the biosynthesis of nucleotide sugar precursors varies based on the particular K-type and based on the arrangement of the repeating sugar unit. The genes encode sugar-specific glycosyl transferases include, *rmlA, wbaP*, *wckA, wcaN, manC, wcaA, wcuD, wcuM,* and *wclH* (Rahn et al. 1999). The flippase enzyme encoded by the *Wzx* gene identifies specific sugar-repeating units and links to undecaprenyl-pyrophosphate (Und-PP), which further aids in the transport of the sugar units to the periplasmic side. The further polymerization process is mediated by Wzy co-polymerase enzyme (Whitfield and Paiment 2003). Finally, the outer-membrane translocon (Wza), the tyrosine autokinase (Wzc), and the phosphatase (Wzb) work in unison to anchor CPS to the outer-membrane protein (Wzi). The genes vary at cps locus, and it is used in differentiating specific K-antigens. Cps cluster gene sequences such as *wzi, wza, wzb, wzc, wcaj, wzx, wzy,* and *wbap* vary according to their K-antigen composition (Whitfield and Paiment 2003).

The capsule types of *K. pneumoniae* display significant variations in their virulence, or disease-causing properties. *K. pneumoniae* is also classified as classical *K. pneumoniae* and hypervirulent *K. pneumoniae* based on their capsular structure. Hypervirulent *K. pneumoniae* (hvKp) isolates are typically restricted to the capsule types K1, K2, K16, K28, K57, and K63. Notably, the K1 and K2 serotypes are associated with approximately 70% of all globally reported hvKp infection cases (Karampatakis et al. 2023). Hence, there is a significant role of the *K. pneumoniae* capsule in providing virulence to the species, and it is emerging as a popular target for potential vaccines and therapeutic interventions against invasive *Klebsiella* infections (Opoku-Temeng et al. 2019).

These capsules have crucial role in enabling the bacteria to evade the immune system and ensure their survival. They achieve this through several mechanisms, such as blocking the immune system's phagocytic clearance by repelling negatively charged phagocytes using similarly charged CPS. They also prevent the exposure of the bacterial cell wall structures and associated proteins to complement-mediated opsonophagocytosis, impersonate host glycans antigenically, or boost the bacteria's resistance to oxidative killing. Cps acts as a physical barrier to protect lipopolysaccharide (Fig. 9.1). However, the precise mechanism by which the capsule aids in bacterial pathogenesis during a *K. pneumoniae* infection remains largely unexplored

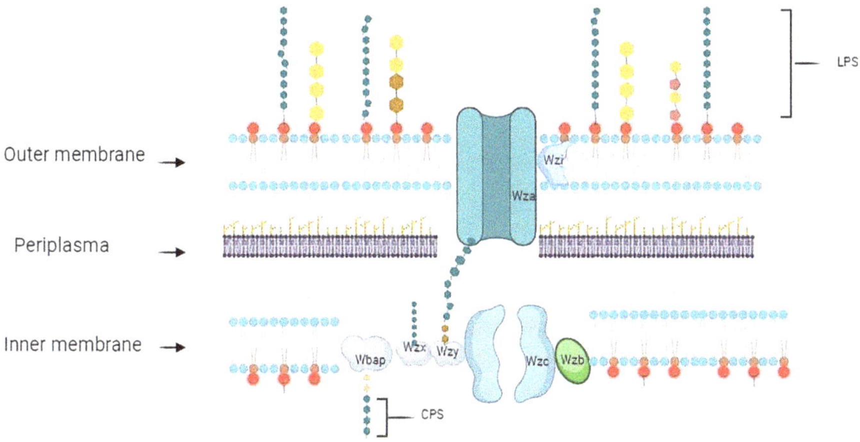

Fig. 9.1 Cell membrane of *K. pneumoniae* with capsule and lipopolysaccharide components

and uncertain (Huang et al. 2022). CPS protects the *Klebsiella species* from complement cascade antimicrobial peptides and from engulfment and phagocytosis by host immune cells.

9.3.2 Lipopolysaccharides

Lipopolysaccharides are glycolipids that evade cationic antimicrobial peptide action, as they enclose the outer membrane of the bacteria. Both CPS and LPS hence provide a counteract to the complement system. The composition of lipopolysaccharides (LPS) includes three key elements: core oligosaccharides, lipid A, and O-antigens. *Klebsiella* species are highly variable by their varying O-antigen composition; hence, it is also used in serotyping of Klebsiella species. There are eleven O-antigen types identified to date of which O1, O2, and O3 are clinically important. The O-antigens found in *Klebsiella* species are comprised of various sugars, including D-galactose, D-mannose, D-galactofuranose, D-ribofuranose, and *N*-acetyl-D-glucosamine. Their specific makeup can vary, resulting in differences in their antigenic properties. Similar to the K-antigen, variations in the O-antigens can be seen in the types of sugar they are composed of, the nature of the glycosidic linkage, the number of repeating units, and the forms of the epimers and enantiomers (Follador et al. 2016; Clarke et al. 2018). The biosynthesis of LPS production begins with nucleotide sugar synthesis, which acts as a building block of the LPS molecule.

The O-antigen biosynthesis is regulated by six genes, which are termed as wb cluster. It includes *wzm, wzt, wbbM, glf, wbbN,* and *wbbO.* Each gene encodes an enzyme required for the biosynthesis pathway (Li et al. 2014). The biosynthesis of LPS is a complex process that takes place at the cytosolic and periplasmic regions of the inner membrane. The production of lipopolysaccharides (LPS) is a multi-step process that includes the biosynthesis of lipid A, the linking of the core-oligosaccharide (core-OS) to lipid A, the translocation of the lipid A-core-OS complex to the periplasmic side, the formation of O-antigen polysaccharide (O-PS) at the cytoplasmic end, the flipping of O-PS to the periplasmic region, and the final attachment of O-PS to the lipid A-core-OS complex within the periplasmic region. The entire process of LPS formation and surface export involves four distinct gene clusters: *lpx*, *waa*, *rfb*, and *lpt*. The products of the *lpx*, *waa*, and *rfb* genes play crucial roles in the biosynthesis of lipid A, core-OS, and O-PS, respectively (Patro and Rathinavelan 2019). Lipopolysaccharide and O-antigen are highly variable and act as a protective barrier against host-mediated complement components of which the O1 serotype is more clinically relevant, and O-antigen acts as a protective barrier against host-mediated complement components (Guan et al. 2001).

9.3.3 Fimbriae and Pili

Adhesins, which are found on the cell surfaces of bacteria, promote adhesion or attachment to host cells or surfaces, playing a crucial role in bacterial infection or colonization. They are considered as promising targets for the prevention or treatment of bacterial infections. In *K. pneumoniae*, there are two types of adhesive structures are present called fimbriae: type 1 and type 3. Type 1 fimbriae are thin and rigid hair-like structures on the cell surface that are assembled via chaperones/usher pathways and are encoded by the *fim* gene cluster. Two specific genes, *fimA* and *fimH*, encode the main structural component and a small adhesin subunit, respectively. Another gene, *fimK*, is also present in *K. pneumoniae* and is important in regulating type 1 fimbriae. The lack of this gene can prevent the expression of these fimbriae. Interestingly, type 1 fimbriae are expressed in the urinary tract but not in the gastrointestinal tract or lungs. This allows them to invade bladder cells and create a biofilm there (Gerlach et al. 1989). Type 3 fimbriae, on the other hand, are encoded by the *mrkABCD* gene cluster and can bind to host cells. These fimbriae are significant for bacterial virulence, leading to colonization and disease. The situation becomes particularly concerning when multi-drug-resistant, and even carbapenem-resistant (CR), hvKP strains appear. This presents a significant public health threat (Riwu et al. 2022).

In terms of biofilm formation and adhesion to host tissues, these are supported by the fimbriae of *K. pneumoniae*. They can bind to abiotic surfaces (like medical devices) and to the extracellular matrix. *K. pneumoniae*, especially the hypermucoid variety, can cause invasive disease in a range of species and is often responsible for mastitis in dairy farms. It can survive in various hosts and environments, including water and soil. Even though hypermucoviscous isolates carry the *fimH* and *mrkD* genes that encode type 1 and type 3 fimbriae, these isolates have been reported to exhibit reduced adhesion due to the presence of an abundant capsule that conceals these fimbriae (Kot et al. 2023). Type 1 and type 3 fimbriae found in *K. pneumoniae* are equipped with adhesins, playing a crucial role in adhering to epithelial cells and promoting immune evasion, and they are beneficial for attaching to inanimate or non-living surfaces. The gene associated with fimbriae is shown in Table 9.1.

Table 9.1 Genes associated with fimbriae

Sl. no	Gene cluster	Gene	Function	References
1.	*fim* gene cluster	*fimA* and *fimH fimK*	Regulating type 1 fimbriae Present in *K. pneumoniae* only	Rifaat and Ghaima (2023); Venkitapathi et al. (2022)
2.	*mrkABCD* gene cluster	*mrkD*	Significant for bacterial virulence, leading to colonization and disease	Shen et al. (2022)

9.3.4 Iron Acquisition Systems (Siderophores)

Siderophores are small, iron-chelating molecules secreted by a wide variety of microorganisms that are critical for providing virulence in many Gram-negative bacteria. *K. pneumoniae* needs to secrete siderophores (iron uptaking system), which are small, iron-binding molecules, for its replication, and to exhibit its full harmful effects. Siderophores are critical for bacterial growth and replication. The specific mix of siderophores that *K. pneumoniae* releases during an infection can influence where in the body the bacteria localize, how they spread throughout the system, and, ultimately, the survival in the host (Holden et al. 2016).

K. pneumoniae is able to infect both healthy and individuals with compromised immune systems. In comparing the hypervirulent (HV) strain and the classic *K. pneumoniae* (cKP) strain, it was found that both almost universally express enterobactin, a primary iron uptake system in *K. pneumoniae*. The *irp* gene codes for proteins necessary for the synthesis of yersiniabactin, while the *fyu* and *ybt* genes encode siderophores, and the *ybtQ* gene codes for receptors, responsible for the uptake of siderophores (Choby et al. 2020).

The hypervirulent *K. pneumoniae* (hvKP) strain is notably more virulent and pathogenic, and it causes different cardiovascular diseases compared to cKP. In the absence of a protein called lipocalin-2, enterobactin can promote pulmonary infiltration and the spread of the bacteria. However, if lipocalin-2 is present, *K. pneumoniae* strains that only produce this siderophore are outcompeted. Salmochelin is a modified form of enterobactin that enables iron transport, and this alteration prevents its binding to lipocalin-2, thereby avoiding neutralization of the siderophore and preventing lipocalin-2-dependent inflammation (Xiao et al. 2017). Aerobactin, another siderophore made up of hydroxamate and citrate, is occasionally expressed by clinical isolates of classic nosocomial *K. pneumoniae*, and its presence is always linked with hypercapsulation, but not all hypercapsulated strains express aerobactin.

9.3.5 Toxin

Pathogenic *Klebsiella pneumoniae* produces endotoxin, colibactin. Colibactin is a polyketide hybrid genotoxin that was first described in *E. coli* (Nougayrède et al. 2006). Colibactin is a genotoxic secondary metabolite produced by certain strains of *Klebsiella pneumoniae*. The specific genotoxic activity of colibactin is linked to its ability to cause DNA damage, which could lead to mutations and cancer development (Strakova et al. 2021). The colibactin is encoded by 19 genes enclosed in pks (polyketide synthase) island consisting of colibactin genes *clbC, clbI clbO, clbN, clbH, clbJ, clbB, clbK, and clbD* amidase *clbL* and *clbQ*. Colibactin is produced as an inactive molecule that is exported in the periplasm by the efflux pump ClbM. The ClbS is a hydrolase enzyme that inactivates colibactin and binds to and protects DNA (Bossuet-Greif et al. 2016). *ClbQ* affects the flux of colibactin production, while *ClbR* is the main transcriptional activator and is known for efficient regulation of colibactin production. Colibactin acts as cytotoxic necrotizing factors and cycle-inhibiting factors modulating functions such as cell differentiation, apoptosis, and proliferation. It promotes carcinogenic process, by effector proteins that damage DNA in host cells (Dziubańska-Kusibab et al. 2020).

9.4 Biofilm Formation by *Klebsiella pneumoniae*

Biofilms are essentially structured communities of bacteria, consisting of single or more species, which are enveloped within an extracellular matrix constructed from polysaccharides, proteins, and DNA. The durability and antibiotic resistance of *Klebsiella pneumoniae*, however, is largely attributed to the complex biofilm, where cells are encased in a self-generated extracellular matrix. The capacity of *K. pneumoniae* to form biofilms contributes to its virulence and resistance traits (Guerra et al. 2022). The primary stages in biofilm development entail adhering to the colonization surface, facilitated by fimbriae and flagella. This is followed by the formation of microcolonies through the production of a self-generated extracellular polymeric substance (EPS), which is composed of polysaccharides, proteins, lipids, and nucleic acids. The secretion of this EPS persists all through the maturation stage, giving the biofilm its distinctive three-dimensional structure. Finally, the detachments of planktonic cells occur, and these cells can spread and colonize or infect nearby areas or at distant sites (Fig. 9.2).

9.4.1 Factors Contributing Biofilm Formation

There are several key elements involved in the formation of biofilms by *K. pneumoniae*. These include the bacterial polysaccharide capsule, tiny appendages such as fimbriae and pili, the management of iron, and the interaction with various other types of bacteria.

Capsule influences the initial surface adhesion and maturation. Research assessing the formation of biofilms among strains of *K. pneumoniae* causing bloodstream infection discovered a correlation between the level of biofilm formation and the activity of the *wcaG, treC,* and *sugE* virulence gene, which plays a significant role in the creation of the bacterial capsule. The capsular polysaccharides produced by *K. pneumoniae* also possess anti-biofilm properties against other bacterial species.

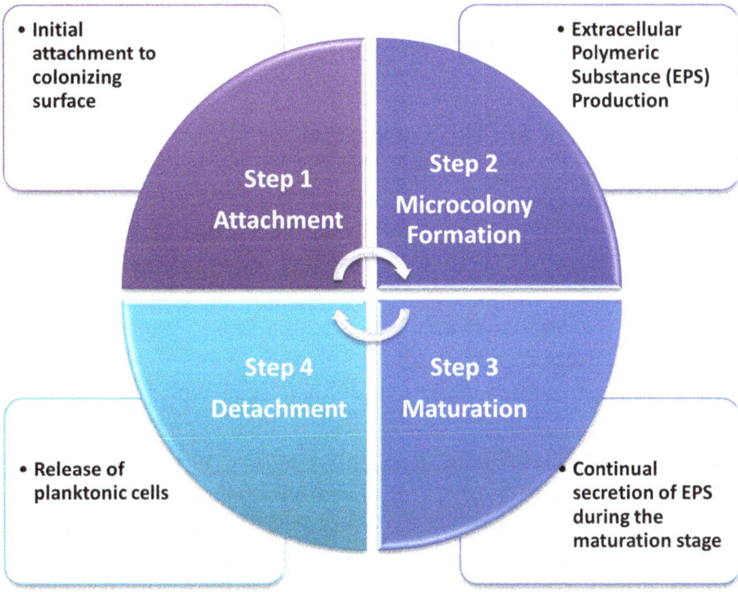

Fig. 9.2 Steps involved in biofilm formation

This gives *K. pneumoniae* a competitive edge when interacting in environments with a mixture of different bacteria (Goncalves Mdos et al. 2014).

Lipopolysaccharides (LPS) also play a crucial role in the early stages of biofilm formation by facilitating the primary adhesion of *K. pneumoniae* on inanimate surfaces. Hence, LPS is a critical element in initiating the process of biofilm creation. The involvement of *wzm* and *wbbM* genes involved in the biosynthesis of LPS also played a role in biofilm production by *K. pneumoniae.* This has been demonstrated by observing a notable increase in the activity of these genes (Vuotto et al. 2017).

The fimbriae in *K. pneumoniae* also serve critical functions in biofilm formation, the specific of which can change based on the infection site or non-living surface. Their roles go beyond just the primary adherence of the bacteria. The expression of fimbriae is a complicated procedure, as it is managed by various regulators. These regulators respond to a range of environmental signals in different host environments. The fact that type III and type I fimbriae are not co-expressed suggests a co-regulation mechanism where they mutually inhibit each other. Both type I and type III fimbriae are essential in biofilm creation, particularly during the initial stages of bacterial attachment, which pave the way for biofilm expansion. Various factors, including iron availability, oxidative stress, and specific DNA-binding proteins, intricately dictate the expression of these fimbriae. Research has been conducted on the gene clusters, *kpf* and *ecp*, associated with fimbriae, to understand their contributions to biofilm development and how bacteria stick to surfaces. The ecp operon contains the *ecpR* genes and is responsible for producing ECP. Meanwhile, the *kpf* cluster, which includes the *kpfR* genes, produces type I-like fimbriae (Sources: Guerra et al. 2022; Shadkam et al. 2021).

9.4.2 Regulation of Biofilm Formation

1. **Quorum Sensing (QS)**

   QS is an advanced communication mechanism that bacteria employ, both within their own species and among different species cohabiting in a similar environment. This communication is regulated by the creation, discharge, and recognition of specific molecules called autoinducers. When these molecules accumulate and exceed a certain level in the surrounding environment, the bacteria detect this and subsequently modify their gene expression. This results in changes in their phenotype characteristics, their ability to cause disease, resistance to acidity, and the way they form biofilms. Additionally, quorum sensing is crucial for the upkeep of existing biofilms.

   Gram-negative bacteria primarily utilize two major quorum-sensing system: type 1 and type 2 (Miller and Bassler 2001). The first type, type 1, is mainly for communication within the same species. It employs molecules termed acyl-homoserine lactones (AHL) or type 1 autoinducers (AI-1). When AI-1 links with its distinct receptor, *LuxR*, it governs the transcription of specific genes. Interestingly, *K. pneumoniae* has an encoded receptor called *SdiA* that responds to AHLs produced by different bacterial species, a fact highlighted by Pacheco and colleagues in their study (Pacheco et al. 2021). Conversely, the type 2 system fosters communication both internally among the same species and externally with different species. It leverages molecules called cyclic furanone compounds or type 2 autoinducers (AI-2). The production of AI-2 molecules is facilitated by an enzyme named LuxS synthase. Notably, a gene resembling luxS was discovered in *K. pneumoniae* genome. Quorum sensing regulates the biofilm development of *Klebsiella pneumoniae* by producing AI-2 autoinducers that communicate between species (De Araujo et al. 2010). Shadkam and associates (Shadkam et al. 2021) emphasized the significance of quorum sensing in coordinating bacterial activities within biofilms, whether they are composed of one or multiple microbial species.

2. **Iron Metabolism**

   Like other bacteria, *K. pneumoniae* iron balance is managed by a transcriptional regulator known as *Fur*, according to Gomes et al. (2021). *Fur* can either activate or repress the transcription of target genes by binding with its cofactor, iron, or existing in its apo-form (without its cofactor). Being a global transcriptional regulator, the *Fur gene* influences the expression of genes associated with iron metabolism and a variety of genes related to virulence factors, some of which are involved in the development of biofilm. For instance, *Fur* plays an important role in biofilm formation by controlling the expression of type I (Gomes et al. 2021) and type III (Wu et al. 2012; Wu et al. 2014) fimbrial genes, as well as the expression of fimbriae transcriptional regulators, in the presence of iron.

9.5 Importance of Biofilm Formation in *Klebsiella pneumoniae* Pathogenesis

Biofilms are intricate communities of microorganisms with a high degree of organization, which exhibit enhanced resistance against antimicrobial substances and host defense mechanisms, such as the complement system, phagocytosis, and antimicrobial proteins. Biofilms serve as a bacterial reserve during the colonization of a host and also contribute to their attachment to non-living surfaces. As such, they play a significant role in the disease-causing ability of numerous bacterial species. Biofilm formation provides heightened resistance to exogenous stressors and other antimicrobial factors. A study conducted in Iran shows a strong correlation of AMR genes and biofilm-forming genes (Mirzaei et al. 2023). In a similar study, a comparison of genetic factors of AMR and biofilm has demonstrated the same result suggesting the importance of biofilm screening for the effective control of nosocomial infections. Several genes like *wabG* and *treC, bsA, pgaC, and luxS* were common in all clades of MDR strains (Devanga Ragupathi et al. 2020). Table 9.2 shows a list of various genes involved in the biofilm formation of *Klebsiella pneumoniae*. These genes play critical roles in the adhesion, production of extracellular matrix, regulation, and communication during biofilm formation in *Klebsiella pneumoniae* leading to its pathogenesis.

Table 9.2 List of various genes involved in biofilm formation

Sl. no.	Genes involved	Function	References
1.	*wcaG, treC,* and *sugE*	Virulence gene related to biofilm formation	Guerra et al. (2022); Shadkam et al. (2021)
2.	*wbbM* and *wzm*	Biosynthesis of LPS	
3.	*fim, ecp, mrkABCDF,* and *kpa* to *kpg* gene clusters	Fimbriae for adhesion	
4.	*fim H*	Adhesion to host cells and abiotic surfaces	Rifaat and Ghaima (2023)
5.	*Fur*	Transcriptional regulator for iron balance	Gomes et al. (2021)
6.	*rmpA/rmpA2*	Synthesis of capsular polysaccharides	Wei et al. (2021)
7.	*wcaG/wza/wzc/wzb*	Synthesis of EPS	Wei et al. (2021); Shadkam et al. (2021)
8.	*pgaABCD*	Involves in glucosamine synthesis cell-to-cell adhesion	Shadkam et al. (2021)
9.	*adrA*	Capsule and biofilm formation	Lamey et al. (2023)
10.	*luxS*	Quorum sensing	Shadkam et al. (2021)

9.6 Function and Role of *Klebsiella pneumoniae* Virulence Factors and Biofilm Components

In *Klebsiella pneumoniae*, pili, capsules, lipopolysaccharide (LPS), and iron carriers have all been identified as pathogenic factors (Shon et al. 2013). Type 1 and type 3 adhesins of *K. pneumoniae* facilitate bacterial adherence to epithelial, immunological, and abiotic surfaces. While the composition of the capsule is pivotal in providing protection to *K. pneumoniae* from the host's immune system, it is plausible that other elements also play a role in enhancing the pathogen's ability to cause disease. In fact, some strains of *K. pneumoniae* may alter LPS to a degree that the host cells do not recognize, while other strains may employ capsules to hide LPS from Toll-like receptor (TLR4) receptor identification (Merino et al. 1992). These changes decrease bacterial clearance and attenuate the inflammatory response. As the *K. pneumoniae* strains are able to create biofilm especially on the surfaces of catheters and implanted medical devices, it is essential to monitor the host-pathogen interactions, immune evasion strategies, and impact on disease progression and severity.

9.7 Host-Pathogen Interactions

The dynamic process of host-pathogen interaction exposes not only the nature of the relationship between the two, but also the persistence of the pathogens during pathogenesis. The initial line of defense against pathogen attack is the host's innate immune response (Pranavathiyani et al. 2020). Over the past two decades, there has been extensive research into the involvement of cytokine production in the host's immune response against lung infections caused by *K. pneumoniae* (Standiford et al. 1999). In both developed and developing nations, bacterial pneumonia is a major source of morbidity and mortality, and according to study reports, fatality rates from human pneumococcal lung infection that is exacerbated by bacteremia are increased by two to three times (Mannes et al. 1991). For the purpose of researching bacterial infection and host interaction, the use of various animal models has become increasingly crucial (Wand et al. 2013). Several animal models such as *Galleria mellonella* (wax moth larvae) and murine models have been implemented for the effective understanding of the host-pathogen interaction for the past two decades.

9.8 Immune Evasion Strategies

To evade the host's immune defenses, *K. pneumoniae* employs diverse surface features such as capsule polysaccharide (CPS), lipopolysaccharide (LPS), OmpA, OmpK36, and AcrAB. Consequently, the bacteria are capable of resisting complement-triggered elimination, countering the impact of antimicrobial peptides from the host, and avoiding engulfment by epithelial cells, macrophages, neutrophils, and dendritic cells (DCs) (Fig. 9.3).

These enable them to decrease the production of the antimicrobial peptides hBD2 and hBD3, as well as the pro-inflammatory cytokine IL-8, and the intracellular killing of ingested bacteria by neutrophils. They also stop DCs from developing further (Li et al. 2014). Two types of interactions can be seen as evading strategies:

1. **As Stealth Pathogen**

   Here, the strategy works by preventing complement mediated bactericidal effect and opsonization where bacteria can evade the complement system's bactericidal effect and opsonization by limiting the deposition of complement component C3b. This is achieved through factors such as capsular polysaccharides (CPS) and lipopolysaccharide (LPS) *O*-polysaccharide (Álvarez et al. 2000; Merino et al. 1992; de Astorza et al. 2004). Another mechanism is by limiting the antimicrobial activity of collectins where bacteria can blunt the interaction of collectins (e.g., SP-A and SP-D) with their surfaces, thereby reducing their susceptibility to these antimicrobial proteins. CPS is one of the factors involved in this mechanism (Ofek et al. 2001; Kabha et al. 1997; Kostina et al. 2005). Another evasion technique used by bacteria to limit the contact of cationic antimicrobial peptides (CAMPs) and polymyxins with their surfaces is the prevent-

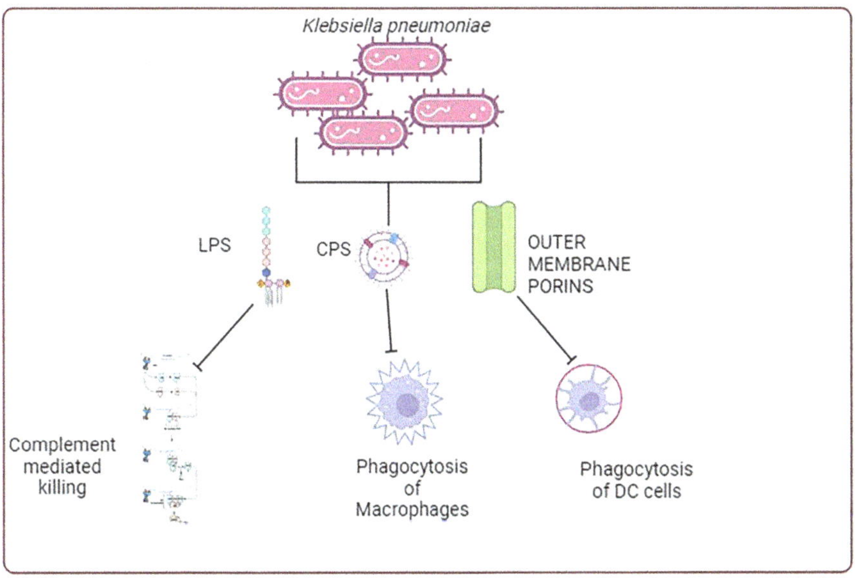

Fig. 9.3 *K. pneumoniae* uses a variety of surface characteristics, such as CPS, LPS, and outer membrane porins, to get beyond the host's immune systems. As a result, the bacterium develops resistance to host-derived antimicrobial peptides, complement-triggered elimination, and dendritic cell, macrophage, neutrophil, and epithelial cell engulfment

ing bactericidal action of CAMPs and polymyxins, which is frequently accomplished through efflux mechanisms. Factors such as CPS, LPS lipid A decorations, and AcrAB contribute to this evasion strategy (Llobet et al. 2011; Kidd et al. 2017; Campos et al. 2004; Mills et al. 2017; Padilla et al. 2010). Bacteria can also lessen their association with immune cells by inhibiting epithelial cell engulfment, which involves CPS (Regueiro et al. 2006; Cortés et al. 2002). Attenuation can also be carried out by avoiding phagocytosis by neutrophils and macrophages, through factors like such as CPS, OmpK36, OmpA, and LPS lipid A decorations (Regueiro et al. 2006; Pan et al. 2011). Bacteria can evade host immune recognition by limiting the activation of PRRs, such as Toll-like receptor 4 (TLR4), often targeted through mechanisms involving LPS lipid A 2-hydroxylation.

2. **Subversion of Host Defenses**

   Here, the subversion of host defenses can be achieved through a variety of mechanisms as listed below:

   a. Attenuating Cell-Intrinsic Immunity: Bacteria can control various aspects of cell-intrinsic immunity, including the maturation of dendritic cells, involving factors such as CPS and LPS *O*-polysaccharide (Evrard et al. 2010) and also by manipulating phagosome maturation through the PI3K-AKT-Rab14 axis, although specific factors are not identified (Cano et al. 2015).
   b. TLR-controlled inflammatory responses are eliminated: Bacteria can manipulate Toll-like receptor (TLR) signaling to dampen inflammatory responses, achieved through abolishing TLR signaling using factors such as CPS, LPS *O*-polysaccharide, OmpA, and type 2 secretion system (T2SS) (Frank et al. 2013; March et al. 2011; Tomás et al. 2015).
   d. Increasing deubiquitinase CYLD by targeting NOD1 and EGFR, NF-B signaling is suppressed, whereas CPS and other unidentified mechanisms are also involved (Frank et al. 2013; Regueiro et al. 2011).
   e. Blunting MAPKs by activating the MAPK phosphatase MKP-1 through NOD1 while the exact mechanism is uncertain (Regueiro et al. 2011).
   f. Manipulating Mucosal Immunity: Bacteria can manipulate mucosal immunity by inducing the production of interleukin-10 (IL-10), although the specific factors are not identified (Yoshida et al. 2001; Greenberger et al. 1995).
   g. Nutritional Immunity Reduction: Bacteria can counteract nutritional immunity by secreting siderophores, such as yersiniabactin, salmochelin, and aerobactin, to scavenge iron from the host environment (Bachman et al. 2011; Lawlor et al. 2007; Russo et al. 2011).

9.9 Impact on Disease Progression and Severity

Determining the association between specific bacterial compounds and the onset and development of disease remains a major challenge. It is possible for pathogenic *K. pneumoniae* to cause severe bacterial pneumonia, which is marked by a significant amount of lung inflammation, hemorrhage, and the development of necrotic lesions in the lungs. Bacteremia and sepsis are frequently caused by this condition (Rammaert et al. 2012). In a study conducted by Willingham et al., the reports have shown that after *K. pneumoniae* bacterial infection, the Nlrp12$^{-/-}$ mice would exhibit rapid disease development, lung inflammation, and cytokine production linked to an overactive immune response (Willingham et al. 2009). The body temperature is a potential indicator that can be used to determine the clinical beginning and development of the disease progression as suggested by Allen and coworkers (Allen et al. 2013). A study by Lawlor and coworkers reported the disease progression in pulmonary and systemic diseases by *K. pneumoniae* virulence determinants using an intranasal infection model (Lawlor et al. 2005). In which the significant bacterial growth could be seen in the trachea and lungs indicated by an increased influx of neutrophil, thereby leading to fatal systemic disease. It is also noteworthy to identify the key factors associated with this disease progression, and in a research study by Podschun and Ullmann, it was reported that the few factors whose role in the development of disease, which have been investigated and proven, are the capsule, fimbriae, LPS, siderophores, urease, and efflux pumps (Podschun and Ullmann 1998).

9.10 Inhibition of *K. pneumoniae* Virulence Factors and Biofilm Formation

K. pneumoniae is a formidable bacterial pathogen that exposes a huge threat to clinical healthcare systems owing to its virulence factors and capacity to form robust biofilms. The inhibition of biofilm formation and the prevention of the spreading of virulence factors of the bacteria are essential to manage the effectiveness of treatment options and potentially reduce the risk of *K. pneumoniae* infections. Strategies encompassing quorum-sensing disruption, enzymatic degradation, siderophore interference, and antivirulence compounds offer innovative ways to counteract the bacterial pathogenicity.

9.10.1 Current Approaches and Strategies

Combining antibiotics with different mechanisms of action has been explored for a long time to combat drug-resistant *K. pneumoniae.* This approach aims to enhance bacterial killing and prevent the emergence of resistance. For example, combining antibiotics that target cell wall synthesis (e.g., β-lactams) with those affecting protein synthesis (e.g., aminoglycosides) has shown synergistic effects against drug-resistant strains (Tängdén et al. 2014). Targeting virulence factors, such as CPS and siderophores, has also gained attention. Compounds that interfere with CPS synthesis or disrupt siderophore-mediated iron acquisition could attenuate pathogenicity. Inhibition of CPS has been shown to reduce bacterial survival and enhance immune recognition (Russo and Marr 2019). Interfering with bacterial quorum-sensing systems can disrupt communication and coordination among bacterial populations.

QSIs may hinder biofilm formation and virulence factor production. Synthetic compounds and natural products have been investigated as QSIs for *K. pneumoniae* (Rutherford and Bassler 2012). Biofilms contribute to chronic infections and antibiotic resistance. Enzymes that degrade biofilm matrices, such as DNase, have been studied as potential agents to disrupt biofilms (Lee and Yoon 2017).

Bacteriophages, viruses that infect and kill bacteria, are being explored as an alternative treatment. Phages can specifically target and lyse bacterial cells, potentially offering a precision therapy against *K. pneumoniae* infections (Ferry et al. 2018). Boosting the host immune response against *K. pneumoniae* is another avenue. Monoclonal antibodies targeting specific virulence factors or surface antigens have been investigated for their potential to enhance bacterial clearance (Bengoechea and Sa Pessoa 2019). Nanoparticles have shown potential in inhibiting bacterial growth and biofilm formation. They can enhance antibiotic delivery and disrupt bacterial membranes (Hernandez-Delgadillo et al. 2012). Non-traditional approaches, such as herbal extracts and natural compounds, are being explored for their antimicrobial and anti-biofilm properties against *K. pneumoniae* (Adonizio 2008). Given the complexity of *K. pneumoniae* infections, combining multiple strategies, such as virulence factor inhibition and biofilm disruption, could offer more effective treatment outcomes (Bassetti et al. 2018).

9.10.2 Small Molecule Inhibitors

Increasing multidrug-resistant *Klebsiella* sp. cases and declining antimicrobial medication activity represent a serious clinical problem on a global scale. Beta-lactam antibiotics were once used to treat *Klebsiella* infections, but these treatments have since proven ineffective owing to the rise of multidrug-resistant strains and hypervirulent strains of the bacteria (Doorduijn et al. 2016; Paczosa and Mecsas 2016; Santajit and Indrawattana 2016). Finding new antibiotics along with medication combinations to treat these fatal illnesses is therefore urgently needed. By examining the pathogen's metabolism at the genome level and providing in-depth knowledge to discover more effective treatment targets, Cesur and coworkers evaluated the novel therapeutic strategies to treat *Klebsiella* infections (Cesur et al. 2020). Small molecule compound such as furanone has been studied for their potential to disrupt quorum sensing, a cell-to-cell communication system that regulates virulence factor production and biofilm formation in bacteria. By interfering with quorum sensing, furanones may inhibit the expression of virulence factors and reduce biofilm formation in *K. pneumoniae*. Research suggests that furanones could be used as antivirulence agents to attenuate bacterial pathogenicity (Rasmussen et al. 2005). Curcumin is another compound found in turmeric with known anti-inflammatory and antimicrobial properties. Studies have indicated that curcumin possesses anti-biofilm activity against various bacterial pathogens, including *K. pneumoniae*. It can disrupt biofilm formation and inhibit bacterial adherence, potentially through interference with cell-to-cell signaling and extracellular matrix production (Packiavathy et al. 2021). Cinnamaldehyde is an essential oil component of cinnamon that has been investigated for its potential to inhibit biofilm formation. Research suggests that cinnamaldehyde can interfere with bacterial adherence, disrupt biofilm matrix components, and inhibit the expression of biofilm-related genes (Mahrous et al. 2023). Polyphenolic compounds present in green tea, particularly epigallocatechin gallate (EGCG), have shown inhibitory effects on biofilm formation and virulence factor production in various bacterial pathogens, including *K. pneumoniae*. EGCG may influence biofilm growth by modulating quorum-sensing systems (Yu et al. 2017). The ability of *N*-acetylcysteine (NAC), a mucolytic drug and antioxidant, to reduce biofilm formation and break established biofilms has been investigated. It may function by interfering with bacterial adherence and extracellular matrix components, resulting in a less stable biofilm structure (dos Santos et al. 2020). Enzymes are essential for biofilm development and virulence factor expression. Small molecule inhibitors of these enzymes, such as protease inhibitors, have been investigated for their ability to disrupt biofilm development. Protease inhibition can limit the degradation of extracellular matrix components and interfere with biofilm formation (Rendueles et al. 2014). Brominated chemicals have been studied for their capability to prevent the formation of biofilm and virulence factor production. These chemicals may disrupt quorum sensing and other regulatory processes involved in biofilm formation (Rasmussen et al. 2005). Synthetic peptides that have been designed have shown promise in altering bacterial adhesion and biofilm formation pathways. These peptides have the ability to block early adhesion and interfere with biofilm matrix components, making them promising anti-biofilm agents (Galdiero et al. 2021). Phenylthiazole compounds have been investigated for their capacity to disrupt quorum sensing and other regulatory mechanisms, hence inhibiting biofilm development. These chemicals may regulate bacterial communication and reduce pathogenicity associated with biofilms (Swati 2016). Compounds that release nitric oxide have been investigated as potential anti-biofilm agents. Nitric oxide has the ability to disrupt biofilm formation and increase antibiotic resistance in bacterial pathogens by modifying biofilm matrix components (Barraud et al. 2015). All the above mentioned compounds provide a starting point for exploring the potential of small molecule inhibitors to inhibit *K. pneumoniae* virulence factors and biofilm formation.

9.10.3 Antibodies and Vaccines

Developing vaccines against *K. pneumoniae* is a promising approach. Capsular polysaccharides (CPS) are key virulence factors that contribute to the bacterial pathogenicity. Vaccines targeting specific CPS types have been investigated to prevent infections caused by these strains. The study on the same by Cryz and his coworkers long before had proved the same by investigating the immunogenicity and safety of the CPS of *K. pneumoniae* K1, a hypervirulent strain, in humans (Cryz Jr et al. 1985). Another instance is where a vaccine candidate targeting the same showed efficacy in animal models (Foerster and Bachman 2015). However, the diversity of CPS types presents a challenge in developing a universal vaccine. The

CPS has previously been used as a target for the development of drugs and vaccines, despite the fact that there is not an approved CPS-based vaccination or medication for the treatment of CR *K. pneumoniae* infections. The need to implement immunoprophylactic and immunotherapeutic strategies that have been tested in the past for the treatment of *Klebsiella* infections and potential strategies to encourage the development of CPS-based vaccines and therapies for the prevention and treatment of *K. pneumoniae* infections are crucial to regaining an effective treatment regimen. Rats are protected from experimental *K. pneumoniae*, according to Held and coworkers by a murine monoclonal antibody made against a K2 serotype CPS (Held et al. 1992). Rats treated with this mAb had fewer bacterial counts in the lungs and more normal lung function as compared to BSA-pretreated control rats. Emphasizing the impact humoral immunity has on the emergence of drug resistance, especially for extremely resistant *K. pneumoniae* infections that can contribute toward the prevention of the widespread of infectious diseases.

9.11 Future Directions and Challenges

The therapeutic necessity for doctors treating patients with significant bacterial infections is to provide effective, timely, and adequately administered antibiotic therapy or other alternative techniques while minimizing additional bacterial resistance. Phage therapy has regained prominence as a research subject due to the oversight of warnings concerning the overreliance on antibiotics, which has given rise to the emergence of bacteria like *K. pneumoniae* that are resistant to multiple drugs. Consequently, the advancement of early antimicrobial marketing is no longer sufficient in light of the need for sustained infection control and bacterial containment. It is imperative to consider strategic approaches for employing specific antibacterial remedies. In the realm of phage therapy, these approaches might involve utilizing combination treatments, such as sophisticated combinations of phages or blends of phages and drugs. These combinations should be carefully designed to mitigate the development of bacterial resistance during treatment. In the context of healthcare-associated infection control processes, the primary emphasis should be on prevention. In addition to the development of novel antibacterial medications, it is essential to implement enhanced insurance policies aimed at reducing the likelihood of patients contracting bacterial infections. These measures are crucial for curbing the proliferation of treatment-resistant strains. These protocols may encompass rigorous environmental decontamination, the maintenance of immune system integrity for clinical equipment, prompt removal of medical devices once they are no longer necessary, the implementation of screening and decolonization initiatives, and the prudent use of antimicrobial agents.

9.12 Conclusion

Exploring the future of dealing with *Klebsiella* infections entails both novel approaches and considerable challenges. Alternative medicines such as phage therapy, monoclonal antibodies, and new antimicrobial agents are emerging as viable alternatives as antibiotic resistance spreads. Precision medicine, treatment tailoring based on strain characteristics, and the development of combination medications could all improve efficacy. Vaccination and timely diagnoses are critical preventive interventions. However, obstacles loom big. Antibiotic resistance spreads quickly among *Klebsiella* strains, highlighting the importance of developing new therapies. Ineffective intervention is hampered by a lack of options, necessitating interdisciplinary collaboration among researchers and clinicians. The regulatory complexity involved in creating novel treatments and managing infection control in hospital settings presents considerable challenges. Furthermore, the worldwide nature of drug-resistant *Klebsiella* infections necessitates collaborative efforts and public education to prevent resistance from spreading. To overcome these issues, multiple stakeholders will need to work together in a comprehensive and collaborative manner.

References

Adonizio AL (2008) Anti-quorum sensing agents from South Florida medicinal plants and their attenuation of Pseudomonas aeruginosa pathogenicity. Doctoral dissertation, Florida International University

Álvarez D, Merino S, TomÁs JM, Benedí VJ, Albertí S (2000) Capsular polysaccharide is a major complement resistance factor in lipopolysaccharide O side chain-deficient *Klebsiella pneumoniae* clinical isolates. Infect Immun 68(2):953–955

Allen IC, McElvania-TeKippe E, Wilson JE, Lich JD, Arthur JC, Sullivan JT et al (2013) Characterization of NLRP12 during the in vivo host immune response to Klebsiella pneumoniae and Mycobacterium tuberculosis. PloS one 8(4):e60842. https://doi.org/10.1371/journal.pone.0060842

Bachman MA, Oyler JE, Burns SH, Caza M, Lépine F, Dozois CM, Weiser JN (2011) Klebsiella pneumoniae yersiniabactin promotes respiratory tract infection through evasion of lipocalin 2. Infect Immun 79(8):3309–3316

Barraud N, Kelso MJ, Rice SA, Kjelleberg S (2015) Nitric oxide: a key mediator of biofilm dispersal with applications in infectious diseases. Curr Pharm Des 21(1):31–42

Bassetti MATTEO, Giacobbe DR, Giamarellou H, Viscoli C, Daikos GL, Dimopoulos G, Poulakou G (2018) Management of KPC-producing *Klebsiella pneumoniae* infections. Clin Microbiol Infect 24(2):133–144

Bengoechea JA, Sa Pessoa J (2019) *Klebsiella pneumoniae* infection biology: living to counteract host defences. FEMS Microbiol Rev 43(2):123–144

Bossuet-Greif N, Dubois D, Petit C, Tronnet S, Martin P, Bonnet R et al (2016) Escherichia coli ClbS is a colibactin resistance protein. Mol Microbiol 99(5):897–908. https://doi.org/10.1111/mmi.13272

Campos MA, Vargas MA, Regueiro V, Llompart CM, Albertí S, Bengoechea JA (2004) Capsule polysaccharide mediates bacterial resistance to antimicrobial peptides. Infect Immun 72(12):7107–7114

Cano V, March C, Insua JL, Aguiló N, Llobet E, Moranta D, Bengoechea JA (2015) Klebsiella *pneumoniae* survives within macrophages by avoiding delivery to lysosomes. Cell Microbiol 17(11):1537–1560

Cesur MF, Siraj B, Uddin R, Durmuş S, Çakır T (2020) Network-based metabolism-centered screening of potential drug targets in *Klebsiella pneumoniae* at genome scale. Front Cell Infect Microbiol 9:447

Chang D, Sharma L, Dela Cruz CS, Zhang D (2021) Clinical epidemiology, risk factors, and control strategies of *Klebsiella pneumoniae* infection. Front Microbiol 12:750662

Choby JE, Howard-Anderson J, Weiss DS (2020) Hypervirulent Klebsiella pneumoniae—clinical and molecular perspectives. J Intern Med 287(3):283–300

Clarke BR, Ovchinnikova OG, Kelly SD, Williamson ML, Butler JE, Liu B et al (2018) Molecular basis for the structural diversity in serogroup O2-antigen polysaccharides in Klebsiella pneumoniae. J Biol Chem 293(13):4666–4679. https://doi.org/10.1074/jbc.RA117.000646

Cortés G, Borrell N, de Astorza B, Gómez C, Sauleda J, Albertí S (2002) Molecular analysis of the contribution of the capsular polysaccharide and the lipopolysaccharide O side chain to the virulence of *Klebsiella pneumoniae* in a murine model of pneumonia. Infect Immun 70(5):2583–2590

Cryz SJ Jr, Fürer E, Germanier R (1985) Safety and immunogenicity of *Klebsiella pneumoniae* K1 capsular polysaccharide vaccine in humans. J Infect Dis 151(4):665–671

De Araujo C, Balestrino D, Roth L, Charbonnel N, Forestier C (2010) Quorum sensing affects biofilm formation through lipopolysaccharide synthesis in Klebsiella pneumoniae. Res Microbiol 161(7):595–603

de Astorza B, Cortés G, Crespí C, Saus C, Rojo JM, Albertí S (2004) C3 promotes clearance of *Klebsiella pneumoniae* by A549 epithelial cells. Infect Immun 72(3):1767–1774

Devanga Ragupathi NK, Muthuirulandi Sethuvel DP, Triplicane Dwarakanathan H, Murugan D, Umashankar Y, Monk PN, Veeraraghavan B (2020) The influence of biofilms on carbapenem susceptibility and patient outcome in device associated *K. pneumoniae* infections: insights into phenotype vs genome-wide analysis and correlation. Front Microbiol 11:591679

Dong N, Yang X, Chan EWC, Zhang R, Chen S (2022) *Klebsiella* species: taxonomy, hypervirulence and multidrug resistance. EBioMedicine 79:103998

Doorduijn DJ, Rooijakkers SH, van Schaik W, Bardoel BW (2016) Complement resistance mechanisms of *Klebsiella pneumoniae*. Immunobiology 221(10):1102–1109

dos Santos Goncalves M, Delattre C, Balestrino D, Charbonnel N, Elboutachfaiti R, Wadouachi A et al (2014) Anti-biofilm activity: a function of Klebsiella pneumoniae capsular polysaccharide. PLoS One 9(6):e99995. https://doi.org/10.1371/journal.pone.0099995

Dziubańska-Kusibab PJ, Berger H, Battistini F, Bouwman BA, Iftekhar A, Katainen R, Meyer TF (2020) Colibactin DNA-damage signature indicates mutational impact in colorectal cancer. Nat Med 26(7):1063–1069

Evrard B, Balestrino D, Dosgilbert A, Bouya-Gachancard JL, Charbonnel N, Forestier C, Tridon A (2010) Roles of capsule and lipopolysaccharide O antigen in interactions of human monocyte-derived dendritic cells and *Klebsiella pneumoniae*. Infect Immun 78(1):210–219

Ferry T, Boucher F, Fevre C, Perpoint T, Chateau J, Petitjean C, Laurent F (2018) Innovations for the treatment of a complex bone and joint infection due to XDR Pseudomonas aeruginosa including local application of a selected cocktail of bacteriophages. J Antimicrob Chemother 73(10):2901–2903

Foerster J, Bachman M (2015) Beyond passive immunization: toward a nanoparticle-based IL-17 vaccine as first in class of future immune treatments. Nanomedicine 10(8):1361–1369

Follador R, Heinz E, Wyres KL, Ellington MJ, Kowarik M, Holt KE, Thomson NR (2016) The diversity of *Klebsiella pneumoniae* surface polysaccharides. Microb Genom 2(8). https://doi.org/10.1099/mgen.0.000073

Frank CG, Reguerio V, Rother M, Moranta D, Maeurer AP, Garmendia J, Bengoechea JA (2013) Klebsiella pneumoniae targets an EGF receptor-dependent pathway to subvert inflammation. Cell Microbiol 15(7):1212–1233

Gerlach GF, Clegg STEVEN, Allen B (1989) Identification and characterization of the genes encoding the type 3 and type 1 fimbrial adhesins of Klebsiella pneumoniae. J Bacteriol 171(3):1262–1270. https://doi.org/10.1128/jb.171.3.1262-1270.1989

Galdiero E, Salvatore MM, Maione A, Carraturo F, Galdiero S, Falanga A et al (2021) Impact of the peptide WMR-K on dual-Species biofilm Candida albicans/Klebsiella pneumoniae and on the untargeted metabolomic profile. Pathogens 10(2):214. https://doi.org/10.3390/pathogens10020214

Gomes AÉI, Pacheco T, Santos CDSD, Pereira JA, Ribeiro ML, Darrieux M, Ferraz LFC (2021) Functional insights from KpfR, a new transcriptional regulator of fimbrial expression that is crucial for *Klebsiella pneumoniae* pathogenicity. Front Microbiol 11:601921

Greenberger MJ, Strieter RM, Kunkel SL, Danforth JM, Goodman RE, Standiford TJ (1995) Neutralization of IL-10 increases survival in a murine model of Klebsiella pneumonia. J Immunol (Baltimore, Md.: 1950) 155(2):722–729

Guan S, Clarke AJ, Whitfield C (2001) Functional analysis of the galactosyltransferases required for biosynthesis of D-galactan I, a component of the lipopolysaccharide O1 antigen of *Klebsiella pneumoniae*. J Bacteriol 183(11):3318–3327

Guerra MES, Destro G, Vieira B, Lima AS, Ferraz LFC, Hakansson AP, Converso TR (2022) *Klebsiella pneumoniae* biofilms and their role in disease pathogenesis. Front Cell Infect Microbiol 12:877995

Held TK, Trautmann M, Mielke ME, Neudeck H, Cryz SJ Jr, Cross AS (1992) Monoclonal antibody against *Klebsiella* capsular polysaccharide reduces severity and hematogenic spread of experimental *Klebsiella pneumoniae* pneumonia. Infect Immun 60(5):1771–1778

Hernandez-Delgadillo R, Velasco-Arias D, Diaz D, Arevalo-Niño K, Garza-Enriquez M, De la Garza-Ramos MA, Cabral-Romero C (2012) Zerovalent bismuth nanoparticles inhibit Streptococcus mutans growth and formation of biofilm. Int J Nanomedicine 7:2109–2113

Holden VI, Breen P, Houle S, Dozois CM, Bachman MA (2016) *Klebsiella pneumoniae* siderophores induce inflammation, bacterial dissemination, and HIF-1α stabilization during pneumonia. MBio 7(5):10–1128

Huang X, Li X, An H, Wang J, Ding M, Wang L, Zhang JR (2022) Capsule type defines the capability of *Klebsiella pneumoniae* in evading Kupffer cell capture in the liver. PLoS Pathog 18(8):e1010693

Jadimurthy R, Mayegowda SB, Nayak SC, Mohan CD, Rangappa KS (2022) Escaping mechanisms of ESKAPE pathogens from antibiotics and their targeting by natural compounds. Biotechnol Rep 34:e00728

Kabha KISRA, Schmegner JUTTA, Keisari YONA, Parolis HARALAMBOS, Schlepper-Schaeffer J, Ofek ITZHAK (1997) SP-A enhances phagocytosis of *Klebsiella* by interaction with capsular polysaccharides and alveolar macrophages. Am J Phys Lung Cell Mol Phys 272(2):L344–L352

Karampatakis T, Tsergouli K, Behzadi P (2023) Carbapenem-resistant *Klebsiella pneumoniae*: virulence factors, molecular epidemiology and latest updates in treatment options. Antibiotics 12(2):234

Kidd TJ, Mills G, Sá-Pessoa J, Dumigan A, Frank CG, Insua JL, Bengoechea JA (2017) A *Klebsiella pneumoniae* antibiotic resistance mechanism that subdues host defences and promotes virulence. EMBO Mol Med 9(4):430–447

Kostina E, Ofek I, Crouch E, Friedman R, Sirota L, Klinger G, Keisari Y (2005) Noncapsulated *Klebsiella pneumoniae* bearing mannose-containing O antigens is rapidly eradicated from mouse lung and triggers cytokine production by macrophages following opsonization with surfactant protein D. Infect Immun 73(12):8282–8290

Kot B, Piechota M, Szweda P, Mitrus J, Wicha J, Grużewska A, Witeska M (2023) Virulence analysis and antibiotic resistance of Klebsiella pneumoniae isolates from hospitalised patients in Poland. Sci Rep 13(1):4448

Lamey AE, Selim AO, Atia NM, Mowafy RE (2023) Genes contributed on biofilm forming bacteria incriminated in various disease conditions in cattle. J Adv Vet Res 13(3):312–321

Lawlor MS, Hsu J, Rick PD, Miller VL (2005) Identification of *Klebsiella pneumoniae* virulence determinants using an intranasal infection model. Mol Microbiol 58(4):1054–1073

Lawlor MS, O'connor C, Miller VL (2007) Yersiniabactin is a virulence factor for Klebsiella pneumoniae during pulmonary infection. Infect Immun 75(3):1463–1472

Lee K, Yoon SS (2017) Pseudomonas aeruginosa biofilm, a programmed bacterial life for fitness. J Microbiol Biotechnol 27(6):1053–1064

Li B, Zhao Y, Liu C, Chen Z, Zhou D (2014) Molecular pathogenesis of Klebsiella pneumoniae. Future Microbiol 9(9):1071–1081

Llobet E, Campos MA, Giménez P, Moranta D, Bengoechea JA (2011) Analysis of the networks controlling the antimicrobial-peptide-dependent induction of *Klebsiella pneumoniae* virulence factors. Infect Immun 79(9):3718–3732

Mahrous SH, El-Balkemy FA, Abo-Zeid NZ, El-Mekkawy MF, El Damaty HM, Elsohaby I (2023) Antibacterial and anti-biofilm activities of Cinnamon Oil against multidrug-resistant Klebsiella pneumoniae isolated from Pneumonic Sheep and Goats. Pathogens 12(9):1138. https://doi.org/10.3390/pathogens12091138

Mannes GP, Boersma WG, Baur CH, Postmus PE (1991) Adult respiratory distress syndrome (ARDS) due to bacteraemic pneumococcal pneumonia. Eur Respir J 4(4):503–504

March C, Moranta D, Regueiro V, Llobet E, Tomás A, Garmendia J, Bengoechea JA (2011) Klebsiella pneumoniae outer membrane protein A is required to prevent the activation of airway epithelial cells. J Biol Chem 286(12):9956–9967

Merino S, Camprubí S, Albertí S, Benedí VJ, Tomas J (1992) Mechanisms of *Klebsiella pneumoniae* resistance to complement-mediated killing. Infect Immun 60(6):2529–2535

Miller MB, Bassler BL (2001) Quorum sensing in bacteria. Annu Rev Microbiol 55(1):165–199

Mills G, Dumigan A, Kidd T, Hobley L, Bengoechea JA (2017) Identification and characterization of two *Klebsiella pneumoniae* lpxL lipid A late acyltransferases and their role in virulence. Infect Immun 85(9):10–1128

Mirzaei B, Ebrahimi A, Keshavarzi S, Hydarzadeh S, Badmasti F, Dadar M, Moradi N (2023) Antibiotic susceptibility, biofilm-forming ability, and prevalence of extended-spectrum beta-lactamase (ESBL)-and biofilm-associated genes among *Klebsiella pneumoniae* isolates from hospitalized patients in northwest of Iran. Curr Microbiol 80(5):175

Nougayrède JP, Homburg S, Taieb F, Boury M, Brzuszkiewicz E, Gottschalk G, Oswald E (2006) Escherichia coli induces DNA double-strand breaks in eukaryotic cells. Science 313(5788):848–851

Ofek I, Mesika A, Kalina M, Keisari Y, Podschun R, Sahly H, Crouch E (2001) Surfactant protein D enhances phagocytosis and killing of unencapsulated phase variants of *Klebsiella pneumoniae*. Infect Immun 69(1):24–33

Opoku-Temeng C, Kobayashi SD, DeLeo FR (2019) *Klebsiella pneumoniae* capsule polysaccharide as a target for therapeutics and vaccines. Comput Struct Biotechnol J 17:1360–1366

Pacheco T, Gomes AÉI, Siqueira NMG, Assoni L, Darrieux M, Venter H, Ferraz LFC (2021) SdiA, a quorum-sensing regulator, suppresses fimbriae expression, biofilm formation, and quorum-sensing signaling molecules production in. Front Microbiol 12:597735

Packiavathy IASV, Kannappan A, Thiyagarajan S, Srinivasan R, Jeyapragash D, Paul JBJ, Ravi AV (2021) AHL-Lactonase producing Psychrobacter sp. from Palk Bay sediment mitigates quorum sensing-mediated virulence production in Gram negative bacterial pathogens. Front Microbiol 12:634593

Paczosa MK, Mecsas J (2016) *Klebsiella pneumoniae*: going on the offense with a strong defense. Microbiol Mol Biol Rev 80(3):629–661

Padilla E, Llobet E, Doménech-Sánchez A, Martínez-Martínez L, Bengoechea JA, Albertí S (2010) *Klebsiella pneumoniae* AcrAB efflux pump contributes to antimicrobial resistance and virulence. Antimicrob Agents Chemother 54(1):177–183

Pan YJ, Lin TL, Hsu CR, Wang JT (2011) Use of a Dictyostelium model for isolation of genetic loci associated with phagocytosis and virulence in *Klebsiella pneumoniae*. Infect Immun 79(3):997–1006

Patro LPP, Rathinavelan T (2019) Targeting the sugary armor of *Klebsiella* species. Front Cell Infect Microbiol 9:367

Podschun R, Ullmann U (1998) *Klebsiella* spp. as nosocomial pathogens: epidemiology, taxonomy, typing methods, and pathogenicity factors. Clin Microbiol Rev 11(4):589–603

Pranavathiyani G, Prava J, Rajeev AC, Pan A (2020) Novel target exploration from hypothetical proteins of *Klebsiella pneumoniae* MGH 78578 reveals a protein involved in host-pathogen interaction. Front Cell Infect Microbiol 10:109

Rahn A, Drummelsmith J, Whitfield C (1999) Conserved organization in the cps gene clusters for expression of Escherichia coli group 1 K antigens: relationship to the colanic acid biosynthesis locus and the cps genes from Klebsiella pneumoniae. J Bacteriol 181(7):2307–2313. https://doi.org/10.1128/JB.181.7.2307-2313.1999

Rammaert B, Goyet S, Beauté J, Hem S, Te V, Try PL, Vong S (2012) Klebsiella pneumoniae related community-acquired acute lower respiratory infections in Cambodia: clinical characteristics and treatment. BMC Infect Dis 12(1):1–7

Rasmussen TB, Bjarnsholt T, Skindersoe ME, Hentzer M, Kristoffersen P, Kote M, Givskov M (2005) Screening for quorum-sensing inhibitors (QSI) by use of a novel genetic system, the QSI selector. J Bacteriol 187(5):1799–1814

Regueiro V, Campos MA, Pons J, Albertí S, Bengoechea JA (2006) The uptake of a *Klebsiella pneumoniae* capsule polysaccharide mutant triggers an inflammatory response by human airway epithelial cells. Microbiology 152(2):555–566

Regueiro V, Moranta D, Frank CG, Larrarte E, Margareto J, March C, Bengoechea JA (2011) Klebsiella pneumoniae subverts the activation of inflammatory responses in a NOD1-dependent manner. Cell Microbiol 13(1):135–153

Rendueles O, Beloin C, Latour-Lambert P, Ghigo JM (2014) A new biofilm-associated colicin with increased efficiency against biofilm bacteria. ISME J 8(6):1275–1288

Rifaat RM, Ghaima KK (2023) Detection of type1 fimbriae genes (fimH and fimA) of *Klebsiella pneumoniae* among Iraqi Patients with catheter-associated urinary tract infections. Acta Biomed 94(2):e2023064

Riwu KHP, Effendi MH, Rantam FA, Khairullah AR, Widodo A (2022) A review: virulence factors of *Klebsiella* pneumonia as emerging infection on the food chain. Vet World 15(9):2172

Russo TA, Marr CM (2019) Hypervirulent *Klebsiella pneumoniae*. Clin Microbiol Rev 32(3):10–1128

Russo TA, Shon AS, Beanan JM, Olson R, MacDonald U, Pomakov AO, Visitacion MP (2011) Hypervirulent K. pneumoniae secretes more and more active iron-acquisition molecules than "classical" K. pneumoniae thereby enhancing its virulence. PLoS One 6(10):e26734

Rutherford ST, Bassler BL (2012) Bacterial quorum sensing: its role in virulence and possibilities for its control. Cold Spring Harb Perspect Med 2(11):a012427

Sachdeva S, Palur RV, Sudhakar KU, Rathinavelan T (2017) E. coli group 1 capsular polysaccharide exportation nanomachinary as a plausible antivirulence target in the perspective of emerging antimicrobial resistance. Front Microbiol 8:70. https://doi.org/10.3389/fmicb.2017.00070

Santajit S, Indrawattana N (2016) Mechanisms of antimicrobial resistance in ESKAPE pathogens. Biomed Res Int 2016:2475067

Santos PBDRED, Avila DDS, Ramos LDP, Yu AR, Santos CEDR, Berretta AA, Oliveira LDD (2020) Effects of Brazilian green propolis extract on planktonic cells and biofilms of multidrug-resistant strains of *Klebsiella pneumoniae* and Pseudomonas aeruginosa. Biofouling 36(7):834–845

Shadkam S, Goli HR, Mirzaei B, Gholami M, Ahanjan M (2021) Correlation between antimicrobial resistance and biofilm formation capability among *Klebsiella pneumoniae* strains isolated from hospitalized patients in Iran. Ann Clin Microbiol Antimicrob 20:1–7

Shen L, Zhang J, Xue J, Du L, Yuan L, Nie H, Li Y (2022) Regulation of ECP fimbriae-related genes by the transcriptional regulator RcsAB in *Klebsiella pneumoniae* NTUH-K2044. J Basic Microbiol 62(5):593–603

Shon AS, Bajwa RP, Russo TA (2013) Hypervirulent (hypermucoviscous) *Klebsiella pneumoniae*: a new and dangerous breed. Virulence 4(2):107–118

Standiford TJ, Wilkowski JM, Sisson TH, Hattori N, Mehrad B, Bucknell KA, Moore TA (1999) Intrapulmonary tumor necrosis factor gene therapy increases bacterial clearance and survival in murine gram-negative pneumonia. Hum Gene Ther 10(6):899–909

Strakova N, Korena K, Karpiskova R (2021) *Klebsiella pneumoniae* producing bacterial toxin colibactin as a risk of colorectal cancer development—a systematic review. Toxicon 197:126–135

Swati S (2016) A study on microbiological profile of symptomatic catheter associated urinary tract infection in an intensive care unit setup in a Tertiary care hospital. Doctoral dissertation, Madras Medical College, Chennai

Tacconelli E (2017) Global priority list of antibiotic-resistant bacteria to guide research, discovery, and development. WHO pp 1–7. https://policycommons.net/artifacts/1818147/global-priority-list-of-antibiotic-resistant-bacteria-to-guide-researchdiscovery-and-development/2555608/

Tängdén T, Hickman RA, Forsberg P, Lagerbäck P, Giske CG, Cars O (2014) Evaluation of double- and triple-antibiotic combinations for VIM-and NDM-producing *Klebsiella pneumoniae* by in vitro time-kill experiments. Antimicrob Agents Chemother 58(3):1757–1762

Tesfa T, Mitiku H, Edae M, Assefa N (2022) Prevalence and incidence of carbapenem-resistant *K. pneumoniae* colonization: systematic review and meta-analysis. Syst Rev 11(1):1–15

Thorpe HA, Booton R, Kallonen T, Gibbon MJ, Couto N, Passet V, Feil EJ (2022) A large-scale genomic snapshot of *Klebsiella* spp. isolates in Northern Italy reveals limited transmission between clinical and non-clinical settings. Nat Microbiol 7(12):2054–2067

Tomás A, Lery L, Regueiro V, Pérez-Gutiérrez C, Martínez V, Moranta D, Bengoechea JA (2015) Functional genomic screen identifies Klebsiella pneumoniae factors implicated in blocking nuclear factor κB (NF-κB) signaling. J Biol Chem 290(27):16678–16697

Tsai CC, Lin JC, Chen PC, Liu EYM, Tsai YK, Yu CP, Siu LK (2023) A 20-year study of capsular polysaccharide seroepidemiology, susceptibility profiles, and virulence determinants of *Klebsiella pneumoniae* from bacteremia patients in Taiwan. Microbiol Spectr 11(3):e00359–e00323

Venkitapathi S, Wijesundara YH, Cornelius SA, Herbert FC, Gassensmith JJ, Zimmern PE, De Nisco NJ (2022) Conserved FimK truncation coincides with increased expression of type 3 fimbriae and cultured bladder epithelial cell association in *Klebsiella quasipneumoniae*. J Bacteriol 204(9):e00172–e00122

Vuotto C, Longo F, Pascolini C, Donelli G, Balice MP, Libori MF et al (2017) Biofilm formation and antibiotic resistance in *Klebsiella pneumoniae* urinary strains. J Appl Microbiol 123(4):1003–1018. https://doi.org/10.1111/jam.13533

Wand ME, McCowen JW, Nugent PG, Sutton JM (2013) Complex interactions of *Klebsiella pneumoniae* with the host immune system in a Galleria mellonella infection model. J Med Microbiol 62(12):1790–1798

Wang G, Zhao G, Chao X, Xie L, Wang H (2020) The characteristic of virulence, biofilm and antibiotic resistance of *Klebsiella pneumoniae*. Int J Environ Res Public Health 17(17):6278

Wei DD, Xiong XS, Mei YF, Du FL, Wan LG, Liu Y (2021) Microbiological and clinical characteristics of *Klebsiella pneumoniae* isolates of K57 capsular serotype in China. Microb Drug Resist 27(3):391–400

Whitfield C, Paiment A (2003) Biosynthesis and assembly of Group 1 capsular polysaccharides in *Escherichia coli* and related extracellular polysaccharides in other bacteria. Carbohydr Res 338(23):2491–2502

Willingham SB, Allen IC, Bergstralh DT, Brickey WJ, Huang MTH, Taxman DJ, Ting JPY (2009) NLRP3 (NALP3, Cryopyrin) facilitates in vivo caspase-1 activation, necrosis, and HMGB1 release via inflammasome-dependent and-independent pathways. J Immunol 183(3):2008–2015

Wu CC, Lin CT, Cheng WY, Huang CJ, Wang ZC, Peng HL (2012) Fur-dependent MrkHI regulation of type 3 fimbriae in Klebsiella pneumoniae CG43. Microbiology 158(4):1045–1056. https://doi.org/10.1099/mic.0.053801-0

Wu C-C, Wang C-K, Chen Y-C, Lin T-H, Jinn T-R, Lin C-T (2014) IscR regulation of capsular polysaccharide biosynthesis and iron-acquisition systems in klebsiella pneumoniae CG43. PLoS ONE 9(9):e107812. https://doi.org/10.1371/journal.pone.0107812

Xiao X, Yeoh BS, Vijay-Kumar M (2017) Lipocalin 2: an emerging player in iron homeostasis and inflammation. Annu Rev Nutr 37:103–130

Yang MT (1974) Structural studies of *Klebsiella* capsular polysaccharides. Doctoral dissertation, University of British Columbia

Yoshida K, Matsumoto T, Tateda K, Uchida K, Tsujimoto S, Iwakura Y, Yamaguchi K (2001) Protection against pulmonary infection with Klebsiella pneumoniae in mice by interferon-γ through activation of phagocytic cells and stimulation of production of other cytokines. J Med Microbiol 50(11):959–964

Yu HH, Zhang L, Yu F, Li F, Liu ZY, Chen JH (2017) Epigallocatechin-3-gallate and Epigallocatechin-3-O-(3-O-methyl)-gallate Enhance the Bonding Stability of an Etch-and-Rinse Adhesive to Dentin. Materials 10(2):183

Acinetobacter baumannii Virulence Factors and Biofilm Components: Synthesis, Structure, Function, and Inhibitors

10

Pitchaipillai Sankar Ganesh, Pathoor Naji Naseef, Raman Muthusamy, Sathish Sankar, Rajesh Kanna Gopal, and Esaki Muthu Shankar

Abstract

Acinetobacter baumannii (*A. baumannii*) is a Gram-negative opportunistic human pathogen, and the drastic dissemination of multidrug resistance in *A. baumannii* is presently posing a serious concern in healthcare settings. This bacterium has the ability to multiply and produce various virulence factors and biofilm formation on the surface of host cells. Due to the production of virulence factors and biofilm formation, the bacteria's ability contributes to chronic or persistent infections and resistance to various antibiotic substance. In this chapter, we mainly focus on the mechanisms of quorum sensing-controlled virulence factors, drug resistance, and biofilm formation of *A. baumannii* and a brief account of quorum sensing (QS) inhibitors and anti-biofilm compounds as an alternative therapeutic treatment for multidrug-resistant *A. baumannii*.

Keywords

Acinetobacter baumannii · Virulence · Quorum sensing · Inhibitors · Alternative medicine

P. S. Ganesh (✉) · P. Naji Naseef · R. Muthusamy · S. Sankar · R. K. Gopal
Department of Microbiology, Centre for Infectious Diseases, Saveetha Dental College and Hospitals, Saveetha Institute of Medical and Technical Sciences (SIMATS) (Deemed to be University), Chennai, India

E. M. Shankar
Department of Biotechnology, Infection and Inflammation, Central University of Tamil Nadu, Thiruvarur, Tamil Nadu, India

S. Busi, R. Prasad (eds.), *ESKAPE Pathogens*,
https://doi.org/10.1007/978-981-99-8799-3_10

10.1 Introduction

Acinetobacter baumannii (*A. baumannii*) is Gram-negative pathogenic bacterium, recognized as an important opportunistic human pathogen and primarily associated with hospital-acquired infections. It is aerobic, pleomorphic, non-motile, non-fermenting, and coccobacilli (Howard et al. 2012). Being a group 1 priority pathogen (WHO 2016), *A. baumannii* has been considered one of the most critical pathogens associated with nosocomial infections, especially among the immunocompromised due to reports of increased rates of antimicrobial resistance and its ability to form biofilm on both biotic and abiotic surfaces (Chen 2020; Antunes et al. 2014; Yang et al. 2019). The route of infection is mainly via moist tissues, such as mucosa or exposed skin, caused by an accident or injury, leading to respiratory infections, bacteremia, sepsis, etc. (Howard et al. 2012). The prevalence of *A. baumannii* infection is relatively higher in males when compared the females, possibly due to age factor and changing lifestyle choices such as smoking, alcohol, and drugs, as well as comorbidities such as obstructive pulmonary diseases, diabetics, and renal disease (Falagas and Rafailidis 2007). The infection and/or colonization of *A. baumannii* depends on several specific and nonspecific virulence factors, which mainly play critical roles in adhesion, cytotoxicity, immune evasion, microbial interaction, genetic re-arrangements, and, most importantly, QS-controlled virulence factors and biofilm formation (Gaddy and Actis 2009; Kon et al. 2020). Some of these predisposing virulence factors in *A. baumannii* determine the formation of extracellular exopolysaccharide (EPS), two-component systems such as Bfm/S and BfmR, outer membrane protein A (OmpA, bacterial biofilm-associated protein (Bap)), PER-1, chaperon-usher pilus assembly system of pili (CsuBABCDE), and the QS system that are involved in the infection process and downstream pathobiologic events, viz. activation of immune responses, cellular damage, serum resistance, and bacterial adherence to epithelial cells (Kon et al. 2020; Aly et al. 2016; Ryu et al. 2017). Of the several factors, the biofilm-associated protein (bap) is crucial for bacterial cell accumulation, cell (intercellular) adhesion, and the development of biofilms (Fattahian et al. 2011).

Quorum sensing (QS) is a complex cell-to-cell communication system among bacteria by their ability to monitor the changes in the surrounding microenvironment to maintain population density through small diffusible signal molecules called auto-inducers (AIs) (Zhong and He 2021). AI molecules are mainly released by bacteria during the growth phase, whose secretion is directly proportional to the bacterial cell density population. By virtue of secretion and receipt of these signal molecules, the QS system regulates gene expression aiding bacteria to acclimatize to adverse external environmental conditions (Zhong and He 2021). As a result, QS interference by QS inhibitors or quorum quenchers (QQs) is a probable novel target for antimicrobial therapy since bacterial biofilm is one of the major underlying causes of multidrug resistance (Zhong and He 2021; Rao et al. 2008). There are several QS systems exist among pathogenic bacteria such as *Escherichia coli, Vibrio cholerae, A. baumannii, Pseudomonas aeruginosa, and* phytopathogens *Erwinia* and *Ralstonia,* which have been extensively studied (Sibanda et al. 2018). QS regulation in biofilm formation is an indispensable factor for the development of strategies to control biofilms (Subhadra et al. 2018).

Biofilms are multicellular bacterial conglomerates that mainly adhere to a surface and are highly submerged into a self-secreted polymeric substance called exopolysaccharide (EPS). Biofilm mainly involved various virulence factors such as adhesion of collagen, aggregation of substances, siderophore-mediated iron acquisition, expression of pili, secretion of necessary virulence factors, and, most importantly, communication between surrounding cells (Longo et al. 2014). Biofilm confers resistance against many conventional antibiotics due to a thick exopolysaccharide matrix that acts as a protective shield against antibiotics. *A. baumannii* isolates carrying plcN and lasB exhibit higher levels of virulence as compared to other strains (Aliramezani et al. 2019). Bacteria capable of producing biofilm show a strong degree of resistance against a wide range of conventional antibiotics, including gentamicin, piperacillin, ceftazidime, and ticarcillin (Yang et al. 2019).

Antimicrobial resistance (AMR) represents one of the most serious threats to public health, warranting a search for newer anti-infective agents against biofilm formation, preferably from natural resources potentially exhibiting anti-QS activities (Goel et al. 2021). The World Health Organization (WHO) estimates that ~80% of the global population is dependent on traditional remedies for ailments, and >35,000 plant species are being utilized globally in traditional medications (WHO 2022). These medicinal plants serve as potential sources for alternative drugs to alleviate the challenges linked to the emergence of AMR and biofilm formation (Vaou et al. 2021). Specific compounds, such as rosmarinic acid, limonoid, changing, quercetin, catechin, and apigenin, possess protective properties against infections by biofilm-producing microbes (Asfour 2018; Martin 2009; Bravo 2009; Liu 2004). The 3-HBA is a naturally occurring active molecule that has antimicrobial, anti-inflammatory, antioxidant, and anti-carcinogenic naturally found in *Taxus baccata* (Kalinowska et al. 2021). It serves as an intermediate in the production of pharmaceuticals, plasticizers, and resins. 3-HBA is commonly seen in fruits such as *Citrus paradisi*, *Olea europaea*, and *Mespilus germanica* (Bendini et al. 2007).

10.2 *Acinetobacter baumannii* Infection

A. baumannii is an important pathogen in a hospital environment, and its ability to upregulate the antimicrobial determinants and the marked ability to increase mortality in critically ill subjects uplift *A. baumannii,* formerly known as *Acinetobacter calcoaceticus,* into the forefront of pathogenic microbiology research. Likewise, the bacteria and its ability to survive in the hospital and healthcare environment for a prolonged duration increase its threat of nosocomial spread (Peleg et al. 2008). The infection rate of A. *baumannii* is very high in critically admitted patients connected to different lifesaving equipment that causes breaches into the skin or airway passages and affects mainly organs with high fluid contents such as the peritoneal cavity, respiratory, and urinary tracts. It is challenging to distinguish between an infection and colonization attributed to *A. baumannii* (Peleg et al. 2008; Jung and Park 2015).

A. baumannii causes a broad range of infections both in the hospital and in the community, like infections involving urinary tract infections (UTIs), skin and soft tissue, bacterial meningitis, bacteremia, and pneumonia, with the latter being the most subsequently reported in both the aforementioned settings. The organism is frequently associated with colonization or infection associated with catheters. *A. baumannii* rarely causes complicated UTIs among outpatients. One study showed that it accounts for ~1.6% of the ICU-acquired UTIs (Peleg et al. 2008). However, the most frequent reasons for *A. baumannii* bacteremia are intravascular and respiratory tract catheters. Endocarditis can have uncommon origins viz. from surgical incisions, burns, and UTIs. About 21–70% of cases have an unknown source of bacteremia (Cisneros and Rodríguez-Baño 2002). More importantly, multidrug-resistant *A. baumannii*, which causes nosocomial post-neurosurgical meningitis, is a burning concern in modern days. Acinetobacters are attributed to deliberate greater than 10% of Gram-negative bacillary and about 4% of all the nosocomial meningitides in adults (Doughari et al. 2011).

10.3 *A. baumannii*: An Emerging Hospital-Associated Pathogen and Colonizer

A. baumannii is a most important nosocomial pathogen in a hospital environment, and due to its ability to pose a significant risk to healthcare professionals, it is classified as a "red alert" bacterial pathogen largely owing to resist a wide range of conventional antibiotics (Cerqueira and Peleg 2011). Over the course of 17 years (from 1986 to 2003), the National Nosocomial Infections Surveillance System found that the proportion of *A. baumannii* among all pathogenic Gram-negative bacteria has increased significantly (Weinstein et al. 2005). Infections due to acinetobacters can have varying degrees of severity, from mild to moderate to severe.

A. baumannii is one of three closely related species collectively referred to as the ACB (*Acinetobacter calcoaceticus-baumannii*) complex (along with three genomic species and 13TU) that are primarily associated with human infections and are difficult to distinguish from one another in typical clinical settings (Bouvet and Grimont 1987; Bergogne-Bérézin and Towner 1996).

A. baumannii has been found to coexist with humans, and its "colonization attributes" have been extensively studied. Various existing research publications emphasize that the frequency of probable transition from colonization to infection is more likely higher as compared to infections acquired from distant environmental sources (Towner 2009).

Recently, the complex virulence factors are mainly contributed to the colonization and pathogenicity of *A. baumannii,* which has been extensively scrutinized, and several models of the same have been proposed. These models emphasize various characteristics such as the presence and activity of exoproteases and exopolysaccharides (which enable the bacterium to form biofilms), as well as resistance to iron acquisition, serum, and desiccation. Other attributes, such as adherence and colonization, invasion of epithelial cells, and the ability to acquire the foreign particle via lateral transfer, are also described as important virulence factors (Cerqueira and Peleg 2011; Antunes et al. 2011). *A. baumannii* can persist for extended periods in both living and non-living dry surfaces and can cause outbreaks under suitable circumstances. The precise natural habitats of various *Acinetobacter* sp. are not yet completely comprehended and may warrant extensive investigations.

10.4 Antimicrobial Resistance of *A. baumannii*

A. baumannii has increasingly become a growing concern in healthcare environments because it can develop resistance against various antimicrobial drugs. AMR is attributed to the presence of AMR genes, which is considered to be one of its primary mechanisms. *A. baumannii* possesses various resistance mechanisms against different antibiotic classes, (multidrug efflux pumps aminoglycoside-modifying enzymes, permeability issues, β-lactamases, and target site) modifications (Gordon and Wareham 2010). A vast majority of these mechanisms of resistance can downgrade the effect of various groups of antibiotics although multiple mechanisms can synergistically collaborate to create resistance toward a specific group of antibiotics. *A. baumannii* has developed significant resistance to carbapenem via the production of β-lactamases that can hydrolyze carbapenems. In addition to carbapenem resistance, *A. baumannii* may also be resistant to other classes of antibiotics due to a range of other resistance mechanisms, such as efflux pumps and modification of the bacterial cell wall (Perez et al. 2007). Higher production of efflux pumps and β-lactamases work together to enhance antibiotic resistance (Cecchini et al. 2018), which means that the AdeRS two-component system is responsible for regulating the expression of the efflux pump. If there are point mutations in the AdeRS operon, it can result in higher levels of being expressed, which can then cause antibiotic resistance (Marchand et al. 2004; Leus et al. 2018). Penicillin-binding proteins (PBPs) are a class of enzymes that catalyze the polymerization of peptidoglycan molecules, leading to the assembly of the cell wall. Their role is key to ensuring the proper insertion of peptidoglycan into the cell wall (Sauvage et al. 2008). Since β-lactams mimic their substrate, they bind to PBPs. When PBPs are blocked by β-lactams, it causes unevenness in cell wall metabolism, which ultimately leads to cell death (Zapun et al. 2008). While the significance of this resistance mechanism seems to be minor, it should not be disregarded entirely. However, resistance to imipenem in a specific clone of *A. baumannii* is due to a complicated alteration in the PBPs. Additionally, it has been documented that an MDR *A. baumannii* strain (with a MIC of imipenem >120 μg/mL) exhibited an increase in PBP expression as a result of exposure to imipenem (Harding et al. 2018).

The drastic dissemination of multidrug resistance in *A. baumannii* is presently posing a serious concern in healthcare settings. *A. baumannii* is ability to multiply and produce various virulence factors which disiminate and counteract host immune system and also it can damage host cells. Due to the production of virulence factors and biofilm formation, the bacteria's ability contributes to chronic/persistent infections and resistance to various antibiotic substance. Hence, QS-controlled virulence factors of exopolysaccharide that normally surrounds bacteria allow them to withstand even extremely harsh environments and also prevent the entry of various conventional antibiotics. As a result, most of the antibiotics now available for treating *A. baumannii* QS-controlled virulence factors and biofilm-associated infections are mostly infective.

10.5 Quorum Sensing

Communication between the bacterial cells, known as QS, that allows bacteria to acquire information about the solidity of the cell and species gathering of the adjacent communal and modify their gene expression patterns accordingly with the production and release of small signaling molecules (autoinducers (AIs)) is called QS system (Bassler and Losick 2006). The number of AIs produced is directly proportional to the bacterial population density. The production of various virulence traits, bioluminescence, and formation of biofilm are some of the characteristic features regulated by the QS system. As such, targeting the QS could be a promising approach for developing antimicrobial strategies. Different QS systems are used by both Gram-negative bacteria and Gram-positive bacteria. Auto inducer peptides (AIPs) are signaling molecules, which is mainly used by Gram-positive bacteria. QS system in Gram-positive bacteria contains a two-component system such as membrane-bound sensor kinase receptors and cytoplasmic transcription factors that help in the modification of their gene expression (Rutherford and Bassler 2012). In contrast, Gram-negative bacteria have the ability to communicate and coordinate using small signaling molecules such as *N*-acyl homoserine lactones (AHLs), which control the expression of relevant genes (Whitehead et al. 2001).

10.5.1 Quorum Sensing in Gram-Negative Bacteria

Gram-negative bacterial QS system has been well-studied, and it was first described in a marine species called *Vibrio fischeri* (*V. fischeri*), which can release a class of signaling factors called AHL into the environment. *V. fischeri* bioluminescence gene cluster includes eight lux genes (luxAe, luxG, luxI, and luxR) organized in two bi-directionally transcribed operons (Engebrecht et al. 1983; Swartzman et al. 1990). The byproducts of the luxI and luxR genes act as bioluminescence regulators (Engebrecht and Silverman 1984). Luxl is an autoinducer synthase molecule that is responsible for AHL production (*S*-adenosyl methionine). LuxR stimulates the transcription of structural operon luxCDABE by binding with cognate autoinducer molecules. The regulatory protein homologs aid in the regulation of various arsenal virulence factors and biofilm production (Lupp et al. 2003).

10.5.2 Quorum-Sensing System in *A. baumannii*

A complicated system of sensors governs the activation of virulence factors in *Acinetobacter* sp. by integrating signals from both inside and outside of the cell (Harding et al. 2018). The QS system of *A. baumannii* has two components, i.e., the *AbaI* inducer and its cognate to receptor *AbaR*, which is similar to the typical *LuxI/ LuxR* QS systems found in other Gram-negative bacteria. The bar gene produces the sensor protein AbaI, which serves as an AI synthase and generates AHL signal molecules, and AbaR serves as a protein receptor for AHL, which on binding to AbaR sets off a chain of events. Only AbaI synthases and acyl transferases may be responsible for the structural diversity of AHLs (Niu et al. 2008). The production of the EPS poly-*N*-acetylglucosamine (PNAG), which is necessary for adhesion and aggregation, is a significant ancillary factor that aids *A. baumannii* in forming biofilm (Choi et al. 2009).

10.5.3 Biofilm Formation in *A. baumannii*

In the current scenario of infection, AMR is becoming a huge threat worldwide in preventing and treating persistent infections. AMR in a microorganism can be due to many intrinsic or extrinsic factors (Dadgostar 2019). One such factor is the formation of biofilms, which are homogenous or heterogeneous microbial communities covered in a self-formed sheath of EPS where bacteria remain metabolically inactive in stage (Ghosh et al. 2020). Biofilms are a serious threat, as they remain the major cause of life-threatening diseases involving the lungs, urinary tract, or even maybe nosocomial (Goel et al. 2021). Biofilm enhances the bacteria's ability to cause diseases and make them survive in a wide range of environmental stress conditions (Yin et al. 2019). *A. baumannii* release various arsenal virulence factors and forms biofilms on a variety of surfaces, such as indwelling medical devices, catheters, and wound dressings. The process of biofilm formation in *A. baumannii* is seemingly complex and involves several steps (Gaddy and Actis 2009).

10.5.4 Biofilm Development

During the biofilm development the bacteria's ability to attach biotic and abiotic surface, and steps involved attachment cell surface, formation of micro-colony, maturation, and detachment of the biofilm, which are predominantly controlled by the QS system (Fig. 10.1). The first step is the attachment of a bacterium to a surface, which is mediated by adhesion. Upon adhesion, the bacterium begins to produce EPS, which forms a matrix that holds the biofilm together. EPS production is regulated by a complex network of genes and environmental signals (Limoli et al. 2015). As the biofilm grows, it becomes more complex, with distinct layers of bacteria and EPS. Within the biofilm, bacteria may undergo drastic changes in the

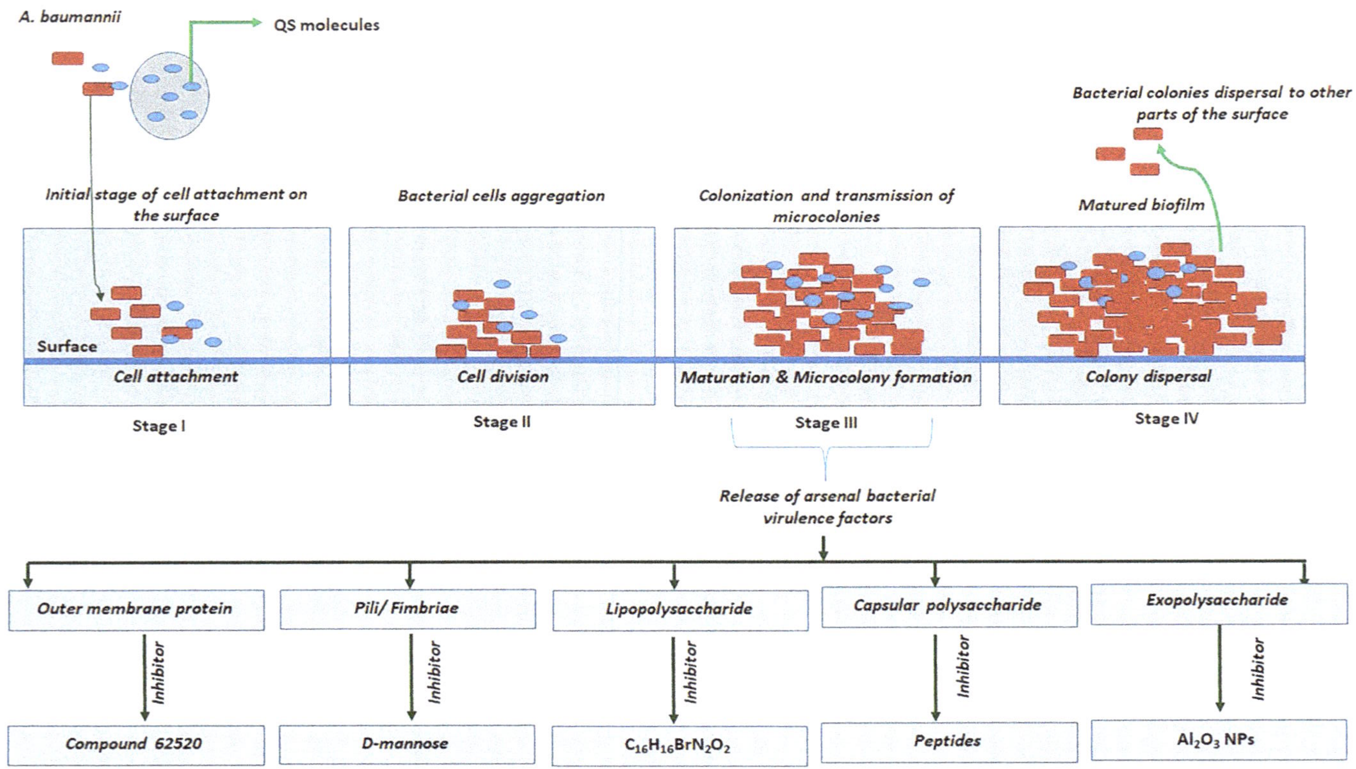

Fig. 10.1 Life cycle of biofilm formation and production of QS-controlled arsenal virulence factors of *Acinetobacter baumannii* and natural/synthetic QS inhibitors

expression of genes and physiological properties, which can render them more resistant to antibiotics and other external stresses.

Biofilms can be challenging to treat, as they are resistant to percolation by many conventional antimicrobial agents. Therefore, preventing biofilm formation is an important strategy for controlling infections caused by *A. baumannii*. Some approaches to preventing biofilm formation include the use of antibacterial coatings on medical equipment, developing novel antimicrobial agents that target biofilm formation, and improvement of infection control practices across healthcare settings (Roy et al. 2018).

10.5.5 Arsenal of QS-Controlled Virulence Factors Deployed by *A. baumannii*

A. baumannii has potent several QS-controlled arsenals of virulence factors that allow the establishment of infection at various degree levels of pathogenesis. It is also used for hiding the host immune exposure to attachment, internalization, and apoptosis of host cells (Tiku 2022). A set of major arsenal virulence traits are involved in the formation of biofilm and preventing the entry of various antibiotic substances. There are different types of virulence factors that are elucidated by *A. baumannii* such as outer membrane proteins, porin proteins, efflux pumps, phospholipases, and capsular polysaccharides (Gedefie et al. 2021). These arsenal virulence traits are involved in the establishment of pathogenesis on the host cells (Fig. 10.1) (Lee et al. 2017).

10.5.5.1 Outer Membrane Proteins (OMPs) and Inhibitors

Outer membrane proteins (OMPs) are one of the major key virulence factors that mainly help in attachment to the host epithelial cell layer. Outer membrane protein A (OmpA) is a major component of OMPs in Gram-negative bacteria (Shadan et al. 2023). This protein (OmpA) involves the formation of biofilm on the surface of the host cells and the establishment of infection in eukaryotic cells. Similarly, its resistance to various antibiotics (nalidixic acid, imipenem, chloramphenicol, aztreonam, and meropenem) activates host immunomodulation. The apoptosis process begins through the secretion of host cell apoptotic factors advancing to cell death (Nie et al. 2020). Besides their diverse role in antibiotic resistance, OMPs confer arsenal virulence factors to *A. baumannii.*

A. baumannii is mainly attributed to adhering to the surface of host cells and persisting in host lung epithelial cells via direct attachment with fibronectin on the cell surfaces (Zoghlami et al. 2023). It induces a pro-inflammatory response, which leads to cellular (metastasis) invasion and cell death via the triggering of caspase-3 as OmpA, and it has been detected to localize in nuclei and mitochondria (McIlwain et al. 2013). In *A. baumannii* OmpA has been abundantly found in outer membrane vesicle (OMV). OmpA has been abundantly found in outer membrane vesicles (OMV), which are mainly released by pathogenic bacteria of *A. baumannii,* conferring upon its cytotoxic effects (Tiku 2022). OmpA protein also inhibits the degradation of caspase-1 and promotes the activation of the NLRP3 inflammasome, which triggers the secretion of pro-inflammatory responses such as interleukin-18 (IL-18) and IL-1 β, and finally, it leads to serious inflammatory injuries to the host cells (Li et al. 2021). Na et al. have reported that the sub-inhibitory concentration level of compound 62520 has potentially inhibited the OmpA expression and sessile formation of biofilm in *A. baumannii* ATCC 17978 (Na et al. 2021). In silico study revealed that the phyto-compounds such as isosakuranetin, aloe emodin, and pinocembrin possess considerable binding affinity toward the selected receptor site of OmpA and OmpW, which indicated that some of the natural compounds potentially inhibited the biofilm formation in a dose-dependent manner (Shahryari et al. 2021). Therefore, any natural inhibitors with the ability to suppress the functions of outer membrane proteins (OmpA and OmpW) could be possible to treat the infection caused by multidrug resistance *A. baumannii.* Similarly, Vila-Farres et al. have reported that in virtual screening cyclic hexapeptide AOA-2 could inhibit the adhesion of *A. baumannii* biofilm formation (Vila-Farrés et al. 2017).

10.5.5.2 Biofilm-Associated Arsenal Virulence Factors and Inhibitors

Pili/Fimbriae and Inhibitors

Pili are tiny, hair-like proteinaceous appendages that are normally found on prokaryotic bacteria. It has a major role in the colonization of specific host tissues on both biotic and abiotic surfaces, twisting motility, biofilm, cell adhesion, conjugation, horizontal gene transfer, and cell adhesion (Ronish et al. 2019; Upmanyu et al. 2022). All those factors contribute to the cytotoxic effects of *A. baumannii. A. baumannii* expresses type IV pili, which is mainly used for twisting motility and natural competence and contributes to host cell adherence and biofilm formation (Ronish et al. 2019). The establishment of biofilms is stimulated by Csu system (type 1 pili). Csu pili is composed of four subunits of *CsuA/B* (pilin subunit), *CsuA, CsuB, and CsuE* (tip subunit) assembled via the chaperone-usher (CU) pathway (Pakharukova et al. 2018). *CsuC* and *CsuD* are well known as transport proteins, which are extremely conserved in the formation of biofilm in *A. baumannii* and more perilous for adherence to abiotic surface (Eijkelkamp et al. 2014). CsuE protein is located at the pilus tip, and the expression of *CsuE* is part of the *CsuA/BABCED* chaperone-usher assembly system of pili, which is essential for surface adherence to the biotic and abiotic substrate during the biofilm development in host cells (De Breij et al. 2009). Kishii et al. have reported that the Csu operon positive strain of *A. baumannii* formed biofilm very thickly when compared to the Csu operon negative strain of *A. baumannii* (Kishii et al. 2020). Chen et al. reported that D-mannose effectively suppresses the Csu pilus of *A. baumannii* (Chen et al. 2022). Eugenol and geraniol have potentially inhibited biofilm formation and exopolysaccharide production, reduced the expression of Csu genes, and prevented the assembly of mature pilus (Choudhary et al. 2022).

Lipopolysaccharide

Biosynthesis of Lipopolysaccharide and Inhibitors

Lipopolysaccharide (LPS) is the major integral component of the outer membrane (OM) of Gram-negative bacterial pathogen that contribute to shielding the bacterium from the external harsh environment (Bertani and Ruiz 2018). The main advantages are to maintain the structural integrity and permeability barrier to prevent the entry of small molecules from the external environment (Nikaido 2003). LPS is a complex glycolipid that mainly comprises three major structural components such as lipid A, glycosylated with core oligosaccharide, and O-antigen (Raetz and Whitfield 2002). Lipid A consists of a complex glycolipid that compresses two phosphorylated *N*-acetylglucosamine (GlcNAc), and various acyl chains are linked to the two sugars of lipid tails (14–24 carbon atoms) that are directly joined to the external leaflet of the OM. Biosynthesis of LPS pathway comprises sequences of reactions catalyzed and involves nine enzymes *LpxA, LpxB, LpxC, LpxD, LpxH, LpxK, LpxL KdtA,* and LpxM. *LpsB* glycosyltransferase is a vital enzyme required for the synthesis of LPS in *A. baumannii* (Luke et al. 2010). Similarly, *LpxA, LpxC, LpxD, LpxH, and LpxK* enzymes are also important in Gram-negative bacteria. All these enzymes together catalyze the diacylation of uridine diphosphate *N*-acetylglucosamine (UDP-GlcNAc), following the concentration of two molecules, and phosphorylation yielding lipid IVA. Other enzymes such as KdtA, LpxL, and LpxM are majorly involved in the further formation of glycosylation and acylation into Kdo-lipid A and the final product to which core oligosaccharide and O-antigen are added. However, these downstream enzyme (KdtA, LpxL, and LpxM) complexes in O-antigen assembly are recognized to stimulate the membrane permeability, antibiotic resistance, virulence factors, etc. (Romano and Hung 2023). LPS has been proven to contribute both to bacterial evasion of host immune response (innate and acquired immune response) and to initiation of an overwhelming host pro-inflammatory response that significantly increases with the mortality and mortality of infected patients (Chen 2020).

There are some inhibitors that predominantly inhibit the synthesis of LPS in *A. baumannii* and reduce bacterial resistance. LpxC inhibitor (PF-5081090) inhibits lipid A biosynthesis and increases the susceptibility to conventional antibiotics (García-Quintanilla et al. 2016). Based on the virtual screening and in vitro study revealed that *N*-[(2S)-3-amino-1-(hydroxyamino)-1-oxopropan-2-yl]-4-(4-bromophenyl) benzamide (CS250) (chemical formula: $C_{16}H_{16}BrN_2O_2$) tightly intact with LpxC enzyme and reduced the bacterial resistance (Zoghlami et al. 2023).

Capsular Polysaccharides and Exopolysaccharide

A baumannii produces capsular polysaccharide (CPS), which encloses the OM. It consists of tightly filled repeating oligosaccharide subunits (K units) (Singh et al. 2019). CPS develops a separate layer on the bacterial cell surface and is considered an important virulence factor in *A. baumannii.* It protects from external environmental stress and counteracts the host defense mechanism and also its resistance to various conventional antibiotics through CPS (Geisinger and Isberg 2015). Similarly,

CPS plays a pivotal role in the development of bacterial biofilm and surface colonization (Antunes et al. 2011). In *A. baumannii,* the synthesis of CPS assembly and export occurs through a Wzy CPS pathway. Wza is an OM protein encoded by the wza gene, which is responsible for transporting CPS from the periplasmic space to the surface of the bacterium, and it contributes to the formation of bacterial capsule (Singh et al. 2019). Bacterial biofilms are microbial communities adhering to living and non-living surfaces enclosed in a self-secreted extracellular polymeric substance (EPS) that holds microbial consortia together onto the surface and environ of the bacterial population (Geisinger and Isberg 2015; García-Quintanilla et al. 2013). This substance provides protection from the harsh environments. Al_2O_3 NPs drastically suppressed the biofilm formation and exopolysaccharide (EPS) production at the lowest concentration level of 120 μg/mL against multidrug-resistant *A. baumannii* (Muzammil et al. 2020).

10.6 Pathogenesis of *Acinetobacter baumannii* Infections

A. baumannii produces a range of surface molecules, including lipopolysaccharides (LPS), capsular polysaccharides (CPS), and pili, which contribute to its pathogenesis (Monem et al. 2020). LPS is an essential component of the OM of Gram-negative bacteria and plays a pivotal role in bacterial survival and immune evasion. The LPS of *A. baumannii* contains lipid A modifications that contribute to its resistance to antimicrobial peptides (Maldonado et al. 2016). CPS is the capsule that surrounds the bacterial cell and provides protection against host defenses. The CPS of *A. baumannii* is highly variable and has been associated with virulence and antibiotic resistance (Tickner et al. 2021). Pili are hair-like appendages that extend from the surface and are involved in adhesion to host cells and formation of biofilm in *A. baumannii*. *A. baumannii* produces a variety of pili, type I, type IV, and chaperone-usher pili.

A. baumannii also produces a range of enzymes, including proteases, lipases, and phospholipases, which contribute to its pathogenesis. These enzymes can degrade host tissues and facilitate bacterial dissemination (Pakharukova et al. 2018). Another critical aspect of *A. baumannii* pathogenesis is its ability to acquire and maintain drug resistance genes. *A. baumannii* is inherently resistant to many drugs due to its impermeable outer membrane and efflux pumps that can pump out antibiotics. However, it can also acquire additional resistance genes through horizontal gene transfer, which can occur through plasmids, transposons, or integrons (Abdi et al. 2020). Overall, the molecular pathogenesis of *A. baumannii* involves a complex interplay of virulence factors and antibiotic resistance mechanisms. Understanding their role is critical for the development of newer intervention strategies against *A. baumannii* infections.

10.7 Quorum-Sensing Inhibitors

There are several pathological events relevant to the QS system. There are different pathways, such as AI synthesis inhibition, AI secretion or transport inhibition, receptor binding downstream target inhibition, AI degradation using enzymes, antibodies that can cover and block receptors of AIs, and antagonistic AI receptors (Asfour 2018). Interference of cell communication without altering cell growth is the principal theme of these attributes. The characteristics of QS inhibitors include chemical stability, high effectiveness in low molecular mass molecules, and a high degree of specificity.

Quorum quenching (QQ) refers to any activity that disrupts the process of QS (Sikdar and Elias 2020). Examples of QQ compounds that are widely recognized include lactonases/acylases, which break down AIs known as *N*-(3-oxoctanoyl)-homoserine lactone (HSL), inhibitors of synthase such as analogs of anthranilic acid that prevent the synthesis of quinolone signals, and receptor inhibitors such as brominated furanone (Lesic et al. 2007). Niclosamide is a type of QQ compound found in many drugs that have been approved for clinical use. It has been shown to decrease surface motility, the formation of biofilm, and the production of various virulence factors (Imperi et al. 2013). Several studies have reported the identification of QSIs in *A. baumannii*. Earlier experiments screened an organic compound 3-HBA and identified that this compound inhibited QS-regulated phenotypes in *A. baumannii*. Studies also signify the role of 3-HBA in inhibiting the QS of *Staphylococcus aureus* (Ganesh et al. 2022). The greater the number of hydroxyl groups, the greater will be their effectiveness against microorganisms. Licoricone and glycerin derived from the family Fabaceae/Leguminosae also appear to inhibit bacterial QS (Bhargava et al. 2015) (Table 10.1).

10.8 Conclusion

A. baumannii is a highly resistant and invasive bacterial pathogen mostly acquired through the existing healthcare system. Hence, it is very pertinent to study the various divergent-resistant mechanisms of this nosocomial infectious agent. With its unique ability to survive using both phenotypic and genotypic characteristics, a combination of strategies had to be evaluated in managing this multidrug-resistant bacterial pathogen. In this chapter, the various mechanisms of its biofilm development, the virulence factors, and other resistant mechanisms existing in situ with potential solutions for combating such mechanisms of this hospital-acquired infectious agent have been dealt with in detail. It also covered the various intrinsic and extrinsic futuristic solutions that can be exploited for preventing and controlling this most dangerous bacterial pathogen.

Table 10.1 List of virulence factors and biofilm inhibitors against *A. baumannii*

S. no.	Inhibitors	Target	Reference
1.	Compound 62520	Outer membrane protein	Na et al. (2021)
2.	Isosakuranetin, aloe-emodin, and pinocembrin	OmpA and OmpW	Shahryari et al. (2021)
3.	Cyclic hexapeptide AOA-2	Outer membrane and biofilm	Vila-Farrés et al. (2017)
4.	D-mannose	Csu pilus	Chen et al. (2022)
5.	Eugenol and geraniol	Csu, biofilm, exopolysaccharide	Choudhary et al. (2022)
6.	LpxC inhibitor PF-5081090	Lipid A	García-Quintanilla et al. (2016)
7.	*N*-[(2S)-3-amino-1-(hydroxyamino)-1-oxopropan-2-yl]-4-(4-bromophenyl) benzamide	LpxC enzyme	Zoghlami et al. (2023)
8.	Al_2O_3 NPs	Exopolysaccharide	Muzammil et al. (2020)
9.	Anthranilic acid	Quinolone signal	Lesic et al. (2007)
10	Furanone	Receptor inhibitor	Lesic et al. (2007)
11.	Niclosamide	Inhibit motility, biofilm formation	Imperi et al. (2013)
12.	3-Hydroxybenzoic acid	Biofilm, virulence	Ganesh et al. (2022)
13.	Licoricone and glycerin	Virulence and biofilm	Bhargava et al. (2015)
14.	Cec4 peptide	Outer membrane, efflux pump genes	Liu et al. (2020)
15.	Siphonocholin (Syph-1)	Biofilm and pellicle formation	Alam et al. (2020)
16.	Palmitoleic acid (POA) and myristic acid (MOA)	Biofilm formation	Nicol et al. (2018)
17.	Zinc lactate, stannous fluoride	Biofilm formation	Peng et al. (2020)
18.	Octopromycin	Quorum sensing (QS) and biofilm	Rajapaksha et al. (2023)
19.	Phenothiazine	IV pilus-dependent surface motility and biofilm formation	Vo et al. (2023)
20.	Nothoscordum bivalve	Biofilm and virulence factors	Alejandro et al. (2018)

References

Abdi SN et al (2020) *Acinetobacter baumannii* efflux pumps and antibiotic resistance. Infect Drug Resist 13:423–434

Alam P, Alqahtani AS, Mabood Husain F, Tabish Rehman M, Alajmi MF, Noman OM (2020) Siphonocholin isolated from red sea sponge Siphonochalina siphonella attenuates quorum sensing controlled virulence and biofilm formation. Saudi Pharm J 28:1383–1391. https://doi.org/10.1016/j.jsps.2020.09.002

Alejandro H, Humberto M, Fidel G, Antonio G, Eduardo S (2018) Inhibition of *Acinetobacter baumannii* biofilm formation by methanolic extract of Nothoscordum bivalve. Adv Microbiol 8:422–438. https://doi.org/10.4236/aim.2018.85028

Aliramezani A, Soleimani M, Fard RMN, Nojoomi F (2019) Virulence determinants and biofilm formation of *Acinetobacter baumannii* isolated from hospitalized patients. Germs 9:148–153

Aly MM, Abu Alsoud NM, Elrobh MS, Al Johani SM, Balkhy HH (2016) High prevalence of the PER-1 gene among carbapenem-resistant *Acinetobacter baumannii* in Riyadh, Saudi Arabia. Eur J Clin Microbiol Infect Dis 35:1759–1766

Antunes EM et al (2011) Identification and in vitro anti-esophageal cancer activity of a series of halogenated monoterpenes isolated from the South African seaweeds Plocamium suhrii and Plocamium cornutum. Phytochemistry 72:769–772

Antunes LCS, Visca P, Towner KJ (2014) *Acinetobacter baumannii*: evolution of a global pathogen. Pathog Dis 71:292–301

Asfour H (2018) Anti-quorum sensing natural compounds. J Microsc Ultrastruct 6:1

Bassler BL, Losick R (2006) Bacterially speaking. Cell 125:237–246

Bendini A et al (2007) Phenolic molecules in virgin olive oils: a survey of their sensory properties, health effects, antioxidant activity and analytical methods. An overview of the last decade Alessandra. Molecules 12:1679–1719

Bergogne-Bérézin E, Towner KJ (1996) Acinetobacter spp. as nosocomial pathogens: microbiological, clinical, and epidemiological features. Clin Microbiol Rev 9:148–165

Bertani B, Ruiz N (2018) Function and biogenesis of lipopolysaccharides. EcoSal Plus 8:ecosalplus.ESP-0001-2018

Bhargava N, Singh SP, Sharma A, Sharma P, Capalash N (2015) Attenuation of quorum sensing-mediated virulence of *Acinetobacter baumannii* by *Glycyrrhiza glabra* flavonoids. Future Microbiol 10:1953–1968

Bouvet PJM, Grimont PAD (1987) Identification and biotyping of clinical isolates of Acinetobacter. Ann Inst Pasteur Microbiol 138:569–578

Bravo L (2009) Polyphenols: chemistry, dietary sources, metabolism, and nutritional significance. Nutr Rev 56:317–333

Cecchini T et al (2018) Deciphering multifactorial resistance phenotypes in *Acinetobacter baumannii* by genomics and targeted label-free proteomics. Mol Cell Proteomics 17:442–456

Cerqueira GM, Peleg AY (2011) Insights into *Acinetobacter baumannii* pathogenicity. IUBMB Life 63:1055–1060

Chen W (2020) Host innate immune responses to *Acinetobacter baumannii* infection. Front Cell Infect Microbiol 10:486

Chen C-L et al (2022) d-mannose-sensitive pilus of *Acinetobacter baumannii* is linked to biofilm formation and adherence onto respiratory tract epithelial cells. J Microbiol Immunol Infect 55:69–79

Choi AHK, Slamti L, Avci FY, Pier GB, Maira-Litrán T (2009) The *pgaABCD* locus of *Acinetobacter baumannii* encodes the production of poly-β-1-6-*N*-acetylglucosamine, which is critical for biofilm formation. J Bacteriol 191:5953–5963

Choudhary M, Shrivastava R, Vashistt J (2022) Eugenol and geraniol impede Csu-pilus assembly and evades multidrug-resistant *Acinetobacter baumannii* biofilms: in-vitro and in-silico evidence. Biochem Biophys Res Commun 636:10–17

Cisneros JM, Rodríguez-Baño J (2002) Nosocomial bacteremia due to *Acinetobacter baumannii*: epidemiology, clinical features and treatment. Clin Microbiol Infect 8:687–693
Dadgostar P (2019) Antimicrobial resistance: implications and costs. Infect Drug Resist 12:3903–3910
De Breij A et al (2009) CsuA/BABCDE-dependent pili are not involved in the adherence of *Acinetobacter baumannii* ATCC19606T to human airway epithelial cells and their inflammatory response. Res Microbiol 160:213–218
Doughari HJ, Ndakidemi PA, Human IS, Benade S (2011) The ecology, biology and pathogenesis of Acinetobacter spp.: an overview. Microbes Environ 26:101–112
Eijkelkamp BA, Stroeher UH, Hassan KA, Paulsen IT, Brown MH (2014) Comparative analysis of surface-exposed virulence factors of *Acinetobacter baumannii*. BMC Genomics 15:1020
Engebrecht J, Silverman M (1984) Identification of genes and gene products necessary for bacterial bioluminescence. Proc Natl Acad Sci 81:4154–4158
Engebrecht J, Nealson K, Silverman M (1983) Bacterial bioluminescence: isolation and genetic analysis of functions from Vibrio fischeri. Cell 32:773–781
Falagas ME, Rafailidis PI (2007) Attributable mortality of *Acinetobacter baumannii*: no longer a controversial issue. Crit Care 11:134
Fattahian Y et al (2011) Protection against *Acinetobacter baumannii* infection via its functional deprivation of biofilm associated protein (Bap). Microb Pathog 51:402–406
Gaddy JA, Actis LA (2009) Regulation of *Acinetobacter baumannii* biofilm formation. Future Microbiol 4:273–278
Ganesh PS et al (2022) Biofilm-associated Agr and Sar quorum sensing systems of *Staphylococcus aureus* are inhibited by 3-hydroxybenzoic acid derived from *Illicium verum*. ACS Omega 7:14653–14665
García-Quintanilla M, Pulido MR, López-Rojas R, Pachón J, McConnell MJ (2013) Emerging therapies for multidrug resistant *Acinetobacter baumannii*. Trends Microbiol 21:157–163
García-Quintanilla M et al (2016) Inhibition of LpxC increases antibiotic susceptibility in *Acinetobacter baumannii*. Antimicrob Agents Chemother 60:5076–5079
Gedefie A et al (2021) *Acinetobacter baumannii* biofilm formation and its role in disease pathogenesis: a review. Infect Drug Resist 14:3711–3719
Geisinger E, Isberg RR (2015) Antibiotic modulation of capsular exopolysaccharide and virulence in *Acinetobacter baumannii*. PLoS Pathog 11:e1004691
Ghosh A, Jayaraman N, Chatterji D (2020) Small-molecule inhibition of bacterial biofilm. ACS Omega 5:3108–3115
Goel N, Fatima SW, Kumar S, Sinha R, Khare SK (2021) Antimicrobial resistance in biofilms: exploring marine actinobacteria as a potential source of antibiotics and biofilm inhibitors. Biotechnol Rep 30:e00613
Gordon NC, Wareham DW (2010) Multidrug-resistant *Acinetobacter baumannii*: mechanisms of virulence and resistance. Int J Antimicrob Agents 35:219–226
Harding CM, Hennon SW, Feldman MF (2018) Uncovering the mechanisms of *Acinetobacter baumannii* virulence. Nat Rev Microbiol 16:91–102
Howard A, O'Donoghue M, Feeney A, Sleator RD (2012) *Acinetobacter baumannii*: an emerging opportunistic pathogen. Virulence 3:243–250
Imperi F et al (2013) New life for an old drug: the anthelmintic drug niclosamide inhibits Pseudomonas aeruginosa quorum sensing. Antimicrob Agents Chemother 57:996–1005
Jung J, Park W (2015) Acinetobacter species as model microorganisms in environmental microbiology: current state and perspectives. Appl Microbiol Biotechnol 99:2533–2548
Kalinowska M et al (2021) Plant-derived and dietary hydroxybenzoic acids—a comprehensive study of structural, anti-/pro-oxidant, lipophilic, antimicrobial, and cytotoxic activity in MDA-MB-231 and MCF-7 cell lines. Nutrients 13:3107
Kishii K et al (2020) Differences in biofilm formation and transcription of biofilm-associated genes among *Acinetobacter baumannii* clinical strains belonging to the international clone II lineage. J Infect Chemother 26:693–698

Kon H, Schwartz D, Temkin E, Carmeli Y, Lellouche J (2020) Rapid identification of capsulated *Acinetobacter baumannii* using a density-dependent gradient test. BMC Microbiol 20:285
Lee C-R et al (2017) Biology of *Acinetobacter baumannii*: pathogenesis, antibiotic resistance mechanisms, and prospective treatment options. Front Cell Infect Microbiol 7
Lesic B et al (2007) Inhibitors of pathogen intercellular signals as selective anti-infective compounds. PLoS Pathog 3:e126
Leus IV et al (2018) Substrate specificities and efflux efficiencies of RND efflux pumps of *Acinetobacter baumannii*. J Bacteriol 200:e00049
Li Y et al (2021) Outer membrane protein A inhibits the degradation of caspase-1 to regulate NLRP3 inflammasome activation and exacerbate the *Acinetobacter baumannii* pulmonary inflammation. Microb Pathog 153:104788
Limoli DH, Jones CJ, Wozniak DJ (2015) Bacterial extracellular polysaccharides in biofilm formation and function. Microbiol Spectr 3:3.3.29
Liu RH (2004) Potential synergy of phytochemicals in cancer prevention: mechanism of action. J Nutr 134:3479S–3485S
Liu W, Wu Z, Mao C, Guo G, Zeng Z, Fei Y, Wan S, Peng J, Wu J (2020) Antimicrobial peptide Cec4 eradicates the bacteria of clinical carbapenem-resistant *Acinetobacter baumannii* biofilm. Front Microbiol 11:1532. https://doi.org/10.3389/fmicb.2020.01532
Longo F, Vuotto C, Donelli G (2014) Biofilm formation in *Acinetobacter baumannii*. New Microbiol 37:119–127
Luke NR et al (2010) Identification and characterization of a glycosyltransferase involved in *Acinetobacter baumannii* lipopolysaccharide core biosynthesis. Infect Immun 78:2017–2023
Lupp C, Urbanowski M, Greenberg EP, Ruby EG (2003) The Vibrio fischeri quorum-sensing systems ain and lux sequentially induce luminescence gene expression and are important for persistence in the squid host: Vibrio fischeri ainS. Mol Microbiol 50:319–331
Maldonado RF, Sá-Correia I, Valvano MA (2016) Lipopolysaccharide modification in Gram-negative bacteria during chronic infection. FEMS Microbiol Rev 40:480–493
Marchand I, Damier-Piolle L, Courvalin P, Lambert T (2004) Expression of the RND-type efflux pump AdeABC in *Acinetobacter baumannii* is regulated by the AdeRS two-component system. Antimicrob Agents Chemother 48:3298–3304
Martin KR (2009) Polyphenols as dietary supplements: a double-edged sword. Nutr Diet Suppl 2010:1–12. https://doi.org/10.2147/NDS.S6422
McIlwain DR, Berger T, Mak TW (2013) Caspase functions in cell death and disease. Cold Spring Harb Perspect Biol 5:a008656–a008656
Monem S et al (2020) Mechanisms protecting *Acinetobacter baumannii* against multiple stresses triggered by the host immune response, antibiotics and outside-host environment. Int J Mol Sci 21:5498
Muzammil S et al (2020) Aluminium oxide nanoparticles inhibit EPS production, adhesion and biofilm formation by multidrug resistant *Acinetobacter baumannii*. Biofouling 36:492–504
Na S-H et al (2021) Therapeutic effects of inhibitor of ompA expression against carbapenem-resistant *Acinetobacter baumannii* strains. Int J Mol Sci 22:12257
Nicol M, Alexandre S, Luizet JB, Skogman M, Jouenne T, Salcedo SP et al (2018) Unsaturated fatty acids affect quorum sensing communication system and inhibit motility and biofilm formation of *Acinetobacter baumannii*. Int J Mol Sci 19:214. https://doi.org/10.3390/ijms19010214
Nie D et al (2020) Outer membrane protein A (OmpA) as a potential therapeutic target for *Acinetobacter baumannii* infection. J Biomed Sci 27:26
Nikaido H (2003) Molecular basis of bacterial outer membrane permeability revisited. Microbiol Mol Biol Rev 67:593–656
Niu C, Clemmer KM, Bonomo RA, Rather PN (2008) Isolation and characterization of an autoinducer synthase from *Acinetobacter baumannii*. J Bacteriol 190:3386–3392
Pakharukova N et al (2018) Structural basis for *Acinetobacter baumannii* biofilm formation. Proc Natl Acad Sci 115:5558–5563
Peleg AY, Seifert H, Paterson DL (2008) *Acinetobacter baumannii*: emergence of a successful pathogen. Clin Microbiol Rev 21:538–582

Peng Q, Lin F, Ling B (2020) In vitro activity of biofilm inhibitors in combination with antibacterial drugs against extensively drug-resistant *Acinetobacter baumannii*. Sci Rep 10:18097. https://doi.org/10.1038/s41598-020-75218-y

Perez F et al (2007) Global challenge of multidrug-resistant *Acinetobacter baumannii*. Antimicrob Agents Chemother 51:3471–3484

Raetz CRH, Whitfield C (2002) Lipopolysaccharide endotoxins. Annu Rev Biochem 71:635–700

Rajapaksha DC, Edirisinghe SL, Nikapitiya C, Whang I, De Zoysa M (2023) The antimicrobial peptide octopromycin suppresses biofilm formation and quorum sensing in *Acinetobacter baumannii*. Antibiotics (Basel) 12(3):623. https://doi.org/10.3390/antibiotics12030623

Rao R et al (2008) Correlation between biofilm production and multiple drug resistance in Imipenem Resistant clinical isolates of *Acinetobacter baumannii*. Indian J Med Microbiol 26:333–337

Romano KP, Hung DT (2023) Targeting LPS biosynthesis and transport in gram-negative bacteria in the era of multi-drug resistance. Biochim Biophys Acta Mol Cell Res 1870:119407

Ronish LA, Lillehoj E, Fields JK, Sundberg EJ, Piepenbrink KH (2019) The structure of PilA from *Acinetobacter baumannii* AB5075 suggests a mechanism for functional specialization in Acinetobacter type IV pili. J Biol Chem 294:218–230

Roy R, Tiwari M, Donelli G, Tiwari V (2018) Strategies for combating bacterial biofilms: a focus on anti-biofilm agents and their mechanisms of action. Virulence 9:522–554

Rutherford ST, Bassler BL (2012) Bacterial quorum sensing: its role in virulence and possibilities for its control. Cold Spring Harb Perspect Med 2:a012427–a012427

Ryu SY, Baek W-K, Kim HA (2017) Association of biofilm production with colonization among clinical isolates of *Acinetobacter baumannii*. Korean J Intern Med 32:345–351

Sauvage E, Kerff F, Terrak M, Ayala JA, Charlier P (2008) The penicillin-binding proteins: structure and role in peptidoglycan biosynthesis. FEMS Microbiol Rev 32:234–258

Shadan A, Pathak A, Ma Y, Pathania R, Singh RP (2023) Deciphering the virulence factors, regulation, and immune response to *Acinetobacter baumannii* infection. Front Cell Infect Microbiol 13:1053968

Shahryari S, Mohammadnejad P, Noghabi KA (2021) Screening of anti-*Acinetobacter baumannii* phytochemicals, based on the potential inhibitory effect on OmpA and OmpW functions. R Soc Open Sci 8:201652

Sibanda S, Moleleki LN, Shyntum DY, Coutinho TA (2018) Quorum sensing in gram-negative plant pathogenic bacteria. In: Kimatu JN (ed) Advances in plant pathology. InTech. https://doi.org/10.5772/intechopen.78003

Sikdar R, Elias M (2020) Quorum quenching enzymes and their effects on virulence, biofilm, and microbiomes: a review of recent advances. Expert Rev Anti-Infect Ther 18:1221–1233

Singh JK, Adams FG, Brown MH (2019) Diversity and function of capsular polysaccharide in *Acinetobacter baumannii*. Front Microbiol 9:3301

Subhadra B, Kim D, Woo K, Surendran S, Choi C (2018) Control of biofilm formation in healthcare: recent advances exploiting quorum-sensing interference strategies and multidrug efflux pump inhibitors. Materials 11:1676

Swartzman A, Kapoor S, Graham AF, Meighen EA (1990) A new Vibrio fischeri lux gene precedes a bidirectional termination site for the lux operon. J Bacteriol 172:6797–6802

Tickner J, Hawas S, Totsika M, Kenyon JJ (2021) The Wzi outer membrane protein mediates assembly of a tight capsular polysaccharide layer on the *Acinetobacter baumannii* cell surface. Sci Rep 11:21741

Tiku V (2022) *Acinetobacter baumannii*: virulence strategies and host defense mechanisms. DNA Cell Biol 41:43–48

Towner KJ (2009) Acinetobacter: an old friend, but a new enemy. J Hosp Infect 73:355–363

Upmanyu K, Haq QMR, Singh R (2022) Factors mediating *Acinetobacter baumannii* biofilm formation: opportunities for developing therapeutics. Curr Res Microb Sci 3:100131

Vaou N, Stavropoulou E, Voidarou C, Tsigalou C, Bezirtzoglou E (2021) Towards advances in medicinal plant antimicrobial activity: a review study on challenges and future perspectives. Microorganisms 9:2041

Vila-Farrés X et al (2017) Combating virulence of Gram-negative bacilli by OmpA inhibition. Sci Rep 7:14683

Vo N, Sidner BS, Yu Y, Piepenbrink KH (2023) Type IV pilus-mediated inhibition of *Acinetobacter baumannii* biofilm formation by phenothiazine compounds. Microbiol Spectr 11(4):e0102323. https://doi.org/10.1128/spectrum.01023-23

Weinstein RA, Gaynes R, Edwards JR, National Nosocomial Infections Surveillance System (2005) Overview of nosocomial infections caused by gram-negative bacilli. Clin Infect Dis 41:848–854

Whitehead NA, Barnard AML, Slater H, Simpson NJL, Salmond GPC (2001) Quorum-sensing in Gram-negative bacteria. FEMS Microbiol Rev 25:365–404

WHO (2016) Central Asian and European surveillance of antimicrobial resistance

WHO (2022) Global Centre for Traditional Medicine

Yang C-H, Su P-W, Moi S-H, Chuang L-Y (2019) Biofilm formation in *Acinetobacter baumannii*: genotype-phenotype correlation. Molecules 24:1849

Yin W, Wang Y, Liu L, He J (2019) Biofilms: the microbial "protective clothing" in extreme environments. Int J Mol Sci 20:3423

Zapun A, Contreras-Martel C, Vernet T (2008) Penicillin-binding proteins and β-lactam resistance. FEMS Microbiol Rev 32:361–385

Zhong S, He S (2021) Quorum sensing inhibition or quenching in *Acinetobacter baumannii*: the novel therapeutic strategies for new drug development. Front Microbiol 12:558003

Zoghlami M, Oueslati M, Basharat Z, Sadfi-Zouaoui N, Messaoudi A (2023) Inhibitor assessment against the LpxC enzyme of antibiotic-resistant *Acinetobacter baumannii* using virtual screening, dynamics simulation, and *in vitro* assays. Mol Inform 42:2200061

Pseudomonas aeruginosa Virulence Factors and Biofilm Components: Synthesis, Structure, Function and Inhibitors

11

Mahima S. Mohan, Simi Asma Salim, Nishel Forgia, and Siddhardha Busi

Abstract

Pseudomonas aeruginosa is a versatile opportunistic bacterium reasons for hospital-acquired infections specifically to the patients suffering from cystic fibrosis, burns, wounds and also immunocompromised individuals. Diverse resistance mechanisms possessed by the bacterium make them to overcome the effect of various antimicrobial agents. *P. aeruginosa* possess a plenty of virulence determinants, specifically biofilm, and triggers resistance, which is controlled by cellular communication pathway called quorum sensing (QS). A deep understanding about the pathogenic determinants such as virulence factor and biofilm, and also the interaction with the host, makes the development of therapeutic approaches easier. This chapter gives an insight into the synthesis, structure, function and inhibitors of virulence phenotypes of *Pseudomonas aeruginosa*.

Keywords

Pseudomonas aeruginosa · Virulence factors · Biofilm · Quorum sensing · Inhibitors

11.1 *Pseudomonas aeruginosa:* An Overview

P. aeruginosa is a predominant potential pathogenic microorganism (PPM) (Garcia-Clemente et al. 2020) containing relatively larger genome ~5.5–7 (Pang et al. 2019) with a remarkable plasticity (Langendonk et al. 2021). In ICU patients, *P. aeruginosa* lead to 7.1–7.3% healthcare-associated infections (HAI). The risk of *P.*

M. S. Mohan · S. A. Salim · N. Forgia · S. Busi (✉)
Department of Microbiology, School of Life Sciences, Pondicherry University, Puducherry, India

S. Busi, R. Prasad (eds.), *ESKAPE Pathogens*,
https://doi.org/10.1007/978-981-99-8799-3_11

aeruginosa infection is high in transplanted inpatients, and it is a critical pathogen in immunocompromised patients, particularly in conditions like neutropenia. Catheter-associated UTI is commonly caused by *P. aeruginosa* biofilms, which show elevated morbidity, mortality, and bacteraemia. In patients with burnt wound infection, MDR *P. aeruginosa* cause sepsis and death in an elevated manner (Reynolds and Kollef 2021). *P. aeruginosa* is a dominant pathogenic microorganism associated with chronic pulmonary diseases like asthma, chronic obstructive pulmonary disease (COPD), and bronchiectasis and cystic fibrosis (CF). In CF patients, organism undergoes mutations and genetic changes to adapt within the favourable ecological niche in the host. The chronic bronchial infection (CBI) caused by *P. aeruginosa* leads to decline in lung function. It is characterized by lower FEV1 levels, a predictive marker of airflow occlusion, compared to the uninfected patients. *P. aeruginosa* show substantial resistance towards innate immune effector cells and antimicrobial compounds. The switch between non-mucoid to mucoid form of biofilm also contributes to the advancement in antimicrobial resistance and low life expectancy. The frequency of *P. aeruginosa* infection is reported high in bronchiectasis and capable of persuading CBI in COPD and people with asthma. The co-occurrence of bronchiectasis and asthma is related with the incidence of chronic bronchial expectoration and purulent phlegm (Garcia-Clemente et al. 2020). Bronchiectasis is characterized by chronic infection and airway inflammation, where non-CF bronchiectasis is more frequent than CF bronchiectasis (Reynolds and Kollef 2021).

P. aeruginosa, a critical pathogen in WHO's priority list, contains a large pangenome comprising core genome and accessory genomes, which allow them to easily adapt to the environment and develop antibiotic resistance (AR) (Botelho et al. 2019). It displays both tolerance and resistance against antibiotics where tolerance relies on physical and physiological features, while resistance rely on mutation and external stimulus (Ciofu and Tolker-Nielsen 2019). Antibiotic-resistant mechanism is an energy consuming process (Hwang and Yoon 2019), where intrinsic, acquired and adaptive resistance are the mechanism shown by the bacterium to counter the bactericidal activity of antibiotics (Pang et al. 2019).

The genes present in core genome is responsible for the presence of inherent features (Langendonk et al. 2021) like low membrane penetrability, presence of efflux pumps and antibiotic inactivation enzymes and is considered as its intrinsic ability to resist antibiotics (Pang et al. 2019). Intrinsic resistance is controlled positively or negatively by one or more than one regulatory mechanism (Moradali et al. 2017). The outer membrane containing phospholipids and LPS embedded with porins acts as a selective barricade to inhibit the entry of antimicrobial drugs (Pang et al. 2019). The principal porin OprF is the foremost nonspecific porin, mostly present in closed state (Langendonk et al. 2021), which leads to higher expression of c-di-GMP and thereby induce the establishment of biofilm (Pang et al. 2019). The downregulation of second major porin OprD leads to carbapenem resistance, whereas the smallest porin OprH is a 21.6 kDa protein, interacts with the LPS and leads to polymyxin B and gentamicin resistance (Langendonk et al. 2021; Pang et al. 2019). Efflux pumps in *P. aeruginosa* are a part of the RND family. Among the

12-efflux pump present in *P. aeruginosa*, four (MexAB-OprM, MexCD-OprJ, MexEF-OprN and MexXY-OprM) were take part in resistance mechanism (Pang et al. 2019). MexAB-OprM is responsible for resistance towards aminoglycosides, triclosan and sodium dodecyl sulphate, thus contributing to the progression of MDR in *P. aeruginosa*. MexCD-OprJ present only in nfxB mutants is negatively regulated NfxB protein that leads to antimicrobial resistance. The mexEF-oprN operon was positively controlled by MexT and influence the resistance against carbapenems (Langendonk et al. 2021). Fosfomycin resistance is caused by the presence of FosA homologues (Botelho et al. 2019). The β-lactamase encoded by inducible gene *AmpC* belongs to ambler class C antibiotic inactivation enzyme, hydrolyses β-lactam ring structures and leads to resistance. Metallo β-lactamases (MBLs) is an ambler group B, and most common extended-spectrum β-lactamases (ESBL) in *P. aeruginosa* are the ambler class A enzyme and also contribute to the resistance (Langendonk et al. 2021). Thus, the decreased expression of OprD and the upregulation of *AmpC* and MexAB-OprM chiefly assist in MDR development (Moradali et al. 2017).

Horizontal gene transfer (HGT) and changes by mutation can lead to acquired resistance. It is irreversible in nature and transferred to progeny in a stable manner. *P. aeruginosa* also acquire resistance by spontaneous mutations and mutational modification of target site (Pang et al. 2019). Plasmid acts as a potential vehicle for HGT (Moradali et al. 2017). The accessory genome pool comprises mobile genetic elements acquired through HGT and harbours a variety of genes encoding antibiotic resistance. Beside plasmid, integrons and transposons also contribute to AR. Targeted mutation in *AmpC* led to the advancement of resistance towards cephalosporin (Botelho et al. 2019). Specific alterations in genes *gyrA-gyrB* and *parC-parE* elevate the production of DNA gyrase and topoisomerase which can chemically alter the antibiotic molecules and lead to AR (Azam and Khan 2019). Carbapenemase-encoding genes (CEGs) and extended-spectrum β-lactamases (ESBLs) are extensively present in *P. aeruginosa* which are allied with class I integrons and also contribute to AR (Botelho et al. 2019). Enzymes encoded by genes present in both plasmid and chromosome can lead to the development of acquired and intrinsic resistance. It is very difficult to treat the infections, in case of presence of intrinsic or acquired resistance (Azam and Khan 2019).

During infection, the frequent exposure to the immune response and antibiotics may bring the expression of diverse set of genes in *P. aeruginosa* which enable them to acclimatize to the settings (Moradali et al. 2017). This type of resistance is reversible in nature and independent of genetic mutations (Pang et al. 2019) which can be removed in the absence of external stimuli (Langendonk et al. 2021). Genomic islands (GI) containing virulence factor-encoding genes existing in the accessory genome of *P. aeruginosa* are known as pathogenicity islands (PAPI), which help in host adaptation and diversification of the pathogen (Botelho et al. 2019). Adaptation to the outside environment is an outstanding character shown by *P. aeruginosa* (Moradali et al. 2017). The development of biofilm and persister cell generation is the major adaptive resistance mechanism shown by them (Pang et al. 2019). The presence of efflux pumps contributes to both adaptive and acquired resistance (Azam and Khan 2019). The alteration of lipid A in LPS is caused by supplementing

4-amino-4-deoxy-L-arabinose (L-Ara4N) group to the phosphate component, which lowers net negative charge of the outer membrane and decreases its interaction with positively charged antimicrobial leading to the resistance towards cationic antimicrobial peptides (CAPs) and polymyxin. Swarming and surfing mode of motility show an important role in adaptive resistance by inhibiting aminoglycosides, tetracyclines and certain β-lactams (Langendonk et al. 2021). CRISPR-Cas system is the only adaptive immune responses that act against certain phage infections in *P. aeruginosa.*

P. aeruginosa can produce both mucoid and non-mucoid biofilm (Malhotra et al. 2018). The switching of motile to sessile lifestyle to form mucoid biofilm is a hallmark mechanism to adapt the AR. Mutations in *AmpR* gene trigger biofilm establishment and alginate synthesis (Moradali et al. 2017). The biofilm structure and resistance rely on the environmental factors (Langendonk et al. 2021). The EPS produced by the preeminent expression of c-di-GMP acts as a diffusion obstacle to antibiotics into the biofilm. Alginate production decreases the antibiotic susceptibility and human defence mechanism against biofilm (Azam and Khan 2019). The greater negative charge of alginate enables the mucoid variants to protect from the action of LL-37, while non-mucoid overproduces catalase enzyme to neutralize H_2O_2 (Malhotra et al. 2018). This chapter deals with various virulence mechanisms adopted by *P. aeruginosa* to cause diseases in the host. Further, this chapter summarizes about the synthesis, structure, functions and inhibitors of several virulence phenotypes and biofilm produced by *P. aeruginosa.*

11.2 Virulence Factors of *P. aeruginosa*

P. aeruginosa possess an arsenal of virulence factors or are produced at the time of infection contributing to their attachment and colonization and dampening of host defence mechanisms and immune escape, thereby playing central roles in their pathogenesis (Liao et al. 2022). Here, we classified the virulence factors broadly into three—the surface virulence components, the secreted virulence elements and bacterial cell-to-cell communication. Type IV pili, flagella and other components of outer membrane like lipopolysaccharides and importantly various secretion apparatus are included in the category—surface virulence components. The category secreted factors/elements comprising exopolysaccharides, siderophores, proteases, toxins, etc. Also, bacteria-bacteria signalling network, that is, quorum sensing (QS) signal cascade and biofilm, enacts vital function in the virulence of *P. aeruginosa.* Below sections described these virulence factors in detail (Fig. 11.1).

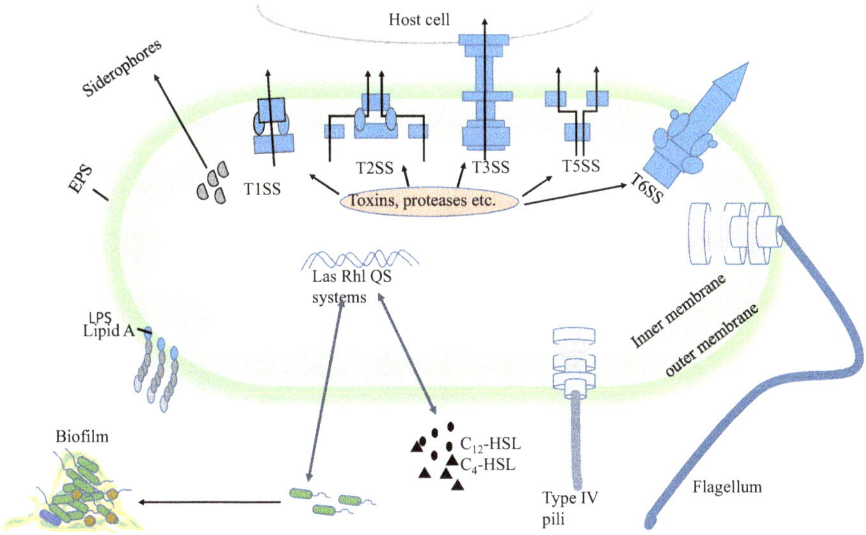

Fig. 11.1 Factors contributing to *Pseudomonas aeruginosa* virulence. Broadly, virulence elements are classified into three: bacterial surface virulence components, secreted virulence elements and bacterial cell-to-cell interaction

11.3 Surface Virulence Components

11.3.1 Type IV Pili (T4P)

P. aeruginosa possesses thin <8-nm-wide fibre-like motorized appendages called type IV pili (T4P). Three subclasses of T4P identified in *P. aeruginosa* are type IV a (T4aP), type IV b (T4bP) and tight adherence pili (Tad). T4aP is comparatively homogenous and broadly distributed group, whereas T4bP is heterogeneous. Tad pili is a monophyletic class of T4bP. Pilin subunits of T4aP and T4bP are ~15–20 kDa, while the Tad pilin subunits (7–8 kDa) are relatively smaller. Regardless of the size difference, all of the pilin subtypes possess a highly preserved, hydrophobic and α-helical N-terminus in which specific type III signal sequence is located. Pilin system consists of major pilus unit along with membrane subcomplex, motor subcomplex and alignment subcomplex. PilA, PilS2 and Flp comprise main pilin subunits of T4aP, T4bP and Tad pili, respectively. PilQ, PilN2 and RcpA are their respective secretins, which forms the channel. Their corresponding lipoprotein, pilotins, which helps in accurate localization and formation of multimers is PilF for T4aP and TadD for Tad pili. Pilotins and secretins are located in the outer membrane. Motor subcomplex is not well characterized, but T4aP motor complex is comprised of PilC, PilB, PilT, PilU and other regulatory proteins based on available studies. The alignment subcomplex is anticipated as the connection between the motor spanning in inner membrane and subcomplexes of outer membrane to guarantee that the flexible budding pilus is placed properly for exit as it commences to outspread through the periplasm from its assembly site. T4aP alignment subcomplex of *P. aeruginosa* include the broadly conserved PilM, PilN, PilO and PilP proteins and the peptidoglycan-binding FimV protein. Twitching is the well-defined type of motility associated with T4aP, whereas T4b and Tad pili are not usually motility allied. Additional T4aP-dependent functions have been ascribed for *P. aeruginosa*, such as swarming, walking and slingshot motilities. Apart from motility, T4P are significant adhesins, facilitating initial adhesion to a range of chemically distinct surfaces. T4aP have role in biofilm architecture and differentiation also. For the reason that T4P aid in establishing initial contact of bacteria with the host, they are considered as significant virulence factors of *P. aeruginosa.* Other possible function of T4P is that it can act as an antenna and respond to in their microenvironment owing to its length of several microns (Burrows 2012).

11.3.2 Flagella

P. aeruginosa is monotrichous, i.e., a single polar flagellum and made of polymerized flagellin subunits (Dasgupta et al. 2003). Apart from its central role in motility, flagella have a great impact on colonization, adhesion and virulence of this pathogen. In *P. aeruginosa*, flagellar development is regulated by a four-tiered hierarchical system. Generally, flagella of Gram-negative bacteria are comprised of three parts: membrane complex, hook and filament. Membrane complex serves as an anchor, hook is a hollow tube connecting the filament and membrane complex, and filament acts as the propeller (Dasgupta et al. 2004; Bouteiller et al. 2021). Membrane complex comprises of the basal body, flagellar type 3 secretion system, stator complex and a connecting rod. C ring present at cytoplasm, MS ring present at inner membrane, P ring present at peptidoglycan layer and L ring located at outer membrane make the basal body. In *Pseudomonas*, the MS ring made by FliF protein subunits, P ring made up of FlgI protein and L ring made of FlgH subunits together these three rings anchor the macromolecular structure in the outer membrane. The cytoplasmic C ring considered as rotor is composed by three different protein subunits—FliG, FliM and FliN, and FliN is the major component and serves as the substrate/chaperon complex-binding site beforehand secretion. FliM attached with chemosensory system and regulates the direction of rotation. The rotor, C ring, is attached to a stator complex comprises of the MotA and MotB proteins and is critical for generating the ion-motive force. FliG subunits of C ring interact with stator complex. Flagellar type 3 secretion system, which has structural similarities with virulence-related type 3 secretion system, has a steady inner-membrane unit consisting of central channel of 1.5 nm made up of FliO, FliP, FliQ, FliR, FlhA, and FlhB proteins t and a dynamic cytoplasmic ATPase unit made of FliI, FliH and FliJ subunits. An 8- to 14-nm-diameter rod structure made up of FlgG, FlgB, FlgC and FlgF proteins connects the inner- and outer-membrane rings and hook of flagella. Hook is made up of 11 FlgE protofilaments. Filament is approximately 10 μm long, made up of ~20,000 subunits of flagellin—FliC monomeric units, and the filament is coupled to hook and basal body through junction proteins—FlgK and FlgL. FliD protein makes up the cap of flagellin. Two discrete forms of FliC flagellin in *P. aeruginosa* are a-type and b-type flagellins, and they differ in their size and antigenic attributes; former one is short and heterogeneous, on the other hand, latter one is homogenous. The carboxy- and amino-terminal areas are similar in both types. PAK strain of *P. aeruginosa* possesses a-type flagellin, whereas the PAO1 strain has b-type flagellin, generating diverse antigenic reactions. Post-translational modification of flagellin such as glycosylation alters its hydrophobic properties, which influences attachment of flagella with different surfaces and thereby plays critical function in *P. aeruginosa* virulence (Bouteiller et al. 2021). Flagellar filament rotation is reliant on chemotaxis and energy production. Chemotaxis is a mechanism common to bacteria that makes them to move towards or away from extracellular molecules by sensing it. The PAO1 strain of *P. aeruginosa* possesses two distinct chemosensory signal transduction complexes, which are confined to the poles of cell (Bouteiller et al. 2021).

11.3.3 Lipopolysaccharide (LPS)

Lipopolysaccharide being the element of outer membrane is found in all strains of *P. aeruginosa.* LPS consists of lipid A moiety, outer- and inner-core oligosaccharides and *O*-antigens. Structure of LPS in the pathogen of concern is as that of the classic Gram-negative bacterial LPS, where lipid A moiety containing a backbone of *N*-acylated and *O*-acylated diglucosamine bisphosphate with differences in 1° acyl group number and the kinds of fatty acids replacing the 1°and 2° acyl groups. Majority of laboratory-adapted *P. aeruginosa* strains synthesize a penta-acylated LPS, with small fraction (25%) of hexa-acylated LPS. Inner core comprises of two D-manno-oct-2-ulosonic acid units and two L-glycero-D-manno-heptose units. The outer oligosaccharide core of the *P. aeruginosa* LPS is generally produced as two glycoform types by a single strain. Both of them consist of a galactosamine residue with *N*-alanylation, three D-glucose units and a L-rhamnose unit, and variation occurs in the site of L-rhamnose residue. *O*-antigen, which determines the serogroup, is linked to the glycoform 2. Most of the LPS-smooth strains have glycoform 2 entirely substituted with *O*-antigen, whereas the glycoform 2 of LPS-rough strains is not substituted with any *O*-antigen. *N*-acyl derivatives of various amino sugars together with rhamnose constitute the sugars within the *P. aeruginosa* *O*-antigen (Pier 2007). LPS aids as a major feature in *P. aeruginosa* virulence mechanisms and related host reactions, but the LPS influences the pathogenesis, and immunity differs based on the susceptibility to infection by individuals, the LPS isoform, princi pally the lipid A element and structural difference in the *O*-antigen side chain. Because LPS functions like microbe-associated molecular pattern (MAMP), it can activate host defence mechanisms through several signal transduction pathways including cystic fibrosis transmembrane conductance regulator (CFTR) and the toll-like receptor 4 (TLR4) (Liao et al. 2022).

11.3.4 Outer-Membrane Vesicles (OMVs)

P. aeruginosa and some similar Gram-negative bacteria continuously secrete a 20- to 200-nm-diameter bilayered proteolipid vesicles termed outer-membrane vesicles (OMVs). Proteins and lipids of outer membrane, periplasmic proteins, LPS, DNA, RNA and other virulence-related biomolecules make up these vesicles. *P. aeruginosa*-derived OMVs encompass numerous secreted virulence factors like phospholipase C, alkaline phosphatase, proelastase, hemolysin and antibacterial factors including murein hydrolases for destruction of host cell and other competitors. Additionally, packaging of QS molecules in *P. aeruginosa* OMVs was also found correlated to the transfer of antibiotic resistance and trafficking signal (Lee et al. 2008).

11.4 Secretion Systems

Protein secretory systems can be described as "molecular nanomachines" adapted by Gram-negative bacteria to survive in their atmosphere. Six different classes of secretory systems, type I to type IV, have been recognized in *P. aeruginosa* (Filloux 2011) and are explained in detail below.

11.4.1 Type I Secretion System (T1SS)

T1SS, the simplest bacterial secretory system, in general involves an outer-membrane protein and ABC transporter spanning in the inner membrane. An adapter protein connects these two components. The substrates for T1SS machinery are targeted with the aid of a non-cleavable C-terminal secretion signal. The unfolded substrate proteins interact briefly with both ABC transporter and adapter protein. Secretion signal interaction with the ABC transporter nucleotide binding domain will lead to alteration of protein's confirmation and ATP binding, leading to ATP hydrolysis. This energy-driven process causes movement of the substrate protein through the machinery further (Bleves et al. 2010).

Two T1SS has been explored intensively in *P. aeruginosa*. The first one, Apr system, comprises AprD, the ABC transporter, AprE, the adapter and AprF, the outer-membrane factor implicated in secretion of alkaline protease AprA extracellularly, which is a foremost virulence element of *P. aeruginosa* and AprX. The second T1SS is *has* gene involved iron utilization system. Here, HasD is the ABC transporter, HasE is the adapter, and HasF is the outer-membrane factor. The secreted protein HasAp plays relevant function in iron accession of *P. aeruginosa* in host environment and is essential all through early stages of infection. HasAp is a haemophore binds to haem from haemoglobin, and then, the HasAp-haem complex binds to HasR which is the outer-membrane receptor (Bleves et al. 2010).

11.4.2 Type II Secretion System (T2SS)

This complex machinery has the unique property of translocating large multimeric proteins folded in periplasm to the external environment. Generally, the T2SS consists of 11–16 widely conserved proteins of Gram-negative bacteria termed GspC-M, GspAB, GspN, GspO and GspS. T2SS structure can be categorized into three parts—the inner-membrane spanning part consisting of the components GspC, E, F, L and M. Gsp E acts as the motor and has ATPase activity. The channel portion spanning in the outer membrane made up of the secretin, GspD. The remaining part—fibrillar structure—called pseudopilus is made up of the major pseudopilin, GspG and capped with minor pseudopilins, GspI, GspJ and GspK.

Working Mechanism: Pseudopilus acts as a piston, assembly of which is triggered by GspE motor, and pushes the secreted protein through the secretin channel to extracellular space.

The genes arranged in two operons *xcpP-Q* and *xcpR-Z* encode for the *P. aeruginosa* T2SS, Xcp. Another gene-*xcpA/pilD* which does not belong to these operons is also involved in T2SS machinery. The ATPase motor protein is the XcpR, and it interacts with other proteins such as XcpY, XcpZ and XcpS to form an inner-membrane complex. Another inner-membrane protein XcpP can transiently with outer-membrane secretin XcpQ. The channel is made of multimers of 12 subunits of XcpQ and is arranged as a ring structure forming an inner large hollow tube of 90 Å. The major pseudopilins, XcpT, and minor ones, XcpU-X, form the pseudopilus and its cap, respectively.

The exoproteins secreted by *P. aeruginosa* Xcp system have crucial role in infections. The exoproteins include the proteolytic enzymes such as LasB elastase (degrades elastin, major lung tissue component), LasA staphylolysin, the aminopeptidase PaAP and protease IV. The other exoprotein family secreted by T2SS is the lipid metabolism enzyme catalysts LipA and LipC lipases and phospholipase C—PlcH, PlcN and PlcB. PhoA is the alkaline phosphatase produced by Xcp. The exotoxin A, also called ToxA, is an AB toxin secreted through T2SS of *P. aeruginosa*.

Another T2SS found in *P. aeruginosa* which only works under phosphate-limiting conditions is the Hxc system. Structurally, the Hxc system is similar to Xcp even though gene organization is different. The only one secretory protein of Hxc system is LapA or low molecular weight alkaline phosphatase (Bleves et al. 2010).

11.4.3 Type III Secretion System (T3SS)

T3SS, also called injectisome, has components and structure similar to flagellum. It is one among the principal virulence factor of *P. aeruginosa*. Generally, T3SS consists of five components which include needle complex, translocation apparatus, regulatory proteins, chaperones and effectors. In *P. aeruginosa*, the structural and regulatory components are encoded by genes located in five distinct operons, whereas the genes for effectors and chaperons are distributed around other parts of chromosome. The needle-like filament is made of PscF proteins and is a hollow tube having length of 60–80 nm and width of 7 nm. PscE and PscG are the chaperons of PscF preventing them from premature polymerization in cytoplasm. The membrane component spans in the inner and outer membrane and separated as basal body and export apparatus. A sequence of rings comprise the basal body and are made up of proteins PscC (outer rings) and PscJ and PscD (inner rings). An internal rod connecting cytoplasmic rings with needle-like filament will be formed by the ring channels coated by PscI. The export apparatus located in the inner membrane is consisted of PscU-T and PcrD, and it interacts with the cytoplasmic ring (C ring). PscU regulates the secretion switch, and PcrD links the ATPase complex to the secretion pore. The C ring composed of PscQ protein and ATPase complex is located beneath basal body and export apparatus, and together, it forms the sorting platform which regulates protein export. The ATPase complex is formed of the four proteins—PscN, PscO, PscL and PscK, which are ATPase, positive ATPase regulator, negative ATPase regulator and ATPase cofactor, respectively.

Translocation apparatus is responsible for injecting effectors to the host cell to form a 1–4 nm pore and is located at needle tip. It is composed of the proteins—PopB, PopD and PcrV, where the first two proteins are involved in pore formation. PcrV helps in the formation of bridge from needle to pore enabling effector transportation and injection to host cell. Four regulatory genes regulate the T3SS complex, and general regulator is the ExsA. The chaperones in T3SS are small proteins which help effectors storage and maintenance and avoid premature interactions or aggregations. Chaperones are also thought to be implicated in secretion hierarchy. Four major T3SS effector proteins identified in *P. aeruginosa* are ExoS, ExoT, ExoU and ExoY. Other two—PemA and PemB—were identified, but only little information is available (Bleves et al. 2010: Horna and Ruiz 2021). The effectors are described in detail in the later section.

11.4.4 Type V Secretory System (T5SS)

It is one of the simple secretion systems consisting of a Sec export system and outer-membrane channel made of β-barrel protein. Two types T5SS are described—autotransporters and the two-partner secretion. Autotransporter (AT or T5Ass and T5cSS), secreted proteins comprises of an N-terminal signal sequence, C-terminal β-domain and in between "passenger" domain carrying the catalytic portion. The signal peptide allows them to export out of inner membrane, and it gets cleaved. On reaching periplasm, the C-terminus domain gets inserted into the outer membrane forming a β-barrel in such a way either exposing the passenger domain at the bacterial periphery or releasing it extracellularly after autoproteolytic cleavage or by a dedicated protease. The only well-studied AT in *P. aeruginosa* is EstA and is found to be involved in rhamnolipid production. The two-partner secretion (TPS orT5bSS) also has similar mechanism like AT, excluding TpsA, the passenger domain, and the TpsB, the β-domain proteins, which are independent. TpsA encompasses the TPS secretion and catalytic domain, whereas the TpsB contains one or more polypeptide transport-associated (POTRA) domain apart from β-domain. These proteins are transported across the inner membrane via Sec export pathway. TpsA is synthesized as a pre-pro-protein, which is transformed into pro-protein after signal peptide maturation. The TpsB introduced into the outer membrane employs the TpsA via one of its POTRA domains that collaborate with the TpsA TPS motif. In the outer membrane, the pro-protein TpsA becomes mature protein by undergoing one more cleavage, and the mature protein may either stay anchored on the bacterial surface or is liberated into the extracellular setting similarly to AT proteins (Bleves et al. 2010).

11.4.5 Type VI Secretion System (T6SS)

The T6SS is highly complex machinery, with 12–25 accessory proteins. Proteins which depend on T6SS for secretion is not yet well elucidated. Hcp1 is generally considered as a constituent of the T6SS rather than an actual substrate. Structure of the *P. aeruginosa* Hcp1 protein elucidated by crystallography demonstrated a hexameric-ring structure. It polymerizes readily in solution forming nanotubes of 9 nm internal diameter. Other T6SS components characterized are a AAA$^+$ family protein—ClpV ATPase, a fork head-associated (FHA) regulatory domain which is the target for several Ser/Thr kinase and phosphatase, DotU-and IcmF-like proteins homologous to components of T4SS and an outer-membrane lipoprotein. The Hcp1 as a T6SS substrate in *P. aeruginosa* is identified in the sputum samples of individuals with cystic fibrosis. The T6SS components of *P. aeruginosa* PAO1 strain is encoded by three loci called HSI-I, HSI-II and HSI-III, each of which contain 15–20 genes. Plant and animal infection studies proved shown that the latter two loci have *P. aeruginosa* virulence role (Bleves et al. 2010).

11.5 Secreted Virulence Phenotypes

11.5.1 Exopolysaccharides (EPS)

P. aeruginosa secrete macromolecules and polysaccharides for improving their survival in harsh conditions like desiccation, oxidizing agents, and host defence mechanisms. It is one among the major component of biofilm extracellular polymeric substances. Apart from their role in structure they also play role in adhesion. Three exopolysaccharides revealed in *P. aeruginosa* include alginate, Psl and Pel. Strains isolated from cystic fibrosis patients extensively secrete alginate, whereas Psl and Pel are mainly secreted by other environmental strains (Liao et al. 2022). **Alginate**, main constituent of mucoid *P. aeruginosa* biofilms, is the most studied EPS. It is a polymer consisting of the monomers and D-mannuronic and L-guluronic acids which are linked by β1-4 linkages and is of high molecular weight. The enzymes essential for alginate synthesis is encoded by *algD* operon. Alginate helps in biofilm architecture, maturation and stability. It can attach to tracheobronchial mucin. Owing to their acetate groups, these macromolecules are highly viscous and can retain water and nutrient. Furthermore, alginate has significant association with bacterial persistence by guarding *P. aeruginosa* from phagocytosis by host immune cells in the lungs and protecting them by quenching ROS generated by activated macrophages and neutrophils, etc. (Jurado-Martín et al. 2021). **Psl**, the mannose- and galactose-rich polysaccharide, is function for adhesion initially and forming mature biofilm. Psl is synthesized at the time of planktonic growth, aids in adhesion to various surfaces and thereby contributes to microcolony formation. **Pel**, cellulose-like polymer rich in glucose, and is encompassed in the development of pellicle at the air-liquid interface (Ghafoor et al. 2011).

11.5.2 Cytotoxins

The production of a plethora of cytotoxins, thereby inducing host cell cytotoxicity in the course of infection, serves a major virulence function in *P. aeruginosa*. The apoptosis inducing cytotoxins of *P. aeruginosa* explained is exotoxin A (Exo A), C_{12} HSL, azurin and pyocyanin. As we earlier discussed in secretory system, **exotoxin A** or toxin A is an AB toxin and is aT2SS substrate. With the help of B component, enzymatic A component crosses cytoplasmic membrane of host cell and ribosylates eEF-2, a eukaryotic elongation factor, thereby blocking protein synthesis followed by apoptotic cell death. **Pyocyanin** is another key T2SS-secreted virulence factor. It is a water soluble, phenazine-derived pigment metabolite having oxidizing and reducing properties and generates reactive oxygen species (ROS). In vivo experiments in mice have been shown that pyocyanin has relevant function in developing chronic infection and in causing lung damage by *P. aeruginosa*. It has impact on host immune response. **C_{12}-HSL** or *N*-3-oxododecanoyl homoserine lactone is a signalling molecule (autoinducer) involved in the *Las* QS system in *P. aeruginosa*. Apart from its part in the QS, various studies on different cell lines have been proved that C_{12}-HSL has apoptosis inducing properties. **Azurin,** which is a cupredoxin-like protein engaged in electron, transfers at the time of the denitrification process in *P. aeruginosa*. Like C_{12}-HSL, azurin also demonstrated apoptosis in different cancer cell lines by plateauing tumour suppressor p53. The membrane-associated cytotoxins are lipopolysaccharides, flagella, rhamnolipids, porins, etc. We already discussed lipopolysaccharides and flagella in detail in the above sections. **Porins** are outer-membrane proteins which help in transport of substances across cell, and *P. aeruginosa* have been found to possess more than 20 porins. These proteins have crucial role in uptake of nutrients and adhesion, etc. Various mutation studies have shown that deletion of different porin genes decreased the virulence in *P. aeruginosa*. **Rhamnolipids** are glycolipid biosurfactants comprised of mono- or di-rhamnose linked to 3-hydroxy-fatty acids of varying lengths. These molecules have different roles in imparting virulence of *P. aeruginosa*. Rhamnolipid-producing strain and purified rhamnolipids were shown to disrupt tight junction and thus can easily invade human respiratory epithelium. In vivo experiments in tracheal rings of guinea pigs shown that rhamnolipids hamper ciliary beating and mucociliary elimination of *P. aeruginosa*. Furthermore, rhamnolipid also facilitates swarming motility in *P. aeruginosa*. This biosurfactant has also been revealed for its role in development of biofilm, and they also have role in modulating membrane composition of *P. aeruginosa* by suppressing LPS and porin membrane proteins, thus efficiently conducive to their renowned inherent antibiotic resistance. Moreover, rhamnolipids can stimulate cytotoxicity in host epithelial cells, and it can trigger and fight host immune responses (Wood et al. 2023). **ExoS** is a 48.3 kDa bifunctional protein comprising 453 amino acid residues. N-terminal region is the GTPase-activating domain that inactivates small Ras-like GTPases such as Rho, Rac and Cdc45, which have crucial role in actin cytoskeleton. Subversion of this system will cause cell rounding, thereby disturbing cell migration and division. The C-terminal ADRPT domain of ExoS has many targets, including Ras, Rab and Rho protein family, as well as ezrin, radixin, myosin, vimentin and cyclophilin. ADP ribosylation of target proteins by ADPRT domain with the help of 14-3-3 host protein as the cofactor will cause of cytoskeleton disruption, endocytosis and cell-to-cell adhesion, as well as DNA synthesis inhibition and finally apoptosis. **ExoU** is another T3SS toxin that is a 73.9 kDa protein having 687 amino acids and needs more than two eukaryotic activators to become active. ExoU is thought to be the foremost driver of the cytotoxic phenotype as it possesses significant cytotoxic potential. It has a patatin-like domain possessing phospholipase A2 activity and can aim various lipids. **ExoT** is a 457 amino acid containing protein having dual functions similar to ExoS, i.e., with GAP and ADPRT activity. It has key role in preventing bacterial phagocytosis by macrophages. **ExoY** is a 41.7 KDa protein having 378 amino acids. In contrast to other toxins, no chaperones were found in association with this toxin. ExoY is a nucleotidyl cyclase, involved in production of cyclic nucleotidyl monophosphates, especially cAMPs, and requires the eukaryotic filamentous actin (F-actin) for its activity. It is supposed that ExoY involved in the development of oedema owing to the key role of cAMP in the endothelial barrier disruption. Moreover, ExoY aids in immune evasion of *P. aeruginosa* by downregulating transforming growth factor β (TGF-β)-activated kinase 1 (TAK1) and thereby contributes in increasing infection severity (Wood et al. 2023; Horna and Ruiz 2021).

11.5.2.1 Proteases

Various proteases secreted by *P. aeruginosa* have significant role in the pathogenic interaction with its host. **Elastase A** (LasA) also called staphylolysin is a zinc metalloprotease and is one among the most secreted endopeptidase by *P. aeruginosa*. The *lasA* gene encodes this protease and is secreted by T2SS. LasA possesses high staphylolytic efficacy and also little elastolytic activity. Moreover, it can enhance the elastolytic property of LasB elastase, thereby enhancing the pathogen's virulence. It also augments the elastolytic property of numerous host elastolytic proteases. Breaking the pentaglycine bridges of *S. aureus* cell wall peptidoglycan will lead to lysis of the competitors of *P. aeruginosa* and will help to outperform *S. aureus* in the course of colonization of the CF lung. Other proteases including LasB or the alkaline protease stimulates and enhances the LasA-induced lysis of *Staphylococci*. The **elastase B** (LasB) encoded by *lasB* gene is another metalloprotease. It has crucial role in successful colonization in lungs of CF patients by degrading the protein elastin, which is a biopolymer forming protein in organs and tissues of vertebrates that contribute them elastic nature. Elastase B can also degrade other biopolymers such as collagen, specifically collagen types III and IV. Also, it interacts with and thereby degrades human immune defence system proteins. The zinc-dependent metalloendopeptidase **alkaline protease** (AprA), also called aeruginolysin, is secreted by T1SS as mentioned earlier. Owing to their laminin degradation property, alkaline protease has plausible direct association with invasion and haemorrhagic tissue necrosis in *P. aeruginosa* infections. Both elastase B and alkaline protease were found to inactivate human γ-IF and TNF-α, thereby disturbing human immune system. Lysyl endopeptidase, also called **protease IV** or iron-regulated protein PrpL, is a serine protease which has the ability to degrade a wide variability of host proteins including fibrinogen, plasminogen, immunoglobulin G (IgG) and the complement components 3 and C1q and contributes to virulence and pathogenicity by *P. aeruginosa* (Hoge et al. 2010). Other proteases contributing to virulence include *P. aeruginosa* small protease (PASP), large exoprotease A (LepA) and MucD. PASP is an 18.5 kDa zinc-dependent leucine aminopeptidase, expression of which is largely found during ocular infection because of their properties to degrade proteins involved in conserving corneal structure including collagens, fibrinogen, complement C3 and antimicrobial peptide LL-37. **Large exoprotease A** (LepA) is approximately a 100 kDa exoprotease involved in cleaving human protease-activated receptors (PARs) 1, 2 and 4. **MucD** is a serine endoprotease present in periplasmic space and modulates host immune responses (Galdino et al. 2017).

11.5.2.2 Siderophores

Bacteria mostly depend on an insoluble ferric iron form for their metabolic needs. In aerobic and neutral pH conditions, Fe yields an insoluble Fe^{3+} oxide hydrate which is not readily accessible to bacteria. So, bacteria produce siderophores, which are iron-chelating molecules (Fe^{3+} particularly) under the conditions of limited iron, to compensate for this limited bioavailability. *P. aeruginosa* assembles two forms of virulent siderophores called pyoverdine and pyochelin (Amin et al. 2009; Cornelis and Dingemans 2013).

Pyoverdine is a class of chromophore linked to an oligopeptide of about 6–12 amino acid residues, where chromophore part helps in chelation of Fe^{3+}. On the basis of configuration, quantity and content of amino acids within oligopeptide part, different types of pyoverdine have been explained. *P. aeruginosa* synthesizes pyoverdines with diverse peptide chains and structural characteristics (Braud et al. 2009: Mislin and Schalk 2014).

Virulence of *P. aeruginosa* has association to pyoverdine. *P. aeruginosa* pathogenesis in many mammals is due to pyoverdine and is involved in acute illness and the maturation process of biofilms (Jeong et al. 2023). Pyoverdine-dependent signal transduction pathways are involved in the synthesis of key virulence components, namely exotoxin A, PrpL and AprA proteases (Visca et al. 2007).

Pyochelin is another siderophore synthesized by all *P. aeruginosa* strains. However, compared to pyoverdine, it has less affinity to Fe^{3+} (Cornelis and Dingemans 2013). The pyochelin structure comprises of a hydroxyphenyl-thiazoline group and a carboxylic acid moiety. Pyochelin promotes bacterial growth, which is correlated to its virulence character. With pyocyanin, pyochelin causes oxidative damage and inflammation, and this process makes hydroxyl radicals, which creates detrimental consequences on host tissues (Jeong et al. 2023).

11.6 Biofilm Formation

Biofilm progression is a sequence of harmonized phenomenon, where the bacterial cells aggregate, and forms an extracellular polymeric matrix through quorum sensing regulatory networks. Biofilms are dynamic in nature; some fraction of the communities detaches and develops as new biofilm communities (Kulasekara et al. 2005).

Biofilm formation is divided into several stages. In the initial phase, using flagella, *P. aeruginosa* attach to surfaces through physical interactions. In next phase, cells get irreversibly attached over the surface and the extra polymeric substances ensheath the cells, where it forms an aggregate of cells. In this phase, microcolonies were formed. Next phase is the maturation stage, in which the microcolonies develop to form mushroom-like structure, where the cells are encapsulated with EPS matrix. In this maturation phase, flagellar motility is reduced and polysaccharide secretion was increased. The establishment of mushroom stalk like framework is driven by type IV pili, observed in glucose rich medium. The ultimate stage is the dispersal of biofilm. The planktonic forms were released into the milieu by lysis of the EPS layer as planktonic forms or as a fraction of biofilm communities (Vetrivel et al. 2021; Häußler 2004).

11.7 Biofilm Matrix Components

Biofilm matrix is comprised of an extracellular polymeric substance, contains polysaccharides, protein, extracellular DNA, lipids and biosurfactants and helps in initial adhesion and stabilizes the biofilm framework (Azam and Khan 2019).

11.7.1 Polysaccharides

The exopolysaccharides produced by the bacteria helps in the initial adhesion to the surfaces during biofilm establishment. As the biofilm matures, the production of polysaccharides increases and forms three-dimensional architecture. Polysaccharides function in the development and shield the bacterial cells which include (1) adhesion, (2) bacterial aggregation to sustain biofilm structure, (3) nourishment of biofilm by protecting the cells from antibiotics and external stress and (4) protection from host immune responses (Chung et al., 202). Polysaccharides synthesized by *P. aeruginosa* are Psl, Pel and alginate (Wei and Ma 2013).

11.7.1.1 Psl Polysaccharide

Polysaccharide synthesis locus or Psl is a neutral polysaccharide comprised of pentasaccharide reiterating units of D-mannose, L-rhamnose and D-glucose. Psl is involved in the biofilm development by providing structural strength and cellular adhesion between surface and the cells (Billings et al. 2013). It triggers the generation of c-di-GMP, a secondary messenger, responsible for the biofilm development in Gram-negative organisms (Dragoš and Kovács 2017). Psl synthesis is controlled by 15 transcribed genes in the polysaccharide synthesis locus (*psl*) such as *pslA* to *pslO*. Out of 15, 11 genes are essential for the *psl* synthesis for biofilm development in non-mucoid bacteria, which do not depend on alginate for biofilm establishment (Byrd et al. 2009; Skariyachan et al. 2018). During biofilm maturation, psl cross-links with the extracellular adhesin, CdrA and e DNA, which maintains structural stability of biofilm structure (Chung et al. 2023).

Synthesis of polysaccharides is mainly depended upon the precursor molecules, which are integrated into the growing polysaccharide chain directly or after modification (Byrd et al. 2009). The biosynthetic mechanism behind psl production was poorly understood. Some evidences suggested that psl synthesis was similar to isoprenoid lipid-based biosynthesis of capsular and extracellular polysaccharide in *E. coli* (Franklin et al. 2011). Psl synthesis can confer resistance towards antibiotics includes colistin, ciprofloxacin and tobramycin. In addition, psl blocks the action of complement protein, opsonization and phagocytic killing of bacterial cells (Morris et al. 2021).

Lectins produced by *P. aeruginosa* such as LecA (PAI-L) and LecB (PAII-L) are able to interact with polysaccharides. Binding affinity studies demonstrated that LecA binds with galactose moieties, whereas LecB attaches with fucose, mannose and mannose-containing sugar units (up to 10 moieties). Lectins mediate the bacterial adhesion to host during infection and also correlated with biofilm development on abiotic surfaces (Passos da Silva et al. 2019). Studies revealed that Psl stained with lectins depicted the helical arrangement of polysaccharide around the periphery of the cells. In another study, *P. aeruginosa* PAO1 biofilm under fluorescence and confocal laser scanning microscopy showed the structure of psl arranged as mat like matrix that links the biofilm (Franklin et al. 2011). Helical arrangement of psl on the bacteria surface helps in the interaction of psl produced by the adjacent cells. Interaction with the psl will create matrix and will upregulate the cellular interaction among the bacterial cells (Ma et al. 2009).

11.7.1.2 Pel Polysaccharide

Besides psl, pel is another polysaccharide formed by *P. aeruginosa*, especially used in biofilm establishment in non-mucoid strains (Colvin et al. 2012). Pel is a cationic polymer mainly associated with pellicle layer development at air-liquid interface. The genes encode for pel biosynthesis were located in the *pel* gene cluster. Pel is basically comprised of galactosamine and *N*-acetyl galactosamine moieties connected partially by de-*N*-acetylated α 1-4 glycosidic bond (Chung et al. 2023; Balducci et al. 2023).

structure in higher density. Studies reported that stalk-forming bacterial population secretes eDNA through PQS signalling pathway, and type IV pili of the motile bacteria binds to the eDNA and initiates the cap development in mushroom structure (Harmsen et al. 2010; Allesen-Holm et al. 2006).

eDNA portrays a decisive role in genetics, especially in the development of competence by the uptake of DNA from the milieu by bacteria. Studies reported that through natural transformation gene exchange happens between bacteria in the biofilm forms. eDNA is able to capture cations and makes a cation-deficient environment in the host and causes disturbances in the membrane potential and cell death. Furthermore, eDNA can enhance antimicrobial resistance in bacteria. In *P. aeruginosa* biofilm, eDNA functions as a key component of biofilm, and it also involved in signalling. Signalling mechanism is due to the acidic domains of eDNA, which lowers the pH of the medium and finally triggers resistance mechanism. For instance, incubation of *P. aeruginosa* cells with eDNA can enhance the minimum inhibitory concentration of aminoglycoside antibiotics. In addition, studies proposed that eDNA facilitates in the migration of cells by serving as an attachment site for type IV pili (Dragoš and Kovács 2017). eDNA serves as carbon source and causes biofilm dispersal (Karygianni et al. 2020).

Release of eDNA was regulated by phenazine molecule, pyocyanin. Mutants (mutation in the phenazine gene) produce less eDNA than wild-type strains. Pyocyanin-e DNA interaction improves the cell surface hydrophobicity, and therefore, it impacts the surface components and causes nonspecific interactions. Studies shown that the removal of eDNA decreases the biofilm thickness around 40%, while phenazine-deficient strains produce biofilm with 40% thickness and 80% biofilm biomass (Skariyachan et al. 2018).

11.7.3 Proteins

Proteins are one of the structural features of biofilm matrix in *P. aeruginosa.* Cellular contacts within the bacterial communities, and the host tissues are maintained by surface proteins. Surface appendages such as type IV pili, flagella and fimbriae primarily function as a structural support for the biofilm development through attachment. Flagella involved in the motility, especially swimming and swarming motility, which helps in the initial cell-surface interaction (Wei and Ma 2013).

CdrA matrix proteins are involved in the aggregation, localization and retention of Psl exopolysaccharides of biofilm. Some other proteins take part in nutrient accession to protection, especially from stress due to oxidation. For instance, ecotin, a serine protease inhibitor, was recently described as matrix protein that bounds to polysaccharide, Psl. Psl-bounded ecotin protects microorganism from host immune response by host immune protease neutrophil elastase (Reichhardt and Parsek 2019).

P. aeruginosa consists of three fimbria cup such as *cupA, cupB* and *cupC* that are necessary for the synthesis of adhesins. These proteins play an important function in biofilm development. Studies revealed that a regulatory protein encoded by *mvaT* functions as a repressor for CupA fimbriae. MvaT regulates numerous surface proteins intended for adhesion which includes *lecA* that encodes for PA-IL lectin and PA4306 to PA4297 that encode for fimbrial determinant Tad (Kulasekara et al. 2005).

Proteins involved in the stress resistance were reported to be present in the matrix throughout biofilm formation. Heat shock proteins produced in different bacterial species promote biofilm development and were abundant in biofilm forms than the planktonic forms (Zhang et al. 2015).

11.8 Virulence Factors and Biofilm Regulatory Mechanisms

11.8.1 Quorum Sensing

Bacterial cell-to-cell signalling pathways portray major function in the advancement of *P. aeruginosa* biofilms. Pseudomonal virulence factors are synchronized by signalling mechanism known as quorum sensing (QS). Bacterial cells talk each other by producing small signalling molecules called autoinducers (AI), which are a high-density population-dependent process. These molecules interact with the cognate receptor proteins and induce the expression of numerous virulence phenotypes and biofilm establishment. This signalling system portrays crucial role in pathogenesis, host colonization and invasion, host immune evasion and resistance towards antibiotics (Tuon et al. 2022).

The foremost QS systems reported in *P. aeruginosa* are las and rhl systems. They are acyl homoserine lactone-dependent system. In las system, *lasI* encodes for autoinducer synthase gene, which produces *N*-(3-oxododecanoyl) homoserine lactone ($3OC_{12}$-HSL), and *lasR* encodes for transcriptional activator protein. Las system regulates the secretion of extracellular virulence toxins such as LasA, LasB and ToxA proteases. Rhl system includes *rhlR* that encodes for the transcriptional activator protein and *rhlI* that encodes for autoinducer synthase gene and produces autoinducer molecule, *N*-butyryl homoserine lactone (C_4-HSL). Rhl system monitors the expression of *rhlAB* operon that encodes for rhamnosyltransferase, necessary for the formation of rhamnolipids. Rhamnolipids are crucial for the production of mushroom cap-like structure during biofilm formation, open channels, microcolonies and dispersal (Sharma et al. 2014).

The third most QS system reported in *P. aeruginosa* is Pseudomonas quinolone signal (PQS), where the signal molecules are quinolone molecule, 2-heptyl-3-hydroxy-4-quinolone. Another QS signalling system is called IQS, where the AI molecules are synthesized from *ambBCDE* gene cluster. This system regulates the PQS and Rhl system and, moreover, regulates the expression of virulence phenotypes such as pyocyanin, elastase and rhamnolipids (Sharma et al. 2014; Tuon et al. 2022).

11.8.2 c-di-GMP

Biofilm matrix formation is coordinated by bis-(3′-5′)-cyclic dimeric guanosine monophosphate (c-di-GMP), an intracellular secondary messenger extensively present in bacteria. It induces the synthesis of adhesins and polysaccharide for the biofilm development. Rise in c-di-GMP levels stimulates the production of adhesins and polysaccharides, while lower levels downregulate matrix components synthesis, resulting in biofilm dispersal. These molecules block the bacterial motility by regulating the evolution between planktonic and biofilm forms. c-di-GMP binds to its cognate receptors or effector molecules, which results in the transduction of signals to phenotypic regulation. *P. aeruginosa* produces mainly four types of effector proteins and includes FleQ, PelD, Alg44 and PilZ. FleQ is involved in the flagella biosynthesis activation, and PelD regulates the pel biosynthesis, which is a part of *pel* operon. Alg44 controls the alginate biosynthesis, and Pilz is type IV fimbria synthesis protein that regulates the twitching motilities upon binding with messenger molecules. FleQ is characterized as repressor of *pel* operon, FleQ-c-di-GMP binding causes protein dissociation from *pel* A promoter region, but FelQ can act as both activator and repressor of *pel* transcription (Fazli et al. 2014; Wei and Ma 2013). It controls the biosynthesis of matrix components which include alginate, CdrA adhesins, Psl and Pel. In addition, small regulatory RNAs (sRNA) regulate the development of biofilm in several bacterial species (Ghosh et al. 2022).

11.8.3 Two-Component System

Studies on *pel* and *psl* gene regulation help in the discovery of regulatory system accountable for the planktonic and biofilm lifestyles in *P. aeruginosa.* GacS/GacA two-component system was known for the regulation of virulence factors production. These system function through the phosphorylation, stimulate conformational modification in the regulatory protein and finally induce gene expression. Global activator of antibiotics and cyanide synthesis or Gac system is best studied among the more than 60 two-component systems. This system portrays a crucial part in acute to chronic infections. GacS is a membrane-bound sensor kinase containing cytoplasmic HAMP domain. GacA is a response regulator that governs the transcription of two small RNAs RsmY and RsmZ, resulting in the upregulation or downregulation of translation of *pel* and *psl* operon. Gac system negatively regulates the synthesis of AIs such as 3-O-C_{12}-HSL and C_4-HSL and virulence factors controlled by AHL system via RsmA. RetS and LadS are two histidine kinases that are able to interact with other sensors in *P. aeruginosa* through Gac system. As a result of RetS/GacS interface, GacS cannot auto-phosphorylate and thereby unable to activate response regulator, GacA. Such kind of heterodimerization phenomenon by RetS is an unusual signalling mechanism in bacteria. Lost adhesion sensor or LadS acts similar to GacS by positively regulating *pel* expression, thereby enhancing biofilm formation and downregulating the expression of T3SS (Mikkelsen et al. 2011; Wei and Ma 2013; Nadal Jimenez et al. 2012).

11.9 Inhibitors of Virulence Factors and Biofilm

P. aeruginosa is a nosocomial pathogen that causes infections to immune compromised individuals, which is considered as a public threat. Current strategies to combat infections rely on the antimicrobial drugs, make them more resistant and the therapy became ineffective due to the expression of several virulence elements and biofilm. Deep insights into the virulence and QS mechanism help in the development of novel strategies to overcome such resistance. Some of the novel therapeutic methods to regulate virulence factor production and biofilm were described below (Table 11.1).

11.9.1 Phytochemicals

Phytochemicals were reported for the antivirulence and antibiofilm activities against *P. aeruginosa.* These molecules act as a quorum quenching molecule, as it shares similar heterocyclic structure with that of signalling molecules. Hence, the molecules can maintain a stable interaction with the receptor protein and can degrade the QS receptors. Phytochemicals such as curcumin, allicin and ajoene have similar structural conformation to that of AHL molecules but having different oxygenation levels. In case of 6-gingerol, zingerone, eugenol and so on having an aromatic ring instead of lactone ring, baicalein, naringin, quercetin, vitexin, etc., have polycyclic structures (Chadha et al. 2022). Ajoene, sulphur-containing QS inhibitor obtained from garlic, shows downregulation of rhamnolipid expression in a concentration-dependent manner, and also, it shows synergism with tobramycin antibiotics against *P. aeruginosa* biofilms (del Mar Cendra and Torrents 2021).

11.9.2 Small-Molecule Inhibitors

Design and synthesis of inhibitor molecules against QS of *P. aeruginosa* are another way to deal with spread of resistance. Some sulphur analogues show more effective against *P. aeruginosa* biofilm formation when combined with ciprofloxacin than the individual antibiotic. These analogues target the Las and Rhl QS system of *P. aeruginosa.* Phosphoramidon and its analogues were reported to be effective against LasB proteins. Hydroxamic acid derivative were found to act as an agonist for PqsR, which decreases the pyocyanin secretion in *P. aeruginosa* (Sharma et al. 2014; Camberlein et al. 2022).

11.9.3 Antimicrobial Peptides

Antimicrobial peptides or AMPs perform a critical function in the innate immune responses. Innate immunity protects the host from pathogens and is considered as the first line of host defence mechanism. A cyclic peptide, ZY4, reported to

Table 11.1 Several therapeutic approaches used to treat *Pseudomonas aeruginosa* biofilm and anti-virulence mechanisms

Sl. no.	Therapeutic strategy	Activity	Targets	References
Plant/phytochemical based				
1	Chloroform and methanolic extracts of *Andrographis paniculata*	Anti-virulence Antibiofilm	Swarming motility, adhesion, decreased production of pyocyanin, protease and rhamnolipids	Silva et al. (2023)
2	Ethyl acetate fraction of *Centella asiatica* leaves	Anti-virulence Antibiofilm	Reduced biofilm biomass and swimming motility	Silva et al. (2023)
3	Ethanol extract of *Cinnamomum verum* J. Presl leaves	Anti-virulence Antibiofilm	Reduced biofilm adhesion, elastase, pyocyanin, swarming motility and gene expression	Silva et al. (2023)
4	Methanolic extract of *Terminalia bellirica* leaves	Anti-QS Antibiofilm	Inhibit biofilm formation, pyocyanin and EPS production	Silva et al. (2023)
5	Naringenin Taxifolin	Antibiofilm	Reduces QS-controlled gene expression	Yadav et al. (2021)
6	Protocatechuic acid + sulfadiazine Sulfadiazine + quercetin	Antibacterial	Bactericidal	Sakharkar et al. (2009)
Synthetic molecules				
1	Phenoxyacetamide, MBX 1641	Anti-virulence	Blocks T3SS secretion of ExoS	Aiello et al. (2010)
2	Hydroxamic acid derivative	Anti-QS	Agonist for PqsR and decrease pyocyanin secretion	Sharma et al. (2014)
Enzymes				
1	Acylase from *Aspergillus melleus*	Antibiofilm	Used as coating catheter to prevent biofilm formation	Yadav et al. (2021)
2	Paraoxonases	Anti-QS	Degradation of C12-HSL molecules	Yadav et al. (2021)
3	DNase 1L2	Anti-biofilm	Biofilm formation inhibition	Tetz et al. (2009)
4	BpiB09 oxidoreductases	Anti-virulence Anti-biofilm	Reduction in motility, biofilm formation pyocyanin production	Bijtenhoorn et al. (2011)
5	Lactonase enzymes, SsPox-W263I	Anti-QS	Inhibit protease, elastase, pyocyanin production Degrades AHLs	Mion et al. (2019)
6	Alginate lyase, AlyP1400	Anti-biofilm	Reduces biofilm formation by targeting polysaccharides	Daboor et al. (2019)
Bacteriophage therapy				
1	AZ1 phage	Anti-biofilm	Killing both planktonic and biofilm forms	Jamal et al. (2017)

(continued)

Table 11.1 (continued)

Sl. no.	Therapeutic strategy	Activity	Targets	References
2	LysPA26, phage endolysin	Anti-biofilm	Effect on planktonic and biofilm states, especially on MDR strains	Guo et al. (2017)
3	MAG1 and MAG4	Anti-biofilm	Effect on carbapenem resistant *P. aeruginosa* biofilm formation	Kwiatek et al. (2017)
4	Vb-Pa4 and vB-Pa5	Anti-biofilm	Prevent biofilm formation and pre-formed biofilm	Chegini et al. (2020)
Nanoparticles				
1	*Acacia nilotica* AgNPs	Anti-biofilm	Biofilm inhibition	Arya et al. (2018)
2	Fucoidan-stabilized gold nanoparticles	Anti-biofilm	Eradicates mature biofilm	Khan et al. (2019)
3	Conjugation of copper nanoparticles, nanofibers of poly-D, L-lactide and poly (ethylene oxide)	Anti-biofilm	Prevent biofilm formation	Ahire et al. (2016)
4	CuO-NPs coated with poly-acrylic acid	Anti-biofilm	Inhibition of biofilm matrix formation, motility behaviour, lipoprotein synthesis and DNA replication Affect pre-formed biofilm	Singh et al. (2019)
5	*Nigella sativa* seed extract ZnNPs	Anti-QS	Inhibition of QS and downregulation of genes encodes for protease and elastase production	Al-Shabib et al. (2016)
6	Chitosan-oligosaccharide loaded with streptomycin	Anti-biofilm	Inhibition of biofilm development Suppression of MexX-MexY efflux pump Downregulation of polysaccharide synthesis genes	Li et al. (2019)
Antimicrobial peptides (AMPs)				
1	Peptoids 1 and 1-C13	Anti-biofilm	Significant reduction in biofilm biomass and cell viability	Kapoor et al. (2011)
2	LL-37, cathelicidin by humans and peptide enantiomer D-LL-37	Anti-biofilm	Inhibit attachment and reduce pre-established biofilm	Dean et al. (2011)
Photodynamic therapy				
1	GD11, boron-dipyrromethene (BODIPY) dye family	Anti-biofilm	Impact on biofilm structure and inhibit biofilm growth	Orlandi et al. (2014)

(continued)

Table 11.1 (continued)

Sl. no.	Therapeutic strategy	Activity	Targets	References
2	Glycosylated PS, pGEMA-1 derived from BODIPY	Anti-biofilm	Selective binding to bacteria and biofilm inhibition	Zhao et al. (2018)
3	Chitosan NP + Methylene blue	Anti-biofilm	Biofilm disruption	Darabpour et al. (2016)

permeabilize and disorganize the *P. aeruginosa* bacterial membrane ultimately leads to immediate killing of bacteria. It also disrupts the planktonic and the biofilm development in *P. aeruginosa.* Furthermore, it also shows activity against persister cells (Mwangi et al. 2019). Cationic polypeptide and poly-L-lysine (pLK) showed anti-pseudomonal property by lysis of the cells and also inhibit biofilm formation (Guillon et al. 2018).

11.9.4 Bacteriophage Therapy

Bacteriophage cocktails are able to reduce bacterial load and show higher decrease in *P. aeruginosa* infections. These cocktails invade biofilm matrix and trigger the production of polysaccharide depolymerase enzyme. These enzymes can degrade carbohydrate moieties of bacterial envelope. Bacteriophage, M-1, retards the growth of *P. aeruginosa* and reduces biofilm formation after 6 h of treatment. Studies reported that PB1-like, phiKZ-like and LUZ24-like phages are able to inhibit biofilm formation in multidrug-resistant *P. aeruginosa.* Phages such as Pa193, Pa204, Pa222 and Pa223 are efficient in hindering biofilm formation in *P. aeruginosa* isolates from chronic rhinosinusitis person. Phages in single dose or cocktail mode can reduce the biofilm formation. Another study reported that ΦMR299-2 and ΦNH-4 phages are able to kill clinical isolates of *P. aeruginosa* (Chegini et al. 2020; del Mar Cendra and Torrents 2021).

11.9.5 Photodynamic Therapy (PDT)

PDT is an oxygen-dependent photochemical reaction which uses certain wavelength of low power lasers to lyse the microbial cells with photosensitizer molecules. After laser exposure, the photosensitizer gets activated and produces reactive oxygen species. Antimicrobial photodynamic therapy using polycationic photosensitizer such as Tetra-Py^{+}-Me reported to aim polysaccharides of biofilm matrix and inactivates *P. aeruginosa* (Sharma et al. 2014; Hu et al. 2018). In this approach, mutually planktonic and biofilm of microorganism can be destroyed. It is a non-antibiotic therapy; there is no development of resistance mechanism.

11.9.6 Nanoparticles

Nanoparticles are reported for the antibacterial and antibiofilm activity. Silver nanoparticles showed antimicrobial activity with less toxicity against *P. aeruginosa.* The adsorption of NPs to the cell surface causes cellular damage owing to the generation of reactive oxygen species. NPs attachment in the EPS matrix causes decrease in biofilm formation. Selenium nanoparticles are reported to disrupt biofilm formation in *P. aeruginosa.* NPs can function as drug delivery vehicle by enhancing the effect of antibiotics. For instance, polymeric nanoparticles combined with tobramycin antibiotics are used against *P. aeruginosa* biofilm-linked pulmonary infections. Combination effect of AgNPs and aztreonam antibiotic decreases the biofilm thickness by 50% and biomass reduction up to 98%. Several gold nanoparticles prepared from plant extracts showed activity against *P. aeruginosa* biofilm and planktonic cells (Thi et al. 2020; Yadav et al. 2021).

11.10 Conclusion

Pseudomonas aeruginosa is an opportunistic nosocomial pathogenic bacterium, which accounts for various infections to immunocompromised individuals including cystic fibrosis patients. *P. aeruginosa* produces several virulence factors and biofilm, which are governed by cell-cell communication network known as quorum sensing. The production of virulence phenotypes results in pathogenicity and causes higher mortality rates. Some of the crucial virulence phenotypes secreted by *P. aeruginosa* include proteases, flagella, type IV pili, LPS, toxins and rhamnolipids. Biofilm is another crucial virulence mechanism of *P. aeruginosa* consisting of an extra polymeric matrix made up of polysaccharides, proteins and extracellular DNA. Matrix components stabilize the biofilm structure and help in the surface attachment of the cells. These forms are highly resistant to various antimicrobial drugs and make the treatment ineffective. In order to overcome such resistance mechanism, researchers explore various strategies to control or inhibit the virulence factor production and biofilm. Several methods are employed to inhibit *P. aeruginosa* virulence behaviours which include the use of phytochemicals, synthesis of QS inhibitors, antimicrobial peptides, bacteriophage therapy, PDT and nanoparticles. Further studies and research are essential in the advancement of more accurate therapeutic approaches to disrupt virulence mechanisms in *P. aeruginosa.*

References

Ahire JJ, Hattingh M, Neveling DP, Dicks LM (2016) Copper-containing anti-biofilm nanofiber scaffolds as a wound dressing material. PLoS One 11(3):e0152755

Aiello D, Williams JD, Majgier-Baranowska H, Patel I, Peet NP, Huang J, Lory S, Bowlin TL, Moir DT (2010) Discovery and characterization of inhibitors of Pseudomonas aeruginosa type III secretion. Antimicrob Agents Chemother 54(5):1988–1999

Allesen-Holm M, Barken KB, Yang L, Klausen M, Webb JS, Kjelleberg S, Molin S, Givskov M, Tolker-Nielsen T (2006) A characterization of DNA release in Pseudomonas aeruginosa cultures and biofilms. Mol Microbiol 59(4):1114–1128

Al-Shabib NA, Husain FM, Ahmed F, Khan RA, Ahmad I, Alsharaeh E, Khan MS, Hussain A, Rehman MT, Yusuf M, Hassan I (2016) Biogenic synthesis of Zinc oxide nanostructures from *Nigella sativa* seed: prospective role as food packaging material inhibiting broad-spectrum quorum sensing and biofilm. Sci Rep 6(1):36761

Amin SA, Green DH, Küpper FC, Carrano CJ (2009) Vibrioferrin, an unusual marine siderophore: iron binding, photochemistry, and biological implications. Inorg Chem 48(23):11451–11458

Arya G, Kumari RM, Sharma N, Chatterjee S, Gupta N, Kumar A, Nimesh S (2018) Evaluation of antibiofilm and catalytic activity of biogenic silver nanoparticles synthesized from *Acacia nilotica* leaf extract. Adv Nat Sci Nanosci Nanotechnol 9(4):045003

Azam MW, Khan AU (2019) Updates on the pathogenicity status of *Pseudomonas aeruginosa*. Drug Discov Today 24(1):350–359

Balducci E, Papi F, Capialbi DE, Del Bino L (2023) Polysaccharides' structures and functions in biofilm architecture of antimicrobial-resistant (AMR) pathogens. Int J Mol Sci 24(4):4030

Bijtenhoorn P, Mayerhofer H, Müller-Dieckmann J, Utpatel C, Schipper C, Hornung C, Szesny M, Grond S, Thürmer A, Brzuszkiewicz E, Daniel R (2011) A novel metagenomic short-chain dehydrogenase/reductase attenuates *Pseudomonas aeruginosa* biofilm formation and virulence on *Caenorhabditis elegans*. PLoS One 6(10):e26278

Billings N, Ramirez Millan M, Caldara M, Rusconi R, Tarasova Y, Stocker R, Ribbeck K (2013) The extracellular matrix component Psl provides fast-acting antibiotic defense in *Pseudomonas aeruginosa* biofilms. PLoS Pathog 9(8):e1003526

Bleves S, Viarre V, Salacha R, Michel GP, Filloux A, Voulhoux R (2010) Protein secretion systems in *Pseudomonas aeruginosa*: a wealth of pathogenic weapons. Int J Med Microbiol 300(8):534–543

Botelho J, Grosso F, Peixe L (2019) Antibiotic resistance in *Pseudomonas aeruginosa*—mechanisms, epidemiology and evolution. Drug Resist Updat 44:100640

Bouteiller M, Dupont C, Bourigault Y, Latour X, Barbey C, Konto-Ghiorghi Y, Merieau A (2021) Pseudomonas flagella: generalities and specificities. Int J Mol Sci 22(7):3337

Braud A, Hoegy F, Jezequel K, Lebeau T, Schalk IJ (2009) New insights into the metal specificity of the Pseudomonas aeruginosa pyoverdine–iron uptake pathway. Environ Microbiol 11(5):1079–1091

Burrows LL (2012) *Pseudomonas aeruginosa* twitching motility: type IV pili in action. Annu Rev Microbiol 66:493–520

Byrd MS, Sadovskaya I, Vinogradov E, Lu H, Sprinkle AB, Richardson SH, Ma L, Ralston B, Parsek MR, Anderson EM, Lam JS (2009) Genetic and biochemical analyses of the *Pseudomonas aeruginosa* Psl exopolysaccharide reveal overlapping roles for polysaccharide synthesis enzymes in Psl and LPS production. Mol Microbiol 73(4):622–638

Camberlein V, Jézéquel G, Haupenthal J, Hirsch AK (2022) The structures and binding modes of small-molecule inhibitors of *Pseudomonas aeruginosa* elastase LasB. Antibiotics 11(8):1060

Chadha J, Harjai K, Chhibber S (2022) Repurposing phytochemicals as anti-virulent agents to attenuate quorum sensing-regulated virulence factors and biofilm formation in *Pseudomonas aeruginosa*. Microb Biotechnol 15(6):1695–1718

Chegini Z, Khoshbayan A, Taati Moghadam M, Farahani I, Jazireian P, Shariati A (2020) Bacteriophage therapy against *Pseudomonas aeruginosa* biofilms: a review. Ann Clin Microbiol Antimicrob 19:1–17

Chung J, Eisha S, Park S, Morris AJ, Martin I (2023) How three self-secreted biofilm exopolysaccharides of *Pseudomonas aeruginosa*, Psl, Pel, and alginate, can each be exploited for antibiotic adjuvant effects in cystic fibrosis lung infection. Int J Mol Sci 24(10):8709

Ciofu O, Tolker-Nielsen T (2019) Tolerance and resistance of *Pseudomonas aeruginosa* biofilms to antimicrobial agents—how *P. aeruginosa* can escape antibiotics. Front Microbiol 10:913

Colvin KM, Irie Y, Tart CS, Urbano R, Whitney JC, Ryder C, Howell PL, Wozniak DJ, Parsek MR (2012) The Pel and Psl polysaccharides provide *Pseudomonas aeruginosa* structural redundancy within the biofilm matrix. Environ Microbiol 14(8):1913–1928

Cornelis P, Dingemans J (2013) *Pseudomonas aeruginosa* adapts its iron uptake strategies in function of the type of infections. Front Cell Infect Microbiol 3:75

Daboor SM, Raudonis R, Cohen A, Rohde JR, Cheng Z (2019) Marine bacteria, a source for alginolytic enzyme to disrupt *Pseudomonas aeruginosa* biofilms. Mar Drugs 17(5):307

Darabpour E, Kashef N, Mashayekhan S (2016) Chitosan nanoparticles enhance the efficiency of methylene blue-mediated antimicrobial photodynamic inactivation of bacterial biofilms: an in vitro study. Photodiagn Photodyn Ther 14:211–217

Dasgupta N, Wolfgang MC, Goodman AL, Arora SK, Jyot J, Lory S, Ramphal R (2003) A four-tiered transcriptional regulatory circuit controls flagellar biogenesis in *Pseudomonas aeruginosa*. Mol Microbiol 50(3):809–824

Dasgupta N, Arora SK, Ramphal R (2004) The flagellar system of *Pseudomonas aeruginosa*. In: Ramos JL (ed) Pseudomonas. Springer, Boston, MA, pp 675–698

Dean SN, Bishop BM, Van Hoek ML (2011) Susceptibility of Pseudomonas aeruginosa biofilm to alpha-helical peptides: D-enantiomer of LL-37. Front Microbiol 2:128

del Mar Cendra M, Torrents E (2021) *Pseudomonas aeruginosa* biofilms and their partners in crime. Biotechnol Adv 49:107734

Dragoš A, Kovács ÁT (2017) The peculiar functions of the bacterial extracellular matrix. Trends Microbiol 25(4):257–266

Fazli M, Almblad H, Rybtke ML, Givskov M, Eberl L, Tolker-Nielsen T (2014) Regulation of biofilm formation in *Pseudomonas* and *Burkholderia* species. Environ Microbiol 16(7):1961–1981

Filloux A (2011) Protein secretion systems in *Pseudomonas aeruginosa*: an essay on diversity, evolution, and function. Front Microbiol 2:155

Franklin MJ, Nivens DE, Weadge JT, Howell PL (2011) Biosynthesis of the *Pseudomonas aeruginosa* extracellular polysaccharides, alginate, Pel, and Psl. Front Microbiol 2:167

Galdino ACM, Branquinha MH, Santos ALS, Viganor L (2017) *Pseudomonas aeruginosa* and its arsenal of proteases: weapons to battle the host. In: Chakraborti S, Dhalla N (eds) Pathophysiological aspects of proteases. Springer, Singapore, pp 381–397

Garcia-Clemente M, de la Rosa D, Máiz L, Girón R, Blanco M, Olveira C, Canton R, Martinez-García MA (2020) Impact of *Pseudomonas aeruginosa* infection on patients with chronic inflammatory airway diseases. J Clin Med 9(12):3800

Ghafoor A, Hay ID, Rehm BH (2011) Role of exopolysaccharides in Pseudomonas aeruginosa biofilm formation and architecture. Appl Environ Microbiol 77(15):5238–5246

Gheorghita AA, Li YE, Kitova EN, Bui DT, Pfoh R, Low KE, Whitfield GB, Walvoort MT, Zhang Q, Codée JD, Klassen JS (2022) Structure of the AlgKX modification and secretion complex required for alginate production and biofilm attachment in *Pseudomonas aeruginosa*. Nat Commun 13(1):7631

Ghosh S, Saha I, Dey A, Lahiri D, Nag M, Sarkar T, Pati S, Rebezov M, Shariati MA, Thiruvengadam M, Ray RR (2022) Natural compounds underpinning the genetic regulation of biofilm formation: an overview. S Afr J Bot 151:92–106

Guillon A, Fouquenet D, Morello E, Henry C, Georgeault S, Si-Tahar M, Hervé V (2018) Treatment of *Pseudomonas aeruginosa* biofilm present in endotracheal tubes by poly-L-lysine. Antimicrob Agents Chemother 62(11):10–1128

Guo M, Feng C, Ren J, Zhuang X, Zhang Y, Zhu Y, Dong K, He P, Guo X, Qin J (2017) A novel antimicrobial endolysin, LysPA26, against *Pseudomonas aeruginosa*. Front Microbiol 8:293

Harmsen M, Yang L, Pamp SJ, Tolker-Nielsen T (2010) An update on *Pseudomonas aeruginosa* biofilm formation, tolerance, and dispersal. FEMS Immunol Med Microbiol 59(3):253–268

Häußler S (2004) Biofilm formation by the small colony variant phenotype of *Pseudomonas aeruginosa*. Environ Microbiol 6(6):546–551

Hoge R, Pelzer A, Rosenau F, Wilhelm S (2010) Weapons of a pathogen: proteases and their role in virulence of *Pseudomonas aeruginosa*. In: Mendez-Vilas A (ed) Current research, technology and education topics in applied microbiology and microbial biotechnology, vol 2, pp 383–395

Horna G, Ruiz J (2021) Type 3 secretion system of *Pseudomonas aeruginosa*. Microbiol Res 246:126719
Hu X, Huang YY, Wang Y, Wang X, Hamblin MR (2018) Antimicrobial photodynamic therapy to control clinically relevant biofilm infections. Front Microbiol 9:1299
Hwang W, Yoon SS (2019) Virulence characteristics and an action mode of antibiotic resistance in multidrug-resistant *Pseudomonas aeruginosa*. Sci Rep 9(1):487
Jamal M, Andleeb S, Jalil F, Imran M, Nawaz MA, Hussain T, Ali M, Das CR (2017) Isolation and characterization of a bacteriophage and its utilization against multi-drug resistant *Pseudomonas aeruginosa*-2995. Life Sci 190:21–28
Jeong GJ, Khan F, Khan S, Tabassum N, Mehta S, Kim YM (2023) *Pseudomonas aeruginosa* virulence attenuation by inhibiting siderophore functions. Appl Microbiol Biotechnol 107(4):1019–1038
Jurado-Martín I, Sainz-Mejías M, McClean S (2021) *Pseudomonas aeruginosa*: an audacious pathogen with an adaptable arsenal of virulence factors. Int J Mol Sci 22(6):3128
Kapoor R, Wadman MW, Dohm MT, Czyzewski AM, Spormann AM, Barron AE (2011) Antimicrobial peptoids are effective against *Pseudomonas aeruginosa* biofilms. Antimicrob Agents Chemother 55(6):3054–3057
Karygianni L, Ren Z, Koo H, Thurnheer T (2020) Biofilm matrixome: extracellular components in structured microbial communities. Trends Microbiol 28(8):668–681
Khan F, Manivasagan P, Lee JW, Pham DTN, Oh J, Kim YM (2019) Fucoidan-stabilized gold nanoparticle-mediated biofilm inhibition, attenuation of virulence and motility properties in *Pseudomonas aeruginosa* PAO1. Mar Drugs 17(4):208
Kulasekara HD, Ventre I, Kulasekara BR, Lazdunski A, Filloux A, Lory S (2005) A novel two-component system controls the expression of *Pseudomonas aeruginosa* fimbrial cup genes. Mol Microbiol 55(2):368–380
Kwiatek M, Parasion S, Rutyna P, Mizak L, Gryko R, Niemcewicz M, Olender A, Łobocka M (2017) Isolation of bacteriophages and their application to control Pseudomonas aeruginosa in planktonic and biofilm models. Res Microbiol 168(3):194–207
Langendonk RF, Neill DR, Fothergill JL (2021) The building blocks of antimicrobial resistance in *Pseudomonas aeruginosa*: implications for current resistance-breaking therapies. Front Cell Infect Microbiol 11:665759
Lee EY, Choi DS, Kim KP, Gho YS (2008) Proteomics in gram-negative bacterial outer membrane vesicles. Mass Spectrom Rev 27(6):535–555
Li R, Yuan X, Wei J, Zhang X, Cheng G, Wang ZA, Du Y (2019) Synthesis and evaluation of a chitosan oligosaccharide-streptomycin conjugate against *Pseudomonas aeruginosa* biofilms. Mar Drugs 17(1):43
Liao C, Huang X, Wang Q, Yao D, Lu W (2022) Virulence factors of *Pseudomonas aeruginosa* and antivirulence strategies to combat its drug resistance. Front Cell Infect Microbiol 12:926758
Ma L, Conover M, Lu H, Parsek MR, Bayles K, Wozniak DJ (2009) Assembly and development of the *Pseudomonas aeruginosa* biofilm matrix. PLoS Pathog 5(3):e1000354
Ma L, Wang J, Wang S, Anderson EM, Lam JS, Parsek MR, Wozniak DJ (2012) Synthesis of multiple *Pseudomonas aeruginosa* biofilm matrix exopolysaccharides is post-transcriptionally regulated. Environ Microbiol 14(8):1995–2005
Malhotra S, Limoli DH, English AE, Parsek MR, Wozniak DJ (2018) Mixed communities of mucoid and nonmucoid *Pseudomonas aeruginosa* exhibit enhanced resistance to host antimicrobials. MBio 9(2)
Mikkelsen H, Sivaneson M, Filloux A (2011) Key two-component regulatory systems that control biofilm formation in *Pseudomonas aeruginosa*. Environ Microbiol 13(7):1666–1681
Mion S, Rémy B, Plener L, Brégeon F, Chabrière E, Daudé D (2019) Quorum quenching lactonase strengthens bacteriophage and antibiotic arsenal against *Pseudomonas aeruginosa* clinical isolates. Front Microbiol 10:2049
Mislin GL, Schalk IJ (2014) Siderophore-dependent iron uptake systems as gates for antibiotic Trojan horse strategies against *Pseudomonas aeruginosa*. Metallomics 6(3):408–420

Moradali MF, Ghods S, Rehm BHA (2017) *Pseudomonas aeruginosa* lifestyle: a paradigm for adaptation, survival, and persistence. Front Cell Infect Microbiol 7:39

Morris AJ, Jackson L, Yau YCW, Reichhardt C, Beaudoin T, Uwumarenogie S, Guttman KM, Lynne Howell P, Parsek MR, Hoffman LR, Nguyen D (2021) The role of Psl in the failure to eradicate *Pseudomonas aeruginosa* biofilms in children with cystic fibrosis. npj Biofilms Microbiomes 7(1):63

Mwangi J, Yin Y, Wang G, Yang M, Li Y, Zhang Z, Lai R (2019) The antimicrobial peptide ZY4 combats multidrug-resistant *Pseudomonas aeruginosa* and *Acinetobacter baumannii* infection. Proc Natl Acad Sci 116(52):26516–26522

Nadal Jimenez P, Koch G, Thompson JA, Xavier KB, Cool RH, Quax WJ (2012) The multiple signaling systems regulating virulence in *Pseudomonas aeruginosa*. Microbiol Mol Biol Rev 76(1):46–65

Orlandi VT, Rybtke M, Caruso E, Banfi S, Tolker-Nielsen T, Barbieri P (2014) Antimicrobial and anti-biofilm effect of a novel BODIPY photosensitizer against *Pseudomonas aeruginosa* PAO1. Biofouling 30(8):883–891

Pang Z, Raudonis R, Glick BR, Lin T-J, Cheng Z (2019) Antibiotic resistance in *Pseudomonas aeruginosa*: mechanisms and alternative therapeutic strategies. Biotechnol Adv 37(1):177–192

Passos da Silva D, Matwichuk ML, Townsend DO, Reichhardt C, Lamba D, Wozniak DJ, Parsek MR (2019) The *Pseudomonas aeruginosa* lectin LecB binds to the exopolysaccharide Psl and stabilizes the biofilm matrix. Nat Commun 10(1):2183

Pier GB (2007) *Pseudomonas aeruginosa* lipopolysaccharide: a major virulence factor, initiator of inflammation and target for effective immunity. Int J Med Microbiol 297(5):277–295

Reichhardt C, Parsek MR (2019) Confocal laser scanning microscopy for analysis of *Pseudomonas aeruginosa* biofilm architecture and matrix localization. Front Microbiol 10:677

Reynolds D, Kollef M (2021) The epidemiology and pathogenesis and treatment of *Pseudomonas aeruginosa* infections: an update. Drugs 81(18):2117–2131

Sakharkar MK, Jayaraman P, Soe WM, Chow VT, Sing LC, Sakharkar KR (2009) *In vitro* combinations of antibiotics and phytochemicals against *Pseudomonas aeruginosa*. J Microbiol Immunol Infect 42(5):364–370

Sharma G, Rao S, Bansal A, Dang S, Gupta S, Gabrani R (2014) *Pseudomonas aeruginosa* biofilm: potential therapeutic targets. Biologicals 42(1):1–7

Silva E, Teixeira JA, Pereira MO, Rocha CM, Sousa AM (2023) Evolving biofilm inhibition and eradication in clinical settings through plant-based antibiofilm agents. Phytomedicine 119:154973

Singh N, Paknikar KM, Rajwade J (2019) Gene expression is influenced due to 'nano' and 'ionic' copper in pre-formed *Pseudomonas aeruginosa* biofilms. Environ Res 175:367–375

Skariyachan S, Sridhar VS, Packirisamy S, Kumargowda ST, Challapilli SB (2018) Recent perspectives on the molecular basis of biofilm formation by *Pseudomonas aeruginosa* and approaches for treatment and biofilm dispersal. Folia Microbiol 63:413–432

Tetz GV, Artemenko NK, Tetz VV (2009) Effect of DNase and antibiotics on biofilm characteristics. Antimicrob Agents Chemother 53(3):1204–1209

Thi MTT, Wibowo D, Rehm BH (2020) *Pseudomonas aeruginosa* biofilms. Int J Mol Sci 21(22):8671

Tuon FF, Dantas LR, Suss PH, Tasca Ribeiro VS (2022) Pathogenesis of the *Pseudomonas aeruginosa* biofilm: a review. Pathogens 11(3):300

Vetrivel A, Ramasamy M, Vetrivel P, Natchimuthu S, Arunachalam S, Kim GS, Murugesan R (2021) *Pseudomonas aeruginosa* biofilm formation and its control. Biologics 1(3):312–336

Visca P, Imperi F, Lamont IL (2007) Pyoverdine siderophores: from biogenesis to biosignificance. Trends Microbiol 15(1):22–30

Wei Q, Ma LZ (2013) Biofilm matrix and its regulation in Pseudomonas aeruginosa. Int J Mol Sci 14(10):20983–21005

Wood SJ, Goldufsky JW, Seu MY, Dorafshar AH, Shafikhani SH (2023) *Pseudomonas aeruginosa* cytotoxins: mechanisms of cytotoxicity and impact on inflammatory responses. Cells 12(1):195

Yadav J, Kumari RM, Verma V, Nimesh S (2021) Recent development in therapeutic strategies targeting Pseudomonas aeruginosa biofilms—a review. Mater Today Proc 46:2359–2373
Zhang W, Sun J, Ding W, Lin J, Tian R, Lu L, Liu X, Shen X, Qian PY (2015) Extracellular matrix-associated proteins form an integral and dynamic system during *Pseudomonas aeruginosa* biofilm development. Front Cell Infect Microbiol 5:40
Zhao Y, Lu Z, Dai X, Wei X, Yu Y, Chen X, Zhang X, Li C (2018) Glycomimetic-conjugated photosensitizer for specific *Pseudomonas aeruginosa* recognition and targeted photodynamic therapy. Bioconjug Chem 29(9):3222–3230

Enterobacter spp. Virulence Factors and Biofilm Components: Synthesis, Structure, Function, and Inhibitors

12

Srujana Kathi

Abstract

Enterobacter species are multidrug-resistant opportunistic pathogens that produce a variety of virulence factors (VF) to cause disease. They are Gram-negative bacteria widely distributed in the environment, particularly in soil, water, and the gastrointestinal tract of animals and humans. These opportunistic pathogens are considered as ESKAPE pathogens due to their ability to escape the effects of many antimicrobial agents and cause a range of infections that are difficult to treat. The prevalence of *Enterobacter*-associated infections has increased in recent years in healthcare settings, and many strains of *Enterobacter* have become resistant to multiple antibiotics, including carbapenems, which are often considered a last resort for treating multidrug-resistant infections. Antibiotic-resistant *Enterobacter* strains are a major public health issue because they can cause serious infections that are difficult to cure. These organisms produce a variety of VFs, including antibiotic resistance genes, biofilms, and toxins, which enable them to survive in hostile environments and evade host immune defenses. They have a mobile genetic makeup that allows them to acquire and exchange antibiotic resistance genes, which favor antimicrobial resistance. Therefore, understanding the mechanisms by which these bacteria cause disease is essential for the development of effective treatments and prevention strategies. Their synthesis, structure, function, and inhibitors play a significant role in virulence and biofilm formation. One of the VFs of *Enterobacter* spp. is the production of various enzymes, such as beta-lactamases, that can degrade antibiotics, making the bacteria resistant to treatment. These organisms are also capable of forming biofilms, which provide a protective environment for the bacteria to survive in hostile conditions, during exposure to antibiotics and host immune defenses. These

S. Kathi (✉)
UGC-Malaviya Mission Teacher Training Centre, Pondicherry University, Puducherry, India

S. Busi, R. Prasad (eds.), *ESKAPE Pathogens*,
https://doi.org/10.1007/978-981-99-8799-3_12

substances form a matrix that surrounds the bacterial cells and protects them from environmental stressors. *Enterobacter* spp. can produce a variety of extracellular polymeric substances (EPS) components, including cellulose, curli fibers, and fimbriae, which promote biofilm formation and provide structural stability to the matrix. Other VFs include the ability to adhere to host cells, invasion into host tissues, and produce toxins that damage host cells. The bacteria can also evade host immune defenses by modifying their surface structures, such as lipopolysaccharides, to avoid detection by host immune cells. The function of VFs produced by *Enterobacter* spp. is critical to their pathogenicity. Inhibitors of VFs are potential therapeutics for combating infections caused by these bacteria.

Keywords

Antimicrobial · Biofilm · Inhibitor · *Enterobacter* spp. · Virulence factors

12.1 Introduction

The rise of drug-resistant microorganisms is a global health threat, and multidrug-resistant bacterial pathogens are a major concern. These pathogens are becoming increasingly difficult to treat, and they pose a serious risk to public health (Jadimurthy et al. 2022). Antimicrobial resistance (AMR) is caused by the rapid mutation of ESKAPE pathogens, being increasingly common in both developed and developing countries. Nonadherence to antibiotic regimens also contributes to AMR. Furthermore, WHO reports rapid increase in mortality rates due to drug-resistant strains of ESKAPE pathogens (Mukhopadhyay et al. 2020; Kathi 2020). AMR can spread quickly between different environments, and a multitude of factors contribute to this, including the movement of people and animals, the runoff of surface water, and the exchange of agricultural products (Allel et al. 2023). Many of these pathogens acquire antibiotic resistance genes (ARGs) due to selective pressure from antibiotics used in agriculture. This selective pressure leads to the evolution of ARG-carrying pathogens that are resistant to most available antibiotics posing a serious threat to human and environmental health, as well as food safety (Iwu and Okoh 2019). Livestock-associated bacteria, which are often resistant to multiple antibiotics, can be released into the environment from wastewater treatment plants and process water from slaughterhouses. This poses a risk to public health, especially for slaughterhouse workers who are occupationally exposed to these bacteria (Savin et al. 2020).

ESKAPE pathogens are distinguished by their higher antibiotic resistance. This resistance is brought on by a number of factors, such as the bacterial genome mutations, the development of biofilms, and the acquisition of resistance genes (Ma et al. 2020; Denissen et al. 2022). *Enterococcus faecium, Staphylococcus aureus, Klebsiella pneumoniae, Acinetobacter baumannii, Pseudomonas aeruginosa*, and *Enterobacter* spp. are the six nosocomial pathogens represented by the acronym ESKAPE (Mulani et al. 2019). ESKAPE pathogens have developed resistance mechanisms to oxazolidinones, lipopeptides, macrolides, fluoroquinolones,

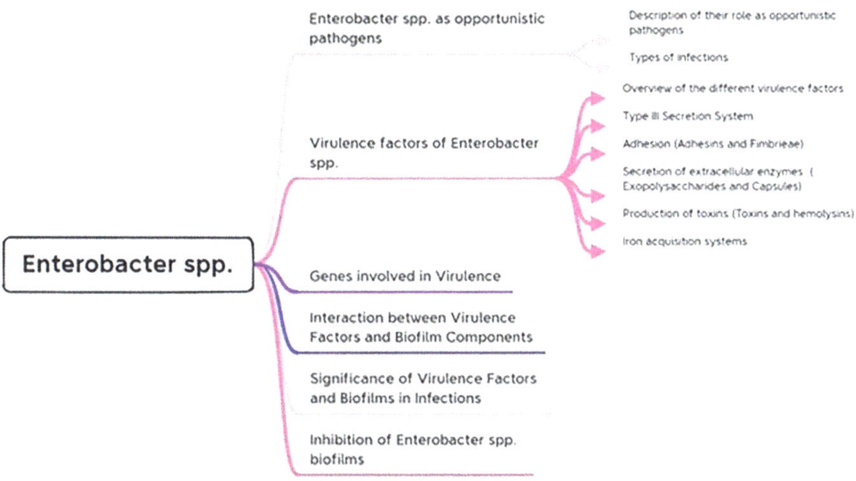

Fig. 12.1 A study on *Enterobacter* spp. as opportunistic pathogen

tetracyclines, -lactams, -lactam--lactamase inhibitor combinations, and antibiotics used as a last resort, such as carbapenems, glycopeptides, and clinically unfavorable polymyxins (De Oliveira et al. 2020). The Enterobacterales order includes both beneficial and harmful microbes. The pathogenic members of the order produce VFs that make them more harmful to their hosts and can also develop resistance to antimicrobial agents. This makes them a major threat to public health (Azam et al. 2023). To infect and injure their hosts, these bacteria employ a range of VFs, including enzymes, toxins, lipids, flagella, exopolysaccharides, and quorum sensing mechanisms. Signaling is a major determinant of pathogenicity, as it can alter the conformation of receptors for ligands, which can have a profound impact on the mechanisms of disease (Meena et al. 2019). Recent developments in nanotechnological techniques, such as solid lipid nanoparticles, fiber-based nanoparticles, dendrimers, cyclodextrin-based delivery, metallic nanoparticles, functionalized metallic nanoparticles, polymer drug conjugates, and liposomes, could provide effective technology-based remedies for infectious conditions (Kumar et al. 2023). This chapter provides an overview of the VFs, biofilm components, genes involved, and inhibitors of *Enterobacter* spp. (Fig. 12.1).

12.2 *Enterobacter* spp. as Opportunistic Pathogens

Enterobacter spp. is an opportunistic Gram-negative bacillus belonging to ESKAPE group. These disease causing pathogens are extensively distributed in the environment causing the most resistant bacterial infections (Zagui et al. 2023; Davin-Regli et al. 2019). Numerous species in this genus are opportunistic pathogens linked to nosocomial infections in immunocompromised individuals (Mohammad et al. 2020). Opportunists are bacteria that only cause disease in people who are already sick or immunocompromised. True pathogens are more likely to be reported than opportunists because they are more likely to cause serious illness (Watkin et al. 2023). They are capable of inflicting a variety of illnesses, especially in people with reduced immune systems, such as newborns, the elderly, and those with immunodeficiency, such as pneumonia, UTIs, bloodstream infections, and sepsis (Gkitsaki et al. 2023). Several *Enterobacter* species are common clinical pathogens. The majority of them have built-in resistance to both older and few newer antibacterial substances. They have a natural tendency to acquire resistance to new drugs (Anju et al. 2020). The capability of these microbes to thrive in diverse environments, from hospitals to the environment, and to infect variety of hosts, from plants to humans, makes it important to understand their population structure, their resistance mechanisms, and how these mechanisms might function in natural ecosystems. This understanding is essential for preventing the spread of antibiotic resistance (Sanz-García et al. 2021).

12.3 Virulence Factors of *Enterobacter* spp.

Urinary pathogens have VFs that allow them to adhere to mucosal surfaces and cause infections. Multidrug-resistant Enterobacteriaceae, especially those that produce ESBLs and carbapenemases, are a major public health menace in developing countries with poor hygiene and antibiotic overuse (Dougnon et al. 2020). These bacteria possess multiple VFs, including toxins and biofilm formation, which enable them to successfully colonize their hosts. The increasing antibiotic resistance of these pathogens necessitates the advance of novel strategies to combat them (de M Campos et al. 2020). Their type III secretion system (T3SS) is a key VF. *SpuE* is a virulence factor that is shared by many ESKAPE pathogens. It is a spermidine-binding protein that belongs to the ATP-binding cassette transporter family and regulates the T3SS in *Pseudomonas aeruginosa* (Zhang et al. 2019). T3SS is a well-regulated secretion system used by many pathogenic Gram-negative bacteria to export VF. The molecular architecture of two essential components of the T3SS was recently described, along with the interface between them, which is necessary for effector secretion. It can endure significant contractions while preserving virulence, and that the make-up of the effector network influences host adaptability (Jensen et al. 2020). The genome-wide sequencing of *Enterobacter cloacae* (Ecc) strains reveals information about the spread and development of antibiotic resistance and pathogenicity, together with the type VI secretion system (T6SS), efflux system, and environmental stress-related genes. T6SS expression is high in ASM medium, indicating that it is important for Ecc survival in a variety of settings. Ecc genome sequence data may be utilized in real time to predict antibiotic resistance genes, which is useful for patient treatment (Mustafa et al. 2020). Enterobacterial uropathogens in cancer patients pose a high risk of morbidity and therapeutic failures due to the combination of VFs, antimicrobial resistance, and biofilm formation. The current findings may aid in the development of antibiotic treatment strategies for cancer patients with urinary tract infections (Adenipekun et al. 2021).

Adhesins are the key VFs that allow bacteria to initiate a urinary tract infection. They can also facilitate bacterial invasion, biofilm formation, host cell signaling, and vaccine development. Later in the infection, toxins produced by the bacteria can make the infection more severe (Oelschlaeger et al. 2002). Type 1 fimbriae (T1F) are long, thin adhesive structures that are composed mainly of FimA proteins. Many Enterobacteriaceae members recognize and bind high-mannose oligosaccharides on host cell surfaces via FimH adhesin. The long, thin structures, which are primarily composed of FimA proteins, are responsible for identifying and binding to high-mannose oligosaccharides that are carried by various glycoproteins and generated on the host cell surface by the FimH adhesin located at the top of T1F (Kolenda et al. 2019).

Microorganisms secrete extracellular polymeric substances (EPS), which are a complex mixture of macromolecular substances. EPS play an important role in the extracellular adsorption of heavy metals (HMs) by microbes. EPS are mainly composed of polysaccharides, humic-like substances, proteins, nucleic acids, lipids, and other molecules (Li et al. 2021). Additionally, Enterobacteriaceae have been linked to advantageous traits including the capacity to manufacture peptides and proteins, which can play a part in the organization of microbial communities. Higher molecular mass antimicrobial compounds are known as colicins, while smaller molecular mass antimicrobial molecules are known as microcins (Marković et al. 2022). *E. hormaechei* capsule is an important immune evasion factor, and it is a potential target for prophylactics and therapies to tackle these increasingly multidrug-resistant human infections (St John et al. 2023). The capsule is made of an anionic and hydrophilic acidic, high molecular-weight polysaccharide. Cyclomodulins are bacterial toxins and effectors that disrupt the eukaryotic cell cycle (Moxley 2022). The Rcs (regulator of capsule synthesis) system is a crucial signal transduction pathway that is present in a large number of Enterobacteriaceae members. This system can control gene expression, integrate environmental inputs, and modify bacterial physiological behavior. The Rcs system offers a special paradigm for investigating the intricacies of bacterial environmental adaptability. The theoretical foundation for the expansion of bacterial infection, prevention, control, and therapy will be aided by knowledge of the Rcs system (Meng et al. 2021). Exopolysaccharides (EPS), which are produced by bacteria, have developed into a significant class of biopolymers used in bioremediation (Dhanya et al. 2021).

12.4 Biofilm Formation in *Enterobacter* spp.

Bacteria shield themselves from host defense, disinfectants, and antibiotics by building a biofilm. Bacteria in biofilm are far more resistant to antimicrobial agents than bacteria in planktonic forms (Dincer et al. 2020). *Enterobacter* spp., especially *E. cloacae* and *E. faecalis*, are adept biofilm formers that pose risks in clinical and industrial settings. Ramos-Vivas et al. (2019) found that 16% of *E. coli*, 73% of *K. pneumoniae*, and 4% of *Enterobacter* spp. from organ transplant patients formed biofilms. These biofilms are surface-attached communities of bacteria embedded in a self-produced matrix (Kim et al. 2009). Dancer et al. (2009) showed milk components influence *E. sakazakii* biofilm formation. Nyenje et al. (2013) demonstrated temperature, medium, and time affect *E. cloacae* biofilm formation, with BHI and TSB media and 37 °C being optimal. Oleiwis (2021) reported that all tested *E. cloacae* isolates formed biofilms, and 60% were strong producers. It was shown that there is no link between hypermutation and biofilm formation in Enterobacteriaceae from UTIs (Urinary Tract Infections) and BSIs (bloodstream infections). While *Enterobacter* spp. from UTIs formed more biofilms, mutation rates did not differ between isolation sources (Kovács 2013). Peter (2013) reported that *E. faecalis* chiefly formed biofilms, aided by glucose, alkaline pH, and 37 °C (Table 12.1).

Biofilms allow *Enterobacter* spp. to persist in environments like food processing plants and hospitals (Yaron and Römling 2014). Multiple factors influence *Enterobacter* biofilm formation, including growth medium, temperature, and strain (Kovács 2013). Additionally, richer media and higher temperatures promote more robust biofilm formation (Nyenje et al. 2013; Kovács 2013). Some *Enterobacter* spp. can form biofilms even in acidic conditions, indicating their adaptability to harsh environments (Kim et al. 2009). *E. cloacae* and *Cronobacter* spp. are able to adhere to and form biofilms on surfaces like polystyrene and polyethylene (Kim et al. 2009; Nyenje et al. 2013). *E. cloacae* biofilm formation is associated with cell surface properties like coaggregation, or the ability of cells to stick to one another (Nyenje et al. 2013). Milk components also influence *Enterobacter* biofilm formation. For example, lactose promotes *Cronobacter* spp. biofilm formation, while bovine serum albumin inhibits it (Dancer et al. 2009). Clinical Enterobacteriaceae isolates from urinary tract infections tend to form more biofilms than bloodstream isolates, though hypermutability does not directly correlate with increased biofilm formation in these isolates (Kovács 2013). Some *Enterobacter* spp. interacts with plants through biofilm formation, which aids in plant colonization and persistence (Yaron and Römling 2014). *Enterobacter* spp. are adept biofilm formers that use biofilms to persist in environments and interact with surfaces like food processing equipment, medical devices, and plants. Their biofilm formation abilities are complex and influenced by many factors, though richer media, higher temperatures, and certain strains tend to promote more robust biofilms. Controlling Enterobacter biofilms may help limit foodborne illness and hospital-acquired infections.

Cyclic di-guanosine monophosphate is crucial for the development of bacterial biofilms. Peptide-based therapies that sequester intracellular c-di-GMP have shown promise in the treatment of antibiotic-resistant bacteria that form biofilms (Poulin and Kuperman 2021). The mature biofilms produced by *S. aureus* and *S. epidermidis* strains were dissociated (>85%) and biofilm development was prevented (>65%) by the cell-free supernatant from sponge-associated *Enterobacter* strain 84.3. The detection of novel antimicrobial agents is hampered by the rising antimicrobial resistance, which is partially due to the capacity to build biofilms (de Oliveira Nunes et al. 2021). Carvacrol oil efficiently prevented *E. cloacae* from forming biofilms and from producing EPS, which is helpful for enhancing the inactivation of *E. cloacae* biofilms by other antimicrobial agents (Liu et al. 2021). Shahid et al. (2021), showed that the inhibitory effect of pesticides on the biofilm-formation ability of

Table 12.1 Structural components and functions of biofilms in *Enterobacter* spp.

Structure component	Contribution	Reference
DNA	The contribution of extracellular DNA to the formation and stability of biofilms	Gilan and Sivan (2013)
Protein factors	The influence of protein factors and secreted polysaccharides on biofilm formation was examined. Proteinase K treatment inhibited the ability of Sp-s to stimulate biofilm formation	Vacheva et al. (2011)
Flagella	Flagellar motility is essential for the initial attachment of bacteria to surfaces and for subsequent biofilm formation	Misra et al. (2022)
Outer membrane proteins	Outer membrane proteins have the potential to form amyloids, which are protein fibrils with a highly ordered beta-sheet structure	Belousov et al. (2022)
	Outer Membrane Vesicles in Enterobacter cloacae are membrane-enclosed vesicles that play a role in a variety of bacterial functions, including virulence, antibiotic resistance, and communication with other bacteria	Bhar et al. (2021)
Quorum sensing molecules	EasR is a LuxR-type transcriptional regulator that is involved in quorum sensing in *Enterobacter asburiae*. When EasR is activated by the quorum sensing signal *N*-butanoyl homoserine lactone (C4-HSL), it binds to DNA and regulates the expression of genes involved in biofilm formation and virulence	Lau et al. (2020)
	In *Enterococcus faecalis,* gelatinase biosynthesis-activating pheromone (GBAP) binds to FsrB and stabilizes its secondary structure which is important for the activation of FsrB and the subsequent regulation of gene expression. The findings of this study provide new insights into the mechanism of action of GBAP in quorum sensing	Littlewood et al. (2020)
Enterococcal surface protein, gelatinase, aggregation substance, and capsule formation	New insights into the role of Esp, a surface protein in biofilm formation by *E. faecalis*. These fibers could provide a structural framework for the biofilm and help to protect the bacteria from the host immune system	Taglialegna et al. (2020)
Extracellular Polymeric Substances (EPS)	Biofilm formation was stronger on the surfaces corresponding to cellular density and EPS production	Misra et al. (2022)
Adhesins	AtaA is a trimeric autotransporter adhesin that could be used to immobilize *E. aerogenes* on polyurethane foam. The immobilized cells were more resistant to detachment than the free cells, and they were also able to produce more biohydrogen. The use of adhesins in the immobilization of bacteria is a promising new technology	Nakatani et al. (2018)

(continued)

Table 12.1 (continued)

Structure component	Contribution	Reference
Lipopolysaccharides (LPS)	Peptidomimetic antibiotics can disrupt the lipopolysaccharide transport bridge (Lpt) of drug-resistant Enterobacteriaceae. The Lpt is a protein complex that is essential for the transport of lipopolysaccharide (LPS), a major component of the outer membrane of Gram-negative bacteria. This study provides new insights into the mechanism of action of peptidomimetic antibiotics and their potential for the treatment of drug-resistant infections	Schuster et al. (2023)

strain EAM 35 of *Enterobacter cloacae* was dose-dependent. The *E. cloacae* strain EAM 35 was subjected to four distinct OCPs in the study: endrin (ES), dieldrin (DE), chlorpyrifos (CP), and lindane (BHC). All four OCPs suppressed the development of *E. cloacae strain EAM 35* in a dose-dependent manner, according to the findings. ES showed the greatest inhibition, followed by DE, CP, and BHC. Additionally, biofilm was composed of a heterogeneous mixture of biomolecules, which include proteins, carbohydrates, lipids, and nucleic acids. The composition of the biofilm changed over time, with the relative abundance of different biomolecules varying at different stages of its formation. Such a biofilm formed by *E. cloacae SBP-8* was associated with the formation of nanotube-like structures between neighboring cells. These nanotubes may play a role in the communication and coordination of the cells within the biofilm (Misra et al. 2022).

12.5 Genes Involved in the Virulence Factor Production

The whole genome and plasmid were sequenced and analyzed in *Enterobacter bugandensis*, a highly pathogenic *Enterobacter* species linked to neonatal sepsis, where the results showed that the chromosome is linked to virulence traits, whereas a 299 kb *IncHI* plasmid is found exclusively where antibiotic-resistance genes are present. The fact that all antibiotic resistance genes are found on plasmids that can be shared between bacteria is a stark warning of how quickly and easily pathogenic bacteria can evolve to become resistant to our most effective treatments, presenting a significant hazard to public health (Pati et al. 2018). The clinical isolates of *Enterobacter aerogenes* and *Enterobacter cloacae* from Brazil had a high prevalence of VF genes. The majority of the isolates also had β-lactamase encoding genes, including *AmpC, CTX-M-1*, and *KPC* genes. The isolates with more virulence factors were more likely to be resistant to antibiotics. This suggests that VF might play a role in the antibiotic resistance of *Enterobacter aerogenes* and *Enterobacter cloacae* (Azevedo et al. 2018). Acinetobacter, Enterobacteriaceae, and Aeromonas all acquire AMR from environmental species by the transfer of miniature inverted-repeat transposable elements (Baquero et al. 2021). Soares (2014) linked biofilm formation in *Enterococcus* spp. to the g*elE*, *esp*, and *agg* genes. A highly drug-resistant and virulent *Enterobacter hormaechei* strain ST133 with genes producing siderophore and *mcr-9.1* has emerged. Four main clades were formed from them. The 178 ST133 *E. hormaechei* strains in the database and C210017's comparative genomic analysis revealed that all 178 strains belonged to serotype O3, and the majority (77.5%) of them possessed the *IncHI2 "superplasmids"* that might be related to the resistance, virulence, and adaptability (Huang et al. 2023). A new species of Enterobacter, namely *Enterobacter pseudoroggenkampii* sp. *nov.*, is reported and described using phenotypic and genomic analysis techniques. These Enterobacter sps have a number of VF, including *iroN* and *iucABCD-iutA*, which encode salmochelin and aerobactin, respectively. It is widely known that Enterobacter's pathogenicity is related to their association. This unique species serves as a reservoir for *qnrE*, a crucial antibiotic resistance gene (Wu et al. 2023). Carbapenem resistance in Enterobacter spp. can develop through mutations in the *ampC* gene, which leads to overexpression of the *ampC* enzyme. This enzyme can break down carbapenems, making them ineffective against the bacteria. Carbapenem resistance can also develop through the acquisition of mobile genetic elements that carry carbapenemase genes. These genes encode enzymes that can also break down carbapenems (Finney et al. 2022).

E. cloacae cells with mutations in the *rpoS* gene were less virulent than wild-type cells. This suggests that the *rpoS* gene is an important regulator of virulence in *E. cloacae* (Gao et al. 2022). The production of biofilm and adhesion VFs by Enterobacteriaceae and *Enterococcus* spp. was higher in patients with decompensation and infections, and these VFs were associated with a higher risk of death and hospitalization. The genes involved in virulence factor production may be important targets for the development of new treatments for infections in patients with cirrhosis (Bajaj et al. 2021). *OmpA* gene, S fimbriae genes, Type III secretion system genes, Capsule genes, and VF regulator genes are involved in the production of VFs by *Enterobacter*. Clinical isolates of *Enterobacter* had more genes involved in VF production than environmental isolates. This suggests that clinical isolates are more likely to cause infections than environmental isolates (Mishra et al. 2020). Hydrocinnamic acid (HCA), a compound produced by *Enterobacter xiangfangensis*, can inhibit the production of AIs by *P. aeruginosa*. This inhibition of AHL-based quorum sensing leads to a reduction in the production of virulence factors such as pyocyanin and biofilm formation (Sharma et al. 2019).

12.6 Interaction Between VFs and Biofilm Components

The interaction between biofilm and cellular immunity factors is a complex and dynamic process that is not fully understood. Many of the defensive mechanisms produced by the immune system and fixed evolutionarily become available to bacteria in biofilms (Shlepotina et al. 2020). In persistent infections, quorum-sensing-mediated cell signaling predominantly controls the development of biofilms and is crucial for promoting virulence (Warrier et al. 2021). Further, marine saprotrophic bacteria could acquire VF from the pathogenic enterobacteria, which could make them more pathogenic. The study assumes significance as they raise the possibility that pathogenic enterobacteria could transfer VF to marine saprotrophic bacteria, which could have implications for the environment and human health. Other marine microbes may have their biological characteristics changed by saprophytic marine bacteria that have acquired VF from pathogenic enterobacteria (Eskova et al. 2022). Biofilms have a viscoelastic character that might make it challenging to remove them from surfaces. This may make it challenging to treat illnesses linked to biofilms (Gloag et al. 2020). The Rcs is a two-component regulatory mechanism found in many Enterobacteriaceae members. It is associated with environmental stress, including envelope stress, nutrient limitation, and quorum sensing. The Rcs system helps to regulate the expression of VF by activating genes that encode these proteins (Meng et al. 2021). Most common species of Enterobacteriaceae isolated from UTIs in Benin were *Escherichia coli* (31.07%), *Klebsiella pneumoniae* (24.39%), *Enterobacter cloacae* (14.29%), and *Proteus mirabilis* (10.71%). The study also found that the majority of the isolates were resistant to multiple antibiotics, including ampicillin, ciprofloxacin, and trimethoprim/sulfamethoxazole. The isolates were positive for a number of VFs, including fimbriae, hemolysins, and Extended-spectrum beta-lactamase (Assouma et al. 2023; Soujanya and Banashankari 2023). A new lectin, EclA, in the bacterium *Enterobacter cloacae* has been reported. Lectins are proteins that can bind to carbohydrates and other molecules. *EclA* is thought to play a role in the virulence of *E. cloacae*, as it can bind to host cells and promote their invasion and colonization. The knockout of *eclA* in *E. cloacae* significantly reduces biofilm formation. The present study highlights that EclA was expressed at high levels in *E. cloacae* isolates that were associated with infections. They also found that EclA could bind to a variety of host cell molecules, including glycoproteins, glycolipids, and carbohydrates. They suggest that EclA may be a promising target for the development of new anti-infective agents. They also suggest that EclA could be used as a diagnostic marker for *E. cloacae* infections (Fares et al. 2023).

12.7 Significance of Virulence Factors and Biofilms in Infections

Ragazzo-Sánchez et al. (2016) studied the ability of enterobacteria to form biofilms on the surface of mango fruit (*Mangifera indica* L.) cv Ataulfo. This knowledge could be used to develop new strategies for preventing foodborne infections. Mango fruit is a common food item that can be contaminated with enterobacteria. The formation of biofilms by enterobacteria on mango fruit can make it difficult to remove them and can contribute to the spread of infection. Therefore, it is important to take steps to prevent the contamination of mango fruit with enterobacteria and to remove any contamination that does occur. The pathogenesis of periodontal disease is a complex process that involves the interaction of bacteria, the host immune system, and the environment. The ability of oral bacteria to persist in the inflamed gingival crevice is due to their production of VFs. Targeting VFs could be a way to disrupt the bacteria's ability to survive and cause disease (Dahlen et al. 2019). The complexity of biofilm-related diseases and the diversity of bacteria that can cause them make them difficult to treat. Most microbial diseases are caused by a combination of different types of bacteria, as is the case with cystic fibrosis, otitis media, dental caries, and chronic wound infections. Understanding the mechanisms of polymicrobial interactions and microbial diversity in chronic disorders is extremely beneficial for antibacterial research (Anju et al. 2022). The high abundance of antimicrobial resistance (AMR) genes in *Enterobacter* spp. and their ability to survive in harsh environments pose a serious threat to human health. To comprehend the worldwide distribution of AMR and virulence genes in these bacteria, more genomic investigations are required (Bolourchi et al. 2022).

12.8 Inhibition of *Enterobacter* spp. Biofilms

Given the importance of biofilms in infectious illnesses, there has been an increased focus on developing small compounds that influence bacterial biofilm growth and maintenance (Worthington et al. 2012). Various ways to inhibit biofilm formation include blocking bacterial surface adhesion, interfering with quorum sensing, modulating second nucleotide messenger signaling molecules, chemically inhibiting biofilm maturation, and disrupting mature biofilms (Table 12.2). Specific pathway inhibitors can modify second nucleotide messenger signaling molecules, prevent biofilm production chemically, and break existing biofilms (Ghosh et al. 2020). Bacteriocins have the potential to be a valuable tool for fighting biofilm infections. They are relatively safe and can be used in combination with other treatments. Regrettably, bacteriocins are often not effective against established biofilms and bacteria can become resistant to bacteriocins (Mathur et al. 2018).

Biofilms can be eliminated using a variety of techniques. The approach chosen will be determined by the unique application. In industrial settings, physical procedures such as heat treatment and ultrasound are frequently employed. In medical contexts, chemical techniques such as antibiotics and biocides are more routinely employed. Biological approaches, such as bacteriophages and probiotics, represent an exciting new area of study. Novel techniques, such as nanotechnology and photodynamic therapy, are showing promise in the eradication of biofilms (Zabielska et al. 2017). The combination of either various antibiotics or antibiotics with substances that can disrupt the biofilm matrix is the most promising approach to address the problem of resistance (Li et al. 2020).

Table 12.2 Potential inhibitors of Enterobacter biofilm formation and virulence factors

Virulence factor/biofilm	Test organism	Inhibitor	Reference
A combination of - β lactam antibiotics and - β lactamase inhibitors, which have been used clinically to overcome resistance by inhibiting -lactamases	Enterobacteriaceae members	β lactamase inhibitors	Carcione et al. (2021)
Antimicrobial and anti-biofilm activities	*Enterobacter aerogenes, Enterobacter cloacae, Klebsiella pneumoniae, Staphylococcus aureus, Staphylococcus haemolyticus*	EpTI a Kunitz-type inhibitor	de Barros et al. (2021)
Drug-extrusion and disrupt biofilm formation	ESKAPE pathogens	Efflux pump-inhibitors (EPIs)	Reza et al. (2019)
Eighteen putative VFs related to motility, quorum sensing, biofilm formation, and endotoxin production	*Cronobacter sakazakii*	Citral	Shi et al. (2017)
The virulence genes *cylR1, cylA, gelE*, and *sprE* that are associated with quorum sensing	*Enterococcus faecalis*	By plasma activating water, ultra-low dosage reactive oxygen species are generated	Li et al. (2019)
Quorum sensing	ESKAPE	Phytochemicals QS inhibitors (QSIs)	Ghosh et al. (2022)
Haemolysin and production of protease	*Enterococcus faecalis, Salmonella enterica serovar Typhimurium and Pseudomonas aeruginosa Staphylococcus aureus Escherichia coli and methicillin-resistant Staphylococcus aureus* (MRSA)	Prodigiosin	Yip et al. (2021)

12.9 Conclusion

Enterobacter spp. exhibits a variety of VFs to enhance their pathogenicity and successfully navigate the host surroundings. Biofilm development appears as a critical approach for *Enterobacter* spp., where the creation of an extracellular matrix and the purposeful deployment of structural components contribute to microbial community resilience and protection. The interactions between VFs and biofilm components are especially remarkable, indicating synergistic linkages that boost pathogenic potential while ensuring the microbial consortium's survival. Essentially, it is impossible to overestimate the importance of these virulence components and biofilm matrices in infection scenarios. These characteristics increase the ability of *Enterobacter* spp. to evade human defenses and antimicrobial therapies, hence sustaining persistent infections and complicating clinical consequences. Therefore, the inhibition of *Enterobacter* spp. biofilms necessitates a thorough, multipronged approach involving physical, chemical, and biological techniques that target both biofilm structure and formation mechanisms, offering potential for therapeutic advancements, while acknowledging the wide-ranging clinical significance of these biofilms across medical contexts, highlighting the need for informed diagnostic and treatment approaches. In essence, this chapter delves into the intricacies between *Enterobacter* spp. VFs and biofilm components, unraveling the intertwined strategies that underscore their success as ESKAPE pathogens. There are a number of ways to inhibit *Enterobacter* spp. biofilms, but more research is needed to develop effective and safe methods and improve patient outcomes in the face of these challenging adversaries.

References

Adenipekun EO, Lawal AO, Nwaokorie FO, Ogbonnaya G, Iwalokun BA (2021) Phenotypic assay of virulence factors, biofilms and antibiotic resistance among enterobacterial uropathogens from cancer patients. Univ Lagos J Basic Med Sci 6(1 & 2)

Allel K, Day L, Hamilton A, Lin L, Furuya-Kanamori L, Moore CE et al (2023) Global antimicrobial-resistance drivers: an ecological country-level study at the human–animal interface. Lancet Planet Health 7(4):e291–e303

Anju VT, Siddhardha B, Dyavaiah M (2020) Enterobacter infections and antimicrobial drug resistance. In: Model organisms for microbial pathogenesis, biofilm formation and antimicrobial drug discovery. pp 175–194

Anju VT, Busi S, Imchen M, Kumavath R, Mohan MS, Salim SA et al (2022) Polymicrobial infections and biofilms: clinical significance and eradication strategies. Antibiotics 11(12):1731

Assouma FF, Sina H, Adjobimey T, Noumavo ADP, Socohou A, Boya B et al (2023) Susceptibility and virulence of Enterobacteriaceae isolated from urinary tract infections in Benin. Microorganisms 11(1):213

Azam MW, Zarrilli R, Khan AU (2023) Updates on the virulence factors produced by multidrug-resistant Enterobacterales and strategies to control their infections. Microorganisms 11(8):1901

Azevedo PAA, Furlan JPR, Oliveira-Silva M, Nakamura-Silva R, Gomes CN, Costa KRC et al (2018) Detection of virulence and β-lactamase encoding genes in Enterobacter aerogenes and Enterobacter cloacae clinical isolates from Brazil. Braz J Microbiol 49:224–228

Bajaj JS, Shamsaddini A, Acharya C, Fagan A, Sikaroodi M, Gavis E et al (2021) Multiple bacterial virulence factors focused on adherence and biofilm formation associate with outcomes in cirrhosis. Gut Microbes 13(1):1993584

Baquero F, Martinez JL, Lanza VF, Rodríguez-Beltrán J, Galán JC, San Millán A et al (2021) Evolutionary pathways and trajectories in antibiotic resistance. Clin Microbiol Rev 34(4):e00050–e00019

Belousov MV, Kosolapova AO, Sulatskaya AI, Sulatsky MI, Antonets KS, Nizhnikov AA (2022) OmpF porins of Enterobacteriaceae possess amyloidogenic properties. FASEB J 36

Bhar S, Edelmann MJ, Jones MK (2021) Characterization and proteomic analysis of outer membrane vesicles from a commensal microbe, Enterobacter cloacae. J Proteome 231:103994

Bolourchi N, Giske CG, Nematzadeh S, Mirzaie A, Abhari SS, Solgi H, Badmasti F (2022) Comparative resistome and virulome analysis of clinical NDM-1–producing carbapenem-resistant Enterobacter cloacae complex. J Glob Antimicrob Resist 28:254–263
Carcione D, Siracusa C, Sulejmani A, Leoni V, Intra J (2021) Old and new beta-lactamase inhibitors: Molecular structure, mechanism of action, and clinical Use. Antibiotics 10(8):995
Dahlen G, Basic A, Bylund J (2019) Importance of virulence factors for the persistence of oral bacteria in the inflamed gingival crevice and in the pathogenesis of periodontal disease. J Clin Med 8(9):1339
Dancer GI, Mah JH, Kang DH (2009) Influences of milk components on biofilm formation of Cronobacter spp. (Enterobacter sakazakii). Lett Appl Microbiol 48(6):718–725
Davin-Regli A, Lavigne JP, Pagès JM (2019) Enterobacter spp.: update on taxonomy, clinical aspects, and emerging antimicrobial resistance. Clin Microbiol Rev 32(4):10–1128
de Barros P, Pimentel RD, Rossoni MT, Garcia V, de Lima Kaminski, Flávio Vieira Loures, Beth Burgwyn Fuchs, Eleftherios Mylonakis, and Juliana Campos Junqueira. (2021) "The anti-biofilm efficacy of caffeic acid phenethyl ester (CAPE) in vitro and a murine model of oral candidiasis." Frontiers in Cellular and Infection. Microbiology 11:700305
de M Campos JC, Antunes LC, Ferreira RB (2020) Global priority pathogens: virulence, antimicrobial resistance and prospective treatment options. Future Microbiol 15(8):649–677
de Oliveira Nunes S, da Silva Rosa H, Canellas ALB, Romanos MTV, Dos Santos KR, Muricy G et al (2021) High reduction of staphylococcal biofilm by aqueous extract from marine sponge-isolated Enterobacter sp. Res Microbiol 172(1):103787
De Oliveira DM, Forde BM, Kidd TJ, Harris PN, Schembri MA, Beatson SA et al (2020) Antimicrobial resistance in ESKAPE pathogens. Clin Microbiol Rev 33(3):10–1128
Denissen J, Reyneke B, Waso-Reyneke M, Havenga B, Barnard T, Khan S, Khan W (2022) Prevalence of ESKAPE pathogens in the environment: antibiotic resistance status, community-acquired infection and risk to human health. Int J Hyg Environ Health 244:114006
Dhanya BE, Athmika, Rekha PD (2021) Characterization of an exopolysaccharide produced by Enterobacter sp. YU16-RN5 and its potential to alleviate cadmium induced cytotoxicity in vitro. 3 Biotech 11:1–11
Dincer S, Uslu FM, Delik A (2020) Antibiotic resistance in biofilm. In: Bacterial biofilms. IntechOpen
Dougnon V, Assogba P, Anago E, Déguénon E, Dapuliga C, Agbankpè J et al (2020) Enterobacteria responsible for urinary infections: a review about pathogenicity, virulence factors and epidemiology. J Appl Biol Biotechnol 8(1):117–124
Eskova AI, Andryukov BG, Yakovlev AA, Kim AV, Ponomareva AL, Obuhova VS (2022) Horizontal transfer of virulence factors by pathogenic enterobacteria to marine saprotrophic bacteria during co-cultivation in biofilm. BioTech 11(2):17
Fares M, Wagner S, Zemlin M, Bischoff M, Titz A (2023) The newly discovered EclA lectin: an emerging player in Enterobacter cloacae infections. https://www.eurocarb2023.com
Finney AG, Perry JM, Evans DR, Westbrook KJ, McElheny CL, Iovleva A et al (2022) Isolation and characterization of lytic bacteriophages targeting diverse Enterobacter spp. clinical isolates. Ther Appl Res 3(1):50–58
Gao X, Qian Q, Zhu Y, Chen Z, Xu J, Xu W et al (2022) Transcriptomic and phenotype analysis revealed the role of rpoS in stress resistance and virulence of pathogenic Enterobacter cloacae from Macrobrachium rosenbergii. Front Microbiol 13:1030955
Ghosh A, Jayaraman N, Chatterji D (2020) Small-molecule inhibition of bacterial biofilm. ACS Omega 5(7):3108–3115
Ghosh S, Lahiri D, Nag M, Dey A, Pandit S, Sarkar T et al (2022) Phytocompound mediated blockage of quorum sensing cascade in ESKAPE pathogens. Antibiotics 11(1):61
Gilan I, Sivan A (2013) Extracellular DNA plays an important structural role in the biofilm of the plastic degrading actinomycete Rhodo-coccus ruber. Adv Microbiol 3(8):543–551
Gkitsaki I, Papachristoforou A, Michailidou S, Karamvalis N, Iliadis I, Graikini D et al (2023) The transmittable through stinging microbiota differs between honeybees and wasps: a potentially greater microbial risk of the wasp sting for humans. Int Microbiol 26(3):663–674

Gloag ES, Fabbri S, Wozniak DJ, Stoodley P (2020) Biofilm mechanics: implications in infection and survival. Biofilms 2:100017

Huang Y, Wu Y, Cai C, Zhang R, Chen G, Dong N (2023) Phenotypic and genomic characterization of ST133 siderophore-encoding extensively drug-resistant Enterobacter hormaechei. Antimicrob Agents Chemother 67(4):e01737–e01722

Iwu CD, Okoh AI (2019) Preharvest transmission routes of fresh produce associated bacterial pathogens with outbreak potentials: a review. Int J Environ Res Public Health 16(22):4407

Jadimurthy R, Mayegowda SB, Nayak SC, Mohan CD, Rangappa KS (2022) Escaping mechanisms of ESKAPE pathogens from antibiotics and their targeting by natural compounds. Biotechnol Rep 34:e00728

Jensen JL, Yamini S, Rietsch A, Spiller BW (2020) The structure of the Type III secretion system export gate with CdsO, an ATPase lever arm. PLoS Pathog 16(10):e1008923

Kathi S (2020) Modeling the spread of antibiotics and AMR/ARGs in soil. In: Antibiotics and antimicrobial resistance genes in the environment. Elsevier, pp 189–194

Kim SA, Lee YM, Oh SW, Gwak HS, Hwang IG, Kang DH et al (2009) Biofilm formation and low pH viability of Cronobacter spp. (Enterobacter sakazakii) isolated from powdered infant formula and infant foods in Korea. Korean J Food Sci Anim Resour 29(6):702–708

Kolenda R, Ugorski M, Grzymajlo K (2019) Everything you always wanted to know about Salmonella type 1 fimbriae, but were afraid to ask. Front Microbiol 10:1017

Kumar L, Bisen M, Harjai K, Chhibber S, Azizov S, Lalhlenmawia H, Kumar D (2023) Advances in nanotechnology for biofilm inhibition. ACS Omega 8(24):21391–21409

Lau YY, How KY, Yin WF, Chan KG (2020) Functional characterization of quorum sensing LuxR-type transcriptional regulator, EasR in Enterobacter asburiae strain L1. PeerJ 8:e10068

Li Y, Pan J, Wu D, Tian Y, Zhang J, Fang J (2019) Regulation of Enterococcus faecalis biofilm formation and quorum sensing related virulence factors with ultra-low dose reactive species produced by plasma activated water. Plasma Chem Plasma Process 39:35–49

Li Y, Xiao P, Wang Y, Hao Y (2020) Mechanisms and control measures of mature biofilm resistance to antimicrobial agents in the clinical context. ACS Omega 5(36):22684–22690

Li Y, Xin M, Xie D, Fan S, Ma J, Liu K, Yu F (2021) Variation in extracellular polymeric substances from Enterobacter sp. and their Pb2+ adsorption behaviors. ACS Omega 6(14):9617–9628

Littlewood S, Tattersall H, Hughes CS, Hussain R, Ma P, Harding SE et al (2020) The gelatinase biosynthesis-activating pheromone binds and stabilises the FsrB membrane protein in Enterococcus faecalis quorum sensing. FEBS Lett 594(3):553–563

Liu F, Jin P, Sun Z, Du L, Wang D, Zhao T, Doyle MP (2021) Carvacrol oil inhibits biofilm formation and exopolysaccharide production of Enterobacter cloacae. Food Control 119:107473

Ma YX, Wang CY, Li YY, Li J, Wan QQ, Chen JH et al (2020) Considerations and caveats in combating ESKAPE pathogens against nosocomial infections. Adv Sci 7(1):1901872

Mathur H, Field D, Rea MC, Cotter PD, Hill C, Paul Ross R (2018) Fighting biofilms with lantibiotics and other groups of bacteriocins. npj Biofilms and Microbiomes 4(1):9

Marković KG, Grujović MŽ, Koraćević MG, Nikodijević DD, Milutinović MG, Semedo-Lemsaddek T, Djilas MD (2022) Colicins and microcins produced by Enterobacteriaceae: characterization, mode of action, and putative applications. Int J Environ Res Public Health 19(18):11825

Meena M, Swapnil P, Zehra A, Aamir M, Dubey MK, Patel CB, Upadhyay RS (2019) Virulence factors and their associated genes in microbes. In: New and future developments in microbial biotechnology and bioengineering. Elsevier, pp 181–208

Meng J, Young G, Chen J (2021) The Rcs system in Enterobacteriaceae: envelope stress responses and virulence regulation. Front Microbiol 12:627104

Mishra M, Panda S, Barik S, Sarkar A, Singh DV, Mohapatra H (2020) Antibiotic resistance profile, outer membrane proteins, virulence factors and genome sequence analysis reveal clinical isolates of Enterobacter are potential pathogens compared to environmental isolates. Front Cell Infect Microbiol 10:54

Misra T, Tare M, Jha PN (2022) Insights into the dynamics and composition of biofilm formed by environmental isolate of Enterobacter cloacae. Front Microbiol 13:877060

Mohammad MK, Dari WA, Shatti ZO (2020) Review of Enterobacter spp.: taxonomy, pathogenisity and antibiotic resistance. Biochem Cell Arch 20(2)
Moxley RA (2022) Enterobacteriaceae: Escherichia. In: Veterinary microbiology. pp 56–74
Mukhopadhyay S, Bharath Prasad AS, Mehta CH, Nayak UY (2020) Antimicrobial peptide polymers: no escape to ESKAPE pathogens—a review. World J Microbiol Biotechnol 36:1–14
Mulani MS, Kamble EE, Kumkar SN, Tawre MS, Pardesi KR (2019) Emerging strategies to combat ESKAPE pathogens in the era of antimicrobial resistance: a review. Front Microbiol 10:539
Mustafa A, Ibrahim M, Rasheed MA, Kanwal S, Hussain A, Sami A et al (2020) Genome-wide analysis of four *Enterobacter cloacae* complex type strains: insights into virulence and niche adaptation. Sci Rep 10(1):8150
Nakatani H, Ding N, Ohara Y, Hori K (2018) Immobilization of Enterobacter aerogenes by a trimeric autotransporter adhesin, AtaA, and its application to biohydrogen production. Catalysts 8(4):159
Nyenje ME, Green E, Ndip RN (2013) Evaluation of the effect of different growth media and temperature on the suitability of biofilm formation by *Enterobacter cloacae* strains isolated from food samples in South Africa. Molecules 18(8):9582–9593
Oelschlaeger TA, Dobrindt U, Hacker J (2002) Virulence factors of uropathogens. Curr Opin Urol 12(1):33–38
Oleiwis SR, Najim SS, Radif HM (2021) Morphological and Molecular Study of Biofilm Formation by Enterobacter cloacae. Ann Trop Med Public Health 24:176–186
Pati NB, Doijad SP, Schultze T, Mannala GK, Yao Y, Jaiswal S et al (2018) Enterobacter bugandensis: a novel enterobacterial species associated with severe clinical infection. Sci Rep 8(1):5392
Poulin MB, Kuperman LL (2021) Regulation of biofilm exopolysaccharide production by cyclic di-guanosine monophosphate. Front Microbiol 12:730980
Ragazzo-Sánchez JA, González-Estrada RR, Santana-Martínez MA, del Carmen Wacher-Rodarte M, Eslava-Campos CA, Calderón-Santoyo M (2016) Data on enterobacteria activity on biofilm formation at surface mango fruit (Mangifera indica L.) cv Ataulfo. Data Brief 9:746–748
Ramos-Vivas J, Chapartegui-González I, Fernández-Martínez M, González-Rico C, Fortún J, Escudero R et al (2019) Biofilm formation by multidrug resistant Enterobacteriaceae strains isolated from solid organ transplant recipients. Sci Rep 9(1):8928
Reza A, Sutton JM, Rahman KM (2019) Effectiveness of efflux pump inhibitors as biofilm disruptors and resistance breakers in gram-negative (ESKAPEE) bacteria. Antibiotics 8(4):229
Sanz-García F, Gil-Gil T, Laborda P, Ochoa-Sánchez LE, Martínez JL, Hernando-Amado S (2021) Coming from the wild: multidrug resistant opportunistic pathogens presenting a primary, not human-linked, environmental habitat. Int J Mol Sci 22(15):8080
Savin M, Bierbaum G, Hammerl JA, Heinemann C, Parcina M, Sib E et al (2020) ESKAPE bacteria and extended-spectrum-β-lactamase-producing Escherichia coli isolated from wastewater and process water from German poultry slaughterhouses. Appl Environ Microbiol 86(8):e02748–e02719
Schuster M, Brabet E, Oi KK, Desjonquères N, Moehle K, Le Poupon K et al (2023) Peptidomimetic antibiotics disrupt the lipopolysaccharide transport bridge of drug-resistant Enterobacteriaceae. *Science*. Advances 9(21):eadg3683
Shahid M, Manoharadas S, Altaf M, Alrefaei AF (2021) Organochlorine pesticides negatively influenced the cellular growth, morphostructure, cell viability, and biofilm-formation and phosphate-solubilization activities of Enterobacter cloacae strain EAM 35. ACS Omega 6(8):5548–5559
Sharma S, Gopu V, Sivasankar C, Shetty PH (2019) Hydrocinnamic acid produced by Enterobacter xiangfangensis impairs AHL-based quorum sensing and biofilm formation in Pseudomonas aeruginosa. RSC Adv 9(49):28678–28687
Shi C, Sun Y, Zhiyuan Liu D, Guo HS, Sun Z, Chen S et al (2017) Inhibition of Cronobacter sakazakii virulence factors by citral. Sci Rep 7(1):43243
Shlepotina NM, Peshikova MV, Kolesnikov OL, Shishkova YS (2020) Modern conceptions about the mechanisms of interaction between biofilm and cellular immunity factors. J Microbiol Epidemiol Immunobiol 97(1):83–90

Soares RO, Fedi AC, Reiter KC, Caierão J, d'Azevedo PA (2014) Correlation between biofilm formation and gelE, esp, and agg genes in Enterococcus spp. clinical isolates. Virulence 5(5):634–637

Soujanya BR, Banashankari GS (2023) Phenotypic detection of virulence factors of uropathogenic Enterobacteriaceae. J Pure Appl Microbiol 17(2)

St John A, Perault AI, Giacometti SI, Sommerfield AG, DuMont AL, Lacey KA et al (2023) Capsular polysaccharide is essential for the virulence of the antimicrobial-resistant pathogen Enterobacter hormaechei. mBio 14(2):e02590–e02522

Taglialegna A, Matilla-Cuenca L, Dorado-Morales P, Navarro S, Ventura S, Garnett JA et al (2020) The biofilm-associated surface protein Esp of Enterococcus faecalis forms amyloid-like fibers. npj Biofilms Microbiomes 6(1):15

Vacheva A, Ivanova R, Stoitsova S (2011) Stimulation of sessile growth of a biofilm-deficient strain Escherichia coli K-12 by protein (s) secreted by Enterobacteriaceae species. Biotechnol Biotechnol Equip 25(Suppl 1):88–91

Warrier A, Satyamoorthy K, Murali TS (2021) Quorum-sensing regulation of virulence factors in bacterial biofilm. Future Microbiol 16(13):1003–1021

Watkin S, Ciric L, Kiernan M, Cloutman-Green E (2023) A proposed classification system for opportunistic pathogens for improved healthcare infection prevention and control risk assessments. J Hosp Infect 135:206–207

Worthington RJ, Richards JJ, Melander C (2012) Small molecule control of bacterial biofilms. Org Biomol Chem 10(37):7457–7474

Wu S, He Y, Feng Y, Zong Z (2023) Enterobacter pseudoroggenkampii sp. nov. carrying quinolone-resistant gene qnrE recovered from clinical samples in China. Antonie Van Leeuwenhoek 116(7):643–651

Yaron S, Römling U (2014) Biofilm formation by enteric pathogens and its role in plant colonization and persistence. Microb Biotechnol 7(6):496–516

Yip C-H, Mahalingam S, Wan K-L, Nathan S (2021) Prodigiosin inhibits bacterial growth and virulence factors as a potential physiological response to interspecies competition. PLoS One 16(6):e0253445

Zabielska J, Tyfa A, Kunicka-Styczyńska A (2017) Methods for eradication of the biofilms formed by opportunistic pathogens using novel techniques—a review. Folia Biol Oecol 12:26–37

Zagui GS, Moreira NC, Santos DV, Paschoalato CFPR, Sierra J, Nadal M et al (2023) Multidrug-resistant Enterobacter spp. in wastewater and surface water: molecular characterization of β-lactam resistance and metal tolerance genes. Environ Res 233:116443

Zhang Y, Sun X, Qian Y, Yi H, Song K, Zhu H et al (2019) A potent anti-SpuE antibody allosterically inhibits type III secretion system and attenuates virulence of Pseudomonas aeruginosa. J Mol Biol 431(24):4882–4896

Antibiotic Adjuvants and Their Synergistic Activity Against ESKAPE Pathogens

13

V. T. Anju, Siddhardha Busi, and Madhu Dyavaiah

Abstract

ESAKPE pathogens are acquiring resistance to the conventional and modern therapeutic strategies, becoming difficult to eradicate. The development of antibiotic resistant strains and their evolving escape process are one of the major global health concerns. The multidrug resistance mechanism and virulence pathways exhibited by ESKAPE pathogens urge the medical system to find effective alternatives to fight these pathogens. These pathogens exhibit resistance towards antibiotics by different mechanisms; drug modifications, drug efflux, target modifications, and decreased penetration through cell membrane. To combat antibiotic resistance, new antibiotics and antibacterial agents are explored from plant and bacterial sources. Recently, ESKAPE pathogens exhibited resistance towards these novel compounds or therapeutics. Alternative and effective therapeutics such as antibiotic combinations or in combination with adjuvants are in clinical trials. In recent years, antibiotic adjuvants used against ESKPE pathogens showed promising results as the development of resistance towards multiple agents is less. Also, the toxicity of drugs used in single therapy can be minimized when synergistic effect of drugs is displayed. In addition, more activity is expected in synergy in comparison to the monotherapy. Also, adjuvants especially from plant sources help to make the ineffective drugs to effective against the pathogens.

V. T. Anju · M. Dyavaiah
Department of Biochemistry and Molecular Biology, School of Life Sciences, Pondicherry University, Pondicherry, India

S. Busi (✉)
Department of Microbiology, School of Life Sciences, Pondicherry University, Puducherry, India

S. Busi, R. Prasad (eds.), *ESKAPE Pathogens*,
https://doi.org/10.1007/978-981-99-8799-3_13

Keywords

Antibiotic resistance · Virulence · Therapeutics · Adjuvants · Synergy

13.1 The Emerging Problem of Antibiotic Resistance

Antibiotic resistance (AMR) crisis is a scenario where bacteria develop different mechanisms to escape from drugs and evolve into different resistant strains capable of causing infections that are difficult to treat by traditional antibiotics. At present, World Health Organization (WHO) has warned health care system and scientists regarding the post-antibiotic era and the situation where all of us are running out of necessary antibiotics. The heightening peak of AMR is always a critical health challenge failing to control infectious diseases (Chinemerem Nwobodo et al. 2022). According to WHO, antibiotic resistance is one the leading worldwide health challenge which needs urgent attention. Approximately, a rise in global economy to $100 trillion in the year 2050 is predicted as death rate elevate by 10 million/year (Strathdee et al. 2020). AMR genes are found in the environment; however, redundant and inappropriate use of antibiotics leads to AMR crisis (Akova 2016).

A list of pathogens was studied in 2017 by WHO requiring urgent attention in the field of treatment. This list contains ESKAPE group: *Enterococcus faecium*, *Staphylococcus aureus*, *Klebsiella pneumoniae*, *Acinetobacter baumannii*, *Pseudomonas aeruginosa*, and *Enterobacter* sp. (Rice 2008). This list of priority pathogens develops many life threatening infections in hospitalized and immunocompromised persons (Rice 2010). The spread of AMR can be owing to (1) bacterial genetic mutations, (2) inappropriate use of antibiotics during the period of intake, (3) overuse, (4) taking drugs at lower doses than prescribed, (5) antibiotic residues containing dairy products or other animal products, and (6) fertilizers with antibiotic residues for farming (Helmy et al. 2023). These pathogens are major contributors to the elevated global mortality and morbidity rates. And they are associated with the incidence and persistence of infections such as urinary tract diseases, pneumonia, bacteraemia, wound-related infections, meningitis, etc. (Navidinia et al. 2017). ESKAPE pathogens developed resistance towards all commonly available antibiotics including beta lactam antibiotics, macrolides, tetracyclines, lipopeptides, oxazolidinones, and even last generation drugs such as carbapenems, polymyxins, and glycolipids (Denissen et al. 2022).

13.2 Mechanism of AMR in ESKAPE Pathogens

This list of pathogens is associated with high death rate and health care cost owing to the antibiotic resistance crisis (Founou et al. 2017). A list of 12 bacterial pathogens involving ESKAPE which require urgent intervention methods including novel antibiotics is reported by WHO. These pathogens fall in the category of critical, high, and medium priority list as per the therapeutics they require for the eradication. The critical priority pathogens are carbapenem antibiotic-resistant *P. aeruginosa* and *A. baumannii*¸ extended spectrum beta-lactamase-producing, or carbapenem-resistant *K. pneumoniae* and *Enterobacter* sp. High priority pathogens are listed as vancomycin-resistant *E. faecium* and *S. aureus* and methicillin-resistant *S. aureus* (Mulani et al. 2019).

In order to counteract the attack of antibiotics, bacteria evolve with different survival methods. The mechanism of antibiotic resistance includes (Peterson and Kaur 2018),

1. Reducing the penetrative ability of antibiotic to reach the target in bacteria.
2. Extruding antibiotics from cell through different efflux pumps present on cell membrane.
3. Inactivation of antibiotic and inability to attach to the target site.
4. Modifying the antibiotic target site.

Table 13.1 provides details of different antibiotic-resistant mechanisms exhibited by the ESKAPE pathogens (Table 13.1).

The specialized protected form of growth exhibited by pathogens with the help of exopolymer matrix are biofilms. These biofilms are population of cells growing under the exopolymeric substances and protected from outside harsh environments. Biofilms escape from the attack of host defense mechanisms and antibiotics to survive and adapt to the long-term exposure of different antibiotics. Dormant persister cells formed inside biofilms protect themselves from the attack of antibiotics and can evolve after the release of antibiotic pressure causing recalcitrant infections (Lewis 2007). Clinically, antibiotic resistance is expressed in terms of the minimum inhibitory dose (MIC) of drug in μg/mL unit. This is the lowest dose of drug required to inhibit the growth of pathogen. Bacteria are observed as resistant when they can grow and cause infection even after applying MIC. Susceptible (S) bacteria is treated with a standard drug regimen, whereas a resistant (R) one requires very high dose than MIC as reported by European Committee on Antimicrobial Susceptibility Testing (EUCAST) (Kalpana et al. 2023).

13.3 Need for New Drug Discovery

ESKAPE group is also recognized by several governmental and non-governmental research agencies and groups to improve the research and development in the area of novel antibiotics against them. However, there is a lacuna in the research and development process owing to the uncoordinated assessment of AMR around the globe. It has been reported that increase in the rate of AMR occurs along with increased antibiotic usage, simple procedures to obtain different antibiotics, inappropriate use of antibiotics, and guidelines related to its application. In addition, AMR frequency is found increasing owing to the insufficient effective second and third line of antibiotics, procurement of poor quality or false antibiotics, and not having sufficient hygienic techniques (Kalpana et al. 2023).

Table 13.1 Details of different antibiotic-resistant mechanisms present in ESKAPE pathogens

Antibiotic-resistant mechanisms	Pathogen	Enzymes/mechanism	Antibiotics
Chemical modification/ breakdown of the drug	*S. aureus*	*blaZ*–encoded penicillinases	Penicillins
	K. pneumoniae	Carbapenemase	β-lactams, including carbapenems
	Enterobacter sp., *P. aeruginosa*, and *Acinetobacter* sp.	Cephalosporinases	Cephalosporins Plus aztreonam
	S. aureus and *Enterococcus* sp.	Aminoglycoside Phosphotransferases	Amikacin
	K. pneumoniae and *S. aureus*	Aminoglycoside nucleotidyltransferases	4, 6-di-substituted aminoglycosides, kanamycin A/B/C, gentamicin A, amikacin, tobramycin, and neomycin B/C
Prevention of antibiotic influx	*A. baumannii*	OprD channel	Panipenem/imipenem
	P. aeruginosa	OprD channel	Imipenem
Expulsion through efflux pumps	*P. aeruginosa*	MexXY (OprM), MexCD-OprJ, MexEF-OprN, MexXY, and MuxBC–OpmB	Beta lactams, tigecycline, aminoglycosides, chloramphenicol, fluoroquinolones, macrolides, quinolones, tetracycline, and novobiocin
Modification of antibiotic target	Methicillin-resistant *S. aureus*	*mecA* gene encoding penicillin-binding protein 2a	Methicillin and other Beta-lactam antibiotics
	E. faecium	Penicillin-binding protein 5	Penicillin, ampicillin, and other Beta-lactam drugs
	A. baumannii	Penicillin-binding protein	Carbapenem
	K. pneumoniae and *Enterobacter* sp.	Plasmid-mediated quinolone resistance (PMQR) by Qnr-family proteins	Fluoroquinolones
	S. aureus and *Enterococcus* sp.	*erm*-encoded rRNA methyltransferases	Macrolide, lincosamide, and streptogramin B

In the past years, novel antimicrobial agents are discovered and isolated from natural sources. Secondary metabolites isolated from plants, microorganisms (fungi, bacteria), and marine vertebrates are found to be potent antibacterial agents against variety of human diseases. There are more than potent 13,000 bioactive molecules isolated from different bacterial sources. In this, 70% of compounds are isolated from actinomycetes group itself. Approximately, 1% of bacterial sources are

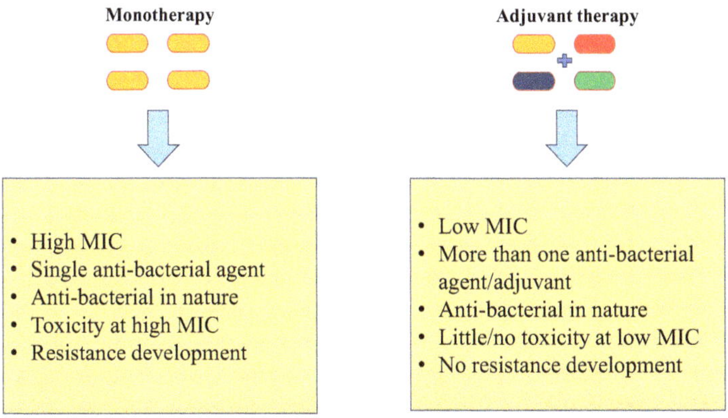

Fig. 13.1 Comparison of merits and demerits of monotherapy and adjuvant therapy

unculturable and need special techniques involving whole genome sequencing to unleash the hidden potential (Jadimurthy et al. 2022).

Novel antimicrobial agents developed so far failed to eradicate AMR strains and associated infections completely. Thus, combining one or multiple agents of antimicrobial activity and/or non-antimicrobial activity exists as a potent strategy to tackle AMR. As monotherapy using single antibiotic is not fully effective in treating bacterial illness and antibiotic-resistant strains, adjuvants are combined along with antibiotic to improve the result (De Oliveira et al. 2020) (Fig. 13.1).

13.4 Antibiotic Adjuvants and Synergy

Antibiotic adjuvants are molecules which may not have specific target, but enhance the activity of antibiotics that act on specific target. These adjuvants improve the life of people by potentiating already available antibiotics and are active against antibiotic-resistant pathogens (Gill et al. 2015). In another view, adjuvants do not exhibit antimicrobial properties by themselves, but exhibit synergistic activity with the existing antibiotics or inhibit the emergence of resistance. Antibiotic activity is restored or conserved once resistance is blocked by adjuvant molecules. There are different types of antibiotic adjuvants developed so far which include beta-lactamase inhibitors, efflux pump inhibitors, and membrane permeabilizers (Dhanda et al. 2023) (Fig. 13.2). In antibiotic adjuvant concept, physical combinations of two antimicrobial agents are designed to exert combination effect. Combination of two antibiotics is already an available concept and widely used in clinical practice (Kerantzas and Jacobs 2017).

These adjuvants work by exhibiting synergistic activity, enhance the activity, and block the resistance mechanism. These molecules focus on either active or passive resistance mechanism of bacteria. Synergistic mechanism of adjuvants is assessed using checkerboard and isobologram analysis. Fractional inhibitory index (FICI)

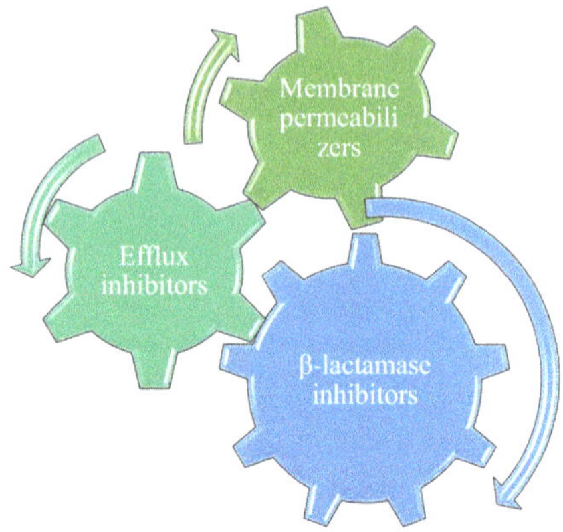

Fig. 13.2 Diagram showing different types of antibiotic adjuvants studied in the treatment of ESKAPE pathogens

and potentiation factors are different parameters studied using the above assays (Dhanda et al. 2023). According to the checkerboard study, Minimum Inhibitory Concentration (MIC) of compound get reduced in the presence of an adjuvant. The synergy interactions are assessed based on the FICI (Rishi et al. 2018). FICI index is calculated by comparing the individual MICs and combination MICs of compounds. The interaction is considered as synergy when index is ≤0.5, additive for a FICI in between 0.5 and 2 and FICI of ≥2 is considered as antagonistic combination (Yap et al. 2021). FICI is calculated as follows:

$$\begin{aligned} \text{FICI} &= \left(\text{MIC of compound a in combination} \div \text{MIC of a alone}\right) \\ &+ \left(\text{MIC of compound b in combination} \div \text{MIC of b alone}\right) \end{aligned}$$

There are direct and indirect resistance breakers used in adjuvant therapy. The direct resistance breakers are inhibitors of enzyme and efflux pumps. Indirect breakers include those acting on teichoic acid synthesis, membrane targeting compounds, and enzymes of cellular processes (Dhanda et al. 2023).

13.4.1 Beta-Lactamase Inhibitors

Beta-lactamases are enzymes found in antibiotic-resistant bacteria which hydrolyze beta lactam antibiotics (penicillins, cephalosporins, monobactams, and carbapenems). There are A, B, C, and D classes of beta-lactamases. The most common inhibitors are non-beta lactam such as boron-based inhibitors, 1,6-diazabicyclo [3,2,1] octanes, thiol derivatives, and dicarboxylate derivatives (González-Bello et al. 2020). Two examples of boron-based inhibitor in the development, QPX7728 and VNRX-7145, are used as inhibitors of metallo-β-lactamases present in multidrug-resistant *A. baumannii*, *P. aeruginosa*, and Enterobacteriaceae (González-Bello et al. 2020).

Ceftriaxone + sulbactam + adjuvant disodium edetate is a well-approved adjuvant identity approved and used widely in clinical system. This combination along with beta-lactamase inhibitor antibiotic was used to cure sepsis in patients. Good clinical outcome was observed in sepsis patients with Gram-negative infections of *E. coli*, *Klebsiella* sp., *Acinetobacter* sp., and *Pseudomonas* sp. (Verma 2015). Clavulanic acid as an inhibitor along with penicillins and cephalosporins was successfully employed for the nosocomial infections caused by *P. aeruginosa*, *E. coli*, and *K. Pneumoniae* (Blush and Brandford 2016). The potential of plant molecule, stigmasterol, as an adjuvant to beta lactam antibiotic, ampicillin, was studied by Yenn and coworkers. The combination of inhibitor and antibiotic exhibited synergistic antibacterial property against *E. coli*, *S. aureus*, and *P. aeruginosa* (Yenn et al. 2017). Aspergillomarasmine is a fungal-derived inhibitor combined with meropenem to regain the antibacterial mechanism against Enterobacteriaceae, *Acinetobacter* sp., and *Pseudomonas* sp. (Dolgin 2016).

13.4.2 Inhibitors of Efflux Pumps (EPIs)

EPs are proteins present in Gram-negative and positive bacterial cell membrane that contribute to the elimination of harmful molecules from cells to outside system. There are primary and secondary transporters based on energy requirements. Primary transporters work by energy utilized by ATP. Secondary transporters work based on proton or sodium ion potential across the membrane. Various efflux pumps are major facilitator superfamily (MFS), resistance nodulation cell division (RND), ATP-binding cassette (ATP), multidrug and toxic compound extrusion, and small multidrug resistance (Tegos et al. 2012; Sharma et al. 2019; Willers et al. 2017).

A plant-based EP inhibitor, reserpine, is found to act synergistically with tetracycline and repressed the clearance of antibiotic from methicillin-resistant *S. aureus*. In *E. coli*, ciprofloxacin activity is enhanced when combined with reserpine (Tariq et al. 2019). Microbially isolated, compound ethyl 4-bromopyrrole-2-carboxylate potentiated antibiotics against bacterial strains of *E. coli* and *P. aeruginosa*. Compound acted on the mechanism of RNA efflux pumps. Efflux pump inhibitor showed synergistic activity with tetracycline in *E. coli* and levofloxacin in *P. aeruginosa* (Tambat et al. 2022). Phe-Arg-beta-naphthylamide is a well-studied EPI as well as virulence inhibitor of *P. aeruginosa* which affects the production of different RND pumps such as MexAB-OprM, MexCD-OprJ, and MexEF-OprN18. This inhibitor works synergistically with the fluoroquinolone antibiotic, levofloxacin (Rampioni et al. 2017). In another study, EPI along with a permeability increasing drug exhibited synergy interaction involving multidrug-resistant antibiotics. Polymyxin B nonapeptide enhanced the activity of antibiotics (azithromycin and doxycycline) along with inhibitors, phenylalanine-arginine beta-naphthylamide) and 1-(1-naphthylmethyl)-piperazine against AMR *P. aeruginosa* (Ferrer-Espada et al. 2019). EPI, 1-(1-naphthylmethyl)-piperazine is also found active against *E. coli*, *A. baumannii*, and *K. pneumoniae* in combination with tetracyclines, fluoroquinolones, macrolides, penicillins, and rifampicin (Yang et al. 2017). A lipopeptide was used in combination against Gram-negative ESKAPE (*K. pneumoniae*, *P. aeruginosa*, and *A. baumannii*) pathogens. The adjuvant was found acting on outer membrane permeability and efflux pump mechanism and exhibited synergy with rifampin and erythromycin. The adjuvant compound, C10BBc12B, also protected mice from infections with enhanced survival and reduced bacterial load in different organs (Meir et al. 2022).

13.4.3 Membrane Permeabilizers

Permeability of cell membrane is always a problem for the entry of sufficient antibiotics to the bacteria. Thus, improving the permeability in bacteria, especially Gram-negative bacteria, enhances antibiotic mechanism. Lipopolysaccharide layer of outer membrane present in Gram-negative bacteria requires membrane permeabilizers to improve antibiotic efficiency (Vaara 2019; Schmid et al. 2019). In *A. baumannii*, activity of azithromycin was enhanced when combined with colistin which weakens the cell membrane (Lin et al. 2015). In another work, when colistin was combined with different antibiotics against carbapenem-resistant *A. baumannii*, synergistic interaction was observed. A maximum of synergy was studied with colistin-vancomycin combinations and then by colistin-minocycline (Sertcelik et al. 2020). Antibacterial activity of rifampin against *E. coli* and *P. aeruginosa* was enhanced by polymyxin B nonapeptide which permeabilized the cell membrane (Vaara 1992). Phenylalanine-arginine beta-naphthylamide is one of most discussed membrane permeabilizers against *P. aeruginosa* (González-Bello 2017). Recently, nanoparticles are employed as antibacterial agents against broad spectrum pathogenic bacteria. In a recent study, ability of cerium oxide nanoparticle to enhance the efficacy of beta-lactam drugs in *K. pneumoniae* was studied. Interestingly, nanoparticle-antibiotics combinations exhibited synergistic interactions as cerium oxide nanoparticles can act on cell membrane. Activity of cefotaxime, amoxicillin, amoxicillin/clavulanate, and imipenem was improved in presence of nanoparticle (Bellio et al. 2018). 4-Hexylresorcinol was used as antibiotic adjuvant with polymyxin antibiotic and efficiently eradicated mice infected with *K. pneumoniae*. The combination which works on cell membrane protected 75% of infected animal from sepsis and death with respect to antibiotic alone (Nikolaev et al. 2020).

13.5 Antibiotic Adjuvants in Biofilm and Anti-Quorum Sensing Therapy

Antibiotic adjuvants are being explored widely in the therapy targeting biofilm formation and quorum sensing mechanisms of pathogens. As biofilm establishment and quorum sensing are also related to the emergence of antibiotic resistance, inhibitors or adjuvants used along with conventional antibiotics serve as an excellent therapeutic strategy (Sionov and Steinberg 2022). Virulence of bacteria in host is also impaired by inhibiting the biofilms and quorum sensing. Quercetin is a plant flavonoid observed as an adjuvant along with antibiotics against broad spectrum bacteria including *S. aureus*, *K. pneumoniae*, *P. aeruginosa*, *E. coli*, and *A. baumannii*. Quercetin and amikacin and tobramycin combinations exhibited synergy interaction against clinical isolates of *P. aeruginosa*. Biofilm development and metabolic activity of pathogenic bacteria were reduced with the synergy association of antibiotics and flavonoid (Vipin et al. 2020). In a similar work, biofilms of *P. aeruginosa* were inhibited by using a molecule acting on quorum sensing in combination with ciprofloxacin. A novel, quorum sensing inhibitor, (R)-2-(4-(3-(6-chloro-4-oxoquinazolin-3(4H)yl)-2-hydroxypropoxy)phenyl)acetonitrile coated with polymer, potentiated the activity of ciprofloxacin and impaired biofilm formation (Soukarieh et al. 2023). Plant-derived quorum sensing inhibitor, baicalein in association with beta lactam antibiotics, exhibited synergy against methicillin-resistant *S. aureus*. It was detected that the MIC of tetracycline against test pathogen was reduced significantly when combined with adjuvant (Fujita et al. 2005).

Anti-biofilm agent, 2-aminoimidazole/triazole conjugate, exhibited synergistic effect with conventional antibiotics against antibiotic-resistant *S. aureus* and *A. baumannii*. Colistin and novobiocin in combination with anti-biofilm agent reduced MIC of drugs against *A. baumannii* and *S. aureus*, respectively (Rogers et al. 2010). At present, nanoparticles are best utilized as potential anti-bacterial and anti-biofilm agents against broad spectrum pathogenic bacteria. Silver nanoparticles were employed in the therapy which target biofilm and quorum sensing in *S. aureus*, *E. coli*, and *P. aeruginosa*. Also, antibiofilm and anti-quorum sensing silver nanoparticles improved the anti-bacterial properties of different antibiotics such as amoxycillin, ampicillin, and methicillin (Kumar et al. 2013). In a similar study, silver nanoparticle synthesized through fungus enhanced the mechanism of antibiotics (kanamycin, oxytetracycline, and streptomycin) against *S. aureus*, *E. coli*, and *P. aeruginosa* (Barapatre et al. 2016).

13.6 Clinical Improvements in Adjuvant Therapy

There are several combinations of antibiotics or adjuvants studied in different stages of clinical trials. It is imperative to evaluate the in vitro studied drug-dose regimes, mechanisms, pharmacodynamics, toxicity, and stability of combinations in in vivo conditions also. It is always found that the effectiveness of drugs varies in in vivo conditions. Most of the clinically approved and widely used antibiotic adjuvants are from beta-lactamase inhibitors. The very first approved and widely practiced beta-lactamase inhibitor and beta lactam antibiotic was sold under the name Augmentin. It is the combination of clavulanic acid and amoxicillin. An effective antibacterial combination of avibactam and ceftazidime available in market is known as Avycaz (Dhanda et al. 2023).

A clinical trial has been conducted for beta-lactamase inhibitors to restore the efficacy of vaborbactam against carbapenemase producing *K. pneumoniae*. The trial included 41 subjects and confirmed the safety of application among humans. Metal ion chelators such as EDTA, deferasirox, and deferoxamine as antibiotic adjuvants for imipenem, tobramycin, and vancomycin were tested in murine infection models of *S. aureus* and *P. aeruginosa*. The results showed reduced bacterial load in model system (Mulani et al. 2019). In addition, some clinical studies reported toxicity and ineffectiveness of combinations against several pathogens. Combinations of two antibiotics, rifampin and colistin, were clinically proven as ineffective against extensively drug-resistant *A. baumannii* and also exhibited hepatotoxicity (Durante-Mangoni et al. 2013). In another study, metal chelators used as adjuvant sequestered ions from host cells also rendering ineffectiveness (Yoshizumi et al. 2013).

13.7 Conclusions

The crisis of AMR among ESKPAE group is one of the global health challenges requiring utmost attention in the design of new therapeutic agents. Often, the process of novel drug design and development and associated research get stalled by several factors including the strict guidelines of drug design. In such scenario, non-antibiotic compounds are developed as antibiotic adjuvants able to improve the efficacy of antibiotics against pathogens. Antibiotic adjuvants are often used in combination with traditional antibiotics to control antibiotic resistance problem and to improve the antibiotic regimes. Synergistic property of antibiotic and adjuvant combinations is studied in vitro and in vivo models in recent years. To address the problem of insufficient and inferior anti-bacterial drugs, several classes of adjuvants targeting the beta-lactamases, efflux pumps, and cell membrane were explored and studied. To date, few antibiotic adjuvants are clinically approved and more candidates are in the queue to get approval. Thus, the gap observed in the development of different adjuvants needs to be discussed further.

References

Akova M (2016) Epidemiology of antimicrobial resistance in bloodstream infections. Virulence 7:252–266

Barapatre A, Aadil KR, Jha H (2016) Synergistic antibacterial and antibiofilm activity of silver nanoparticles biosynthesized by lignin-degrading fungus. Bioresour Bioprocess 3:8

Bellio P, Luzi C, Mancini A, Cracchiolo S, Passacantando M, Di Pietro L et al (2018) Cerium oxide nanoparticles as potential antibiotic adjuvant. Effects of CeO 2 nanoparticles on bacterial outer membrane permeability. Biochim Biophys Acta Biomembr 1860(11):2428–2435

Blush K, Brandford PA (2016) Beta-lactam and beta-lactamase inhibitors: an overview. Cold Spring Harb Perspect Med 6(8):a025247

Chinemerem Nwobodo D, Ugwu MC, Oliseloke Anie C, Al-Ouqaili MTS, Chinedu Ikem J, Victor Chigozie U et al (2022) Antibiotic resistance: the challenges and some emerging strategies for tackling a global menace. J Clin Lab Anal 36(9):e24655

De Oliveira DMP, Forde BM, Kidd TJ, Harris PNA, Schembri MA, Beatson SA et al (2020) Antimicrobial resistance in ESKAPE pathogens. Clin Microbiol Rev 33(3):e00181–e00119

Denissen J, Reyneke B, Waso-Reyneke M, Havenga B, Barnard T, Khan S et al (2022) Prevalence of ESKAPE pathogens in the environment: antibiotic resistance status, community-acquired infection and risk to human health. Int J Hyg Environ Health 244:114006

Dhanda G, Acharya Y, Haldar J (2023) Antibiotic adjuvants: a versatile approach to combat antibiotic resistance. ACS Omega 8(12):10757–10783

Dolgin E (2016) Combating antibiotic resistance from the ground up. Proc Natl Acad Sci U S A 113(42):11642–11643

Durante-Mangoni E, Signoriello G, Andini R, Mattei A, De Cristoforo M, Murino P et al (2013) Colistin and rifampicin compared with colistin alone for the treatment of serious infections due to extensively drug-resistant *Acinetobacter baumannii*: a multicenter, randomized clinical trial. Clin Infect Dis 57:349–358

Ferrer-Espada R, Shahrour H, Pitts B, Stewart PS, Sánchez-Gómez S, Martínez-de-Tejada G (2019) A permeability-increasing drug synergizes with bacterial efflux pump inhibitors and restores susceptibility to antibiotics in multi-drug resistant *Pseudomonas aeruginosa* strains. Sci Rep 9:3452

Founou RC, Founou LL, Essack SY (2017) Clinical and economic impact of antibiotic resistance in developing countries: a systematic review and meta-analysis. PLoS One 12(12):e0189621

Fujita M, Shiota S, Kuroda T, Hatano T, Yoshida T, Mizushima T, Tsuchiya T (2005) Remarkable synergies between baicalein and tetracycline, and baicalein and beta-lactams against methicillinresistant Staphylococcus aureus. Microbiol Immunol 49(4):391–396. https://doi.org/10.1111/j.1348-0421.2005.tb03732.x. PMID: 15840965

Gill EE, Franco OL, Hancock REW (2015) Antibiotic adjuvants: diverse strategies for controlling drug-resistant pathogens. Chem Biol Drug Des 85(1):56–78

González-Bello C (2017) Antibiotic adjuvants—a strategy to unlock bacterial resistance to antibiotics. Bioorg Med Chem Lett 27(18):4221–4228

González-Bello C, Rodríguez D, Pernas M, Rodríguez Á, Colchón E (2020) β-Lactamase inhibitors to restore the efficacy of antibiotics against superbugs. J Med Chem 63:1859–1881

Helmy YA, Taha-Abdelaziz K, Hawwas HAEH, Ghosh S, AlKafaas SS, Moawad MMM et al (2023) Antimicrobial resistance and recent alternatives to antibiotics for the control of bacterial pathogens with an emphasis on foodborne pathogens. Antibiotics 12(2):274

Jadimurthy R, Mayegowda SB, Nayak SC, Mohan CD, Rangappa KS (2022) Escaping mechanisms of ESKAPE pathogens from antibiotics and their targeting by natural compounds. Biotechnol Rep 34:e00728

Kalpana S, Lin W-Y, Wang Y-C, Fu Y, Lakshmi A, Wang H-Y (2023) Antibiotic resistance diagnosis in ESKAPE pathogens—a review on proteomic perspective. Diagnostics (Basel) 13:1014

Kerantzas CA, Jacobs WR (2017) Origins of combination therapy for tuberculosis: lessons for future antimicrobial development and application. mBio. 8(2):e01586–e01516

Kumar MA, Mahesh N, Kumar SB, Kumar RS, Priyadharshni S (2013) Antiquorum sensing and antibacterial activity of silver nanoparticles synthesized by mutant *Klebsiella pneumoniae* MTCC 3354. Asian J Chem 25(17):9961–9964

Lewis K (2007) Persister cells, dormancy and infectious disease. Nat Rev Microbiol 5:48–56

Lin L, Nonejuie P, Munguia J, Hollands A, Olson J, Dam Q et al (2015) Azithromycin synergizes with cationic antimicrobial peptides to exert bactericidal and therapeutic activity against highly multidrug-resistant gram-negative bacterial pathogens. EBioMedicine 2(7):690–698

Meir O, Zaknoon F, Mor A (2022) An efflux-susceptible antibiotic-adjuvant with systemic efficacy against mouse infections. Sci Rep 12(1):17673

Mulani MS, Kamble EE, Kumkar SN, Tawre MS, Pardesi KR (2019) Emerging strategies to combat ESKAPE pathogens in the era of antimicrobial resistance: a review. Front Microbiol 10:539

Navidinia M, Goudarzi M, Rameshe SM, Farajollahi Z, Asl PE, Khosravi SZ et al (2017) Molecular characterization of resistance genes in MDR-ESKAPE pathogens. J Pure Appl Microbiol 11(2):779

Nikolaev YA, Tutel'yan AV, Loiko NG, Buck J, Sidorenko SV, Lazareva I et al (2020) The use of 4-Hexylresorcinol as antibiotic adjuvant. PLoS One 15(9):e0239147

Peterson E, Kaur P (2018) Antibiotic resistance mechanisms in bacteria: relationships between resistance determinants of antibiotic producers, environmental bacteria, and clinical pathogens. Front Microbiol 9:2928

Rampioni G, Pillai CR, Longo F, Bondì R, Baldelli V, Messina M et al (2017) Effect of efflux pump inhibition on *Pseudomonas aeruginosa* transcriptome and virulence. Sci Rep 7:11392

Rice LB (2008) Federal funding for the study of antimicrobial resistance in nosocomial pathogens: no ESKAPE. J Infect Dis 197(8):1079–1081

Rice LB (2010) Progress and challenges in implementing the research on ESKAPE pathogens. Infect Control Hosp Epidemiol 31:S7–S10

Rishi P, Vij S, Maurya IK, Kaur UJ, Bharati S, Tewari R (2018) Peptides as adjuvants for ampicillin and oxacillin against methicillin-resistant *Staphylococcus aureus* (MRSA). Microb Pathog 124:11–20

Rogers SA, Huigens RW, Cavanagh J, Melander C (2010) Synergistic effects between conventional antibiotics and 2-aminoimidazole-derived antibiofilm agents. Antimicrob Agents Chemother 54(5):2112–2118

Schmid A, Wolfensberger A, Nemeth J, Schreiber PW, Sax H, Kuster SP (2019) Monotherapy versus combination therapy for multidrug-resistant Gram-negative infections: systematic review and meta-analysis. Sci Rep 9:15290

Sertcelik A, Baran I, Akinci E, Mumcuoglu I, Bodur H (2020) Synergistic activities of colistin combinations with meropenem, sulbactam, minocycline, disodium Fosfomycin, or vancomycin against different clones of carbapenem-resistant *Acinetobacter baumannii* strains. Microb Drug Resist 26:429–433

Sharma A, Gupta VK, Pathania R (2019) Efflux pump inhibitors for bacterial pathogens: from bench to bedside. Indian J Med Res 149:129–145

Sionov RV, Steinberg D (2022) Targeting the holy triangle of quorum sensing, biofilm formation, and antibiotic resistance in pathogenic bacteria. Microorganisms 10:1239

Soukarieh F, Gurnani P, Romero M, Halliday N, Stocks M, Alexander C et al (2023) Design of quorum sensing inhibitor-polymer conjugates to penetrate *Pseudomonas aeruginosa* biofilms. ACS Macro Lett 12:314–319

Strathdee SA, Davies SC, Marcelin JR (2020) Confronting antimicrobial resistance beyond the COVID-19 pandemic and the 2020 US election. Lancet 396:P1050–P1053

Tambat R, Mahey N, Chandal N, Verma DK, Jangra M, Thakur KG et al (2022) A microbe-derived efflux pump inhibitor of the resistance-nodulation-cell division protein restores antibiotic susceptibility in *Escherichia coli* and *Pseudomonas aeruginosa*. ACS Infect Dis 8:255–270

Tariq A, Sana M, Shaheen A, Ismat F, Mahboob S, Rauf W et al (2019) Restraining the multidrug efflux transporter STY4874 of *Salmonella Typhi* by reserpine and plant extracts. Lett Appl Microbiol 69(3):161–167. https://doi.org/10.1111/lam.13196

Tegos GP, Haynes M, Jacob Strouse J, Md T Khan M, Bologa CG, Oprea TI et al (2012) Microbial efflux pump inhibition: tactics and strategies. Curr Pharm Des 17(13):1291–1302

Vaara M (1992) Agents that increase the permeability of the outer membrane. Microbiol Rev 56(3):395–411

Vaara M (2019) Polymyxins and their potential next generation as therapeutic antibiotics. Front Microbiol 10:1689

Verma S (2015) A retrospective study to evaluate the efficacy of a new antibiotic adjuvant entity (β-lactam/β-lactamase inhibitor/adjuvant disodium edetate combination) for management of sepsis. Res J Infect Dis 3:3

Vipin C, Saptami K, Fida F, Mujeeburahiman M, Rao SS, Athmika et al (2020) Potential synergistic activity of quercetin with antibiotics against multidrug-resistant clinical strains of *Pseudomonas aeruginosa*. PLoS One 15(11):e0241304

Willers C, Wentzel JF, du Plessis LH, Gouws C, Hamman JH (2017) Efflux as a mechanism of antimicrobial drug resistance in clinical relevant microorganisms: the role of efflux inhibitors. Expert Opin Ther Targets 21(1):23–36

Yang X, Goswami S, Gorityala BK, Domalaon R, Lyu Y, Kumar A et al (2017) A tobramycin vector enhances synergy and efficacy of efflux pump inhibitors against multidrug-resistant gram-negative bacteria. J Med Chem 60:3913–3932

Yap JKY, Tan SYY, Tang SQ, Thien VK, Chan EWL (2021) Synergistic antibacterial activity between 1,4-naphthoquinone and β-lactam antibiotics against methicillin-resistant *Staphylococcus aureus*. Microb Drug Resist 27:234–240

Yenn TW, Arslan Khan M, Amiera Syuhada N, Chean Ring L, Ibrahim D, Tan WN (2017) Stigmasterol: an adjuvant for beta lactam antibiotics against beta-lactamase positive clinical isolates. Steroids 128:68–71

Yoshizumi A, Ishii Y, Livermore DM, Woodford N, Kimura S, Saga T et al (2013) Efficacies of calcium-EDTA in combination with imipenem in a murine model of sepsis caused by *Escherichia coli* with NDM-1 β-lactamase. J Infect Chemother 19:992–995

Phytochemicals as Potential Antibacterial Agents Against ESKAPE Pathogens

14

Subhaswaraj Pattnaik, Monika Mishra, and Pradeep Kumar Naik

Abstract

The alarming increase in antimicrobial resistance (AMR) phenomena becomes a serious global public health threat across the globe. Irrational use, misuse, and overprescription of antibiotics with non-specific purpose are the primary causes of AMR in pathogens. According to WHO, the priority pathogens, particularly ESKAPE pathogens (*Enterococcus faecium*, *Staphylococcus aureus*, *Klebsiella pneumoniae*, *Acinetobacter baumannii*, *Pseudomonas aeruginosa*, and *Enterobacter* sp.), pose extensive resistance to several antibiotics leading to global health hazards. These pathogenic bacteria are associated with several chronic health conditions by producing pathogenic elements, development of biofilm dynamics, expressions of efflux pumps, and inference of AMR. Owing to the importance of bacterial virulence and biofilm dynamics in the occurrence of resistance to antibiotics in clinical healthcare settings, it is imperative to look for alternative therapeutic agents to mitigate bacterial pathogenesis and biofilm dynamics. In search of novel antibacterial and antibiofilm agents, several synthetic and semi-synthetic drugs are being considered for efficient management of chronic microbial diseases. Nevertheless, several factors such as biodegradability issues, toxicity, and less absorptivity limit their candidature as promising therapeutic agents against chronic bacterial infections. In this context, pharmacologically important medicinal plants with high-throughput values in folkloric practices could invariably be developed as therapeutic agents for the treatment of bacterial infections associated with ESKAPE pathogens. In recent years, several reports have emphasized the use of medicinal plants and plant-derived secondary metabolites against bacterial pathogenesis, biofilm mechanics, and drug resis-

S. Pattnaik · M. Mishra · P. K. Naik (✉)
Centre of Excellence in Natural Products and Therapeutics, Department of Biotechnology and Bioinformatics, Sambalpur University, Sambalpur, Odisha, India
e-mail: pknaik1973@suniv.ac.in

S. Busi, R. Prasad (eds.), *ESKAPE Pathogens*,
https://doi.org/10.1007/978-981-99-8799-3_14

tance severity. The present chapter, thus, emphasizes the therapeutic role of phytochemicals in the development of antibacterial agents, inhibitors of quorum sensing- (QS) mediated bacterial pathogenesis, inhibitors of efflux machinery, and mitigators of biofilm matrices against ESKAPE group of microorganisms. The chapter also critically evaluates the FDA-approved phytochemical drugs and their formulations for therapeutic management of bacterial drug resistance. An interactive discussion also emphasizes the repurposing of FDA-approved phytochemicals towards the mitigation of bacterial infections and biofilms, which are being used earlier for other therapeutic roles. The chapter also focuses on combinatorial tactics using phytochemicals with existing antibiotics and phytochemicals-based nanoformulations to mitigate the severity of bacterial infections in ESKAPE pathogens.

Keywords

Antimicrobial resistance (AMR) · Antibacterial agent · Biofilms · Efflux pump · ESKAPE · Medicinal plants · Phytochemicals · Quorum sensing

14.1 Introduction

Antibiotic discovery in 1920s gained considerable interest as promising therapeutic medications against chronic microbial infections. Antibiotics have emerged as "Wonder Drugs" with widespread abilities to mitigate the severity of microbial infections at varied level of targets with special reference to clinical settings (Liu et al. 2020; Nwobodo et al. 2022). Since the discovery of first of its kind of antibiotic, in a span of just over 100 years, antibiotics have completely revolutionized the concept of modern medicines to treat microbial infections by critically extending the average lifespan of human beings. The period from the 1940s to 1960s is considered the "Golden age of antibiotics" as the majority of antibiotics in clinical practices today were discovered during this period (Hutchings et al. 2019). A large number of clinically relevant antibiotics are derived from natural products with Actinomycetes being the highest stakeholders of producing antibiotics. Since the antibiotics were discovered from several biological and synthetic sources, it is imperative that the diverse class of antibiotics exhibits widespread functional attributes. For example, the antibiotics available in the market specifically target one or more cellular functions including synthesis of cell wall and interfering with different phases of Central Dogma (replication, transcription, translation of RNA to protein) and energy metabolism (Hutchings et al. 2019; Matsui et al. 2020). In addition, the combinatorial effect of several groups of antibiotics could be employed against the pathogenic microorganisms causing severe chronic infections for improved therapeutic outcome (Olsson et al. 2020).

14.1.1 Antibiotics Against MDR Bacteria

Antibiotics are regarded as therapeutic drugs to treat both acute and chronic diseases caused by pathogenic bacteria. The discovery of antibiotics has proven to be an integral part of our therapeutic regimens in the twentieth century providing a number of advantages over conventional strategies (Botelho et al. 2019). Several groups of antibiotics (i.e. Aminoglycosides, Phenicols, Sulfonamides, Tetracyclines, Macrolides, Cephalosporins, Carbapenems, Quinolones, Rifamycins, Lincosamides, etc.) covering several generations have been discovered with a widespread mechanism of action against wide range of pathogenic microorganisms (Coates et al. 2011). Similar to the diversity of antibiotic classes, the mechanisms of antibiotic action on target microbial pathogens are also highly diverse starting from cell wall synthesis to nucleotide synthesis and protein synthesis. For example, Carbapenems, Cephalosporins, Penicillins, etc. target the cell wall synthesis pathway for targeting the bacterial pathophysiological mechanisms. Similarly, Fluoroquinolones and Sulfonamides group of antibiotics specifically target the nucleic acid synthesis mechanism. Meanwhile, Aminoglycosides, Tetracyclines, Macrolides, Chloramphenicol, etc. specifically target the protein synthesis pathway (Kapoor et al. 2017; O'Rourke et al. 2020; Parmanik et al. 2022). The fourth generation antibiotics also initially showed an inhibitory effect on ESKAPE pathogens (*Enterococcus faecium*, *Staphylococcus aureus*, *Klebsiella pneumoniae*, *Acinetobacter baumannii*, *Pseudomonas aeruginosa*, and *Enterobacter* sp.). However, irrational uses and production of specific β-Lactamases by these priority pathogens counteract the action of the antibiotics used, thereby decreasing the sensitivity towards the antibiotics (Jadimurthy et al. 2022). However, a combination of different antibiotics targeting several pathophysiological processes could be an interesting aspect in handling and mitigation of ESKAPE pathogens-associated health hazards.

14.1.2 Antibiotic Resistance: A Major Threat

No doubt, the antibiotic discovery has revolutionized the therapeutic settings for effective regulation of microbial pathogenesis; its unregulated extensive use results in the advent of resistance to antimicrobial agents. The wrath of antibiotic resistance becomes a potential menace to public health with approximately 0.7 million deaths every year globally. As per the estimates, if we could not address this global menace, it could result in 10 million death per year globally by 2050 which ultimately create selective pressure on the global economy costing around $100 trillion (Nwobodo et al. 2022; Yuan et al. 2022). Even developed countries like USA also witness the wrath of drug resistance with more than 2 million infections and more than 23,000 deaths per year as per CDC report (Rather et al. 2021). WHO has categorized the ESKAPE group of pathogens into high priority pathogens as these MDR pathogens require urgent attention as they infer extensive resistance to next-generation antibiotics (Abdul-Jabbar et al. 2022). Despite the launch of the Global

Action Plan to tackle antimicrobial resistance (AMR) by practicing evidence-based prescription of antibiotics and development of low-cost diagnostic tools, the ever-increasing global antibiotic consumption rate with an alarming increase in Daily Defined Doses (DDDs) in developing countries led to resistance at the community level (Malik and Bhattacharyya 2019). ESKAPE pathogens play pivotal influence on the majority of hospital-acquired nosocomial infections and are resistant to several groups of conventional antibiotics, leading to severe health ailments and socio-economic downfall. The increased incidence of antibiotic resistance, if unattended, could lead to catastrophic effects on public health, agricultural settings, food security, animal husbandry, and the global economy (Murray et al. 2015; Nicolas et al. 2019). Though the wrath of AMR critically controls the socio-economic values at the global platform, countries with comparatively lower GDP values received the highest-burden as evidenced by the AMR-associated deaths in Asian and African countries as compared to European and American countries (Ayobami et al. 2022). As per the estimates, the antimicrobial resistance, if unaddressed, could lead to a severe economic crisis with approximately 24 million people forced to below the poverty level and a drastic fall in the Gross domestic product (GDP) of both LICs and LMICs (Zhen et al. 2019).

The antibiotic resistance profile of ESKAPE pathogens could be classified into intrinsic (i.e. alteration in permeability of bacterial membrane, activation of antibiotic-specific degrading enzymes, and increased expression of efflux pump), acquired (i.e. horizontal gene transfer and mutation), and adaptive resistance (i.e. transformation into biofilm mode). Apart from that, the tendency of a small bacterial sub-population within the bacterial community to switch towards metabolic dormancy leading to persistent behaviour also critically influences antibiotic failure against ESKAPE pathogens (Pang et al. 2019). Mutation has a critical role to play in inferring resistance profile in pathogenic microorganisms apart from the inadvertent exposure of microorganisms to high doses of antibiotics (Ciofu and Tolker-Nielsen 2019). In terms of physiological responses, bacterial acquisition and the activation of drug efflux transporters also critically minimize the sensitivity of traditional antibiotics of interest towards pathogenic microorganisms (Miesel et al. 2003). Besides, environmental wastes such as hospital wastewater, pharmaceutical effluents, agricultural wastes, run-off from livestock, etc. also critically create a selective pressure on microorganisms and influence the upsurging of AMR (Danner et al. 2019; Monahan et al. 2022). Based upon the severity of antibiotic resistance in hospital-acquired infections and simultaneous failure of current therapeutic regimens, we are facing the state of "Post-Antibiotic era", where we need to develop alternative therapeutics on an urgent basis bypassing the antibiotic therapies (Terreni et al. 2021).

14.1.3 AMR Profile in ESKAPE Pathogens

As mentioned earlier, ESKAPE group of pathogens as listed by WHO is highly resistant to conventional antibiotics including the next-generation antibiotics of several class. As evident from research and development in the field of AMR in

ESKAPE pathogens concerned, several noted mechanisms are responsible for the emergence of AMR. These mechanisms include drug inactivation by bacterial enzymes, modification of drug binding targets by gene mutation, reduction in intracellular drug permeability through modification in membrane dynamics via activation of efflux pump transporters, activation of alternative metabolic pathways, evolution of recalcitrant biofilm matrix, evolution of bacterial tolerance to transient exposure to antibiotics, and bacterial persistence (Fig. 14.1) (Blair et al. 2015; Santajit and Indrawattana 2016; De Oliveira et al. 2020; Parmanik et al. 2022). The resistance-encoding genes in ESKAPE pathogens causing resistance patterns are further transferred to other pathogens through genetic recombination, particularly through horizontal gene transfer (Ashkenazi 2013). The acquisition of resistant mobile genetic elements by ESKAPE pathogens provides new dimensions to the resistance patterns against antibiotics including next-generation Carbapenems (De Oliveira et al. 2020).

As per the received trends, Vancomycin-resistant ESKAPE pathogen, *E. faecium,* exhibited resistance towards Cephalosporins, Ampicillin, Linezolid, etc. through altered target binding, activation of efflux pump transporters, low uptake capacity, and mutation (Arias and Murray 2012; Remschmidt et al. 2018). Meanwhile, Methicillin-resistant *S. aureus* showed increased resistance towards Macrolides, β-lactams, Aminoglycosides, Fluoroquinolones, and Chloramphenicol (Harkins et al. 2017; Turner et al. 2019). Pneumonia and soft tissue infections causing ESKAPE pathogen, *K. pneumonia,* displayed resistance against Carbapenems, third generation Cephalosporins, Fluoroquinolones, Aminoglycosides, etc. (Zowawi et al. 2015). The priority ESKAPE pathogen, *A. baumannii,* inferred resistance against β-lactams, Carbapenems, fourth generation Cephalosporins, etc. (Xie et al. 2018). Similarly, opportunistic ESKAPE pathogen, *P. aeruginosa,* exhibited resistance against second generation Cephalosporins, β-lactams, Carbapenems, Polymixins, and Piperacillin-Tazobactams (Yayan et al. 2015). Meanwhile, fourth generation Cephalosporins, β-lactams, Carbapenems antibiotics showed the least sensitivity against neonatal pneumonia-causing pathogen, *Enterobacter* sp. (Thiolas et al. 2005).

In clinical healthcare settings, Gram-negative ESKAPE pathogens have the tendency to produce several enzymes, which specifically target the β-lactam group of antibiotics. Hence, the production of these enzymes in ESKAPE pathogens provides an added dimension to eventually degrade the β-lactams and is thus influential in the extended MDR phenomenon (Pandey et al. 2021). The ESKAPE pathogens have the tendency to produce several post-translational modifications (PTMs) which tend to modulate the host cellular pathway, thereby inferring resistance against host immunity and antibiotic treatment (Tiwari 2019).

Biofilms infer the microbial aggregates that harbour within the matrix of extracellular materials secreted by the embedding microbes with polysaccharides and proteins constituting the major proportions of the matrix. The EPS also constitutes the extracellular lipids and nucleic acids. As compared to their planktonic state, the biofilm mode of life style significantly promotes greater survival instincts against stress environment including tolerance towards antibiotic exposure (Flemming

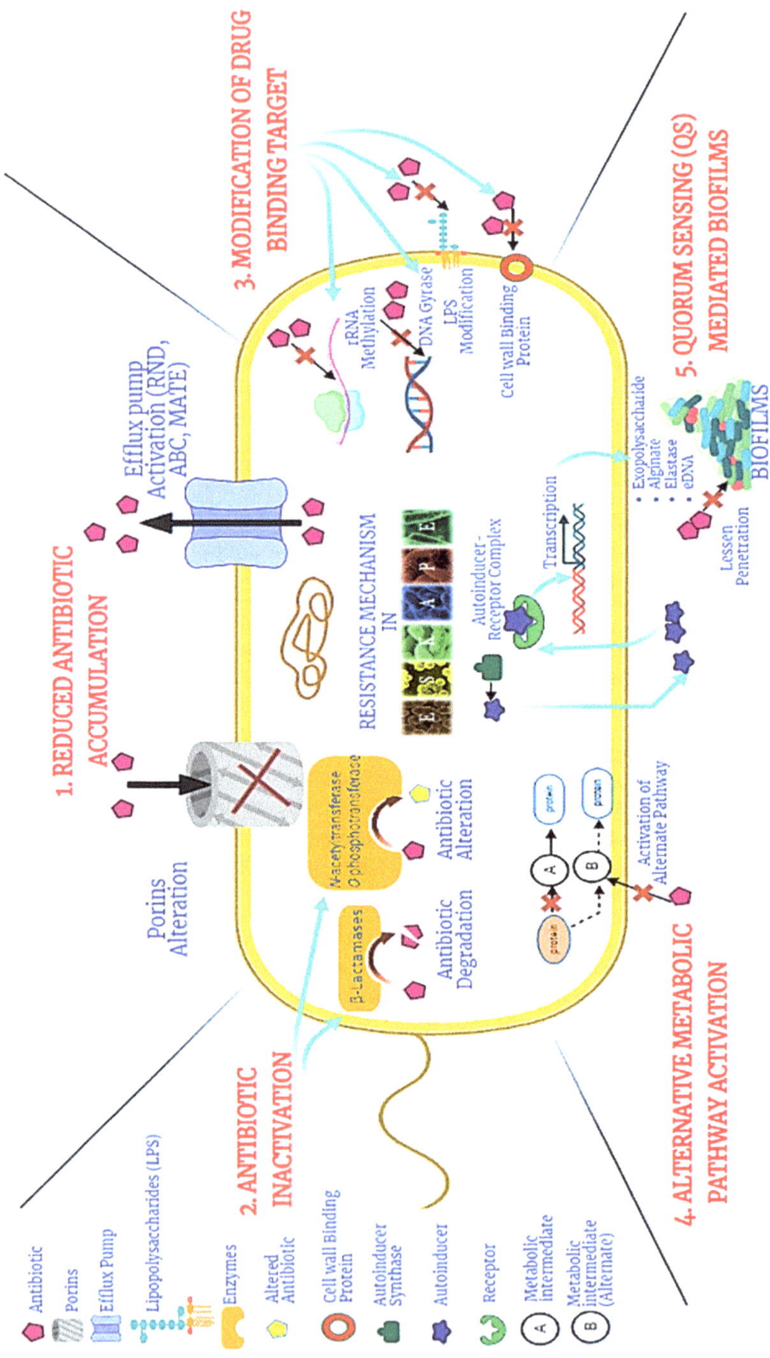

Fig. 14.1 Schematic representation of the mechanism of antibiotic resistance in ESKAPE pathogens

et al. 2016). The embedding biofilm matrix provides multifaceted advantages such as resource capture and retention ability, improved social cooperation and interactions, increased pathogenicity, and providing a stable platform for the dissemination of resistance genes through horizontal gene transfer (Zhang et al. 2015; Flemming et al. 2016). Hence, biofilm matrices are considered the hotspots of antibiotic resistance by inhabiting the drug-resistant genes and mobile genetic elements (Balcázar et al. 2015). Biofilm formation and its implications in drug resistance also have a critical influence on several food-borne outbreaks with the MDR *S. aureus* (Farha et al. 2020).

14.1.4 Biofilm-Related Drug Resistance in ESKAPE Pathogens

The biofilm matrix not only maintains the architecture of biofilms, but also promotes innate tolerance towards the penetration of antimicrobial agents by forming a mechanical barrier. Hence, bacterial pathogens with a biofilm mode of life style are approximately 10,000 times more resistant than that of their free-living planktonic stage (Rima et al. 2022). Basically, the ESKAPE pathogens utilize biofilm mechanics as a protective tool not only to evade the effect of antibiotic exposure, but also to escape the wrath of host defence mechanisms (Thornton et al. 2021). As per the report published by National Institute of Health (NIH), biofilms contribute significantly to the majority of chronic microbial infections with a global share of 65–80% (Patil et al. 2021). The dynamic changes in the environment (i.e. pH, temperature, the concentration of chemical moieties, etc.) critically affect the physiological response of embedded microbial communities within the biofilm matrix. As a result of physiological and biochemical changes within the bacterial community, the microorganisms attain the state of resistance to the stress conditions including antibiotic exposure (Zhao et al. 2020). Since biofilms are the reservoir of genetic diversity which provides an added advantage of adaptation, evolution, and survival in a hostile environment, the biofilm lifestyle suits the concerned bacterial community against the wrath of antibiotics (Dincer et al. 2020). The transition of pathogenic microorganisms from planktonic to sessile stage is controlled by Gac/Rsm pathway or the involvement of cyclic diguanosine monophosphate (c-di-GMP) signalling pathway (Moscoso et al. 2014). For instance, the binding of transcription factor, FleQ (flagellar gene expression), to c-di-GMP in *P. aeruginosa* coherently regulates transition between planktonic to biofilm state (Baraquet and Harwood 2016). In addition, ESKAPE pathogens produce several types of adhesins, which are associated with the establishment of biofilms and are influential for its regulation (Patil et al. 2021).

The biofilm state of pathogenic microorganisms invariably generates gradients of nutrient dispersion which ultimately generate differential metabolic states at a given point of time, resulting in the development of tolerance to antibiotic exposure and bacterial persistence (Uruén et al. 2021). The pathophysiological responses recorded during acute and chronic bacterial infections and their recalcitrance to antibiotics are the results of the evolution of biofilm mechanics. The sessile biofilms formed over the biomedical devices become catastrophic influence on immuno compromised patients. The etiological response of biofilm-forming ESKAPE pathogens in Ventilator-associated pneumonia is one of the prime examples of the extensive catastrophic health hazards in different healthcare settings (Baidya et al. 2021). Apart from that, pathogenic bacteria form extracellular matrices on biomedical indwelling catheters, orthopaedic implants, prosthetic organs, contact lenses, and endotracheal tubes, which confer an uphill challenge to clinical healthcare settings (Yadav et al. 2020). Since biofilms act as physiological barrier against the host immunity and antibiotic treatment, it is deliberately used as a strategic display of resistance (Karami et al. 2020). Traditionally, biofilm-associated infections in ESKAPE pathogens are controlled by several strategies such as bacteriophage therapy, physicochemical methods for removal of matrix, biophysical tools to facilitate matrix dissemination, anti-virulence therapeutics, and use of matrix destabilizing agents (Lazar et al. 2021).

14.1.5 Quorum Sensing (QS) in ESKAPE Pathogens

In the later phase of twentieth century, it was discovered that a highly defined cascade of events collectively termed "quorum sensing" (QS) in ESKAPE pathogens controls the fate of bacterial virulence, biofilm formation, and AMR. QS refers to cell density-dependent communication network, which infers the synthesis of diffusible autoinducers (AIs), release into the environment, and accumulation followed by detection through species-specific signaling receptors. These cascades of events are stringently associated with the bacterial cell density (S. Mukherjee et al. 2017; Mukherjee and Bassler 2019; Choi et al. 2022a). At higher cell density, i.e. above the critical threshold level, upon the detection of AIs to their cognate receptors, the QS signalling network undergoes activation which in turn activates upregulation of genes associated with bacterial pathogenesis and biofilm mechanics. The diversity of AIs associated with QS regulatory pathways could be observed from the fact that Gram-negative ESKAPE pathogens use acyl-homoserine lactones- (AHL) based AIs, whereas the Gram-positive ESKAPE pathogens utilize oligopeptides as the AIs (L. Kumar et al. 2022; Wang et al. 2022). Apart from these AIs, these pathogens also tend to utilize inter-species signalling autoinducers, i.e. AI-2 signals for bacterial communication and pathogenesis (Li et al. 2015a; Shukla et al. 2022).

QS signaling network controls the expression of several virulence phenotypes of pathophysiological importance, biofilm determinants, extracellular enzymes, cytotoxic elements, and motility factors. The synchronous expression of these virulence phenotypes upon QS activation in ESKAPE pathogens critically regulates the bacterial resistance profile and pathophysiological response (Quecan et al. 2019; Santajit et al. 2022). The successful establishment of persistent infections and inferring multidrug resistance patterns against several antibiotics are controlled by this highly intricate and complex cascade of QS hierarchy. The success of QS signaling cascade could be attributed to the production of several pathophysiological factors (e.g. EPS, alginate, rhamnolipids, eDNA, etc.), adhesion factors, exoenzymes (e.g.

Elastases, Proteases, etc.), and toxins (i.e. exotoxin A, lipid A, Alpha-toxins, Haemolycins, etc.) which critically modulate the biofilm development in ESKAPE pathogens leading to antibiotic resistance (Lazar et al. 2021; Santajit et al. 2022). The interplay between QS signaling cascade, biofilm mechanics, and inference of antibiotic resistance in ESKAPE pathogens critically provides an arsenal not only against the hostile environment, but also against antibiotic exposure (Sionov and Steinberg 2022). Hence, targeting QS and biofilm mechanics in ESKAPE bacteria seems to be an intriguing therapeutic approach for the mitigation of chronic bacterial infections.

14.1.6 Multidrug-Resistant Efflux Pump in ESKAPE Pathogens

Among the different factors responsible for inferring drug resistance in ESKAPE pathogens, multidrug efflux transporters contribution remains relevant in the emergence and maintenance of MDR phenomena. The presence of an efflux pump inherently extrudes the toxic substances including secondary metabolites and antimicrobial agents out of the bacterial cells into the surrounding medium, thereby minimizing the bioactivity of the drugs/antibiotics used against the bacterial infections (Mangiaterra et al. 2017). The efflux transporters in ESKAPE pathogens critically modulate the bacterial pathophysiology through the efficient management of QS, biofilm mechanics, pathogenic profile, surface motility, and host-microbe association and infer drug resistance (Verma et al. 2022). Basically, the efflux pumps have been categorized into five major families such as (1) ATP binding cassette (ABC) transporter, (2) Resistance nodulation cell division (RND) family, (3) Multidrug and toxin extrusion (MATE) superfamily, (4) Major facilitator superfamily (MFS) transporter, and (5) Small multidrug resistance (SMR) family. The efflux transporters generally extrude the substrates from intracellular space with lower concentration to extracellular space with higher concentration and hence these efflux transporters are highly energy-dependent. Among these major superfamilies, only ABC transporters belong to primary efflux transporter (i.e. gained energy by active hydrolysis of ATP), whereas the other efflux transporter families are categorized as 2° active transporters (i.e. gained energy from Proton motive force of ion gradients) (A Sharma et al. 2019; De Gaetano et al. 2023). The efflux transporters-mediated MDR in Gram-positive ESKAPE pathogens are basically cytoplasmic membrane-localized transporters. Meanwhile, the efflux transporters-mediated MDR in Gram-negative ESKAPE pathogens are categorically tripartite protein channels and much more complex due to the complex membrane system (Li et al. 2015b; Schindler and Kaatz 2016; Sharma et al. 2019). The presence of multiple promiscuous RND efflux transporters in ESKAPE pathogens critically modulates the active efflux of several classes of antibiotics out of the bacterial system and thereby minimizing the sensitivity of antibiotics leading to MDR (Cunrath et al. 2019). The relative manifestation of genes encoding efflux transporters is considered to be responsible for the acquisition of genes pertaining to resistance profile (Lorusso et al. 2022). Based on the intensity of several pathophysiological factors in inferring the ESKAPE pathogens resistance to conventional antibiotics, understanding the key concept of resistance profile is necessary to design strategic therapeutic measures.

14.1.7 Current Therapeutic Approaches Against ESKAPE Pathogens

With the increased incidence of MDR and extensive drug resistance (XDR) in pathogenic microorganisms, the pathogens have developed quintessential strategies to escape from conventional antibiotic treatment. Hence, therapeutic modalities are being explored to tackle the issues pertaining to antibiotic resistance. In addition to conventional antibiotic therapies, several advanced therapeutic approaches such as Bacteriophage therapy, the development of anti-virulence strategies to overcome bacterial pathogenicity, antimicrobial Photodynamic therapy (aPDT), antimicrobial peptides (AMPs), and essential oils (EOs) targeting the key regulators of bacterial infections have been employed for AMR mitigation (Mulani et al. 2019; Motiwala et al. 2022). In addition, a combination of more than one antibiotic, and/or synergistic activities of plant-derived phytochemicals with conventional antibiotics could also be undertaken to improve the therapeutic spectrum of used drug moieties against the drug-resistant ESKAPE pathogens (Mulani et al. 2019). The development of antiresistant potentiators by restoring the capacity of currently used antibiotics proved to be an alternative measure to mitigate bacterial infections and drug resistance phenomena. The use of β-Lactamse and efflux pump inhibitors is considered antiresistant drugs which have the tendency to block the resistance mechanism or potentiate the available antibiotics (Ma et al. 2020). In last decade, nanomaterials-based therapeutics are also employed as effective measures in controlling biofilm-mediated recurrent infections of clinical importance (Mukherjee et al. 2023). In this context, the bioactive secondary metabolites from natural sources with special reference to plant-derived phytocompounds and plant-derived EOs are considered for their role against bacterial survival and virulence mechanisms owing to their widespread folkloric practices (Jadimurthy et al. 2022; Panda et al. 2022).

14.2 Synthetic Drugs as Regulators of Bacterial Pathogenesis and Biofilm Mechanics

To develop innovative therapeutic modalities to disarm the pathogenic profile of ESKAPE pathogens and to complement the failure of antibiotics, synthetic antimicrobial and antibiofilm peptides have proven their candidature as promising agents in the management of MDR pathogens. For example, spatially designed antimicrobial peptide, SAAP-148, exhibited encouraging results in inferring bacteriocidal effect with biofilm modulatory properties against *S. aureus* and *A. baumannii* (van Gent et al. 2022). Recently, WLBU2, synthetic peptide, also exhibited promising inhibitory properties against the recalcitrant biofilm mechanics by modulating QS network (Masihzadeh et al. 2023). Similarly, a peculiar hybridization technique was employed to synthesize the Quinolone scaffolds and Sulfonamide moiety to improve the biofilm inhibitory activities against drug-resistant *P. aeruginosa* (Ghorab et al. 2023). The antimicrobial and antibiofilm potential of halogenated Benzyl-derivatives against nosocomial infections causing the pathogen, *P. aeruginosa, was* also critically established (Aboagye et al. 2023). The synthetic Chalcones and their derivatives also received significant attention as promising antagonists to bacterial growth and biofilm dynamics due to the ease of synthesis and their improved bioactivity. Naturally occurring chalcones, the precursor of flavonoids, also showed promising antibacterial and antibiofilm potential (Uchil et al. 2021). Though synthetic chemistry-derived antibacterial and antibiofilm agents are effective in the management of bacterial infections, limitations such as bioavailability and biodegradability profile hinder their widespread occurrence. Hence, in several countries, the research community has a keen interest to explore pharmacologically relevant natural products, i.e. primarily from plants followed by microbes and animals in dismantling bacterial pathogenesis (Qadri et al. 2022). Besides, natural products with special emphasis on phytochemicals could invariably be used as effective modules in the drug development pipelines against several life-threatening disorders and infectious diseases including chronic bacterial infections (Atanasov et al. 2021).

14.3 Antibacterial Properties of Phytochemicals

Owing to the alarming situation of drug resistance and slow progress on the development of potent antibacterial drugs in pharmaceutical industries, it is imperative to develop an integrated approach to utilize the age-old pharmacologically relevant natural products as therapeutic modules against ESKAPE pathogens (Ahmad et al. 2019). Different parts of plants and their Ayurvedic formulations are being actively used in folkloric practices against several diseases and disorders including microbial infections from several ages. The highly defined inhibitory potential of different plant parts against microbial infections could be attributed to the presence of flavanones, flavanols, polyphenols, coumarin derivatives, terpenes, terpenoids, alkaloids, organosulfur compounds, etc. (Khameneh et al. 2019; Sharma et al. 2023). Based upon the structural diversity of phytochemicals, the antibacterial actions of these phytochemicals also greatly vary with several potential therapeutic targets. For instance, phytochemicals such as Gallic acid and Pyrogallol exhibited synergistic activities with the conventional antibiotics to act upon the ESKAPE pathogen, *S. aureus* (Lima et al. 2016). The antibacterial activities of phenolic acids (4-Hydroxycinnamic acid and Chlorogenic acid) against ESKAPE pathogens, *S. aureus* and *K. pneumoniae* were attributed to the membrane damage (Lou et al. 2012; Bajko et al. 2016; Hochma et al. 2021).

Similarly, naturally derived alkaloids (e.g. Berberine, and Sanguinarine) showed antibacterial properties against ESKAPE pathogens (Yan et al. 2021). The flavonoids class of phytochemicals (e.g. Kaemferol, Quercetin, Naringenin, etc.) also been widely reported for antibacterial properties against ESKAPE pathogens by modulating the cytoplasmic membrane dynamics, nucleic acid synthesis, and metabolic activities (Shamsudin et al. 2022). The prenylated flavonoids, α-Mangostin, and Isobavachalcone showed a significant bactericidal effect on Methicillin-resistant Gram-positive bacterium, *S. aureus,* by disrupting the membrane rigidity which concomitantly hampers the bacterial homeostasis (Song et al. 2021). The widespread occurrence of terpenes and terpenoids (e.g. Terpineol, Terpinen-4-ol, Eugenol, etc.) in pharmacologically important plant-derived essential oils critically influences the antibacterial properties against MDR ESKAPE pathogens (Guimarães et al. 2019; Cordeiro et al. 2020; Cheruvanachari et al. 2023a, b). Besides, they also proved to be effective against the molecular determinants of antibiotic resistance such as bacterial signaling cascade (i.e. QS), biofilm mechanics, and efflux pump activation. Based upon the encouraging results of plant-derived phytocompounds from the preliminary investigations, translational mechanistic studies could be undertaken to control the severity of AMR phenomena (Khare et al. 2021). Table 14.1 depicted the list of bioactive phytochemicals derived from medicinal plants for their potential antibacterial properties against ESKAPE pathogens (Table 14.1).

14.3.1 Mechanism of Antibacterial Properties

The widespread antibacterial properties of phytoconstituents against ESKAPE pathogens could be attributed to several therapeutic targets. The mechanistic insight into the mechanism of antibacterial actions of plant-derived phytoconstituents suggested their influence on cell wall synthesis, modulation of bacterial membrane dynamics, alteration in cell permeability, modulation of bacterial physiology (i.e. nucleotide and protein synthesis), inhibition of bacterial enzymes, modulation of antibiotic susceptibility, inhibition of bacterial growth by modulating the bacterial ATP synthase production, management of biofilm, and attenuation of bacterial virulence by interfering the QS signaling network (Barbieri et al. 2017; AlSheikh et al. 2020; Jubair et al. 2021; Khameneh et al. 2021). For example, type II fatty acid (FAS-II) biosynthetic pathway catalysed by species-specific β-Hydroxyacyl-acyl carrier protein (FabZ) is considered as therapeutic target for antibacterial actions. Since FAS-II is devoid in humans, targeting the FAS-II pathway in microbial pathogens could promote effective inhibitory agents for bacterial pathogenesis (Rajarathinam et al. 2018).

14.4 QS Modulation Mechanism by Phytochemicals

Since QS controls the expression of virulence phenotypes of pathophysiological importance, competence, and sporulation, regulates the biofilm mechanics, and promotes the survival of the pathogens under harsh environmental conditions including antibiotic exposure, interference of QS network is considered as an alternative therapeutic regimen against chronic microbial infections. Besides, QS inhibition

Table 14.1 List of pharmacologically important Plant-derived phytochemicals for their potential antibacterial properties against ESKAPE pathogens

Sl. no.	Phytochemical class	Phytochemicals	Source plants	Target ESKAPE pathogen	MIC	Mechanism of action	References
1	Alkaloids	Chelerythrine	*Chelidonium majus*	*Pseudomonas aeruginosa*	0.019 mg/mL	–	Zielińska et al. (2019)
		Sanguinarine	*C. majus*	*Staphylococcus aureus*	0.019 mg/mL	–	Zielińska et al. (2019)
		Sanguinarine chloride hydrate	*Macleaya cordata*	*S. aureus*	0.128 mg/mL	Interferes with integrity and permeability of bacterial cell wall and membrane by producing reactive oxygen species (ROS)	Gu et al. (2023)
		Nigritanine	*Strychnos nigritana*	*S. aureus* ATCC 25923	128 μM	–	Casciaro et al. (2020)
		Berberine	*Berberis aristata*	*S. aureus*	0.2 mg/mL	Exhibits promising antibacterial activity and shows synergy with chitosan nanoparticles	Dash et al. (2020)
		Dihydrochelerythrine	*Zanthoxylum tingoassuiba*	*S. aureus* ATCC 25923	85.8 μM	Shows promising antibacterial activities	Costa et al. (2017)

(continued)

Table 14.1 (continued)

Sl. no.	Phytochemical class	Phytochemicals	Source plants	Target ESKAPE pathogen	MIC	Mechanism of action	References
2	Flavonoids	Naringenin	*Garcinia multiflora*	Colistin-resistant *Acinetobacter baumannii*, *Klebsiella pneumoniae*	>0.512 mg/mL	Exhibits promising antibacterial activities and shows synergistic activities with Colistin	Xu et al. (2022)
		Quercetin	Pure	*S. aureus* NCTC 5655	IC_{50}: 65 µM	Inhibits bacterial growth by modulating the cell membrane permeability	Veiko et al. (2023)
		Baicalein	Pure	*A. baumannii* AB145, *K. pneumoniae* KP1	>0.125 mg/mL	Exhibits synergistic activity with Doxycycline for improved antibacterial properties	Wang et al. (2023)
		Propolin D (Prenylated flavanones)	*Macaranga tanarius*	*S. aureus*	10–50 µg/mL	Exhibits antibacterial properties with specific impact on biofilm dynamics	Lee et al. (2019)
		Hesperetin	*Citrus* sp.	*S. aureus*	125 µg/mL	Shows antibacterial activities by producing ROS	Choi et al. (2022b)
		Kuwanon G	*Morus alba*	Methicillin resistant *S. aureus*	2 µg/mL	Shows antibacterial activities by modulating the membrane dynamics	Wu et al. (2019)
3	Terpenes and terpenoids	Terpinen-4-ol	*Melaleuca alternifolia*	*S. aureus*	0.25% (v/v)	Bactericidal properties	Cordeiro et al. (2020)
		Farnesal	*Eremophila lucida*	*S. aureus* ATCC 29213, ATCC 25923	65 µg/mL	Shows antibacterial properties by disruption of cell membrane	Biva et al. (2019)
		α-Pinene, Linalool	Pure	*S. aureus*	0.420 mg/mL	Bactericidal properties	Wang et al. (2019)
		Syzygiumursanolide D	*Syzygium szemaoense*	*S. aureus*	50 µg/mL	Disrupts the cell membrane integrity	Xu et al. (2020)
		(+)-Nootkatone	Plant-derived essential oil	*S. aureus* ATCC 25923	200 µg/mL	Shows both bactericidal and bacteriostatic activities	Farha et al. (2020)
		2-Phenyl ethyl methyl ether (PEME)	*Pandanus odorifer*	*S. aureus*	50 mM	Shows antibacterial properties by down-regulation of virulence genes	Cheruvanachari et al. (2023a, b)

strategies as compared to conventional antibiotic therapy put lesser selective pressure upon the microorganisms as it affects the bacterial pathogenicity without interfering the bacterial growth (Shaaban et al. 2019; Hemmati et al. 2020). QS inhibition therapy vows to potentiate the efficacy of conventional antimicrobial therapies using antibiotics. Thus, targeting QS cascade in the fight against microbial infections received considerable interest in the last few decades with special emphasis on the development of novel therapeutic agents as well as improving the efficacy of traditional antibiotics (Azuama et al. 2020). In the fight against QS-mediated pathogenicity in ESKAPE pathogens, several synthetic compounds and their derivatives have been reported to exhibit promising attenuation potential against the highly complex QS signaling cascade. Basically, derivatives of Furanone, Benzimidazoles, Benzothiazoles, Quinolines, Pyranones, Pyrroles, Pyridine, Pyrimidines, Indoles, Esters, Thiazoles, etc. were reported for their role in QS inhibition (Vashistha et al. 2023). No doubt, these synthetic derivatives showed characteristic features in QS inhibition, thereby promoting the reduction in antibiotic resistance; the bioavailability and biodegradability of these synthetic chemical moieties and the toxic by-products limit their role in QS attenuation. Hence, in the mechanistic regulation of QS, a significant therapeutic shift has been observed in finding novel alternatives to these synthetic derivatives by looking into the natural sources of pharmacological importance.

In recent years, due to the immense folkloric practices in Ayurvedic treatment for infectious diseases, medicinal plants of ethnopharmacological importance are considered effective alternative therapeutics against QS controlled bacterial virulence and biofilm mechanics (Banarjee et al. 2017; Díaz et al. 2020). In the quest for novel antivirulence therapeutics against QS-mediated pathogenicity and drug resistance, it is important to consider certain key points such as the putative drug candidates should not hamper the host tissues and should maintain the integrity of endogenous microbiome during and/or after the dosage administration. Since natural products are being actively used for several folkloric practices, plants and plant-derived compounds should be considered for antivirulence/anti-infective drug development (Zhang et al. 2020). The bioactive phytochemicals like phenylpropanoids (e.g. *trans*-cinnamaldehyde, Eugenol, etc.), Coumarin derivatives (e.g. Aesculetin, Umbelliferone, etc.), Phenolic acids (e.g. Gallic acid, Coumaric acid), flavonoids (e.g. Apigenin, Quercetin, Naringenin, etc.), Tannins (e.g. Proanthocyanidins, Punicalagin, etc.), Diarylheptanoids (e.g. Curcumin, Hirsutenone, etc.), Benzoic acid derivatives (e.g. Vanillin, Vanillic acid, etc.), Monoterpenes (e.g. Thymol, Limonene, etc.), Stilbenes (e.g. Resveratrol), and Sulfur containing phytocompounds (e.g. Iberin, Diallyl disulphide, etc.) were reported to attenuate QS-mediated virulence in ESKAPE pathogens (Deryabin et al. 2019; Lal et al. 2021). The list of medicinal plants and their bioactive phytoconstituents reported to attenuate QS regulatory pathways and their by-products were presented in Table 14.2.

Table 14.2 Effect of plant-derived phytochemicals on the quorum sensing- (QS) regulated virulence and biofilm formation in drug-resistant ESKAPE pathogens

Sl. no.	Chemical class	Phytocompounds	Source of plants	Target microorganism	Concentration (IC_{50}/MIC)	Effects	Reference
1	Alkaloids	Berberine	*Berberis* sp.	*Pseudomonas aeruginosa* PAO1	1.25 mg/mL	Modulation of biofilm dynamics	Aswathanarayan and Vittal (2018)
		Reserpine	*Rauwolfia serpentina*	*P. aeruginosa* PAO1	0.8 mg/mL	Significantly inhibits biofilm formation, attenuates production of QS-mediated virulence phenotypes, proteases, pyocyanin, rhamnolipids	Parai et al. (2018)
2	Anthocyanins	Carboxypyranocyanidin-3-*O*-glucoside	–	*Staphylococcus aureus*, *P. aeruginosa*	>0.512 mg/mL	Controls the regulatory expression of QS-mediated virulence genes	Coelho et al. (2021)
		Petunidin	Pure	*Klebsiella pneumoniae*	0.2 mg/mL	Potential inhibitor of QS signalling molecules towards *LasR* protein activity	Gopu et al. (2016)
3	Anthraquinone	Emodin	*Rheum palmatum*	*Staphylococcus aureus*	8 µg/mL	Down-regulation of *cidA*, *icaA*, *dltB*, *agrA*, *sortaseA*, and *sarA* gene	Yan et al. (2017)
		Rhein	*Cassia fistula*	*P. aeruginosa*	>0.15 mg/mL	Regulates the expression profiles of QS controlled genes, *lasI*, *lasR*, *rhlI*, and *rhlR*	Peerzada et al. (2022)

Sl. no.	Chemical class	Phytocompounds	Source of plants	Target microorganism	Concentration (IC_{50}/MIC)	Effects	Reference
4	Carotenoids	Lutein	–	*P. aeruginosa*	22 μM	Exhibit promising biofilm inhibitory activities without affecting cell viability, improves antibiotic susceptibility when used in combination	Mahavy et al. (2022)
5	Coumarins	4-Farnesyloxycoumarin	Pure	*P. aeruginosa* PAO1	>0.2 mg/mL	Controls the expression profile of *pqsR*, mitigates biofilm formation, and shows synergistic potential with Tobramycin	Tajani et al. (2023)
		6-Methylcoumarin	Pure	*P. aeruginosa* PAO1	>0.25 mg/mL	Shows promising antibiofilm potential by improving the survival rate of *Caenorhabditis elegans*	Bajire et al. (2021)

(continued)

Table 14.2 (continued)

Sl. no.	Chemical class	Phytocompounds	Source of plants	Target microorganism	Concentration (IC_{50}/MIC)	Effects	Reference
6	Flavonoids	Naringenin	*Pure*	*P. aeruginosa*	>1 µM	Interferes with the QS regulatory behaviour by competitively inhibiting the binding of natural ligand, C12-Homoserine lactone onto LasR	Hernando-Amato et al. (2020)
		Morin	*Psidium guajava*	*S. aureus*	0.4 mg/mL	Significantly inhibits bacterial motility, biofilm dynamics by targeting SarA	Chemmugil et al. (2019)
		Methoxyisoflavan	*Trigonella stellata*	*P. aeruginosa*	0.512 mg/mL	Significantly inhibits the expression profiles of QS regulatory genes, *lasI*, *lasR*, *rhlI*, and *rhlR*	Naga et al. (2022)
		Epigallocatechin-3-gallate	*Camellia sinesis*	*P. aeruginosa*	0.512 mg/mL	Significant inhibition in QS-regulated virulence phenotypes, biofilm formation, and down-regulation of QS regulatory genes (i.e. *las*, *rhl*, and *pqs*)	Hao et al. (2021)
		Calycopterin	*Marcetia latifolia*	*P. aeruginosa* PA14	>32 µM	Modulation of QS-mediated behaviours, motility, and biofilm dynamics	Froes et al. (2020)
7	Furan derivatives	5-Hydroxymethylfurfural	Pure	*P. aeruginosa* PAO1	2.5 µL/mL; 400 µg/mL	Biofilm inhibition and down-regulation of QS regulatory genes (*lasI* and *rhlI*, *lasR*, and *rhlR*)	Rajkumari et al. (2019); Vijayakumar and Ramanathan (2020)

Sl. no.	Chemical class	Phytocompounds	Source of plants	Target microorganism	Concentration (IC_{50}/MIC)	Effects	Reference
8	Organosulfur compounds	Allicin	*Alium sativum*	*P. aeruginosa*	8 μM	Down-regulation of QS pathway by interference of Rhl- and PQS system	Xu et al. (2019)
9	Phenolic acids	Salicylic acid	*Salix* sp.	*P. aeruginosa*	3.62 mM	Intereference in QS regulatory network with inhibition at transcriptional and extracellular level	Ahmed et al. (2019)
10	Phloroglucinol	Pulverulentone A	*Callistemon citrinus*	Methicillin-resistant *S. aureus, P. aeruginosa*	–	Inhibit Staphyloxanthin synthesis, mitigates biofilm formation; Interferes on the synthesis of pyocyanin and exopolysachharides in *P. aeruginosa*	Shehabeldine et al. (2020); Ismail et al. (2021)
11	Phenylpropanoids	Methyl gallate	*Guiera senegalensis*	*P. aeruginosa*	25 μg/mL	Attenuates QS-regulated behaviour with reduction in pyocyanin synthesis	Ouedraogo et al. (2019)
		Curcumin	*Curcuma longa*	*Acinetobacter baumannii*	50 μg/mL	Down-regulation of *bfmR* gene	Raorane et al. (2019)
12	Tannins	Tannic acid	Pure	*S. aureus*	0.08 mg/mL	Shows promising biofilm inhibitory activities	Dong et al. (2018)

(continued)

Table 14.2 (continued)

Sl. no.	Chemical class	Phytocompounds	Source of plants	Target microorganism	Concentration (IC_{50}/MIC)	Effects	Reference
13	Terpenes and terpenoids	Carvacrol	*Origanum vulgare*	*P. aeruginosa*	3.8 mM	Attenuation of QS regulatory genes expression, and biofilm inhibition	Tapia-Rodriguez et al. (2019)
		Myrtenol	*Myrtus communis*	Methicillin-resistant *S. aureus* (MRSA) ATCC33591	0.6 mg/mL	Concentration-dependent reduction in the production of extracellular lipase and hemolysins; inhibition of Staphyloxanthin, and mitigation of biofilms	Selvaraj et al. (2019)
		Glycyrrhetinic acid	*Glycyrrhiza glabra*	*P. aeruginosa* ATCC 25619	0.08 mg/mL	Inhibition of QS-controlled production of pathogenic determinants and biofilm formation	Kannan et al. (2019)
		Terpinen-4-ol	Plant-derived essential oils	*P. aeruginosa* MTCC 3541	0.5% (v/v)	Down-regulates the expression of QS regulatory genes, exhibits synergistic activity with Ciprofloxacin	Bose et al. (2020a, b)
		Terpinen-4-ol	Plant-derived essential oils	*S. aureus* ATCC 25923	0.25% (v/v)	Inhibition of QS-controlled biofilm formation	Cordeiro et al. (2020)
		Nerolidol	*Myrocarpus frondosus*	*P. aeruginosa, K. pneumoniae, S. aureus*	0.5–1 mg/mL	Mitigates biofilm formation	de Moura et al. (2021)

14.4.1 Potential Therapeutic Targets for Quorum Sensing (QS) Inhibition

As discussed earlier, QS signaling network in ESKAPE pathogens controls the secretion of pathophysiological elements like pyocyanin, rhamnolipids, EPS, alginate, proteases, elastases, chitinases, exotoxins, etc. Hence, these pathophysiological factors could be considered as potential therapeutic targets to regulate QS signaling cascade (Ghosh et al. 2022). The QS also controls the production of several pathophysiologically relevant biofilm phenotypes (e.g. EPS, rhamnolipids, alginates, etc.), which play a pivotal role in establishing biofilm-associated infections. Hence, in the therapeutic strategies against QS cascade, targeting these biofilm phenotypes seems to be a promising alternative (Jiang et al. 2019). Besides, the QS regulatory genes (e.g. *lasI*, *lasR*, *rhlR*, *luxI* and *luxR* analogs) responsible for the activation and transcription of several pathophysiological factors during QS signaling pathway are also considered as potential target sites for therapeutic regimens to develop inhibitors of QS network (Ahmed et al. 2019). During the *P. aeruginosa* infections to the host, QS plays a pivotal role in the transcription of several pathophysiologically important genes including *lasB*, encoding the elastase B. LasB elastase has the inherent potential to hydrolyse proteins of interest in hosts as well as in other pathogens. Besides, LasB elastase also critically disrupts the host immune responses upon the establishment of bacterial infection in the host. Based upon the pathophysiological relevance of LasB elastase, it is considered a prospective therapeutic target to modulate QS-regulated pathogenesis (Everett and Davies 2021). Similarly, the QS circuit in *P. aeruginosa* controls the production of several exotoxins, lytic enzymes, and host colonization factors required for infection severity and hence, these products could be considered as potential targets for QS attenuation (Chadha et al. 2022b).

14.4.2 Quorum Sensing (QS) Inhibition Mechanism

Several strategies are considered for the attenuation of QS-regulated behaviours. Basically, three approaches such as inhibition of AI biosynthesis, degradation of AI signals by specific enzymes, and inactivation of signal molecule detection by cognate receptors by interfering the interaction of AI to its cognate receptors are reported for QS inhibition in ESKAPE pathogens (Fig. 14.2) (Rekha et al. 2016; D'Almeida et al. 2017; Hossain et al. 2017; Naga et al. 2023). The interference in the QS signaling pathway concomitantly promotes the down-regulation of virulence factors production which ultimately controls biofilm dynamics. The regulation of biofilm dynamics further promotes the regulation of resistance patterns in ESKAPE pathogens (Zhong and He 2021). Since the QS signaling pathway promotes the extension of the repertoire of potential therapeutic targets, the QS inhibition strategies seem to be invariably advantageous over the conventional antimicrobial targets (Haque et al. 2019).

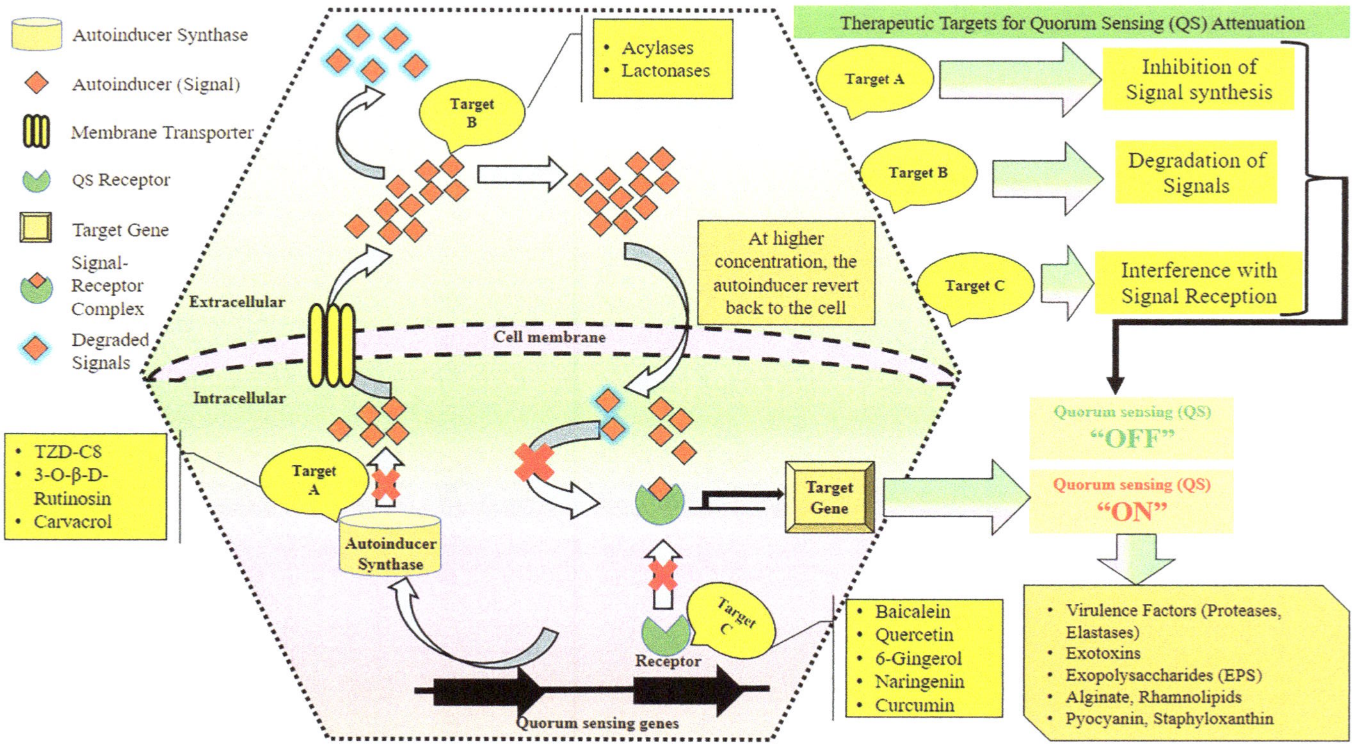

Fig. 14.2 An overview of therapeutic strategies employed to attenuate quorum sensing- (QS) regulated virulence and biofilm mechanics in ESKAPE pathogens

Plant-derived monoterpenoids act as potent inhibitors of QS signaling network by not only acting as inhibitors of signal biosynthesis by interaction with signal synthase, but also interfering with the signal reception system by modulating the signal receptors (Deryabin et al. 2019). Similarly, 5-Hydroxymethylfurfural binds to the QS cognate receptors of ESKAPE pathogen, *P. aeruginosa,* and competitively blocks the binding of AIs to its respective transcriptional receptors, thereby blocking the regulation of QS signaling resulting in the reduction in the production of several pathophysiological factors associated with biofilms and drug resistance (Rajkumari et al. 2019). Enzymes such as Lactonases and Acylases are classic examples of mitigators of QS signaling pathways by degradation of signaling molecules with high specificity in Gram-negative ESKAPE pathogens (Ahmed et al. 2019). The revolutionary trends received to develop inhibitors of QS network provide an extensive platform to regulate QS-mediated drug resistance phenomena. However, in vivo work on these QS inhibitors needs to be optimized and an in-depth clinical investigation needs to be carried out to establish their efficacy as promising alternatives to conventional antibiotics (Piewngam et al. 2020). Based upon the promising potential of QS inhibitors to regulate pathogenic profile of ESKAPE pathogens, mechanistic insight studies could be undertaken to develop potential therapeutic regimens (Lu et al. 2022).

14.5 Regulatory Role of Phytochemicals on Biofilm Dynamics in ESKAPE Pathogens

Based upon the pharmacological evidence and folkloric practices, the major classes of plant-derived phytochemicals reported for promising antibiofilm applications are Alkaloids, polyphenols, Terpenoids, Lectins, and Polypeptides. These phytochemicals inhibit biofilms by mediating deprivation of substrates, disruption of the membrane, disruption of pre-formed biofilm matrices, down-regulation of QS-assisted biofilm phenotypes, and binding to adhesion complex (Parrino et al. 2019; Mishra et al. 2020). Phytol, an important diterpene, exhibited promising antibacterial and biofilm inhibitory potential against MDR *K. pneumoniae*. Phytol not only alters the biofilm architecture, but also critically influences the adhesion capabilities of the highly resistant ESKAPE pathogen, *K. pneumoniae* (Adeosun et al. 2022). Similarly, the highly profound flavonol in plants, Quercetin, also critically inhibits the recalcitrant biofilm matrices in *S.* aureus, *E. faecalis*, and *P. aeruginosa* by an array of therapeutic interventions including prevention of bacterial adhesion, attenuation of QS signaling, blockage of efflux pump transporters, and membrane disruption (Memariani et al. 2019). Phenylpropanoids class of phytochemicals (e.g. *Trans*-Cinnamaldehyde) obtained from different plant-based products also critically down-regulate the biofilm mechanics in nosocomial pathogens by interfering with biofilm mechanics (Firmino et al. 2018; Subhaswaraj et al. 2018; Neelam et al. 2020). Similarly, other flavonoids such as flavones (e.g. Baicalein) and Flavanones (e.g. Propolin D) significantly inhibited biofilm formation in ESKAPE pathogen, *S. aureus,* by mitigating the cell aggregation and down-regulation of biofilm regulatory genes (Lee et al. 2019; Matilla-Cuenca et al. 2020). Recently, members of Coumarins (e.g. 4-Farnesyloxycoumarin and Farnesifrol B) also critically regulate the biofilm mechanics in *P. aeruginosa* PAO1 (Tajani et al. 2023). Tannic acid, naturally derived Tannins, also reported to exhibit antibiofilm properties against MRSA by altering the cell wall integrity (Dong et al. 2018) (Table 14.2).

14.5.1 Understanding the Mechanism of Biofilm Inhibition

Since biofilm mechanics in ESKAPE pathogens critically affects the bacterial pathogenesis and drug resistance patterns, inhibition of biofilm matrices proved to be an influential therapeutic regimen. The biofilm mitigation strategies include mechanical disruption of biofilm matrices, utilization of specific hydrolytic enzymes targeting the biofilm matrix, and down-regulation of pathophysiological factors linked to QS network for biofilm inhibition (Guzzo et al. 2020). Basically, the antibiofilm strategies could be designed to target the different stages of biofilm formation and development. At distinct phases of biofilm development, antibiofilm agents could be designed to (1) prevent adhesion, and interference on the transition from planktonic to sessile form, (2) interfere with the production of EPS, (3) disrupt the sessile community with interference on physiological dormancy, and (4) activate the dispersal mechanism (Parrino et al. 2019). Besides, phytochemicals-based therapeutic approaches target several pathophysiological mechanisms associated with biofilm formation and development. For example, modulation of microbial attachment to the surface by inhibiting surface attachment proteins, microbial adhesion by altering adhesion proteins (Sortase), alteration in coaggregation mechanism by inhibiting EPS production, and attenuation of QS signaling cascade by targeting signaling pathways and associated pathophysiological factors are the possible therapeutic targets for combating biofilm mechanics in ESKAPE pathogens (Lu et al. 2019).

The regulation in EPS secretion vows to direct the biofilm mitigation strategies. In a recent report, Quercetin critically inhibits the EPS production by the down-regulation of biofilm-associated genes with special reference to polysaccharide intracellular adhesion (Mu et al. 2021). Several metal ions such as Calcium, magnesium, Zinc, etc. play pivotal roles in the maintenance of bacterial biofilm architecture by providing stability and cohesiveness to the biofilm matrix. Hence, targeting the metal ions associated with the stability of the biofilm matrix by using specific metal ion chelators could provide novel avenues to fight against the biofilm-associated drug resistance. Apart from synthetic metal chelators, phytochemicals such as polyphenols, flavonoids, and phenolic acids have inherent metal ion chelating properties and hence could be taken into consideration for biofilm inhibition and eradication strategies (Borges et al. 2016).

14.6 Phytochemicals as Inhibitors of Efflux Pump in ESKAPE Pathogens

The activation of efflux pump transporters in pathogenic bacteria is an important strategy to extrude the antibiotics into the surrounding environment and infer potential resistance to therapeutic drugs. Hence, targeting the efflux pump mechanism is an attractive approach in the rational management of drug resistance. As per the recent trends, the efflux pump inhibition could be accomplished by several mechanisms such as (1) down-regulation of efflux pump-associated genes, (2) dismantling the efflux pump assembly, (3) blocking the efflux pump to alter the binding of substrates to the active sites, and (4) inhibition of energy mechanism associated with efflux pump activation (Sharma et al. 2019). Several plant-derived phytochemicals such as Reserpine, Capsaicin, Palmatine, Berberine, Epigallocatechin gallate, etc. belonging to different phytochemical classes were reported to inhibit the NorA efflux pump (Kumar and Tudu 2023). In addition, several phytochemicals such as Carvacrol, α-Bisabolol, Estragole, Limonene, etc. belonging to Terpenes also critically inhibit the NorA and MepA efflux pumps suggesting their role in the management of drug resistance (da Cruz et al. 2020; Freitas et al. 2020; da Costa et al. 2021; dos Santos Barbosa et al. 2021; De Oliveira Dias et al. 2022).

In an earlier study, phytochemicals such as Lanatoside C and Diadzein critically accumulate ethidium bromide intracellularly, thereby increasing the intracellular level of drug and concomitantly decreasing the MexB efflux transport protein mechanism in *P. aeruginosa* (Aparna et al. 2014). The phytochemicals-based therapeutic agents act upon the efflux pump transporters by either reversal of drug resistance phenomena or resensitizing the available antibiotics against the pathogens (Shriram et al. 2018).

14.7 Phytochemicals-Based Nanoformulations for Antibacterial and Antibiofilm Applications

The intervention of nanotechnology in the field of biomedicine has revolutionized the therapeutic modalities of bioactive phytochemicals. Since the encapsulation of bioactive phytochemicals to the biocompatible nanoparticles critically improves the solubility and bioavailability issues in controlled and sustainable manner, the spatially designed nanoparticles could be the efficient therapeutic modalities against drug-resistant bacterial infections and associated health ailments (Ilk et al. 2017). One of the most advantageous properties of using nano-based carriers for effective delivery of bioactive phytochemicals is that the amount and frequency of dosage could be easily regulated with effective drug delivery at the infected site (Abed and Couvreur 2014). The phytochemicals-based nanoformulation critically improves the solubility issues associated with the majority of phytochemicals and also prolonged the therapeutic modalities of the phytochemicals at the target site in a controlled manner. In addition, phytochemicals-encapsulated nanomaterials also improved the antibacterial activities by a concomitant multi-fold decrease in the minimum inhibitory concentrations (MICs) against the test pathogens (Li et al. 2019).

In recent times, plant-derived essential oils and their bioactive phytoconstituents are being actively engaged in the nanomaterial synthesis and investigated for their promising applications in tackling AMR crisis (Manju et al. 2016). For example, α-Terpineol-loaded nanostructured lipid carriers (NLCs) significantly improved the QS inhibitory effect and increased biofilm penetration capacity of α-Terpineol with prolonged therapeutic stability and improved bioavailability against ESKAPE pathogen, *P. aeruginosa* (Bose et al. 2020a, b). The concept of combinatorial study in the drug discovery pipelines also critically influences the antibacterial action and regulation of biofilm mechanics. Basically, silver nanoparticles (AgNPs) and bioactive phytochemicals, curcumin, are well known for their promising antibacterial and antibiofilm potential. When the AgNPs and the curcumin-based nanoparticles were combined and evaluated, it was observed that the spatially designed nanoformulations critically regulate biofilm matrix formation in ESKAPE pathogens (Loo et al. 2016).

More recently, curcumin was encapsulated with AgNPs and copper nanoparticles (CuNPs) and were evaluated for their role in biofilm mitigation. The curcumin-AgNPs and curcumin-CuNPs work synergistically in the mitigation of biofilms in nosocomial pathogenic bacteria, *P. aeruginosa,* with improved entrapment efficacy and sustained release profile (Targhi et al. 2021). Recently, spatially designed Graphene nanoplatelets-tannic acid-silver (GNP-TA-Ag) nanocomposite showed improved inhibitory potential against MRSA (Singhal et al. 2021). Kaempferitrin is employed to engineer AgNPs, which exhibited potential mitigation of biofilm matrices in Methicillin-resistant *S. aureus* by modulating membrane dynamics (Shamprasad et al. 2022) (Fig. 14.3).

14.8 Recent Trends and Future Perspectives

In the quest for novel therapeutic modalities as an alternative to conventional antibiotic therapies, drug moieties with the potential ability to attenuate QS, biofilms, and efflux pumps in ESKAPE pathogens received considerable attention. However, the translational shift of scientific evidence from in vitro studies to clinical settings and further validation needs to be taken into consideration with concerted scientific efforts (Abdelhamid and Yousef 2023). No doubt, the advent of QS mitigation strategies has revolutionized the scientific validations in curbing the severity of chronic bacterial infections; the frequent mutations in the genetic level in the ESKAPE pathogens lead to the failure of these QS inhibitors. In this context, the putative QS inhibitors could be strategically used along with antibiotics to act upon the pathogenic bacteria synergistically with aided efficacy (Li et al. 2021). Owing to the extended AMR, phytochemicals with improved pharmacological relevance could also be used to promote increased sensitivity of the traditional antibiotics against the pathogenic bacterial infections. In this regard, Hordenine, an important plant-derived alkaloid, improved the susceptibility of several aminoglycosides against

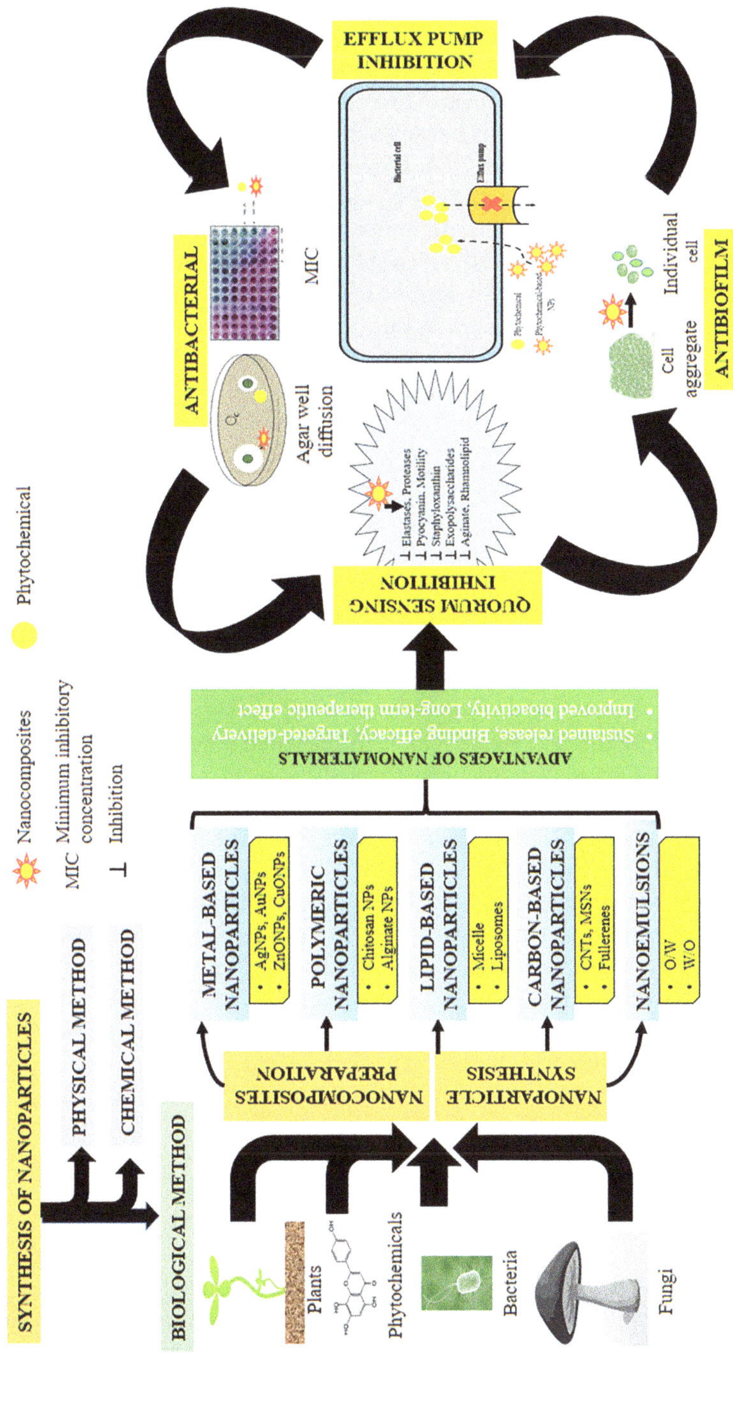

Fig. 14.3 Schematic overview of the phytochemicals-based nanomaterials in combating biofilm mechanics in ESKAPE pathogens

P. aeruginosa PAO1. Thus, Hordenine could be used synergistically with these aminoglycoside antibiotics for efficient management of bacterial resistance (Zhou et al. 2018).

More recently, plant-derived isoquinoline alkaloid, Berberine, also acts synergistically with Azithromycin to attenuate QS-controlled production of virulence phenotypes, biofilm mitigation by down-regulation of alginate production, and bacterial motility (Zhao et al. 2022). Apart from alkaloids, naturally derived Quercetin also exhibited a promising influence on the regulation of QS and biofilms in *P. aeruginosa* when used in combination with aminoglycoside antibiotics. The improved synergistic activities are associated with a substantial decrease in the infection rate and thus provide novel avenues for chronic infection control measures (Vipin et al. 2020). Based upon the promising synergistic potential of plant-derived phytochemicals with conventional antibiotics to control QS and biofilm mechanics, the synergistic potential could further be explored in the rational design of putative drug candidates against the severity of the ESKAPE infections.

Drug repurposing could be instrumental in the development of potential drug candidates to treat QS and biofilm-mediated resistance. Since the drug repurposing approach hovers around the use of clinically approved drugs/bioactive phytochemicals, though originally applied against other pathophysiological conditions, these drug moieties could be screened against chronic microbial infections against ESKAPE pathogens with improved efficacy and reduced cost and time as compared to conventional drug discovery pipelines (D'Angelo et al. 2018; Chadha et al. 2022a). FDA-approved drugs with immense biomedical and pharmaceutical relevance could also be repurposed towards the down-regulation of coordinated expression of QS signaling cascade. Hence, clinically approved drugs could be considered as therapeutic modules to treat drug resistance phenomena.

The computational approaches further ease the cost associated with the screening of the FDA-approved drugs to tackle bacterial virulence and AMR controlled by QS and biofilm dynamics (Mellini et al. 2019). The computational approaches critically accelerate the screening procedure to identify the potent drug candidates. Based on the predictions generated through in silico tools, further experimental studies should be undertaken to develop the potential drug candidates against the MDR patterns in ESKAPE pathogens (Krżyzek 2019; Mellini et al. 2019). The computational tools-based virtual screening and lead discovery critically improve the cost and time required for conventional drug identification for intended therapeutic applications, thereby providing novel avenues for the selection of putative drug molecules for therapeutic use.

14.9 Conclusion

In recent years, ESKAPE pathogens developed protective modules by promoting QS signaling, biofilm matrix formation, and activation of efflux pumps to exhibit resistance to several classes of antibiotics. Plant-based phytochemicals have received considerable attention to tackle QS-regulated virulence, biofilm dynamics, and efflux transporters-mediated resistance and are regarded as potential alternatives. Several groups of phytochemicals have exerted promising influence on the mechanics of QS signaling cascade, biofilm formation and development, and efflux pumps by targeting several pathophysiological targets of clinical importance. Besides, these phytochemicals also have the privilege to work synergistically with conventional antibiotics for effective clearance of drug resistance. The phytochemical-based nanoformulations also critically enhance the therapeutic potency by increasing the target-based therapies. The concept of repurposing by exploring clinically tested phytochemicals towards the mitigation of biofilms and drug resistance also received wide acceptance across the globe. The computational approaches-based screening of phytochemicals further improved the conventional drug screening pipelines to mitigate chronic bacterial infections. No doubt, phytochemicals have shown promising therapeutic effects on ESKAPE pathogens; more in-depth studies should be undertaken with improved therapeutic modalities to fight the AMR in the post-antibiotic era.

Acknowledgement The authors greatly acknowledge OHEPEE, Govt. of Odisha, for providing financial support. The author, Subhaswaraj Pattnaik, also gratefully acknowledges the Science and Engineering Research Board (SERB), Govt. of India, for the award of N-PDF (Reference No. PDF/2021/001260).

References

Abdelhamid AG, Yousef AE (2023) Combating bacterial biofilms: current and emerging antibiofilm strategies for treating persistent infections. Antibiotics 12:1005. https://doi.org/10.3390/antibiotics12061005

Abdul-Jabbar AM, Hussian NN, Mohammed HA, Aljarbou A, Akhter N, Khan RA (2022) Combined anti-bacterial actions of Lincomycin and freshly prepared silver nanoparticles: overcoming the resistance to antibiotics and enhancement of the bioactivity. Antibiotics 11:1791. https://doi.org/10.3390/antibiotics11121791

Abed N, Couvreur P (2014) Nanocarriers for antibiotics: a promising solution to treat intracellular bacterial infections. Int J Antimicrob Agent 43:485–496. https://doi.org/10.1016/j.ijantimicag.2014.02.009

Aboagye CI, Ampomah GB, Mensah JO, Mensah CN, Nartey D, Gasu EN, Borquaye LS (2023) N-benzylimidazoles as potential antimicrobial and antibiofilm agents-Syntheses, bioactivity, and molecular docking studies. Sci Afr 19:e01529. https://doi.org/10.1016/j.sciaf.2022.e01529

Adeosun IJ, Baloyi IT, Cosa S (2022) Anti-biofilm and associated anti-virulence activities of selected phytochemical compounds against *Klebsiella pneumoniae*. Plants (Basel) 11(11):1429. https://doi.org/10.3390/plants11111429

Ahmad I, Qais FA, Samreen, Abulreesh HH, Ahmad S, Rumbaugh KP (2019) Antibacterial drug discovery: perspective insights. In: Ahmad I, Ahmad S, Rumbaugh KP (eds) Antibacterial drug discovery to combat MDR: natural compounds, nanotechnology and novel synthetic sources. Springer Nature, pp 1–21. https://doi.org/10.1007/978-981-13-9871-1_1

Ahmed SAKS, Rudden M, Smyth TJ, Dooley JSG, Marchant R, Banat IM (2019) Natural quorum sensing inhibitors effectively down regulate gene expression of *Pseudomonas aeruginosa* virulence factors. Appl Microbiol Biotechnol 103:3521–3535. https://doi.org/10.1007/s00253-019-09618-0

AlSheikh HMA, Sultan I, Kumar V, Rather IA, Al-Sheikh H, Jan AT, Haq QMR (2020) Plant-based phytochemicals as possible alternative to antibiotics in combating bacterial drug resistance. Antibiotics 9(8):480. https://doi.org/10.3390/antibiotics9080480

Aparna V, Dineshkumar K, Mohanalakshmi N, Velmurugan D, Hopper W (2014) Identification of natural compound inhibitors for multidrug efflux pumps of *Escherichia coli* and *Pseudomonas aeruginosa* using *in silico* high-throughput virtual screening and *in vitro* validation. PLoS One 9(7):e101840. https://doi.org/10.1371/journal.pone.0101840

Arias CA, Murray BE (2012) The rise of the *Enterococcus*: beyond Vancomycin resistance. Nat Rev Microbiol 10:266–278. https://doi.org/10.1038/nrmicro2761

Ashkenazi S (2013) Beginning and possibly the end of the antibiotic era. J Paediatr Child Health 49:E179–E182. https://doi.org/10.1111/jpc.12032

Aswathanarayan JB, Vittal RR (2018) Inhibition of biofilm formation and quorum sensing mediated phenotypes by berberine in *Pseudomonas aeruginosa* and *Salmonella typhimurium*. RSC Adv 8:36133–36141. https://doi.org/10.1039/c8ra06413

Atanasov AG, Zotchev SB, Dirsch VM, International Natural Product Sciences Taskforces, Supuran CT (2021) Natural products in drug discovery: advances and opportunities. Nat Rev Drug Dis 20:200–216. https://doi.org/10.1038/s41573-020-00114-z

Ayobami O, Brinkwirth S, Eckmanns T, Markwart R (2022) Antibiotic resistance in hospital-acquired ESKAPE-E infections in low- and lower-middle-income countries: a systematic review and meta-analysis. Emerg Microbes Infect 11(1):443–451. https://doi.org/10.1080/22221751.2022.2030196

Azuama OC, Ortiz S, Quirós-Guerrero L, Bouffartigues E, Tortuel D, Maillot O, Feuilloley M, Cornelis P, Lesouhaitier O, Grougnet R, Boutefnouchet S, Wolfender JL, Chevalier S, Tahrioui A (2020) Tackling *Pseudomonas aeruginosa* virulence by Mulinane-like diterpenoids from *Azorella atacamensis*. Biomolecules. 10:1626. https://doi.org/10.3390/biom10121626

Baidya S, Sharma S, Mishra SK, Kattel HP, Parajuli K, Sherchand JB (2021) Biofilm formation by pathogens causing Ventilator-associated pneumonia at intensive care units in a tertiary care hospital: an armor for refuge. Biomed Res Int 2021:8817700. https://doi.org/10.1155/2021/8817700

Bajire SK, Jain S, Johnson RP, Shastry RP (2021) 6-Methylcoumarin attenuates quorum sensing and biofilm formation in *Pseudomonas aeruginosa* PAO1 and its applications on solid surface coatings with polyurethane. Appl Microbiol Biotechnol 105:8647–8661. https://doi.org/10.1007/s00253-021-11637-9

Bajko E, Kalinowska M, Borowski P, Siergiejczyk L, Lewandowski W (2016) 5-*O*-Caffeoylquinic acid: a spectroscopic study and biological screening for antimicrobial activity. LWT Food Sci Technol 65:471–479. https://doi.org/10.1016/j.lwt.2015.08.024

Balcázar JL, Subirats J, Borrego CM (2015) The role of biofilms as environmental reservoirs of antibiotic resistance. Front Microbiol 6:1216. https://doi.org/10.3389/fmicb.2015.01216

Banarjee M, Moulick S, Bhattacharya KK, Parai D, Chattopadhyay S, Mukherjee SK (2017) Attenuation of *Pseudomonas aeruginosa* quorum sensing, virulence, and biofilm formation by extracts of *Andrographis paniculata*. Microb Pathog 113:85–93. https://doi.org/10.1016/j.micpath.2017.10.023

Baraquet C, Harwood CS (2016) FleQ DNA binding consensus sequence revealed by studies of FleQ-dependent regulation of biofilm gene expression in *Pseudomonas aeruginosa*. J Bacteriol 198(1):178–186. https://doi.org/10.1128/JB.00539-15

Barbieri R, Coppo E, Marchese A, Daglia M, Sobarzo-Sánchez E, Nabavi SF, Nabavi SM (2017) Phytochemicals for human disease: an update on plant-derived compounds antibacterial activity. Microbiol Res 196:44–68. https://doi.org/10.1016/j.micres.2016.12.003

Biva IJ, Ndi CP, Semple SJ, Griesser HJ (2019) Antibacterial performance of terpenoids from the Australian plant *Eremophila lucida*. Antibiotics 8:63. https://doi.org/10.3390/antibiotics8020063

Blair JMA, Webber MA, Baylay AJ, Ogbolu DO, Piddock LJV (2015) Molecular mechanisms of antibiotic resistance. Nat Rev Microbiol 13:42–51. https://doi.org/10.1038/nrmicro3380

Borges A, Abreu AC, Dias C, Saavedra MJ, Borges F, Simões M (2016) New perspectives on the use of phytochemicals as an emergent strategy to control bacterial infections including biofilms. Molecules 21(7):877. https://doi.org/10.3390/molecules21070877

Bose SK, Chauhan M, Dhingra N, Chhibber S, Harjai K (2020a) Terpinen-4-ol attenuates quorum sensing regulated virulence factors and biofilm formation in *Pseudomonas aeruginosa*. Future Microbiol 15(2):127–142. https://doi.org/10.2217/fmb-2019-0204

Bose SK, Nirbhavane P, Batra M, Chhibber S, Harjai K (2020b) Nanolipoidal α-Terpineol modulates quorum sensing regulated virulence and biofilm formation in *Pseudomonas aeruginosa*. Nanomedicine 15:1743. https://doi.org/10.2217/nnm-2020-0134

Botelho J, Grosso F, Peixe L (2019) Antibiotic resistance in *Pseudomonas aeruginosa*—mechanisms, epidemiology and evolution. Drug Resist Update 44:26–47. https://doi.org/10.1016/j.drup.2019.07.002

Casciaro B, Calcaterra A, Cappiello F, Mori M, Loffredo MR, Ghirga F et al (2020) Nigritanine as a new potential antimicrobial alkaloid for the treatment of *Staphylococcus aureus*-induced infections. Toxins 11(9):511. https://doi.org/10.3390/toxins11090511

Chadha J, Harjai K, Chhibber S (2022a) Revisiting the virulence hallmarks of *Pseudomonas aeruginosa*: a chronicle through the perspective of quorum sensing. Environ Microbiol 24(6):2630–2656. https://doi.org/10.1111/1462-2920.15784

Chadha J, Harjai K, Chhibber S (2022b) Repurposing phytochemicals as anti-virulent agents to attenuate quorum sensing-regulated virulence factors and biofilm formation in *Pseudomonas aeruginosa*. Microb Biotechnol 15(6):1695–1718. https://doi.org/10.1111/1751-7915.13981

Chemmugil P, Lakshmi PTV, Annamalai A (2019) Exploring Morin as an anti-quorum sensing agent (anti-QSA) against resistant strains of *Staphylococcus aureus*. Microb Pathog 127:304–315. https://doi.org/10.1016/j.micpath.2018.12.007

Cheruvanachari P, Mishra M, Pattnaik S, Naik PK (2023a) Determination of antibacterial and antibiofilm potential of Kewda essential oil against *Staphylococcus aureus* and *Klebsiella pneumoniae*. Biologia 78:2253. https://doi.org/10.1007/s11756-023-01360-3

Cheruvanachari P, Pattnaik S, Mishra M, Pragyandipta P, Pattnaik A, Naik PK (2023b) Deciphering the antibiofilm potential of 2-Phenylethyl methyl ether (PEME), a bioactive compound of Kewda essential oil against *Staphylococcus aureus*. Microb Pathog 179:106093. https://doi.org/10.1016/j.micpath.2023.106093

Choi HY, Le DD, Kim WG (2022a) Curvularin isolated from *Phoma macrostoma* is an antagonist of RhlR quorum sensing in *Pseudomonas aeruginosa*. Front Microbiol 13:913882. https://doi.org/10.3389/fmicb.2022.913882

Choi SS, Lee SH, Lee KA (2022b) A comparative study of hesperetin, hesperidin and hesperidin glucoside: antioxidant, antiinflammatory, and antibacterial activities *in vitro*. Antioxidants 11:1618. https://doi.org/10.3390/antiox11081618

Ciofu O, Tolker-Nielsen T (2019) Tolerance and resistance of *Pseudomonas aeruginosa* biofilms to antimicrobial agents—how *P. aeruginosa* can escape antibiotics. Front Microbiol 10:913. https://doi.org/10.3389/fmicb.2019.00913

Coates ARM, Halls G, Hu Y (2011) Novel classes of antibiotics or more of the same. Br J Pharmacol 163(1):184–194. https://doi.org/10.1111/j.1476-5381.2011.01250.x

Coelho P, Oliveira J, Fernandes I, Araujo A, Pereira AR, Gameiro P, Bessa LJ (2021) Pyranoanthocyanins interfering with the quorum sensing of *Pseudomonas aeruginosa* and *Staphylococcus aureus*. Int J Mol Sci 22(16):8559. https://doi.org/10.3390/ijms.22168559

Cordeiro L, Figueiredo P, Souza H, Sousa A, Andrade-Júnior F, Medeiros D, Nóbrega J, Silva D, Martins E, Barbosa-Filho J, Lima E (2020) Terpinen-4-ol as an antibacterial and antibiofilm agent against *Staphylococcus aureus*. Int J Mol Sci 21(12):4531. https://doi.org/10.3390/ijms21124531

Costa RS, Lins MO, Hyaric ML, Barros TF, Velozo ES (2017) *In vitro* antibacterial effects of *Zanthoxylum tingoassuiba* root bark extracts and two of its alkaloids against multiresistant *Staphylococcus aureus*. Braz J Pharmacogn 27:195–198. https://doi.org/10.1016/j.bjp.2016.11.001

Cunrath O, Meinel DM, Maturana P, Fanous J, Buyck JM, Auguste PS, Seth-Smith HMB, Körner J, Dehio C, Trebosc V, Kemmer C, Neher R, Egli A, Bumann D (2019) Quantitative contribution of efflux to multi-drug resistance of clinical *Escherichia coli* and *Pseudomonas aeruginosa* strains. EBioMedicine 41:479–487. https://doi.org/10.1016/j.ebiom.2019.02.061

D'Almeida RE, Molina RDI, Viola CM, Luciardi MC, Nieto Peñalver C, Bardón A, Arena ME (2017) Comparison of seven structurally related coumarins on the inhibition of Quorum sensing of *Pseudomonas aeruginosa* and *Chromobacterium violaceum*. Bioorg Chem 73:37–42. https://doi.org/10.1016/j.bioorg.2017.05.011

D'Angelo F, Baldelli V, Halliday N, Pantalone P, Polticelli F, Fiscarelli E, Williams P, Visca P, Leoni L, Rampioni G (2018) Identification of FDA-approved drugs as antivirulence agents targeting the *pqs* quorum-sensing system of *Pseudomonas aeruginosa*. Antimicrob Agents Chemother 62(11):e01296–e01218. https://doi.org/10.1128/AAC.01296-18

da Costa RHS, Rocha JE, de Freitas TS, Pereira RLS, Pereira FN Jr, de Oliveira MRC et al (2021) Evaluation of antibacterial activity and reversal of the NorA and MepA efflux pump of estragole against *Staphylococcus aureus* bacteria. Arch Microbiol 203(6):3551–3555. https://doi.org/10.1007/s00203-021-02347-x

da Cruz RP, de Freitas TS, Costa MS, Dos Santos ATL, Campina FF, Pereira RLS et al (2020) Effect of α-Bisabolol and its β-Cyclodextrin complex as TetK and NorA efflux pump inhibitors in *Staphylococcus aureus* strains. Antibiotics (Basel) 9(1):28. https://doi.org/10.3390/antibiotics9010028

Danner MC, Robertson A, Behrends V, Reiss J (2019) Antibiotic pollution in surface fresh waters: occurrence and effects. Sci Total Environ 664:793–804. https://doi.org/10.1016/j.scitotenv.2019.01.406

Dash S, Kumar M, Pareek N (2020) Enhanced antibacterial potential of berberine via synergism with chitosan nanoparticles. Mater Today Proc 31:640–645. https://doi.org/10.1016/j.matpr.2020.05.506

De Gaetano GV, Lentini G, Famà A, Coppolino F, Beninati C (2023) Antimicrobial resistance: two-component regulatory systems and multidrug efflux pumps. Antibiotics 12(6):965. https://doi.org/10.3390/antibiotics12060965

De Moura DF, Rocha TA, de Mello Barros D, da Silva MM, dos Santos Santana M, Neta BM et al (2021) Evaluation of the antioxidant, antibacterial, and antibiofilm activity of the sesquiterpene nerolidol. Arch Microbiol 203(7):4303–4311. https://doi.org/10.1007/s00203-021-02377-5

De Oliveira Dias KJS, Miranda GM, Bessa JR, de Araújo ACJ, Freitas PR, De Almeida RS, Paulo CLR, de Araújo Neto JB, Coutinho HDM, Ribeiro-Filho J (2022) Terpenes as bacterial efflux pump inhibitors: a systematic review. Front Pharmacol 13:953982. https://doi.org/10.3389/fphar.2022.953982

De Oliveira DMP, Forde BM, Kidd TJ, Harris PNA, Schembri MA, Beatson SA, Paterson DL, Walker MJ (2020) Antimicrobial resistance in ESKAPE pathogens. Clin Microbiol Rev 33:e00181–e00119. https://doi.org/10.1128/CMR.00181-19

Deryabin D, Galadzhieva A, Kosyan D, Duskaev G (2019) Plant-derived inhibitors of AHL-mediated quorum sensing in bacteria: modes of action. Int J Mol Sci 20(22):5588. https://doi.org/10.3390/ijms20225588

Díaz MA, González SN, Alberto MR, Arena ME (2020) Human probiotic bacteria attenuate *Pseudomonas aeruginosa* biofilm and virulence by quorum sensing inhibition. Biofouling 36(5):597–609. https://doi.org/10.1080/08927014.2020.1783253

Dincer S, Uslu FM, Delik A (2020) Antibiotic resistance in biofilm. In: Dincer S (ed) Bacterial biofilms. Intech Open, pp 1–15. https://doi.org/10.5772/intechopen.92388

Dong G, Liu H, Yu X, Zhang X, Lu H, Zhou T, Cao J (2018) Antimicrobial and anti-biofilm activity of tannic acid against *Staphylococcus aureus*. Nat Prod Res 32(18):2225–2228. https://doi.org/10.1080/14786419.2017.1366485

dos Santos Barbosa CR, Scherf JR, de Freitas TS, de Menezes IRA, Pereira RLS, dos Santos JFS et al (2021) Efect of Carvacrol and Thymol on NorA efflux pump inhibition in multidrug-resistant (MDR) *Staphylococcus aureus* strains. J Bioenerg Biomembr 53:489–498. https://doi.org/10.1007/s10863-021-09906-3

Everett MJ, Davies DT (2021) *Pseudomonas aeruginosa* elastase (LasB) as a therapeutic target. Drug Dis Today 26(9):2108–2123. https://doi.org/10.1016/j.drugdis.2021.02.026

Farha AK, Yang QQ, Kim G, Zhang D, Mavumengwana V, Habimana O, Li HB, Corke H, Gan RY (2020) Inhibition of multidrug-resistant foodborne *Staphylococcus aureus* biofilms by a natural terpenoid (+)-nootkatone and related molecular mechanism. Food Cont 112:107154. https://doi.org/10.1016/j.foodcont.2020.107154

Firmino DF, Cavalcante TTA, Gomes GA, Firmino NCS, Rosa LD, de Carvalho MG, Catunda FEA Jr (2018) Antibacterial and antibiofilm activities of *Cinnamomum* sp. essential oil and Cinnamaldehyde: antimicrobial activities. Sci World J 2018:7405736. https://doi.org/10.1155/2018/7405736

Flemming HC, Wingender J, Szewzyk U, Steinberg P, Rice SA, Kjelleberg S (2016) Biofilms: an emergent form bacterial life. Nat Rev Microbiol 14:563–575. https://doi.org/10.1038/nrmicro.2016.94

Freitas PR, de Araújo ACJ, dos Santos Barbosa CR, Muniz DF, da Silva ACA, Rocha JE et al (2020) GC-MS-FID and potentiation of the antibiotic activity of the essential oil of *Baccharis reticulata* (ruiz & pav.) pers. and α-Pinene. Ind Crops Prod 145:112106. https://doi.org/10.1016/j.indcrop.2020.112106

Froes TQ, Nicastro GG, de Oliveira Pereira T, de Oliveira Carneiro K, Reis IMA, Conceição RS et al (2020) Calycopterin, a major flavonoid from *Marcetia latifolia*, modulates virulence-related traits in *Pseudomonas aeruginosa*. Microb Pathog 144:104142. https://doi.org/10.1016/j.micpath.2020.104142

Ghorab MM, Soliman AM, El-Sayyad GS, Abdel-Kader MS, El-Batal AI (2023) Synthesis, antimicrobial, and antibiofilm activities of some novel 7-Methoxyquinolone derivatives bearing Sulfonamide moiety against urinary tract infection-causing pathogenic microbes. Int J Mol Sci 24(10):8933. https://doi.org/10.3390/ijms24108933

Ghosh S, Lahiri D, Nag M, Dey A, Pandit S, Sarkar T, Pati S, Kari ZA, Ishak AR, Edinur HA, Ray RR (2022) Phytocompound mediated blockage of quorum sensing cascade in ESKAPE pathogens. Antibiotics 11:61. 103390/antibiotics11010061

Gopu V, Meena CK, Murali A, Shetty PH (2016) Petunidin as a competitive inhibitor of acylated homoserine lactones in *Klebsiella pneumoniae*. RSC Adv 6(4):2592–2601. https://doi.org/10.1039/C5RA20677D

Gu Y, Dong J, Li J, Luo Q, Dong X, Tang G et al (2023) Antibacterial activity and mechanism of sanguinarine against *Staphylococcus aureus* by interfering with the permeability of the cell wall and membrane and inducing bacterial ROS production. Front Vet Sci 10:1121082. https://doi.org/10.3389/fvets.2023.1121082

Guimarães AC, Meireles LM, Lemos MF, Guimarães MCC, Endringer DC, Fronza M, Scherer R (2019) Antibacterial activity of terpenes and terpenoids present in essential oils. Molecules 24(13):2471. https://doi.org/10.3390/molecules24132471

Guzzo F, Scognamiglio M, Fiorentino A, Buommino E, D'Abrosca B (2020) Plant derived natural products against *Pseudomonas aeruginosa* and *Staphylococcus aureus*: antibiofilm activity and molecular mechanisms. Molecules 25:5024. https://doi.org/10.3390/molecules25215024

Hao S, Yang D, Zhao L, Shi F, Ye G, Fu H et al (2021) EGCG-mediated potential inhibition of biofilm development and quorum sensing in *Pseudomonas aeruginosa*. Int J Mol Sci 22(9):4946. https://doi.org/10.3390/ijms.22094946

Haque S, Yadav DK, Bisht SC, Yadav N, Singh V, Dubey KK, Jawed A, Wahid M, Dar SA (2019) Quorum sensing pathways in Gram-positive and -negative bacteria: potential of their interruption in abating drug resistance. J Chemother 31(4):161–187. https://doi.org/10.1080/1120009X.2019.1599175

Harkins CP, Pichon B, Doumith M, Parkhill J, Westh H, Tomasz A, De Lencastre H, Bentley SD, Kearns AM, Holden MTG (2017) Methicillin resistant *Staphylococcus aureus* emerged long before the introduction of Methicillin into clinical practice. Genome Biol 18(1):130. https://doi.org/10.1186/s13059-017-1252-9

Hemmati F, Salehi R, Ghotaslou R, Kafil HS, Hasani A, Gholizadeh P, Nouri R, Rezaee MA (2020) Quorum quenching: a potential target for antipseudomonal therapy. Infect Drug Resist 13:2989–3005. https://doi.org/10.2147/IDR.S263196

Hernando-Amato S, Alcalde-Rico M, Gil-Gil T, Valverde JR, Martinez JL (2020) Naringenin inhibition of the *Pseudomonas aeruginosa* quorum sensing response is based on its time-dependent competition with N-(3-Oxo-dodecanoyl)-L-homoserine lactone for LasR binding. Front Mol Biosci 7:25. https://doi.org/10.3389/fmolb.2020.00025

Hochma E, Yarmolinsky L, Khalfin B, Nisnevitch M, Ben-Shabat S, Nakonechny F (2021) Antimicrobial effect of phytochemicals from edible plants. Processes 9(11):2089. https://doi.org/10.3390/pr9112089

Hossain MA, Lee SJ, Park NH, Mechesso AF, Birhanu BT, Kang JW, Reza MA, Suh JW, Park SC (2017) Impact of phenolic compounds in the acyl homoserine lactone-mediated quorum sensing regulatory pathways. Sci Rep 7:10618. https://doi.org/10.1038/s41598-017-10997-5

Hutchings MI, Truman AW, Wilkinson B (2019) Antibiotics: past, present and future. Curr Opin Microbiol 51:72–80. https://doi.org/10.1016/j.mib.2019.10.008

Ilk S, Sağlam N, Özgen M, Korkusuz F (2017) Chitosan nanoparticles enhances the anti-quorum sensing activity of kaemferol. Int J Biol Macromol 94:653–662. https://doi.org/10.1016/j.ijbiomac.2016.10.068

Ismail MM, Hassan M, Moawad SS, Okba MM, Ashour RM, Fayek NM, Saber FR (2021) Exploring the antivirulence activity of Pulverulentone A, a phloroglucinol-derivative from *Callistemon citrinus* leaf extract, against multi-drug resistant *Pseudomonas aeruginosa*. Antibiotics 10:907. https://doi.org/10.3390/antibiotics10080907

Jadimurthy R, Mayegowda SB, Nayak SC, Mohan CD, Rangappa KS (2022) Escaping mechanisms of ESKAPE pathogens from antibiotics and their targeting by natural compounds. Biotechnol Rep 34:e00728. https://doi.org/10.1016/j.btre.2022.e00728

Jiang Q, Chen J, Yang C, Yin Y, Yao K (2019) Quorum sensing: a prospective therapeutic target for bacterial diseases. Biomed Res Int 2019:2015978. https://doi.org/10.1155/2019/2015978

Jubair N, Rajagopal M, Chinnappan S, Abdullah NB, Fatima A (2021) Review on the antibacterial mechanism of plant-derived compounds against multidrug-resistant bacteria (MDR). Evid Based Complement Alternat Med 2021:3663315. https://doi.org/10.1155/2021/3663315

Kannan S, Sathasivam G, Marudhamuthu M (2019) Decrease of growth, biofilm and secreted virulence in opportunistic nosocomial *Pseudomonas aeruginosa* ATCC 25619 by glycyrrhetinic acid. Microb Pathog 126:332–342. https://doi.org/10.1016/j.micpath.2018.011.26

Kapoor G, Saigal S, Elongavan A (2017) Action and resistance mechanisms of antibiotics: a guide for clinicians. J Anaesthesiol Clin Pharmacol 33(3):300–305. https://doi.org/10.4103/joacp.JOACP_349_15

Karami P, Khaledi A, Mashoof RY, Yaghoobi MH, Karami M, Dastan D, Alikhani MY (2020) The correlation between biofilm formation capability and antibiotic resistance pattern in *Pseudomonas aeruginosa*. Gene Rep 18:100561. https://doi.org/10.1016/j.genrep.2019.100561

Khameneh B, Iranshahy M, Soheili V, Bazzaz BSF (2019) Review on plant antimicrobials: a mechanistic viewpoint. Antimicrob Resist Infect Cont 8:118. https://doi.org/10.1186/s13756-019-0559-6

Khameneh B, Eskin NAM, Iranshahy M, Bazzaz BSF (2021) Phytochemicals: a promising weapon in the arsenal against antibiotic-resistant bacteria. Antibiotics (Basel) 10(9):1044. https://doi.org/10.3390/antibiotics10091044

Khare T, Anand U, Dey A, Assaraf YG, Chen ZS, Liu Z, Kumar V (2021) Exploring phytochemicals for combating antibiotic resistance in microbial pathogens. Front Pharmacol 12:720726. https://doi.org/10.3389/fphar.2021.720726

Krżyzek P (2019) Challenges and limitations of anti-quorum sensing therapies. Front Microbiol 10:2473. https://doi.org/10.3389/fmicb.2019.02473

Kumar G, Tudu AK (2023) Tackling multidrug-resistant *Staphylococcus aureus* by natural products and their analogues acting as NorA efflux pump inhibitors. Bioorg Med Chem 80:117187. https://doi.org/10.1016/j.bmc.2023.117187

Kumar L, Patel SKS, Kharga K, Kumar R, Kumar P, Pandohee J, Kulshresha S, Harjai K, Chhibber S (2022) Molecular mechanisms and applications of N-Acyl homoserine lactone-mediated quorum sensing in bacteria. Molecules 27:7584. https://doi.org/10.3390/molecules27217584

Lal AF, Singh S, Franco FC Jr, Bhatia S (2021) Potential of polyphenols in curbing quorum sensing and biofilm formation in Gram-negative pathogens. Asian Pac J Trop Biomed 11(6):231–243. https://doi.org/10.4103/2221-1691.314044

Lazar V, Holban AM, Curutiu C, Chifiriuc MC (2021) Modulation of quorum sensing and biofilms in less investigated Gram-negative ESKAPE pathogens. Front Microbiol 12:676510. https://doi.org/10.3389/fmicb.2021.676510

Lee JH, Kim YG, Khadke SK, Yamano A, Woo JT, Lee J (2019) Antimicrobial and antibiofilm activities of prenylated flavanones from *Macaranga tanarius*. Phytomedicine 63:153033. https://doi.org/10.1016/j.phymed.2019.153033

Li H, Li X, Wang Z, Fu Y, Ai Q, Dong Y, Yu J (2015a) Autoinducer-2 regulates *Pseudomonas aeruginosa* PAO1 biofilm formation and virulence production in a dose-dependent manner. BMC Microbiol 15:192. https://doi.org/10.1186/s12866-015-0529-y

Li XZ, Plésiat P, Nikaido H (2015b) The challenge of efflux-mediated antibiotic resistance in Gram-negative bacteria. Clin Microbiol Rev 28(2):337–418. https://doi.org/10.1128/cmr.00117-14

Li CH, Chen X, Landis RF, Geng Y, Makabenta JM, Lemnios W, Gupta A, Rotello VM (2019) Phytochemical-based nanocomposites for the treatment of bacterial biofilms. ACS Infect Dis 5(9):1590–1596. https://doi.org/10.1021/acsinfecdis.9b00134

Li X, Liu Y, Wang Y, Lin Z, Wang D, Sun H (2021) Resistance risk induced by quorum sensing inhibitors and their combined use with antibiotics: mechanism and its relationship with toxicity. Chemosphere 265:129153. https://doi.org/10.1016/j.chemosphere.2020.129153

Lima VN, Oliveira-Tintino CDM, Santos ES, Morais LP, Tintino SR, Freitas TS, Geraldo YS, Pereira RLS, Cruz RP, Menezes IRA, Coutinho HDM (2016) Antimicrobial and enhancement of the antibiotic activity by phenolic compounds: gallic acid, caffeic acid and pyrogallol. Microb Pathog 99:56–61. https://doi.org/10.1016/j.micpath.2016.08.004

Liu Y, Jia Y, Yang K, Li R, Xiao X, Zhu K, Wang Z (2020) Metformin restores tetracyclines susceptibility against multidrug resistant bacteria. Adv Sci 2020:1902227. https://doi.org/10.1002/advs.201902227

Loo CY, Rohanizadeh R, Young PM, Traini D, Cavaliere R, Whitchurch CB, Lee WH (2016) Combination of silver nanoparticles and curcumin nanoparticles for enhanced anti-biofilm activities. J Agric Food Chem 64(12):2513–2522. https://doi.org/10.1021/acs.jafc.5b04559

Lorusso AB, Carrara JA, Barroso CDN, Tuon FF, Faoro H (2022) Role of efflux pumps on antimicrobial resistance in *Pseudomonas aeruginosa*. Int J Mol Sci 23(24):15779. https://doi.org/10.3390/ijms232415779

Lou Z, Wang H, Rao S, Sun J, Ma C, Li J (2012) *p*-Coumaric acid kills bacteria through dual damage mechanisms. Food Cont 25(2):550–554. https://doi.org/10.1016/j.foodcont.2011.11.022

Lu L, Hu W, Tian Z, Yuan D, Yi J, Zhou Y, Cheng Q, Zhu J, Li M (2019) Developing natural products as potential anti-biofilm agents. Chin Med 14:11. https://doi.org/10.1186/s13020-019-0232-2

Lu L, Li M, Yi G, Liao L, Cheng Q, Zhu J, Zhang B, Wang Y, Chen Y, Zeng M (2022) Screening strategies for quorum sensing inhibitors in combating bacterial infections. J Pharm Anal 12:1–14. https://doi.org/10.1016/j.jpha.2021.03.009

Ma YX, Wang CY, Li YY, Li J, Wan QQ, Chen JH, Tay FR, Niu LN (2020) Considerations and caveats in combating ESKAPE pathogens against nosocomial infections. Adv Sci 7:1901872. https://doi.org/10.1002/advs.201901872

Mahavy CE, Mol A, Andrianarisoa B, Duez P, El Jaziri M, Baucher M, Rasamiravaka T (2022) The xanthophyll carotenoid lutein reduces the invasive potential of *Pseudomonas aeruginosa* and increases its susceptibility to Tobramycin. Int J Mol Sci 23:7199. https://doi.org/10.3390/ijms23137199

Malik B, Bhattacharyya S (2019) Antibiotic drug-resistance as a complex system driven by socio-economic growth and antibiotic misuse. Sci Rep 9:9788. https://doi.org/10.1038/s41598-019-46078-y

Mangiaterra G, Laudadio E, Cometti M, Mobbili G, Minnelli C, Massaccesi L, Citterio B, Biavasco F, Galeazzi R (2017) Inhibitors of multidrug efflux pumps of *Pseudomonas aeruginosa* from natural sources: an in silico high-throughput virtual screening and in vitro validation. Med Chem Res 26:414–430. https://doi.org/10.1007/s0004-016-1761-1

Manju S, Malaikozhundan B, Vijayakumar S, Shanthi S, Jaishabanu A, Ekambaram P, Vaseeharan B (2016) Antibacterial, antibiofilm and cytotoxic effects of *Nigella sativa* essential oil coated gold nanoparticles. Microb Pathog 91:129–135. https://doi.org/10.1016/j.micpath.2015.11.021

Masihzadeh S, Amin M, Farshadzadeh Z (2023) *In vitro* and *in vivo* antibiofilm activity of the synthetic antimicrobial peptideWLBU2 against multiple drug resistant *Pseudomonas aeruginosa* strains. BMC Microbiol 23:131. https://doi.org/10.1186/s12866-023-02886-x

Matilla-Cuenca L, Gil C, Cuesta S, Rapún-Araiz B, Žiemytė M, Mira A, Lasa I, Valle J (2020) Antibiofilm activity of flavonoids on Staphylococcal biofilms through targeting BAP amyloids. Sci Rep 10:18968. https://doi.org/10.1038/s41598-020-75929-2

Matsui K, Kan Y, Kikuchi J, Takemura M, Maki H, Kozono I, Ueda T, Minagawa K (2020) Stalobacin: discovery of novel Lipopeptide antibiotics with potent antibacterial activity against multidrug-resistant bacteria. J Med Chem 63(11):6090–6095. https://doi.org/10.1021/acs.jmedchem.0c00295

Mellini M, Di Muzio E, D'Angelo F, Baldelli V, Ferrillo S, Visca P, Leoni L, Polticelli F, Rampioni G (2019) *In silico* selction and experimental validation of FDA-approved drugs as anti-quorum sensing agents. Front Microbiol 10:2355. https://doi.org/10.3389/fmicb.2019.02355

Memariani H, Memariani M, Ghasemian A (2019) An overview of anti-biofilm properties of quercetin against bacterial pathogens. World J Microbiol Biotechnol 35(9):143. https://doi.org/10.1007/s11274-019-2719-5

Miesel L, Greene J, Black TA (2003) Genetic strategies for antibacterial drug discovery. Nat Rev Genet 4:442–456. https://doi.org/10.1038/nrg1086

Mishra R, Panda AK, Mandal SD, Shakeel M, Bisht SS, Khan J (2020) Natural anti-biofilm agents: strategies to control biofilm-forming pathogens. Front Microbiol 11:566325. https://doi.org/10.3389/fmicb.2020.566325

Monahan C, Harris S, Morris D, Cummins E (2022) A comparative risk ranking of antibiotic pollution from human and veterinary antibiotic usage—an Irish case study. Sci Total Environ 826:154008. https://doi.org/10.1016/j.scitotenv.2022.154008

Moscoso JA, Jaegar T, Valentini M, Hul K, Jenal U, Fllloux A (2014) The diguanylate cyclase SadC is a central player in Gac/Rsm-mediated biofilm formation in *Pseudomonas aeruginosa*. J Bacteriol 196(23):4081–4088. https://doi.org/10.1128/JB.01850-14

Motiwala T, Mthethwa Q, Achilonu I, Khoza T (2022) ESKAPE pathogens: looking at Clp ATPases as potential drug targets. Antibiotics 11:1218. https://doi.org/10.3390/antibiotics11091218

Mu Y, Zeng H, Chen W (2021) Quercetin inhibits biofilm formation by decreasing the production of EPS and altering the composition of EPS in *Staphylococcus epidermidis*. Front Microbiol 12:631058. https://doi.org/10.3389/fmicb.2021.631058

Mukherjee S, Bassler BL (2019) Bacterial quorum sensing in complex and dynamically changing environments. Nat Rev Microbiol 17(6):371–382. https://doi.org/10.1038/s41579-019-0186-5

Mukherjee S, Moustafa D, Smith CD, Goldberg JB, Bassler BL (2017) The RhlR quorum-sensing receptor controls *Pseudomonas aeruginosa* pathogenesis and biofilm development independently of its canonical homoserine lactone autoinducer. PLoS Pathog 13(7):e1006504. https://doi.org/10.1371/journal.ppat.1006504

Mukherjee A, Bose S, Shaoo A, Das SK (2023) Nanotechnology based therapeutic approaches: an advanced strategy to target the biofilm of ESKAPE pathogens. Mater Adv 4:2544–2572. https://doi.org/10.1039/d2ma00846g

Mulani MS, Kamble EE, Kumar SN, Tawre MS, Pardesi KR (2019) Emerging strategies to combat ESKAPE pathogens in the era of antimicrobial resistance: a review. Front Microbiol 10:539. https://doi.org/10.3389/fmicb.2019.00539

Murray JL, Kwon T, Marcotte EM, Whiteley M (2015) Intrinsic antimicrobial resistance determinants in the superbug *Pseudomonas aeruginosa*. mBio 6(6):e01603–e01615. https://doi.org/10.1128/mBio.01603-15

Naga NG, Zaki AA, El-Badan DE, Rateb HS, Ghanem KM, Shaaban MI (2022) Methoxyisoflavan derivative from *Trigonella stellata* inhibited quorum sensing and virulence factors of *Pseudomonas aeruginosa*. World J Microbiol Biotechnol 38:156. https://doi.org/10.1007/s11274-022-03337-x

Naga NG, El-Badan DE, Ghanem KM, Shaaban MI (2023) It is the time for quorum sensing inhibition as alternative strategy of antimicrobial therapy. Cell Commun Signal 21:133. https://doi.org/10.1186/s12964-023-01154-9

Neelam, Khatkar A, Sharma KK (2020) Phenylpropanoids and its derivatives: biological activities and its role in food, pharmaceutical and cosmetic industries. Crit Rev Food Sci Nutr 60(16):2655–2675. https://doi.org/10.1080/10408398.2019.1653822

Nicolas I, Bordeau V, Bondon A, Baudy-Floc'h M, Felden B (2019) Novel antibiotics effective against Gram-positive and Gram-negative multi-resistant bacteria with limited resistance. PLoS Biol 17(7):e3000337. https://doi.org/10.1371/journal.pbio.3000337

Nwobodo DC, Ugwu MC, Anie CO, Al-Ouqaili MTS, Ikem JC, Chigozie UV, Saki M (2022) Antibiotic resistance: the challenges and some emerging strategies for tackling a global menace. J Clin Lab Anal 36:e24655. https://doi.org/10.1002/jcla.24655

O'Rourke A, Beyhan S, Choi Y, Morales P, Chan AP, Espinoza JL, Dupont CL, Meyer KJ, Spoering A, Lewis K, Nierman WC, Nelson KE (2020) Mechanism-of-action classification of antibiotics by global transcriptome profiling. Antimicrob Agents Chemother 64(3):e01207–e01219. https://doi.org/10.1128/AAC.01207-19

Olsson A, Wistrand-Yuen P, Nielsen EI, Friberg LE, Sandegren L, Lagerbäck P, Tängdéna T (2020) Efficacy of antibiotic combinations against multidrug-resistant *Pseudomonas aeruginosa* in automated time-lapse microscopy and static time-kill experiments. Antimicrob Agents Chemother 64(6):e02111–e02119. https://doi.org/10.1128/AAC.02111-19

Ouedraogo V, Sombie PAED, Compaore M, Kiendrebeogo M (2019) Anti-quorum quenching activity of methyl gallate isolated from galls of *Guiera senegalensis* J. F. Gmel (Combretaceae). Afr J Microbiol Res 13(16):290–297. https://doi.org/10.5897/AJMR2019.9056

Panda SK, Buroni S, Swain SS, Bonacorsi A, da Fonseca Armorim EA, Kulshrestha M, da Silva LCN, Tiwari V (2022) Recent advances to combat ESKAPE pathogens with special reference to essential oils. Front Microbiol 13:1029098. https://doi.org/10.3389/fmicb.2022.1029098

Pandey R, Mishra SK, Shreshtha A (2021) Characterization of ESKAPE pathogens with special reference to multidrug resistance and biofilm production in a Nepalese hospital. Infect Drug Resist 14:2201–2212. https://doi.org/10.2147/IDR.S306688

Pang Z, Raudonis R, Glick BR, Lin TJ, Cheng Z (2019) Antibiotic resistance in *Pseudomonas aeruginosa*: mechanisms and alternative therapeutic strategies. Biotechnol Adv 37:177–192. https://doi.org/10.1016/j.biotechadv.2018.11.013

Parai D, Banerjee M, Dey P, Chakraborty A, Islam E, Mukherjee SK (2018) Effect of reserpine on *Pseudomonas aeruginosa* quorum sensing mediated virulence factors and biofilm formation. Biofouling 34(3):320–334. https://doi.org/10.1080/08927014.2018.1437910

Parmanik A, Das S, Kar B, Bose A, Dwivedi GR, Pandey MM (2022) Current treatment strategies against multidrug-resistant bacteria: a review. Curr Microbiol 79(12):388. https://doi.org/10.1007/s00284-022-03061-7

Parrino B, Schillaci D, Carnevale I, Giovannetti E, Diana P, Cirrincione G, Cascioferro S (2019) Synthetic small molecules as anti-biofilm agents in the struggle against antibiotic resistance. Eur J Med Chem 161:154–178. https://doi.org/10.1016/j.ejmech.2018.10.036

Patil A, Banerjee R, Kanojiya P, Saroj SD (2021) Foodborne ESKAPE biofilms and antimicrobial resistance: lessons learned from clinical isolates. Pathog Glob Health 115(6):339–356. https://doi.org/10.1080/20477724.2021.1916158

Peerzada Z, Kanhed AM, Desai KB (2022) Effects of active compounds from *Cassia fistula* on quorum sensing mediated virulence and biofilm formation in *Pseudomonas aeruginosa*. RSC Adv 12:15196–15214. https://doi.org/10.1039/d1ra008351a

Piewngam P, Chiou J, Chatterjee P, Otto M (2020) Alternative approaches to treat bacterial infections: targeting quorum-sensing. Expert Rev Anti-Infect Ther 18(6):499–510. https://doi.org/10.1080/14787210.2020.1750951

Qadri H, Shah AH, Ahmad SM, Alshehri B, Almilaibary A, Mir MA (2022) Natural products and their semi-synthetic derivatives against antimicrobial-resistant human pathogenic bacterial and fungi. Saudi J Biol Sci 29(9):103376. https://doi.org/10.1016/j.sjbs.2022.103376

Quecan BXV, Santos JTC, Rivera MLC, Hassimotto NMA, Almeida FA, Pinto UM (2019) Effect of quercetin rich onion extracts on bacterial quorum sensing. Front Microbiol 10:867. https://doi.org/10.3389/fmicb.2019.00867

Rajarathinam G, Sundaramurthi JC, Sarada Dronamraju VL (2018) Antibacterial activity of flavonoid isolated from *Trianthema decandra* against *Pseudomonas aeruginosa* and molecular docking study of FabZ. Microb Pathog 121:87–92. https://doi.org/10.1016/j.micpath.2016.05.016

Rajkumari J, Borkotoky S, Reddy D, Mohanty SK, Kumavath R, Murali A, Suchiang K, Busi S (2019) Anti-quorum sensing and anti-biofilm activity of 5-Hydroxymethylfufural against *Pseudomonas aeruginosa* PAO1: insights from *in vitro*, *in vivo*, and *in silico* studies. Microbiol Res 226:19–26. https://doi.org/10.1016/j.micres.2019.05.001

Raorane CJ, Lee JH, Kim YG, Rajasekharan SK, Garcia-Contreras R, Lee J (2019) Antibiofilm and antivirulence efficacies of flavonoids and curcumin against *Acinetobacter baumannii*. Front Microbiol 10:990. https://doi.org/10.3389/fmicb.2019.00990

Rather MA, Gupta K, Mandal M (2021) Inhibition of biofilm and quorum sensing-regulated virulence factors in *Pseudomonas aeruginosa* by *Cuphea carthagenesis* (Jacq.) J. F. Macbr. Leaf extract: an *in vitro* study. J Ethnopharmacol 269:113699. https://doi.org/10.1016/j.jep.2020.113699

Rekha PD, Vasavi HS, Vipin C, Saptami K, Arun AB (2016) A medicinal herb *Cassia alata* attenuates quorum sensing in *Chromobacter violaceum* and *Pseudomonas aeruginosa*. Lett Appl Microbiol 64:231–238. https://doi.org/10.1111/lam.12710

Remschmidt C, Schröder C, Behnke M, Gastmeier P, Geffers C, Kramer TS (2018) Continuous increase of Vancomycin resistance in *Enterococci* causing nosocomial infections in Germany—10 years of surveillance. Antimicrob Resist Infect Control 7:54. https://doi.org/10.1186/s13756-018-0353-x

Rima M, Trognon J, Latapie L, Chbani A, Roques C, El Garah F (2022) Seaweed extracts: a promising source of antibiofilm agents with distinct mechanisms of action against *Pseudomonas aeruginosa*. Mar Drugs 20:92. https://doi.org/10.3390/md20020092

Santajit S, Indrawattana N (2016) Mechanisms of antimicrobial resistance in ESKAPE pathogens. Biomed Res Int 2016:2475067. https://doi.org/10.1155/2016/2475067

Santajit S, Sookrung N, Indrawattana N (2022) Quorum sensing in ESKAPE bugs: a target for combating antimicrobial resistance and bacterial virulence. Biology 11:1466. https://doi.org/10.3390/biology11101466

Schindler BD, Kaatz GW (2016) Multidrug efflux pumps of Gram-positive bacteria. Drug Resist Updat 27:1–13. https://doi.org/10.1016/j.drup.2016.04.003

Selvaraj A, Jayasree T, Valliammai A, Pandian SK (2019) Myrtenol attenuates MRSA biofilm and virulence by suppressing *sarA* expression dynamism. Front Microbiol 10:2027. https://doi.org/10.3389/fmicb.2019.02027

Shaaban M, Elgaml A, Habib EE (2019) Biotechnological applications of quorum sensing inhibition as novel therapeutic strategies for multidrug resistant pathogens. Microb Pathog 127:138–143. https://doi.org/10.1016/j.micpath.2018.11.043

Shamprasad BR, Lotha R, Nagarajan S, Sivasubramanian A (2022) Metal nanoparticles functionalized with nutraceutical Kaempferitrin from edible *Crotalaria juncea*, exert potent antimicrobial and antibiofilm effects against Methicillin-resistant *Staphylococcus aureus*. Sci Rep 12:7061. https://doi.org/10.1038/s41598-022-11004-2

Shamsudin NF, Ahmed QU, Mahmood S, Shah SAA, Khatib A, Mukhtar S, Alsharif MA, Parveen H, Zakaria ZA (2022) Antibacterial effects of flavonoids and their structure-activity relationship study: a comparative interpretation. Molecules 27(4):1149. https://doi.org/10.3390/molecules27041149

Sharma A, Gupta VK, Pathania R (2019) Efflux pump inhibitors for bacterial pathogens: from bench to bedside. Indian J Med Res 149(2):129–145. https://doi.org/10.4103/ijmr.IJMR_2079_17

Sharma A, Harjai K, Ramniwas S, Singh D (2023) Bioactivity of citral and its nanoparticle in attenuating the pathogenicity of *Pseudomonas aeruginosa* and controlling *Drosophila melanogaster*. J Nanomater 2023:4703650. https://doi.org/10.1155/2023/4703650

Shehabeldine A, Ashour RM, Okba MM, Saber FR (2020) *Callistemon citrinus* bioactive metabolites as new inhibitors of methicillin-resistant *Staphylococcus aureus* biofilm formation. J Ethnopharmacol 254:112669. https://doi.org/10.1016/j.jep.2020.112669

Shriram V, Khare T, Bhagwat R, Shukla R, Kumar V (2018) Inhibiting bacterial drug efflux pumps via phyto-therapeutics to combat threatening antimicrobial resistance. Front Microbiol 9:2990. https://doi.org/10.3389/fmicb.2018.02990

Shukla M, Singh V, Habeeballah H, Alkhanani MF, Lata M, Hussain Y, Mukherjee M, Pasupuleti M, Meena A, Mishra BN, Haque S (2022) Quorum quenching-guided inhibition of mixed bacterial biofilms and virulence properties by protein derived from leaves of *Carissa carandas*. Front Cell Infect Microbiol 12:836819. https://doi.org/10.3389/fcimb.2022.836819

Singhal AV, Malwal D, Thiyagarajan S, Lahiri I (2021) Antimicrobial and antibiofilm activity of GNP-Tannic Acid-Ag nanocomposite and their epoxy-based coatings. Prog Org Coat 159:106421. https://doi.org/10.1016/j.porgcoat.2021.106421

Sionov RV, Steinberg D (2022) Targeting the holy triangle of quorum sensing, biofilm formation, and antibiotic resistance in pathogenic bacteria. Microorganisms 10:1239. https://doi.org/10.3390/microorganisms10061239

Song M, Liu Y, Li T, Liu X, Hao Z, Ding S, Panichayupakaranant P, Zhu K, Shen J (2021) Plant natural flavonoids against multidrug resistant pathogens. Adv Sci 2021:2100749. https://doi.org/10.1002/advs.202100749

Subhaswaraj P, Barik S, Macha C, Chiranjeevi PV, Siddhardha B (2018) Anti quorum sensing and anti biofilm efficacy of cinnamaldehyde encapsulated chitosan nanoparticles against *Pseudomonas aeruginosa* PAO1. LWT Food Sci Technol 97:752–759. https://doi.org/10.1016/j.lwt.2018.08.011

Tajani AS, Tehranizadeh ZA, Pourmohammad A, Pourmohammad A, Iranshahi M, Farhadi F, Soheili V, Bazzaz BSF (2023) Anti-quorum sensing and antibiofilm activity of coumarin derivatives against *Pseudomonas aeruginosa* PAO1: insights from in vitro and in silico studies. Iran J Basic Med Sci 26(4):445–452. https://doi.org/10.22038/IJBMS.2023.69016.15047

Tapia-Rodriguez MR, Bernal-Mercado T, Gutierrez-Pacheco MM, Vazquez-Armenta FJ, Hernandez-Mendoza A, Gonzalez-Aguilar GA et al (2019) Virulence of *Pseudomonas aeruginosa* exposed to carvacrol: alterations of the Quorum sensing at enzymatic and gene levels. J Cell Commun Signal 13:531–537. https://doi.org/10.1007/s12079-019-00516-8

Targhi AA, Moammeri A, Jamshidifar E, Abbaspour K, Sadeghi S, Lamakani L, Akbarzadeh I (2021) Synergistic effect of curcumin-Cu and curcumin-Ag nanoparticle loaded niosome: enhanced antibacterial and anti-biofilm activities. Bioorg Chem 115:105116. https://doi.org/10.1016/j.bioorg.2021.105116

Terreni M, Taccani M, Pregnolato M (2021) New antibiotics for multidrug-resistant bacterial strains: latest research developments and future perspectives. Molecules 26:2671. https://doi.org/10.3390/molecules26092671

Thiolas A, Bollet C, La Scola B, Raoult D, Pagès J-M (2005) Successive emergence of *Enterobacter aerogenes* strains resistant to imipenem and colistin in a patient. Antimicrob Agents Chemother 49:1354–1358. https://doi.org/10.1128/AAC.49.4.1354-1358.2005

Thornton JM, Walker JM, Kadiyam Sundarasivarao PY, Spur BW, Rodriguez A, Yin K (2021) Lipoxin A4 promotes reduction and antibiotic efficacy against *Pseudomonas aeruginosa* biofilm. Prostaglandins Other Lipid Mediat 152:106505. https://doi.org/10.1016/j.prostaglandins.2020.106505

Tiwari V (2019) Post-translational modification of ESKAPE pathogens as a potential target in drug discovery. Drug Dis Today 24(3):814–822. https://doi.org/10.1016/j.drugdis.2018.12.005

Turner NA, Sharma-Kuinkel BK, Maskarinec SA, Eichenberger EM, Shah PP, Carugati M, Holland TL, Fowler VG Jr (2019) Methicillin-resistant *Staphylococcus aureus*: an overview of basic and clinical research. Nat Rev Microbiol 17:203–218. https://doi.org/10.1038/s41579-018-0147-4

Uchil A, Murali TS, Nayak R (2021) Escaping ESKAPE: a chalcone perspective. Result Chem 3:100229. https://doi.org/10.1016/j.rechem.2021.100229

Uruén C, Chopo-Escuin G, Tommassen J, Mainar-Jaime RC, Arenas J (2021) Biofilms as promoters of bacterial antibiotic resistance and tolerance. Antibiotics 10:3. https://doi.org/10.3390/antibiotics10010003

van Gent ME, van der Reijden TJK, Lennard PR, de Visser AW, Schonkeren-Ravensbergen B, Dolezal N, Cordfunke RA, Drijfhout JW, Nibbering PH (2022) Synergism between the synthetic antibacterial and antibiofilm peptide (SAAP)-148 and Halicin. Antibiotics 11(5):673. https://doi.org/10.3390/antibiotics11050673

Vashistha A, Sharma N, Nanaji Y, Kumar D, Singh G, Barnwal RP, Yadav AK (2023) Quorum sensing inhibitors as therapeutics: bacterial biofilm inhibition. Bioorg Chem 136:106551. https://doi.org/10.1016/j.bioorg.2023.106551

Veiko AG, Olchwik-Grabarek E, Sekowski S, Roszkowska A, Lapshina EA, Dobrzynska I et al (2023) Antimicrobial activity of Quercetin, Naringenin and Catechin: flavonoids inhibit *Staphylococcus aureus*-induced hemolysis and modify membranes of bacteria and erythrocytes. Molecules 28:1252. https://doi.org/10.3390/molecules28031252

Verma P, Tiwari M, Tiwari V (2022) Potentiate the activity of current antibiotics by naringin dihydrochalcone targeting the AdeABC efflux pump of multidrug-resistant *Acinetobacter baumannii*. Int J Biol Macromol 217:592–605. https://doi.org/10.1016/j.ijbiomac.2022.07.065

Vijayakumar K, Ramanathan T (2020) *Musa acuminata* and its bioactive metabolite 5-Hydroxymethylfurfural mitigates quorum sensing (las and rhl) mediated biofilm and virulence production of nosocomial pathogen *Pseudomonas aeruginosa in vitro*. J Ethnopharmacol 246:112242. https://doi.org/10.1016/j.jep.2019.112242

Vipin C, Saptami K, Fida F, Mujeeburrahiman M, Rao SS, Athmika et al (2020) Potential synergistic activity of quercetin with antibiotics against multidrug-resistant clinical strains of *Pseudomonas aeruginosa*. PLoS One 15(11):e0241304. https://doi.org/10.1371/journal.pone.0241304

Wang CY, Chen YW, Hou CY (2019) Antioxidant and antibacterial activity of seven predominat terpenoids. Int J Food Prop 22(1):230–238. https://doi.org/10.1080/10942912.2019.1582541

Wang W, Lin X, Yang H, Huang X, Pan L, Wu S, Yang C, Zhang L, Li Y (2022) Anti-quorum sensing evaluation of methyleugenol, the principal bioactive component, from the *Melaleuca bracteata* leaf oil. Front Microbiol 13:970520. https://doi.org/10.3389/fmicb.2022.970520

Wang Y, Su J, Zhou Z, Yang J, Liu W, Zhang Y et al (2023) Baicalein resensitizes multidrug-resistant gram-negative pathogens to doxycycline. Microbiol Spectr 11(3):e0470222. https://doi.org/10.1128/spectrum.04702-22

Wu SC, Han F, Song M, Chen S, Li Q, Zhang Q et al (2019) Natural flavones from *Morus alba* against MRSA via targeting the proton motive force and membrane permeability. J Agric Food Chem 67(36):10222–10234. https://doi.org/10.1021/acs.jafc.9b01795

Xie RQ, Zhang XD, Zhao Q, Peng B, Zheng J (2018) Analysis of global prevalence of antibiotic resistance in *Acinetobacter baumannii* infections disclosed a faster increase in OECD countries. Emerg Microbes Infect 7:31. https://doi.org/10.1038/s41426-018-0038-9

Xu Z, Zhang H, Yu H, Dai Q, Xiong J, Sheng H et al (2019) Allicin inhibits *Pseudomonas aeruginosa* virulence by suppressing the *rhl* and *pqs* quorum-sensing systems. Can J Microbiol 65(8):563–574. https://doi.org/10.1139/cjm-2019-0055

Xu W, Tan J, Mu Y, Zheng D, Huang X, Li L (2020) New antimicrobial terpenoids and phloroglucinol glucosides from *Syzygium szemaoense*. Bioorg Chem 103:104242. https://doi.org/10.1016/j.bioorg.2020.104242

Xu M, Yao Z, Zhao Y, Shi S, Sun Y, Feng L et al (2022) Naringenin restores colistin activation against colistin-resistant gram-negative bacteria *in vitro* and *in vivo*. Front Microbiol 13:916587. https://doi.org/10.3389/fmicb.2022.916587

Yadav MK, Vidal JE, Song JJ (2020) Microbial biofilms on medical indwelling devices. In: Yadav MK, Singh BP (eds) New and future developments in microbial biotechnology and bioengineering: microbial biofilms. Elsevier, pp 15–28. https://doi.org/10.1016/B978-0-444-64279-0.00002-5

Yan X, Gu S, Shi Y, Cui X, Wen S, Ge J (2017) The effect of emodin on *Staphylococcus aureus* strains in planktonic form and biofilm formation *in vitro*. Arch Microbiol 199(9):1267–1275. https://doi.org/10.1007/s00203-017-1396-8

Yan Y, Li X, Zhang C, Lv L, Gao B, Li M (2021) Research progress on antibacterial activities and mechanisms of natural alkaloids: a review. Antibiotics 10(3):318. https://doi.org/10.3390/antibiotics10030318

Yayan J, Ghebremedhin B, Rasche K (2015) Antibiotic resistance of *Pseudomonas aeruginosa* in pneumonia at a single university hospital center in Germany over a 10-year period. PLoS One 10:e0139836. https://doi.org/10.1371/journal.pone.0139836

Yuan Y, Yang X, Zeng Q, Li H, Fu R, Du L, Liu W, Zhang Y, Zhou X, Chu Y, Zhang X, Zhao K (2022) Repurposing Demetridazole and Ribavirin to disarm *Pseudomonas aeruginosa* virulence by targeting the quorum sensing system. Front Microbiol 13:978502. https://doi.org/10.3389/fmicb.2022.978502

Zhang W, Sun J, Ding W, Lin J, Tian R, Lu L, Liu X, Shen X, Qian PY (2015) Extracellular matrix-associated proteins form an integral and dynamic system during *Pseudomonas aeruginosa* biofilm development. Front Cell Infect Microbiol 5:40. https://doi.org/10.3389/fcimb.2015.00040

Zhang P, Wu Q, Chen L, Duan K (2020) Virulence-inhibiting herbal compound Falcarindiol significantly reduced mortality in mice infected with *Pseudomonas aeruginosa*. Antibiotics 9:136. https://doi.org/10.3390/antibiotics9030136

Zhao X, Yu Z, Ding T (2020) Quorum-sensing regulation of antimicrobial resistance in bacteria. Microorganisms 8:425. https://doi.org/10.3390/microorganisms8030425

Zhao Z, Guo M, Xu X, Hu Y, Liu D, Wang C, Liu X, Li Y (2022) In vitro synergistic inhibitory activity of natural alkaloid Berberine combined with Azithromycin against alginate production in *Pseudomonas aeruginosa* PAO1. Oxid Med Cell Longev 2022:3858500. https://doi.org/10.1155/2022/3858500

Zhen X, Lundborg CS, Sun X, Hu X, Dong H (2019) Economic burden of antibiotic resistance in ESKAPE organisms: a systematic review. Antimicrob Resist Infect Control 8:137. https://doi.org/10.1186/s13756-019-0590-7

Zhong S, He S (2021) Quorum sensing inhibition or quenching in *Acinetobacter baumannii*: the novel therapeutic strategies for new drug development. Front Microbiol 12:558003. https://doi.org/10.3389/fmicb.2021.558003

Zhou JW, Hou B, Liu GY, Jiang H, Sun B, Wang ZN et al (2018) Attenuation of *Pseudomonas aeruginosa* biofilm by Hordenine: a combinatorial study with aminoglycoside antibiotics. Appl Microbiol Biotechnol 102:9745–9758. https://doi.org/10.1007/s00253-018-9315-8

Zielińska S, Wojciak-Kosior M, Dziagwa-Becker M, Glensk M, Sowa I, Fijalkowski K et al (2019) The activity of Isoquinoline alkaloids and extracts from *Chelidonium majus* against pathogenic bacteria and *Candida* sp. Toxins 11(7):406. https://doi.org/10.3390/toxins11070406

Zowawi HM, Forde BM, Alfaresi M, Alzarouni A, Farahat Y, Chong TM, Yin WF, Chan KG, Li J, Schembri MA, Beatson SA, Paterson DL (2015) Stepwise evolution of pandrug-resistance in *Klebsiella pneumoniae*. Sci Rep 5:15082. https://doi.org/10.1038/srep15082

Applications of Photodynamic Therapy for the Eradication of ESKAPE Pathogens 15

V. T. Anju, Siddhardha Busi, and Madhu Dyavaiah

Abstract

Antimicrobial photodynamic therapy (aPDT) is one of the efficient and alternate therapeutic strategies to combat planktonic and biofilm cells of ESKAPE pathogens. This therapeutic method requires a photosensitizer, light, and oxygen. ESKAPE superbugs are responsible for most of the nosocomial infections and it needs urgent attention in healthcare system. aPDT is emerged as a safer and effective method against the elimination of resistant ESKPAE pathogens, as the therapy is not involved in the generation of antibiotic-resistant strains. Reactive oxygen species produced through type I and II photoreactions of aPDT has multiple cellular targets in the bacteria, i.e., protein, DNA, and lipids. ROS exerts oxidation of cellular targets where the photosensitizer is localized. There are several classes of photosensitizers tested and used against variety of pathogens, especially ESKAPE such as from the family of chlorins, porphyrins, phthalocyanines, and bacteriochlorins. Additionally, several photosensitizers of plant origin and nanoparticles are found to be efficient against ESKPAE pathogens. In comparison to Gram-negative pathogens, aPDT is more efficient against Gram-positive pathogens owing to their cell wall structure. Cationic and neutral photosensitizers are more efficient in the eradication of Gram-positive pathogens, while photosensitizers are modified for Gram-negative pathogens. Here, we discussed applications and effectiveness of different photosensitizers against pathogenic forms of ESKAPE group.

V. T. Anju · M. Dyavaiah
Department of Biochemistry and Molecular Biology, School of Life Sciences, Pondicherry University, Pondicherry, India

S. Busi (✉)
Department of Microbiology, School of Life Sciences, Pondicherry University, Puducherry, India

S. Busi, R. Prasad (eds.), *ESKAPE Pathogens*,
https://doi.org/10.1007/978-981-99-8799-3_15

Keywords

Antimicrobial photodynamic therapy · Biofilms · Photosensitizer · Reactive oxygen species

15.1 Introduction

Many years back, photodynamic therapy (PDT) was introduced to kill certain microorganisms using harmless light and non-toxic photosensitizers (PS). Different PS and light combinations have been studied to inactivate bacteria, fungi, and viruses (Huang et al. 2010). The unsuitable application of antibiotics caused the global health challenge leading to the emergence of several antibiotic-resistant strains. Antimicrobial photodynamic therapy (aPDT) is used for the elimination of antibiotic-resistant strains and this therapeutic strategy holds the capacity to replace even antibiotics. aPDT inactivates microorganisms using reactive oxygen species (ROS) synthesized by photoactivated PS. PDT exhibited potential results in the treatment of cancer cells along with its antimicrobial therapeutic properties (Anas et al. 2021). There are different types of PS available so far with antimicrobial properties including both natural and synthetic compounds. Many of the PSs exhibited antimicrobial activities towards planktonic and biofilm state of cells. The putative targets of aPDT in microorganism by ROS are their vital biomolecules such as lipids, proteins, and nucleic acids (Cieplik et al. 2018).

ESKAPE pathogens are a group of six bacteria accountable for variety of nosocomial infections and can escape from the attack of antibiotics. Several multidrug-resistant strains are evolved among the pathogens, *Enterococcus faecium*, *Staphylococcus aureus*, *Klebsiella pneumoniae*, *Acinetobacter baumannii*, *Pseudomonas aeruginosa*, and *Enterobacter* sp. owing to their virulence mechanisms (Navidinia 2016; Rice 2008). World Health Organization has categorized ESKAPE pathogens requiring greatest attention owing to the reported risk of mortality rate. These bacteria are named as critical, high, and medium priority pathogens for which traditional antibiotics are failed and novel and effective antibiotics are needed. *A. baumannii* and *P. aeruginosa* with carbapenem resistance and *K. pneumoniae* and *Enterobacter* sp. with carbapenem resistance or extended spectrum beta lactamase enzymes (ESBL) are grouped as critical priority pathogens. High priority pathogens are *E. faecium* showing vancomycin resistance (VRE) and *S. aureus* with methicillin (MRSA) and vancomycin resistance (VRSA) (Tacconelli et al. 2018).

ESKAPE pathogens exhibit different resistance mechanisms. These include inactivation of drugs, antibiotic target site modification, and efflux out of antibiotic from cell. Drugs are inactivated by some enzymes produced by the organisms which cleave them in an irreversible manner. Modification target site where binding of drug occurs so that drug is no longer able to bind to the target site for its action. Presence of efflux pumps or porin proteins on cell membrane reduces the accumulation of drug inside the cell. They can also develop into biofilm growth state during unfavorable conditions of antibiotic therapy. Biofilms protect the cells from antibiotic exposure by the production of exopolymeric matrix and prevent the entry of antibiotic to the deeper layers of cells. Biofilms can also reach to a state of dormancy where formation of persister cells occurs when treated with high concentrations of antibiotic and can revive the cell growth once the antibiotic pressure is released (Santajit and Indrawattana 2016; Lewis 2007).

These pathogens are evolved with multidrug-resistant (MDR), pandrug-resistant, and extensively drug-resistant (XDR) strains which cause several clinical infections and are difficult to treat/manage by traditional therapies. Potential therapeutics used against these pathogens include antimicrobial peptides, bacteriophage-coded proteins, natural products, etc. (Panda et al. 2022). New antibiotics targeting cell wall synthesis, outer membrane, and DNA synthesis are developed and some are in clinical development process against ESKAPE pathogens. Moreover, there are effective anti-resistance potentiators, nanoparticles as anti-microbial agents, molecules targeting biofilm formation stages, efflux pump action, quorum sensing, and its virulence gene expressions. Recent therapeutic strategies include bacteriophage therapy, nanoparticle-based drug delivery systems, and light-mediated anti-microbial therapy, i.e., aPDT (Ma et al. 2020). In this chapter, antimicrobial photodynamic inactivation of ESKAPE pathogens is discussed.

15.2 What Is an Antimicrobial Photodynamic Therapy (aPDT)

aPDT is developed as an effective therapeutic method to target the global threat of antibiotic resistance. It is widely accepted for the inactivation of broad-spectrum microorganisms such as Gram-negative and positive bacteria, fungi, viruses, and parasites. In this method of antimicrobial killing, ROS produced by type I or II mechanism such as superoxide ions, hydrogen peroxides, hydroxyl ions, and singlet oxygen are cytotoxic to cell components. These radical ions cause irreversible damage to the biomolecules and mediate microbial death. As aPDT acts on multiple targets of bacteria and causes irreversible damage to the vital molecules, there is no selective pressure to develop resistance as observed in traditional therapeutics (Almeida 2020).

15.2.1 aPDT Mechanism and ROS Production

The whole process of aPDT uses a non-toxic PS, light of suitable wavelength and oxygen. PS is transformed from its ground state of having low energy to singlet excited form when light is illuminated on PS. Later, it can return to lower energy state through fluorescence or convert to triplet state with high energy. Then, triplet state interacts with molecules either through type I or II pathway. Free radicals are produced through energy transfer in type I pathway and these radicals are converted to ROS when react with oxygen. Type I pathway-generated ROS interfere with cell membrane stability. Electron transfer mechanism in type II pathway produces extremely cytotoxic, singlet oxygen. Singlet oxygen causes most of the cell damage as it produces localized effect (Rajesh et al. 2011).

There are some ideal properties for effective aPDT. PS having cationic charge and hydrophilicity property is ideal for Gram-negative pathogens. And an ideal PS should have good photosensitivity by producing enough amount of ROS upon illumination. Light used for illumination should not exert toxicity or other side effects to the surrounding host cells along with good light transmittance property (Cieplik et al. 2018).

15.2.2 Types of Photosensitizers

There are different types of antimicrobial photosensitizers available so far. They are classified according to their structure, origin, and function. There are PSs such as synthetic dyes, tetra-pyrrole structures, natural compounds, and nanocompounds. There are first, second, and third generation PSs. First-generation phenothiazinium dyes are available such as methylene blue, toluidine blue, and Rose Bengal. Among these, except Rose Bengal all are cationically charged PS. Rose Bengal, eosin Y, and erythrosine are anionic PSs. Derivatives of methylene blue such as new methylene blue, dimethyl methylene blue, etc. are also studied previously. Examples of natural PSs are hypericin, flavin derivatives, and curcumin. Tetra-pyrrole structures include neutral PSs such as phthalocyanine, and chlorine and cationic, zinc phthalocyanine derivatives, and porphyrin (Ghorbani et al. 2018). The most used natural PS classes include curcuminoids (curcumin), alkaloids (pterin-6-carboxylic acid), perylenequinones (hypericin, hypocrellin), anthraquinones (parietin, anthraquinone, aloe emodin), flavins (riboflavin), chlorin-type compounds (chlorophyllin sodium salt), and porphyrin precursor (5-aminolevulinic acid) (Polat and Kang 2021).

15.2.3 Molecular Targets of aPDT

The major targets of aPDT in bacteria are proteins, lipids, polysaccharides, DNA, and/or RNA. In Gram-negative bacteria, most of the PS are phototoxic owing to their cell wall structure and charge, whereas for Gram-positive bacteria cationic or neutral PS combined with membrane breaking molecules are required (Youf et al. 2021).

Cell membrane and associated biomolecules are potential cellular targets. Damaged cell membrane caused by photoactivated PS cause leakage of cellular materials. Along with metabolite leakage from cells, oxidative damage of proteins and lipid bilayers of cell membrane also occurs. This in turn inactivates aerobic respiratory complexes and interfere with the ATP production. Enzymes present in cytosol which participate in ATP production are also targeted by aPDT. Thus, the phototoxicity of PS on microbial cell leads to the deficiency of ATP, NADH, and NADPH. This negotiates the cell defense mechanisms against ROS (Awad et al. 2016) (Fig. 15.1).

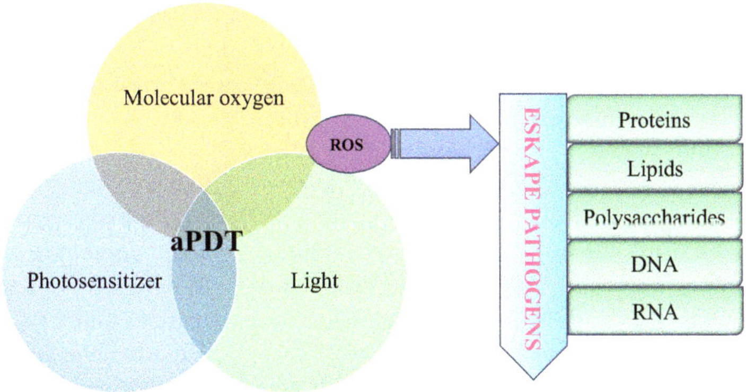

Fig. 15.1 Different targets of antimicrobial photodynamic therapy in ESKAPE pathogens

DNA damage by photoactivated PS causes alteration or modulation of nucleic acid bases. Single or double strand breaks and loss of supercoiled plasmid DNA are observed. Moreover, some PS cleaves DNA by reducing guanine bases through electrostatic bonds. In biofilms, PS affects biofilm matrix by affecting extracellular DNA and polysaccharides. In addition, biofilm formation is affected by PS by inhibiting or quenching quorum sensing mechanism and virulence gene expression profiles (Youf et al. 2021).

15.3 aPDT for Gram-Positive ESKAPE Pathogens

Gram-positive ESKAPE pathogens, *Enterococcus faecium* and *Staphylococcus aureus,* were tested with several PSs for their photoinactivation properties. Antimicrobial photodynamic inactivation of bacteria has showed increased susceptibility in pathogens. As described by Tomb et al. (2017), aPDT using blue light photodynamically inactivated methicillin-resistant and sensitive *S. aureus*. Around, 5 $\log_{10}$ reduction was observed for sensitive culture under 216 J/cm^2 light treatment. In addition, sensitivity to several antibiotics (fusidic acid, mupirocin, and rifampicin) was altered after aPDT (Tomb et al. 2017). *S. aureus* biofilms were photoinactivated by methylene blue combined with an efflux pump inhibitor, verapamil. The combination of methylene blue (200 μg/mL) and pump inhibitor (312 μg/mL) exhibited 3.65 $\log_{10}$ reductions under 22 J/cm^2 light. Also, 80% of metabolic activity was inhibited in biofilms of *S. aureus* (De Aguiar Coletti et al. 2017). In a combination therapy, methylene blue-mediated aPDT along with cationic antimicrobial peptide aurein 1.2 (16 μM) reduced planktonic cells of *S. aureus* by 6 $\log_{10}$ CFU/mL (De Freitas et al. 2018). A new-PS nano emulsion was developed to kill biofilms of *S. aureus*. Porphyrin containing nanoemulsion was prepared by encapsulating with food grade oil core for the efficient delivery of PS to biofilms. There was an high number of cell reduction by 6 $\log_{10}$ when incubated the new PS (100 μM) for long time under the conditions of 60 J/cm^2 (Buzzá et al. 2022). Clinical isolates of

S. aureus isolated from wound patients were photoirradiated using toluidine blue dye. The viable bacterial count studied through plate count method showed decreased number on different time intervals and with a laser of output power, 7.5 mW. The reduction was achieved as 100% killing upon exposure to 50 μg/mL of toluidune blue dye for 15 min (Hajim et al. 2010).

E. faecium causes several nosocomial infections and is resistant to traditional antibiotic therapy. Several in vitro aPDT studies explored killing of resistant planktonic and biofilm cells of *E. faecium*. Green light of 6.4 J/cm^2 combined with Rose Bengal dye at low doses (0.1 μM) reduced cells of *E. faecium* by 2.5 $\log_{10}$ CFU/mL. Similarly, a second PS, fullerene, exhibited 5 log reductions under similar light conditions and 0.5 μM concentration. In aPDT-synergy experiments, both the PSs exhibited combination or synergy effect with gentamycin, streptomycin, tigecycline, doxycycline, or daptomycin antibiotics (Woźniak et al. 2021). In a study, insect model inoculated with *E. faecium* was inhibited by methylene blue-mediated aPDT and subsequent antibiotic therapy. In this study, vancomycin-resistant bacteria-infected caterpillars (*Galleria mellonella*) exhibited improved survival after aPDT and vancomycin treatment. This is because aPDT enhanced the susceptibility of resistant strains to vancomycin and protected model system from infection and death (Chibebe Junior et al. 2013). Two different wavelengths of 450 (Blue light) and 405 (Violet light) nm irradiation caused reduction of *E. faecium* cells by 3.26 and 3.06 $\log_{10}$ CFU/mL with an energy fluence rate of 1250 and 825 J/cm^2, respectively (Hoenes et al. 2021). More than 5 $\log_{10}$ decrease exhibited for planktonic cells of *E. faecium* caused by methylene blue mediated photodynamic inactivation under a light treatment of 40 J/cm^2 (Sabino et al. 2020).

15.4 aPDT for Gram-Negative ESKAPE Pathogens

As Gram-negative pathogens possess thick outer membrane, anionic and neutral PS encounter permeability problems. Membrane disorganizing agents such as EDTA and cationic charges on PS are used to enhance the permeability of PSs. Poly-L-lysine, polyethylenimine, and polymyxin B nonapeptide (PMBN) are some cationic groups linked to anionic or neutral PS to enhance aPDT of Gram-negative pathogens. aPDT of Gram-negative bacteria with anionic porphyrins exhibited reduced permeability to PSs. For example, *P. aeruginosa* was not inactivated by anionic, hematoporphyrin derivative, or deuteroporphyrin. However, pretreatment of *P. aeruginosa* with EDTA caused efficient photodynamic inactivation by deuteroporphyrin due to the altered membrane permeability (Sperandio et al. 2013).

Methylene blue-dependent aPDT is well explored for various ESKAPE pathogens. Methylene blue-mediated photodynamic inactivation of *Klebsiella pneumoniae* was studied by Songsantiphap and coworkers. Multidrug-resistant clinical isolate of *K. pneumoniae* exhibited phototoxicity in the presence of 50 mg/L of methylene blue. Around 2.82 and 2.94 $\log_{10}$ reduction at 80 J/cm^2 energy fluence rate was observed (Songsantiphap et al. 2022). In an interesting study, antimicrobial blue light-mediated photodynamic inactivation of *K. pneumoniae* under a light fluence of 61.2 J/cm^2 was observed. The study enabled 2.2 $\log_{10}$ CFU/mL reduction on fifth day of consecutive treatments (Rapacka-Zdonczyk et al. 2021). Phenothiazinium dyes photoinactivated both planktonic cells and biofilms of *K. pneumoniae*. Three dyes, toluidine blue, azure A, and new methylene blue, caused about 3 $\log_{10}$ reductions after 300 s of irradiation time. In addition, more than 42% of biofilm reduction was observed with all the 3 dyes. Further, they observed cell membrane and DNA as the major cellular targets (Misba et al. 2017). In another work, 5-aminolevulinic acid and 5-aminolevulinic acid methyl ester significantly affected the viability of planktonic cells and biofilms of *K. pneumoniae*. In a comparative study, extended spectrum beta lactamase producing pathogen exhibited photodynamic reduction of 3.20 and 4.52 $\log_{10}$ reduction when treated with 5-aminolevulinic acid and 5-aminolevulinic acid methyl ester, respectively. The mechanism of action discovered the consequence on DNA and cell envelope. Also, denaturation of cytoplasmic contents, aggregation of ribosomes, and fast leakage of biopolymers were reported by the authors (Liu et al. 2016).

Natural PSs, riboflavin, and chlorophyllin were used for planktonic cells and biofilms of *Acinetobacter baumannii*. Around 6.7 $\log_{10}$ and 5.7 $\log_{10}$ reduction of planktonic cells was observed with riboflavin and chlorophyllin under blue light irradiation. Light and dark toxicity was not observed in the study. Similarly, decrease of 5.6 $\log_{10}$ and 2.3 $\log_{10}$ biofilm cells were observed with riboflavin and chlorophyllin, respectively. More antibiofilm effect by riboflavin was correlated to higher amount of ROS produced by the PS (Buchovec et al. 2023). Nosocomial infection causing XDR *A. baumannii* strains isolated from Thailand was subjected to methylene blue-mediated aPDT under red light. It caused more than 2 $\log_{10}$ reduction at 50 mg/L of methylene blue dye (Songsantiphap et al. 2022). Phenothiazinium dyes, toluidine blue O, methylene blue, 1,9-dimethylmethylene blue, and new methylene blue exhibit aPDT against MDR *A. baumannii*. Under a light treatment of 22.5 J/cm^2, all the dyes exhibited efficient photodynamic toxicity against the bacteria. Maximum activity was observed with new methylene blue with a reduction of 6 $\log_{10}$. Other dyes exhibited only 2–3 $\log_{10}$ reduction under similar light conditions (Ragàs et al. 2010).

Methylene blue-mediated photodynamic inactivation of *Enterobacter aerogenes* impaired carbapenemase enzyme activity and improved their susceptibility to carbapenem antibiotics. Methylene blue at dose of 10 μg/mL when irradiated by 660 nm of laser was able to inactivate periplasmic carbapenemases and also improved the efficacy of imipenem antibiotic. The results suggest the inactivation of antibiotic-resistant enzymes by MB-aPDT (Feng et al. 2020). Chlorin e6 trisodium salt was successfully used for the photodynamic inactivation of antibiotic-resistant *E. cloacae*. Photosensitizer (50 μmol/L) irradiated under the blue light (80 J/cm^2) caused a reduction of 7 * 10^4 CFU in vitro (Kustov et al. 2023). aPDT on ESKAPE pathogens and various light parameters are summarized in Table 15.1.

The results of aPDT on *P. aeruginosa* were studied previously by several groups. In a study, aPDT effect of aminolevulinic acid on biofilms of *P. aeruginosa* was explored. PS inhibited biofilm development in a dose-dependent method to 42% and 61% at 10 and 20 mM concentration. aPDT results showed a reduced synthesis

Table 15.1 Antimicrobial photodynamic inactivation of Gram-negative and positive ESKAPE pathogens

ESKAPE pathogen	PS class	Light conditions	Activity	References
Enterococcus faecium	Rose Bengal	0.1 μM, green light (6.4 J/cm^2)	2.5 log_{10} CFU/mL	Woźniak et al. (2021)
	Fullerene	0.15 μM to 0.5 mM, green light (6.4 J/cm^2)	5 log_{10} CFU/mL	Woźniak et al. (2021)
Staphylococcus aureus	Chlorin e6 (Ce6)	4 μg/mL, 660 nm light	100%	Zhang et al. (2019)
	Hypocrellin	3.12 μmol/L, orange LED	4 log	Malacrida et al. (2020)
	Curcumin	460–465 nm, 25 μM	5 log_{10} CFU/mL	Li et al. (2020)
Klebsiella pneumoniae	Porphyrin PS + silver nanoparticles	LED light of 10 or 20 J/cm^2, 1.56–50 μM of PS and 3.38 mg/L of nanoparticles	5 log_{10} CFU/mL	Malá et al. (2021)
	Methylene blue + carboxypterin	Ultraviolet and visible light, 100 μM of antibiotic, and 2.5–10 μM of PS	3 log_{10} CFU/mL	Tosato et al. (2020)
	Isomeric tetra-cationic porphyrin	White LED light of 180 J/cm^2, 10^{-4} M	More than 3 log_{10}	da Silveira et al. (2020)
Acinetobacter baumannii	Ce6	40 μg/mL, 660 nm light	100%	Zhang et al. (2019)
	Methylene blue	12.5 and 25 mg/mL, 638 nm	3.10 ± 0.2 log_{10}	Maliszewska and Goldeman (2023)
	Zinc oxide nanoparticles	Blue light of 5.4 J/cm^2, 0.5–1 mg/mL	100%	Yang et al. (2018)
Pseudomonas aeruginosa	Polyethylene terephthalate discs functionalized with a cationic porphyrin	White LED light	1.51 ± 0.03 log reduction	Shatila et al. (2023)
	BODIPY	400–800 nm, 800 μg/mL	90%	Hao et al. (2019)
	Toluidine blue	5 mg, Red LED	>99.9999%	Moore et al. (2023)

(continued)

Table 15.1 (continued)

ESKAPE pathogen	PS class	Light conditions	Activity	References
Enterobacter sp (*E. cloacae*)	Rose Bengal	Green light of 57.2 J/cm^2, 10 and 20 μM	≥3 $\log_{10}$ CFU/mL	Woźniak et al. (2022)
Enterobacter sp. (*E. aerogenes*)	Methylene blue + Ceftriaxone Sodium	LED of 660 nm and 10 and 25 J/cm^2, 100 μg/mL of methylene blue, and 32 μg/mL of Ceftriaxone Sodium	3.14 log	Costa Magacho et al. (2020)

of various virulence factors of bacteria (pyocyanin and elastase). The antibacterial activity was confirmed with downregulated expression of various quorum sensing and virulence genes (Tan et al. 2018). Songsantiphap and coworkers also reported effective methylene blue-mediated aPDT of XDR *P. aeruginosa* under red light at a fluence rate of 80 J/cm^2. The viable cell number reductions were recorded for 2 strains of XDR as 3.13 and 3.17 $\log_{10}$ (Songsantiphap et al. 2022). A new boron-dipyrromethene (BODIPY) class PS, GD11 was employed for the killing of *P. aeruginosa* PAO1. Both planktonic cells and biofilms of this bacteria was photoinactivated at 2.5 μM concentration of GD11. Around 7 log reductions were observed when treated with photoactivated dye (Orlandi et al. 2014).

15.5 Animal Models to Study aPDT Against ESKAPE Pathogens

There are several PS and its combinations studied in vitro against ESKAPE and other important pathogens. Still, successful application in translational medicine is required to fully uncover the pharmacokinetics and pharmacodynamics properties of PS. Studies involving animal models provide large amount of information on the suitability of aPDT for the removal of planktonic and biofilm cells.

Burn wound infection model of *S. aureus* was employed to evaluate the effect of porphyrins in aPDT. Two porphyrins, hemin and deuteroporphyrin, in combination were used as anti-Staphylococcal agent in guinea pigs. Almost 99% of bacterial reduction was recorded in the infected model system. The above study employed two 100 W incandescent lamps for irradiation procedures (Orenstein et al. 1997). Another burn wound mice infection model was used for the study of antimicrobial photodynamic inactivation of *A. baumannii* using decacationic monoadducts and bisadducts of fullerene (C_{70}). Bacterial count was reduced in infected burn wound after the light treatment of 48 and 72 J/cm^2 even from the first day of treatment to day 6. Interestingly, PS not showed any toxicity in dark conditions (Huang et al. 2014). Hypocrellin B with lanthanide ions as PS and its effect on burn wound mice model were studied previously. In an in vivo model, bacterial load was reduced to 6 $\log_{10}$ when applied blue and red light. Bacteremia was reduced and maintained bacterial load in blood 2–3 log lower than the control infection group. Also, the infected mice with treatment and under light irradiation were survived up to 48 h of infection irrespective of control group (Hashimoto et al. 2012). In an interesting study, aPDT using methylene blue (0.01%) was used for infectious stomatitis in snakes. The inflammatory signs and caseous material in oral cavity were reduced in snakes followed by aPDT under red light of 280 J/cm^2. The reduction of bacterial isolates, *P. aeruginosa*, and *K. pneumoniae* are reported in the above study (Grego et al. 2017).

15.6 Future Perspectives and Conclusions

Antimicrobial photodynamic inactivation of ESKAPE pathogens is one of the promising therapeutic strategies to eliminate antibiotic-resistant biofilms and planktonic cells. As conventional therapeutic strategies fail to eliminate antibiotic-resistant strains, novel therapeutics are the need of hour. There are several PSs of natural origin and synthetic compounds used against planktonic cells as well as biofilms. To improve the efficacy of PS, several nanoparticles or combination of antibiotics are provided. Though plenty of evidences are available for in vitro activity of these PSs against pathogens, studies about their clinical trials and translation into medicine are scarce. The in vitro studies provide information of dose of PSs, light intensity, time of incubation of PS, exposure time, and dark toxicity. However, these parameters can be altered in various in vivo models and clinical practice. The emerging studies available so far including both in vitro and in vivo conditions specify the successful application of aPDT for the replacement of antimicrobials.

References

Almeida A (2020) Photodynamic therapy in the inactivation of microorganisms. Antibiotics 9:138

Anas A, Sobhanan J, Sulfiya KM, Jasmin C, Sreelakshmi PK, Biju V (2021) Advances in photodynamic antimicrobial chemotherapy. J Photochem Photobiol C: Photochem Rev 49:100452

Awad MM, Tovmasyan A, Craik JD, Batinic-Haberle I, Benov LT (2016) Important cellular targets for antimicrobial photodynamic therapy. Appl Microbiol Biotechnol 100:7679–7688

Buchovec I, Vyčaitė E, Badokas K, Sužiedelienė E, Bagdonas S (2023) Application of antimicrobial photodynamic therapy for inactivation of *Acinetobacter baumannii* biofilms. Int J Mol Sci 24:722

Buzzá HH, Alves F, Tomé AJB, Chen J, Kassab G, Bu J et al (2022) Porphyrin nanoemulsion for antimicrobial photodynamic therapy: effective delivery to inactivate biofilm-related infections. Proc Natl Acad Sci 119:e2216239119

Chibebe Junior J, Fuchs BB, Sabino CP, Junqueira JC, Jorge AOC, Ribeiro MS et al (2013) Photodynamic and antibiotic therapy impair the pathogenesis of *Enterococcus faecium* in a whole animal insect model. PLoS One 8:e55926

Cieplik F, Deng D, Crielaard W, Buchalla W, Hellwig E, Al-Ahmad A et al (2018) Antimicrobial photodynamic therapy—what we know and what we don't. Crit Rev Microbiol 44:571–589

Costa Magacho C, Guerra Pinto J, Müller Nunes Souza B, Correia Pereira AH, Ferreira-Strixino J (2020) Comparison of photodynamic therapy with methylene blue associated with ceftriaxone in gram-negative bacteria; an in vitro study. Photodiagn Photodyn Ther 30:101691

da Silveira CH, Vieceli V, Clerici DJ, Santos RCV, Iglesias BA (2020) Investigation of isomeric tetra-cationic porphyrin activity with peripheral [Pd(bpy)Cl]+ units by antimicrobial photodynamic therapy. Photodiagn Photodyn Ther 31:101920

De Aguiar Coletti TMSF, De Freitas LM, Almeida AMF, Fontana CR (2017) Optimization of antimicrobial photodynamic therapy in biofilms by inhibiting efflux pump. Photomed Laser Surg 35:378–385

De Freitas LM, Lorenzón EN, Santos-Filho NA, Zago LHDP, Uliana MP, De Oliveira KT et al (2018) Antimicrobial photodynamic therapy enhanced by the peptide aurein. Sci Rep 8:4212

Feng Y, Palanisami A, Ashraf S, Bhayana B, Hasan T (2020) Photodynamic inactivation of bacterial carbapenemases restores bacterial carbapenem susceptibility and enhances carbapenem antibiotic effectiveness. Photodiagn Photodyn Ther 30:101693

Ghorbani J, Rahban D, Aghamiri S, Teymouri A, Bahador A (2018) Photosensitizers in antibacterial photodynamic therapy: an overview. Laser Ther 27:293–302

Grego KF, de Carvalho MPN, Cunha MPV, Knöbl T, Pogliani FC, Catão-Dias JL et al (2017) Antimicrobial photodynamic therapy for infectious stomatitis in snakes: clinical views and microbiological findings. Photodiagn Photodyn Ther 20:196–200

Hajim KI, Salih DS, Rassam YZ (2010) Laser light combined with a photosensitizer may eliminate methicillin-resistant strains of *Staphylococcus aureus*. Lasers Med Sci 25:743–748

Hao J, Lu ZS, Li CM, Xu LQ (2019) A maltoheptaose-decorated BODIPY photosensitizer for photodynamic inactivation of Gram-positive bacteria. New J Chem 43:15057–15065

Hashimoto MCE, Prates RA, Kato IT, Núñez SC, Courrol LC, Ribeiro MS (2012) Antimicrobial photodynamic therapy on drug-resistant *Pseudomonas aeruginosa*-induced infection. An *in vivo* study. Photochem Photobiol 88:590–595

Hoenes K, Bauer R, Meurle T, Spellerberg B, Hessling M (2021) Inactivation effect of violet and blue light on ESKAPE pathogens and closely related non-pathogenic bacterial species—a promising tool against antibiotic-sensitive and antibiotic-resistant microorganisms. Front Microbiol 11:612367

Huang L, Dai T, Hamblin MR (2010) Antimicrobial photodynamic inactivation and photodynamic therapy for infections. Methods Mol Biol 635:155–173

Huang L, Wang M, Dai T, Sperandio FF, Huang YY, Xuan Y et al (2014) Antimicrobial photodynamic therapy with decacationic monoadducts and bisadducts of [70]fullerene: in vitro and in vivo studies. Nanomedicine 9:253–266

Kustov AV, Berezin DB, Zorin VP, Morshnev PK, Kukushkina NV, Krestyaninov MA et al (2023) Monocationic chlorin as a promising photosensitizer for antitumor and antimicrobial photodynamic therapy. Pharmaceutics 15:61

Lewis K (2007) Persister cells, dormancy and infectious disease. Nat Rev Microbiol 5:48–56

Li T, Zhao Y, Matthews K, Gao J, Hao J, Wang S et al (2020) Antibacterial activity against *Staphylococcus aureus* of curcumin-loaded chitosan spray coupled with photodynamic treatment. LWT 134:110073

Liu C, Zhou Y, Wang L, Han L, Lei J, Ishaq HM et al (2016) Photodynamic inactivation of *Klebsiella pneumoniae* biofilms and planktonic cells by 5-aminolevulinic acid and 5-aminolevulinic acid methyl ester. Lasers Med Sci 31:557–565

Ma Y, Wang C, Li Y, Li J, Wan Q, Chen J et al (2020) Considerations and caveats in combating ESKAPE pathogens against nosocomial infections. Adv Sci 7:1901872

Malá Z, Žárská L, Bajgar R, Bogdanová K, Kolář M, Panáček A et al (2021) The application of antimicrobial photodynamic inactivation on methicillin-resistant *S. aureus* and ESBL-producing *K. pneumoniae* using porphyrin photosensitizer in combination with silver nanoparticles. Photodiagn Photodyn Ther 33:102140

Malacrida AM, Dias VHC, Silva AF, dos Santos AR, Cesar GB, Bona E et al (2020) Hypericin-mediated photoinactivation of polymeric nanoparticles against *Staphylococcus aureus*. Photodiagn Photodyn Ther 30:101737

Maliszewska I, Goldeman W (2023) Increasing photoeradication's efficiency of *Acinetobacter baumannii* by polyphosphonic chelating agents. Photodiagn Photodyn Ther 43:103672

Misba L, Zaidi S, Khan AU (2017) A comparison of antibacterial and antibiofilm efficacy of phenothiazinium dyes between Gram positive and Gram negative bacterial biofilm. Photodiagn Photodyn Ther 18:24–33

Moore JV, Wylie MP, Andrews GP, McCoy CP (2023) Photosensitiser-incorporated microparticles for photodynamic inactivation of bacteria. J Photochem Photobiol B Biol 241:112671

Navidinia M (2016) The clinical importance of emerging ESKAPE pathogens in nosocomial infections. Arch Adv Biosci 7:43–57

Orenstein A, Klein D, Kopolovic J, Winkler E, Malik Z, Keller N et al (1997) The use of porphyrins for eradication of *Staphylococcus aureus* in burn wound infections. FEMS Immunol Med Microbiol 19:307–314

Orlandi VT, Rybtke M, Caruso E, Banfi S, Tolker-Nielsen T, Barbieri P (2014) Antimicrobial and anti-biofilm effect of a novel BODIPY photosensitizer against *Pseudomonas aeruginosa* PAO1. Biofouling 30:883–891

Panda SK, Buroni S, Swain SS, Bonacorsi A, da Fonseca Amorim EA, Kulshrestha M et al (2022) Recent advances to combat ESKAPE pathogens with special reference to essential oils. Front Microbiol 13:1029098

Polat E, Kang K (2021) Natural photosensitizers in antimicrobial photodynamic therapy. Biomedicine 9:584

Ragàs X, Dai T, Tegos GP, Agut M, Nonell S, Hamblin MR (2010) Photodynamic inactivation of *Acinetobacter baumannii* using phenothiazinium dyes: in vitro and in vivo studies. Lasers Surg Med 42:384–390

Rajesh S, Koshi E, Philip K, Mohan A (2011) Antimicrobial photodynamic therapy: an overview. J Indian Soc Periodontol 15:323–327

Rapacka-Zdonczyk A, Wozniak A, Kruszewska B, Waleron K, Grinholc M (2021) Can gram-negative bacteria develop resistance to antimicrobial blue light treatment? Int J Mol Sci 22:11579

Rice LB (2008) Federal funding for the study of antimicrobial resistance in nosocomial pathogens: no ESKAPE. J Infect Dis 197:1079–1081

Sabino CP, Wainwright M, Ribeiro MS, Sellera FP, dos Anjos C, da Silva Baptista M et al (2020) Global priority multidrug-resistant pathogens do not resist photodynamic therapy. J Photochem Photobiol B Biol 208:111893

Santajit S, Indrawattana N (2016) Mechanisms of antimicrobial resistance in ESKAPE pathogens. Biomed Res Int 2016:2475067

Shatila F, Tieman GMO, Musolino SF, Wulff JE, Buckley HL (2023) Antimicrobial photodynamic inactivation of planktonic and biofilm cells by covalently immobilized porphyrin on polyethylene terephthalate surface. Int Biodeterior Biodegradation 178:105567

Songsantiphap C, Vanichanan J, Chatsuwan T, Asawanonda P, Boontaveeyuwat E (2022) Methylene blue–mediated antimicrobial photodynamic therapy against clinical isolates of extensively drug resistant Gram-negative bacteria causing nosocomial infections in Thailand, an in vitro study. Front Cell Infect Microbiol 12:929242

Sperandio F, Huang Y-Y, Hamblin M (2013) Antimicrobial photodynamic therapy to kill Gram-negative bacteria. Recent Pat Antiinfect Drug Discov 8:108–120

Tacconelli E, Carrara E, Savoldi A, Harbarth S, Mendelson M, Monnet DL et al (2018) Discovery, research, and development of new antibiotics: the WHO priority list of antibiotic-resistant bacteria and tuberculosis. Lancet Infect Dis 18:318–327

Tan Y, Cheng Q, Yang H, Li H, Gong N, Liu D et al (2018) Effects of ALA-PDT on biofilm structure, virulence factor secretion, and QS in *Pseudomonas aeruginosa*. Photodiagn Photodyn Ther 24:88–94

Tomb RM, Maclean M, Coia JE, MacGregor SJ, Anderson JG (2017) Assessment of the potential for resistance to antimicrobial violet-blue light in *Staphylococcus aureus*. Antimicrob Resist Infect Control 6:100

Tosato MG, Schilardi P, Lorenzo de Mele MF, Thomas AH, Lorente C, Miñán A (2020) Synergistic effect of carboxypterin and methylene blue applied to antimicrobial photodynamic therapy against mature biofilm of *Klebsiella pneumoniae*. Heliyon 6:e03522

Woźniak A, Kruszewska B, Pierański MK, Rychłowski M, Grinholc M (2021) Antimicrobial photodynamic inactivation affects the antibiotic susceptibility of *Enterococcus* spp. clinical isolates in biofilm and planktonic cultures. Biomol Ther 11(5):693

Woźniak A, Burzyńska N, Zybała I, Empel J, Grinholc M (2022) Priming effect with photoinactivation against extensively drug-resistant *Enterobacter cloacae* and *Klebsiella pneumoniae*. J Photochem Photobiol B Biol 235:112554

Yang MY, Chang KC, Chen LY, Wang PC, Chou CC, Wu ZB et al (2018) Blue light irradiation triggers the antimicrobial potential of ZnO nanoparticles on drug-resistant *Acinetobacter baumannii*. J Photochem Photobiol B Biol 180:235–242

Youf R, Müller M, Balasini A, Thétiot F, Müller M, Hascoët A et al (2021) Antimicrobial photodynamic therapy: latest developments with a focus on combinatory strategies. Pharmaceutics 13:1–56

Zhang R, Li Y, Zhou M, Wang C, Feng P, Miao W et al (2019) Photodynamic chitosan nano-assembly as a potent alternative candidate for combating antibiotic-resistant bacteria. ACS Appl Mater Interfaces 11:26711–26721

Antimicrobial Peptides and Antibacterial Antibodies for the Elimination of ESKAPE Pathogens

16

Hemavathy Nagarajan, Sampathkumar Ranganathan, Jeyakanthan Jeyaraman, and Srujana Chitipothu

Abstract

ESKAPE pathogens, responsible for hospital-acquired infections, pose a global threat due to their multi-drug resistance attained by excessive and inappropriate usage of antimicrobial products. Therefore, the unearthing of a novel therapy for the drug resistance infections acquired by ESKAPE pathogens has been painstaking. To date, both the anti-microbial peptides (AMPs) and antibacterial antibodies have been promising therapeutics against the ESKAPE pathogens more than the conventional broad-spectrum antibiotics that develop resistance. AMPs are the first line of innate immune mechanisms towards various microbes. AMPs act as potential therapeutics due to their unique mechanism of action, which involves disrupting bacterial membranes and lower resistance development compared to conventional antibiotics. On the other side, antibacterial antibodies are

H. Nagarajan
Centre for Bioinformatics, Kamalnayan Bajaj Institute for Research in Vision and Ophthalmology, Vision Research Foundation, Chennai, Tamil Nadu, India

Structural Biology and Bio-Computing Lab, Department of Bioinformatics, Science Block, Alagappa University, Karaikudi, Tamil Nadu, India

S. Ranganathan
Department of Chemical Engineering, Konkuk University, Seoul, Republic of Korea

J. Jeyaraman
Structural Biology and Bio-Computing Lab, Department of Bioinformatics, Science Block, Alagappa University, Karaikudi, Tamil Nadu, India

S. Chitipothu (✉)
Centre for Bioinformatics, Kamalnayan Bajaj Institute for Research in Vision and Ophthalmology, Vision Research Foundation, Chennai, Tamil Nadu, India

Central Research Instrumentation Facility, Kamalnayan Bajaj Institute for Research in Vision and Ophthalmology, Vision Research Foundation, Chennai, Tamil Nadu, India
e-mail: drsrujana@snmail.org

S. Busi, R. Prasad (eds.), *ESKAPE Pathogens*,
https://doi.org/10.1007/978-981-99-8799-3_16

proteins that can precisely target and neutralize bacterial pathogens. Antimicrobial antibodies have the potential to recognize and bind with specific bacterial antigens, which trigger opsonization or complement activation, which can facilitate bacterial clearance from the body. Moreover, antimicrobial agents can prevent bacterial attachment to the host cells or inhibit bacterial virulence factors. Several recent studies of AMPs (AMP LL-37) combined with antimicrobial antibodies (vancomycin) showed the synergistic potential against *Staphylococcus* MRSA strains. Similarly, the recent patented study of a monoclonal antibody targeting *Klebsiella pneumoniae* was shown to be effective in reducing the bacterial burden and improving survival in mouse models. Despite both AMPs and antibacterial antibodies being promising approaches for the elimination of these pathogens, further in-depth studies are needed to optimize and improve their pharmacokinetics profile toward increased efficacy and safety for clinical usage. In conclusion, the development of novel AMPs is indispensable to fend off the threat posed by ESKAPE pathogens.

Keywords

ESKAPE pathogens · Multidrug-resistant bacteria · Anti-microbial peptides · Antimicrobial antibodies · Anti-microbial agents

16.1 Introduction

Antimicrobial peptides (AMPs) have attained significant consideration in recent years as highly preferred contenders to address antibiotic resistance (Aslam et al. 2018). AMPs are natural defense molecules found in multicellular organisms that directly kill bacteria or inhibit their functions, leading to cell death. They showcase broad-spectrum activity, low toxicity to host cells, and low induction of resistance and possess several advantages over regular antibiotics. Most AMPs are cationic and amphipathic, with a hydrophilic face that interacts with microbial cell membranes that are negatively charged while the hydrophobic face disrupts the membrane which leads to cell death. Certain AMPs can also act on intracellular targets (Erdem Büyükkiraz and Kesmen 2022). Certain peptide-based antibiotics, such as vancomycin and daptomycin, have been approved by regulatory authorities and are in clinical use. Pharmaceutical companies are also exploring the development of AMPs as commercially available medications (Kamaruzzaman et al. 2019). Evolution of multidrug-resistant (MDR) pathogens, such as ESKAPE, has underscored the need for alternative antimicrobial strategies (Tenover 2006; Watson et al. 2019) as they frequently demonstrate resistance to multiple antibiotics and are associated with a high mortality rate.

The ESKAPE pathogens, including *Enterococcus faecium, Staphylococcus aureus, Klebsiella pneumoniae, Acinetobacter baumannii, Pseudomonas aeruginosa,* and *Enterobacter* species, exhibit multidrug resistance through several mechanisms. These mechanisms highlight their innate adaptive nature and also their capability to acquire resistance towards antibiotics (Dixit et al. 2019). Therefore, understanding these mechanisms is crucial for the unveiling of innovative antimicrobial strategies and novel agents that can effectively fight against drug-resistant ESKAPE pathogens (Gruenheid and Le Moual 2012). The major mechanisms that invoke drug resistance are summarized as follows (Capelo-Martínez and Igrejas 2019; Ma et al. 2020; Olesen et al. 2018; Piddock 2006; Rice 2008; Santajit and Indrawattana 2016):

- **Antibiotic Inactivation/Modification:** Bacteria can produce enzymes that chemically modify or break down antibiotics, rendering them ineffective. For example, β-lactamase hydrolyzes β-lactam antibiotics (penicillins and cephalosporins), leading to their inactivation. Many ESKAPE pathogens produce β-lactamases, contributing to their resistance to β-lactam antibiotics (Capelo-Martínez and Igrejas 2019; Da Costa de Souza et al. 2022).
- **Target site modification:** Bacteria can alter the target sites of antibiotics, such as bacterial enzymes or proteins, through mutations or acquisition of resistance genes. This modification reduces the binding affinity of antibiotics to their targets, rendering them ineffective. For instance, methicillin-resistant *Staphylococcus aureus* (MRSA) acquired an altered gene PBP2a for penicillin-binding protein leading to resistance against methicillin and other β-lactam antibiotics (Dixit et al. 2019).
- **Biofilm formation:** ESKAPE pathogens possess the ability to form biofilms where groups of bacteria attach to a structure lodged in an autogenic extracellular matrix. Biofilms provide a protective milieu for bacteria, making them highly resistant to antibiotics. Thus, the slow growth rate of bacteria within biofilms further contributes to antibiotic resistance (Davis et al. 2006).
- **Acquisition of resistance genes:** Bacteria can obtain resistance genes via horizontal gene transfer, permitting them to acquire resistance to more than one antibiotic. These genes can exist in the form of plasmids, transposons, and integrons and they can easily shuffle between different bacteria by genetic transformation. The transfer of resistance genes contributes to the unfold of antibiotic resistance amongst one-of-a-kind bacterial species (Prestinaci et al. 2015).
- **Reduced Permeability:** Some bacteria can develop mechanisms to reduce the entry of antibiotics into their cells. They may alter the structure of their cell wall or outer membrane, making it less permeable to drugs and limiting their effectiveness.
- **Efflux pumps:** Bacteria can have efflux pumps that actively pump out antibiotics from the bacterial cell, preventing them from reaching their target sites and maintaining high intracellular antibiotic concentrations. This mechanism allows bacteria to expel a wide range of antibiotics, making them resistant to multiple drug classes.

These mechanisms collectively contribute to the multidrug resistance observed in ESKAPE pathogens, making them challenging to treat with conventional antibiotics (Tenover 2006). Developing new antibiotics or alternative treatment strategies that can overcome these mechanisms is crucial to combat hospital-acquired infections caused by these pathogens (Ma et al. 2020). These difficult-to-treat pathogens pose a significant challenge in clinical settings, as they have developed resistance mechanisms against many conventional antibiotics. AMPs with their broad-spectrum and unique mechanisms of action offer a promising alternative. In addition, their potentiality to target the cell membrane and intracellular components exhibits their effectiveness against drug-resistant strains (Mahlapuu et al. 2016).

AMPs are widely available across organisms, functioning as primary defense agents against pathogens. AMPs are ancient host defense molecules found in eukaryotes, offering advantages over acquired immunity (Nicolas and Mor 1995). They can be found in higher organisms such as plants, fish, amphibians, reptiles, birds, mammals, and invertebrates and also predominantly in microorganisms. Nisin, the first discovered bacterial AMP (Hazam et al. 2022), competes for nutrients. AMPs are also found in fungi (e.g., Copsin) (Tabbene et al. 2015) and plants (e.g., thionins, defensins, and cyclotides). Invertebrates rely on AMPs like cecropins and drosocin due to their innate immune systems (Mendelson et al. 2016; Zasloff 2019).

Mammals, along with inorganic substances and antibacterial proteins, possess AMPs as a component of their innate immune system, such as cathelicidins and defensins (Hassan et al. 2012). Cathelicidins have a conserved cathelin domain and exhibit broad-spectrum antibacterial activity. Fish also have hepcidins and piscidins as AMPs. Cathelicidins are stored in neutrophils and macrophages in an inactive form and become activated upon leukocyte activation. Amphibians, such as frogs, have skin as a significant source of AMPs, including cathelicidin-PV. Defensins are another major class of AMPs seen in different organisms, including humans, but they have varying lengths, amino acid composition, and structures. They contribute to the innate immune system's antimicrobial defense in vertebrates.

16.2 Peptide-Based Antibiotics

Peptide-based antibiotics have attained increased approval from the FDA and are now available on the market (Mendelson et al. 2016). Examples of approved peptide-based antibiotics include vancomycin (isolated from the fungus *Streptomyces orientalis*), to treat Gram-positive bacterial infections specifically MRSA. Another approved peptide-based antibiotic is daptomycin against Gram-positive bacterial infections. Apart from these, telavancin, dalbavancin, and oritavancin have also been approved to combat *S. aureus* infections (Hancock 1997). These peptide-based antibiotics provide further treatment alternatives against infections caused by this bacterium. Pharmaceutical companies are also actively exploring the use of antimicrobial peptides (AMPs) as commercially available medications. Companies such as Novabiotics and Lytix Biopharma are developing AMPs in preclinical and clinical stages, respectively. These efforts highlight the potential of AMPs as future therapeutics for various infections, as well as MRSA (Vlieghe et al. 2010). However, there are certain limitations despite their advantages. For instance, hemolytic activity, protease instability, sensitivity to salt, and high synthesis costs. Overcoming these limitations is crucial to maximizing the therapeutic potential of AMPs. Henceforth, to improve/enhance AMPs, several approaches are adapted such as optimization and truncation of sequences, and hybrid analog modification. These approaches aim to enhance the efficacy, stability, and production feasibility of AMPs, expediting their clinical development and application. The continuous AMPs development and optimization unveil favorable alternatives to address the challenges of antibiotic resistance and expand effective treatments against various microbial infections.

16.2.1 Overview of AMP Properties

Characterizing AMPs is crucial for enhancing their utilization as therapeutic agents. They exhibit bacteriostatic and bactericidal effects with lower resistance development than conventional antibiotics. The composition of amino acids in AMPs determines their properties which in turn adopts their selective action against microbes. AMPs rely on electrostatic interactions with anionic phospholipids in bacterial membranes and cationic AMPs to facilitate their antimicrobial activity. In contrast, the outer membrane primarily consists of neutral phospholipids, sphingomyelins, and cholesterol, leading to less cytotoxicity toward eukaryotic cells. Also, cholesterol in the membrane decreases its binding affinity to mammalian cell membranes.

AMPs possess an amphipathic structure, with distinct hydrophobic and hydrophilic regions. The hydrophilic cationic residues, facilitates electrostatic interactions with microbial cell membranes. The hydrophobic amino acids (Tryptophan, Alanine, Glycine, and Leucine) contribute to AMPs lipophilicity and enable microbial membrane disruption. This disruption can lead to pore formation and cell death. Most AMPs are cationic and contain arginine (R) and lysine (K) residues, which contribute to their antimicrobial properties (Hancock 2001; Ladokhin et al. 1997; Schibli et al. 2002). Tryptophan, being hydrophobic, participates in cation-π interactions through its aromatic indole side chain. The cationic arginine residues can form noncovalent bonds with the aromatic π face of tryptophan, complementing each other effectively in AMPs. Thus, AMPs exert their antimicrobial activity by interacting with membranes through electrostatic and hydrophobic interactions.

16.2.2 Mechanisms of Action

Based on their secondary structure, AMPs are classified as linear, α-helical, β-sheet, linear extended, and peptides containing both α-helix and β-sheet structures (Powers and Hancock 2003) (Fig. 16.1). They primarily kill cells by permeabilizing microbial cytoplasmic membranes. However, α-helical peptides permeate and destroy the microbial membranes either by the carpet or barrel-stave pore model, while β-sheet peptides prevent the formation of cell walls or by binding onto the lipid components. Elongated AMPs are rich in amino acids that diffuse through the pathogen membranes and interact with cytoplasmic proteins.

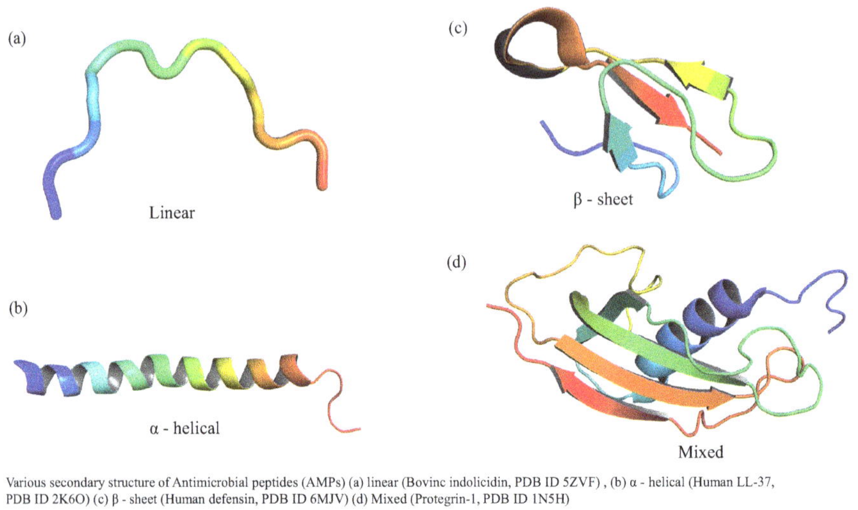

Fig. 16.1 Various secondary structures of antimicrobial peptides (AMPs) (**a**) linear (bovine indolicidin, PDB ID 5ZVF), (**b**) α-helical (human LL-37, PDB ID 2K6O), (**c**) β–β-sheet (human defensin, PDB ID 6MJV), and (**d**) mixed (Protegrin-1, PDB ID 1N5H)

The mechanism of AMP of action can be broadly classified into direct killing (targeting membranes or interfering with cellular processes) and immunological regulation. Examples of AMPs include defensin, LL37, melittin, pleurocidin, pyrrhocidin, and mersacidin. AMPs can disrupt membrane integrity and induce cell death through receptor-mediated or non-receptor-mediated mechanisms. In the case of receptor-mediated, AMPs have receptor-binding or pore-forming domains, allowing them to interact with specific membrane components and exhibit antimicrobial effects. Non-receptor-mediated AMPs interact with membrane components, such as LPS (Gram-negative bacteria) and teichoic acid (Gram-positive bacteria), through electrostatic attraction. These interactions cause membrane disruption, release of intracellular components, and eventual cell death. Non-membrane targeting mechanisms of action in AMPs can be divided into those that target intracellular organelles and those that target the bacterial cell wall (Chertov et al. 1996; Goldman et al. 1997; Zhang et al. 2021). They selectively act against microbes due to their positive charge and have weaker interactions with mammalian membranes (Fig. 16.2).

16.2.2.1 Membrane Targeting Mechanisms

- **Membrane Disruption:** AMPs have a high affinity for microbial membranes due to their positive charge and amphipathic nature. This interaction disrupts the membrane integrity leading to the leakage of intracellular components and subsequent cell death. Selective targeting of microbial membranes and effective elimination of microbial pathogens by the AMPs is the highlight of this mechanism.

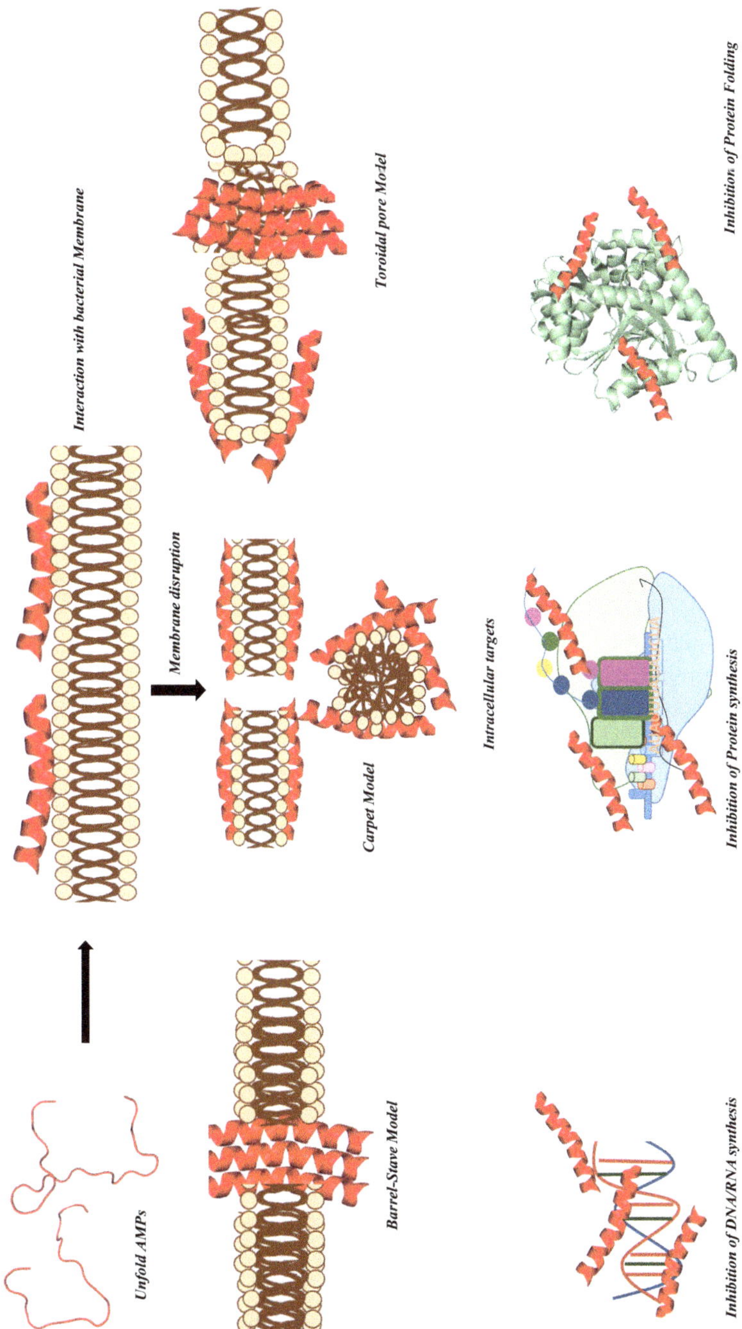

Fig. 16.2 Mechanism of action of AMPs

- **Pore Formation:** In addition to membrane disruption, some AMPs exert their antimicrobial effect by implanting into the microbial membrane. These AMPs form pores and ion channels in microbial membranes. They insert themselves into the lipid bilayer, disrupting the electrochemical gradient across the membrane and leading to cell death (Brogden 2005).

16.2.2.2 Non-membrane Targeting Mechanisms

- **Intracellular Targeting:** Some AMPs can penetrate microbial cell membranes without causing significant membrane disruption. Buforin II and Indolicidin are examples of such AMPs. These peptides enter the microbial cell and specifically target intracellular molecules such as DNA, RNA, and proteins. By interacting with these vital intracellular molecules, AMPs obstruct essential cellular functions leading to the eventual death of microbial cells. This intracellular targeting mechanism allows AMPs to exert its antimicrobial effects by disrupting key cellular processes compromising the survival and integrity of pathogens (Le et al. 2017; Shah et al. 2016).
- **Binding to Intracellular Targets:** Another mechanism that AMPs follow includes binding to the intracellular targets with in the microbial cells. These targets include enzymes and other essential proteins necessary for the survival and proliferation of microbial cells. AMPs bind to these intracellular targets and disrupt key metabolic pathways and enzymatic processes crucial for microbial viability and ultimately lead to cell death. By targeting intracellular components and disrupting vital cellular functions, AMPs effectively eliminate microbial pathogens and contribute to their antimicrobial activity (Le et al. 2017).
- AMPs not only possess direct microbial cell death mechanisms but also exhibit immunomodulatory effects by modulating the immune response. It can also stimulate the production of pro-inflammatory cytokines, chemokines, and antimicrobial factors, thereby enhancing the immune response against invading pathogens (Bals 2000). Hence apart from solely acting as antimicrobial agents, AMPs also play a role in enhancing the body's capability to fight against pathogens. Immunological cells, such as macrophages and neutrophils, are responsible for AMPs production (Mahlapuu et al. 2016). AMPs in mammals activate the immune system through various mechanisms, including T cell activation, Toll-like receptor stimulation, enhancement of phagocytosis, activation of dendritic cells, and chemoattraction of neutrophils. AMPs are produced by various cells in the body, including epithelial cells, lymphocytes, phagocytes, neutrophils, and keratinocytes. They are found in locations such as the lymphatic system, genitourinary tract, gastrointestinal tract, and immune systems. AMP production can occur constitutively or in response to inflammation. Neutrophils and macrophages constitutively generate AMPs, while epithelial cells produce them upon mucosal surface stimulation. Certain AMPs, like α-defensins, are frequently generated, while others, such as β-defensins, are induced. Human AMPs, such as LL-37 (Grönberg et al. 2014) and β-defensins (Goldman et al. 1997), can attract immune cells such as leukocytes, dendritic cells, and mast cells. Furthermore, AMPs may play a role in wound healing and tissue regeneration processes, high-

lighting their potential contribution to these physiological processes. AMPs exhibit a multifaceted function in the immune system, acting not only as antimicrobial agents but also as modulators of immune responses and facilitators of tissue repair.

Cationic antimicrobial peptides (CAMPs) are attractive candidates for novel antibiotic development due to their limited bacterial resistance and exhibit broad-spectrum antimicrobial activity (Hancock 2001). CAMPs are typically amphipathic, having both cationic and hydrophobic surfaces. The main mode of action of CAMPs is via membrane destruction; however, the mechanisms of this interaction are complex and not fully explored. Many models have been proposed to explain membrane disruption by CAMPs, like the barrel-stave model, toroidal pore wormhole model, carpet model, sinking raft model, electroporation, and ATP-dependent cellular uptake model.

In the **barrel-stave model**, CAMPs initially interact with the membrane through electrostatic interactions. They undergo conformational change and acquire an amphipathic structure. The peptides (such as Alamethicin and Gramicidin S) insert themselves between the lipid bilayer, and through lateral interactions form an ion channel or pore. The formation of a "barrel" pore leads to the exudation of internal cell contents leading to cell death.

Peptides in the **toroidal pore wormhole model** attach in a parallel orientation to the membrane and induce a puncture in the hydrophobic part of the membrane. This rupture, along with membrane thinning, weakens the membrane's structure, making it susceptible to the disruptive action of AMPs. When the peptide concentration extents a critical threshold, self-aggregates are formed creating a toroidal pore complex perpendicular to the bilayer membrane surface. AMPs such as magainin 2, lacticin Q, and melittin are examples that act through the toroidal pore model.

In the **carpet model**, cationic AMPs initially interacted with the phospholipids and then oriented parallel to the bilayer surface. As peptides hoard to a critical concentration, they form a "carpet" leading to membrane cracks and micelle formation. Examples of AMPs that act through the carpet model include cecropin and aurein.

Other models propose mechanisms such as the sinking raft model, which involves the formation of unstable holes in the membrane, and electroporation, which alters the charge on both sides of the membrane to create holes. Additionally, ATP-dependent cellular uptake mechanisms, like macropinocytosis, involve the folding of the plasma membrane along with the peptide to generate vesicles called macropinosomes. The AMPs inside vesicles are then released into the cytoplasm, exerting their antibacterial effects (Fig. 16.3).

16.3 Databases of Antimicrobial Peptides

Antimicrobial peptide databases are specialized repositories that store and provide access to various information about AMPs. The databases collect, organize, and present data on AMPs, such as sequence information, its structure, and respective

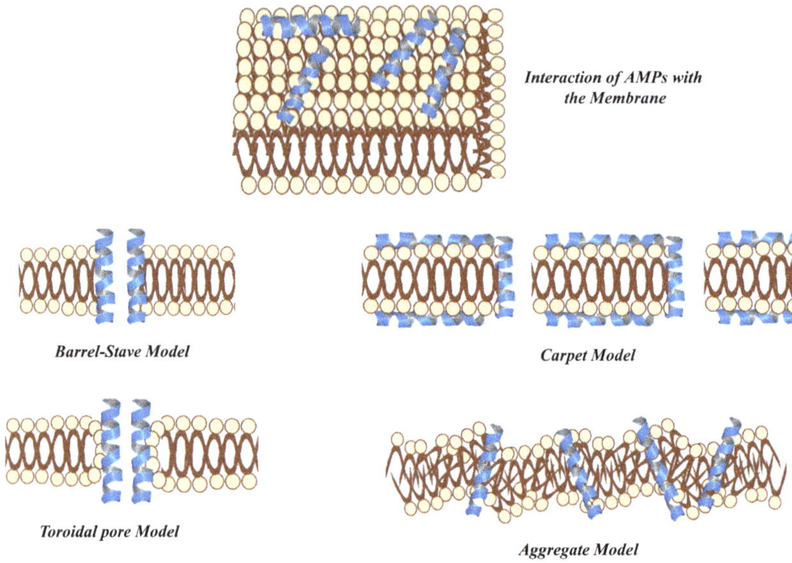

Fig. 16.3 Models of antibacterial mechanisms of AMPs

activities. In addition, the other modifications, and relevant properties. Hence, databases serve as valuable resources for researchers, scientists, and bioinformaticians working in the field of antimicrobial research. Few predominant AMP databases are dbAMP (Jhong et al. 2022), DBAASP (Pirtskhalava et al. 2021), LAMP (Zhao et al. 2013), DRAMP (Kang et al. 2019), CAMP (Waghu et al. 2016), APD (Wang et al. 2016), Hemolytik (Gautam et al. 2014), BACTIBASE (Hammami et al. 2010), DefensinsKB (Seebah et al. 2007), PhytAMP (Hammami et al. 2009), BAGEL4 (Van Heel et al. 2018), CyBase (Wang et al. 2008), and T3DB (Lim et al. 2010). Antimicrobial peptide databases play a role, in speeding up research and development efforts in therapy. These databases bring together a wealth of information on peptides (AMPs) making it easier to discover, design, and optimize AMPs with powerful antimicrobial properties and reduced resistance potential. There are methods and tools available for predicting AMPs and their functional activities. These tools utilize machine learning (ML) techniques. Incorporate features such as amino acid composition, physicochemical properties, sequence, net charge, and secondary structure tendencies. Some notable examples include the Ensemble AMPPred (Lertampaiporn et al. 2021), Deep AmPEP30 (Yan et al. 2020), AntiCP (Agrawal et al. 2021), and AmpGram (Burdukiewicz et al. 2020). By leveraging these features, computational approaches can accurately predict AMP activity offering a cost-high throughput classification of AMPs with accuracy. It is important to highlight that while there are tools for predicting AMPs ongoing research is focused on developing reliable prediction models specifically targeting peptides, with particular activities. However, it is highly essential to constantly refine existing techniques to improve the precision and dependability of AMP prediction. These tools assist in identifying AMP candidates optimizing their characteristics and contributing to the progress of AMP research and therapeutic uses. They provide support, for researchers, in this field speeding up the discovery of AMPs and addressing the increasing problem of antimicrobial resistance.

16.4 Host Defense Peptides

Host defense peptides (HDPs) are conserved peptides of the human innate immune system against microbial infections (Maiti et al. 2014). Also, HDPs feature both immunomodulatory and antibiofilm activities. HDPs are composed of an increased range of positively charged hydrophobic amino acids since its immense hydrophobic nature reduces antimicrobial activity and increases hemolytic potential. This unique feature of HPDs aids in both interaction and translocation across the bacterial and host membranes, thereby facilitating its functionality (Yeung et al. 2011).

Amongst the HDPs, cathelicidins are defined by the conserved pro-region (Dürr et al. 2006) which exhibits direct activity and act as "alarmins" to coordinate immune responses. It also plays a crucial role in modulating inflammation to prevent host damage which is implicated in wound healing, angiogenesis, and the elimination of abnormal cells. For instance, LL-37, a short helical peptide (37 amino acids) derived from human cathelicidin hCAP18 exhibits both cytotoxic and immunomodulatory effects. Hence, its therapeutic potential is being explored, particularly for treating systemic infections and prolonged wounds. However, challenges such as protease degradation limit the usage of therapeutic peptides. Researchers have developed LL-37 variants and analogs to improve their stability and efficacy (Rodríguez-Martínez et al. 2008).

16.5 α-Helical Peptides

To date, α-helical peptides of membrane-active AMPs have been studied extensively (Malmsten 2016; Nguyen et al. 2011). A few α-helical peptides have exhibited tremendous breakthroughs against antibiotic resistance. The first was a hybrid peptide PA-13, with potent efficacy against a broad range of microbes (Klubthawee et al. 2020). PA-13 adopts an α-helical nature and interacts with the membrane, in specific the tryptophan's bulky side chain aids in rapid penetration across the membrane, eventually leading to its disruption, pore formation, and leakage of intracellular contents, resulting in bacterial cell death. PA-13 also exhibits anti-inflammatory activity by binding and neutralizing LPS an outer membrane component of Gram-negative bacteria. However, further studies, including effectiveness, and safety evaluations in vivo, are warranted to explore its clinical potential.

16.6 Antibiofilm Peptides

In recent years, the development of synthetic antibiofilm peptides based on the structure of HDPs (Strempel et al. 2015) is on the rise due to its promising antibiofilm activity. Studies have shown significant differences between naturally occurring peptides, synthetic peptides, and conventional antibiotics in terms of their antibiofilm activity. Some synthetic peptides derived from LL-37 and other cathelicidins have demonstrated antibiofilm activity against bacteria (Gram-positive and Gram-negative) and fungi (Overhage et al. 2008). Rational designing methods have allowed advancements in synthetic peptides with improved biological functions, including antibiofilm activity. These peptides have the potential to overcome antibiotic resistance, as biofilms are highly resistant to conventional antimicrobials.

16.7 Other Strategies

Owing to the increasing resistance of pathogens to various antibiotics and AMPs, there is a need for alternative treatments. Mechanisms of resistance to AMPs vary among bacterial species, and understanding these mechanisms is crucial for developing effective AMPs. However, there is limited information available on the resistance mechanisms in certain pathogenic species, such as *Enterococcus faecalis*. For instance, understanding the resistance and cross-resistance mechanisms of *E. faecalis* against pore-forming AMPs, specifically alamethicin, and pediocin. Alamethicin is a natural peptide produced by the fungus *Trichoderma viride*, while pediocins are AMPs synthesized by lactic acid bacteria. Both AMPs disrupt the cell membrane by forming pores. Mehla and Sood (2011) study highlights the protective power of cell aggregation and higher cell densities in withstanding exposure to AMPs. It also suggests that understanding resistance mechanisms, such as cell aggregation and changes in cell surface properties, aids in designing AMPs and developing effective treatment strategies against resistant bacterial strains (Mehla and Sood 2011).

Thereby, to optimize the properties of peptides, various physicochemical modifications have been employed to the existing AMPs, namely, amino acid deletion/substitution, cyclization, retro-inverso design, D-enantiomer, truncations, and hybrid assembly, have been employed to optimize the peptides' properties.

Polymyxin B, another potentiator, has been highly used for synergist activity along with many antibiotics against drug-resistant bacteria (Trimble et al. 2016). The mechanism of synergism involves increased bacterial membrane permeabilization caused by polymyxin B (Berditsch et al. 2015).

AMPs, such as DP7 and CLP-19, have also shown synergistic effects with antibiotics against antibiotic-resistant strains by reducing cell wall proteins and disrupting cell walls, respectively. They can reduce antibiotic-induced release of LPS from Gram-negative bacteria. Vancomycin, which is typically ineffective against them due to the LPS outer leaflet, can be potentiated by β-naphthyl alanine end-tagged AMPs to fight against similar infections. Thus, future research aims to design peptides that selectively target specific pathogens while sparing beneficial microbiota.

Additionally, antibiofilm peptides have the potential to enhance the action of available antibiotics, reducing the required antibiotic dosage and minimizing the development of resistance. In the next few sections, example studies of AMPs targeting specific bacteria and their therapeutic efficacy are discussed.

16.8 *Enterococcus faecium*-Specific AMPs

The short peptides, namely, 5L and 6L extracted from bovine lactoferrin (LfcinB6) (Mishra et al. 2022) have exerted their antimicrobial activity by depolarizing the *E. faecium* cell membrane that leads to membrane disruption. They also affected bacterial metabolic pathways, including the tricarboxylic acid cycle, glutathione, and purine metabolism. Further, peptides exhibited antibiofilm activity by downregulating genes responsible for biofilm adherence. They showed increased hydrophobicity compared to LfcinB6, resulting in enhanced antimicrobial potency. Further, the downregulation of Peptide 5L, in particular, acts as a narrow-spectrum antimicrobial agent against *E. faecium*. Further investigations are needed to advance these peptides for potential anti-infective drug development.

Another AMP, buwchitin, discovered from the rumen microbiome is a cationic peptide with 71 amino acids and demonstrated antimicrobial activity against multidrug-resistant *Enterococcus faecalis* (Oyama et al. 2017). It exhibited a bacteriostatic effect without significant changes in cell density. Buwchitin's minimal inhibitory concentration against *E. faecalis* ranged from 100 to 200 μg/mL. The peptide interacted with the cell envelope rather than causing direct membrane disruption. The discovery of buwchitin highlights the potential of exploring microbial communities for new antimicrobial compounds to combat antibiotic-resistant infections.

The combination of LL-37 and endolysins exhibited synergistic antibacterial and antibiofilm activities against *E. faecalis* including against vancomycin-resistant strains (Zhang et al. 2023). The Ply2660 + LL-37 combination likely enhances LL-37's activity by degrading the cell wall and increasing membrane accessibility. It also shows higher antibiofilm activity than individual treatments, targeting both planktonic and dormant cells within the biofilm. This combination offers the potential to address antibiotic resistance and biofilm-related challenges posed by *E. faecalis*.

16.9 *Acinetobacter baumannii*-Specific AMPs

A. baumannii infections have been known to enhance mortality risk, although the direct attribution of mortality to these infections is still debated (Avershina et al. 2021). MDR strains of *A. baumannii*, resistant to carbapenem antibiotics, are prevalent worldwide, with different mechanisms of resistance, including the production of carbapenemases (Bonomo et al. 2018). Carbapenem antibiotics are the primary treatment for Gram-negative bacterial infections, but the emergence of CRAB (carbapenem-resistant *A. baumannii*) poses a global health threat (Poirel et al. 2017). Various AMPs extracted from frogs and insects including cathepsins, defensins, etc., have shown effectiveness against *A. baumannii*. Combination therapies involving AMPs and antibiotics have also shown promise in improving treatment efficacy against microbial infections. Alternative therapies such as combination therapies with both β-lactam and β-lactamase inhibitors have been explored, yet the efficacy and toxicity of existing treatments remain problematic (Da Costa de Souza et al. 2022).

Recent research studies investigating the synergistic effect of AMP Esc (1–21) (frog-skin AMP esculentin-1a) with colistin inhibited the growth and killed colistin-resistant *A. baumannii* strains (Casciaro et al. 2020). Esc(1–21) enhanced the membrane permeability of colistin, leading to microbial death at lower concentrations than individual compounds (Elham and Fawzia 2019; Sacco et al. 2022; Vila-Farres et al. 2012).

Interestingly, Cec4, a peptide derived from the housefly, demonstrated antimicrobial activity by altering both membrane potential and fluidity and also by activating efflux pathways (Mao et al. 2022). Furthermore, Cec4 treatment upregulated genes associated with macrolide efflux pumps in *A. baumannii*, suggesting interactions with the bacterial secretion pathway proteins that negatively affect bacterial cells (Liu et al. 2020). Additionally, Cec4 exhibited therapeutic potential in treating *A. baumannii* infections. Combination therapy of Cec4 with the antibiotic polymyxin B was also explored, revealing potential synergistic or additive effects. In conclusion, it exhibited minimal toxicity to human cells and showed therapeutic efficacy in treating skin and sepsis infections (Peng et al. 2023). Further research is needed to refine and optimize Cec4 and explore its potential in clinical settings.

Two cationic antimicrobial peptides, Mt6 and D-Mt6, were developed based on MAF-1, a peptide isolated from housefly larvae (Kong et al. 2022; Zhou et al. 2016). These peptides disrupt membrane integrity leading to cell death. The anti-inflammatory effect of these peptides is attributed to their ability to bind LPS and inhibit LPS-activated MAPK pathways in macrophages. The peptides have the potential to complement combination therapy with traditional antibiotics. This dual antimicrobial and anti-inflammatory potential makes D-Mt6 a promising candidate against *A. baumannii* infections. Overall, Kong et al. (2022) research highlights potential AMPs, such as D-Mt6, in combating antibiotic-resistant infections and addressing the challenges posed by *A. baumannii* infections.

16.10 *Klebsiella pneumonia*-Specific AMPs

Klebsiella pneumoniae is a bacterium that infects humans and animals, specifically intestines, genitals, and respiratory tracts. Amongst them, the hypervirulent strains of *K. pneumoniae* have emerged in Southeast Asia and have become a global health concern (Russo and Marr 2019). These strains spread from liver abscesses to other tissues through bacteremia, leading to infections for example, endophthalmitis, meningitis, etc. (Siu et al. 2012). One of the key features of hypervirulent strains is the presence of a thick capsule and the ability to form iron-siderophores. The capsule is linked to plasmids carrying the *rmpA* gene, and gene *iucA* (encodes for siderophore aerobactin), plays a vital role in the virulence of these strains. Carbapenem-resistant *K. pneumoniae* is listed by the WHO as one of the three priority pathogens requiring the development of new antimicrobial agents (Bassetti et al. 2016).

Pneumonia is a foremost cause of death and economic burden in the European Union, with Gram-negative bacteria accounting for the majority of healthcare-associated pneumonia cases (Marshall et al. 2018). The upsurge of resistance and the lack of effective drugs pose a great challenge for treatment regimens against pneumonia. Despite their potential, AMPs have been underutilized in pneumonia treatment due to limitations such as toxic side effects, short half-life, and minimal efficacy. However, the enhancement of new AMPs and innovative delivery systems, such as nanoparticle-based formulations, offer promising solutions to overcome these challenges. Nanoparticles can improve the stability, bioavailability, and sustained release of AMPs, thereby enhancing their therapeutic potential (Zazo et al. 2016). In the PneumoNP research consortium, funded by the European Union, AMP nanomedicines were developed and investigated for the treatment of pneumonia. However, various patents have been related to the development of AMPs against polymyxin-resistant *Klebsiella pneumoniae*. It is important to note that most of the reviewed patents are in the experimental or preclinical phase, and further development is required to progress them to clinical trials and market availability.

One of the AMPs studied was AA139, derived from marine lugworm *Arenicola marina* (Van der Weide et al. 2020). AA139 was modified to decrease toxicity and enhance its antimicrobial properties. It functions by binding directly to membrane phospholipids and destroying phospholipid transportation, leading to bacterial cell death. Two nanomedicine formulations, AA139-PNP (polymeric nanoparticles) and AA139-MCL (lipid-core micelles), were created to encapsulate AA139. These formulations exhibited comparable antimicrobial activity to free AA139, suggesting that the nanocarriers did not compromise its efficacy. Importantly, both formulations showed improved lung residence time compared to free AA139, indicating their potential for sustained drug release and reduced toxicity.

Cationic proline-rich antimicrobial peptides (PrAMPs) are effective against various Gram-negative pathogens (Li et al. 2014; Roy et al. 2015). PrAMPs like apidaecins and oncocins penetrate the bacterial outer membrane and promote peptide unfolding which binds to the bacteria's ribosome. This in turn inhibits translation, suggesting its potential use in treating invasive *K. pneumoniae* infections. For instance, Krieger et al. (2021) study demonstrated the bactericidal effects of PrAMPs, specifically Api137, against *K. pneumoniae* in porcine blood. The results also suggest the pig model provides a valuable tool for further studies on AMPs against systemic infections, and it also reflects host-pathogen interaction observed in humans.

16.11 *Pseudomonas aeruginosa*-Specific AMPs

The indiscriminate usage of antibiotics led to the increased multidrug-resistant strains of *P. aeruginosa. P. aeruginosa* possesses multifaceted regulatory networks and forms biofilms, contributing to its ability to thrive in different environments (Balasubramanian et al. 2012). Quorum sensing (QS) is a communication system of *P. aeruginosa* to regulate population density and the production of virulence factors in response to varied bacterial density. The studies of Brackman and Coenye (2015) and Shang et al. (2021) found that the peptides affected bacterial swarming motility and decreased the activity of the antibiotic efflux pump MexAB/OprM. The Trp-containing peptides also demonstrated synergistic effects when combined with certain antibiotics, such as ceftazidime and piperacillin, inhibiting the production of beta-lactamase and reducing antibiotic efflux from the bacterial cells (Shang et al. 2021).

Chronic lung infections caused by *Pseudomonas aeruginosa* are a major cause of death in cystic fibrosis (CF) patients (Saint-Criq and Gray 2017). *P. aeruginosa* obtains protection from the immune system and antimicrobial agents by various resistance mechanisms and by forming biofilms (Ciofu and Tolker-Nielsen 2019). Recently, a de novo peptide family called D, L-K6L9 was designed with enhanced antimicrobial activity, stability, and reduced toxicity (Ben Hur et al. 2022). These peptides effectively combat *P. aeruginosa*, including antibiotic-resistant strains, and inhibit biofilm formation. They maintain their antibiofilm activity and do not induce constitutive resistance. These peptides offer advantages over natural AMPs for treating CF-associated lung infections. Combination therapy with AMPs and antibiotics or rotational medication may help prevent resistance and enhance therapeutic potential. Thus, synthetic peptides like D, L-K6L9 are suggested to have efficient antimicrobial activity, resistance to sputum proteases, and potential for treating CF lung infections.

The potential of P-113D, an antimicrobial peptide, as an inhalant therapy for CF patients (Sajjan et al. 2001), has shown potent activity against important CF pathogens. However, P-113D is unstable in sputum, which presents a challenge for its use in CF patients. Nevertheless, P-113D remained stable and active in sputum, effectively targeting endogenous microorganisms present in CF patients' sputum. The peptide's activity was observed despite the presence of divalent cations, which are abundant in sputum. The use of rhDNase to liquefy sputum and disrupt biofilms enhanced the access of P-113D to bacteria. It is important to note that not all sputum samples will respond equally to P-113D due to patient-specific factors and sample composition variability. P-113D, despite its small size, retains potent activity and has advantages in terms of cost-effective synthesis and minimal immune response elicitation. Importantly, no evidence of cross-resistance was known between antibiotics and antimicrobial peptides, making P-113D a promising candidate for inhalant therapy. Additionally, antimicrobial peptides like P-113D have the potential to neutralize bacterial LPS, which contributes to the inflammatory response and lung damage in CF patients. Therefore, P-113D holds promise for chronic suppressive therapy in CF patients, offering activity against CF pathogens, avoidance of antibiotic resistance, and potential anti-inflammatory effects.

The ideal cyclic peptide, ZY4, was known to be highly effective in rapidly killing bacteria (*Pseudomonas aeruginosa* and *Acinetobacter baumannii*) while exhibiting a low likelihood of inducing resistance (Mwangi et al. 2019). ZY4 hinders both the growth and formation of biofilms and also exhibits biofilm eradication properties. Unlike some antibiotics that struggle to eliminate slow-growing and dormant persister cells within the biofilm, ZY4's targeting of the cell membrane is effective against dormant cells. Thus, AMPs composed of tryptophan and arginine, like ZY4, shall be potential contenders in the battle against antibiotic resistance with better antibiofilm and antipersister activities. In animal models, ZY4 demonstrated its potential as a drug candidate by suppressing bacterial growth, cytokine release, and inflammation in target organs and blood, offering antibacterial and anti-inflammatory properties.

16.12 *Staphylococcus aureus*-Specific AMPs

S. aureus is known to cause infectious diseases, and the emergence of multidrug-resistant strains has posed a significant challenge to treating clinicians. AMPs such as MPX an antimicrobial peptide derived from wasp venom have gained attention as potential alternatives to antibiotics. MPX, specifically, has shown promise in protecting mice against *Actinobacillus pleuropneumoniae*, a Gram-negative bacterium, in previous experiments.

The study by Zhu et al. (2022) demonstrated strong bactericidal activity of MPX by inhibiting biofilm formation, disrupting membrane integrity, and also by releasing cellular contents. Moreover, it also exhibited stability in various conditions, including salt ion solutions, and maintains its antimicrobial activity. MPX-based ointment against wound infections was also observed to reduce wound colonization, and inflammation, and promote wound healing.

A novel AMP called S2, which combines a targeting domain derived from the quorum-sensing signal pheromone of *staphylococci,* has been designed to specifically target *S. aureus* (Shang et al. 2020). The peptide's mode of action involves binding to lipoteichoic acid (LTA) on the membrane, inducing membrane depolarization, and causing outflow of intracellular content. The study by Shang et al. (2020) also found that S2 shows selectivity towards bacterial membranes over mammalian cells, indicating its potential safety and therapeutic use. Thus, AMP S2 acts potent as an effective and selective antimicrobial agent against *S. aureus* infections.

Amphibians, such as frogs, are considered natural sources of HDPs/AMPs, and brevinins, including brevinin-1E-OG9, have been found to possess various bioactivities, making them valuable skin infection treatments. The loop structure of OG9 plays a crucial role in its antibacterial activity, likely attributed to cationic residues within the loop. Among the analogs, OG9c-De-NH2 demonstrated that its strong binding to both LTA and LPS induces membrane permeabilization. Although it did not exhibit rapid bacterial lysis, its increased membrane permeability contributed to its efficacy (Fei et al. 2023).

The amassed problem of antimicrobial resistance, particularly the ESKAPE pathogens, has led to untreatable systemic infections and high mortality rates. Novel plasmid-encoded antibiotic resistance genes further exacerbate the crisis by conferring resistance to last-resort antibiotics. Biofilm formation and bacterial metabolic changes contribute to reduced antibiotic susceptibility and the persistence of infections. Synthetic AMPs or modified AMPs have potent activity against multidrug-resistant pathogens. In this regard, Shi et al. (2022) study, revealed a new synthetic antibacterial peptide called LI14 using rational drug design and demonstrated broad-spectrum bactericidal activity including MRSA and *E. coli* strains harboring resistant genes. Mechanistic studies revealed that LI14 targeted membranes cause its disruption, thereby perturbing its metabolism which accumulates toxic products. LI14 specifically targets LPS of the outer membrane and peptidoglycan of the inner membrane, resulting in membrane impairment. The disruption of the proton motive force (PMF) played a crucial role in LI14's lethality, impairing basic bacterial functions and accelerating bacterial death. LI14 also inhibited biofilm formation and eliminated mature biofilms. Overall, LI14 established broad-spectrum activity, anti-biofilm and persister cell eradication abilities, low resistance development, stability, and efficacy in animal models emphasizing its therapeutic potential.

Cyclic AMP, Bac21 is a promising alternative for antibiotic-resistant infections specifically vancomycin-resistant *enterococci* and MRSA. Yet, further research has to be enforced to assess its efficiency, toxicity, and potential clinical applications. Monitoring the emergence of resistance to bacteriocins like Bac21 is also important for long-term effectiveness.

16.13 AMPs Targeting *Enterobacter Species*

The challenges in developing effective drugs for sepsis treatment particularly focus on both antiseptic and anti-inflammatory properties (Fleischmann et al. 2016). Specifically, practical applications complexity, and limitations of current approaches, such as using anti-TNF antibodies. The direction of drug development has shifted towards targeting organ dysfunction and immune response disruption. HDPs have been a potential solution, but their systemic application has been limited by low efficacy and high toxicity. Amongst the available antibiotics, AMPR-11 is suggested to have advantages over chemical antibiotics, including broad-spectrum activity, lower likelihood of resistance development, and minimal environmental pollution (Lee et al. 2020). Currently, combination therapy with AMPR-11 and other antibiotics is also proposed as a promising strategy.

SAAP-148 is a highly effective peptide against multidrug-resistant bacteria, comprising those resistant to last-resort antibiotics like colistin (de Breij et al. 2018). SAAP-148 demonstrates broad-spectrum bactericidal activity and is more effective than most antimicrobial peptides (AMPs) tested in clinical trials, especially under physiological conditions. The peptide also has efficacy against

Clostridium difficile and can prevent and eradicate biofilms and persister cells, which are known to contribute to chronic and recurrent infections. Investigations reveal that SAAP-148 interacts with bacterial membranes, permeabilizing them and leading to bacterial death. The upswing in antibiotic resistance necessitates the development of effective treatments for wound infections, and SAAP-148 incorporated into a hypromellose gel has shown high efficacy in eradicating MRSA and *Acinetobacter baumannii* (de Breij et al. 2018). The ointment's safety profile and efficacy against various wound infections make it a promising treatment option; however, further studies, including deeper wound infections and tailored drug delivery systems, are needed to overcome potential challenges.

BKR1 is a hybrid peptide composed of Esculentin-1a and melittin designed with enhanced physicochemical properties, increased positive net charge, and increased helicity (Al Tall et al. 2023). BKR1 is effective against bacteria while its potency is not affected by the degree of resistance. It exhibits bactericidal behavior without hemolytic effects at effective inhibitory concentrations, indicating improved selectivity compared to its parent peptides. In addition, BKR1 shows synergistic effects with conventional antibiotics, enhancing their activity against resistant strains of bacteria. The best synergistic combinations are observed with antibiotics such as chloramphenicol and rifampicin suggesting that BKR1 and antibiotic combinations could be potential candidates for combating infections (Al Tall et al. 2023). Yet, its exact mechanism of action and its potential applications need to be explored.

To overcome limitations, AMPs can be designed or modified based on the key parameters that influence the activity of AMPs. Variation in any of the parameter severely alters the others hence these modifications have to be curated properly. Different strategies can be implemented for modifications such as C-terminal truncation and D-amino acid substitution. Both these modifications have the capability of enhancing AMP activity, selectivity, and stability. Based on these limitations, many AMPs have been designed, one amongst them is a novel hybrid peptide called AP19, derived from α-helical cathelicidin and aurein. In this peptide, the C-terminal was truncated to increase its antibacterial activity and reduce hemolytic activity (Jariyarattanarach et al. 2022). Furthermore, to improve the stability of existing truncated AP19 (AMP), D-amino acid substitution was employed which resulted in another novel peptide D-AP19. The modified D-AP19 has immense activity against both MDR and XDR specifically the clinical isolates of *A. baumannii*. Amongst these modified AMPs, despite increased anti-bacterial efficacy by both AP19 and D-AP19 with high therapeutic index, only D-AP19 withstands the proteolytic cleavage and, hence, is a suitable candidate for clinical use. These findings contribute to the growth of novel peptide-based drugs against multidrug-resistant pathogens.

Cecropins are a type of cationic AMP (37 residues) found in Diptera and Lepidoptera insects (Brady et al. 2019). Cecropins penetrate bacterial membranes through its amphipathic α-helices leading to membrane disruption. This disruption depletes the electrochemical gradients, resulting in either cell lysis or cell death. One challenge in using peptides as antibiotics is their susceptibility to proteases. Many Cecropins contain amino acid sequences recognized by proteases, making them susceptible to degradation. Designing stable peptides has economic and technological value for their therapeutic and industrial applications.

OMN6, an engineered cyclic AMP derived from Cecropin A, has enhanced stability and reduced proteolytic degradation (Mandel et al. 2021). OMN6 exhibits its antimicrobial activity via membrane disruption, which distinguishes it from small-molecule agents that target specific enzymes or sites. OMN6 displays effectiveness against both sensitive and MDR clinical isolates of *Acinetobacter baumannii*, a major threat in healthcare-associated infections. Importantly, OMN6 has a significantly lower likelihood of mutation-driven resistance development. However, Gram-positive bacteria, with their unique membrane features and components, may be less susceptible to OMN6 due to factors such as the presence of peptidoglycan cell walls. Otherwise, OMN6 is a stable non-toxic antimicrobial agent that shall be considered for future therapeutic applications.

16.14 Future Perspective

The increased prevalence and emergence of new multidrug-resistant ESKAPE pathogens have emphasized the need for alternative antimicrobial strategies. Particularly, their acquired resistance through mechanisms such as antibiotic modification, target site modification, biofilm formation, acquisition of resistance genes, reduced permeability, and efflux pumps limits the current treatment strategies. Hence, antimicrobial peptides (AMP) therapy has been uncovered as a potential candidate for combating antibiotic resistance and developed to address the challenge of multidrug-resistant ESKAPE pathogens. The added advantage of AMPs includes broad-spectrum activity, rapid production, and unique mechanisms of action that target the cell membrane and intracellular components of bacteria. Future perspectives involve optimizing AMP-based therapies by further enhancing their efficacy, stability, and safety. Strategies to overcome resistance mechanisms, improve delivery methods, and minimize potential toxicities will be important modalities that need to be addressed as listed below.

1. **Clinical Development:** Further studies are needed to evaluate the safety, pharmacokinetics, and efficacy of AMPs in clinical settings. Preclinical and clinical trials will help determine the optimal dosage, administration route, and treatment regimens for AMPs. Additionally, investigating potential drug interactions and assessing their efficacy in combination with existing antibiotics could be explored.
2. **Resistance Development:** Understanding the potential mechanisms of resistance and developing novel strategies to mitigate them will be important to preserve the effectiveness of AMPs. Hence, continued surveillance and monitoring of resistance development to AMPs will be crucial.
3. **Combination Therapies:** Exploring combination therapies could provide synergistic effects and enhance treatment outcomes. However, they aid in overcoming resistance mechanisms and also improve treatment efficacy.

4. **Formulation and Delivery:** Both these are essential for the clinical application of AMPs. Strategies such as encapsulation in nanoparticles, liposomes, or other drug delivery vehicles can improve stability, bioavailability, and targeted delivery to infection sites.
5. **Biofilm Eradication:** Further investigations into the ability of AMPs to eradicate biofilms, including complex and antibiotic-resistant biofilms, are warranted. Combination approaches that integrate LJ14 with biofilm-disrupting agents or adjuvants could enhance biofilm eradication and prevent recurrent infections.
6. **Synthetic Peptide Design:** Continued advancements in computer-aided design and chemical modification techniques can further enhance the potency, stability, and specificity of synthetic AMPs. Rational design approaches can be employed to create novel peptide scaffolds with improved antimicrobial properties and reduced toxicity.
7. **Alternative Applications:** Apart from their antimicrobial properties, AMPs may have additional therapeutic applications. Exploring their immunomodulatory effects, wound-healing properties, and potential anticancer activities could expand their clinical utility.
8. **Resistance Reversal:** Investigating whether AMPs can restore susceptibility to conventional antibiotics in drug-resistant bacteria is an intriguing avenue. AMPs have been shown to synergize with antibiotics and reverse resistance mechanisms, potentially revitalizing the effectiveness of existing antibiotics.
9. **Environmental Impact:** Assessing the potential environmental impact of synthetic AMPs is important to understand any ecological consequences associated with their use. Evaluating their biodegradability and potential effects on non-target organisms will contribute to responsible development and application.
10. **Cost and Accessibility:** Considering the cost-effectiveness and accessibility of AMPs and synthetic AMPs will be crucial for their widespread adoption. Development strategies that optimize production processes, minimize manufacturing costs, and ensure affordability for patients and healthcare systems will be essential.

These efforts towards unlocking the most optimal AMPs as therapy for ESKAPE pathogens will contribute to address the global challenge of antimicrobial resistance and providing effective treatments against drug-resistant bacterial infections. Exploring new sources of AMPs and understanding their specific modes of action would also contribute to the development of more effective antimicrobial strategies. Overall, continued exploration of AMPs offers promising solutions for addressing multidrug-resistant pathogens like the ESKAPE group, thereby addressing the challenge of antibiotic resistance and improving outcomes in infectious diseases.

References

Agrawal P, Bhagat D, Mahalwal M, Sharma N, Raghava GPS (2021) Anticp 2.0: an updated model for predicting anticancer peptides. Brief Bioinform 22(3):bbaa153. https://doi.org/10.1093/bib/bbaa153

Al Tall Y, Al-Nassar B, Abualhaijaa A, Sabi SH, Almaaytah A (2023) The design and functional characterization of a novel hybrid antimicrobial peptide from Esculentin-1a and melittin. Pharmacia 70(1):161–170. https://doi.org/10.3897/PHARMACIA.70.E97116

Aslam B, Wang W, Arshad MI, Khurshid M, Muzammil S, Rasool MH et al (2018) Antibiotic resistance: a rundown of a global crisis. Infect Drug Resist 11:1645–1658. https://doi.org/10.2147/IDR.S173867

Avershina E, Shapovalova V, Shipulin G (2021) Fighting antibiotic resistance in hospital-acquired infections: current state and emerging technologies in disease prevention, diagnostics and therapy. Front Microbiol 12:707330. https://doi.org/10.3389/fmicb.2021.707330

Balasubramanian D, Schneper L, Merighi M, Smith R, Narasimhan G, Lory S, Mathee K (2012) The regulatory repertoire of Pseudomonas aeruginosa AmpC ß-lactamase regulator AmpR includes virulence genes. PLoS One 7(3):e34067. https://doi.org/10.1371/journal.pone.0034067

Bals R (2000) Epithelial antimicrobial peptides in host defense against infection. Respir Res 1(3):141–150. https://doi.org/10.1186/rr25

Bassetti M, Welte T, Wunderink RG (2016) Treatment of Gram-negative pneumonia in the critical care setting: is the beta-lactam antibiotic backbone broken beyond repair? Crit Care 20:19. https://doi.org/10.1186/s13054-016-1197-5

Ben Hur D, Kapach G, Wani NA, Kiper E, Ashkenazi M, Smollan G et al (2022) Antimicrobial peptides against multidrug-resistant Pseudomonas aeruginosa biofilm from cystic fibrosis patients. J Med Chem 65(13):9050–9062. https://doi.org/10.1021/acs.jmedchem.2c00270

Berditsch M, Jäger T, Strempel N, Schwartz T, Overhage J, Ulrich AS (2015) Synergistic effect of membrane-active peptides polymyxin B and gramicidin S on multidrug-resistant strains and biofilms of Pseudomonas aeruginosa. Antimicrob Agents Chemother 59(9):5288–5296. https://doi.org/10.1128/AAC.00682-15

Bonomo RA, Burd EM, Conly J, Limbago BM, Poirel L, Segre JA, Westblade LF (2018) Carbapenemase-producing organisms: a global scourge. Clin Infect Dis 66(8):1290–1297. https://doi.org/10.1093/cid/cix893

Brackman G, Coenye T (2015) Quorum sensing inhibitors as anti-biofilm agents. Curr Pharm Des 21(1):5–11. https://doi.org/10.2174/1381612820666140905114627

Brady D, Grapputo A, Romoli O, Sandrelli F (2019) Insect cecropins, antimicrobial peptides with potential therapeutic applications. Int J Mol Sci 20(23):5862. https://doi.org/10.3390/ijms20235862

Brogden KA (2005) Antimicrobial peptides: pore formers or metabolic inhibitors in bacteria? Nat Rev Microbiol 3(3):238–250. https://doi.org/10.1038/nrmicro1098

Burdukiewicz M, Sidorczuk K, Rafacz D, Pietluch F, Chilimoniuk J, Rödiger S, Gagat P (2020) Proteomic screening for prediction and design of antimicrobial peptides with AmpGram. Int J Mol Sci 21(12):4310. https://doi.org/10.3390/ijms21124310

Capelo-Martínez J-L, Igrejas G (2019) Antibiotic drug resistance. Wiley, New York. https://doi.org/10.1002/9781119282549

Casciaro B, Cappiello F, Loffredo MR, Ghirga F, Mangoni ML (2020) The potential of frog skin peptides for anti-infective therapies: the case of esculentin-1a(1-21)NH2. Curr Med Chem 27(9):1405–1419. https://doi.org/10.2174/0929867326666190722095408

Chertov O, Michiel DF, Xu L, Wang JM, Tani K, Murphy WJ et al (1996) Identification of defensin-1, defensin-2, and CAP37/azurocidin as T-cell chemoattractant proteins released from interleukin-8-stimulated neutrophils. J Biol Chem 271(6):2935–2940. https://doi.org/10.1074/jbc.271.6.2935

Ciofu O, Tolker-Nielsen T (2019) Tolerance and resistance of Pseudomonas aeruginosa biofilms to antimicrobial agents—how P. aeruginosa can escape antibiotics. Front Microbiol 10:913. https://doi.org/10.3389/fmicb.2019.00913

Da Costa de Souza G, Roque-Borda CA, Pavan FR (2022) Beta-lactam resistance and the effectiveness of antimicrobial peptides against KPC-producing bacteria. Drug Dev Res 83(7):1534–1554. https://doi.org/10.1002/ddr.21990

Davis SC, Martinez L, Kirsner R (2006) The diabetic foot: the importance of biofilms and wound bed preparation. Curr Diab Rep 6(6):439–445. https://doi.org/10.1007/s11892-006-0076-x

de Breij A, Riool M, Cordfunke RA, Malanovic N, de Boer L, Koning RI et al (2018) The antimicrobial peptide SAAP-148 combats drug-resistant bacteria and biofilms. Sci Transl Med 10(423):eaan4044. https://doi.org/10.1126/scitranslmed.aan4044

Dixit A, Kumar N, Kumar S, Trigun V (2019) Antimicrobial resistance: progress in the decade since emergence of New Delhi metallo-β-lactamase in India. Indian J Community Med 44(1):4–8. https://doi.org/10.4103/ijcm.IJCM_217_18

Dürr UHN, Sudheendra US, Ramamoorthy A (2006) Ll-37, the only human member of the cathelicidin family of antimicrobial peptides. Biochim Biophys Acta 1758(9):1408–1425. https://doi.org/10.1016/j.bbamem.2006.03.030

Elham B, Fawzia A (2019) Colistin resistance in Acinetobacter baumannii isolated from critically ill patients: clinical characteristics, antimicrobial susceptibility and outcome. Afr Health Sci 19(3):2400–2406. https://doi.org/10.4314/ahs.v19i3.13

Erdem Büyükkiraz M, Kesmen Z (2022) Antimicrobial peptides (AMPs): a promising class of antimicrobial compounds. J Appl Microbiol 132(3):1573–1596. https://doi.org/10.1111/jam.15314

Fei F, Wang T, Jiang Y, Chen X, Ma C, Zhou M et al (2023) A frog-derived antimicrobial peptide as a potential anti-biofilm agent in combating Staphylococcus aureus skin infection. J Cell Mol Med 27(11):1565–1579. https://doi.org/10.1111/jcmm.17785

Fleischmann C, Scherag A, Adhikari NKJ, Hartog CS, Tsaganos T, Schlattmann P et al (2016) Assessment of global incidence and mortality of hospital-treated sepsis. Current estimates and limitations. Am J Respir Crit Care Med 193(3):259–272. https://doi.org/10.1164/rccm.201504-0781OC

Gautam A, Chaudhary K, Singh S, Joshi A, Anand P, Tuknait A et al (2014) Hemolytik: a database of experimentally determined hemolytic and non-hemolytic peptides. Nucleic Acids Res 42(Database issue):D444–D449. https://doi.org/10.1093/nar/gkt1008

Goldman MJ, Anderson GM, Stolzenberg ED, Kari UP, Zasloff M, Wilson JM (1997) Human beta-defensin-1 is a salt-sensitive antibiotic in lung that is inactivated in cystic fibrosis. Cell 88(4):553–560. https://doi.org/10.1016/S0092-8674(00)81895-4

Grönberg A, Mahlapuu M, Ståhle M, Whately-Smith C, Rollman O (2014) Treatment with LL-37 is safe and effective in enhancing healing of hard-to-heal venous leg ulcers: a randomized, placebo-controlled clinical trial. Wound Repair Regen 22(5):613–621. https://doi.org/10.1111/wrr.12211

Gruenheid S, Le Moual H (2012) Resistance to antimicrobial peptides in Gram-negative bacteria. FEMS Microbiol Lett 330(2):81–89. https://doi.org/10.1111/j.1574-6968.2012.02528.x

Hammami R, Ben Hamida J, Vergoten G, Fliss I (2009) Phytamp: a database dedicated to antimicrobial plant peptides. Nucleic Acids Res 37(Database issue):D963–D968. https://doi.org/10.1093/nar/gkn655

Hammami R, Zouhir A, Le Lay C, Ben Hamida J, Fliss I (2010) Bactibase second release: a database and tool platform for bacteriocin characterization. BMC Microbiol 10:22. https://doi.org/10.1186/1471-2180-10-22

Hancock RE (1997) Peptide antibiotics. Lancet 349(9049):418–422. https://doi.org/10.1016/S0140-6736(97)80051-7

Hancock RE (2001) Cationic peptides: effectors in innate immunity and novel antimicrobials. Lancet Infect Dis 1(3):156–164. https://doi.org/10.1016/S1473-3099(01)00092-5

Hassan M, Kjos M, Nes IF, Diep DB, Lotfipour F (2012) Natural antimicrobial peptides from bacteria: characteristics and potential applications to fight against antibiotic resistance. J Appl Microbiol 113(4):723–736. https://doi.org/10.1111/j.1365-2672.2012.05338.x

Hazam PK, Cheng C-C, Hsieh C-Y, Lin W-C, Hsu P-H, Chen T-L et al (2022) Development of bactericidal peptides against multidrug-resistant Acinetobacter baumannii with enhanced stability and low toxicity. Int J Mol Sci 23(4):2191. https://doi.org/10.3390/ijms23042191

Jariyarattanarach P, Klubthawee N, Wongchai M, Roytrakul S, Aunpad R (2022) Novel D-form of hybrid peptide (D-AP19) rapidly kills Acinetobacter baumannii while tolerating proteolytic enzymes. Sci Rep 12(1):15852. https://doi.org/10.1038/s41598-022-20236-1

Jhong J-H, Yao L, Pang Y, Li Z, Chung C-R, Wang R et al (2022) Dbamp 2.0: updated resource for antimicrobial peptides with an enhanced scanning method for genomic and proteomic data. Nucleic Acids Res 50(D1):D460–D470. https://doi.org/10.1093/nar/gkab1080

Kamaruzzaman NF, Li Tan P, Hamdan RH, Choong SS, Wong WK, Gibson AJ et al (2019) Antimicrobial polymers: the potential replacement of existing antibiotics? Int J Mol Sci 20(11):2747. https://doi.org/10.3390/ijms20112747

Kang X, Dong F, Shi C, Liu S, Sun J, Chen J et al (2019) Dramp 2.0, an updated data repository of antimicrobial peptides. Sci Data 6(1):148. https://doi.org/10.1038/s41597-019-0154-y

Klubthawee N, Adisakwattana P, Hanpithakpong W, Somsri S, Aunpad R (2020) A novel, rationally designed, hybrid antimicrobial peptide, inspired by cathelicidin and aurein, exhibits membrane-active mechanisms against Pseudomonas aeruginosa. Sci Rep 10(1):9117. https://doi.org/10.1038/s41598-020-65688-5

Kong D, Hua X, Zhou R, Cui J, Wang T, Kong F et al (2022) Antimicrobial and anti-inflammatory activities of MAF-1-derived antimicrobial peptide Mt6 and its D-enantiomer D-Mt6 against Acinetobacter baumannii by targeting cell membranes and lipopolysaccharide interaction. Microbiol Spectr 10(5):e0131222. https://doi.org/10.1128/spectrum.01312-22

Krieger A-K, Knappe D, Öhlmann S, Mayer L, Eder IB, Köller G et al (2021) Proline-rich antimicrobial peptide Api137 is bactericidal in porcine blood infected ex vivo with a porcine or human Klebsiella pneumoniae strain. J Glob Antimicrob Resist 24:127–135. https://doi.org/10.1016/j.jgar.2020.12.012

Ladokhin AS, Selsted ME, White SH (1997) Bilayer interactions of indolicidin, a small antimicrobial peptide rich in tryptophan, proline, and basic amino acids. Biophys J 72(2 Pt 1):794–805. https://doi.org/10.1016/S0006-3495(97)78713-7

Le C-F, Fang C-M, Sekaran SD (2017) Intracellular targeting mechanisms by antimicrobial peptides. Antimicrob Agents Chemother 61(4):e02340-16. https://doi.org/10.1128/AAC.02340-16

Lee H-R, You D-G, Kim HK, Sohn JW, Kim MJ, Park JK et al (2020) Romo1-derived antimicrobial peptide is a new antimicrobial agent against multidrug-resistant bacteria in a murine model of sepsis. mBio 11(2):e03258-19. https://doi.org/10.1128/mBio.03258-19

Lertampaiporn S, Vorapreeda T, Hongsthong A, Thammarongtham C (2021) Ensemble-AMPPred: robust AMP prediction and recognition using the ensemble learning method with a new hybrid feature for differentiating AMPs. Genes 12(2):137. https://doi.org/10.3390/genes12020137

Li W, Tailhades J, O'Brien-Simpson NM, Separovic F, Otvos L, Hossain MA, Wade JD (2014) Proline-rich antimicrobial peptides: potential therapeutics against antibiotic-resistant bacteria. Amino Acids 46(10):2287–2294. https://doi.org/10.1007/s00726-014-1820-1

Lim E, Pon A, Djoumbou Y, Knox C, Shrivastava S, Guo AC et al (2010) T3db: a comprehensively annotated database of common toxins and their targets. Nucleic Acids Res 38(Database issue):D781–D786. https://doi.org/10.1093/nar/gkp934

Liu W, Wu Z, Mao C, Guo G, Zeng Z, Fei Y et al (2020) Antimicrobial peptide Cec4 eradicates the bacteria of clinical carbapenem-resistant Acinetobacter baumannii biofilm. Front Microbiol 11:1532. https://doi.org/10.3389/fmicb.2020.01532

Ma Y-X, Wang C-Y, Li Y-Y, Li J, Wan Q-Q, Chen J-H et al (2020) Considerations and caveats in combating ESKAPE pathogens against nosocomial infections. Adv Sci (Weinh) 7(1):1901872. https://doi.org/10.1002/advs.201901872

Mahlapuu M, Håkansson J, Ringstad L, Björn C (2016) Antimicrobial peptides: an emerging category of therapeutic agents. Front Cell Infect Microbiol 6:194. https://doi.org/10.3389/fcimb.2016.00194

Maiti S, Patro S, Purohit S, Jain S, Senapati S, Dey N (2014) Effective control of Salmonella infections by employing combinations of recombinant antimicrobial human β-defensins hBD-1 and hBD-2. Antimicrob Agents Chemother 58(11):6896–6903. https://doi.org/10.1128/AAC.03628-14

Malmsten M (2016) Interactions of antimicrobial peptides with bacterial membranes and membrane components. Curr Top Med Chem 16(1):16–24. https://doi.org/10.2174/1568026615666150703121518

Mandel S, Michaeli J, Nur N, Erbetti I, Zazoun J, Ferrari L et al (2021) Omn6 a novel bioengineered peptide for the treatment of multidrug resistant Gram negative bacteria. Sci Rep 11(1):6603. https://doi.org/10.1038/s41598-021-86155-9

Mao C, Wang Y, Yang Y, Li L, Yuan K, Cao H et al (2022) Cec4-derived peptide inhibits planktonic and biofilm-associated methicillin resistant Staphylococcus epidermidis. Microbiol Spectr 10(6):e0240922. https://doi.org/10.1128/spectrum.02409-22

Marshall DC, Goodson RJ, Xu Y, Komorowski M, Shalhoub J, Maruthappu M, Salciccioli JD (2018) Trends in mortality from pneumonia in the Europe union: a temporal analysis of the European detailed mortality database between 2001 and 2014. Respir Res 19(1):81. https://doi.org/10.1186/s12931-018-0781-4

Mehla J, Sood SK (2011) Substantiation in Enterococcus faecalis of dose-dependent resistance and cross-resistance to pore-forming antimicrobial peptides by use of a polydiacetylene-based colorimetric assay. Appl Environ Microbiol 77(3):786–793. https://doi.org/10.1128/AEM.01496-10

Mendelson M, Røttingen J-A, Gopinathan U, Hamer DH, Wertheim H, Basnyat B et al (2016) Maximising access to achieve appropriate human antimicrobial use in low-income and middle-income countries. Lancet 387(10014):188–198. https://doi.org/10.1016/S0140-6736(15)00547-4

Mishra B, Felix L, Basu A, Kollala SS, Chhonker YS, Ganesan N et al (2022) Design and evaluation of short bovine lactoferrin-derived antimicrobial peptides against multidrug-resistant Enterococcus faecium. Antibiotics (Basel) 11(8):1085. https://doi.org/10.3390/antibiotics11081085

Mwangi J, Yin Y, Wang G, Yang M, Li Y, Zhang Z, Lai R (2019) The antimicrobial peptide ZY4 combats multidrug-resistant Pseudomonas aeruginosa and Acinetobacter baumannii infection. Proc Natl Acad Sci U S A 116(52):26516–26522. https://doi.org/10.1073/pnas.1909585117

Nguyen LT, Haney EF, Vogel HJ (2011) The expanding scope of antimicrobial peptide structures and their modes of action. Trends Biotechnol 29(9):464–472. https://doi.org/10.1016/j.tibtech.2011.05.001

Nicolas P, Mor A (1995) Peptides as weapons against microorganisms in the chemical defense system of vertebrates. Annu Rev Microbiol 49:277–304. https://doi.org/10.1146/annurev.mi.49.100195.001425

Olesen SW, Barnett ML, MacFadden DR, Brownstein JS, Hernández-Díaz S, Lipsitch M, Grad YH (2018) The distribution of antibiotic use and its association with antibiotic resistance. elife 7:e39435. https://doi.org/10.7554/eLife.39435

Overhage J, Campisano A, Bains M, Torfs ECW, Rehm BHA, Hancock REW (2008) Human host defense peptide LL-37 prevents bacterial biofilm formation. Infect Immun 76(9):4176–4182. https://doi.org/10.1128/IAI.00318-08

Oyama LB, Crochet J-A, Edwards JE, Girdwood SE, Cookson AR, Fernandez-Fuentes N et al (2017) Buwchitin: a ruminal peptide with antimicrobial potential against Enterococcus faecalis. Front Chem 5:51. https://doi.org/10.3389/fchem.2017.00051

Peng J, Wang Y, Wu Z, Mao C, Li L, Cao H et al (2023) Antimicrobial peptide Cec4 eradicates multidrug-resistant Acinetobacter baumannii in vitro and in vivo. Drug Des Devel Ther 17:977–992. https://doi.org/10.2147/DDDT.S405579

Piddock LJV (2006) Clinically relevant chromosomally encoded multidrug resistance efflux pumps in bacteria. Clin Microbiol Rev 19(2):382–402. https://doi.org/10.1128/CMR.19.2.382-402.2006

Pirtskhalava M, Amstrong AA, Grigolava M, Chubinidze M, Alimbarashvili E, Vishnepolsky B et al (2021) Dbaasp v3: database of antimicrobial/cytotoxic activity and structure of peptides as a resource for development of new therapeutics. Nucleic Acids Res 49(D1):D288–D297. https://doi.org/10.1093/nar/gkaa991

Poirel L, Jayol A, Nordmann P (2017) Polymyxins: antibacterial activity, susceptibility testing, and resistance mechanisms encoded by plasmids or chromosomes. Clin Microbiol Rev 30(2):557–596. https://doi.org/10.1128/CMR.00064-16

Powers J-PS, Hancock REW (2003) The relationship between peptide structure and antibacterial activity. Peptides 24(11):1681–1691. https://doi.org/10.1016/j.peptides.2003.08.023

Prestinaci F, Pezzotti P, Pantosti A (2015) Antimicrobial resistance: a global multifaceted phenomenon. Pathog Glob Health 109(7):309–318. https://doi.org/10.1179/2047773215Y.0000000030

Rice LB (2008) Federal funding for the study of antimicrobial resistance in nosocomial pathogens: no ESKAPE. J Infect Dis 197(8):1079–1081. https://doi.org/10.1086/533452

Rodríguez-Martínez S, Cancino-Diaz JC, Vargas-Zuñiga LM, Cancino-Diaz ME (2008) Ll-37 regulates the overexpression of vascular endothelial growth factor (VEGF) and c-IAP-2 in human keratinocytes. Int J Dermatol 47(5):457–462. https://doi.org/10.1111/j.1365-4632.2008.03340.x

Roy RN, Lomakin IB, Gagnon MG, Steitz TA (2015) The mechanism of inhibition of protein synthesis by the proline-rich peptide oncocin. Nat Struct Mol Biol 22(6):466–469. https://doi.org/10.1038/nsmb.3031

Russo TA, Marr CM (2019) Hypervirulent Klebsiella pneumoniae. Clin Microbiol Rev 32(3):e00001-19. https://doi.org/10.1128/CMR.00001-19

Sacco F, Bitossi C, Casciaro B, Loffredo MR, Fabiano G, Torrini L et al (2022) The antimicrobial peptide esc(1-21) synergizes with colistin in inhibiting the growth and in killing multidrug resistant Acinetobacter baumannii strains. Antibiotics (Basel) 11(2):234. https://doi.org/10.3390/antibiotics11020234

Saint-Criq V, Gray MA (2017) Role of CFTR in epithelial physiology. Cell Mol Life Sci 74(1):93–115. https://doi.org/10.1007/s00018-016-2391-y

Sajjan US, Tran LT, Sole N, Rovaldi C, Akiyama A, Friden PM et al (2001) P-113D, an antimicrobial peptide active against Pseudomonas aeruginosa, retains activity in the presence of sputum from cystic fibrosis patients. Antimicrob Agents Chemother 45(12):3437–3444. https://doi.org/10.1128/AAC.45.12.3437-3444.2001

Santajit S, Indrawattana N (2016) Mechanisms of antimicrobial resistance in ESKAPE pathogens. Biomed Res Int 2016:2475067. https://doi.org/10.1155/2016/2475067

Schibli DJ, Epand RF, Vogel HJ, Epand RM (2002) Tryptophan-rich antimicrobial peptides: comparative properties and membrane interactions. Biochem Cell Biol 80(5):667–677. https://doi.org/10.1139/o02-147

Seebah S, Suresh A, Zhuo S, Choong YH, Chua H, Chuon D et al (2007) Defensins knowledgebase: a manually curated database and information source focused on the defensins family of antimicrobial peptides. Nucleic Acids Res 35(Database issue):D265–D268. https://doi.org/10.1093/nar/gkl866

Shah P, Hsiao FS-H, Ho Y-H, Chen C-S (2016) The proteome targets of intracellular targeting antimicrobial peptides. Proteomics 16(8):1225–1237. https://doi.org/10.1002/pmic.201500380

Shang L, Li J, Song C, Nina Z, Li Q, Chou S et al (2020) Hybrid antimicrobial peptide targeting Staphylococcus aureus and displaying anti-infective activity in a murine model. Front Microbiol 11:1767. https://doi.org/10.3389/fmicb.2020.01767

Shang D, Han X, Du W, Kou Z, Jiang F (2021) Trp-containing antibacterial peptides impair quorum sensing and biofilm development in multidrug-resistant Pseudomonas aeruginosa and exhibit synergistic effects with antibiotics. Front Microbiol 12:611009. https://doi.org/10.3389/fmicb.2021.611009

Shi J, Chen C, Wang D, Wang Z, Liu Y (2022) The antimicrobial peptide LI14 combats multidrug-resistant bacterial infections. Commun Biol 5(1):926. https://doi.org/10.1038/s42003-022-03899-4

Siu LK, Yeh K-M, Lin J-C, Fung C-P, Chang F-Y (2012) Klebsiella pneumoniae liver abscess: a new invasive syndrome. Lancet Infect Dis 12(11):881–887. https://doi.org/10.1016/S1473-3099(12)70205-0

Strempel N, Strehmel J, Overhage J (2015) Potential application of antimicrobial peptides in the treatment of bacterial biofilm infections. Curr Pharm Des 21(1):67–84. https://doi.org/10.2174/1381612820666140905124312

Tabbene O, Di Grazia A, Azaiez S, Ben Slimene I, Elkahoui S, Alfeddy MN et al (2015) Synergistic fungicidal activity of the lipopeptide bacillomycin D with amphotericin B against pathogenic Candida species. FEMS Yeast Res 15(4):fov022. https://doi.org/10.1093/femsyr/fov022

Tenover FC (2006) Mechanisms of antimicrobial resistance in bacteria. Am J Med 119(6 Suppl 1):S3–S10. https://doi.org/10.1016/j.amjmed.2006.03.011; discussion S62–70

Trimble MJ, Mlynárčik P, Kolář M, Hancock REW (2016) Polymyxin: alternative mechanisms of action and resistance. Cold Spring Harb Perspect Med 6(10):a025288. https://doi.org/10.1101/cshperspect.a025288

Van der Weide H, Cossío U, Gracia R, Te Welscher YM, ten Kate MT, van der Meijden A et al (2020) Therapeutic efficacy of novel antimicrobial peptide AA139-nanomedicines in a multidrug-resistant Klebsiella pneumoniae pneumonia-septicemia model in rats. Antimicrob Agents Chemother 64(9):e00517-20. https://doi.org/10.1128/AAC.00517-20

Van Heel AJ, de Jong A, Song C, Viel JH, Kok J, Kuipers OP (2018) Bagel4: a user-friendly web server to thoroughly mine RiPPs and bacteriocins. Nucleic Acids Res 46(W1):W278–W281. https://doi.org/10.1093/nar/gky383

Vila-Farres X, Garcia de la Maria C, López-Rojas R, Pachón J, Giralt E, Vila J (2012) In vitro activity of several antimicrobial peptides against colistin-susceptible and colistin-resistant Acinetobacter baumannii. Clin Microbiol Infect 18(4):383–387. https://doi.org/10.1111/j.1469-0691.2011.03581.x

Vlieghe P, Lisowski V, Martinez J, Khrestchatisky M (2010) Synthetic therapeutic peptides: science and market. Drug Discov Today 15(1–2):40–56. https://doi.org/10.1016/j.drudis.2009.10.009

Waghu FH, Barai RS, Gurung P, Idicula-Thomas S (2016) Campr3: a database on sequences, structures and signatures of antimicrobial peptides. Nucleic Acids Res 44(D1):D1094–D1097. https://doi.org/10.1093/nar/gkv1051

Wang CKL, Kaas Q, Chiche L, Craik DJ (2008) Cybase: a database of cyclic protein sequences and structures, with applications in protein discovery and engineering. Nucleic Acids Res 36(Database issue):D206–D210. https://doi.org/10.1093/nar/gkm953

Wang G, Li X, Wang Z (2016) Apd3: the antimicrobial peptide database as a tool for research and education. Nucleic Acids Res 44(D1):D1087–D1093. https://doi.org/10.1093/nar/gkv1278

Watson D, Spaulding AB, Dreyfus J (2019) Risk-set matching to assess the impact of hospital-acquired bloodstream infections. Am J Epidemiol 188(2):461–466. https://doi.org/10.1093/aje/kwy252

Yan J, Bhadra P, Li A, Sethiya P, Qin L, Tai HK et al (2020) Deep-AmPEP30: improve short antimicrobial peptides prediction with deep learning. Mol Ther Nucleic Acids 20:882–894. https://doi.org/10.1016/j.omtn.2020.05.006

Yeung ATY, Gellatly SL, Hancock REW (2011) Multifunctional cationic host defence peptides and their clinical applications. Cell Mol Life Sci 68(13):2161–2176. https://doi.org/10.1007/s00018-011-0710-x

Zasloff M (2019) Antimicrobial peptides of multicellular organisms: my perspective. Adv Exp Med Biol 1117:3–6. https://doi.org/10.1007/978-981-13-3588-4_1

Zazo H, Colino CI, Lanao JM (2016) Current applications of nanoparticles in infectious diseases. J Control Release 224:86–102. https://doi.org/10.1016/j.jconrel.2016.01.008

Zhang Q-Y, Yan Z-B, Meng Y-M, Hong X-Y, Shao G, Ma J-J et al (2021) Antimicrobial peptides: mechanism of action, activity and clinical potential. Mil Med Res 8(1):48. https://doi.org/10.1186/s40779-021-00343-2

Zhang H, Zhang X, Liang S, Wang J, Zhu Y, Zhang W et al (2023) Bactericidal synergism between phage endolysin Ply2660 and cathelicidin LL-37 against vancomycin-resistant Enterococcus faecalis biofilms. NPJ Biofilms Microbiomes 9(1):16. https://doi.org/10.1038/s41522-023-00385-5
Zhao X, Wu H, Lu H, Li G, Huang Q (2013) Lamp: a database linking antimicrobial peptides. PLoS One 8(6):e66557. https://doi.org/10.1371/journal.pone.0066557
Zhou J, Kong L, Fang N, Mao B, Ai H (2016) Synthesis and functional characterization of MAF-1A peptide derived from the larvae of housefly, Musca domestica (Diptera: Muscidae). J Med Entomol 53(6):1467–1472. https://doi.org/10.1093/jme/tjw110
Zhu C, Zhao Y, Zhao X, Liu S, Xia X, Zhang S et al (2022) The antimicrobial peptide MPX can kill Staphylococcus aureus, reduce biofilm formation, and effectively treat bacterial skin infections in mice. Front Vet Sci 9:819921. https://doi.org/10.3389/fvets.2022.819921

Antimicrobial Activity of Nanomaterials and Nanocomposites Against ESKAPE Pathogens

17

Sudhakar Pola

Abstract

The rise of multidrug-resistant infections is now a critical concern on a worldwide scale. The ESKAPE group, consisting of *Enterococcus faecium, Staphylococcus aureus, Klebsiella pneumoniae, Acinetobacter baumannii, Pseudomonas aeruginosa,* and *Enterobacter species*, is of particular alarm due to its potential to profoundly impact public health. To combat these infections, nanotechnology has gained popularity as a viable strategy. Nanomaterials and nanocomposites provide distinctive properties that can improve antibacterial activity and overcome the drawbacks of traditional antibiotics. This chapter examines the mechanisms through which nanomaterials and nanocomposites exert their antimicrobial effects, including, Physical Interactions and Membrane Disruption and Generation of reactive oxygen species. Moreover, the evaluation of antimicrobial activity is discussed, encompassing in vitro and in vivo assessment methods. The synergistic approaches of combining nanomaterials with conventional antimicrobial agents are explored, highlighting their potential to overcome drug resistance. Nevertheless, safety considerations regarding the potential cytotoxicity and adverse effects of nanomaterials are essential in their clinical translation. Regulatory aspects, including standardized testing and risk assessments, are vital to ensure the safe development and utilization of nanomaterial-based antimicrobial therapies. Furthermore, this chapter discusses future directions and challenges, such as Combination Therapies, and Targeted Delivery. Ultimately, the understanding of the "antimicrobial activity" of nanomaterials and nanocomposites Hostile to "ESKAPE pathogens" holds great promise in revolutionizing antimicrobial therapies and advancing public health outcomes.

S. Pola (✉)
Department of Biotechnology, College of Science and Technology, Andhra University, Visakhapatnam, Andhra Pradesh, India
e-mail: sudhakar@andhrauniversity.edu.in

S. Busi, R. Prasad (eds.), *ESKAPE Pathogens*,
https://doi.org/10.1007/978-981-99-8799-3_17

Keywords

Multidrug-resistant infections · Nanotechnology · Nanomaterials · Antimicrobial activity · Drug resistance and targeted delivery

17.1 Introduction

"ESKAPE pathogens" are a group of microorganisms that have gained considerable attention due to their remarkable capability to "escape" the effects of antimicrobial treatments. The acronym ESKAPE (Fig. 17.1) represents six bacterial species: "*Enterococcus faecium*, *Staphylococcus aureus*, *Klebsiella pneumoniae*, *Acinetobacter baumannii*, *Pseudomonas aeruginosa,* and *Enterobacter*" species (Boucher et al. 2009). These pathogens are known for their high levels of antimicrobial resistance, causing a significant burden on healthcare systems worldwide.

The role of nanomaterials and nanocomposites in combating ESKAPE pathogens is a rapidly evolving field with immense potential for addressing the challenges posed by antimicrobial resistance. Nanotechnology offers unique advantages in terms of targeted delivery, enhanced antimicrobial activity, and the ability to overcome traditional antibiotic limitations.

17.2 Nanomaterials and Nanocomposites

Nanomaterials and nanocomposites have captured the attention of experts across diverse fields, owing to their distinct characteristics and versatile applications. At the nanoscale, materials exhibit novel characteristics and behaviors, distinct from their bulk counterparts, making them attractive for numerous technological advancements. This introduction provides an overview of nanomaterials and nanocomposites, highlighting their properties, synthesis methods, and potential applications.

Nanomaterials are materials with dimensions ranging from 1 to 100 nm in at least one dimension. They can be categorized into various groups based on their composition, particularly metals, semiconductors, ceramics, polymers, and carbon-based materials (Alwarappan and Joshi 2016). These materials often exhibit unique physical, chemical, and "biological properties," particularly increased proportion of external area to internal volume, quantum confinement effects, enhanced mechanical strength, and tailored optical and electrical properties (Pinto et al. 2016).

Nanocomposites, on the other hand, are composite materials consisting of a combination of nanoscale materials, such as nanoparticles, nanofibers, or nanosheets, embedded within a matrix material. The matrix material can be organic (polymer-based) or inorganic (ceramic or metallic) (Balázsi et al. 2012). The incorporation of nanoscale components into the matrix allows for the manipulation of characteristics, including mechanical toughness, ability to conduct electricity, and thermal stability, and chemical reactivity (Wang et al. 2020).

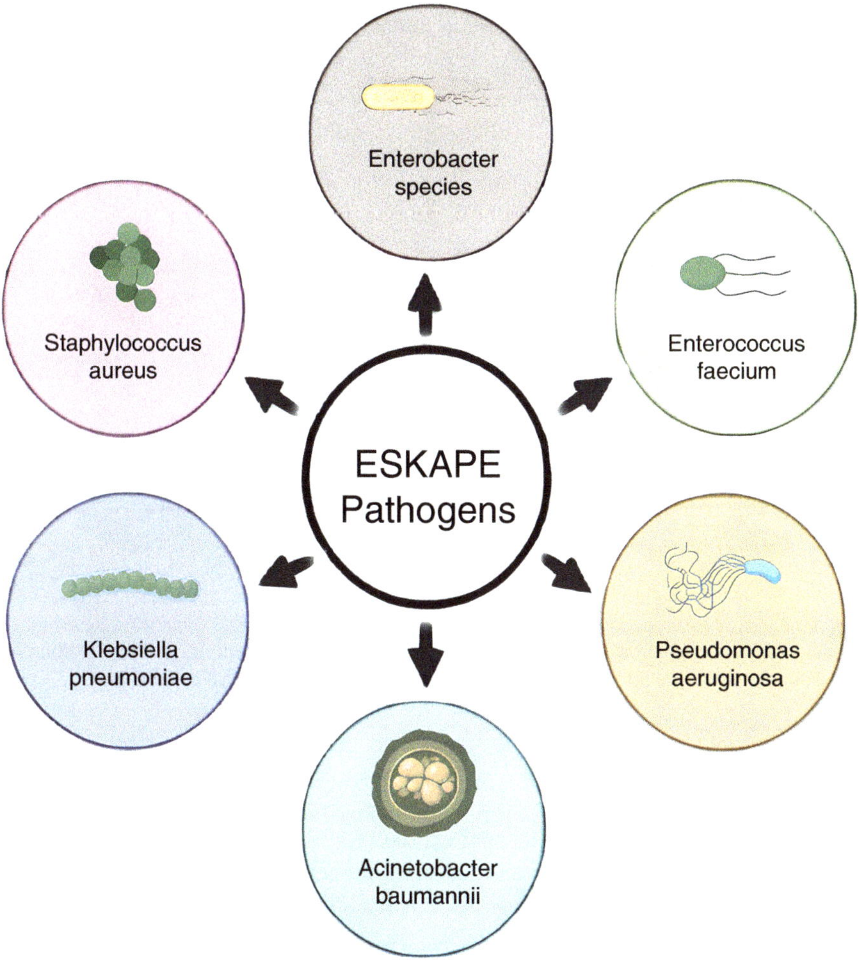

Fig. 17.1 ESKAPE pathogens. This figure showcases the six multidrug-resistant ESKAPE pathogens, which comprise *Enterococcus faecium, Staphylococcus aureus, Klebsiella pneumoniae, Acinetobacter baumannii, Pseudomonas aeruginosa,* and *Enterobacter species*. The pathogens' clinical impact and antibiotic resistance underscore the urgency for innovative therapeutic approaches

The production of nanoparticles and nanocomposites (Fig. 17.2) can be achieved through various methods, including from the grassroots and higher echelons approaches. From the grassroots, approaches involve building materials atom by atom or molecule by molecule, particularly gas-phase accumulation, solution-gelation synthesis, and self-assembly techniques (Leid et al. 2010). Top-down approaches, on the other hand, involve the reduction of larger materials to the nanoscale, such as ball milling, lithography, and template synthesis.

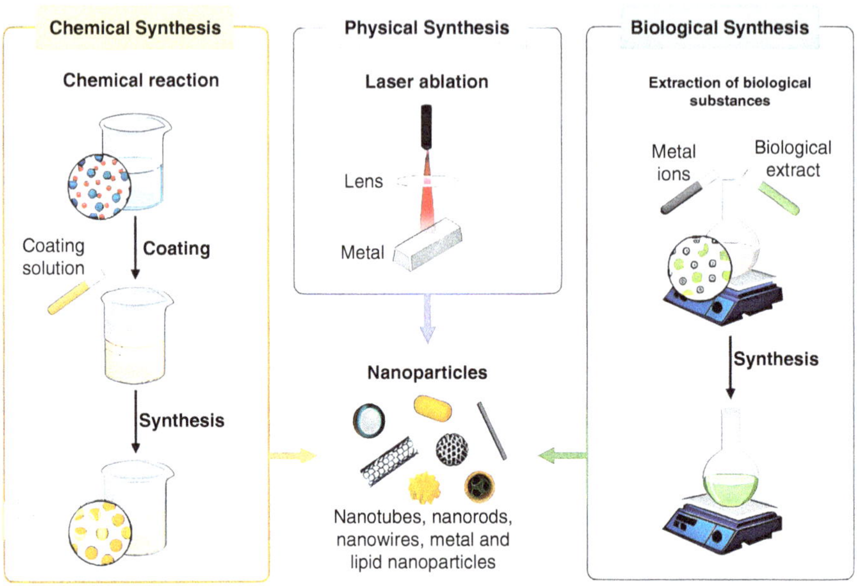

Fig. 17.2 Basic methods of nanoparticle synthesis. This figure outlines key nanoparticle synthesis methods, including chemical reduction, sol-gel, and physical vapor deposition. These techniques yield diverse nanoparticles with distinct properties, shaping their potential applications

The special characteristics of nanomaterials and nanocomposites have brought about their utilization in a variety of applications, such as electronics, power storage and conversion, Catalyst-driven reaction, sensors, biomedical devices, environmental remediation, and antimicrobial applications (Raliya and Tarafdar 2019; Balandin et al. 2008). In particular, their antimicrobial properties have garnered significant interest in combating microbial infections and addressing the challenges of antimicrobial resistance (Rai et al. 2014).

The antimicrobial activity of nanomaterials and nanocomposites arises from their interactions with microbial cells, which can include damaging cell membranes, disrupting cellular processes, inhibiting biofilm formation, and inducing oxidative stress. The use of nanomaterials and nanocomposites as antimicrobial agents offers potential advantages over traditional antibiotics, such as broader spectrum activity, reduced resistance development, and the possibility of targeted delivery (Hajipour et al. 2012; Ventola 2015).

17.2.1 Types of Nanomaterials and Nanocomposites Used in Antimicrobial Applications

Nanomaterials and nanocomposites have emerged as promising tools in combating microbial infections due to their unique properties and interactions with pathogens.

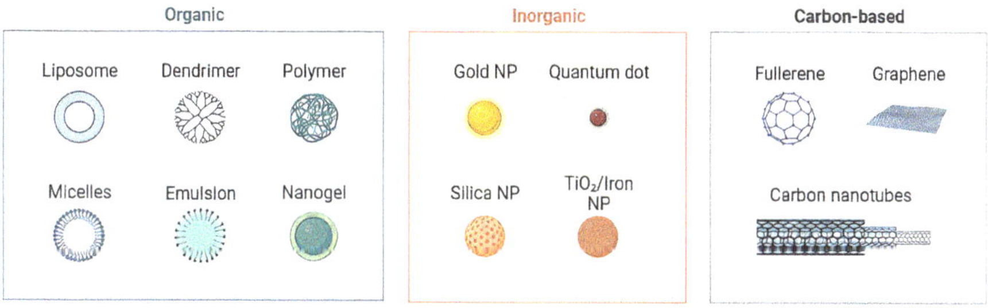

Fig. 17.3 Different types of nanoparticles. This figure visually presents antimicrobial nanoparticles such as silver, copper, zinc oxide, carbon nanotubes, and graphene oxide. The illustration highlights their unique structural attributes, offering versatility for combating ESKAPE pathogens

Various types of nanomaterials (Fig. 17.3) have been explored for their antimicrobial applications, offering different mechanisms of action and advantages.

Metal Nanoparticles: A lot of research has been done on the potent antibacterial capabilities of metal nanoparticles like silver (Ag), copper (Cu), and zinc oxide (ZnO) Rai et al. 2009; Jemal et al. (2017); Safawo et al. (2018). These nanoparticles can interact with bacterial cell membranes, penetrate biofilms, and induce oxidative stress, leading to microbial inactivation (Panáček et al. 2018a, b). Silver nanoparticles, in particular, have undergone thorough research for broad-spectrum antimicrobial involvement with various bacteria, including ESKAPE pathogens (Morones et al. 2005). Copper nanoparticles have also shown significant antimicrobial efficacy, with potential applications in reducing healthcare-associated infections (Durán et al. 2005). Additionally, ZnO nanoparticles have shown to have antibacterial properties against both Gram-positive and Gram-negative bacteria (Sirelkhatim et al. 2015).

17.2.2 Carbon-Based Nanomaterials

Nanomaterials made of carbon, such as graphene and carbon nanotubes (CNTs), have shown promise in antimicrobial applications. CNTs possess high surface area and mechanical strength, allowing for effective physical interaction with bacterial membranes and subsequent disruption (Kang et al. 2008). Graphene and its derivatives exhibit comprehensive antimicrobial action because of their capability to cause bacterial membrane damage and oxidative stress (Hu et al. 2010). These carbon-based nanomaterials offer potential advantages in terms of stability, biocompatibility, and large-scale production (Zhang et al. 2007).

17.2.3 Polymer Nanomaterials

Polymer-based nanomaterials, including polymer nanoparticles and nanofibers, have undergone investigation due to their antimicrobial properties (Pandey et al.

2020). Chitosan, polyvinyl alcohol (PVA), and poly(lactic-*co*-glycolic acid) (PLGA) exhibit significant antimicrobial properties against a broad spectrum of pathogens. Chitosan, derived from chitin, possesses inherent antimicrobial properties and can disrupt bacterial membranes (Rabea et al. 2003). PVA and PLGA nanoparticles have been investigated for their potential in controlled drug release and antimicrobial applications (Bricháč et al. 2013).

17.2.4 Mesoporous Silica Nanoparticles (MSNs)

Mesoporous silica nanoparticles (MSNs) are a type of nanomaterial that is made up an irregular silica framework with well-arranged porous molecular sieves. MSNs possess several characteristics that render them promising in support of antimicrobial implementations, including its extensive expanse, consistent pore size and biocompatibility (Lakshmi and Pola 2020).

17.2.5 Advantages of Using Nanomaterials and Nanocomposites for Antimicrobial Activity

The utilization of nanomaterials and nanocomposites for antimicrobial activity offers several advantages and opportunities in combating microbial infections.

Enhanced Antimicrobial Activity: Nanomaterials and nanocomposites often exhibit enhanced antimicrobial activity compared to traditional antimicrobial agents. Their individual physicochemical traits, exemplified by a notable surface area-to-volume ratio and tailored surface functionalities, allow for effective interactions with microbial cells, leading to improved antimicrobial efficacy (Rai et al. 2009).

Broad-Spectrum Activity: Many nanomaterials and nanocomposites demonstrate comprehensive antimicrobial properties, being versatile in targeting different microorganisms, including fungi and viruses. This broad activity makes them attractive for applications where multiple types of pathogens need to be targeted.

Reduced Antimicrobial Resistance Development: Nanomaterials and nanocomposites offer potential approaches to overcome microbial resistance to drugs. Their strategies of action such as cell membrane disruption, oxidative stress induction, and inhibition of bacterial adhesion, differ from traditional antibiotics, reducing the likelihood of resistance development (Panáček 2018).

Targeted Delivery: Nanomaterials can be functionalized or engineered to enable targeted delivery of antimicrobial agents to specific sites of infection. This targeted approach enhances treatment efficacy, reduces systemic toxicity, and minimizes the impact on beneficial microbiota (Sun et al. 2020).

17.3 Mechanisms of Antimicrobial Activity

17.3.1 Modes of Action of Nanomaterials and Nanocomposites Against ESKAPE Pathogens

Nanomaterials and nanocomposites exhibit various modes of action against ESKAPE pathogens, contributing to their antimicrobial activity. These mechanisms involve physical interactions, cellular membrane disruption, oxidative stress induction, and inhibition of essential cellular processes (Table 17.1). Understanding the modes of action is crucial for the effective utilization of nanomaterials and nanocomposites in combating microbial infections.

Physical Interactions and Membrane Disruption: Nanomaterials can physically interact with microbial cells and disrupt their cellular membranes. These materials can penetrate the outer membrane of the microbial cell and cause leakage of intracellular components, leading to death of the microorganism (Raghunath and Perumal 2017). For example, silver nanoparticles interact with bacterial cell membranes, causing structural damage and compromising membrane integrity, leading to cell lysis (Sondi and Salopek-Sondi 2004). Carbon nanotubes and graphene can

Table 17.1 Mechanism of antimicrobial activity exhibited by nanomaterials and nanocomposites against ESKAPE pathogens

Mechanism	Description	Examples of nanomaterials	References
Cell membrane disruption	Interaction with cell membranes, leading to leakage and cell death	Silver nanoparticles, graphene oxide	Rai et al. (2009), Li et al. (2010)
Reactive oxygen species (ROS) generation	Induction of ROS, causing oxidative stress and cellular damage	Titanium dioxide nanoparticles, zinc oxide nanoparticles	Murugan et al. (2016), Zhang et al. (2007)
Intracellular process interference	Disruption of key cellular processes or organelles	Carbon nanotubes, chitosan nanoparticles	Jiang et al. (2009), Li et al. (2012)
Metal ion release	Release of metal ions that disrupt microbial growth	Copper nanoparticles, zinc oxide nanoparticles	Ruparelia et al. (2008), Ravindran et al. (2013)
Quorum sensing inhibition	Interference with microbial communication and biofilm formation	Gold nanoparticles, silver nanoparticles	Fayaz et al. (2010), Li et al. (2010)
Nanoparticle-induced stress	Generation of mechanical or physical stress within cells	Magnetic nanoparticles, silica nanoparticles	Raffi et al. (2010), Sotiriou and Pratsinis (2010)
DNA/RNA interaction	Binding to microbial genetic material, inhibiting replication	Quantum dots, carbon nanotubes	Lok et al. (2007), Kang (2008)
Nanoparticle internalization	Uptake of nanoparticles by cells, leading to intracellular damage	Polymeric nanoparticles, liposomes	Hu et al. (2006), Gupta and Saleh (2011)

pierce bacterial cell walls, physically damaging the cell membrane and resulting in bacterial death (Liu et al. 2009).

Oxidative Stress Induction: Many nanomaterials and nanocomposites exert antimicrobial activity by inducing oxidative stress within microbial cells. Upon contact, these materials can generate "reactive oxygen species" (ROS), such as superoxide radicals and hydrogen peroxide, which overwhelm the antioxidant defense systems of microbial cells (Li et al. 2010). The accumulation of ROS leads to oxidative damage to cellular components ultimately causing cell death (Arokiyaraj et al. 2014). Metal nanoparticles, including Ag and Cu nanoparticles, are noted for its capacity to provoke oxidative stress within microbial cells, contributing to their antimicrobial activity (Morones et al. 2005).

Inhibition of Essential Cellular Processes: Nanomaterials and nanocomposites can interfere with essential cellular processes of ESKAPE pathogens, impairing their survival and growth. For instance, some nanomaterials can disrupt bacterial adhesion, inhibiting the formation of biofilms, which are important for bacterial colonization and persistence (Li et al. 2006). Others can disrupt essential enzymatic processes or interfere with microbial signaling pathways, leading to cellular dysfunction and inhibition of microbial growth (Zodrow et al. 2009a, b). These inhibitory effects contribute to the antimicrobial pursuit associated with nanomaterials and nanocomposites against ESKAPE pathogens.

Release of Antimicrobial Agents: In certain cases, nanocomposites can be designed to release antimicrobial agents, such as antibiotics or other bioactive compounds, in a controlled manner. The incorporation of antimicrobial agents within the nanocomposite matrix allows for sustained release over time, ensuring a prolonged antimicrobial effect (Nguyen et al. 2017). This approach enhances the effectiveness of the antimicrobial treatment and minimizes the potential for resistance development.

17.3.2 Interaction of Nanomaterials and Nanocomposites with Bacterial Cells

Nanomaterials and nanocomposites exhibit unique interactions with bacterial cells that are indispensable in their antimicrobial activity. These encounters can occur at the cell surface, cell membrane, and intracellular levels. Understanding the specific interactions is essential for designing effective nanomaterials and nanocomposites for antimicrobial applications.

Cell Surface Interactions: Nanomaterials and nanocomposites can interact with bacterial cells at the surface level, leading to various effects. For example, nanomaterials such as silver nanoparticles can adhere the outer layer through electrostatic forces and form aggregates, leading to the clustering of bacteria and hindering their growth. Similarly, graphene oxide can adsorb onto the microbial surface, disrupting membrane integrity and restraining bacterial growth (Liu et al. 2009).

17.3.3 Cell Membrane Interactions

The intersection of nanomaterials and nanocomposites with bacterial outer membranes is crucial for their antimicrobial activity. Metal nanomaterials can interface with cell membranes, leading to organelle damage and increased membrane porosity (Padmavathy and Vijayaraghavan 2008). This breakdown of the membrane integrity results in leakage of cellular components and ultimately cell death. Carbon-based nanomaterials can physically penetrate the cell membrane, causing mechanical damage and compromising membrane integrity (Liu et al. 2011).

Intracellular Interactions: Certain nanomaterials and nanocomposites can enter bacterial cells, leading to intracellular interactions that affect cellular processes. For example, some metal oxide nanoparticles, such as zinc oxide (ZnO) nanoparticles, can penetrate bacterial cells and generate reactive oxygen species (ROS) within the cytoplasm, inducing oxidative stress and damaging cellular components (Li et al. 2010). This oxidative stress can disrupt essential cellular processes, impairing bacterial growth and survival.

17.4 Evaluation of Antimicrobial Activity

17.4.1 In Vitro Assessment Methods for Antimicrobial Activity

In vitro assessment methods are crucial for evaluating the antimicrobial activity of nanomaterials and nanocomposites against pathogens. These methods provide valuable information on the effectiveness of these materials in inhibiting the growth and survival of microorganisms (Table 17.2).

Agar Dilution Method: The agar dilution method involves preparing agar plates with different concentrations of the antimicrobial agent (nanomaterial or nanocomposite). Microbial suspensions are then inoculated onto the plates, and the plates are incubated. The minimum inhibitory concentration (MIC) is determined as the lowest concentration of the antimicrobial agent that inhibits visible microbial growth. This method provides quantitative information on the antimicrobial activity of the test samples (Andrews 2001).

Broth Dilution Method: The broth dilution method involves preparing serial dilutions of the antimicrobial agent in a liquid growth medium. Microbial suspensions are added to the dilutions, and the tubes or microplates are incubated. The MIC is determined as the lowest concentration of the antimicrobial agent that prevents visible growth of the microorganisms. This method is useful for assessing the antimicrobial activity of nanomaterials and nanocomposites in a liquid medium (CLSI 2018).

Disk Diffusion Method: The disk diffusion method, also known as the Kirby-Bauer method, involves impregnating paper disks with the antimicrobial agent (nanomaterial or nanocomposite). The disks are then placed on agar plates inoculated with the test microorganism. After incubation, the diameter of the zone of inhibition surrounding the disk is measured. A larger zone of inhibition indicates

Table 17.2 Methods for the evaluation of antimicrobial activity of nanomaterials and nanocomposites against ESKAPE pathogens

Evaluation method	Description	Advantages	Limitations	References
Agar diffusion assay	Measurement of inhibition zone on agar plates	Simple, cost-effective	Limited quantitative information	Andrews et al. (2001)
Minimum inhibitory concentration (MIC)	Lowest concentration inhibiting bacterial growth	Quantitative assessment	Time-consuming, endpoint measurement	CLSI (2018)
Minimum bactericidal concentration (MBC)	Lowest concentration killing bacteria	Assessment of bactericidal effect	May not correlate with clinical efficacy	Kaplan (2011)
Time-kill kinetics	Bacterial viability over time at different concentrations	Dynamic antimicrobial profile	Labor-intensive, requires serial sampling	Aljeboree et al. (2011)
Biofilm inhibition/ destruction assays	Disruption or prevention of biofilm formation	Relevant for biofilm-associated infections	Complex assay, may not mimic in vivo	O'Toole (2011)
Cell viability assays	Impact on mammalian cell viability	Evaluation of cytotoxicity	Doesn't directly measure microbial activity	Borenfreund and Puerner (1985)
Quantitative PCR (qPCR)	Changes in microbial gene expression	Molecular-level assessment	Requires advanced equipment and expertise	Bustin (2009)
Flow cytometry	Microbial cell population analysis	Single-cell analysis	Limited information on mechanism of action	Wang et al. (2017)

stronger antimicrobial activity. This method provides a qualitative assessment of the antimicrobial efficacy of the test samples (Bauer et al. 1966).

Time-Kill Kinetics Assay: The time-kill kinetics assay involves exposing microbial cultures to a fixed concentration of the antimicrobial agent over a specific period. At designated time intervals, samples are taken, diluted, and plated onto agar plates. The viable microbial count is determined by colony counting. This assay provides information on the bactericidal or bacteriostatic effects of the antimicrobial agent over time (Mataraci and Dosler 2005).

Biofilm Inhibition and Eradication Assays: Biofilm inhibition and eradication assays are specifically designed to assess the antimicrobial activity of nanomaterials and nanocomposites against biofilms. These assays involve the formation of biofilms on various substrates, such as microtiter plates or catheters. The antimicrobial agent is then applied, and the inhibition or eradication of biofilms is evaluated using microscopy, crystal violet staining, or quantification of viable cells.

17.4.2 In Vivo Evaluation of Nanomaterials and Nanocomposites Against ESKAPE Pathogens

Animal Infection Models: Animal infection models, such as mice, rats, or rabbits, are commonly used to evaluate the antimicrobial efficacy of nanomaterials and nanocomposites against ESKAPE pathogens. These models involve infecting the animals with the target pathogen and subsequently treating them with the nanomaterial or nanocomposite of interest. The efficacy of the treatment is assessed by monitoring the survival rate, bacterial load, histopathological changes, and other relevant parameters (Zhang et al. 2018).

Wound Infection Models: Wound infection models are specifically designed to evaluate the antimicrobial activity of nanomaterials and nanocomposites in the context of wound healing and infection. These models involve creating a controlled wound and infecting it with ESKAPE pathogens. The nanomaterial or nanocomposite is applied to the wound site, and the progression of infection, wound healing, bacterial load, and tissue response are assessed (Sheikholeslami et al. 2018).

17.5 Synergistic Approaches

Combination therapies involving the use of nanomaterials and nanocomposites have gained significant attention in the field of antimicrobial research. These therapies offer the potential to increase the potency of antimicrobial regimens, overcome drug resistance, and minimize side effects.

Antibiotic-Nanomaterial Combinations: Combining antibiotics with nanomaterials or nanocomposites can enhance their antimicrobial activity and overcome bacterial resistance. The nanomaterials act as carriers or delivery systems for antibiotics, improving their bioavailability and targeting specific sites of infection. For example, the co-administration of antibiotics with silver nanoparticles or carbon nanotubes has demonstrated synergistic effects against drug-resistant bacteria (Rai et al. 2012).

Photothermal Therapy-Nanomaterial Combinations: Photothermal therapy (PTT) involves the use of light-absorbing nanomaterials to generate localized heat and induce bacterial cell death. Combining PTT with nanomaterials, such as gold nanorods or carbon nanotubes, enhances the therapeutic efficacy. The heat generated by these nanomaterials effectively destroys bacterial cells and disrupts biofilms, providing a complementary approach for antimicrobial treatment (Gao et al. 2017). Photodynamic Therapy-Nanomaterial Combinations: Photodynamic therapy (PDT) involves the use of photosensitizers and photons to generate reactive oxygen species that can eradicate microorganisms. Combining PDT with nanomaterials, such as zinc oxide nanoparticles or graphene oxide, can enhance the delivery and targeting of photosensitizers, leading to improved antimicrobial efficacy. These combination therapies have shown promise in combating drug-resistant bacteria and biofilms (Xue et al. 2017).

17.5.1 Enhanced Antimicrobial Activity Through Functionalization and Surface Modifications

Enhancing the antimicrobial activity of nanomaterials and nanocomposites can be achieved through functionalization and surface modifications. These strategies involve modifying the surface properties of the materials to improve their interactions with microorganisms, enhance stability, and promote targeted antimicrobial effects.

17.5.2 Surface Charge Modification

Modifying the exterior charge of nanomaterials can influence their interactions with microbial cells. For example, cationic surfaces have the ability to come into contact with anionic charged microbial cell membranes, leading to enhanced antimicrobial activity. Functionalization with cationic polymers or molecules, such as chitosan or quaternary ammonium compounds, can impart positive charges and improve antimicrobial efficacy (Hu et al. 2015).

Surface Coating with Antimicrobial Agents: Coating the surface of nanomaterials with antimicrobial agents can provide a sustained release of active compounds, leading to enhanced antimicrobial activity. The coatings can be composed of natural or chemically synthesized antimicrobial compounds such as silver nanoparticles, quaternary ammonium compounds, or antimicrobial peptides. These coatings release antimicrobial agents over time, ensuring prolonged effectiveness (Cui et al. 2013).

17.5.3 Incorporation of Antibiotics

Incorporating antibiotics into nanomaterials or nanocomposites can enhance their antimicrobial capacity and eliminate bacterial immunity. The encapsulation or loading of antibiotics within nanomaterials can improve their stability, bioavailability, and targeted delivery. Functionalization strategies, such as mesoporous structures or polymer matrices, enable controlled release of antibiotics and improve their efficacy (Jafarzadeh et al. 2017).

The combination of nanomaterials and nanocomposites with conventional antimicrobial agents has shown promising results in enhancing antimicrobial activity, overcoming drug resistance, and reducing the required dosage of antibiotics.

Synergy with Antibiotics: Combining nanomaterials or nanocomposites with antibiotics can result in synergistic effects, leading to enhanced antimicrobial activity. The nanomaterials can act as carriers or delivery systems for antibiotics, improving their bioavailability and targeted delivery. Synergy with antibiotics has been demonstrated using various nanomaterials, and nanoparticles, in combination with antibiotics like penicillin, tetracycline, and vancomycin (Rai et al. 2009).

17.6 Safety and Toxicity Considerations

Potential cytotoxicity and adverse effects of nanomaterials: While nanomaterials hold great promise for antimicrobial applications, their potential cytotoxicity and unwanted reactions on individual physical condition and the ecosystem need to be carefully evaluated.

Nanoparticles often acquire significant toxicity, posing considerable risks to both the environment and human health. The primary factor responsible for these adverse effects is the size distribution of nanoparticles, typically ranging from 1 to 100 nm. Their ability to enter ecological food chains through microorganisms has been a straightforward process, disrupting the delicate ecological balance. This disturbance further aggravates the impact on both the environment and the health system (Sanapala and Pola 2020).

Cytotoxicity: Some nanomaterials may exhibit cytotoxic effects on mammalian cells, including cell death, oxidative stress, inflammation, and DNA damage. The cytotoxicity can vary depending on factors such as nanoparticle size, shape, surface charge, surface functionalization, and dose. It is important to assess the cytotoxicity of nanomaterials using appropriate cell culture models and evaluation methods, such as cell viability assays, oxidative stress measurements, and genotoxicity assays (Fadeel et al. 2018a, b).

Immunotoxicity: Nanomaterials can interact with the immune system, potentially leading to immunotoxic effects. This includes activation of inflammatory responses, release of pro-inflammatory cytokines, modulation of immune cell function, and interference with immune signaling pathways. Immunotoxicity assessments should consider immune cell viability, cytokine profiling, phagocytic activity, and antibody production, among other parameters (Pacheco et al. 2020).

Genotoxicity: Some nanomaterials have the potential to induce genetic damage in cells, which can lead to mutations and genomic instability. Genotoxicity studies evaluate the ability of nanomaterials to cause DNA damage, chromosomal aberrations, or mutations using techniques such as comet assay, micronucleus assay, and DNA damage markers. Understanding the genotoxic potential of nanomaterials is critical for their safe use (Singh et al. 2010).

Strategies for Mitigating Nanomaterial Toxicity Mitigating the potential toxicity of nanomaterials is crucial for their safe and sustainable use. Several strategies can be employed to reduce the adverse effects associated with nanomaterials.

Rational Nanomaterial Design: By carefully designing nanomaterial characteristics such as dimensions, structure, exterior charge, and surface functionalization, it is possible to optimize their biocompatibility and reduce potential toxicity. Rational design strategies aim to minimize interactions with biological systems that may lead to adverse effects. For example, surface modifications with biocompatible coatings or polymers can improve nanomaterial stability and reduce cytotoxicity (Fadeel et al. 2018a, b).

Encapsulation and Controlled Release: Encapsulating nanomaterials within protective matrices or coatings can minimize their direct contact with biological systems, reducing potential toxicity. Controlled release systems allow for the gradual and targeted release of nanomaterials, limiting their concentration and exposure to sensitive tissues or cells. Encapsulation strategies include liposomes, micelles, and polymeric nanoparticles, which can enhance biocompatibility and reduce toxicity (Chou et al. 2014).

Surface Functionalization: Modifying the surface of nanomaterials with biomolecules, such as proteins or polymers, can improve their biocompatibility and reduce toxicity. Functionalization can provide a protective barrier, reduce protein corona formation, and improve stability in biological environments. Surface functionalization can be tailored to specific applications, aiming to minimize cytotoxicity and enhance targeted delivery (Wu et al. 2005).

Environmental Impact of Nanomaterials and Nanocomposites The environmental impact of nanomaterials and nanocomposites is an important consideration due to their potential Emission into the environment during production, utilization, and dumping stages.

Ecotoxicity and Bioaccumulation: Nanomaterials released into the environment can cause harm to ecosystems and organisms. They can build up in diverse environmental compartments, such as ground, liquid, and particulates, and may pose a risk to aquatic and terrestrial organisms. Studies have shown that certain nanomaterials, such as silver nanoparticles or titanium dioxide nanoparticles, can cause toxicity and impair the growth, reproduction, and behavior of various organisms.

Interactions with Biota: Nanomaterials can interact with different organisms, including microorganisms, plants, and animals, which can have cascading effects on ecological processes. For example, nanomaterials can influence microbial community structures, nutrient cycling, and plant growth. The impacts can vary depending on factors such as nanomaterial properties, exposure concentrations, and environmental conditions (Dimkpa et al. 2013).

Environmental Fate and Transport: Understanding the movement and distribution of nanomaterials in the environment is crucial for assessing their potential environmental impact. Nanomaterials can undergo transformations and aggregate, affecting their mobility and distribution. Elements like particle dimensions, exterior features, and environmental conditions influence their behavior in soil, water, and air. Research has focused on studying the environmental destiny and migration of nanomaterials to assess their potential for dispersion and accumulation (Keller and Lazareva 2014).

17.7 Future Directions and Challenges

Future directions in the development of antimicrobial nanomaterials are driven by the need to address emerging challenges in combating microbial infections, including the rise of antibiotic resilient germs and the increasing occurrence regarding healthcare-associated illnesses. Researchers are exploring innovative approaches to enhance the effectiveness and specificity of antimicrobial nanomaterials while ensuring their safety and environmental sustainability.

Combination Therapies: Combining various antimicrobial nanomaterials, including metal nanoparticles, carbon derived nanomaterials and nanocomposites with conventional antibiotics or other antimicrobial agents can lead to synergistic effects and overcome drug resistance. Researchers are investigating the optimal combinations and ratios of nanomaterials to improve antimicrobial efficacy and reduce the risk of resistance development.

Targeted Delivery: Developing nanomaterials with specific targeting capabilities can enhance their selectivity and reduce off-target effects. Functionalizing nanomaterials with ligands or antibodies that recognize microbial surface markers allows for targeted delivery to infected sites, increasing their antimicrobial activity while minimizing damage to healthy tissues (Singh et al. 2018).

Biodegradable Nanomaterials: The design of biodegradable antimicrobial nanomaterials that can be broken down into non-toxic components after fulfilling their antimicrobial function is gaining attention. Biodegradable nanomaterials reduce the risk of long-term environmental accumulation and promote sustainable antimicrobial solutions (Dhand et al. 2016).

Immunomodulatory Nanomaterials: Researchers are exploring nanomaterials that can modulate the host immune response to enhance antimicrobial defense mechanisms. Immunomodulatory nanomaterials can boost the immune system's ability to fight infections and promote tissue repair, complementing their direct antimicrobial activity.

17.8 Conclusion

The remarkable potential of nanomaterials and nanocomposites in combating ESKAPE pathogens offers a promising avenue for addressing the escalating challenges posed by multidrug-resistant infections. Extensive research has demonstrated their efficacy in exerting antimicrobial effects through diverse mechanisms, ranging from cell membrane disruption to reactive oxygen species generation. This versatility positions nanomaterials as valuable candidates for innovative therapeutic interventions.

As these nanomaterials transition from laboratories to clinical applications, regulatory considerations become paramount. Regulatory agencies mandate comprehensive safety assessments and risk evaluations to ensure the responsible use of nanomaterial-based products. Rigorous testing for potential toxicity, environmental impact, and long-term effects is essential to establish their safety profiles. Adhering to these risk assessment protocols and guidelines provides a robust framework for the secure integration of nanomaterials into medical and consumer realms, enhancing public health and safety.

Acknowledgements I, Sudhakar Pola, wish to extend my heartfelt thanks to authorities of Andhra University for their support in conducting this study.

Conflict of Interest I, Sudhakar Pola, hereby state that, I have no competing interests related to this research.

References

Aljeboree AM, Alshirifi AN, Alkaim AF (2011) Kinetics and mechanism of silver ions adsorption on to almond shell. Desalination 276(1–3):27–36

Alwarappan S, Joshi RK (2016) Nanomaterials: classification, fabrication methods, and applications. In: Joshi RK, Kumar A (eds) Nanoscience in food and agriculture 2. Springer, Berlin, pp 1–20

Andrews JM (2001) Determination of minimum inhibitory concentrations. J Antimicrob Chemother 48(Suppl 1):5–16

Arokiyaraj S, Vincent S, Saravanan M (2014) Biosynthesis of copper oxide nanoparticles using Bacillus subtilis shows potent antibacterial activity. Orient J Chem 30(3):1401–1408

Balandin AA, Ghosh S, Bao W, Calizo I, Teweldebrhan D, Miao F et al (2008) Superior thermal conductivity of single-layer graphene. Nano Lett 8(3):902–907

Balázsi C, Eckert J, Stoica M, Balázsi K (2012) Nanocomposites: synthesis, structure, properties and new application opportunities. J Nanomater 2012:1–14

Bauer AW, Kirby WMM, Sherris JC, Turck M (1966) Antibiotic susceptibility testing by a standardized single disk method. Am J Clin Pathol 45(4):493–496

Borenfreund E, Puerner JA (1985) Toxicity determined in vitro by morphological alterations and neutral red absorption. Toxicol Lett 24(2–3):119–124

Boucher HW, Talbot GH, Bradley JS, Edwards JE, Gilbert D, Rice LB, Scheld M, Spellberg B, Bartlett J (2009) Bad bugs, no drugs: no ESKAPE! An update from the Infectious Diseases Society of America. Clin Infect Dis 48(1):1–12

Bricháč J, Čejková J, Štěpánek F, Hrušková I, Cépe K, Rezek B (2013) Antibacterial effect of polyvinyl alcohol and poly(vinyl alcohol)/poly(acrylic acid) electrospun nanofibers containing mupirocin. Eur J Pharm Sci 50(2):201–207

Bustin SA, Benes V, Garson JA et al (2009) The MIQE guidelines: minimum information for publication of quantitative real-time PCR experiments. Clin Chem 55:611–622

Chou LY, Zagorovsky K, Chan WC (2014) DNA assembly of nanoparticles. ACS Nano 8(1):65–76

CLSI (2018) Methods for dilution antimicrobial susceptibility tests for bacteria that grow aerobically, 11th edn. Clinical and Laboratory Standards Institute, Berwyn

Cui H, Bai M, Li T, Li C, Wang Y, Guan Y et al (2013) Antibacterial activity of chitosan-coated silver nanoparticles on Campylobacter jejuni. Food Control 34(1):117–122

Dhand C, Dwivedi N, Loh XJ, Ying AN, Verma NK, Beuerman RW et al (2016) Methods to access biodegradable nanoparticles for drug delivery. Curr Med Chem 23(40):4717–4736

Dimkpa CO, McLean JE, Martineau N, Britt DW, Haverkamp R, Anderson AJ (2013) Silver nanoparticles disrupt wheat (Triticum aestivum L.) growth in a sand matrix. Environ Sci Technol 47(17):9453–9460

Durán N, Marcato PD, De Souza G, Alves OL, Esposito E (2005) Mechanistic aspects of biosynthesis of silver nanoparticles by several Fusarium oxysporum strains. J Nanobiotechnol 3(1):8

Fadeel B, Farcal L, Hardy B, Vázquez-Campos S (2018a) Advanced tools for the safety assessment of nanomaterials. Nat Nanotechnol 13(6):537–543

Fadeel B, Feliu N, Vogt C, Abdelmonem AM, Parak WJ (2018b) Bridge over troubled waters: understanding the synthetic and biological identity of nanomaterials. Wiley Interdiscip Rev Nanomed Nanobiotechnol 10(2):e1489

Fayaz AM, Balaji K, Girilal M, Yadav R, Kalaichelvan PT, Venketesan R (2010) Biogenic synthesis of silver nanoparticles and their synergistic effect with antibiotics: a study against gram-positive and gram-negative bacteria. Nanomedicine 6(1):103–109

Gao M, Liang C, Song X, Chen Q, Jin Q, Wang C et al (2017) Hydrogenated black TiO2 nanocrystals: bridging the gap between oxygen vacancies and photothermal conversion for cancer therapy. Angew Chem Int Ed Engl 56(49):15272–15276

Gupta A, Saleh NM (2011) Synthesis and characterization of magnetic poly(styrene-co-acrylic acid) nanocomposites for removal of heavy metal ions from aqueous solution. J Hazard Mater 185(1):329–340

Hajipour MJ, Fromm KM, Akbar Ashkarran A, Jimenez de Aberasturi D, de Larramendi IR, Rojo T et al (2012) Antibacterial properties of nanoparticles. Trends Biotechnol 30(10):499–511

Hu J, Chen M, Li ZY, Au L, Hartland GV, Li XD, Marquez M, Xia Y (2006) Gold nanostructures: engineering their plasmonic properties for biomedical applications. Chem Soc Rev 35(11):1084–1094

Hu W, Peng C, Luo W, Lv M, Li X, Li D et al (2010) Graphene-based antibacterial paper. ACS Nano 4(7):4317–4323

Hu Y, Zuo P, Shang L, Liu J, Xu Q, Tang J (2015) Fabrication of antibacterial biomimetic surfaces by coating a layer of cell membrane onto nanofibrous mats. ACS Appl Mater Interfaces 7(33):18653–18661

Jafarzadeh S, Nemati M, Shakerian A, Nemati N, Ardalan S (2017) Antibiotic-loaded nanofibrous scaffolds prepared by electrospinning: a review. Nanomedicine (Lond) 12(12):1353–1370

Jemal K, Sandeep BV, Pola S (2017) Synthesis, characterization, and evaluation of the antibacterial activity of Allophylus serratus leaf and leaf derived callus extracts mediated silver nanoparticles. J Nanomater 2017:4213275. https://doi.org/10.1155/2017/4213275. 11 pages

Jiang W, Mashayekhi H, Xing B (2009) Bacterial toxicity comparison between nano- and micro-scaled oxide particles. Environ Pollut 157(5):1619–1625

Kang S, Herzberg M, Rodrigues DF, Elimelech M (2008) Antibacterial effects of carbon nanotubes: size does matter! Langmuir 24(13):6409–6413

Kaplan JB (2011) Antibiotic-induced biofilm formation. Int J Artif Organs 34(9):737–751

Keller AA, Lazareva A (2014) Predicted releases of engineered nanomaterials: from global to regional to local. Environ Sci Technol Lett 1(1):65–70

Lakshmi P, Pola S (2020) Mesoporous silica nanomaterials as antibacterial and antibiofilm agents. In: Prasad R, Siddhardha B, Dyavaiah M (eds) Nanostructures for antimicrobial and antibiofilm applications. Nanotechnology in the life sciences. Springer, Cham, pp 375–397. https://doi.org/10.1007/978-3-030-40337-9_16

Leid JG, Ditto AJ, Knapp A, Shah PN, Wright BD, Blust R et al (2010) Inactivation of bacterial and viral biothreat agents on metallic copper surfaces. J Appl Microbiol 109(2):695–701

Li Y, Leung P, Yao L, Song QW, Newton E (2006) Antimicrobial effect of surgical masks coated with nanoparticles. J Hosp Infect 62(1):58–63

Li WR, Xie XB, Shi QS, Zeng HY, Ou-Yang YS, Chen YB (2010) Antibacterial activity and mechanism of silver nanoparticles on Escherichia coli. Appl Microbiol Biotechnol 85(4):1115–1122

Li Y, Zhang W, Niu J, Chen Y (2012) Mechanism of photogenerated reactive oxygen species and correlation with the antibacterial properties of engineered metaloxide nanoparticles. ACS Nano 6(6):5164–5173

Liu S, Zeng TH, Hofmann M, Burcombe E, Wei J, Jiang R et al (2011) Antibacterial activity of graphite, graphite oxide, graphene oxide, and reduced graphene oxide: membrane and oxidative stress. ACS Nano 5(9):6971–6980

Mataraci E, Dosler S (2005) In vitro antibacterial activity of propolis samples from different geographical origins against certain oral pathogens. Anaerobe 11(5):372–378

Morones JR, Elechiguerra JL, Camacho A, Holt K, Kouri JB, Ramírez JT et al (2005) The bactericidal effect of silver nanoparticles. Nanotechnology 16(10):2346–2353

Murugan K, Senthilkumar B, Senbagam D, Al-Sohaibani S, Chandrasekaran N, Benelli G (2016) Green synthesis of titanium dioxide nanoparticles using Psidium guajava extract and its antibacterial and antioxidant properties. Asian Pac J Trop Med 9(4):386–391

Nguyen TA, Nguyen HT, Vu HA, Le VT, Hoang Thi TT, Dao TK et al (2017) Biocompatible silver nanoparticle/chitosan-coated poly(lactic-co-glycolic acid) nanocomposites for enhanced antibacterial and wound healing applications. Int J Nanomedicine 12:1953–1967

O'Toole GA (2011) Microtiter dish biofilm formation assay. J Vis Exp 47:2437

Pacheco KA, André LC, de Araújo MV, Fraceto LF (2020) Toxicity assessment of nanoparticles in the immune system. Int J Environ Res Public Health 17(21):8022

Padmavathy N, Vijayaraghavan R (2008) Enhanced bioactivity of ZnO nanoparticles—an antimicrobial study. Sci Technol Adv Mater 9(3):035004

Panáček A, Kvítek L, Smékalová M, Večeřová R, Kolář M, Röderová M et al (2018a) Bacterial sensitivity to silver nanoparticles and ionic silver assessed by universal stress protein responses. Nanotoxicology 12(3):318–328

Panáček A, Kvítek L, Smékalová M, Večeřová R, Kolář M, Röderová M et al (2018b) Bacterial resistance to silver nanoparticles and how to overcome it. Nat Nanotechnol 13(1):65–71

Pandey A, Nikam AN, Mutalik SP, Fernandes G, Shreya AB, Padya BS et al (2020) Architectured therapeutic and diagnostic nanoplatforms for combating SARS-CoV-2: role of inorganic, organic, and radioactive materials. ACS Biomater Sci Eng 7(1):31–54

Pinto AM, Gonçalves IC, Magalhães FD, Ferreira RA, Segundo MA, Costa Lima SA (2016) Nanomaterials: advantages and disadvantages in nanomedicine. In: Grumezescu AM (ed) Nanobiomaterials in drug delivery: applications of nanobiomaterials. Elsevier, Amsterdam, pp 55–92

Rabea EI, Badawy ME-T, Stevens CV, Smagghe G, Steurbaut W (2003) Chitosan as antimicrobial agent: applications and mode of action. Biomacromolecules 4(6):1457–1465

Raffi M, Mehrwan S, Bhatti TM, Akhter JI, Hameed A, Yawar W, Ul Hasan MM (2010) Investigations into the antibacterial behavior of copper nanoparticles against Escherichia coli. Ann Microbiol 60(1):75–80

Raghunath A, Perumal E (2017) Metal oxide nanoparticles as antimicrobial agents. In: Thakur VK, Thakur MK, Raghavan P (eds) Handbook of composites from renewable materials, volume 4: functionalization. Wiley, New York, pp 263–289

Rai M, Yadav A, Gade A (2009) Silver nanoparticles as a new generation of antimicrobials. Biotechnol Adv 27(1):76–83

Rai M, Deshmukh S, Ingle A, Gade A (2012) Silver nanoparticles: the powerful nanoweapon against multidrug-resistant bacteria. J Appl Microbiol 112(5):841–852

Rai M, Kon K, Ingle A, Duran N, Galdiero S, Galdiero M (2014) Broad-spectrum bioactivities of silver nanoparticles: the emerging trends and future prospects. Appl Microbiol Biotechnol 98(5):1951–1961

Raliya R, Tarafdar JC (2019) Nanocomposites and their potential applications in agricultural biotechnology. In: Nandwani V (ed) Advances in plant breeding strategies: nut and beverage crops. Springer, Berlin, pp 445–464

Ravindran A, Chandran P, Khan SS (2013) Biofunctionalized silver nanoparticles: advances and prospects. Colloids Surf B Biointerfaces 105:342–352

Ruparelia JP, Chatterjee AK, Duttagupta SP, Mukherji S (2008) Strain specificity in antimicrobial activity of silver and copper nanoparticles. Acta Biomater 4(3):707–716

Safawo T, Sandeep BV, Pola S, Tadesse A (2018) Synthesis and characterization of zinc oxide nanoparticles using tuber extract of anchote (Coccinia abyssinica (lam.) Cong.) for antimicrobial and antioxidant activity assessment. OpenNano 3:56–63

Sanapala P, Pola S (2020) Toxicological evaluation of nanoparticles using prokaryotic model organisms. In: Siddhardha B, Dyavaiah M, Kasinathan K (eds) Model organisms to study biological activities and toxicity of nanoparticles. Springer, Singapore, pp 277–296. https://doi.org/10.1007/978-981-15-1702-0_14

Sheikholeslami S, Mirrahimi M, Kafil HS, Amini M, Moghadam MF (2018) The in vivo antibiofilm effect of gold nanoparticles. Nanomedicine 13(1):89–98
Singh N, Jenkins GJ, Asadi R, Doak SH (2010) Potential toxicity of superparamagnetic iron oxide nanoparticles (SPION). Nano Rev 1(1):5358
Singh AV, Vyas V, Patil R, Sharma V, Scopelliti PE, Bongiovanni R et al (2018) Gold nanoparticles in diagnostics and therapeutics for human cancer. Int J Mol Sci 19(7):1979
Sirelkhatim A, Mahmud S, Seeni A, Kaus NHM, Ann LC, Bakhori SKM et al (2015) Review on zinc oxide nanoparticles: antibacterial activity and toxicity mechanism. Nano-Micro Lett 7(3):219–242
Sondi I, Salopek-Sondi B (2004) Silver nanoparticles as antimicrobial agent: a case study on E. coli as a model for Gram-negative bacteria. J Colloid Interface Sci 275(1):177–182
Sotiriou GA, Pratsinis SE (2010) Antibacterial activity of nanosilver ions and particles. Environ Sci Technol 44(14):5649–5654
Sun B, Long P, Liu D, Zhang X, Zhang L, Zhou J et al (2020) Nanomaterials for regulating the dental biofilm: current understanding and future perspectives. Front Chem 8:834
Ventola CL (2015) The antibiotic resistance crisis: part 1: causes and threats. P T 40(4):277–283
Wang X, Yang L, Li L, Yang Z, Zhang Z, Wu S et al (2017) Investigation of the antibacterial activity and mechanism of action of graphene oxide. J Nanobiotechnol 15(1):34
Wang M, Yu Y, Shao H, Cong H (2020) Mechanical properties of nanocomposites. In: Pielichowski K, Majka TM (eds) Polymer composites with functionalized nanoparticles: synthesis, properties, and applications. Elsevier, Amsterdam, pp 61–98
Wu W, Wieckowski S, Pastorin G, Benincasa M, Klumpp C, Briand JP et al (2005) Targeted delivery of amphotericin B to cells by using functionalized carbon nanotubes. Angew Chem Int Ed Engl 44(39):6358–6362
Xue X, Wu J, Liu S, Liu H, Liu Q, Zhou Y et al (2017) Antibacterial activities of zinc oxide nanoparticles against Escherichia coli O157:H7. J Appl Microbiol 122(1):161–170
Zhang L, Jiang Y, Ding Y, Povey M, York D (2007) Investigation into the antibacterial behaviour of suspensions of ZnO nanoparticles (ZnO nanofluids). J Nanopart Res 9(3):479–489
Zhang Y, Kang S, Arslan Z, Mahmoudi M (2018) Insights into the interaction of nanomaterials with blood in the circulatory system. Trends Pharmacol Sci 39(11):1064–1082
Zodrow K, Brunet L, Mahendra S, Li D, Zhang A, Li Q et al (2009a) Polysulfone ultrafiltration membranes impregnated with silver nanoparticles show improved biofouling resistance and virus removal. Water Res 43(3):715–723
Zodrow KR, Schiffman JD, Elimelech M (2009b) Biodegradable polymers decorated with silver nanoparticles as a platform for antimicrobial applications. Nano Lett 9(3):1225–1229

Bacteriophage Therapy to Combat ESKAPE Pathogens

18

Sayak Bhattacharya

Abstract

Anti-microbial resistant ESKAPE pathogens are global threat to human due to their pathogenecity. These pathogens cause a substantial percentage of nosocomial infections in modern hospitals. However, these pathogens show antibiotics resistant property and thus it is a challenging task to eliminate them. Alternative therapy like bacteriophage therapy has drawn attention to treat these pathogens. Bacteriophages are natural killer of bacteria, and they destroy the host cell completely. They are abundant in nature and also effective against drug-resistant pathogens. It had been observed by animal model that either treatment with them or with their enzymes or synergistically with other therapeutic agents shows activity against pathogens. In this chapter, we will review the process of bacteriophage therapy and how they have been used to treat against ESKAPE pathogens.

Keywords

Pathogen · Bacteriophage · Antimicrobial · Nasocomial · Infection

18.1 Introduction

In recent times, 'ESKAPE pathogens' (*Enterococcus faecium*, *Staphylococcus aureus*, *Klebsiella pneumoniae*, *Acinetobacter baumannii*, *Pseudomonas aeruginosa* and *Enterobacter* spp.) drew attention for its escaping property from cidal effect of antibiotics (Rice 2008). They are designated as they 'escape' from the action of antibiotics (Rice 2008). ICUs all around the world have been suffering

S. Bhattacharya (✉)
Department of Microbiology, Bijoy krishna Girls' College, Howrah, West Bengal, India

S. Busi, R. Prasad (eds.), *ESKAPE Pathogens*,
https://doi.org/10.1007/978-981-99-8799-3_18

from crisis as ESKAPE pathogens are untreatable with antibiotics (Pendleton et al. 2013). They cause deadly infections among immune deficient and clinically weak patients (Founou et al. 2018). ESKAPE pathogens are responsible for critical infections and death (Okwu et al. 2019; Taha Yassin et al. 2020). Their multidrug-resistant property (MDRO) has been emerging due to overuse of antibiotics in agriculture and health sector and new antimicrobials have been establishing slowly (Marshall and Levy 2011). Actually, Gram-negative microbes (ESKAPE pathogens) could not be treated with antibiotics as they contain double membrane that protects from antibiotics and toxic compounds (Aitken et al. 2016). It focuses on new paradigms on disease formation and its resistance. Gradually increasing research on mechanism behind disease progression of these pathogens leads to advanced study on new antimicrobial options.

ESKAPE pathogens are very common microorganisms that cause nosocomial infections (Rice 2008). There is high probability of death due to ESKAPE pathogens and thus costs are increasing (Founou et al. 2017). Estimation by the U.S. Centers for Disease Control and Prevention indicated that these antibiotic-resistant microbes cause more than 2 million infections in every year in the U.S., leading to at least 23,000 deaths (CDC 2014).

When microbes encounter antibiotics, naturally antimicrobial resistance property develops. Overuse in environment, agriculture, and health sector causes this resistant (Holmes et al. 2016). Lack of knowledge on how to consume antibiotic also leads to the antimicrobial resistance (Prestinaci et al. 2015). Antimicrobial-resistant microbes lead to around 25,000 deaths every year in the EU and 700,000 worldwide (Cecchini et al. 2015). Due to continuous emergence of antibiotics resistance property, ECDC in Europe and CDC in the USA classified bacteria in three groups like multidrug-resistant, extensively drug-resistant, and pan drug-resistant (Abbas et al. 2017). These pathogens exhibit continuous antibiotic resistance property (Ma et al. 2020). In spite of overuse of antibiotics is a problem, still practice has been continuing and antibiotics were used abruptly. It could be thought that antibiotics use has been drastically reduced by national and international health organizations (Shapiro et al. 2014). ESKAPE pathogens adopt several mechanisms against antibiotics. They modify target site, inactivate the drugs, and reduce drug accumulation (Santajit and Indrawattana 2016). Additionally, these bacteria exhibit challenges to the drug therapy beside their resistance property (Garai et al. 2019). Although these pathogens were recognized as exclusively extracellular bacteria, they now survived in intracellular environment like in macrophages (Parra-Millán et al. 2018). *S. aureus, K. pneumoniae, A. baumannii,* and *P. aeruginosa* often form biofilm on ventilators. This leads to ventilator-associated pneumonia (Jones 2010; Park 2005). ESKAPE pathogens are able to form biofilm and antibiotics could not reach effectively. Moreover, they protect dormant cells that are tolerant to antibiotics (Lewis 2007).

The emergence of pathogens resisting antibiotics is happening due to unlimited use of antibiotics, transfer of horizontal genes and obviously for bacterial evolution. Thus, it is making a worse global health problem (Walsh 2003). Thus, new generation of antimicrobials is required against these pathogens; however, the scope is limited (Walsh 2003). Some effective antibiotics are in process to treat ESKAPE pathogens for clinical usage (Bassetti and Righu 2015). The main threat to the global health is antimicrobial resistance (AMR). These microbes acquire several techniques to do horizontal gene transfer (Giedraitienė et al. 2011). Thus, ESKAPE pathogens have devastating effect on human.

A bacteriophage can target one particular bacterial strain, and, thus it has been used for targeted treatment with minimal side effect (Dufour et al. 2017). There are estimated 10^{32} phages present on Earth, and thus, they are the most diversified biological group (Bobay and Ochman 2018). The use of phages gained momentum in Eastern Europe since their introduction in health care system but limited to rigorous scientific standards (Villarroel et al. 2017). An alternative method is the use of bacteriophage against MDRO infections (Loc-Carrillo and Abedon 2011). Phages are similar to antibiotics in terms of production and mode of action (Regeimbal et al. 2016).

18.2 Pathogenicity of ESKAPE Pathogens

The Infectious Diseases Society of America indicated that the ESKAPE pathogens cause nosocomial infections. One of the major advantages is their multidrug resistance (MDR) property. It can cause a fatal condition in immunocompromised patients (Santajit and Indrawattana 2016). According to the Infectious Diseases Society of America, the dearth of new antibiotics makes it impossible to treat ESKAPE pathogens and advised innovations in 2009 (Boucher et al. 2009). Many new approaches have been tried like antimicrobial peptides, bacteriophages, new antibiotics, and nanoparticles for last 10 years (Mulani et al. 2019). Light also modulates the pathogenicity of these microbes (Tuttobene et al. 2021).

Pathogenicity of each pathogenesis a real concern to humankind. *Klebsiella pneumonia,* one of the pathogens in this group, is of public health concern due to its multidrug-resistant property (WHO 2017). Wyres et al. (2020) reviewed the genomic framework of this pathogen to understand the disease surveillance and management of this pathogen. In Singapore, 20–26% of population showed mortality infected with *K. pneumoniae* (Chew et al. 2017). In China, 11.9% of population admitted in ICU suffered from pneumonia (Zhang et al. 2014). After getting entry into the host, this pathogen stimulate different immune mediators (Piperaki et al. 2017). Cheung et al. (2021) extensively described the pathogenicity of *Staphylococcus aureus*. The main problem associated with *S. aureus* infections is it could not be treated by antibiotics and methicillin-resistant *S. aureus* (MRSA) is most worried clinically (Turner et al. 2019). *S. aureus* can exacerbate skin infections, and it had been shown that after eradication of original pathogen, complication arises (Yeboah-Manu et al. 2013). Kwiecinski and Horswill (2020) reviewed the mechanisms deployed by *S. aureus* for invasive infection. *Acinetobacter baumannii*, another important pathogen, deserve attention by WHO due to its severe pathogenicity (Tacconelli et al. 2018). *A. baumannii* showed antibiotics-resistant property, and thus, it persists in the environment for the long time (De Oliveira et al. 2020). *Enterococcus* is crucial part of human microbiome and they thrive in the GE tract (Rowland et al. 2018). The two most vital *Enterococcus* species are multidrug-resistant *E. faecium* and *E. faecalis* (Zhen et al. 2019). Between these pathogens listed drug-resistant *E. faecium* demands alternative therapies (Santajit and Indrawattana 2016). Another rod-shaped pigment producer of Gram-negative bacteria is *Pseudomonas aeruginosa* which is naturally occurring microbe (Curran et al. 2018; Moradali et al. 2017). The main concern is the deadly disease pneumonia caused by *P. aeruginosa* (Guillamet et al. 2016), sepsis (Britt et al. 2018) and UTI (Jones et al. 2017). Serious complication could be happened in lungs due to the inflammatory response in cystic fibrosis patients (Malhotra et al. 2019). *Enterobacter* spp., another anaerobe rod-shaped bacteria, which causes nosocomial infections like bacteremia, sepsis, pneumonia, urinary tract infection, meningitis, and endocarditis (Davin-Regli et al. 2019). *Enterobacter* isolates produce extended-spectrum β-lactamases (ESBL) as well as carbapenems and that leads to emergence of antibiotic resistance (De Florio et al. 2018). The most common in the hospital environment are *E. cloacae*, *E. aerogenes*, and *E. hormaechei* (Davin-Regli et al. 2019; De Florio et al. 2018). Each of the mentioned microbes has the ability to form biofilm and cause serious infection.

18.3 Novel Treatments Against ESKAPE Pathogens

There are many reports on how to trat the diseases caused by ESKAPE pathogens. Nakonieczna et al. (2019) described the antimicrobial photodynamic inactivation as a therapeutic option against the pathogens. Similarly, it had been observed that natural antimicrobial peptides showed bacteriolytic activity against these pathogens (Mukhopadhyay et al. 2020). Non-thermal plasma treatment has been used against both planktonic and biofilm forms of ESKAPE pathogens and reviewed by Scholtz et al. (2021). Quercetin, a flavonoid isolated from plants, showed anti quorumsensing activity against ESKAPE pathogens (Ghosh et al. 2022). Saha et al. (2023) reviewed the use of CRISPR to treat ESKAPE pathogens. Behroozian et al. (2016) showed the antibacterial action of Kisameet Clay against the ESKAPE pathogens

18.4 Use of Bacteriophages

Bacteriophages attack and kill bacteria. After launch of the new *Bacteriophage* journal in early 2011, Alexander Sulakvelidze described bacteriophages as the most common organism on Earth and it helps to maintain microbial balance on this earth (Sulakvelidze 2011). The continuing emergence of antibiotic-resistant pathogens poses a serious threat like MRSA infection causes 20,000 deaths in 2005 in the US (Klevens et al. 2007). Phage vaccine was also used against *Plasmodium falciparum* in a rabbit model in 1988 (De la Cruz et al. 1988). Recently described bacteriophage therapy is going to be popular due to antibiotic-resistant property by microbes (Gordillo Altamirano and Barr 2019). Phage therapy is an alternative method against MRSA infection (Kutter et al. 2010). Bacteriophages are widely used in Eastern

European countries and antibiotics replaces them in Western countries (Kutter et al. 2010). Phage therapy is not only an alternative to antibiotics due to its cidal property, but also it modify bacterial cells chemically and physically (Kaur et al. 2021). The success of phage therapy was determined by abrupt use of antibiotics. Synergistic effects of phage-antibiotics therapy showed anti-bacterial effect (Jeon and Ahn 2020; Luong et al. 2020; Jennes et al. 2017; Luscher et al. 2020). Phages were used as a therapeutic agent to curb the *E. coli* infections around 1980s (Smith and Huggins 1982; Smith and Huggins 1983; Smith et al. 1987). The use of phages as an antimicrobials is not a new phenomenon and the phages were used as a therapeutic agent in Eastern Europe and the former Soviet Union for a long time (Jeon and Yong 2019). Recently, phages are used to treat the untreatable infections (Monteiro et al. 2019). Phages can self-replicate at the infection site, and they are unlike antibiotics and thus resistant property does not happen (Chan et al. 2018). To avoid resistant property, cocktail is constituted with different phages to treat the infections (Khawaldeh et al. 2011; Schooley et al. 2017). Some minimum regulatory conditions like strict lytic behavior of phages, antimicrobial activity on host microbes and elimination of contaminations are required for effective bacteriophage therapy (Young and Gill 2015). Additionally, knowledge regarding phage receptor should be known clearly as it gives clue on phage resistance and evolution trade off (Young and Gill 2015). Now a days, modified phages have been discovered to enhance the circulation of phages in newly infected mice (Kim et al. 2008). Host ranges of phages have been increased by genetic engineering process (Pouillot et al. 2010). However, a framework is needed for phage therapy and to gain widespread acceptance, it has to pass regulatory hurdles.

Phage therapy has several advantages. One of the major point is auto dose (Abedon and Thomas-Abedon 2010; Capparelli et al. 2010). Phages control themselves, their number depends on number of the hosts, and thus, it is called auto dose as they decide the dose (Abedon and Thomas-Abedon 2010).

As ESKAPE pathogens could not be treated effectively by the antibiotics, and thus the most abundant bacteriophages are needed in the treatment process (Srinivasiah et al. 2008).

18.4.1 Bacteriophage Therapy Against *Enterococcus faecium*

The eradication of *E. faecalis* poses a serious challenge in past few days (Guzman Prieto et al. 2016). The emergence of antibiotic-resistant bacterial strains lead to the failure of conventional treatments, and thus, alternative methods are necessitated. It was observed that *E. faecalis* lytic phage EFDG1 genome does not possess any virulent genes like other phages (Li et al. 2014). It had been concluded that phages if used safely could be best alternative specially when biofilm formed by multidrug-resistant microorganisms (Knoll and Mylonakis 2014). It had been observed that genetically engineered fEf11/fFL1C(D36,)^{PnisA} was effective against *Enterococcus* which is vancomycin resistant (Tinoco et al. 2016). Two bacteriophages, SSsP-1 bacteriophage belonged to the *Saphexavirus* genus, and the GVEsP-1 bacteriophage belonged to the *Schiekvirus* genus, carry putative components of anti-CRISPR systems and were effective against *Enterococcus* (Tkachev et al. 2022).

18.4.2 Bacteriophage Therapy Against *Acinetobacter baumannii*

Acinetobacter baumannii strains show resistance property against several drugs during infections, and thus, it creates a global health problem (Karageorgopoulos and Falagas 2008). Due to multidrug-resistant property, it is really a challenging job to treat the infections and thus it ultimately leads to death (Landman et al. 2008). The application of therapeutics in animal models narrow down the connection between in vitro and clinical studies (Kusradze et al. 2016). *A. baumannii* showed antibiotics resistant property against carbapenem (Queenan and Bush 2007). Tigecycline, a potential antibiotics, its therapy caused superinfection and emergence of resistance (Karageorgopoulos et al. 2008). Bacteriophage therapy sheds the light on the difficult treatment conditions of multidrug-resistant bacterial infections (Burrowes et al. 2011; Lyon 2017). Wang et al. (2018) demonstrated the efficiency of bacteriophage against multidrug-resistant *A. baumannii* in mouse model. It had been shown that lytic phage SH-Ab15599 was capable to infect carbapenem resistant *A. baumannii* in mouse model (Hua et al. 2018). Phages were applied to COVID 19 infected patients to treat secondary drug-resistant bacterial infection (Wu et al. 2021).

18.4.3 Bacteriophage Therapy Against *Klebsiella pneumonia*

Klebsiella pneumoniae is a most important multidrug-resistant nosocomial pathogen due to its biofilm making capability, to escape from immune system, and encoded antibiotic degrading enzymes like carbapenemases (De Oliveira et al. 2020). It had been observed that antibiotic therapy along with treatment by phages showed more efficacy against *K. pneumonia* (Eskenazi et al. 2022). Five phages showed efficacy against *K. pneumoniae* in infected mouse (Kumari et al. 2009). Dhungana et al. (2021) showed the antibacterial potential of locally isolated lytic phages against *K. pneumoniae* in mouse. When two phages are administered together, it showed better result than single phage in mice infected with *K. pneumoniae* ST258 (Hesse et al. 2021).

18.4.4 Bacteriophage Therapy Against *Pseudomonas aeruginosa*

Huge number of *P. aeruginosa* infecting bacteriophages were notified and all the genomes are stored in the NCBI databases (Ackermann 2007). By bioluminescent bacteria, authors compared several phages and classified to use as a therapeutic agent (Debarbieux et al. 2010). Chegini et al. (2020) reviewed bacteriophage therapy against *P. aeruginosa*. It had been observed that bacteriophages and antibiotics together lead to destroy *P. aeruginosa* PA14 biofilm (Chaudhry et al. 2017). Similarly, when phages were delivered with ciprofloxacin, bacterial load was reduced 10,000-fold but not when treated separately (Oechslin et al. 2017). Moreover, it had been noted that these effect on *P. aeruginosa* occurred synergistically (Oechslin et al. 2017). The treatment of MDR *P. aeruginosa* joint infection was carried out with the combination of systemic antibiotics and local phage (Tkhilaishvili et al. 2020). Phage cocktail with antibiotics showed effectiveness against pandrug-resistant *P. aeruginosa* spinal abscess (Ferry et al. 2022). Phage cocktails (Psu1, Psu2 and Psu3) showed its activity against colistin resistant *P. aeruginosa* and totally destroyed them (Shokri et al. 2017).

18.4.5 Bacteriophage Therapy Against *Staphylococcus aureus*

Multidrug-resistance property of *Staphylococcus* makes it challenging to treat staphylococcal infection (Raju et al. 2010). Joseph Maisin and Richard Bruynoghe used bacteriophage against staphylococcal skin infections and they observed improvement within 24–48 h (Lavigne and Robben 2012). Sunagar et al. (2010) showed that phages show effectiveness against Staphylococcal infection in mice without antibiotics. Myoviridae bacteriophages (AB-SA01) was injected in 13 patients to treat Staphylococcal infection and no adverse effect has been reported though further research is required to check the efficacy (Fabijan et al. 2020). Phage cocktail was also used to treat Staphylococcal infection (Alves et al. 2014) (Table 18.1).

Table 18.1 List of bacteriophages identified against various bacterial pathogens

Sr. no.	Pathogen	Bacteriophage
1	*Enterococcus faecium*	Lytic phage EFDG1 (Li et al. 2014)
2	*Staphylococcus aureus*	Phage S25-3 and S13′ (Takemura-Uchiyama et al. 2013)
3	*Staphylococcus aureus*	Phages φMR-5 and φMR-10 (Chhibber et al. 2018)
4	*Pseudomonas aeruginosa*	Phage M-1 (Adnan et al. 2020)
5	*Acinetobacterbaumanni*	Lytic phage IsfAB78 (Ebrahimi et al. 2021)
6	*Acinetobacter baumannii*	Phage ϕAB2 (Lin et al. 2010)
7	*Staphylococcus aureus*	Phage DRA88 and Phage K (Alves et al. 2014)
8	*Staphylococcus aureus*	Bacteriophages AB-SA01 (Fabijan et al. 2020)
9	*Klebsiella pneumoniae*	Bacteriophage KpV74 (Volozhantsev et al. 2022)
10	*Klebsiella pneumoniae*	Bacteriophage vB_Kpn_3 (Habibinava et al. 2022)
11	*Klebsiella pneumoniae*	Phage IME184 (Li et al. 2022)
12	*Klebsiella pneumoniae*	Bacteriophage IME268 (Nazir et al. 2022)

18.5 Adverse Effects of Bacteriophage Therapy

The main disadvantage during lysis of bacterial cell is the release of endotoxins as well as superantigens that may cause inflammatory response and as a result morbidity could happen. So, phages could be genetically engineered, so that it will be non-replicative (Hagens and Blasi 2003). Genetically engineered phages contain restriction endonucleases and holins that express during infection in *E. coli* (Hagens and Blasi 2003). These phages secrete very low amount of endotoxins and but cell lysis did not happen though they were toxic (Hagens et al. 2004).

18.6 The Hurdle of Bacteriophage Delivery

The main challenge of phage delivery is the immune system. Thus, non-systemic diseases drew attentions (Sulakvelidze et al. 2001). Jerne (1956) described phages are immunogenic and thus antibody prevents them to show anti-bacterial activity. The efficiency of phages is dependent on immune response against the administered phages (Aslam et al. 2020). Phages are allowed to passage serially through animals and mutants which are present for longer periods in vivo will be isolated (Merril et al. 1996). Mutants escape from the immune system due to its mutations in the head protein (Geier et al. 1973). Several technologies have been deployed to phage delivery and among them polyethylene glycol was used to conjugate with the phages (Kim et al. 2008). Labrie et al. (2010) already reviewed on how microbes evolve antiviral strategies against phages. The animal model could be used to make bridge between in vivo and in vitro study of phage therapy but therapy is limited to chronic infections whereas animal models are used to examine on acute infections (Kortright et al. 2019). While phage therapy has been used now a days, the main challenging matter is delivery of high titers of phages to the lungs and maintenance of its morphology and functionality. The failure of burn treatment by phages highlights the requirement for a bio material based platform and stability of phages is required at the infection site (Kortright et al. 2019). Liquid formation of phages by nebulizer resulted in reduction of phage titres (Jault et al. 2019).

Some phages infect a broad range of host cells whereas other phages infect narrow range of host cells and the specificity of phage receptors determines the host range. The main problem is when phages show narrow host range and it will be difficult to treat polymicrobial biofilms, which are formed on medical devices (Donlan 2009; Donlan and Costerton 2002). Thus, it will be better to choose single phage which destroys extracellular polymer and act against polymicrobial biofilms.

Phages interact with some components reversibly except cell wall component. Phages cannot get the entry if cell wall are absent but if capsule or teichoic acid is present, it can get entry into the cells (Pires et al. 2017; Rakhuba et al. 2010). Hu et al. (2010) studied and suggested the nature of phages is similar to antibiotics and both they penetrate less in densely packed biofilm forming cells.

Lytic phages are the only option for therapy (Górski et al. 2009). The standard method to prepare phage cocktail exists but there is no clear guidelines available

(Pirnay et al. 2011). Another difficulty in the preparation of a stable virion is the damage caused by various external and physical factors (Jończyk et al. 2011). Phage lytic proteins (endolysin) could be used as a new antimicrobials (Fischetti 2006).

Another factor that creates problem is the transduction method where phages can transfer apart of genome from one microorganism to another and thus new pathogen could be created (Brabban et al. 2005; Maiques et al. 2007). The maximum part of phage genome is unknown or unidentified (Hatfull 2008).

As phages are totally dependent on host cells, it is really a major challenge to treat biofilm forming cells as these types of cells showed less metabolic activity due to under nutrient conditions (Yonezawa et al. 2015; Azeredo and Sutherland 2008; Łoś et al. 2007). Phages infect efficiently in planktonic state of bacteria than biofilm state because in planktonic state, phages attack directly on them (Sillankorva et al. 2004) (Fig. 18.1). Slower lysis of bacterial cells is observed in biofilm forming cells than planktonic cells (Cerca et al. 2007). Actually, cells show more sensitivity in exponential growth phase unlike the stationary growth phase in planktonic cells (Cerca et al. 2007).

Another challenge is the lack of clinical trials. There are several unique properties like dosage of phages should be placed into consideration (Payne and Janser 2003). Sometimes, immunity is generated against phages and thus more studies are needed to understand it fully like generation of adverse effects (Krut and Bekeredjian-Ding 2018).

It could be considered that humans encounter phages, and thus, humans could not show adverse reactions against phages. However, there are several factors that need to be considered like purity of phage preparation. Parracho et al. (2012) already set the quality factors that should be recommended to produce phages. Another crucial factor is toxic shock effect as a result of bactericidal effect of phages (Speck and Smithyman 2016). A advanced understanding is needed to examine the phage distribution to maximize its effect (Levin and Bull 2004; Abedon 2014).

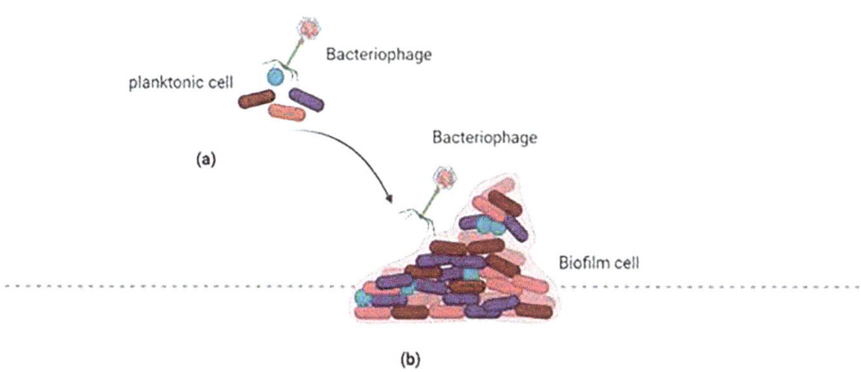

Fig. 18.1 Bacteriophage behavior with (a) planktonic cells (b) biofilm forming cells. From the above picture, it is certainly clear that phages could not get access easily into the cells of biofilm due to polymeric substance

Some problems remain during phage therapy, and these problems need to be solved: (a) phages could be inactivated by neutralizing antibodies and allergic reaction could be generated, (b) mutations could be raised and phages will be inactivated, and (c) phages can capture toxic genes generally present in pathogens (Matsuzaki et al. 2005). Thus, phages should be engineered such a way that they cannot package extra host DNA (Górski et al. 2009).

18.7 Future of Bacteriophage Therapy

As there are many advantages of bacteriophage therapy against the failing antibiotics, scientists worldwide are using phages (Khalifa et al. 2016). Phage therapy was first coined in France by Felix d'Herelle in 1919 (Summers 1999) during the treatment of children suffering from acute dysentery. Afterwards, many nations adopt phage therapy (Kutter et al. 2015). The standardization of key components of therapy by phages is carried out and published (Fauconnier 2018), and it was already suggested to use three to five phages at high titre values with unique properties, are most effective and guaranteed (Chan et al. 2013; Carlson 2005). It was noted that alone phages could not function efficiently rather than they require antibiotics (Kortright et al. 2019). It had been reported that phages showed effectiveness synergistically against planktonic cells (Nouraldin et al. 2016; Jansen et al. 2018; Yazdi et al. 2018) and old biofilms (Bedi et al. 2009; Chaudhry et al. 2017; Akturk et al. 2019), whereas limited success was observed in the individual treatments. Though phages can interact with bacteria directly, their study is deprived due to lack of bridge between phenotypic and genotypic data (Lamy-Besnier et al. 2021). The regulatory framework of phage products and delivery should be controlled strictly for safe outcome (Chan et al. 2013).

Phages could be applied to develop vaccine and to target cancer cells as well as delivery agents of drugs (Aghebati-Maleki et al. 2016; Hess and Jewell 2020). Phages could be used in other fields like agriculture, animal husbandry, and aquaculture but still additional study is needed for effective application and outcome (Mutalik and Arkin 2022). Phage DNA act as a protective antigens in the form of DNA vaccine. Cancer cells could be treated by phages as an anti-cancer agent and bioplanning method is required to select effective phages (Gray and Brown 2014; Abbineni et al. 2010). Intravenous delivery of phages has been successful to treat Staphylococcal infection and septic shock and no adverse effect was observed. Thus, phages could be administered in an effective way (Fabijan et al. 2020). Adhikari and Acharya (2020) reviewed the efficacy of phages against antibiotic resistant microbial strains but their effectiveness could be studied as it depends on many factors like host cell and environment. Though the actual pathways still need to be established, pre-clinical and clinical trials (Duan et al. 2019) could be carried out to determine effectiveness of phage therapy to specifically target *E. faecalis* and *K. pneumoniae* in the gut microbiome to reduce liver disease progression (Hsu et al. 2021; U.S. National Library of Medicine 2022).

To increase the activity, enzymes could be delivered with phages like depolymerases, to improve their anti-biofilm activity (Gutiérrez et al. 2015). Genetic engineering approach could be adopted to develop phage therapy outcomes (Kilcher et al. 2018). A major goal is to control of a phage's host range in phage therapy. Phages should be with desirable properties. This property makes *Pseudomonas* OMKO1 as a suitable therapeutic interest. Recombinant LysSS along with an outer membrane permeabiliser showed activity against MDR microbes (Kim et al. 2020). In conclu sion, by genetic engineering, antimicrobial properties could be improved. There are more commercial values in the phages which are genetically engineered than naturally available phages (Kortright et al. 2022).

The availability of phage therapy worldwide depends on a global plan. Collaboration between countries is required to overcome the regulatory hurdles. Similarly, collaboration between scientists and researchers are very much needed for advances in this field (Pires et al. 2020). The regulation in each country of Europe is different but recently it has been trying to bring more systematic framework among countries (Verbeken and Pirnay 2022). There is no standard protocol available for phage therapy in the human body except Gorgia where phage therapy available in the healthcare systems (Kutateladze 2015). Some institutes specifically George Eliva Institute has phage preparation, and they supply to medical practitioners (Kutateladze 2015). However, phages still are under a unique regulatory agenda.

References

Abbas M, Paul M, Huttner A (2017) New and improved? A review of novel antibiotics for Gram-positive bacteria. Clin Microbiol Infect 23(10):697–703

Abbineni G, Modali S, Safiejko-Mroczka B, Petrenko VA, Mao C (2010) Evolutionary selection of new breast cancer cell-targeting peptides and phages with the cell-targeting peptides fully displayed on the major coat and their effects on actin dynamics during cell internalization. Mol Pharm 7(5):1629–1642

Abedon ST (2014) Phage therapy: eco-physiological pharmacology. Scientifica (Cairo) 2014:581639

Abedon ST, Thomas-Abedon C (2010) Phage therapy pharmacology. Curr Pharm Biotechnol 11(1):28–47

Ackermann HW (2007) 5500 Phages examined in the electron microscope. Arch Virol 152(2):227–243

Adhikari N, Acharya KP (2020) Effectiveness of bacteriophage therapy in field conditions and possible future applications. Curr Pharm Biotechnol 21(5):364–373

Adnan M, Shah MRA, Jamal M, Jalil F, Andleeb S, Nawaz MA, Pervez S, Hussain T, Shah I, Imran M (2020) Isolation and characterization of bacteriophage to control multidrug-resistant *Pseudomonas aeruginosa* planktonic cells and biofilm. Biologicals 63:89–96

Aghebati-Maleki L, Bakhshinejad B, Baradaran B, Motallebnezhad M, Aghebati-Maleki A, Nickho H, Yousefi M, Majidi J (2016) Phage display as a promising approach for vaccine development. J Biomed Sci 23(1):66

Aitken SL, Tarrand JJ, Deshpande LM, Tverdek FP, Jones AL, Shelburne SA, Prince RA, Bhatti MM, Rolston KVI, Jones RN, Castanheira M, Chemaly RF (2016) High rates of nonsusceptibility to ceftazidime-avibactam and identification of New Delhi metallo-β-lactamase production in enterobacteriaceae bloodstream infections at a major cancer center. Clin Infect Dis 63(7):954–958

Akturk E, Oliveira H, Santos SB, Costa S, Kuyumcu S, Melo LDR, Azeredo J (2019) Synergistic action of phage and antibiotics: parameters to enhance the killing efficacy against mono and dual-species biofilms. Antibiotics 8(3):103

Alves DR, Gaudion A, Bean JE, Perez Esteban P, Arnot TC, Harper DR, Kot W, Hansen LH, Enright MC, Jenkins ATA (2014) Combined use of bacteriophage K and a novel bacteriophage to reduce Staphylococcus aureus biofilm formation. Appl Environ Microbiol 80(21):6694–6703

Aslam S, Lampley E, Wooten D, Karris M, Benson C, Strathdee S, Schooley RT (2020, September) Lessons learned from the first 10 consecutive cases of intravenous bacteriophage therapy to treat multidrugresistant bacterial infections at a single center in the United States. In Open forum infectious diseases, vol. 7, No. 9. Oxford University Press, US, p ofaa389

Azeredo J, Sutherland IW (2008) The use of phages for the removal of infectious biofilms. Curr Pharm Biotechnol 9(4):261–266

Bassetti M, Righu E (2015) Development of novel antibacterial drugs to combat multiple resistant organisms. Langenbecks ArchSurg 400:153

Bedi MS, Verma V, Chhibber S (2009) Amoxicillin and specific bacteriophage can be used together for eradication of biofilm of *Klebsiella pneumoniae* B5055. World J Microbiol Biotechnol 25:1145–1151

Behroozian S, Svensson SL, Davies J (2016) Kisameet clay exhibits potent antibacterial activity against the ESKAPE pathogens. mBio 7(1):e01842-15

Bobay LM, Ochman H (2018) Biological species in the viral world. Proc Natl Acad Sci U S A 115:6040–6045

Boucher HW, Talbot GH, Bradley JS, Edwards JE, Gilbert D, Rice LB, Scheld M, Spellberg B, Bartlett J (2009) Bad bugs, no drugs: no ESKAPE! An update from the Infectious Diseases Society of America. Clin Infect Dis 48(1):1–12

Brabban AD, Hite E, Callaway TR (2005) Evolution of foodborne pathogens via temperate bacteriophage-mediated gene transfer. Foodborne Pathog Dis 2(4):287–303

Britt NS, Ritchie DJ, Kollef MH, Burnham CA, Durkin MJ, Hampton NB, Micek ST (2018) Importance of site of infection and antibiotic selection in the treatment of carbapenem-resistant *Pseudomonas aeruginosa* sepsis. Antimicrob Agents Chemother 62(4):e02400-17

Burrowes B, Harper DR, Anderson J, Mcconville M, Enright MC (2011) Bacteriophage therapy: potential uses in the control of antibiotic-resistant pathogens. Expert Rev Anti-Infect Ther 9(9):775–785

Capparelli R, Nocerino N, Iannaccone M, Ercolini D, Parlato M, Chiara M et al (2010) Bacteriophage therapy of *Salmonella enterica*: a fresh appraisal of bacteriophage therapy. J Infect Dis 201(1):52–61

Carlson K (2005) Appendix: working with bacteriophages: common techniques and methodological approaches. In: Kutter E, Sulakvelidze A (eds) Bacteriophages: biology and applications. CRC Press, Boca Raton

Cecchini M, Langer J, Slawomirski L (2015) Antimicrobial resistance in G7 countries and beyond: economic issues, policies and options for action. Organisation for Economic Co-operation and Development, Paris

Centers for Disease Control and Prevention (2014) US Department of Health and Human Services. Antibiotic resistance threats in the United States, 2013. Atlanta CDC and HHS

Cerca N, Oliveira R, Azeredo J (2007) Susceptibility of Staphylococcus epidermidis planktonic cells and biofilms to the lytic action of *staphylococcus* bacteriophage K. Lett Appl Microbiol 45(3):313–317

Chan BK, Abedon ST, Loc-Carrillo C (2013) Phage cocktails and the future of phage therapy. Future Microbiol 8(6):769–783

Chan BK, Turner PE, Kim S, Mojibian HR, Elefteriades JA, Narayan D (2018) Phage treatment of an aortic graft infected with *Pseudomonas aeruginosa*. Evol Med Public Health 2018(1):60–66

Chaudhry WN, Concepcion-Acevedo J, Park T, Andleeb S, Bull JJ, Levin BR (2017) Synergy and order effects of antibiotics and phages in killing Pseudomonas aeruginosa biofilms. PLoS One 12(1):e0168615

Chegini Z, Khoshbayan A, Taati Moghadam M, Farahani I, Jazireian P, Shariati A (2020) Bacteriophage therapy against *Pseudomonas aeruginosa* biofilms: a review. Ann Clin Microbiol Antimicrob 19(1):45

Cheung GYC, Bae JS, Otto M (2021) Pathogenicity and virulence of *Staphylococcus aureus*. Virulence 12(1):547–569

Chew KL, Lin RTP, Teo JWP (2017) *Klebsiella pneumoniae* in Singapore: hypervirulent infections and the carbapenemase threat. Front Cell Infect Microbiol 7:515

Chhibber S, Kaur J, Kaur S (2018) Liposome entrapment of bacteriophages improves wound healing in a diabetic mouse MRSA infection. Front Microbiol 9:561

Curran CS, Bolig T, Torabi-Parizi P (2018) Mechanisms and targeted therapies for *Pseudomonas aeruginosa* lung infection. Am J Respir Crit Care Med 197(6):708–727

Davin-Regli A, Lavigne JP, Pagès JM (2019) *Enterobacter* spp.: update on taxonomy, clinical aspects, and emerging antimicrobial resistance. Clin Microbiol Rev 32(4):e00002-19

De Florio L, Riva E, Giona A, Dedej E, Fogolari M, Cella E, Spoto S, Lai A, Zehender G, Ciccozzi M, Angeletti S (2018) MALDI-TOF MS identification and clustering applied to *Enterobacter* species in nosocomial setting. Front Microbiol 9:1885

De la Cruz V, Lal A, McCutchan TF (1988) Immunogenicity and epitope mapping of foreign sequences via genetically engineered filamentous phage. J Biol Chem 263(9):4318–4322

De Oliveira DMP, Forde BM, Kidd TJ, Harris PNA, Schembri MA, Beatson SA, Paterson DL, Walker MJ (2020) Antimicrobial resistance in ESKAPE pathogens. Clin Microbiol Rev 33(3):e00181-19

Debarbieux L, Leduc D, Maura D, Morello E, Criscuolo A, Grossi O, Balloy V, Touqui L (2010) Bacteriophages can treat and prevent *Pseudomonas aeruginosa* lung infections. J Infect Dis 201(7):1096–1104

Dhungana G, Regmi M, Paudel P, Parajuli A, Upadhyay E, Gyawali I, Upreti H, Nepal R, Pradhan P, Malla R (2021) Therapeutic efficacy of bacteriophage therapy to treat carbapenem resistant *Klebsiella pneumoniae* in mouse model. J Nepal Health Res Counc 19(1):76–82

Donlan RM (2009) Preventing biofilms of clinically relevant organisms using bacteriophage. Trends Microbiol 17(2):66–72

Donlan RM, Costerton JW (2002) Biofilms: survival mechanisms of clinically relevant microorganisms. Clin Microbiol Rev 15(2):167–193

Duan Y, Llorente C, Lang S, Brandl K, Chu H, Jiang L, White RC, Clarke TH, Nguyen K, Torralba M et al (2019) Bacteriophage targeting of gut bacterium attenuates alcoholic liver disease. Nature 575(7783):505–511

Dufour N, Delattre R, Ricard JD, Debarbieux L (2017) The lysis of pathogenic *Escherichia coli* by bacteriophages releases less endotoxin than by β-lactams. Clin Infect Dis 64(11):1582

Ebrahimi S, Sisakhtpour B, Mirzaei A, Karbasizadeh V, Moghim S (2021) Efficacy of isolated bacteriophage against biofilm embedded colistin-resistant Acinetobacter baumannii. Gene Rep 22:100984

Eskenazi A, Lood C, Wubbolts J, Hites M, Balarjishvili N, Leshkasheli L, Askilashvili L, Kvachadze L, van Noort V, Wagemans J, Jayankura M, Chanishvili N, de Boer M, Nibbering P, Kutateladze M, Lavigne R, Merabishvili M, Pirnay JP (2022) Combination of pre-adapted bacteriophage therapy and antibiotics for treatment of fracture-related infection due to pandrug-resistant *Klebsiella pneumoniae*. Nat Commun 13(1):302

Fabijan AP, Khalid A, Maddocks S, Ho J, Gilbey T, Sandaradura I et al (2020) Phage therapy for severe bacterial infections: a narrative review. Medical J Aust 212(6):279

Fauconnier A (2018) Guidelines for bacteriophage product certification. Methods Mol Biol 1693:253–268

Ferry T, Kolenda C, Laurent F, Leboucher G, Merabischvilli M, Djebara S, Gustave CA, Perpoint T, Barrey C, Pirnay JP, Resch G (2022) Personalized bacteriophage therapy to treat pandrug-resistant spinal *Pseudomonas aeruginosa* infection. Nat Commun 13(1):4239

Fischetti VA (2006) Using phage lytic enzymes to control pathogenic bacteria. BMC Oral Health 6:S16

Founou RC, Founou LL, Essack SY (2017) Clinical and economic impact of antibiotic resistance in developing countries: a systematic review and meta-analysis. PloS one 12(12):e0189621
Founou RC, Founou LL, Essack SY (2018) Extended spectrum beta-lactamase mediated resistance in carriage and clinical Gram-negative ESKAPE bacteria: a comparative study between a district and tertiary hospital in South Africa. Antimicrob Resist Infect Control 7:134
Garai P, Berry L, Moussouni M, Bleves S, Blanc-Potard A-B (2019) Killing from the inside: intracellular role of T3SS in the fate of *Pseudomonas aeruginosa* within macrophages revealed by mgtC and oprF mutants. PLoS Pathog 15:e1007812
Geier MR, Trigg ME, Merril CR (1973) Fate of bacteriophage lambda in non-immune germ-free mice. Nature 246:221–223
Ghosh S, Lahiri D, Nag M, Dey A, Pandit S, Sarkar T et al (2022) Phytocompound mediated blockage of quorum sensing cascade in ESKAPE pathogens. Antibiotics 11(1):61
Giedraitienė A, Vitkauskienė A, Naginienė R, Pavilonis A (2011) Antibiotic resistance mechanisms of clinically important bacteria. Medicina 47:137–146
Gordillo Altamirano FL, Barr JJ (2019) Phage therapy in the postantibiotic era. Clin Microbiol Rev 32(2):e00066-18
Górski A, Miedzybrodzki R, Borysowski J, Weber-Dabrowska B, Lobocka M, Fortuna W, Letkiewicz S, Zimecki M, Filby G (2009) Bacteriophage therapy for the treatment of infections. Curr Opin Investig Drugs 10:766–774
Gray BP, Brown KC (2014) Combinatorial peptide libraries: mining for cell-binding peptides. Chem Rev 114:1020–1081
Guillamet CV, Vazquez R, Noe J, Micek ST, Kollef MH (2016) A cohort study of bacteremic pneumonia: The importance of antibiotic resistance and appropriate initial therapy? Medicine 95(35)
Gutiérrez D, Briers Y, Rodríguez-Rubio L et al (2015) Role of the pre-neck appendage protein (Dpo7) from phage vB_SepiS-phiIPLA7 as an anti-biofilm agent in staphylococcal species. Front Microbiol 6:1315
Guzman Prieto AM, van Schaik W, Rogers MR, Coque TM, Baquero F, Corander J et al (2016) Global emergence and dissemination of *enterococci* as nosocomial pathogens: attack of the clones? Front Microbiol 7:788
Habibinava F, Soleimani M, Sabouri S, Zargar M, Zolfaghari MR (2022) Isolating and sequencing vB_Kpn_3, a lytic bacteriophage against multidrug-resistant *Klebsiella pneumoniae*. Future Microbiol 17:235–249
Hagens S, Blasi U (2003) Genetically modified filamentous phage as bactericidal agents: a pilot study. Lett Appl Microbiol 37:318–323
Hagens S, Habel A, von Ahsen U, von Gabain A, Blasi U (2004) Therapy of experimental pseudomonas infections with a nonreplicating genetically modified phage. Antimicrob Agents Chemother 48:3817–3822
Hatfull GF (2008) Bacteriophage genomics. Curr Opin Microbiol 11:447–453
Hess KL, Jewell CM (2020) Phage display as a tool for vaccine and immunotherapy development. Bioeng Transl Med 5:e10142
Hesse S, Malachowa N, Porter AR, Freedman B, Kobayashi SD, Gardner DJ, Scott DP, Adhya S, DeLeo FR (2021) Bacteriophage treatment rescues mice infected with multidrug-resistant *Klebsiella pneumoniae* ST258. mBio 12(1):e00034-21
Holmes AH, Moore LS, Sundsfjord A, Steinbakk M, Regmi S, Karkey A, Guerin PJ, Piddock LJ (2016) Understanding the mechanisms and drivers of antimicrobial resistance. Lancet 387(10014):176–187
Hsu CL, Duan Y, Fouts DE, Schnabl B (2021) Intestinal virome and therapeutic potential of bacteriophages in liver disease. J Hepatol 75:1465–1475
Hu J, Miyanaga K, Tanji Y (2010) Difusion properties of bacteriophages through agarose gel membrane. Biotechnol Prog 26:1213–1221
Hua Y, Luo T, Yang Y, Dong D, Wang R, Wang Y et al (2018) Phage therapy as a promising new treatment for lung infection caused by carbapenem-resistant Acinetobacter baumannii in mice. Front Microbiol 8:2659

Jansen M, Wahida A, Latz S et al (2018) Enhanced antibacterial effect of the novel T4-like bacteriophage KARL-1 in combination with antibiotics against multi-drug resistant Acinetobacter baumannii. Sci Rep 8:1–12

Jault P, Leclerc T, Jennes S, Pirnay JP, Que Y-A, Resch G, Rousseau AF, Ravat F, Carsin H, Le Floch R, Schaal JV, Soler C, Fevre C, Arnaud I, Bretaudeau L, Gabard J (2019) Efficacy and tolerability of a cocktail of bacteriophages to treat burn wounds infected by Pseudomonas aeruginosa (PhagoBurn): a randomised, controlled, double-blind phase 1/2 trial. Lancet Infect Dis 19:35

Jennes S, Merabishvili M, Soentjens P, Pang KW, Rose T, Keersebilck E, Soete O, François P-M, Teodorescu S, Verween G et al (2017) Use of bacteriophages in the treatment of colistin-only-sensitive *Pseudomonas aeruginosa* septicaemia in a patient with acute kidney injury—a case report. Crit Care 21:129

Jeon G, Ahn J (2020) Assessment of phage-mediated inhibition of Salmonella Typhimurium treated with sublethal concentrations of ceftriaxone and ciprofloxacin. FEMS Microbiol Lett 367:fnaa159

Jeon J, Yong D (2019) Two novel bacteriophages improve survival in Galleria mellonella infection and mouse acute pneumonia models infected with extensively drug-resistant Pseudomonas aeruginosa. Appl Environ Microbiol 85:e02900-02918

Jerne NK (1956) The presence in normal serum of specific antibody against bacteriophage T4 and its increase during the earliest stages of immunization. J Immunol 76:209–216

Jończyk E, Kłak M, Międzybrodzki R, Górski A (2011) The influence of external factors on bacteriophages—review. Folia Microbiol (Praha) 56:191–200

Jones RN (2010) Microbial etiologies of hospital-acquired bacterial pneumonia and ventilator-associated bacterial pneumonia. Clin Infect Dis 51:S81–S87

Jones BM, Smith B, Bland CM (2017) Use of continuous-infusion ceftolozane/tazobactam in a multidrug-resistant *Pseudomonas aeruginosa* urinary tract infection in the outpatient setting. Ann Pharmacother 51:715–716

Karageorgopoulos DE, Falagas ME (2008) Current control and treatment of multidrug-resistant *Acinetobacter baumannii* infections. Lancet Infect Dis 8(12):751–762

Karageorgopoulos DE, Kelesidis T, Kelesidis I, Falagas ME (2008) Tigecycline for the treatment of multidrug-resistant (including carbapenem-resistant) *Acinetobacter* infections: a review of the scientific evidence. J Antimicrob Chemother 62(1):45–55

Kaur G, Agarwal R, Sharma RK (2021) Bacteriophage therapy for critical and high-priority antibiotic-resistant bacteria and phage cocktail-antibiotic formulation perspective. Food Environ Virol 13:433–446

Khalifa L, Shlezinger M, Beyth S, Houri-Haddad Y, Coppenhagen-Glazer S, Beyth N, Hazan R (2016) Phage therapy against *Enterococcus faecalis* in dental root canals. J Oral Microbiol 8(1):32157

Khawaldeh A, Morales S, Dillon B, Alavidze Z, Ginn AN, Thomas L, Chapman SJ, Dublanchet A, Smithyman A, Iredell JR (2011) Bacteriophage therapy for refractory *Pseudomonas aeruginosa* urinary tract infection. J Med Microbiol 60(Pt 11):1697–1700

Kilcher S, Studer P, Muessner C, Klumpp J, Loessner MJ (2018) Cross-genus rebooting of custom-made, synthetic bacteriophage genomes in L-form bacteria. Proc Natl Acad Sci U S A 115(3):567–572

Kim KP, Cha JD, Jang EH, Klumpp J, Hagens S, Hardt WD, Lee KY, Loessner MJ (2008) PEGylation of bacteriophages increases blood circulation time and reduces T-helper type 1 immune response. Microb Biotechnol 1(3):247–257

Kim S, Lee DW, Jin JS, Kim J (2020) Antimicrobial activity of LysSS, a novel phage endolysin, against *Acinetobacter baumannii* and *Pseudomonas aeruginosa*. J Glob Antimicrob Resist 22:32–39

Klevens RM, Morrison MA, Nadle J, Petit S, Gershman K, Ray S, Harrison LH, Lynfield R, Dumyati G, Townes JM et al (2007) Invasive methicillin-resistant *Staphylococcus aureus* infections in the United States. JAMA 298:1763–1771

Knoll BM, Mylonakis E (2014) Antibacterial bioagents based on principles of bacteriophage biology: an overview. Clin Infect Dis 58:528–534

Kortright KE, Chan BK, Koff JL, Turner PE (2019) Phage therapy: a renewed approach to combat antibiotic-resistant bacteria. Cell Host Microbe 25(2):219–232

Kortright KE, Chan BK, Evans BR, Turner PE (2022) Arms race and fluctuating selection dynamics in *Pseudomonas aeruginosa* bacteria coevolving with phage OMKO1. J Evol Biol 35:1475–1487

Krut O, Bekeredjian-Ding I (2018) Contribution of the immune response to phage therapy. J Immunol 200(9):3037–3044

Kumari S, Harjai K, Chhibber S (2009) Efficacy of bacteriophage treatment in murine burn wound infection induced by *Klebsiella pneumoniae*. J Microbiol Biotechnol 19(6):622–628

Kusradze I, Karumidze N, Rigvava S, Dvalidze T, Katsitadze M, Amiranashvili I, Goderdzishvili M (2016) Characterization and testing the efficiency of Acinetobacter baumannii phage vB-GEC_Ab-M-G7 as an antibacterial agent. Frontiers Microbiol 7:1590

Kutateladze M (2015) Experience of the Eliava Institute in bacteriophage therapy. Virol Sin 30(1):80–81

Kutter E, De Vos D, Gvasalia G, Alavidze Z, Gogokhia L, Kuhl S, Abedon ST (2010) Phage therapy in clinical practice: treatment of human infections. Curr Pharm Biotechnol 11(1):69–86

Kutter EM, Kuhl SJ, Abedon ST (2015) Re-establishing a place for phage therapy in Western medicine. Future Microbiol 10(5):685–688

Kwiecinski JM, Horswill AR (2020) *Staphylococcus aureus* bloodstream infections: pathogenesis and regulatory mechanisms. Curr Opin Microbiol 53:51–60

Labrie S, Samson J, Moineau S (2010) Bacteriophage resistance mechanisms. Nat Rev Microbiol 8:317–327

Lamy-Besnier Q, Brancotte B, Ménager H, Debarbieux L (2021) Viral host range database, an online tool for recording, analyzing and disseminating virus-host interactions. Bioinformatics 37:2798–2801

Landman D, Georgescu C, Martin DA, Quale J (2008) Polymyxins revisited. Clin Microbiol Rev 21(3):449–465

Lavigne R, Robben J (2012) Professor Dr. Richard Bruynoghe: a 1951 overview of his bacteriophage research spanning three decades. Bacteriophage 2:1–4

Levin BR, Bull JJ (2004) Population and evolutionary dynamics of phage therapy. Nat Rev Microbiol 2(2):166–173

Lewis K (2007) Persister cells, dormancy and infectious disease. Nat Rev Microbiol 5(1):48–56

Li X, Ding P, Han C, Fan H, Wang Y, Mi Z, Feng F, Tong Y (2014) Genome analysis of *Enterococcus faecalis* bacteriophage IME-EF3 harboring a putative metallo-beta-lactamase gene. Virus Genes 49(1):145–151

Li F, Tian F, Nazir A, Sui S, Li M, Cheng D, Nong S, Ali A, KaKar MU, Li L, Feng Q, Tong Y (2022) Isolation and genomic characterization of a novel Autographiviridae bacteriophage IME184 with lytic activity against *Klebsiella pneumoniae*. Virus Res 319:198873

Lin NT, Chiou PY, Chang KC, Chen LK, Lai MJ (2010) Isolation and characterization of ϕAB2: a novel bacteriophage of *Acinetobacter baumannii*. Res Microbiol 161(4):308–314

Loc-Carrillo C, Abedon ST (2011) Pros and cons of phage therapy. Bacteriophage 1(2):111–114

Łoś M, Golec P, Łoś JM, Węglewska-Jurkiewicz A, Czyż A, Węgrzyn A, Węgrzyn G, Neubauer P (2007) Effective inhibition of lytic development of bacteriophages λ, P1 and T4 by starvation of their host, *Escherichia coli*. BMC Biotechnol 7:13

Luong T, Salabarria A-C, Roach DR (2020) Phage therapy in the resistance era: where do we stand and where are we going? Clin Ther 42(9):1659–1680

Luscher A, Simonin J, Falconnet L, Valot B, Hocquet D, Chanson M, Resch G, Köhler T, Van Delden C (2020) Combined bacteriophage and antibiotic treatment prevents *Pseudomonas aeruginosa* infection of wild type and cftr-epithelial cells. Front Microbiol 11:1947

Lyon J (2017) Phage therapy's role in combating antibiotic-resistant pathogens. JAMA 318(18):1746–1748

Ma Y, Wang C, Li Y, Li J, Wan Q, Chen J, Tay FR, Niu LN (2020) Considerations and caveats in combating ESKAPE pathogens against nosocomial infections. Adv Sci 7(1):1901872

Maiques E, Ubeda C, Tormo MA, Ferrer MD, Lasa I, Novick RP, Penadés JR (2007) Role of staphylococcal phage and SaPI integrase in intra- and interspecies SaPI transfer. J Bacteriol 189(15):5608–5616

Malhotra S, Hayes D Jr, Wozniak DJ (2019) Cystic fibrosis and Pseudomonas aeruginosa: the hostmicrobe interface. Clin Microbiol Rev 32(3):10–1128

Marshall BM, Levy SB (2011) Food animals and antimicrobials: impacts on human health. Clin Microbiol Rev 24(4):718–733

Matsuzaki S, Rashel M, Uchiyama J, Sakurai S, Ujihara T, Kuroda M, Ikeuchi M, Tani T, Fujieda M, Wakiguchi H, Imai S (2005) Bacteriophage therapy: a revitalized therapy against bacterial infectious diseases. J Infect Chemother 11(5):211–219

Merril CR, Biswas B, Carlton R, Jensen NC, Creed GJ, Zullo S, Adhya S (1996) Long-circulating bacteriophage as antibacterial agents. Proc Natl Acad Sci U S A 93(8):3188–3192

Monteiro R, Pires DP, Costa AR, Azeredo J (2019) Phage therapy: going temperate? Trends Microbiol 27(4):368–378

Moradali MF, Ghods S, Rehm BH (2017) *Pseudomonas aeruginosa* lifestyle: a paradigm for adaptation, survival, and persistence. Front Cell Infect Microbiol 7:39

Mukhopadhyay S, Bharath Prasad AS, Mehta CH, Nayak UY (2020) Antimicrobial peptide polymers: no escape to ESKAPE pathogens—a review. World J Microbiol Biotechnol 36(9):1–14

Mulani MS, Kamble EE, Kumkar SN, Tawre MS, Pardesi KR (2019) Emerging strategies to combat ESKAPE pathogens in the era of antimicrobial resistance: a review. Front Microbiol 10:539–562

Mutalik VK, Arkin AP (2022) A phage foundry framework to systematically develop viral countermeasures to combat antibiotic-resistant using an ensemble machine learning approach. Viruses 25(4):342

Nakonieczna J, Wozniak A, Pieranski M, Rapacka-Zdonczyk A, Ogonowska P, Grinholc M (2019) Photoinactivation of ESKAPE pathogens: overview of novel therapeutic strategy. Future Med Chem 11(5):443–461

Nazir A, Qi C, Shi N, Gao X, Feng Q, Qing H, Li F, Tong Y (2022) Characterization and genomic analysis of a novel Drexlervirial bacteriophage IME268 with lytic activity against *Klebsiella pneumoniae*. Infect Drug Resist 15:1533–1546

Nouraldin AAM, Baddour MM, Harfoush RAH, Essa SAM (2016) Bacteriophage-antibiotic synergism to control planktonic and biofilm producing clinical isolates of *Pseudomonas aeruginosa*. Alexandria J Med 52(2):99–105

Oechslin F, Piccardi P, Mancini S, Gabard J, Moreillon P, Entenza JM, Resch G, Que YA (2017) Synergistic interaction between phage therapy and antibiotics clears *Pseudomonas aeruginosa* infection in endocarditis and reduces virulence. J Infect Dis 215(5):703–712

Okwu MU, Olley M, Akpoka AO, Izevbuwa OE (2019) Methicillin-resistant *Staphylococcus aureus* (MRSA) and anti-MRSA activities of extracts of some medicinal plants: a brief review. AIMS Microbiol 5(2):117–137

Park DR (2005) The microbiology of ventilator-associated pneumonia. Respir Care 50(6):742–765

Parracho HM, Burrowes BH, Enright MC, Mcconville ML, Harper DR (2012) The role of regulated clinical trials in the development of bacteriophage therapeutics. J Mol Genet Med 6:279–286

Parra-Millán R, Guerrero-Gómez D, Ayerbe-Algaba R, Pachón-Ibáñez ME, Miranda-Vizuete A, Pachón J, Smani Y (2018) Intracellular trafficking and persistence of *Acinetobacter baumannii* requires transcription factor EB. mSphere 3(2):e00106-18

Payne RJ, Jansen VA (2003) Pharmacokinetic principles of bacteriophage therapy. Clin pharmacokinet 42:315–325

Pendleton JN, Gorman SP, Gilmore BF (2013) Clinical relevance of the ESKAPE pathogens. Expert Rev Anti-Infect Ther 11(3):297–308

Piperaki ET, Syrogiannopoulos GA, Tzouvelekis LS, Daikos GL (2017) *Klebsiella pneumoniae*: virulence, biofilm and antimicrobial resistance. Pediatr Infect Dis J 36(10):1002–1005

Pires DP, Melo LD, Boas DV, Sillankorva S, Azeredo J (2017) Phage therapy as an alternative or complementary strategy to prevent and control biofilm-related infections. Curr Opin Microbiol 39:48–56
Pires DP, Costa AR, Pinto G, Meneses L, Azeredo J (2020) Current challenges and future opportunities of phage therapy. FEMS Microbiol Rev 44(6):684–700
Pirnay JP, De Vos D, Verbeken G, Merabishvili M, Chanishvili N, Vaneechoutte M, Zizi M, Laire G, Lavigne R, Huys I, Van den Mooter G, Buckling A, Debarbieux L, Pouillot F, Azeredo J, Kutter E, Dublanchet A, Górski A, Adamia R (2011) The phage therapy paradigm: prêt-à-porter or Sur-mesure? Pharm Res 28(4):934–937
Pouillot F, Blois H, Iris F (2010) Genetically engineered virulent phage banks in the detection and control of emergent pathogenic bacteria. Biosecur Bioterror 8(2):155–169
Prestinaci F, Pezzotti P, Pantosti A (2015) Antimicrobial resistance: a global multifaceted phenomenon. Pathog Glob Health 109(7):309–318
Queenan AM, Bush K (2007) Carbapenemases: the versatile beta-lactamases. Clin Microbiol Rev 20(3):440–458
Raju S, Oli AK, Patil SA, Kelmani Chandrakanth R (2010) Prevalnce of multi-drug resistant Staphylococcus aureus in diabetics clinical samples. World J Microbiol Biotechnol 26:171–176
Rakhuba D, Kolomiets E, Dey ES, Novik G (2010) Bacteriophage receptors, mechanisms of phage adsorption and penetration into host cell. Pol J Microbiol 59(3):145–155
Regeimbal JM, Jacobs AC, Corey BW, Henry MS, Thompson MG, Pavlicek RL, Quinones J, Hannah RM, Ghebremedhin M, Crane NJ, Zurawski DV, Teneza-Mora NC, Biswas B, Hall ER (2016) Personalized therapeutic cocktail of wild environmental phages rescues mice from *Acinetobacter baumannii* wound infections. Antimicrob Agents Chemother 60(10):5806–5816
Rice LB (2008) Federal funding for the study of antimicrobial resistance in nosocomial pathogens: no ESKAPE. J Infect Dis 197(8):1079–1081
Rowland I, Gibson G, Heinken A, Scott K, Swann J, Thiele I, Tuohy K (2018) Gut microbiota functions: metabolism of nutrients and other food components. Eur J Nutr 57(1):1–24
Saha U, Gondi R, Patil A, Saroj SD (2023) CRISPR in modulating antibiotic resistance of ESKAPE pathogens. Mol Biotechnol 65(1):1–16
Santajit S, Indrawattana N (2016) Mechanisms of antimicrobial resistance in ESKAPE pathogens. Biomed Res Int 2016:2475067
Scholtz V, Vaňková E, Kašparová P, Premanath R, Karunasagar I, Julák J (2021) Non-thermal plasma treatment of ESKAPE pathogens: a review. Front Microbiol 12:737635
Schooley RT, Biswas B, Gill JJ, Hernandez-Morales A, Lancaster J, Lessor L, Barr JJ, Reed SL, Rohwer F, Benler S, Segall AM, Taplitz R, Smith DM, Kerr K, Kumaraswamy M, Nizet V, Lin L, McCauley MD, Strathdee SA, Benson CA, Pope RK, Leroux BM, Picel AC, Mateczun AJ, Cilwa KE, Regeimbal JM, Estrella LA, Wolfe DM, Henry MS, Quinones J, Salka S, Bishop-Lilly KA, Young R, Hamilton T (2017) Development and use of personalized bacteriophage-based therapeutic cocktails to treat a patient with a disseminated resistant *Acinetobacter baumannii* infection. Antimicrob Agents Chemother 61(10):e00954-17
Shapiro DJ, Hicks LA, Pavia AT, Hersh AL (2014) Antibiotic prescribing for adults in ambulatory care in the USA, 2007-09. J Antimicrob Chemother 69(1):234–240
Shokri D, Soleimani-Delfan A, Fatemi SM (2017) Assessment of phage cocktails with extended host range activity against antibiotic resistant strains of *Pseudomonas aeruginosa*. Comp Clin Pathol 26:417–422
Sillankorva S, Oliveira R, Vieira MJ, Sutherland I, Azeredo J (2004) Bacteriophage Φ S1 infection of *Pseudomonas fluorescens* planktonic cells versus biofilms. Biofouling 20(3):133–138
Smith HW, Huggins MB (1982) Successful treatment of experimental *Escherichia coli* infections in mice using phage: its general superiority over antibiotics. J Gen Microbiol 128(2):307–318
Smith HW, Huggins MB (1983) Effectiveness of phages in treating experimental *Escherichia coli* diarrhoea in calves, piglets, and lambs. J Gen Microbiol 129(8):2659–2675
Smith HW, Huggins MB, Shaw KM (1987) Factors influencing the survival and multiplication of bacteriophages in calves and in their environment. J Gen Microbiol 133(5):1127–1135

Speck P, Smithyman A (2016) Safety and efficacy of phage therapy via the intravenous route. FEMS Microbiol Lett 363(3):fnv242

Srinivasiah S, Bhavsar J, Thapar K, Liles M, Schoenfeld T, Wommack KE (2008) Phages across the biosphere: contrasts of viruses in soil and aquatic environments. Res Microbiol 159(5):349–357

Sulakvelidze A (2011) Bacteriophage: a new journal for the most ubiquitous organisms on earth. Bacteriophage 1:1–2

Sulakvelidze A, Alavidze Z, Morris JG Jr (2001) Bacteriophage therapy. Antimicrob Agents Chemother 45:649–659

Summers WC (1999) Felix d'Herelle and the origins of molecular biology. Yale University Press, New Haven

Sunagar R, Patil SA, Chandrakanth RK (2010) Bacteriophage therapy for *Staphylococcus aureus* bacteremia in streptozotocin-induced diabetic mice. Res Microbiol 161(10):854–860

Tacconelli E, Carrara E, Savoldi A, Harbarth S, Mendelson M, Monnet DL, Pulcini C, Kahlmeter G, Kluytmans J, Carmeli Y, Ouellette M, Outterson K, Patel J, Cavaleri M, Cox EM, Houchens CR, Grayson ML, Hansen P, Singh N, Theuretzbacher U, Magrini N (2018) Discovery, research, and development of new antibiotics: the WHO priority list of antibiotic-resistant bacteria and tuberculosis. Lancet Infect Dis 18(3):318–327

Taha Yassin M, Abdulrahman A-AA, Abdel-Fattah Mostafa A, El-Sheikh MA (2020) Bioactivity of *Syzygium aromaticum* (L.) Merr. & L.M.Perry extracts as potential antimicrobial and anti-cancer agents. J King Saud Univ Sci 32(8):3273–3278

Takemura-Uchiyama I, Uchiyama J, Kato SI, Inoue T, Ujihara T, Ohara N, Daibata M, Matsuzaki S (2013) Evaluating efficacy of bacteriophage therapy against *Staphylococcus aureus* infections using a silkworm larval infection model. FEMS Microbiol Lett 347(1):52–60

Tinoco JM, Buttaro B, Zhang H, Liss N, Sassone L, Stevens R (2016) Effect of a genetically engineered bacteriophage on *Enterococcus faecalis* biofilms. Arch Oral Biol 71:80–88

Tkachev PV, Pchelin IM, Azarov DV, Gorshkov AN, Shamova OV, Dmitriev AV, Goncharov AE (2022) Two novel lytic bacteriophages infecting Enterococcus spp. are promising candidates for targeted antibacterial therapy. Viruses 14(4):831

Tkhilaishvili T, Winkler T, Müller M, Perka C, Trampuz A (2020) Bacteriophages as adjuvant to antibiotics for the treatment of periprosthetic joint infection caused by multidrug-resistant *Pseudomonas aeruginosa*. Antimicrob AgentsChemother 64:e00924-19

Turner NA, Sharma-Kuinkel BK, Maskarinec SA, Eichenberger EM, Shah PP, Carugati M, Holland TL, Fowler VG Jr (2019) Methicillin-resistant *Staphylococcus aureus*: an overview of basic and clinical research. Nat Rev Microbiol 17(4):203–218

Tuttobene MR, Pérez JF, Pavesi ES, Perez Mora B, Biancotti D, Cribb P, Altilio M, Müller GL, Gramajo H, Tamagno G, Ramírez MS, Diacovich L, Mussi MA (2021) Light modulates important pathogenic determinants and virulence in ESKAPE pathogens *Acinetobacter baumannii*, *Pseudomonas aeruginosa*, and *Staphylococcus aureus*. J Bacteriol 203(5):e00566-20

U.S. National Library of Medicine (2022) Safety and efficacy of EcoActive on intestinal adherent invasive *E. coli* in patients with inactive Crohn's disease, January 17, 2019 edition

Verbeken G, Pirnay JP (2022) European regulatory aspects of phage therapy: magistral phage preparations. Curr Opin Virol 52:24–29

Villarroel J, Larsen MV, Kilstrup M, Nielsen M (2017) Metagenomic analysis of therapeutic PYO phage cocktails from 1997 to 2014. Viruses 9:E328

Volozhantsev NV, Borzilov AI, Shpirt AM, Krasilnikova VM, Verevkin VV, Denisenko EA, Kombarova TI, Shashkov AS, Knirel YA, Dyatlov IA (2022) Comparison of the therapeutic potential of bacteriophage KpV74 and phage-derived depolymerase (β-glucosidase) against *Klebsiella pneumoniae* capsular type K2. Virus Res 322:198951

Wang YC, Huang TW, Yang YS, Kuo SC, Chen CT, Liu CP et al (2018) Biofilm formation is not associated with worse outcome in Acinetobacter baumannii bacteraemic pneumonia. Sci Rep 8(1):7289

Walsh C (2003) Where will new antibiotics come from? Nat Rev Microbiol 1(1):65–70

World Health Organization (2017) Global priority list of antibiotic-resistant bacteria to guide research, discovery, and development of new antibiotics. WHO, Geneva

Wu N, Dai J, Guo M et al (2021) Pre-optimized phage therapy on secondary *Acinetobacter baumannii* infection in four critical COVID-19 patients. Emerg Microbes Infect 10(1):612–618

Wyres KL, Lam MMC, Holt KE (2020) Population genomics of *Klebsiella pneumoniae*. Nat Rev Microbiol 18:344–359

Yazdi M, Bouzari M, Ghaemi EA (2018) Isolation and characterization of a lytic bacteriophage (vB-PmiS-TH) and its application in combination with ampicillin against planktonic and biofilm forms of *Proteus mirabilis* isolated from urinary tract infection. J Mol Microbiol Biotechnol 28:37–46

Yeboah-Manu D, Kpeli GS, Ruf MT et al (2013) Secondary bacterial infections of buruli ulcer lesions before and after chemotherapy with streptomycin and rifampicin. PLoS Negl Trop Dis 7(5):e219

Yonezawa H, Osaki T, Kamiya S (2015) Biofilm formation by *Helicobacter pylori* and its involvement for antibiotic resistance. Biomed Res Int 2015:914791

Young R, Gill JJ (2015) Phage therapy redux—what is to be done? Science 350(6265):1163–1164

Zhang Y, Yao Z, Zhan S, Yang Z, Wei D, Zhang J, Li J, Kyaw MH (2014) Disease burden of intensive care unit-acquired pneumonia in China: a systematic review and meta-analysis. Int J Infect Dis 29:84–90

Zhen X, Lundborg CS, Sun X, Hu X, Dong H (2019) Economic burden of antibiotic resistance in ESKAPE organisms: a systematic review. Antimicrobial Resist Infect Control 8:1–23

Computational Approaches for the Inhibition of ESKAPE Pathogens

19

Subhaswaraj Pattnaik, Monika Mishra, and Pradeep Kumar Naik

Abstract

According to the WHO priority pathogens list, the six nosocomial pathogens that exhibit resistance to several antibiotics are named as ESKAPE, including *Enterococcus faecium*, *Staphylococcus aureus*, *Klebsiella pneumoniae*, *Acinetobacter baumannii*, *Pseudomonas aeruginosa*, and *Enterobacter* sp. Since most antibiotics are ineffective against ESKAPE group of microorganisms, it creates selective pressure on the clinical healthcare settings. Hence, it is imperative to design and develop novel therapeutic agents of synthetic and natural origin against these multidrug-resistant (MDR) bacteria. However, the limitations associated with conventional drug discovery pipelines have urged the scientific community to develop reverse pharmacology-based approaches to discover potential therapeutic drug candidates of interest. With the advancement of next generation computational approaches, i.e., computer-aided drug designing (CADD), quantitative structure affinity relationship (QSAR) studies, pharmacophore modelling, and pharmacokinetic profiling using ADMET (Absorption-Digestion-Metabolism-Excretion-Toxicity) predictions, the drug discovery pipelines received quintessential growth in last few years. In silico tools, particularly, molecular docking, MD Simulation, 3D-QSAR, ADMET profiling, etc. are being extensively explored in providing a platform for identifying and developing potential therapeutic agents against several diseases of economic importance. The advanced computational tools and prediction software not only reduced the manpower, time, and selection of drug development processes but also created a favourable atmosphere to design and develop potential drug candidates against chronic microbial infections linked to ESKAPE pathogens. The in silico tools

S. Pattnaik · M. Mishra · P. K. Naik (✉)
Centre of Excellence in Natural Products and Therapeutics, Department of Biotechnology and Bioinformatics, Sambalpur University, Sambalpur, Odisha, India
e-mail: pknaik1973@suniv.ac.in

S. Busi, R. Prasad (eds.), *ESKAPE Pathogens*,
https://doi.org/10.1007/978-981-99-8799-3_19

also provide an integrated platform for repurposing of therapeutic drugs toward the mitigation of bacterial infections and biofilms mediated resistance in bacterial pathogens. The present chapter emphasizes the use of advanced computational tools to identify potential drug candidates against ESKAPE pathogens. Besides, CADD-based approaches for FDA approved drugs with potential therapeutic values repurposed towards antibacterial and biofilm inhibition was also critically discussed. This chapter will thus provide new avenues to design and develop potential drug candidates with high throughput therapeutic values in the regulation of bacterial virulence and drug resistance patterns in ESKAPE pathogens in the near future.

Keywords

ADMET · Biofilms · CADD · ESKAPE · MDR · QSAR

19.1 Introduction

In spite of the revolutionary advancements in the biomedicines, biotechnological, and pharmaceutical sectors in the later phases of the twentieth century and twenty-first century, pathogenic microorganisms concomitantly develop strategic arsenal to withstand the therapeutic regimens and this adaptive approach leads to antimicrobial resistance (AMR). The irrational and indiscriminate use of conventional antibiotics against pathogenic microorganisms led to antibiotic resistance with severe implications on healthcare, pharmaceutical, agricultural, and most importantly economical sectors across the globe with special reference to low- and middle-income countries (Dadgostar 2019; Ikhimiukor et al. 2022). The increased tendency of AMR becomes a global public health issue with an increasing rate of disease prognosis and mortality, and if not addressed properly, it could result in ten million death across the globe every year by 2050 (Liu et al. 2022). Since the insurgence of SARS-CoV-2, no doubt public health become severely shattered but continues to challenge public health life by promoting 2° infections with special emphasis upon the immunocompromised individuals with a significant increase in the irrational use of antimicrobial agents leading to global drug resistance (Seethalakshmi et al. 2022). It is evident from several reports that several factors such as environmental, evolutionary, and anthropogenic events are responsible for the emergence of AMR and requires considerable attention to formulate a strategic arsenal to deal with the AMR crisis. For this, it is important to collaborate with the involvement of several stakeholders starting from scientific professionals to the general public to understand the situation and convey a robust strategic approach in the fight against AMR in the post-antibiotic era (Bassetti et al. 2022; Larsson and Flach 2022). In this regard, global health organizations including the World Health Organization (WHO) address this issue with the identification of priority pathogens that are mainly responsible for the insurgence of drug resistance.

19.1.1 A Brief Introduction to ESKAPE Pathogens

As per the WHO, the ESKAPE (*Enterococcus faecium*, *Staphylococcus aureus*, *Klebsiella pneumoniae*, *Acinetobacter baumannii*, *Pseudomonas aeruginosa*, and *Enterobacter* sp.) group of pathogens top the list by inferring a world-wide threat to human fitness, health, and economy. By virtue of resistant genes acquisition, the ESKAPE pathogens critically increased the healthcare burdens, mortality rate, and treatment failure. Based upon the resistance severity inferred by the ESKAPE pathogens, the global initiative for antimicrobial resistance surveillance is the primary focus to design and develop novel antimicrobial agents (De Oliveira et al. 2020). Among the ESKAPE pathogens, particularly the Gram-negative pathogens tend to exhibit resistance to β-Lactams, including Carbapenems by producing β-Lactamase enzymes (Mulani et al. 2019; Vrancianu et al. 2020). The extensive drug resistance in ESKAPE pathogens leading to hospital-acquired and community-acquired infections is mediated by several strategic approaches such as target alteration, decreased drug uptake, activation of efflux pumps, biofilm formation, and alternative metabolic pathway activation (Denissen et al. 2022). Since the ESKAPE group of pathogens causes severe threat to human healthcare settings owing to their widespread resistance profiles, these priority pathogens received considerable attention in the Global Action Plan to challenge the antimicrobial resistance (AMR) (Kalpana et al. 2023).

19.1.2 Chronic Bacterial Infections and Biofilm Dynamics in ESKAPE Pathogens

In pathogenic bacteria, particularly, the ESKAPE pathogens show the tendency to shift from planktonic forms and reorganize into sessile biofilm forms by producing an extracellular matrix as a protective covering. The reorganized biofilm matrix not only protects the embedding bacterial community form environmental stress but also provides a strategic plan to infer a multitude of pathogenesis, especially in chronic infections (Vestby et al. 2020). According to the National Institute of Health (NIH) data, 70–80% of chronic infections are referred to hospital-acquired infections which includes Urinary tract infections (UTIs) and respiratory tract infections are controlled by the biofilm associated infections (BAIs). The persistent chronic infections are further aggravated by the presence of highly complex and intricate system of intracellular signalling system called quorum sensing (QS) which promotes the crosstalk among the bacterial community both at the phenotypic and genotypic level (Lazar et al. 2021; Mohamad et al. 2023). These biofilms in ESKAPE pathogens are considered as the survival strategies of the embedded bacterial community against antibiotic treatment leading to antibiotic resistance. Based upon the severity of antibiotic resistance shown by ESKAPE pathogens by expressing recalcitrant biofilm dynamics, they are categorised as MDR or extensive drug resistant (XDR), and pan-drug resistant (PDR) pathogens (Patel et al. 2021). The BAIs are extremely resistant to conventional antibiotics by promoting several adaptive responses including altered phenotypes, reduced transcriptional and translational activities, and modified pathophysiological metabolism (Mohamad et al. 2023). In the last few decades, several technological advancements and biotechnological progress have revolutionized our scientific efforts in the management of prolonged infections by the progress or advancement of next generation antibiotics. The ESKAPE pathogens also on the other hand critically challenge the therapeutic regimens by continuous mutation and altered metabolism by building a tough challenge to the conventional treatment strategies (de Macedo et al. 2021). With reference to this, it is essential to look for alternative approaches beyond antibiotics to combat chronic microbial infections associated with ESKAPE pathogens.

19.1.3 Therapeutics Against ESKAPE Pathogens

Traditionally, antimicrobial peptides (AMPs), bacteriophage therapy, synthetic drug moieties, and natural products are considered as therapeutic regimens against the ESKAPE biofilms and drug resistance as an alternative to conventional antibiotics. Besides, active and passive immunotherapy are considered as promising approaches for the mitigation of recalcitrant biofilms in ESKAPE pathogens (Gao et al. 2020; Panda et al. 2022). Bacteriophage therapy against ESKAPE pathogens is a promising alternative to antibiotic therapy as the bacteriophages have the advantages of low toxicity, target-based killing of bacterial pathogens without affecting the host's normal microbiota, biofilm degrading ability, and unlikely to infer cross-resistance to antibiotics (El Haddad et al. 2019). The AMPs and the antibiofilm peptides were also found to be influential in treatment of SSTIs, i.e., skin and soft tissue infections, which is triggered by nosocomial ESKAPE pathogens, and the recalcitrant biofilm infections on medical equipment such as catheters, stents, and dentures, suggesting their pivotal part in managing the chronic bacterial infections (Pfalzgraff et al. 2018; Rajput and Kumar 2018). In addition, siRNA could also be used to silence the receptors for bacteria in the host cells, thereby promoting their dissociation with the host cells. Besides, specific chimeric molecules could also be designed which characteristically target the host-pathogen interactions (Sharma et al. 2021b). From the therapeutic alternatives against MDR ESKAPE pathogens, natural products received widespread recognition owing to their widespread pharmacological relevance. The bioactive secondary metabolites from natural sources such as plants and microbes have been explored as potential therapeutic alternatives against several health issues including chronic microbial infections. Among the microbial sources, the vast repertoire of bioactive secondary metabolites in Actinomycetes is considered for therapeutic implications against different healthcare conditions (da Rosa et al. 2020). The ethnopharmacologically relevant plants and their phytochemicals are also being actively engaged in the investigations of their role as effective therapeutic weapons against the QS regulated virulence and biofilms mediated hospital and community-acquired infections caused by ESKAPE pathogens (Schultz et al. 2020).

In the quest for alternative therapeutic approaches, the nanotechnological intervention has revolutionized the scientific investigations to treat BAIs in ESKAPE pathogens. Specifically, the unique physicochemical properties which includes high surface area-to-volume ratio, improved physicochemical properties, improved penetration power etc., biocompatibility, drug loading efficiency, controlled release, and longer therapeutic values are the advantages of spatially designed nanomaterials for the management of recalcitrant biofilm infections. Metallic nanoparticles (e.g., silver and copper nanoparticles), polymeric nanoparticles (e.g., chitosan), lipid-based nanoparticles (e.g., liposomes, and micelle), carbon nanoparticles (e.g., mesoporous silica, carbon nanotubes, graphene oxide, etc.), nanoemulsions, and nanocomposites are designed and developed for antimicrobial and antibiofilm activities against ESKAPE pathogens (Mukherjee et al. 2023). Over the last few years, antimicrobial photodynamic inactivation (aPDI) is used in the regulation of metabolically active ESKAPE pathogens by the production of reactive oxygen species (ROS), which act upon the biological macromolecules leading to the death of the bacterial cells in a targeted manner (Nakonieczna et al. 2019). Based on the current therapeutic methods for the treatment of ESKAPE pathogens, it is important to look beyond the conventional approaches and develop novel and innovative antimicrobial approaches. Though current therapeutic options provide a broad range of molecules as potential drug candidates in pre-clinical phases, the majority of compounds failed to impress during animal studies and clinical trials, and therefore, the success of the approval rate is very minimal. Based upon this, it is imperative to develop strategic therapeutic plans with reduced cost and time and improved efficacy as compared to traditional therapeutics (Stephens et al. 2020). The recent trends in data science and bioinformatics have given a new dimension to the area of structural and system biology, and OMICS studies which correspond to the new aspects of computational approaches in biological studies. The information and data generated through bioinformatics and computational tools provide a novel platform to predict the mechanisms associated with the resistance profile of ESKAPE pathogens. Besides, these computational tools further provide valuable information on the possible therapeutic targets and screening of potent drug candidates against the predicted target sites which play an indispensable role in drug discovery pipelines (Priyamvada et al. 2022).

19.1.4 Drug Discovery and Conventional Drug Development Pipelines

Drug discovery and development is an important phase for translating the preclinical research output to humans for the well-being of society to address scientific queries, operational issues, and medications against several healthcare problems systematically (Mohs and Greig 2017). The traditional and sequential phases of drug discovery and development pipelines start with target recognition followed by hit discovery, generation of lead, development of lead, identification of potential drug candidates, pre-clinical studies, and clinical trials. Conventional drug discovery and development pipeline for the expansion of a potential drug from screening to the approval for medications requires a considerable timeline (approximately more than 10–12 years), and larger costs with approval success rate hovers around 1–5% (Wang et al. 2019). As per the estimates, the cost of research and development from research hypothesis to approval of any drugs and their marketing for public use is approximately US$2.6 billion spanning across 10–15 years of robust scientific research and constant incentives from both Governmental and non-governmental agencies (Alves et al. 2022). The genomics-based precision medicine approach marks its stature in the drug discovery and advancement process by providing an underlying mechanism associated with chronic diseases that concomitantly foster precise medicines at the individual level (Dugger et al. 2018).

19.2 Computational Approaches for Drug Discovery and Development

In the fight against AMR, and community-acquired infections associated with ESKAPE pathogens, computational approaches in the rational design of potential drug candidates targeting the pathophysiological targets with higher efficacy is the new aspect of drug development (Atanaki et al. 2020). The advent of the machine learning (ML) approach has dramatically improved the discovery and precision decision-making in every phase of drug discovery pipelines with specific implications during target validation, identification of therapeutically relevant biomarkers, analysis of research output, and big data mining (Vamathevan et al. 2019). The computational approaches are also observed to be influential in the field of pharmacogenomics and pharmacomicrobiomics by elucidating the effective predictions of the pharmacokinetic resources that includes absorption, distribution, metabolism, and excretion of the assumed or accepted drug molecules of interest. The computational tools also critically predict the toxicity profile of the putative drug candidates, which enables in identifying of potential drug molecules for a specific purpose (Agamah et al. 2020). Based upon the computational predictions, the putative drug candidates intended for specific target, once cleared by the ADMET filters, and could be developed for intended medications. Hence, the computational tools invariably potentiate in identifying potential molecules with improved pharmacokinetic properties (Bergstrom and Larsson 2018). Several computational tools such as Discovery Studio, TOPKAT module of Discovery Studio, admetSAR, ADMET Predictor, ADMETlab, SwissADME, pkCSM, preADMET, etc. are available for comprehensive projection and forecasts of the pharmacokinetic properties of the intended drug molecules of interest (Xiong et al. 2021; Dulsat et al. 2023). These computational prediction tools for ADMET have their separate functional attributes and advantages over the counterpart tools. Based on the information required, standard predictive ADMET tools could be selected and relevant pharmacokinetic parameters could be elucidated. The AI-based development and designing of drug also critically influences the progress of potential drug candidates in the fight against the silent pandemic in the form of antimicrobial resistance (AMR) with improved efficacy and cost efficiency. It is evident from recent reports that the AI-based data mining approach has predicted potential drug candidates, particularly β-Lactamase inhibitors from diverse sources including natural products. Hence, AI could be considered as a potential tool to design and develop therapeutic alternatives for conventional antibiotics against AMR and community-acquired infections (Talat and Khan 2023).

19.2.1 Computer-Aided Drug Designing (CADD)

With the aim of the reduction of the cost, shorten the research investigations schedule, and the risk of failure associated with the conventional method of drug discovery and development along with computer-aided drug design (CADD) is considered as a potential alternative. The CADD approach utilizes several computational tools and databases to identify potential drug targets, screen thousands of possible drug candidates against the potential drug targets, lead validation and optimization, and modelling the interactions of the lead compounds with the selected pathophysiological targets (Ou-yang et al. 2012). The combination and assimilation of CADD in the drug discovery and development procedure has several advantages over conventional high throughput screening with comparatively higher hit identification at a shortened time period when virtual screening is used which ultimately accelerates the drug design and development process (Dubey and Dubey 2020; da Silva et al. 2022). Basically, the CADD approach is classified into either structure-based drug design (SBDD) or ligand-based drug design (LBDD). The SBDD approach primarily focuses on the identification of key target sites (either proteins or RNA) and the prediction of interactions associated with the target sites to design potential drug molecules targeting the biological pathways of pathophysiologic importance. Meanwhile, the LBDD approach hovers around the establishment of a relationship between the physiochemical properties and their possible biological roles of known antibiotics and/or drug molecules followed by the optimization for spatial design with improved activities (Yu and MacKerell 2017).

In terms of the functions of CADD approaches to combat the bacterial drug resistance and microbial infections, the SBDD could be implemented to identify potential drug targets of pathophysiological relevance in ESKAPE pathogens and predict novel drug candidates in the attenuation of pathophysiological responses. Meanwhile, the LBDD approach could be implemented when the 3D design of the target is unknown, and there is an absence of information to build a homology model; in such a scenario, SAR could be developed with the known drug molecules of interest and their prospective derivatives. The LBDD approach utilizes two strategic computational modules such as pharmacophore modeling and quantitative structure-activity relationship (QSAR) for predicting the interactions of ligands to unknown target molecules (Fig. 19.1) (Yu and MacKerell 2017; da Silva et al. 2022). The pharmaceutical sectors and scientific community widely employ the SBDD approach for identifying potential drug molecules of interest targeting the proteins of biomedical importance. Based upon the SBDD, several drug molecules targeting several pathological targets are under clinical investigations and few molecules identified from SBDD were available in the market after getting prior approval from the competent authority. For example, Raltitrexed was successfully

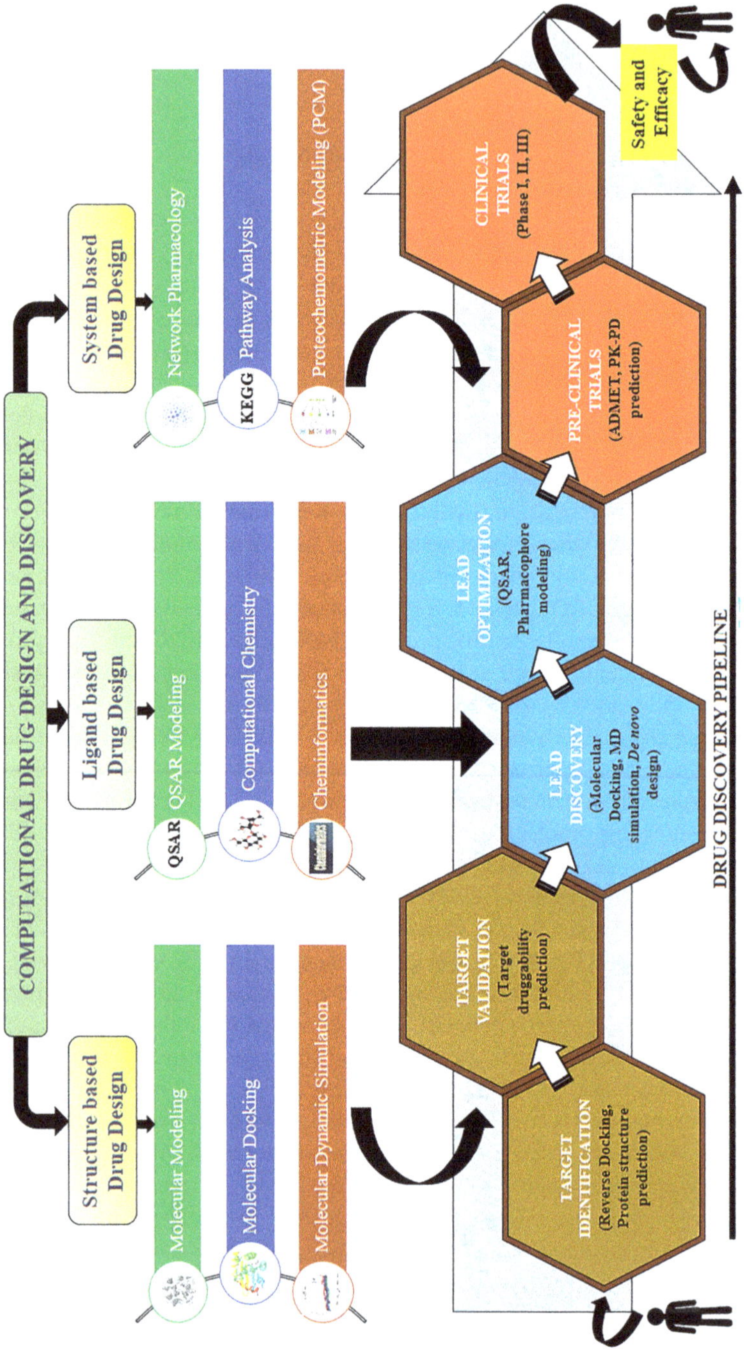

Fig. 19.1 Schematic overview of the computational approaches in drug design and discovery

identified through SBDD and used against the human immunodeficiency virus (HIV) and provides novel avenues to identify and design through SBDD for the management of several diseases including chronic infections caused by ESKAPE pathogens (Batool et al. 2019).

As discussed earlier, the majority of SBDD programmes rely upon the known 3D structure of target proteins of interest, the majority of which are easily obtained from the protein data bank (PDB). However, for new targets, crystal structures are not even available in the PDB, and in that case, the sequence of the target protein could be taken into consideration and Homology Modeling could be employed using several computational tools such as SWISS-MODEL and PHYRE2 to generate a 3D model of the target protein for computational analysis. The generated 3D protein is further optimized and could be used for molecular docking and MD simulation (Cain et al. 2014). The LBDD-based pharmacophore modelling, QSAR analysis, and ADMET (Absorption, Digestion, Metabolism, Excretion, and Toxicity) forecasts were actively employed to determine the inhibitory role of potential ligand molecules against the potential drug target of *S. aureus* with the 3D structure of the target is unknown (Ye et al. 2023). The CADD is being actively employed using the scaffold-hopping approach to identify potential derivatives of clotrimazole and could be used to target pathogenic strains of *S. aureus* (Cortat et al. 2023). In an earlier study, several computational modules of CADD such as pharmacophore modelling, virtual screening, molecular docking, and molecular dynamics simulation (MDS) were incorporated to characterize potential drug candidates from the database against the collagen-binding protein of *E. faecium* (Rasheed et al. 2021).

19.2.1.1 Molecular Docking

Molecular docking is an important computational tool in the development of SBDD, utilizing the virtual screening (VS) mode for screening out thousands of potent drug molecules by study of their conformation and alignment against the binding site of several targets of pathophysiological, biochemical, and molecular targets of interest. Molecular docking is a simple yet fast and economical tool to predict the molecular interactions between the ligands with potential therapeutic targets (Torres et al. 2019). Molecular docking could be used to screen several groups of phytochemicals of pharmacological importance and their analogues to predict their interactions with the potential therapeutic targets of pathophysiological relevance (Abo-Salem et al. 2021). The most significant advantage of using molecular docking tools for virtual screening is that it allows thousands of molecules from synthetic origin to natural sources to screen against the potential therapeutic targets of interest with no aided cost and within the stipulated time period (Liu et al. 2021). The molecular docking is categorized into two sequential events starting with insight exploration of ligand conformations within the binding pocket of the aimed protein followed by the assessment of the quantitative binding energy variations of the predicted conformations of the several ligands used in the predictions (dos Santos et al. 2018). Several molecular docking tools such as GLIDE (Schrödinger), Auto Dock Vina (Scripp Research Institute), LeDock (flexible docking tool), GOLD (based upon genetic algorithm), MOE-Dock, etc. are considered the most widely accepted docking platforms with improved accuracy (Pagadala et al. 2017).

For example, 5-enolpyruvylshikimate-3-phosphate (EPSP) synthase in Shikimate pathway is an important step in the infection process of ESKAPE pathogen, *A. baumannii*. Hence, EPSP could be used as an effective therapeutic target for the mitigation of chronic infection mechanisms, and in this regard, molecular docking analysis utilizing the SBDD approach could be helpful in recognizing potential drug candidates as promising inhibitors of EPSP. Based upon the lead, molecules with best-docked complexes could be explored further to optimize their efficacy in binding to the active site of target molecules and critically alter the infection pathways in *A. baumannii* (Almihyawi et al. 2022). Similarly, molecular docking is used to analyse the interactions of putative drug candidates against the highly complex, intracellular networking system called quorum sensing (QS), which is responsible for producing harmful virulence phenotypes related with biofilm development and the emergence of drug resistance. Based upon the docking scores, further the promising compounds with improved binding affinity with the binding site of the macromolecular target could be employed for in vitro and in vivo studies (Aboagye et al. 2023). The resistance to β-Lactam antibiotics shown by *S. aureus* is primarily activated by the existence of penicillin-binding protein 2a (PBP2a). Hence, PBP2a could be used as probable therapeutic targets for the screening of putative drug molecules and the molecular docking tool facilitates in predicting the binding affinity of the screened molecules with the active binding site of PBP2a. Based on the affinity of the docked complexes, further investigations could be designed to establish the mitigatory role of the important and particular compounds against drug resistance phenomena (Cordeiro et al. 2020a, b).

Molecular docking could also be employed to screen the FDA-approved drugs towards the attenuation of QS and biofilm formation by targeting the ligand binding domain of QS regulatory proteins. This system of repurposing approved drugs towards new therapeutic targets characteristically improves the chance of identifying novel drug molecules with improved efficiency and reduced cost and time (Sadiq et al. 2020).

19.2.1.2 Molecular Dynamics Simulation

Molecular dynamic simulation (MD simulation) is an advanced computational tool which is employed to evaluate and/or predict the strength and stability of the docked complexes formed between the ligand molecules and the target protein achieved from molecular docking analysis. MD simulation by virtue of the enhance computational power associated with running the programme, accurate algorithms in the dynamic environment, and efficiency is considered as potential in silico platform for drug discovery and development with broad spectrum utilities (Palmer et al. 2021). In the computational approach-based drug discovery process, the promising compounds/ligand molecules showing efficient docking complexes with therapeutic targets of interest could be further validated by incorporating them into the MD simulation programme, which not only provides the stability profiles of the docked complexes but also provides an insight into the energy calculations with the docked complexes at the binding site of the target protein. In this aspect, the phytochemicals screened for molecular docking with the potential pathophysiological targets associated with QS and biofilms could further be analyzed using MD simulation to check their stability in the interactions in between the ligand and the protein (Chaieb et al. 2022). The MD simulation basically utilizes force-field parameters for the predictions of the intramolecular interactions and connections between the ligands and the binding site of the aimed protein. Computational tools such as GROMACS, AMBER, CHARMM, and LAMMPS, etc. specifically employ the force-field parameters for the simulation of the docked complexes (Nicolas-Barreales et al. 2021).

No doubt, classical MD simulations have several advantages in validating the stability of the docked complexes and provide new avenues to identify potential drug molecules selectively. However, the electronic polarisation in the classical MD simulation has some limitations as the assigned atoms in the system have a pre-set partial charge and have been maintained during the simulation process. But, in reality, the atoms of the biomolecules are considered to be polarisable as the electronic clouds constantly surrounds around the atoms and alter itself with response to the environment. In this context, the advent of Quantum mechanics, MD simulation (QM MD Simulation) provides a dynamic trajectory of the polarisation effect of the electrons and thus addresses the issues associated with classical MD (Ganesan et al. 2017).

19.2.1.3 De Novo Drug Design

De novo drug design (DNDD) is an effective tool in novel drug discovery programs by providing a novel platform to spatially design ligand molecules with novel chemical entities of interest for the confined pocket of the biological receptor proteins under stationary conditions. To achieve the desired pharmacological properties of the ligands, de novo design addresses three basic queries such as assembly criteria for the ligand molecules, evaluation of their potential values, and the search for the effective binding space (Schneider and Fechner 2005; Mouchlis et al. 2021; Bai et al. 2022). The exploration of larger chemical area for binding, spatial design of ligand molecules with novel physicochemical properties with widespread therapeutic potential, and cost-effective drug design are the trending advantages of DNDD (Mouchlis et al. 2021). In addition, DNDD is considered the therapeutic alternative to the conventional blind virtual screening method in drug design and development (Douguet 2010). Similar to CADD, DNDD is also categorized into either structure-based de novo design (3D structure of the receptor known), or ligand-based de novo design (3D structure of the receptor unknown). HSITE, LUDI, and PRO_LIGAND, etc. are the tools employed for the structure-based de novo design by evaluating several scoring functions including force fields, empirical method, and knowledge-based scoring functions. Meanwhile, the ligand-based de novo design includes TOPAS, SNOPSIS, and DOGS as the empirical tools for the drug design (Mouchlis et al. 2021).

The computational-based DNDD also facilitates the design of novel peptides of interest intending to form the transmembrane channels which act as the inhibitors of

MDR bacterial pathogens, with special emphasis on the ESKAPE pathogens. The designed peptides not only promote bacterial killing even at macromolecular concentrations but also have lower toxicity profiles to the hosts (Deb et al. 2023). The Wasserstein generative adversarial network with gradient penalty (WGAN-GP)-based DNDD developed novel AMPs which characteristically improved the antibacterial potential against the MDR *P. aeruginosa*, and *S. aureus* suggesting their widespread role in the management of drug resistance (Lin et al. 2023). For the generation of energetically favourable binding sites for target ligands, Rosetta Molecular Modeling Suite was used and thereby providing novel avenues of the previously unattained target ligands for widespread biomedical applications (Lucas and Kortemme 2020). DE Novo OPTimization of In/organic Molecules (DENOPTIM), a novel computer-driven de novo design tool, was developed for the virtual screening of functional molecules, with special emphasis on inorganic molecules (Foscato et al. 2019). The Generator User interface (GenUI), an open-source cheminformatics-based computational tool, provides the interface to integrate with de novo molecular generator, DrugEx for improved drug discovery (Sicho et al. 2021). Similarly, REINVENT, an AI triggered de novo design tool, which has revolutionized the conventional DNDD approach for improved drug design and discovery (Blaschke et al. 2020). Since the current scenario of drug resistance in ESKAPE pathogens, and the failure to develop novel drugs against these pathogens have critical implications for human health, it is imperative to focus on the AI-driven antimicrobial design and development through computational de novo approach for providing an impetus to the current research platforms for widespread opportunities (Melo et al. 2021).

19.2.1.4 Sequence-Based Virtual Screening (SVSBI)

In computational approaches-based drug design and discovery, virtual screening (VS) has an important role to play. Particularly, structure-based virtual screening (SBVS) is a robust computational method to forecast the interactions among the ligand and the target protein of interest based on the structure available. The SBVS was further categorised into both ligand-based and target-based VS. For both these tools of SBVS, scoring functions were taken into consideration for predictions of the molecular interactions (Maia et al. 2020). In the drug discovery pipelines, to understand the biomolecular interactions, sequence-based virtual screening (SVSBI) holds rationalized advantages over 3D structure-based virtual screening. To achieve the same, advanced algorithms like NLP (natural language processing) with deep *K*-embedding strategies were developed which subsequently reduced the reliability of 3D structure-oriented virtual screening methods (Shen et al. 2023). Owing to the advanced algorithms, the NLP-based tools such as Autoencoders (AE), DNABERT, Long Short-term Memory (LSTM), and Evolutionary Scale Modeling (ESM) Transformer, etc. accurately predict the biomolecular protein-protein interactions and ligand-protein interactions, DNA-protein, and ligand inhibition of protein-protein relations at the structure-level accuracy. The SBVS infers the deep learning-based non-docking method for virtual screening with the ability to work on both 3D structures as well as 1D sequence information. For example, CarbonAI virtual screening platform was developed which provides the interface to work both on the 3D structure (i.e., Graph Neural Network), and sequence-based BiLSTM (Shen et al. 2023; Wu et al. 2023). Meanwhile, the Reverse Virtual screening method, a robust VS strategy, was also developed to forecast the set of assumed drug targets for any complex with known therapeutic values as an antimicrobial agent (Schottlender et al. 2022).

19.2.1.5 Pharmacophore Modeling

Pharmacophore accounts for the spatial electronic and steric features responsible for the optimized supramolecular interactions throughout the virtual screening of huge databank and is considered an effective tool as compared to molecular docking (Opo et al. 2021). In the computational tools-based designing and discovery of drug, the assemblage of chemical connections among the ligand and protein, their spatial arrangement for interaction patterns, and subsequent de novo drug design infers to the pharmacophore modeling. Among the pharmacophore models established, Pharao (for pharmacophore alignment) and Pharmer are the widely established models for drug design and discovery (Dai et al. 2021). The efficacy of pharmacophore modeling in drug discovery entirely depends upon the grade and class of the established pharmacophore model as the developed model directs the functional attributes of the ligand-receptor binding followed by virtual screening (Giordano et al. 2022). The functional attributes and reliability of the generated pharmacophore models are directly proportional to the authenticated and reliable dataset generated from the in vitro studies (Tyagi et al. 2022). The acceptors and donors of hydrogen bond and hydrophobic areas, the presence of positive and negative ionisable groups, and aromatic groups are the most important criteria for pharmacophore modeling. The pharmacophore models could be generated by either a structure-based approach or a ligand-based methodology. The pharmacophore based on the structure hovers around the interactions of ligands and their targets. Lack of ligand information, the pharmacophore model could be generated exclusively based on the topology of the binding sites as evident from the pharmacophore modeling programs such as Discovery Studio and LigandScout. On the contrary, the ligand-based approach focuses on the alignment of 3D structures of two or more known ligands and the identification of common pharmacophore features, which is considered to be essential for the generation of the model (Kaserer et al. 2015). The other software associated with pharmacophore modeling are MOE, MolSign, UNITY, Quasi, Phase, and Catalyst, etc. (Prachayasittikul et al. 2015).

In recent years, AncPhore, an advanced pharmacophore model, was established, which exhibited promising predictions capabilities in identifying the structurally diverse class of inhibitors aiming Indoleamine/tryptophan 2, Metallo-β-Lactamases, and 3-dioxygenases (Dai et al. 2021). Owing to its suitability in finding mitigators of Metallo-β-Lactamases, the AncPhore model could be instrumental in the expansion of novel therapeutic agents pointing towards the drug resistant pathogens with overexpression of these enzymes. Based upon the pharmacophore model of AMPs, a sequence of small β-Peptidomimetics were generated with improved antibacterial activities against drug resistant ESKAPE pathogen, *S. aureus* suggesting the role of potential pharmacophore model in the management of bacterial pathogenesis (Hansen et al. 2010). Later on, C-glycosyl flavones from *Clinacanthus nutans* showed similar pharmacophore features with higher fit-value to the generated pharmacophore model based upon the selected antibiotics. Based upon the pharmacophore features of C-glycosyl flavones with that of the generated pharmacophore model, these could be instrumental in the widespread antibacterial activities against pathogenic microorganisms (Nyawai et al. 2017). The pharmacophore model-based virtual screening was employed to generate chalcone derivatives analogues of thiol-Michael. The generated chalcone derivatives exhibited promising antibacterial activities against ESKAPE pathogens, MRSA (i.e., methicillin-resistant *S. aureus*) and *Eenterococcus faecalis*. Besides, the one analogue of thiol-Michael also showed bactericidal properties against *S. aureus* (Zhang et al. 2018). In the process of drug design, the 3D pharmacophore modeling creates an ensemble of chemically defined inter connections of the ligands with the target proteins and thus found to be instrumental in virtual screening and mechanistic studies of the proteins of pathophysiological relevance. For example, 3D QSAR-based pharmacophore modeling was developed for the detection of spleen tyrosine kinase inhibitors from the drug-like database to recover the hits associated with novel chemical scaffolds. The advent of machine learning and AI-based 3D pharmacophore modeling further improves its widespread implications in drug discovery pipelines (Kumar et al. 2022; Schaller et al. 2020).

19.2.1.6 Structure-Activity Relationship

The structure-activity relationship (QSAR) model of the CADD approach received considerable attention as it is operative in predicting suitable computational models which is grounded on biological activities. Basically, QSAR explores the suitable molecules of interest with known functional attributes present in databases and offers a predictive model utilizing the molecular structure and associated response. An interesting feature of QSAR models in the assessment of antimicrobial activities is that the QSAR models generate selective requirements based on whether the antimicrobial properties are evaluated against fungi, and even between both Gram-positive, Gram-negative bacteria (de Bruijn et al. 2018). Later on, the QSAR model generated analogues of an arthritis drug, Auranofin, significantly repurposed towards antibacterial activities. The QSAR generated analogues even showed enhanced antibacterial properties than that of Auranofin suggesting the value-addition potential of the employed QSAR model in identifying potent drug candidates against ESKAPE pathogens (Wu et al. 2019). It was evident from the developed QSAR model that characteristically improved the antibacterial potential of novel iminothiadiazolo-pyrimidinone derivatives, cumarin derivatives, and pyrrolidine-based hybrid compounds (Paudel et al. 2013; Qin et al. 2020; Bhat et al. 2023). Hence, QSAR models provide new horizons to design and develop lead compounds against the drug resistant ESKAPE pathogens such as methicillin resistant *S. aureus* and nosocomial infections causing *P. aeruginosa* (Suay-Garcia et al. 2020a).

Several reports have put emphasis on the effective role of antimicrobial peptides (AMPs) as potential antimicrobial and antibiofilm activities against drug resistant

ESKAPE pathogens. In this context, QSAR models could provide a robust computational prediction to design and develop effective AMPs against the pathogenicity and chronic infections associated with ESKAPE pathogens (Cardoso et al. 2020). Machine learning tools such as Tree-based approaches, CART (Classification and Regression Trees), etc. are considered as influential and robust tools for the generation of sophisticated QSAR models in the drug discovery procedure (Suay-Garcia et al. 2020b). The ensemble-based machine learning approaches, particularly Random forest (RF), Linear regression models, and Bayesian Neural networks, are the advanced algorithms with improved predictive ability and robustness that typically improve the predictions made by the QSAR models, thereby improving the intended therapeutic applications (Kwon et al. 2019). Similarly, QSAR-Co, an open source software, was developed based on machine learning algorithms which emphasize the improved robustness of QSAR models with aided validation (Ambure et al. 2019). Hence, QSAR model development, validation, and functional attributes critically enhanced the predictive ability of novel compounds targeting specific targets and could be contributory in rational drug design and discovery pipelines (Neves et al. 2018).

19.3 Computational Tools for Drug Repurposing

The idea of drug repurposing infers the reuse of the clinically accepted drugs originally employed for any health consequences directed toward the new therapeutic applications and thereby improving the therapeutic regimens. Drug repurposing or relocation is an attractive module in the context of extended time frame, irrational cost, and minimal success rate associated with the current scenario of drug discovery processes. Further, it has several advantages such as improved efficacy, shorter timeline for drug development, reduced R&D investment and safety risks, and reduced time for approval as compared to the conventional R&D-based therapeutics (Breijyeh and Karaman 2023; Ma et al. 2023). It also addresses the proper resource management with appropriate utilization of the resources required for the design and advancement of innovative drug molecules targeted towards new therapeutic targets (Gulia et al. 2023). For example, Ticagrelor (originally used for the treatment of atherosclerosis of Cardio vascular diseases) was repurposed towards its ability to hinder the growth of bacterial pathogens. It has been observed that Ticagrelor significantly hindered the growth of ESKAPE pathogen, *S. aureus,* and also mitigated the biofilm dynamics by down regulation of biofilm associated genes (Pant et al. 2022). With regard to this, it is imperative to explore the clinically approved drugs repurposed towards the inhibition of bacterial virulence and mitigation of drug resistance patterns, thereby minimizing the risks associated with AMR (Dubey et al. 2020). However, QS is an attractive approach to target the bacterial virulence mechanisms, thereby providing alternative strategy to combat biofilm-mediated chronic infections. Several pharmacologically important approved drug molecules could be redirected towards the attenuation of QS regulated virulence and biofilms. Based on the results, several drugs such as Chlorpromazine,

Thioridazine, and Mebendazole exhibited promising antibacterial activity against the tested strains, *S. aureus*, and *Enterobacter* sp. Meanwhile, 5-Fluorouracil, Cisplatin, Bleomycin, Chlorpromazine, Thioridazine, etc. also showed significant QS attenuation potential in a dose-dependent manner (Gajdacs and Spengler 2019). In an earlier report, an antihelminthic drug, Niclosamide, could be used in combination with Colistin and successfully repurposed towards Colistin resistant ESKAPE pathogens, *Acinetobacter baumannii*, and *Klebsiella pneumoniae* (Ayerbe-Algaba et al. 2018).

Based on the fundamental principles of the drug repurposing approach (i.e., drugs for a specific pathophysiological condition could also work on other pathophysiological conditions due to interdependence between these conditions, and due to the confounding nature of any drug, it could be used in different pathophysiological conditions), it could be categorized into (a) drug-based approach where information on drugs is available (genome-based and chemical structure profile) and (b) disease-based approach where discovery process hovers around the information on diseases (Phenome-based) (Ashburn and Thor 2004; Jarada et al. 2020). Several data repositories such as genome-based Connectivity Map (CMap), Genome Wide Association Studies (GWAS), Library Integrated Network-based Cellular Signatures (LINCS), Gene Ontology (GO), Kyoto Encyclopedia of Genes and Genomes (KEGG), and Universal Protein Resource (UniProt); Phenome-based Side Effect Resider (SIDER), and ClinicalTrials.gov; Chemical structure-based ChEMBL, DrugBank, Protein Data Bank (PDB), PubChem, and NCGC Pharmaceutical Collection (NPC) deliver innovative possibilities to the scientific society as well as the pharmaceutical sectors to explore these resources for drug repurposing studies with enhanced utility (Hema Sree et al. 2019; Jarada et al. 2020). The drug-based strategy hovers around the information graphics of the drugs of interest with their chemical profiles, biomedical, pharmaceutical, and genomics information, based upon which novel therapeutic modules could be formulated. As per the genome-based tools, functional connections between the drugs of interest, genes, and associated disease profiles could be revealed. Similarly, using the two-dimensional (2D) topological fingerprints and three-dimensional (3D) conformational fingerprints, the structural similarity between the drugs could be measured and concomitantly forms the basis for their modulation of therapeutic targets (Swamidass 2011; Jarada et al. 2020). In addition, several researchers focus on the Inverse Genomic Signature approach utilizing the CMap tool for understanding the complication of genome-wide response (i.e., mRNA expression profiles depicting the higher-level protein interactions) of the hosts at diseased conditions as well as during therapeutic regimens (Law et al. 2013). Meanwhile, the disease-based strategy focuses on detailed information on pathophysiological conditions and phenotypic traits of the disease of interest, and complicacy associated with the drug effect on the disease under consideration (Fig. 19.2).

The technical innovations in the field of data science, bioinformatics, OMICS-based approaches, biological sciences, and most importantly, the advent of high throughput computational approaches further improved the extended horizons of the drug repurposing initiatives (Dotolo et al. 2021). Particularly, Data mining (e.g.,

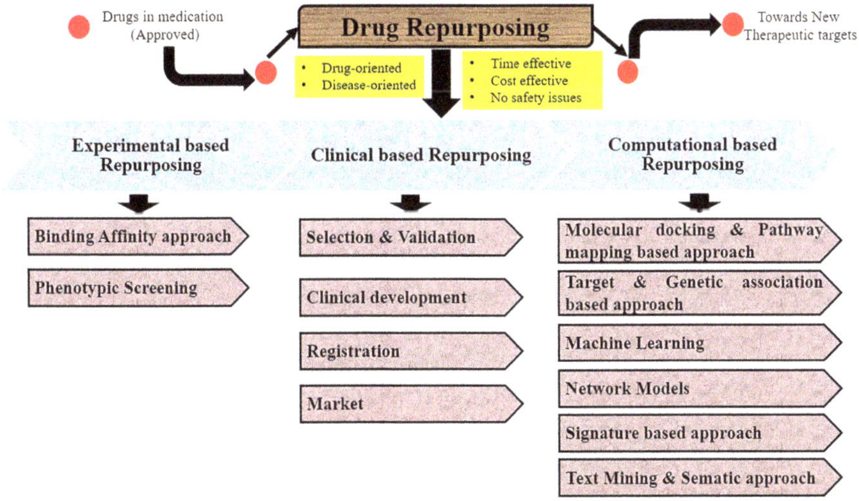

Fig. 19.2 Schematic representation of the computational approaches-based drug discovery in drug repurposing

Text mining and Semantic technologies), Machine Learning (Deep learning), and Network Analysis are the fundamental computational approaches, which are actively explored in drug repurposing strategies. The text-mining approach specifically illustrates the relevant information from the available literature with the help of a computer based program. Besides, cluster algorithms like CLIQUE, STING, DBSCAN, and OPTICS, etc. are explored to find the network clusters determining the relationship between the drug-drug, drug-target, and drug-disease interactions (Jarada et al. 2020). Lomitapide, the Food and Drug Administration approved drug for the treatment of Homozygous familial hypercholesterolemia (HFH), was repurposed as antibacterial and antibiofilm agent against ESKAPE pathogens, *S. aureus*, and *E. faecium* using the protein-protein interaction studies with an ample focus on the down-regulated differentially expressed genes (DEGs) using the STING database (Zhang et al. 2022). Several deep mining approaches from Machine learning tools also critically improve our understanding of the information on drugs, targets, and the disease of interest which effectively provide new horizons for drug repurposing. Several machine learning tools such as Tree Ensemble, Gradient Boosted Trees, Random Forest, Neural Network (NN), and DeepPurpose, etc. are used for systematically integrated approaches for drug repurposing. In this context, machine learning-based predictions on drug repurposing, SperoPredictor, a generic drug repurposing outline could be employed for drug repurposing (Ahmed et al. 2022; Yang et al. 2022). In addition, PREDICT and SPACE, the logistic regression-based Machine Learning framework, also being employed to evaluate the integrated drug-drug and disease-disease similarity which forms the basis for forecasting analogous drugs for analogous diseases by means of logistic regression (Park 2019). The recent advancement in network analysis also comprehensively provides an access to the multiple domains of the data available on drugs, pathophysiological targets, and disease physiology by promoting an interactive biological network between disease-associated genes, and thus promoting new horizons to the drug repurposing strategies (Sagulkoo et al. 2022; Truong et al. 2022). In these biological network models, the network nodes infer gene products or drug, disease whereas the edges infer the interactions between the nodes. The biological network analysis provides a schematic insight into the interactions between drug-target, between the drugs, between drug and disease, and signalling and transcriptional networks (Park 2019).

Thus, the computational approaches-based drug repurposing provides ample opportunities in providing novel therapeutic regimens of the approved drugs and also provide access to the failed drugs towards new therapeutic targets (Jarada et al. 2020). The widespread implications of computational approaches-based drug repurposing could be understood from the fact that the algorithm-based computational tools have the inherent ability to analyse the data sets from diverse resources including OMICS, biomedical associations, literature available, and even from the electronic health records (EHRs) (Zong et al. 2022). In an earlier report, among the clinically-approved compounds with known safety dosages, and suitable pharmacokinetic parameters, diethylene triamine penta-acetic acid (DPTA), a clinically-approved contrasting agent for molecular diagnostic imaging technique, successfully repurposed in the mitigation of alginate biosynthesis, which plays a significant part in the biofilm development in the ESKAPE pathogen, *P. aeruginosa* (Gi et al. 2014). Besides, the advanced machine learning technique-based predictive algorithms could also be employed to develop computational tools to guess the desired biological actions of the putative drug complex, approved drugs, and clinically failed drugs. For example, the machine learning technique-based user-friendly web server program "Anti-Biofilm" was developed, which provides a robust approach to forecast the effectiveness of putative drug molecules in modulating the drug resistance profile of bacterial pathogens with special importance on ESKAPE pathogens (Rajput et al. 2023). No doubt, several natural products, particularly plant-derived phytochemicals exhibited promising inhibitory effects on bacterial biofilms both in vitro and in vivo. However, not a single agent has been permitted by the FDA for clinical uses, which potentiates the usefulness of computational tools to guess the toxicity outline, suitable pharmacokinetic properties, bioavailability, and molecular interactions with the target sites of these phytochemicals. Based on the predictive scores, the concerned drug candidate could be clinically tested and spatially designed against biofilm mechanics in ESKAPE pathogens (Mishra et al. 2020).

19.4 Computational Tools for the Identification of Drugs Targeting ESKAPE Pathogens

In the process of drug discovery, the computational tools have a pivotal role to play as these tools primarily address the evaluation of thousands of molecules of interest, their drug-likeness, pharmacokinetic parameters, and their bioavailability index. Based upon the predicted scores, the selected compounds could further be explored towards desired attributes. For example, several computational tools such as SwissADME (for pharmacokinetic parameters), Osiris (for drug likeness or probability to be a drug), PROTOX (for toxicity prediction), and PASS-Prediction of Activity Spectra for Substances (for prediction of biological activities) were employed for the Cuminaldehyde, an active biocomponent of Cumin seed essential oil. Based on the predicted score, its antimicrobial actions by potentiating the role of Ciprofloxacin against *Staphylococcus aureus* was further estimated in vitro (Montiro-Neto et al. 2020). Similarly, computational tools such as sequence-based Homology modelling, Molecular docking, MD simulation, etc. are widely employed to determine the role of synthetic compounds, plant-derived compounds, their derivatives, and microbial bioactive compounds in the modulation of bacterial pathogenesis with special reference to QS, biofilm-mediated chronic infections, and drug resistance profile in ESKAPE pathogens. These computational tools facilitate in providing an insight into the mechanism of QS and biofilm mitigation by specifically targeting the pathophysiological targets. In the subsequent phase, selected drug candidates are subjected for validations using in vivo, and in vitro experimental analysis. For instance, the FDA-approved drug for clinical trials, e.g., Radezolid, a novel oxazolidinone antibiotic, and Sertindole, an antipsychotic drug were investigated for its antibacterial and biofilm inhibitory potential against *S. aureus*. Based on the biological network analysis by investigating the protein-protein interactions for the DEGs and differentially expressed proteins associated with bacterial metabolism and virulence profiles, both Radezolid, and Sertindole treated *S. aureus* promoted a potential reduction in the development of biofilm and associated infections (Tang et al. 2023; Wang et al. 2023).

The intervention of machine learning-based algorithms in biofilm research and the development of potential drug candidates against recalcitrant biofilms has transformed the drug discovery pipeline. In this regard, machine learning-based Random Forest algorithms were used to develop a webserver named "Molib" for the prediction of putative drug molecules as biofilm inhibitors. Since the developed webserver is user friendly, and a reliable machine learning tool was employed, the webserver could provide novel avenues to predict the biofilm mitigation properties of small molecules by providing an insight into the structural fingerprints, chemical and physical attributes of the small molecules of interest (Srivastava et al. 2020). Similarly, in an earlier study, spatially designed in silico tool-based webserver, dPABBs, was developed to predict the role of antimicrobial peptides (AMPs), and anti-biofilm peptides in controlling the biofilm-associated infections using the Machine learning tools, Support Vector Machine (SVM), and Weka (Sharma et al. 2016). BIPEP, the sequence-based prediction tool based upon the Machine learning approach (i.e., NMR and physicochemical predictors) was also developed to predict the antibiofilm peptides against biofilm-forming bacterial pathogens (Atanaki et al. 2020). In the last few years, revolutionary transformation in the development of computational tools critically facilitates drug development against ESKAPE pathogens. For instance, the STITCH tool is employed for probable target identification, VICMPred is employed for functional analysis of target proteins identified, VirulentPred is the predictive tool to guess the virulence profiles of the aimed proteins, and the PSORTb tool is used to forecast the sub-cellular location of the target proteins. Based on these predictions, the underlying mechanism of antibacterial activities of putative drug molecules against the ESKAPE pathogens could be understood and direct their role as potential therapeutic agents (Girija et al. 2021). The machine learning and artificial intelligence (AI)-based approaches in drug discovery and development critically accelerate the pace of putative drug molecules to enter clinical trials which is approximately 80% faster than that of the conventional approach suggesting the widespread attributes of the in silico tools in the expansion of a potent drug for exact pathophysiological condition (Sharma et al. 2021a). Thus, the state-of-the-art computational approaches critically transform our understanding of AMR in ESKAPE pathogens, target delineation of pathophysiological importance in AMR, identification of putative drug candidates with an insight into their interaction with the target proteins, and the stability of the interactions (Matamoros-Recio et al. 2021).

19.4.1 Phytochemicals as Potent Inhibitors of Quorum Sensing and Biofilms Using Computational Approaches

To combat the QS mediated virulence and development of biofilm layers in ESKAPE pathogens, the advent of computational tools particularly molecular docking approach received widespread attention owing to its user-friendly module, and availability of a free version of the tool. Using the molecular docking tools, several putative drug candidates could be identified based upon their binding affinity with the potential therapeutic targets such as QS machinery, biofilm phenotypes, and other pathophysiological proteins associated with bacterial virulence pathway. In this regard, the phytochemicals identified from plant-derived essential oils were screened against several pathophysiological targets of ESKAPE pathogens. Based on the affinity of docked complexes, suitable molecules of interest could further be explored for their role as promising antibacterial and antibiofilm agents (Noumi et al. 2023). Since, several groups of phytochemicals such as alkaloids, polyphenols, terpenes, terpenoids, anthocyanins, and flavonoids, etc. derived from plant sources were reported as favourable substitute to antibiotics in the attenuation of QS, and biofilm formation in pathogenic microorganisms; these phytochemicals could be explored through advanced in silico tools for the mechanistic insights into the mechanism of QS attenuation, biofilm mitigation, and alteration of drug resistance profiles (Shamim et al. 2023).

In this regard, pharmacologically relevant plant derived flavone, Hispidulin, exhibited promising QS inhibition and biofilm mitigation by exactly pointing on the QS transcriptional regulatory protein, LasR as evident by the molecular docking analysis (Anju et al. 2022). Based upon the molecular docking, MD simulation could be employed to investigate the stability in the docked complexes and further predict the suitability of the ligand molecules as a potent inhibitors of QS and biofilm development in ESKAPE pathogens. For example, several Benzimidazole derivatives were docked to QS regulatory protein, LasR followed by MD simulation to establish the efficacy of the selected derivatives in competitively inhibiting the binding of LasR with its natural ligand and thus altering the QS pathway in *P. aeruginosa*. Thus, based upon the MD simulation analysis, the selected molecules could be employed further to derivatize their role as QS and biofilm inhibiting agents (Abd El-Aleam et al. 2022). Apart from LasR, MvfR is also an important regulatory protein in activating the QS regulatory virulence in *P. aeruginosa* and is used as a potential therapeutic target to attenuate bacterial virulence. In this regard, MD simulation could be employed to elucidate the stability of the screened ligands with the receptor protein, MvfR and the possible mode of interactions (Vieira et al. 2022). Thus, the computational approaches, particularly, the molecular docking tools have been widely used as in silico platforms to understand the interactions between the plant-derived phytochemicals as ligands with the target proteins of interest in the ESKAPE pathogens. Further, in vivo and in vitro studies could be investigated by taking into consideration of the computer generated predictions on antibacterial, QS inhibition, and biofilm mitigation with aided stability and safety profile of the phytochemicals tested (Abishad et al. 2021). The list of plant-derived phytochemicals with anti-quorum sensing and antibiofilm activities against ESKAPE pathogens based on the computational approaches were presented in Table 19.1.

19.4.2 Computational Tools for Identification of Microbial Secondary Metabolites Against ESKAPE Pathogens

Similar to the diversity of phytochemicals from plant origin, microbes (e.g., bacteria and fungi) also have the tendency to produce a highly diverse range of bioactive compounds of pharmacological and pharmaceutical importance. It is evident from the earlier reports that genome mining of microbes displays a rich heritage of secondary metabolites that are essential for widespread potential as medicines. The recent trends in OMICS-based methods have delivered a new horizon to isolate microbial species with rich sources of bioactive metabolites with widespread biomedical and pharmaceutical applications (Dhandapani et al. 2022). The computational tools, particularly the molecular docking approach, facilitate the potential role of these bioactive secondary metabolites in the management of chronic infections pattern in the ESKAPE group of pathogens (Meenambiga and Rajagopal 2018). The list of microbial metabolites with anti-quorum sensing and antibiofilm activities against ESKAPE pathogens based on the computational approaches is presented in Table 19.2.

19.5 Current Trends and Future Prospective

It is evident from the recent trends in drug discovery that computational approaches when used in combination with the in vitro works, the chances of providing novel therapeutic alternatives against several diseases and disorders become a reality. Computational tools offer a vision into the mode of action of the putative drug

Table 19.1 List of plant-derived phytochemicals screened for their binding affinity towards potential therapeutic targets of quorum sensing (QS) regulated virulence and biofilm mechanics in ESKAPE pathogens using computational approaches

Sl. no.	Phytochemicals	Phytochemical class	In silico tools used	Therapeutic targets	Target ESKAPE pathogen	Remarks	References
1.	Berberine	Alkaloid	Homology modeling, molecular docking	QS receptor proteins, LasR and RhlR	*Pseudomonas aeruginosa* PA01	Showed better binding affinity towards RhlR and could modulate the QS regulatory behavior	Aswathanarayan and Vittal (2018)
2.	3,5,7-Trihydroxyflavone	Flavonoids	Molecular docking	Homoserine lactone synthases, LasI and RhlI; cognate receptor, LasR	*P. aeruginosa*	Exhibits binding affinity with the target proteins with highest affinity towards LasR	Abinaya and Gayathri (2019)
3.	Hexadecanoic acid, 1-(hydroxymethyl), 1,2-ethannediylester	Fatty acid	Molecular docking	Acyl homoserine lactone (AHL) synthase	*Acinetobacter baumannii*	Exhibits strong binding affinity with AHL synthase	Namasivayam et al. (2019)
4.	Rutin	Flavonoids	Molecular docking	Outer membrane protein, OprD	*P. aeruginosa*	Exhibits high binding affinity towards the outer membrane protein	Adnan et al. (2020)
5.	Ursolic acid	Terpenoids	Molecular docking	Outer membrane protein, OprD	*P. aeruginosa*	Shows strong interactions with the OprD	Adnan et al. (2020)
6.	Myrtenol	Monoterpenes	Molecular docking and MD simulation	Penicillin binding protein 2 (PBP2)	*Staphylococcus aureus*	Shows higher binding affinity towards PBP2 with stable interactions	Cordeiro et al. (2020a, b)
7.	α-Guaiene, δ-guaiene	Sequiterpenes	Molecular docking	Cognate QS receptor, LasR	*P. aeruginosa*	Shows strong hydrogen binding interactions	Mansuri et al. (2022)

Sl. no.	Phytochemicals	Phytochemical class	In silico tools used	Therapeutic targets	Target ESKAPE pathogen	Remarks	References
8.	Methyl eugenol	Phenyl-propanoids	Molecular docking	QS receptor proteins, LasR, RhlR, and PqsR	*P. aeruginosa*	Exhibits significant QS inhibitory activities by binding to the receptor proteins	Sobieszczanska et al. (2020)
9.	Guanosine	Nucleoside	Molecular docking	Cognate QS receptor, LasR	*P. aeruginosa*	Strong binding interactions with the cognate receptor protein	Baloyi et al. (2021)
10.	Coumarins	Polyphenols	Molecular docking, ADMET profiling	LasR receptor	*P. aeruginosa*	Exhibits attenuation of QS regulatory activities by binding to LasR, ADMET profiling showed the bioactive potential similar to that of ciprofloxacin	Eswaramurthy et al. (2021)
11.	6-Gingerol	Phenolics	Molecular docking, MD simulation	QS cognate receptors, LasR, RhlR, and PhzR	*P. aeruginosa*	Exhibits strong QS attenuation potential by specifically targeting the QS signaling pathway	Shukla et al. (2021)
12.	Pheophorbide, pyropheophorbide	Pheophytin derivative	Molecular docking	Adhesion proteins	*S. aureus*, *P. aeruginosa*	Controls biofilm formation by exhibiting strong affinity towards the adhesion proteins	Awadelkareem et al. (2022)
13.	Vanillin hybrids with eugenol, guaiacol, and cinnamaldehyde	Phenolics	Molecular docking	QS cognate receptors, LasR, RhlR, and PqsR	*P. aeruginosa*	Exhibit strong anti-virulence properties by stable interactions with the QS signaling receptors	Dua et al. (2023)

(continued)

Table 19.1 (continued)

Sl. no.	Phytochemicals	Phytochemical class	In silico tools used	Therapeutic targets	Target ESKAPE pathogen	Remarks	References
14.	Limonene	Terpenes	Molecular docking	QS cognate receptor, LasR	*P. aeruginosa*	Shows strong hydrogen bonding and modulate the virulence pathway	Ghannay et al. (2022)
15.	Thymoquinone	Terpenes	Molecular docking, MD simulation	Transcriptional regulator, qacR	*S. aureus*	Shows stable binding affinity	Qureshi et al. (2022)
16.	Homomangiferin	C-glycosyl xanthone	Molecular docking	QS cognate receptors, LasR, RhlR; biofilm associated proteins, PilY1, PilT	*P. aeruginosa*	Exhibits strong interference on QS regulated virulence and biofilm mechanics	Samreen et al. (2022)
17.	Betulin	Pentacyclic triterpenoid	Molecular docking	Biofilm associated protein, PilY1	*P. aeruginosa*	Inhibits biofilm associated infections by targeting biofilm associated protein	Samreen et al. (2022)
18.	Catechin, epicatechin	Flavan-3-ol (flavonoids)	Molecular docking	Tyrosyl-tRNA synthetases	*S. aureus*	Exhibits strong antibacterial potential with improved biofilm inhibition	Ellafi et al. (2023)
19.	10-Undecenoic acid	Fatty acids	Molecular docking	QS AHL synthase, LasI; cognate receptor, LasR	*P. aeruginosa*	Acts as potential inhibitor of QS regulated virulence by targeting QS circuit	Fernandes et al. (2023)

Sl. no.	Phytochemicals	Phytochemical class	In silico tools used	Therapeutic targets	Target ESKAPE pathogen	Remarks	References
20.	Pinocembrin	Flavanone	Molecular docking, MD simulation	SagS sensor regulator	*P. aeruginosa*	Shows promising antibiofilm potential by providing stable interactions with the sensor regulator	Behera et al. (2022)
21.	Pulverulentone A	Phloroglucinol derivative	Molecular docking	QS response regulators, LasR, PqsR, QscR	*P. aeruginosa*	Exhibits strong antivirulence potential by attenuation of QS signaling pathway	Ismail et al. (2021)
22.	Carvacrol	Phenolic monoterpenoid	Molecular docking	QS regulatory LasI, LasR	*P. aeruginosa*	Inhibits biofilm formation by interfering with the LasI/R pathway	Tapia-Rodriguez et al. (2019)
23.	Mosloflavone	Flavonoids	Molecular docking	QS cognate receptors, LasR, RhlR	*P. aeruginosa* PAO1	Competitively inhibits the binding of natural ligands and alters the QS mediated virulence	Hnamte et al. (2019)

Table 19.2 List of microbes-derived secondary metabolites screened for their binding affinity towards potential therapeutic targets of quorum sensing (QS) regulated virulence and biofilm mechanics in ESKAPE pathogens using computational approaches

Sl. no.	Microbial secondary metabolites	Source microorganism	In silico tools used	Therapeutic targets	Target ESKAPE pathogen	Remarks	References
1.	13*Z*-Octadecenal (fatty acid)	*Streptomyces griseoincarnatus* HK12	Molecular docking	LasI (autoinducer synthase)	*Pseudomonas aeruginosa*	Binds to conserved sites of substrate binding and modulate the QS signalling pathway	Kamarudheen and Rao (2019)
2.	Zeaxanthin (carotenoid)	Pure (lichens)	Molecular docking	LasI, quorum sensing (QS) receptors, LasR, RhlR	*P. aeruginosa*	Shows better binding affinity towards the targets than that of azithromycin	Gökalsın et al. (2017)
3.	Nakinadine B	*Amphimedon* sp. (sponge)	Molecular docking	QS cognate receptor, LasR and PqsE	*P. aeruginosa* PAO1	Inhibits QS mediated virulence by specifically targeting QS cascade	Chaieb et al. (2022)
4.	Hydrocinnamic acid	*Enterobacter xiangfangensis*	Molecular docking	QS cognate receptor, LasR	*P. aeruginosa*	Impairs AHL-based QS mechanism in *P. aeruginosa*	Sharma et al. (2019)
5.	Stigmatellin Y	*Bacillus subtilis* BR4	Molecular docking	QS receptor protein, PqsR	*P. aeruginosa*	Shows antivirulence properties by dismantling the PQS pathway	Boopathi et al. (2017)
6.	Diketopiperazine factor	*Rheinheimera aquimaris* QSI02	Molecular docking, MD simulation	QS cognate receptor, LasR	*P. aeruginosa*	Exhibits stable complex with LasR and interfere the signalling mechanism	Sun et al. (2016)

Sl. no.	Microbial secondary metabolites	Source microorganism	In silico tools used	Therapeutic targets	Target ESKAPE pathogen	Remarks	References
7.	Actinomycin D	*Streptomyces cyaneochromogenes* RC1	Molecular docking	QS cognate receptors, LasR, RhlR	*P. aeruginosa*	Shows greater binding affinity with LasR as compared to the natural ligand suggesting its role in competitive inhibition of binding of natural ligand to LasR	Zeng et al. (2022)
8.	Phenalinolactones A–D, synerazol	Actinomycetes	Molecular docking	QS regulatory protein, AgrA	*S. aureus*	Specifically interact with the ATP-binding site of AgrA and impair the QS signalling cascade	Desouky et al. (2022)
9.	Fenaclon	*Diaporthe phaseolorum* SSP12	Molecular docking	QS cognate receptor protein, LasR	*P. aeruginosa* PAO1	Competitively inhibits the binding of signaling molecule to the cognate LasR protein and modulate QS regulated behaviour	Pattnaik et al. (2018)
10.	2,4-Di-*tert*-butylphenol	*Diaporthe phaseolorum* SSP12	Homology modeling, molecular docking	QS cognate receptor protein, RhlR	*P. aeruginosa* PAO1	Competitively inhibits the binding of signaling molecule to the cognate RhlR protein and modulate QS regulated biofilm dynamics	Pattnaik et al. (2018)

(continued)

Table 19.2 (continued)

Sl. no.	Microbial secondary metabolites	Source microorganism	In silico tools used	Therapeutic targets	Target ESKAPE pathogen	Remarks	References
11.	Tenuazonic acid	*Albophoma* sp. BAPR5	Drug likeness profiling, PASS (prediction of activity spectra for substances) analysis	Membrane permeability factor, membrane integrity factor, GPCR kinase receptor	*Staphylococcus* sp.	Shows promising antibacterial and antibiofilm activities	Rabha et al. (2023)
12.	Avellanin C	*Hamigera ingelheimensis*	Molecular docking, MD simulation	QS regulatory protein, AgrA	*Staphylococcus aureus*	Forms stable complex with AgrA and modulates the QS pathway	Basak et al. (2022)
13.	1-(4-Amino-2-hydroxyphenyl) ethanone (AHE)	*Phomopsis liquidambari* S47	Molecular docking	QS proteins, LasR, RhlR, PqsR	*P. aeruginosa* PAO1	Suppresses the QS mediated virulence and biofilm formation	Zhou et al. (2021)

candidates against possible drug targets, thereby more accurate drug candidates could be selected at a faster rate. The biggest advantage of the computational tools in rational drug design is that the several computational tools could be explored simultaneously for the effective selection of the putative drug molecules of interest (Brogi 2019).

The recent advancement in the group of large data sets associated with cheminformatics databases marked its presence in the part of big data and AI in drug design and development. The vast algorithms associated with AI have provided a platform for big data mining and precise predictions of target-based drug identification through virtual drug screening (Tripathi et al. 2021). The powerful data mining tool of AI with the beginning of machine learning algorithms is presently being explored at various points of drug development. The AI-based data mining approach not only minimizes the R&D costs but also critically influences the increase in drug approval efficacy, thereby promoting the resurgent drug development programmes of pharmaceutical industries (Mak and Pichika 2019; Wang et al. 2019). The resurgence of AI-based bioinformatics tools has been useful in both the target based screening as well as phenotypic screening in the drug discovery process suggesting its wide avenues in the expansion of potent drug molecules (Malandraki-Miller and Riley 2021). The machine learning-based algorithms and AI-based neural networks are also being incorporated in both structure-based as well as ligand-based virtual evaluation, pharmacophore modelling, and QSAR (Quantitative Structure-Activity Relationship) (Gupta et al. 2021) (Fig. 19.3).

The machine learning algorithms and AI-based network explore the sequence-based structures for the generation of peptidic scaffolds for novel AMPs discovery with promising antibacterial potential and could be instrumental in the match against resistance of several drug patterns in ESKAPE pathogens (Agüero-Chapin et al. 2022). Hence, the AI-based approach transforms the computational tools in drug discovery processes by improved predictions of toxicity profiles of drugs, bioactive potential of drug molecules, bioavailability of the drugs screened, and physicochemical characteristics of the putative drug molecules. Once predicted, AI-based models could be instrumental in drug-target interactions, de novo drug design, drug-drug interactions, and binding affinity calculations with improved stability profile generation leading to improved drug discovery applications (Chen et al. 2023). The AI-based approach also concisely facilitates the spatial design and development of approved drug molecules for drug repurposing to improve therapeutic applications (Moingeon et al. 2022).

The advancement in the powerful, faster, versatile, and physiologically relevant mass spectrometry (MS) technique has given a new dimension to the HTS (High Throughput Screening) for the discovery of potential drug molecules from millions of molecules under screening. The amalgamation of the versatile MS technique with HTS also found to be influential in several phases of drug development pipelines such as identification of the target, validation of the target, optimization of lead compounds, and prediction of the compound's cellular mechanism of action (Dueñas et al. 2023). Several factors such as (a) advanced automation in the structural revolution of microcrystallographic techniques to reveal the 3D structures of

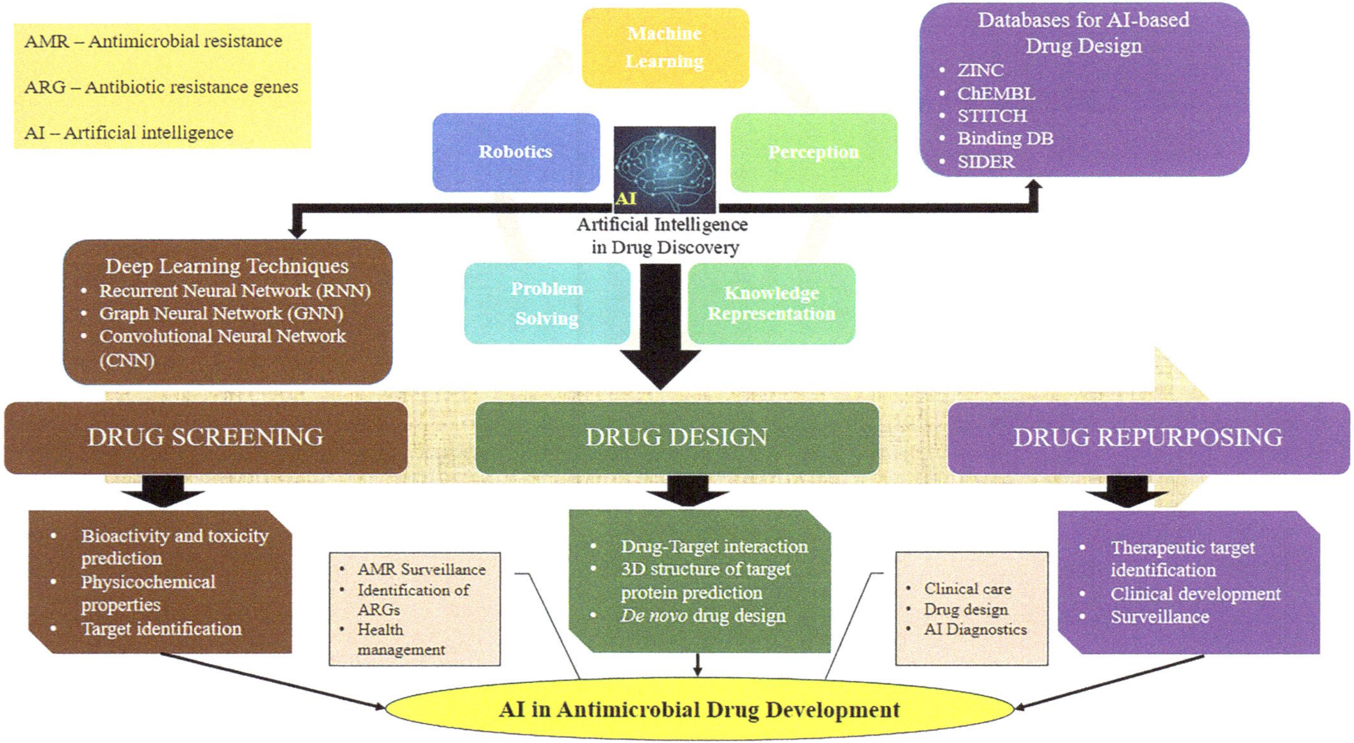

Fig. 19.3 An overview of the artificial intelligence (AI) and machine learning (ML)-based tools in drug screening, design, and repurposing

possible therapeutic targets, (b) fast and robust development of drug-like space with advanced ultra-large virtual libraries housing billions of compounds, and (c) the revolutionary advancement in computational tools to strive upon the ligand molecule abundance are responsible for the revolutionary transformation in the drug discovery and development pipelines (Sadybekov and Katritch 2023). The progress of high throughput computational tools along with the transforming progress in the field of cheminformatics, analytical chemistry, and structural biology has provided the scientific community as well as the pharmaceutical sectors a new horizon to tackle the global issue of AMR by the development of potent anti-infective drugs targeting ESKAPE pathogens (Krell and Matilla 2022).

19.6 Conclusion

In the fight from chronic bacterial infections, and biofilm-related drug resistance in ESKAPE pathogens with the last resort of antibiotics available; it is imperative to enhance our drug design and discovery programmes with an aim to counteract the drug resistant pathogens by providing an alternative to conventional antibiotics. No doubt with the improvement in the pharmaceutical, biotechnological, and biomedical sectors the drug discovery process seems to be accelerated a bit but requires more apprehensions to face the wrath of MDR phenomena. In this regard, it is imperative to look for computational alternatives for a rapid and perfect identification of potent drug molecules against therapeutic targets of pathophysiological importance. The advancement in computational approaches such as MD simulation, molecular docking, QSAR model, de novo drug design, pharmacophore modeling, structure- and sequence-based virtual screening, etc. have transformed our scientific efforts to achieve the identification of potent drug complexes against MDR pathogens. No doubt, the in silico approaches have provided novel avenues to predict potential drug candidates against therapeutic targets associated with drug resistance in ESKAPE pathogens; very few drugs have been approved for marketed use. Hence, the scientific community as well as the pharmaceutical sectors should step up on this opportunity to develop potent computational tools and their widespread applicability for the development of therapeutics with marketed value.

Acknowledgement The author, Subhaswaraj Pattnaik, acknowledges SERB DST, Govt. of India for the National Post-Doctoral Fellowship (NPDF) (Reference No. PDF/2021/001260). The authors would also like to acknowledge OHEPEE, Govt. of Odisha, through the World Bank for giving financial support.

References

Abd El-Aleam RH, Sayed AM, Taha MN, George RF, Georgey HH, Abdel-Rahman HM (2022) Design and synthesis of novel benzimidazole derivatives as potential *Pseudomonas aeruginosa* anti-biofilm agents inhibiting LasR: evidence from comprehensive molecular dynamics simulation and *in vitro* investigation. Eur J Med Chem 241:114629. https://doi.org/10.1016/j.ejmech.2022.114629

Abinaya M, Gayathri M (2019) Inhibition of biofilm formation, quorum sensing activity and molecular docking study of isolated 3,5,7-Trihydroxyflavone from *Alstonia scholaris* leaf against *P. aeruginosa*. Bioorg Chem 87:291–301. https://doi.org/10.1016/j.bioorg.2019.03.050

Abishad P, Niveditha P, Unni V, Vergis J, Kurkure NV, Chaudhari S, Rawool DB, Barbuddhe SB (2021) *In silico* molecular docking and *in vitro* antimicrobial efficacy of phytochemicals against multi-drug-resistant enteroaggregative *Escherichia coli* and non-typhoidal *Salmonella* spp. Gut Pathog 13:46. https://doi.org/10.1186/s13099-021-00443-3

Aboagye CI, Ampomah GB, Mensah JO, Mensah CN, Nartey D, Gasu EN, Borquaye LS (2023) N-Benzylimidazoles as potential antimicrobial and antibiofilm agents—syntheses, bioactivity, and molecular docking studies. Sci Afr 19:e01529. https://doi.org/10.1016/j.sciaf.2022.e01529

Abo-Salem HM, Abd El Salam HA, Abdel-Aziem AM, Abdel-Aziz MS, El-Sawy ER (2021) Synthesis, molecular docking, and biofilm formation inhibitory activity of bis(indolyl)pyridines analogues of the marine alkaloid nortopsentin. Molecules 26(14):4112. https://doi.org/10.3390/molecules26144112

Adnan M, Patel M, Deshpande S, Alreshidi M, Siddiqui AJ, Reddy MN et al (2020) Effect of *Adiantum philippense* extract on biofilm formation, adhesion with its antibacterial activities against foodborne pathogens, and characterization of bioactive metabolites: an *in vitro-in silico* approach. Front Microbiol 11:823. https://doi.org/10.3389/fmicb.2020.00823

Agamah FE, Mazandu GK, Hassan R, Bope CD, Thomford NE, Ghansah A, Chimusa ER (2020) Computational/*in silico* methods in drug target and lead prediction. Brief Bioinform 21(5):1663–1675. https://doi.org/10.1093/bib/bbz103

Agüero-Chapin G, Galpert-Cañizares D, Dominguez-Perez D, Marrero-Ponce Y, Perez-Machado G, Teijeira M, Antunes A (2022) Emerging computational approaches for antimicrobial peptide discovery. Antibiotics. 11(7):936. https://doi.org/10.3390/antibiotics11070936

Ahmed F, Lee JW, Samantasinghar A, Kim YS, Kim KH, Kang IS et al (2022) SperoPredictor: an integrated machine learning and molecular docking-based drug repurposing framework with use case of COVID-19. Front Public Health 10:902123. https://doi.org/10.3389/fpubh.2022.902123

Almihyawi RAH, Naman ZT, Al-Hasani HMH, Muhseen ZT, Zhang S, Chen G (2022) Integrated computer-aided drug design and biophysical simulation approaches to determine natural antibacterial compounds for *Acinetobacter baumannii*. Sci Rep 12:6590. https://doi.org/10.1038/s41598-022-10364-z

Alves VM, Korn D, Pervitsky V, Thieme A, Capuzzi SJ, Baker N et al (2022) Knowledge-based approaches to drug discovery for rare diseases. Drug Discov Today 27(2):490–502. https://doi.org/10.1016/j.drudis.2021.10.014

Ambure P, Halder AK, Diaz HG, Cordeiro MNDS (2019) QSAR-Co: an open source software for developing robust multitasking or multitarget classification-based QSAR models. J Chem Inf Model 59(6):2538–2544. https://doi.org/10.1021/acs.jcim.9b00295

Anju VT, Busi S, Mohan MS, Ranganathan S, Ampasala DR, Kumavath R, Dyavaiah M (2022) *In vivo*, *in vitro* and molecular docking studies reveal the anti-virulence property of hispidulin against *Pseudomonas aeruginosa* through the modulation of quorum sensing. Int Biodeterior Biodegradation 174:105487. https://doi.org/10.1016/j.ibiod.2022.105487

Ashburn TT, Thor KB (2004) Drug repositioning: identifying and developing new uses for existing drugs. Nat Rev Drug Discov 3:673–683. https://doi.org/10.1038/nrd1468

Aswathanarayan JB, Vittal RR (2018) Inhibition of biofilm formation and quorum sensing mediated phenotypes by berberine in *Pseudomonas aeruginosa* and *Salmonella typhimurium*. RSC Adv 8:36133. https://doi.org/10.1039/c8ra06413j

Atanaki FF, Behrouzi S, Ariaeenejad S, Boroomand A, Kavousi K (2020) BIPEP: sequence-based prediction of biofilm inhibitory peptides using a combination of NMR and physicochemical descriptors. ACS Omega 5(13):7290–7297. https://doi.org/10.1021/acsomega.9b04119

Awadelkareem AM, Al-Shammari E, Elkhalifa AO, Adnan M, Siddiqui AJ, Mahmood D et al (2022) Anti-adhesion and antibiofilm activity of *Eruca sativa* Miller extract targeting cell adhesion proteins of food-borne bacteria as a potential mechanism: combined *in vitro-in silico* approach. Plants 11:610. https://doi.org/10.3390/plants11050610

Ayerbe-Algaba R, Gil-Marqués ML, Jiménez-Mejías ME, Sánchez-Encinales V, Parra-Millán R, Pachón-Ibáñez ME, Pachón J, Smani Y (2018) Synergistic activity of niclosamide in combination with colistin against colistin-susceptible and colistin-resistant *Acinetobacter baumannii* and *Klebsiella pneumoniae*. Front Cell Infect Microbiol 8:348. https://doi.org/10.3389/fcimb.2018.00348

Bai Q, Liu S, Tian Y, Banegas-Luna AJ, Pérez-Sánchez H, Huang J, Liu H, Yao X (2022) Application advances of deep learning methods for de novo drug design and molecular dynamics simulation. WIREs Comput Mol Sci 12(3):e1581. https://doi.org/10.1002/wcms.1581

Baloyi IT, Adeosun IJ, Yusuf AA, Cosa S (2021) *In silico* and *in vitro* screening of antipathogenic properties of *Melianthus comosus* (Vahl) against *Pseudomonas aeruginosa*. Antibiotics 10:679. https://doi.org/10.3389/antibiotics10060679

Basak HK, Paswan U, Chatterjee A (2022) A computational approach to explore the binding mechanism of secondary metabolites of *Hamigera ingelheimensis* with AgrA quorum sensing proteins. Vietnam J Chem 60(1):21–36. https://doi.org/10.1002/vjch.202100046

Bassetti S, Tschudin-Sutter S, Egli A, Osthoff M (2022) Optimizing antibiotic therapies to reduce the risk of bacterial resistance. Eur J Intern Med 99:7–12. https://doi.org/10.1016/j.ejim.2022.01.029

Batool M, Ahmad B, Choi S (2019) A structure-based drug discovery paradigm. Int J Mol Sci 20:2783. https://doi.org/10.3390/ijms20112783

Behera SK, Panda AK, Mishra R, Mahanty A, Bisht SS (2022) Structure based virtual screening and molecular dynamics of natural anti-biofilm compounds against SagS response regulator/sensor kinase in *Pseudomonas aeruginosa*. J Biomol Struct Dyn 41:6011. https://doi.org/10.1080/07391102.2022.2100482

Bergstrom CAS, Larsson P (2018) Computational prediction of drug solubility in water-based systems: qualitative and quantitative approaches used in the current drug discovery and development setting. Int J Pharm 540:185–193. https://doi.org/10.1016/j.ijpharm.2018.01.044

Bhat AA, Tandon N, Singh I, Tandon R (2023) Structure-active relationship (SAR) and antibacterial activity of pyrrolidine based hybrids: a review. J Mol Struct 1283:135175. https://doi.org/10.1016/j.molstruc.2023.135175

Blaschke T, Arus-Porus J, Chen H, Margreitter C, Tyrchan C, Engkvist O et al (2020) REINVENT 2.0—an AI tool for de novo drug design. J Chem Inf Model 60(12):5918–5922. https://doi.org/10.1021/acs.jcim.0c00915

Boopathi S, Vashisth R, Manoharan P, Kandasamy R, Sivakumar N (2017) Stigmatellin Y—an anti-biofilm compound from *Bacillus subtilis* BR4 possibly interferes in PQS–PqsR mediated quorum sensing system in *Pseudomonas aeruginosa*. Bioorg Med Chem Lett 27:2113–2118. https://doi.org/10.1016/j.bmcl.2017.03.074

Breijyeh Z, Karaman R (2023) Design and synthesis of novel antimicrobial agents. Antibiotics 12:628. https://doi.org/10.3390/antibiotics12030628

Brogi S (2019) Computational approaches for drug discovery. Molecules 24(17):3061. https://doi.org/10.3390/molecules24173061

Cain R, Narramore S, McPhillie M, Simmons K, Fishwick CWG (2014) Applications of structure-based design to antibacterial drug discovery. Bioorg Chem 55:69–76. https://doi.org/10.1016/j.bioorg.2014.05.008

Cardoso MH, Orozco RQ, Rezende SB, Rodrigues G, Oshiro KGN, Candido ES, Franco OL (2020) Computer-aided design of antimicrobial peptides: are we generating effective drug candidates? Front Microbiol 10:3097. https://doi.org/10.3389/fmicb.2019.03097

Chaieb K, Kouidhi B, Hosawi SB, Baothman OAS, Zamzami MA, Altayeb HN (2022) Computational screening of natural compounds as putative quorum sensing inhibitors targeting drug resistance bacteria: molecular docking and molecular dynamics simulations. Comput Biol Med 145:105517. https://doi.org/10.1016/j.compbiomed.2022.105517

Chen W, Liu X, Zhang S, Chen S (2023) Artificial intelligence for drug discovery: resources, methods, and applications. Mol Ther Nucleic Acid 31:691–702. https://doi.org/10.1016/j.omtn.2023.02.019

Cordeiro L, Figueiredo P, Souza H, Sousa A, Andrade-Junior F, Medeiros D et al (2020a) Terpinen-4-ol as an antibacterial and antibiofilm agent against *Staphylococcus aureus*. Int J Mol Sci 21:4531. https://doi.org/10.3390/ijms21124531

Cordeiro L, Figueiredo P, Souza H, Sousa A, Andrade-Junior F, Barbosa-Filho J, Lima E (2020b) Antibacterial and antibiofilm activity of myrtenol against *Staphylococcus aureus*. Pharmaceuticals 13:133. https://doi.org/10.3390/ph13060133

Cortat Y, Nedyalkova M, Schindler K, Kadakia P, Demirci G, Sovari SN, Crochet A, Salentinig S, Lattuada M, Steiner OM, Zobi F (2023) Computer-aided drug design and synthesis of rhenium Clotrimazole antimicrobial agents. Antibiotics 12(3):619. https://doi.org/10.3390/antibiotics12030619

da Rosa TF, Coelho SS, Foletto VS, Bottega A, Serafin MB, de Souza Machado C, Franco LN, de Paula BR (2020) Alternatives for the treatment of infections caused by ESKAPE pathogens. J Clin Pharm Ther 45(4):863–873. https://doi.org/10.1111/jcpt.13149

da Silva TH, Hachigian TZ, Lee J, King MD (2022) Using computers to ESKAPE the antibiotic resistance crisis. Drug Discov Today 27(2):456–470. https://doi.org/10.1016/j.drudis.2021.10.005

Dadgostar P (2019) Antimicrobial resistance: implications and cost. Infect Drug Resist 12:3903–3910. https://doi.org/10.2147/IDR.S234610

Dai Q, Yan Y, Ning X, Li G, Yu J, Deng J et al (2021) AncPhore: a versatile tool for anchor pharmacophore steered drug discovery with applications in discovery of new inhibitors targeting metallo-β-lactamases and indoleamine/tryptophan 2,3-dioxygenases. Acta Pharm Sin B 11(7):1931–1946. https://doi.org/10.1016/j.apsb.2021.01.018

de Bruijn WJC, Hageman JA, Araya-Cloutier C, Gruppen H, Vincken JP (2018) QSAR of 1,4-benzoxazin-3-one antimicrobials and their drug design perspectives. Bioorg Med Chem 26(23–24):6105–6114. https://doi.org/10.1016/j.bmc.2018.11.016

de Macedo GHR, Costa GDE, Oliveira ER, Damasceno GV, Mendonca JSP, dos Santos SL et al (2021) Interplay between ESKAPE pathogens and immunity in skin infections: an overview of the major determinants of virulence and antibiotic resistance. Pathogens 10(2):148. https://doi.org/10.3390/pathogens10020148

De Oliveira DMP, Forde BM, Kidd TJ, Harris PNA, Schembri MA, Beatson SA, Paterson DL, Walker MJ (2020) Antimicrobial resistance in ESKAPE pathogens. Clin Microbiol Rev 33(3):e00181-19. https://doi.org/10.1128/CMR.00181-19

Deb R, Kabelka I, Pribyi J, Vacha R (2023) *De novo* design of peptides that form transmembrane barrel pores killing antibiotic resistant bacteria. Biophys J 122(3):155a. https://doi.org/10.1016/j.bpj.2022.11.984

Denissen J, Reyneke B, Waso-Reyneke M, Havenga B, Barnard T, Khan S, Khan W (2022) Prevalence of ESKAPE pathogens in the environment: antibiotic resistance status, community-acquired infection and risk to human health. Int J Hyg Environ Health 244:114006. https://doi.org/10.1016/j.ijheh.2022.114006

Desouky SE, Abu-Elghait M, Fayed EA, Selim S, Yousuf B, Igarashi Y et al (2022) Secondary metabolites of Actinomycetales as potent quorum sensing inhibitors targeting Gram-positive pathogens: *in vitro* and *in silico* study. Metabolites 12:246. https://doi.org/10.3390/metabo12030246

Dhandapani R, Thangavelu S, Ragunathan L, Paramasivam R, Velmurugan P, Muthupandian S (2022) Potential bioactive compounds from marine *Streptomyces* sp. and their in vitro antibiofilm and antibacterial activities against antimicrobial-resistant clinical pathogens. Appl Biochem Biotechnol 194:4702–4723. https://doi.org/10.1007/s12010-022-04072-7

dos Santos RN, Ferreira LG, Andricopulo AD (2018) Practices in molecular docking and structure-based virtual screening. In: Gore M, Jagtap UB (eds) Computational drug discovery and design, methods in molecular biology. Springer Nature, Berlin, pp 31–50. https://doi.org/10.1007/978-1-4939-7756-7_3

Dotolo S, Marabotti A, Facchiano A, Tagliaferri R (2021) A review on drug repurposing applicable to COVID-19. Brief Bioinform 22(2):726–741. https://doi.org/10.1093/bib/bbaa288

Douguet D (2010) e-LEA3D: a computational-aided drug design web server. Nucleic Acid Res 38:W615–W621. https://doi.org/10.1093/nar/gkq322

Dua T, Mangal S, Akshita G, Harshdeep AAK, Sharma P et al (2023) Novel vanillin-based hybrids inhibit quorum sensing and silences phenotypical expressions in *Pseudomonas aeruginosa*. Drug Dev Res 84:45–61. https://doi.org/10.1002/ddr.22011

Dubey K, Dubey R (2020) Computational screening of narcissoside a glycosyloxyflavone for potential novel coronavirus 2019 (COVID-19) inhibitor. Biomed J 43:363–367. https://doi.org/10.1016/j.bj.2020.05.002

Dubey KK, Indu, Sharma M (2020) Reprogramming of antibiotics to combat antimicrobial resistance. Arch Pharm 2020:e2000168. https://doi.org/10.1002/ardp.202000168

Duenas ME, Peltier-Heap RE, Leveridge M, Annan RS, Buttner FH, Trost M (2023) Advances in high-throughput mass spectrometry in drug discovery. EMBO Mol Med 15:e14850. https://doi.org/10.15252/emmm.202114850

Dugger SA, Platt A, Goldstein DB (2018) Drug development in the era of precision medicine. Nat Rev Drug Discov 17:183–196. https://doi.org/10.1038/nrd.2017.226

Dulsat J, Lopez-Nieto B, Estrada-Tejedor R, Borrell JI (2023) Evaluation of free online ADMET tools for academic or small biotech environments. Molecules 28(2):776. https://doi.org/10.3390/molecules28020776

El Haddad L, Harb CP, Gebara MA, Stibich MA, Chemaly RF (2019) A systematic and critical review of bacteriophage therapy against multidrug-resistant ESKAPE organisms in humans. Clin Infect Dis 69(1):167–178. https://doi.org/10.1093/cid/ciy947

Ellafi A, Farhat R, Snoussi M, Noumi E, Anouar EH, El May MV et al (2023) Phytochemical profiling, antimicrobial, antibiofilm, insecticidal, and anti-leishmanial properties of aqueous extract from *Juglans regia* L. root bark: *in vitro* and *in silico* approaches. Int J Food Prop 26(1):1079–1097. https://doi.org/10.1080/10942912.2023.2200561

Eswaramurthy R, Hailekiros H, Kedir F, Endale M (2021) *In silico* molecular docking, DFT analysis and ADMET studies of carbazole alkaloid and coumarins from roots of *Clausena anisata*: a potent inhibitor for quorum sensing. Adv Appl Bioinform Chem 14:13–24. https://doi.org/10.2147/AABC.S290912

Fernandes S, Borges A, Gomes IB, Sousa SF, Simoes M (2023) Curcumin and 10-undecenoic acid as natural quorum sensing inhibitors of LuxS/AI-2 of *Bacillus subtilis* and LasI/LasR of *Pseudomonas aeruginosa*. Food Res Int 165:112519. https://doi.org/10.1016/j.foodres.2023.112519

Foscato M, Venkatraman V, Jensen VR (2019) DENOPTIM: software for computational de novo design of organic and inorganic molecules. J Chem Inf Model 59(10):4077–4082. https://doi.org/10.1021/acs.jcim.9b00516

Gajdacs M, Spengler G (2019) The role of drug repurposing in the development of novel antimicrobial drugs: non-antibiotic pharmacological agents as quorum sensing-inhibitors. Antibiotics 8:270. https://doi.org/10.3390/antibiotics8040270

Ganesan A, Coote ML, Barakat K (2017) Molecular dynamics-driven drug discovery: leaping forward with confidence. Drug Discov Today 22(2):249–269. https://doi.org/10.1016/j.drudis.2016.11.001

Gao F, Zhai G, Wang H, Lu L, Xu J, Zhu J, Chen D, Lu H (2020) Protective effects of anti-alginate monoclonal antibody against *Pseudomonas aeruginosa* infection of HeLa cells. Microb Pathog 145:104240. https://doi.org/10.1016/j.micpath.2020.104240

Ghannay S, Aouadi K, Kadri A, Snoussi M (2022) GC-MS profiling, vibriocidal, antioxidant, antibiofilm, and anti-quorum sensing properties of *Carum carvi* L. essential oil: *in vitro* and *in silico* approaches. Plants 11:1072. https://doi.org/10.3390/plants11081072

Gi M, Jeong J, Lee K, Lee KM, Toyofuku M, Yong DE, Yoon SS, Choi JY (2014) A drug-repositioning screening identifies pentetic acid as a potential therapeutic agent for suppressing the elastase-mediated virulence of *Pseudomonas aeruginosa*. Antimicrob Agent Chemother 58(12):7205–7214. https://doi.org/10.1128/AAC.03063-14

Giordano D, Biancaniello C, Argenio MA, Facchiano A (2022) Drug design by pharmacophore and virtual screening approach. Pharmaceuticals 15(5):646. https://doi.org/10.3390/ph15050646

Girija ASS, Gnanendra S, Paramasivam A, Vijayashree Priyadharsini J (2021) Delineating the potential targets of thymoquinone in ESKAPE pathogens using a computational approach. In Silico Pharmacol 9:52. https://doi.org/10.1007/s40203-021-00111-z

Gökalsın B, Aksoydan B, Erman B, Sesal NC (2017) Reducing virulence and biofilm of *Pseudomonas aeruginosa* by potential quorum sensing inhibitor carotenoid: zeaxanthin. Microb Ecol 74:466–473. https://doi.org/10.1007/s00248-017-0949-3

Gulia K, Hassan AHE, Lenhard JR, Farahat AA (2023) Escaping ESKAPE resistance: *in vitro* and *in silico* studies of multifunctional carbamimidoyl-tethered indoles against antibiotic-resistant bacteria. R Soc Open Sci 10:230020. https://doi.org/10.1098/rsos230020

Gupta R, Srivastava D, Sahu M, Tiwari S, Ambasta RK, Kumar P (2021) Artificial intelligence to deep learning: machine intelligence approach for drug discovery. Mol Divers 25:1315–1360. https://doi.org/10.1007/s11030-021-10217-3

Hansen T, Alst T, Havelkova M, Strom MB (2010) Antimicrobial activity of small β-peptidomimetics based on the pharmacophore model of short cationic antimicrobial peptides. J Med Chem 53:595–606. https://doi.org/10.1021/jm901052r

Hema Sree GNS, Saraswathy GR, Murahari M, Krishnamurthy M (2019) An update on drug repurposing: re-written saga of the drug's fate. Biomed Pharmacother 110:700–716. https://doi.org/10.1016/j.biopha.2018.11.127

Hnamte S, Parasuraman P, Ranganathan S, Ampasala DR, Reddy D, Kumavath RN et al (2019) Mosloflavone attenuates the quorum sensing controlled virulence phenotypes and biofilm formation in *Pseudomonas aeruginosa* PAO1: *in vitro*, *in vivo* and *in silico* approach. Microb Pathog 131:128–134. https://doi.org/10.1016/j.micpath.2019.04.005

Ikhimiukor OO, Odih EE, Donado-Godoy P, Okeke IN (2022) A bottom-up view of antimicrobial resistance transmission in developing countries. Nat Microbiol 7:757–765. https://doi.org/10.1038/s41564-022-01124-x

Ismail MM, Hassan M, Moawad SS, Okba MM, Ashour RM, Fayek NM, Saber FR (2021) Exploring the antivirulence activity of Pulverulentone A, a phloroglucinol-derivative from *Callistemon citrinus* leaf extract, against multi-drug resistant *Pseudomonas aeruginosa*. Antibiotics 10:907. https://doi.org/10.3390/antibiotics10080907

Jarada TN, Rokne JG, Alhajj R (2020) A review of computational drug repositioning: strategies, approaches, opportunities, challenges, and directions. J Cheminform 12:46. https://doi.org/10.1186/s13321-020-00450-7

Kalpana S, Lin WY, Wang YC, Fu Y, Lakshmi A, Wang HY (2023) Antibiotic resistance diagnosis in ESKAPE pathogens—a review on proteomic perspective. Diagnostics 13(6):1014. https://doi.org/10.3390/diagnostics13061014

Kamarudheen N, Rao KVB (2019) Fatty acyl compounds from marine *Streptomyces griseoincarnatus* strain HK12 against two major bio-film forming nosocomial pathogens; an *in vitro* and *in silico* approach. Microb Pathog 127:121–130. https://doi.org/10.1016/j.micpath.2018.11.050

Kaserer T, Beck KR, Akram M, Odermatt A, Schuster D (2015) Pharmacophore models and pharmacophore-based virtual screening: concepts and applications exemplified on hydroxysteroid dehydrogenases. Molecules 20(12):22799–22832. https://doi.org/10.3390/molecules.201219880

Krell T, Matilla MA (2022) Antimicrobial resistance: progress and challenges in antibiotic discovery and anti-infective therapy. Microb Biotechnol 15(1):70–78. https://doi.org/10.1111/1751-7915.13945

Kumar V, Parate S, Danishuddin, Zeb A, Singh P, Lee G et al (2022) 3D-QSAR-based pharmacophore modeling, virtual screening, and molecular dynamics simulations for the identification of spleen tyrosine kinase inhibitors. Front Cell Infect Microbiol 12:909111. https://doi.org/10.3389/fcimb.2022.909111

Kwon S, Bae H, Jo J, Yoon S (2019) Comprehensive ensemble in QSAR prediction for drug discovery. BMC Bioinform 20:521. https://doi.org/10.1186/s12859-019-3135-4

Larsson DGJ, Flach CF (2022) Antibiotic resistance in the environment. Nat Rev Microbiol 20:257–269. https://doi.org/10.1038/s41579-021-00649-x

Law GL, Tisoncik-Go J, Korth MJ, Katze MG (2013) Drug repurposing: a better approach for infectious disease drug discovery? Curr Opin Microbiol 25:588–592. https://doi.org/10.1016/j.coi.2013.08.004

Lazar V, Holban AM, Curutiu C, Chifiriuc MC (2021) Modulation of quorum sensing and biofilms in less investigated Gram-negative ESKAPE pathogens. Front Microbiol 12:676510. https://doi.org/10.3389/fmicb.2021.676510

Lin TT, Yang LY, Lin CY, Wang CT, Lai CW, Ko CF, Shih YH, Chen SH (2023) Intelligent *de novo* design of novel antimicrobial peptides against antibiotic-resistant bacteria strains. Int J Mol Sci 24:6788. https://doi.org/10.3390/ijms.24076788

Liu Z, Zhang L, Wang J, Li Y, Chang Y, Huang X et al (2021) Virtual screening and biological evaluation of anti-biofilm agents targeting LuxS in the quorum sensing system. Nat Prod Commun 16(6):1–10. https://doi.org/10.1177/1934578X211019625

Liu G, Thomsen LE, Olsen LE (2022) Antimicrobial-induced horizontal transfer of antimicrobial resistant genes in bacteria: a mini-review. J Antimicrob Ther 77:556–567. https://doi.org/10.1093/jac/dkab450

Lucas JE, Kortemme T (2020) New computational protein design methods for de novo small molecule binding sites. PLoS Comput Biol 16(10):e1008178. https://doi.org/10.1371/journal.pcbi.1008178

Ma Y, Wang YR, He YH, Ding YY, An JX, Zhang ZJ, Zhao WB, Hu YM, Liu YQ (2023) Drug repurposing strategy part 1: from approved drugs to agri-bactericides leads. J Antibiot 76:27–51. https://doi.org/10.1038/s41429-022-00574-y

Maia EHB, Assis LC, de Oliveira TA, da Silva AM, Taranto AG (2020) Structure-based virtual screening: from classical to artificial intelligence. Front Chem 8:343. https://doi.org/10.3389/fchem.2020.00343

Mak KK, Pichika MR (2019) Artificial intelligence in drug development: present status and future prospects. Drug Discov Today 24(3):773–780. https://doi.org/10.1016/j.drudis.2018.11.014

Malandraki-Miller S, Riley PR (2021) Use of artificial intelligence to enhance phenotypic drug discovery. Drug Discov Today 26(4):887–901. https://doi.org/10.1016/j.drudis.2021.01.013

Mansuri A, Lokhande K, Kore S, Gaikwad S, Nawani N, Swamy KV et al (2022) Antioxidant, anti-quorum sensing, biofilm inhibitory activities and chemical composition of Patchouli essential oil: *in vitro* and *in silico* approach. J Biomol Struct Dyn 40(1):154–165. https://doi.org/10.1080/07391102.2020.1810124

Matamoros-Recio A, Franco-Gonzalez JF, Forgione RE, Torres-Mozas A, Silipo A, Martin-Santamaria S (2021) Understanding the antibacterial resistance: computational explorations in bacterial membranes. ACS Omega 6:6041–6054. https://doi.org/10.1021/acsomega.0c05590

Meenambiga SS, Rajagopal K (2018) Antibiofilm activity and molecular docking studies of bioactive secondary metabolites from endophytic fungus *Aspergillus nidulans* on oral *Candida albicans*. J Appl Pharm Sci 8(3):37–45. https://doi.org/10.7324/JAPS.2018.8306

Melo MCR, Maasch JRMA, de la Fuente-Nunez C (2021) Accelarating antibiotic discovery through artificial intelligence. Commun Biol 4:1050. https://doi.org/10.1038/s42003-021-02586-0

Mishra R, Panda AK, Mandal SD, Shakeel M, Bisht SS, Khan J (2020) Natural anti-biofilm agents: strategies to control biofilm-forming pathogens. Front Microbiol 11:566325. https://doi.org/10.3389/fmicb.2020.566325

Mohamad F, Alzahrani RR, Alsaadi A, Alrfaei BM, Yassin AAB, Alkhulafi MM, Halwani M (2023) An explorative review on advanced approaches to overcome bacterial resistance by curbing bacterial biofilm formation. Infect Drug Resist 16:19–49. https://doi.org/10.2147/IDR.5380883

Mohs RC, Greig NH (2017) Drug discovery and development: role of basic biological research. Alzheimers Dement (N Y) 3(4):651–657. https://doi.org/10.1016/j.trci.2017.10.005

Moingeon P, Kuenemann M, Guedj M (2022) Artificial intelligence-enhanced drug design and development: toward a computational precision medicine. Drug Discov Today 27(1):215–222. https://doi.org/10.1016/j.drudis.2021.09.006

Montiro-Neto V, de Souza CD, Gonzaga LF, da Silveira BC, Sousa NCF, Pontes JP et al (2020) Cuminaldehyde potentiates the antimicrobial actions of ciprofloxacin against *Staphylococcus aureus* and *Escherichia coli*. PLoS One 15(5):e0232987. https://doi.org/10.1371/journal.pone.0239287

Mouchlis VD, Afantitis A, Serra A, Fratello M, Papadiamantis AG, Aidinis V et al (2021) Advances in *de novo* drug design: from conventional to machine learning methods. Int J Mol Sci 22(4):1676. https://doi.org/10.3390/ijms.22041676

Mukherjee A, Bose S, Shaoo A, Das SK (2023) Nanotechnology based therapeutic approaches: an advanced strategy to target the biofilm of ESKAPE pathogens. Mater Adv 4:2544–2572. https://doi.org/10.1039/D2MA00846G

Mulani MS, Kamble EE, Kumkar SN, Tawre MS, Pardesi KR (2019) Emerging strategies to combat ESKAPE pathogens in the era of antimicrobial resistance: a review. Front Microbiol 10:539. https://doi.org/10.3389/fmicb.2019.00539

Nakonieczna J, Wozniak A, Pieranski M, Rapacka-Zdonczyk A, Ogonowska P, Grinholc M (2019) Photoinactivation of ESKAPE pathogens: overview of novel therapeutic strategy. Future Med Chem 11(5):443–461. https://doi.org/10.4155/fmc-2018-0329

Namasivayam SKR, Shankar KG, Vivek JM, Nizar M, Sudarsan AV (2019) In silico and in vitro analysis of quorum quenching active phytochemicals from the ethanolic extract of medicinal plants against quorum sensing mediated virulence factors of *Acinetobacter baumannii*. Ind J Biochem Biophys 56:276–286

Neves BJ, Braga RC, Melo-Filho CC, Moreira-Filho JT, Muratov EN, Andrade CH (2018) QSAR-based virtual screening: advances and applications in drug discovery. Front Pharmacol 9:1275. https://doi.org/10.3389/fphar.2018.01275

Nicolas-Barreales G, Sujar A, Sanchez A (2021) A web-based tool for simulating molecular dynamics in cloud environments. Electronics 10(2):185. https://doi.org/10.3390/electronics10020185

Noumi E, Ahmad I, Adnan M, Merghni A, Patel H, Haddaji N et al (2023) GC/MS profiling, antibacterial, anti-quorum sensing, and antibiofilm properties of *Anethum graveolens* L. essential oil: molecular docking study and *in-silico* ADME profiling. Plants 12:1997. https://doi.org/10.3390/plants12101997

Nyawai TN, Asaruddin MR, Rosli MFA, Romli AM, Radhakrishnan SE, Ahmad MN (2017) Pharmacophore modeling of phytochemicals from *Clinacanthus nutans* for antimicrobial activity. Transact Sci Technol 4(4):498–503

Opo FADM, Rahman MM, Ahammad F, Ahmed I, Bhuiyan MA, Asiri AM (2021) Structure based pharmacophore modeling, virtual screening, molecular docking and ADMET approaches for identification of natural anti-cancer agents targeting XIAP protein. Sci Rep 11:4049. https://doi.org/10.1038/s41598-021-83626-x

Ou-yang S, Lu J, Kong X, Liang Z, Luo C, Jiang H (2012) Computational drug discovery. Acta Pharmacol Sin 33:1131–1140. https://doi.org/10.1038/aps.2012.109

Pagadala NS, Syed K, Tuszynski J (2017) Software for molecular docking: a review. Biophys Rev 9:91–102. https://doi.org/10.1007/s12551-016-0247-1

Palmer N, Massch JRMA, Torres MDT, de la Fuente-Nunez C (2021) Molecular dynamics for antimicrobial peptide discovery. Infect Immun 89:e00703-20. https://doi.org/10.1128/IAI.00703-20

Panda SK, Buroni S, Swain SS, Bonacorsi A, da Fonseca Amorim EA, Kulshreshtha M, da Silva LCN, Tiwari V (2022) Recent advances to combat ESKAPE pathogens with special reference to essential oils. Front Microbiol 13:1029098. https://doi.org/10.3389/fmicb.2022.1029098

Pant N, Miranda-Hernandez S, Rush C, Warner J, Eisen DP (2022) Non-antimicrobial adjuvant therapy using Ticagrelor reduced biofilm-related *Staphylococcus aureus* prosthetic joint infection. Front Pharamacol 13:927783. https://doi.org/10.3389/fphar.2022.927783

Park K (2019) A review of computational drug repurposing. Transl Clin Pharmacol 27(2):59–63. https://doi.org/10.12793/tcp.2019.27.2.59

Patel A, Banerji R, Kanojiya P, Saroj SD (2021) Foodborne ESKAPE biofilms and antimicrobial resistance: lessons learned from clinical isolates. Pathog Glob Health 115(6):339–356. https://doi.org/10.1010/20477724.2021.1916158

Pattnaik SS, Ranganathan S, Ampasala DR, Syed A, Ameen F, Busi S (2018) Attenuation of quorum sensing regulated virulence and biofilm development in *Pseudomonas aeruginosa* PAO1 by *Diaporthe phaseolorum* SSP12. Microb Pathog 118:177–189. https://doi.org/10.1016/j.micpath.2018.03.031

Paudel A, Kaneko K, Watanabe A, Shigeki M, Motomu K, Hamamoto H, Sekimizu K (2013) Structure-activity relationship study of novel iminothiadiazolo-pyrimidinone antimicrobial agents. J Antibiot 66:663–667. https://doi.org/10.1038/ja.2013.69

Pfalzgraff A, Brandenburg K, Weindl G (2018) Antimicrobial peptides and their therapeutic potential for bacterial skin infections and wounds. Front Pharmacol 9:281. https://doi.org/10.3389/fphar.2018.00281

Prachayasittikul V, Worachartcheewan A, Shumbuatong W, Songtawee N, Simeon S, Prachayasittikul V, Nantasenamat C (2015) Computer-aided drug design of bioactive natural products. Curr Top Med Chem 15:1780–1800. https://doi.org/10.2174/1568026615666150506151101

Priyamvada P, Debroy R, Anbarasu A, Ramaiah S (2022) A comprehensive review on genomics, systems biology and structural biology approaches for combating antimicrobial resistance in ESKAPE pathogens: computational tools and recent advancements. World J Microbiol Biotechnol 38:153. https://doi.org/10.1007/s11274-022-03343-z

Qin HL, Zhang ZW, Ravindar L, Rakesh KP (2020) Antibacterial activities with the structure-activity relationship of coumarin derivatives. Eur J Med Chem 207:112832. https://doi.org/10.1016/j.ejmech.2020.112832

Qureshi KA, Imtiaz M, Parvez A, Rai PK, Jaremko M, Emwas AH et al (2022) *In vitro* and *in silico* approaches for the evaluation of antimicrobial activity, time-kill kinetics, and anti-biofilm potential of thymoquinone (2-methyl-5-propan-2-ylcyclohexa-2,5-diene-1,4-dione) against selected human pathogens. Antibiotics 11(1):79. https://doi.org/10.3390/antibiotic.11010079

Rabha J, Chetri BK, Das S, Jha DK (2023) *In-vitro* and *in-silico* evaluation of antimicrobial and antibiofilm secondary metabolites of a novel fungal endophyte, *Albophoma* sp. BAPR5. South Afr J Bot 158:347–368. https://doi.org/10.1016/j.sajb.2023.05.033

Rajput A, Kumar M (2018) Anti-biofilm peptides: a new class of quorum quenchers and their prospective therapeutic applications. In: Kalia VC (ed) Biotechnological applications of quorum sensing inhibitors. Springer Nature, Berlin, pp 87–110. https://doi.org/10.1007/978-981-10-9026-4_5

Rajput A, Bhamare KT, Thakur A, Kumar M (2023) Anti-biofilm: machine learning assisted prediction of IC_{50} activity of chemicals against biofilms of microbes causing antimicrobial resistance and implications in drug repurposing. J Mol Biol 435:168115. https://doi.org/10.1016/j.jmb.2023.168115

Rasheed MA, Iqbal MN, Saddick S, Ali I, Khan FS, Kanwal S et al (2021) Identification of lead compounds against Scm (fms10) in *Enterococcus faecium* using computer-aided drug designing. Life 11:77. https://doi.org/10.3390/life11020077

Sadiq S, Rana NF, Zahid MA, Zargaham MK, Tanweer T, Batool A et al (2020) Virtual screening of FDA-approved drugs against LasR of *Pseudomonas aeruginosa* for antibiofilm potential. Molecules 25:3723. https://doi.org/10.3390/molecules25163723

Sadybekov AV, Katritch V (2023) Computational approaches streamlining drug discovery. Nature 616:673–685. https://doi.org/10.1038/s41586-023-05905-z

Sagulkoo P, Chuntakaruk H, Rungrotmongkol T, Suratanee A, Plaimas K (2022) Multi-level biological network analysis and drug repurposing based on leukocyte transcriptomics in severe COVID-19: *in silico* systems biology to precision medicine. J Pers Med 12(7):1030. https://doi.org/10.3390/jpm12071030

Samreen, Qais FA, Ahmad I (2022) Anti-quorum sensing and biofilm inhibitory effect of some medicinal plants against Gram-negative bacterial pathogens: *in vitro* and *in silico* investigations. Heliyon 8:e11113. https://doi.org/10.1016/j.heliyon.2022.e11113

Schaller D, Sribar D, Noonan T, Deng L, Nguyen TN, Pach S et al (2020) Next generation 3D pharmacophore modeling. WIREs Comput Mol Sci 10(4):e1468. https://doi.org/10.1002/wcms.1468

Schneider G, Fechner U (2005) Computer-based *de novo* design of drug-like molecules. Nat Rev Drug Dis 4:649–663. https://doi.org/10.1038/nrd1799

Schottlender G, Prieto JM, Palumbo MC, Castello FA, Serral F, Sossa EJ et al (2022) From drugs to targets: reverse engineering the virtual screening process on a proteomic scale. Front Drug Discov 2:969983. https://doi.org/10.3389/fddsv.2022.969983

Schultz F, Anywar G, Tang H, Chassagne F, Lyles JT, Garbe LA, Quave CL (2020) Targeting ESKAPE pathogens with anti-infective medicinal plants from the Greater Mpigi region in Uganda. Sci Rep 10:11935. https://doi.org/10.1038/s41598-020-67572-8

Seethalakshmi PS, Charity OJ, Giakoumis T, Kiran GS, Sriskandan S, Voulvoulis N, Selvin J (2022) Delineating the impact of COVID-19 on antimicrobial resistance: an Indian perspective. Sci Total Environ 818:151702. https://doi.org/10.1016/j.scitotenv.2021.151702

Shamim A, Ali A, Iqbal Z, Mirza MA, Aqil M, Kawish SM et al (2023) Natural medicine a promising candidate in combating microbial biofilm. Antibiotics 12(2):299. https://doi.org/10.3390/antibiotics12020299

Sharma A, Gupta P, Kumar R, Bhardwaj A (2016) dPABBs: a novel *in silico* approach for predicting and designing anti-biofilm peptides. Sci Rep 6:21839. https://doi.org/10.1038/srep21839

Sharma S, Gopu V, Sivasankar C, Shetty PH (2019) Hydrocinnamic acid produced by *Enterobacter xiangfangensis* impairs AHL-based quorum sensing and biofilm formation in *Pseudomonas aeruginosa*. RSC Adv 9:28678. https://doi.org/10.1039/c9ra05725k

Sharma PP, Bansal M, Sethi A, Poonam PL, Goel VK et al (2021a) Computational methods directed towards drug repurposing for COVID-19: advantages and limitations. RSC Adv 11:36181. https://doi.org/10.1039/d1ra05320e

Sharma S, Tiwari M, Tiwari V (2021b) Therapeutic strategies against authophagic escape by pathogenic bacteria. Drug Discov Today 26(3):704–712. https://doi.org/10.1016/j.drudis.2020.12.002

Shen L, Feng H, Qiu Y, Wei GW (2023) SVSBI: sequence-based virtual screening of biomolecular interactions. Commun Biol 6(1):536. https://doi.org/10.1038/s42003-023-04866-3

Shukla A, Shukla G, Parmar P, Patel B, Goswami D, Saraf M (2021) Exemplifying the next generation of antibiotic susceptibility intensifiers of phytochemicals by LasR-mediated quorum sensing inhibition. Sci Rep 11:22421. https://doi.org/10.1038/s41598-021-01845-8

Sicho M, Liu X, Svozil D, van Westen GJP (2021) GenUI: interactive and extensible open source software platform for de novo molecular generation and cheminformatics. J Cheminform 13:73. https://doi.org/10.1186/s13321-021-00550-y

Sobieszczanska N, Myszka K, Szwengiel A, Majcher M, Grygier A, Wolko L (2020) Tarragon essential oil as a source of bioactive compounds with anti-quorum sensing and anti-proteolytic activity against *Pseudomonas* spp. isolated from fish—*in vitro*, *in silico* and *in situ* approaches. Int J Food Microbiol 331:108732. https://doi.org/10.1016/j.ijfoodmicro.2020.108732

Srivastava GN, Malwe AS, Sharma AK, Shastri V, Hibare K, Sharma VK (2020) Molib: a machine learning based classification tool for the prediction of biofilm inhibitory molecules. Genomics 112:2823–2832. https://doi.org/10.1016/j.ygeno.2020.03.020

Stephens LJ, Werrett MV, Sedgwick AC, Bull SD, Andrews PC (2020) Antimicrobial innovation: a current update and perspective on the antibiotic drug development pipeline. Future Med Chem 12(22):2035–2065. https://doi.org/10.4155/fmc-2020-0225

Suay-Garcia B, Bueso-Bordils JI, Falco A, Perez-Garcia MT, Anton-Fos G, Aleman-Lopez P (2020a) Quantitative structure-activity relationship methods in the discovery and development of antibacterials. WIREs Comput Mol Sci 10(6):e1472. https://doi.org/10.1002/wcms.1472

Suay-Garcia B, Falco A, Bueso-Bordlis JI, Anton-Fos GM, Perez-Garcia MT, Aleman-Lopez PA (2020b) Tree-based QSAR model for drug repurposing in the discovery of new antibacterial compounds against *Escherichia coli*. Pharmaceuticals 13(12):431. https://doi.org/10.3390/ph13120431

Sun S, Dai X, Sun J, Bu X, Weng C, Li H, Zhu H (2016) A diketopiperazine factor from *Rheinheimera aquimaris* QSI02 exhibits anti-quorum sensing activity. Sci Rep 6:39637. https://doi.org/10.1038/srep39637

Swamidass SJ (2011) Mining small-molecule screens to repurpose drugs. Brief Bioinform 12(4):327–335. https://doi.org/10.1093/bib/bbr028

Talat A, Khan AU (2023) Artificial intelligence as a smart approach to develop antimicrobial drug molecules: a paradigm to combat drug-resistant infections. Drug Discov Today 28(4):103491. https://doi.org/10.1016/j.drudis.2023.103491

Tang Y, Zou F, Chen C, Zhang Y, Shen Z, Liu Y et al (2023) Antibacterial and antibiofilm activities of Sertindole and its antibacterial mechanism against *Staphylococcus aureus*. ACS Omega 8:5415–5425. https://doi.org/10.1021/acsomega.2c06569

Tapia-Rodriguez MR, Bernal-Mercado AT, Gutierrez-Pacheco MM, Vazquez-Armenta FJ, Hernandez-Mendoza A, Gonzalez-Aguilar GA et al (2019) Virulence of *Pseudomonas aeruginosa* exposed to carvacrol: alterations of the quorum sensing at enzymatic and gene levels. J Cell Commun Signaling 13:531–537. https://doi.org/10.1007/s12079-019-00516-8

Torres PHM, Sodero ACR, Jofily P, Silva-Jr FP (2019) Key topics in molecular docking for drug design. Int J Mol Sci 20(18):4574. https://doi.org/10.3390/ijms20184574

Tripathi MK, Nath A, Singh TP, Ethayathulla AS, Kaur P (2021) Evolving scenario of big data and artificial intelligence (AI) in drug discovery. Mol Divers 25:1439–1460. https://doi.org/10.1007/s11030-021-10256-w

Truong TTT, Panizzutti B, Kim JH, Walder K (2022) Repurposing drugs via network analysis: opportunities for psychiatric disorders. Pharmaceutics 14(7):1464. https://doi.org/10.3390/pharmaceutics14071464

Tyagi R, Singh A, Chaudhary KK, Yadav MK (2022) Pharmacophore modeling and its applications. In: Singh DB, Pathak RK (eds) Bioinformatics: methods and applications. Academic, New York, pp 269–289. https://doi.org/10.1016/B978-0-323-89775-4.00009-2

Vamathevan J, Clark D, Czodrowski P, Dunham I, Ferran E, Lee G, Li B, Madabhushi A, Shah P, Spitzer M, Zhao S (2019) Applications of machine learning in drug discovery and development. Nat Rev Drug Dis 18:463–477. https://doi.org/10.1038/s41573-019-0024-5

Vestby LK, Grønseth T, Simm R, Nesse LL (2020) Bacterial biofilms and its role in the pathogenesis of disease. Antibiotics 9(2):59. https://doi.org/10.3390/antibiotics9020059

Vieira TF, Magalhaes RP, Simoes M, Sousa SF (2022) Drug repurposing targeting *Pseudomonas aeruginosa* MvfR using docking, virtual screening, molecular dynamics, and free-energy calculations. Antibiotics 11:185. https://doi.org/10.3390/antibiotics11020185

Vrancianu CO, Gheorghe I, Dobre EG, Barbu IC, Cristian RE, Popa M et al (2020) Emerging strategies to combat β-lactamase producing ESKAPE pathogens. Int J Mol Sci 21(22):8527. https://doi.org/10.3390/ijms21228527

Wang L, Ding J, Pan L, Cao D, Jiang H, Ding X (2019) Artificial intelligence facilitates drug design in the big data era. Chemometr Intell Lab Syst 194:103850. https://doi.org/10.1016/j.chemolab.2019.103850

Wang C, Xiong Y, Bao C, Wei Y, Wen Z, Cao X et al (2023) Antibacterial and anti-biofilm activity of radezolid against *Staphylococcus aureus* clinical isolates from China. Front Microbiol 14:1131178. https://doi.org/10.3389/fmicb.2023.1131178

Wu B, Yang X, Yan M (2019) Synthesis and structure-activity relationship study of antimicrobial Auranofin against ESKAPE pathogens. J Med Chem 62(17):7751–7768. https://doi.org/10.1021/acs.jmedchem.9b00550

Wu J, Jin K, Jiao K, Wang X, Li S, Pan L (2023) CarbonAI, a non-docking deep learning based small molecule virtual screening platform. ChemRxiv. https://doi.org/10.26434/chemrxiv-2022-gk3n6-v2

Xiong G, Wu Z, Yi J, Fu L, Yang Z, Hsieh C et al (2021) ADMETlab 2.0: an integrated online platform for accurate and comprehensive predictions of ADMET properties. Nucleic Acid Res 49:W5–W14. https://doi.org/10.1093/nar/gkab255

Yang F, Zhang Q, Ji X, Zhang Y, Li W, Peng S, Xue F (2022) Machine learning applications in drug repurposing. Interdiscip Sci 14(1):15–21. https://doi.org/10.1007/s12539-021-00487-8

Ye J, Yang X, Ma C (2023) Ligand-based drug design of novel antimicrobials against *Staphylococcus aureus* by targeting bacterial transcription. Int J Mol Sci 24(1):339. https://doi.org/10.3390/ijms24010339

Yu W, MacKerell AD Jr (2017) Computer-aided drug design methods. Methods Mol Biol 1520:85–106. https://doi.org/10.1007/978-1-4939-6634-9_5

Zeng YX, Liu JS, Wang YJ, Tang S, Wang DY, Deng SM, Jia AQ (2022) Actinomycin D: a novel *Pseudomonas aeruginosa* quorum sensing inhibitor from the endophyte *Streptomyces cyaneochromogenes* RC1. World J Microbiol Biotechnol 38:170. https://doi.org/10.1007/s11274-022-03360-y

Zhang M, Prior AM, Maddox MM, Shen WJ, Hevener KE, Bruhn DF et al (2018) Pharmacophore modeling, synthesis, and antibacterial evaluation of chalcones and derivatives. ACS Omega 3:18343–18360. https://doi.org/10.1021/acsomega.8b03174

Zhang Y, Zhang Y, Chen C, Cheng H, Deng X, Li D et al (2022) Antibacterial activities and action mode of anti-hyperlipidemic lomitapide against *Staphylococcus aureus*. BMC Microbiol 22:114. https://doi.org/10.1186/s12866-022-02535-9

Zhou JW, Jia AQ, Jiang H, Li PL, Chen H, Tan XJ, Liu EQ (2021) 1-(4-Amino-2-hydroxyphenyl) ethanone from *Phomopsis liquidambari* showed quorum sensing inhibitory activity against *Pseudomonas aeruginosa*. Appl Microbiol Biotechnol 105(1):341–352. https://doi.org/10.1007/s00253-020-11013-z

Zong N, Wen A, Moon S, Fu S, Wang L, Zhao Y et al (2022) Computational drug repurposing based on electronic health records: a scoping review. Npj Digital Med 5:77. https://doi.org/10.1038/s41746-022-00617-6

GPSR Compliance
The European Union's (EU) General Product Safety Regulation (GPSR) is a set of rules that requires consumer products to be safe and our obligations to ensure this.

If you have any concerns about our products, you can contact us on

ProductSafety@springernature.com

In case Publisher is established outside the EU, the EU authorized representative is:

Springer Nature Customer Service Center GmbH
Europaplatz 3
69115 Heidelberg, Germany

www.ingramcontent.com/pod-product-compliance
Ingram Content Group UK Ltd.
Pitfield, Milton Keynes, MK11 3LW, UK
UKHW022217051026
15624UKWH00001B/210
* 9 7 8 9 8 1 9 9 8 7 9 8 6 *